ANATOMY OF THE

MONOCOTYLEDONS

IX

ACORACEAE AND ARACEAE

Anatomy of the Monocotyledons series

(Volumes I–VII edited by C.R. Metcalfe)

Vol. I. Gramineae. By C. R. Metcalfe, 1960.
Vol. II. Palmae. By P. B. Tomlinson, 1961.
Vol. III. Commelinales-Zingiberales. By P. B. Tomlinson, 1969.
Vol. IV. Juncales. By D. F. Cutler, 1969.
Vol. V. Cyperaceae. By C. R. Metcalfe, 1971.
Vol. VI. Dioscoreales. By E. S. Ayensu, 1972.
Vol. VII. Helobieae (Alismatidae) (including the seagrasses). By P. B. Tomlinson, 1982. ISBN 0-19-854502-9.
Vol. VIII. (eds D. F. Cutler and M. Gregory) Iridaceae. By Paula Rudall, 1995. ISBN 0-19-854504-5.
Vol. IX. (eds M. Gregory and D. F. Cutler) Acoraceae and Araceae. By Richard C. Keating. ISBN 0-19-854535-5.

Series editors: D. F. Cutler, M. Gregory and P. Rudall.

Anatomy of the Dicotyledons series

(Volumes I–III do not have editors)

Vol. I. Systematic anatomy of leaf and stem, with a brief history of the subject.
By C. R. Metcalfe and L. Chalk, 1979. ISBN 0-19-854383-2.
Vol. II. Wood structure and conclusion of the general introduction.
By C. R. Metcalfe and L. Chalk, 1983. ISBN 0-19-854383 (*v*. 1).
Vol. III. Magnoliales, Illiciales, and Laurales (*Sensu* Armen Takhtajan).
By C. R. Metcalfe, 1987. ISBN 0-19-854593-2 (*v*. 3).
Vol. IV. (eds D. F. Cutler and M. Gregory) Saxifragales. By Hazel P. Wilkinson, R. J. Gornall,
K. I. A. Al-Shammery and Mary Gregory, 1998. ISBN 0-19-854792-7.
Vol. I. reprinted with corrections in paperback, 1988. ISBN 0-19-854253-4.
Vol. II. reprinted with corrections in paperback, 1989. ISBN 0-19-854594-0.

Series editors: D. F. Cutler, P. Gasson, M. Gregory and D. W. Stevenson.

ANATOMY OF THE MONOCOTYLEDONS

EDITED BY

M. GREGORY and D. F. CUTLER

Honorary Research Fellows,
Jodrell Laboratory, Royal Botanic Gardens,
Kew, Richmond, Surrey.

IX. ACORACEAE AND ARACEAE

RICHARD C. KEATING

Professor Emeritus,
Southern Illinois University at Edwardsville,
and Research Associate,
Missouri Botanical Garden, USA.

CLARENDON PRESS • OXFORD
2002

This book has been printed digitally and produced in a standard specification in order to ensure its continuing availability

OXFORD
UNIVERSITY PRESS

Great Clarendon Street, Oxford OX2 6DP

Oxford University Press is a department of the University of Oxford.
It furthers the University's objective of excellence in research, scholarship,
and education by publishing worldwide in

Oxford New York

Auckland Cape Town Dar es Salaam Hong Kong Karachi
Kuala Lumpur Madrid Melbourne Mexico City Nairobi
New Delhi Shanghai Taipei Toronto

With offices in

Argentina Austria Brazil Chile Czech Republic France Greece
Guatemala Hungary Italy Japan South Korea Poland Portugal
Singapore Switzerland Thailand Turkey Ukraine Vietnam

Published in the United States
by Oxford University Press Inc., New York

Reprinted 2009

ISBN 978-0-19-854535-4

Printed and bound in Great Britain by CPI Antony Rowe, Chippenham and Eastbourne

FOREWORD

This volume by Richard Keating on the anatomy of the Acoraceae and Araceae, the ninth volume in the series Anatomy of the Monocotyledons, is a much awaited and welcome addition. The progress that has been made in recent years in plant anatomy and plant systematics is represented in this volume. For example, it has been the tradition to represent most illustrations as line art rather than as photomicrographs because of the difficulty inherent in low magnification photomicroscopy and thin section techniques. However, the present volume has been able to employ refined technology and there is no line art but rather 855 photomicrographs in 115 plates that are of the highest quality and clarity. This is of great value to the user because it makes for absolute comparisons between the plates and the slides one has at hand and also allows for the understanding of variability and interpretation.

A new classification system for the Araceae is presented based upon anatomical, morphological, and molecular data: a total evidence system. The system consists entirely of monophyletic groups in keeping with modern phylogenetic systematics. The anatomical and systematic treatments are supplemented with discussions on phytochemistry, economic botany, biogeography, and palaeobotany. Thus, the volume is valuable to a wide audience and places a vast amount of information in a broad context. The book is truly eclectic.

An interesting note is that recent evidence indicates that the Acoraceae, Araceae, and Alismatid families along with Tofieldiacae form a monophyletic group. The present volume along with Volume 7 on the Helobiae (Alismatidae) covers 99.9% of the clade representing the most thorough treatments of any monocot higher group.

The structural botany community has patiently awaited the arrival of this volume through its quarter of a century in the making and the wait has definitely been worthwhile. The preceding two volumes in the series, Anatomy of the Monocotyledons, have set new standards and this one by Richard Keating certainly follows that pattern.

DENNIS WM. STEVENSON, FMLS

AUTHOR'S PREFACE

In the late 1970s when I first conceived of undertaking this study of Araceae (then including *Acorus*), no broad comparative studies of the family had been undertaken since the works by Engler in the late 19th and early 20th centuries. During the course of the anatomical survey, the family has emerged as one of the best known flowering plant families. In terms of plant family overviews, few works can compare to the complementary volumes *The Genera of Araceae* by Mayo, Bogner and Boyce (1997) and *Aroids, Plants of the Arum Family* by Bown (2000).

The invitation to do this treatment offered by the late C. R. Metcalfe presented an outstanding opportunity. A particular inducement was the availability of remarkable material to study. Throughout their careers, taxonomists Thomas Croat (St. Louis) and Josef Bogner (Munich) had accumulated comprehensive field-collected living materials, including virtually all of the rare genera. Added to this, their ecumenical spirit and generosity made the undertaking of this study irresistible.

One might assume of this volume, titled unequivocally, that it could be considered the definitive treatment of the included genera. Regardless that such a compendium will not appear again soon, it should really be viewed as a basis or starting point for future investigations. For each genus sampling is limited, and is wide but not deep for the larger genera. Almost nothing is settled here. The volume was planned to provide a broad, comparative and illustrated view of anatomical variation across the family. Parallel generic descriptions will allow the data to serve as a point of departure for meta-analyses that will lead to new ideas.

A second value is a key to literature, much of which is obscure and scattered. Most of the references emphasize structural work but include as well some introduction to systematic and other disciplines. The number of recent titles suggests strong contemporary interest in these genera and the pace of research seems to be accelerating. While this project was underway, the genus *Acorus* was recognized as not belonging among the aroids. Also, cladistic analyses of molecular genetic data have stripped the aroids of their own order, indicating placement within Alismatales. The duckweeds are now recognized as being embedded among subfamilies of Araceae.

An inspection of these data and references will suggest numerous ways to advance the science. Further work will enhance our understanding of the Araceae and provide explanatory power within the general framework of plant structure and function. Consider several examples. One can only be impressed by the unusual display of calcium oxalate crystals, especially raphides, that demonstrate the greatest variety of crystal bundles and enclosing cell structures to be found in the plant kingdom. Systematically, little sense could be made of such variety during this survey. However, numerous easily grown genera can provide material for ontogenetic, chemical, genetic, and functional studies. I am confident that order can be imposed on this superficially chaotic picture.

Second, wider sampling of many anatomically variable genera, especially the larger ones, will allow amplification of the descriptive and interpretive base. Consider the general trends for collenchyma evolution. Many questions remain regarding ontogeny and homologies with ground tissue and sclerenchyma. Those wishing to elucidate the basis of laticifer specialization might well begin with genera of tribe Colocasieae. As a final example, for these mostly net-veined monocots, an adequate understanding of the ontogeny of leaf venation and its evolution remains an open frontier.

Questions, many yet to be formulated, will involve evolution, development, genetics, physiology and ecology and give us ample reason to engage in investigations of these fascinating plants. My hope for this volume is that it will assist this cause.

June 2002 R.C.K.

ACKNOWLEDGMENTS

I am grateful to the late Charles Russell Metcalfe, anatomist and Keeper of the Jodrell Laboratory, Kew, who encouraged my interest in providing an anatomical account of the Araceae, and also to David Cutler and Mary Gregory of the Laboratory who guided the manuscript through its later stages. The bibliographical compilation by Mary Gregory was crucial as the basis for undertaking the literature review. The existence of living collections of two botanists in particular were vital to conceiving and completing this project. Josef Bogner (Munich) and Thomas Croat (Missouri Botanical Garden, St. Louis) both manage extensive, carefully documented and named living collections which are based on extensive field experience. Both of these aroid taxonomists and growers were enthusiastic supporters of this study early on and were always generous with plant materials. Bogner's collection is especially rich in Old World species and Croat's in New World. Between the two are represented most of the world's aroid genera. Both botanists also freely shared experiences and ideas which were invaluable in helping me understand the family. I am also grateful to David Boufford, Wilbert Hetterscheid, and Simon Mayo for important specimens. Early stages of this study were supported by the School of Graduate Studies and Research, Southern Illinois University Edwardsville, and I am grateful to Library and Research Division staffs at the Missouri Botanical Garden, St. Louis, who helped advance this study in many ways. My thanks also to J. C. French and the Trustees of the Royal Botanic Gardens, Kew, for permission to reproduce the concensus tree in Table 6. Finally, the preparation of over 6000 slides of sections and clearings was moved forward in the early years at Southern Illinois University, Edwardsville, by Jeannie Bain, Jerry Dragich, Marina Ferrari, Sumi Fong, Rob Hamm, Leslie Lammert, and Deborah Thompson, student technicians who learned the discipline and satisfaction of research.

CONTENTS

LIST OF ABBREVIATIONS USED IN TEXT AND GLOSSARY OF SOME TISSUE TYPES

MBB: Mayo, Bogner and Boyce (1997) *The genera of Araceae*. Royal Botanic Gardens, Kew.
TS: Transverse section
LS: Longitudinal section
PS: Paradermal section
***l/w*:** length/width ratio

Mesophyll and ground tissue: aerenchyma types as seen in TS (see Histology and character state definitions, p. 36)

Type 1: Spongy mesophyll of irregularly arranged cells, with or without arms, among irregular air cavities, about the same diameter as the cells or smaller (Fig. 4G). Used to describe leaf tissue.
Type 1a: Spongy mesophyll formed of parenchyma cells with five or more short arms attached bluntly by their arm ends to adjacent cell arms. Rounded cavities about same diameter as the cells between cell bodies (Fig. 4H). Used to describe leaf tissue.
Type 2: Spongy mesophyll or ground tissue formed of long-armed parenchyma cells attached by their arm ends to neighbouring cells. Air cavities 2–3 times diameter of cell bodies (Fig. 5A). Can be applied to any organ.
Type 3: Mesophyll or ground tissue with large air cavities separated by uniseriate partitions of several cells that radiate from corner cells (Fig. 5B). Can be applied to any organ.
Type 4: Midrib, petiolar or stem ground tissue distinguished by having large air cavities separated by multiseriate partitions of ground tissue (Fig. 5D).

Collenchyma types as seen in TS (see Histology and character state definitions, p. 36)

Type B: Banded. Tissue forming a continuous multicellular band around the periphery or on one side of an organ (Fig. 5G).
Type Bi: Banded-interrupted. Collenchyma bands frequently interrupted by ribs of chlorenchyma or other ground tissue (Fig. 5F).
Type S: Independent strands, modified as listed below.
Type Sb: Strands circular or irregular, not aligned with vascular bundles but often between them, i.e. on alternate radii (Fig. 5H).
Type Sv: Strands, usually circular, aligned with vascular bundles, external to them and on same radii (Fig. 5I).
Type Svc: Strands of any shape, aligned with vascular bundles, placed close to xylem or phloem strands, resembling fibre caps in some genera.

Vascular bundle types as seen in TS (in midveins, petioles and stems; see Histology and character state definitions, p. 36)

Type I: Broad vascular bundles, often circular in outline, composed of semicircular strands of xylem and phloem separated by a broad, usually straight boundary (Fig. 6A).
Type II: Narrower vascular bundles with a protoxylem lacuna and 1–3 files of xylem separated from ± rounded or laterally narrowed phloem strand by narrow boundary (Fig. 6B).
Type III: Narrow vascular bundles with a single large metaxylem or persistent protoxylem cell or lacuna and 0–2 very small diameter metaxylem cells. Phloem strand usually laterally compressed (Fig. 6C).

INTRODUCTION

VEGETATIVE GROWTH AND MORPHOLOGY

The Araceae occupy a relatively isolated position among flowering plants due to their possession of a number of distinguishing vegetative and reproductive characters (MBB). Within the family a number of distinctive character states occur, especially in the larger or more basal genera, but few are diagnostic in themselves. It remains best to combine as many characters as possible when making determinations. Particularly in the more derived genera, anatomical characters are diminished in variety and must be combined with other morphological and reproductive characters for this purpose.

Growth form and habitat variety exceeds that of any other monocot family. The range includes vines or large herbs in tropical forests and borders, and plants with rhizomes or corms in seasonal dry or cold habitats (Bogner 1987). Hemiepiphytes and epiphytes are not common but occur in thirteen tropical genera (Benzing 1990). A useful bibliography, including anatomical citations for epiphytic species, can be found in Watson *et al.* (1987). Habitat elevation ranges from sea level to greater than 3–4000 m. For a predominantly terrestrial family, aquatic adaptations of several types are widespread. Steenis (1981, 1987) and Cook (1999) have listed numerous observations on the family; together the lists include 29 genera distributed among at least six subfamilies (including the four lemnoid genera), making 26% of the genera having aquatic species. Most aroid aquatics are rooted plants with emergent stems and/or leaves. Some are submerged (*Cryptocoryne, Jasarum*), or are rheophytes (*Aridarum, Bucephalandra, Piptospatha*). Some prefer brackish water (*Aglaodorum* and some *Cryptocoryne*). They may form rosettes (*Anubias, Cryptocoryne, Jasarum*), be free-floating (*Pistia*) or planktonic (lemnoids). Cook proposed that these adaptations have arisen independently in the family many times.

Meusel (1951) reviewed the work by Engler (1877, and see Engler 1990 with translation by Ray and Renner) on the growth forms of aroids and noted that sympodial growth is the principal mode of extension in genera with elongated internodes, as illustrated by eight examples. Holttum (1955) supported these observations and called attention to the precision (exact number of leaves) of branch architecture in the climbing genera. Engler devised the symbol and line diagrams later used by Ray (see below) to detail the shoot architectural patterns in the family.

Since then, the complex architectural habits of this family have been given improved descriptive tools by Ray (1986, 1987a,b,c). Ray (1987c) engaged the complexity of aroid shoot diversity and produced a scheme that allows structural comparisons across taxa supporting evolutionary and systematic analyses. Beginning with the whole shoot, the article (axis) is produced by the activity of a single meristem from origin to termination, abortion, or onset of sexuality. The shoot is a single (unbranched) article. The segment is an internode plus associated leaf (or leaves) and bud.

In aroids, segments can be monopodial or branching, the latter being a junction between two articles. Ray (1987b) investigated shoot organization, leaf types and growth correlations of the segment. The two shoot types found in the Araceae are (1) monopodial, in which the base of the petiole ensheaths the stem, and (2) sympodial where the shoot is continued after a foliage leaf by a sylleptic axillary bud; the shoot apex terminates by differentiating into an inflorescence, and the petiole base does not encircle the stem. Plants with elongated stems are restricted to taxa with predominantly monopodial segments. Ray examined the allometry among the components of the segments and attempted correlations between parts. His conclusions yielded no immediate systematic application but have potential to contribute to understanding the evolution of growth forms in the family.

Ray (1987c) studied shoot organizational diversity of 83 species representing 27 genera and 20 tribes, plus *Acorus*. He noted that flowering is always terminal in aroids, and that sympodial and monopodial axes are distinguished by whether or not the shoot is renewed after flowering. After reviewing the early German and more recent literature, Ray codified terminology and revised Engler's diagrams that use a series of symbols to represent segment parts. They are useful in correlating structure with phylogeny and systematics. His architectural analysis of an unbranched system fits *Chamberlain's Model* in the system of Hallé *et al.* (1978). Ray and Renner (in Engler 1990) noted that leaf divergence (phyllotaxy) is greatly influenced by shoot diameter. Small shoots have a leaf divergence approaching 1 : 2; larger diameters produce a spiral arrangement.

Leaf types within the Araceae vary considerably (cf. Hay and Mabberly 1991) but were divided by

Ray (1987b) into four classes: those associated with (1) initiation of new shoots; (2) continuation axes; (3) resting axes; and (4) termination and replacement of axes. The terms leaf, prophyll, mesophyll, bracteole, mesobracteole, blastophyll, and cataphyll are used conventionally. When combined with modifiers, such as foliar, flagellar, proleptic, sylleptic, etc, the terms are defined and used in a much more precise way than before. Ray has provided tools by which an evolutionary survey of shoot architecture of the family could be attempted.

Germination and growth

The Araceae are found to produce a larger variety of seedling forms than most monocot families, variation well-illustrated by Tillich (1985, 1995). In general, seedlings are said to be endospermless, with reduced hypocotyl, having weak development of the cotyledonary sheath and primary root. While *Acorus* seems easily excluded from the Araceae, it appears we are not at all close to systematic use of seedling variation within the family. For that we will need comparable stages and morphometric data on relative expansion of mass of embryonic organs. Much of the apparently marked variety in germinating embryos is probably due to heterochrony (cf. Gould 1977: 209 ff) and the variation might be organized using an approach that addresses this possibility.

Shoot ontogeny of *Calla* was studied by Scribailo and Tomlinson (1992). The shoot apex was investigated by Haccius and Lakshmanan (1966) in *Pistia* and *Lemna*. Jaeger (1968) reported on vegetative growth in *Remusatia*. A study of meristematic activity of internodal meristems in monocots (Fisher and French 1976, 1978) included several genera of Araceae. Both localized intercalary meristems (interrupted) and uninterrupted internodal meristems are found in the family. *Anthurium* (three species) and *Zantedeschia* (one species) have uninterrupted meristems on reproductive axes as detected using surface marking data. *Aglaonema* (one species) and *Pistia stratiotes* have interrupted meristems on reproductive and vegetative axes, respectively. *Spathiphyllum* (three species) has uninterrupted and interrupted meristems. No systematic value can be imputed as yet.

An intrafascicular cambium (as seen in TS) has been reported within vascular bundles of some Araceae (Nicolas 1916; Arber 1922a). As described by Arber, in immature vascular bundles of *Calla palustris* and *Arum italicum*, a cambium can be seen producing short files of derivatives that mature between the protoxylem and protophloem points. She considered the small resulting cells to be secondary xylem. Arber also reported 'radial seriation' in Alismataceae, Aponogetonaceae, and Hydrocharitaceae. However, as currently understood, it would appear that she was describing the maturation of cells divided within the procambium. All tissue matures at the xylem–phloem boundary as originally blocked out by the procambium; therefore, this interpretation of the observation may not be credible. It is interesting that the pattern of maturation she noted was also found in the apparently neighbouring Alismatales, which share a similar vascular bundle construction (cf. Tomlinson 1982).

REPRODUCTIVE MORPHOLOGY

The family is strikingly set off from other flowering plants by its signature feature, the inflorescence. A terminal branch, the spadix, is subtended by a single leaf-like organ, the spathe. The spadix may be covered with closely packed bisexual flowers, or the plants may be monoecious with gynoecial flowers crowding the base of the spadix and androecial flowers formed distal to them. Distally, a sterile appendix may occur which may be variously coloured. The spathe may be simple and reflexed, or form an erect sheath, variously elaborated into a tube, and often variously pigmented (MBB).

Inflorescences and flowers

Flowers of the family are usually small and prism-shaped due to their packing geometry and usual unextended or sessile form. Because of the inclusion of numerous crystals and/or mucilaginous materials, microtechnique requires unusual care and patience. Therefore, detailed anatomical knowledge has been slow in coming. Eyde *et al.* (1967) made the first survey of floral anatomy since Engler's period, where they described anatomy and vasculature of 18 genera. A series of detailed studies in floral vasculature were made by Barabé (1982, *Symplocarpus*; 1987, neoteny), Barabé and Chrétien (1985a, *Spathicarpa*; 1985b, *Monstera*; 1986, *Spathiphyllum*), Barabé, Chrétien and Forget (1986, *Anchomanes*; 1987, gynoecia), Barabé and Forget (1987, Calloideae; 1988a, *Zamioculcas*; 1988b, *Aglaonema*; 1992, *Culcasia*), Barabé, Forget and Chrétien (1986, 1987, *Symplocarpus*), Barabé and Labrecque (1983, *Calla*; 1984, *Lysichiton*; 1985, *Orontium*), and Barabé, Labrecque and Chrétien (1984, *Anthurium*). Carvell's (1989) large thesis also considered in detail floral anatomy of numerous genera, emphasizing Engler's Pothoideae and Monsteroideae. Its numerous detailed drawings of floral sections of the family and resulting analysis are worth consulting. Mayo (1986, 1989) investigated floral and inflorescence anatomy of *Philodendron*.

Mercado-Noriel and Mercado (1977–8) studied the floral and seed anatomy of *Pistia*. Inflorescence morphology and flower development in *Pistia* were investigated by Buzgó (1994). The inflorescence and flowers are highly reduced in this floating rosette plant. The spathe and spadix show fusion, congenital at the base and postgenital at the apex. Buzgó believed the genus to be closest to Cryptocoryninae and *Pinellia*. Several other authors have made useful observations on the inflorescence (Engler 1884; Manya-Chernej 1978; Chernej 1977; Barahona Carvajal 1977).

French (1986b) made the first comprehensive investigation of stamen vasculature involving observations on 100 genera and 350 spp of Araceae. Stamen traces vary in their capacity for branching and anastomosing, while synandrium construction complicates vascular organization. His findings are systematically arranged. French (1985a,b) also studied the endothecium in a total of 59 genera. Manning (1996) noted that the Araceae show variation on the annular–helical endothecium pattern. *Acorus* with its palmate-baseplate endothecium is distinct. Sampled plants of the Alismatiflorae show more diversity with their U-shaped, helical and baseplate types.

Pollen and pollination

Beginning with Erdtman's (1952) compendium, a number of observations on araceous pollen have been produced; see especially Thanikaimoni (1969) and Grayum (1984, 1992) for comprehensive surveys using light and electron microscopy, respectively. Also, Grayum (1985) has reviewed pollen starch and pollen nuclear number in the family (Grayum 1986a). See Kuprianova and Tarasevich (1984) and Tarasevich (1990) for pollen surveys of Lemnaceae.

Grayum (1986b) noted that pollen morphology is highly variable in Araceae. Sculpturing is more variable than vegetative morphology for advanced genera. Grayum offered correlations between sculpturing and weight with different families of insect pollinators, and concluded that sculpturing is probably most closely correlated with pollinator type. Unusual pollination syndromes of Araceae involve inflorescences acting as phototropic attractants, insect traps, the formation of nectar-like substances through degeneration of cell contents of stigmatic papillae, and the generation of various odours that act as beetle or fly attractants, as reviewed by Faegri and Pijl (1971). Young (1986) made observations on beetle pollination (cantharophily) in *Dieffenbachia longispatha*, supporting arguments that beetle pollination is an early syndrome in angiosperms. Young noted the work of Croat (1983) that included observations on floral biology of *Dieffenbachia*. Beetle visitors have been thought to be 'mess and spoil' pollinators, but are probably a component of a highly co-evolved system that involves specialized morphology, physiology and behaviour of both beetles and beetle-pollinated plants (Young 1986). Pollination phenomena have also been reviewed by Grayum (1986b, 1990) and Bay (1995). Croat's (1998) review updated the efforts in this active field.

Fruit

Pijl (1969) discussed fruit dispersal in the family and Spjut (1994) provided a systematic treatment of fruit types. In Spjut's thorough review, good illustrations are given for each type, including the bacca (berry): *Acorus calamus;* catoclesium (compound fruit of indehiscent fruitlets, enclosed by leaves, bracts or fused perianth parts): *Typhonodorum*; foraminicidal capsule (capsular fruit opening by irregular diverging cracks or slits): *Alocasia macrorrhiza, Lasia spinosa;* sorosus (compound fruit of many succulent pericarpia that developed on a peduncle): *Symplocarpus foetidus*.

Little is known of sex expression variables within the family. Lovett-Doust *et al.* (1986) studied fecundity and size relationships in *Arisaema triphyllum*, a gender-labile woodland herb. In that species, sex expression correlates with the abundance of stored resources in the underground corm. The plants are obligate outcrossers, having a self-incompatibility system, therefore neighbouring male ramets will not affect the pollination success. The larger the corm, the smaller the percentage of successfully matured ovules, suggesting nutritional competition between corm and ovules. Several other significant correlations are discussed. Fruit and seed structure are also discussed by Kulkarni *et al.* (1990) and by Bochenska and Kozlowski (1974). Grubert (1981) included Araceae in a survey of angiosperms for mucilage.

Seeds and embryos

Seubert (1993) made a comprehensive survey of seed structure of 92 genera. Her work included a description of seed and some embryonic tissues with illustrations. The Araceae are notable for presenting a large amount of total monocot seed variation. She produced comparisons of her data against four prominent classifications. Among the conclusions, she noted the genus *Gymnostachys* is very isolated and most resembles Typhales rather than Araceae. Also Lemnaceae should clearly be a tribe within the Araceae. Later, Seubert (1997b) studied seeds of the 10 genera and 18 species of the tribe Lasieae; these seeds are distinguished from those of all other aroids by being kidney- or horseshoe-shaped, with a warty or smooth surface, and with a curved or straight embryo. Because of the variation, morphoclines of these data must be interpreted carefully since the trends are not synchronized or unidirectional. Takhtajan (1985) described the seed anatomy of many genera.

Grayum (1991) surveyed the systematic embryology of the family. Johri *et al.* (1992) reviewed the systematic embryology of the family, as did Seubert (1993), who demonstrated a close relationship between the Araceae and Lemnaceae; the authors provided a comparison with the Helobieae.

PALAEOBOTANICAL RECORD

The most recent review of the aroid fossil record is found in MBB and additional reviews are given in Gregor and Bogner (1984, 1989) and Grayum (1990). The following summary will emphasize several interesting aspects of the history of vegetative and reproductive materials.

On the basis of cpDNA sequences in plants, dicot–monocot divergence has been placed at 200 mya $\pm$ 40 my, a suggestion that is also supported by an analysis of nuclear genes encoding 26S and 18S ribosomal RNAs (Li *et al.* 1993). This would suggest emergence of the angiosperm lineage in Triassic–Jurassic time, which considerably predates its appearance in the fossil record. Much of what we wish to understand of the genesis of this distinct family must have occurred within this early period. No pre-Cenozoic aroid remains are known and no evidence is available from the period when diversification of subfamilies would have taken place. Grayum (1990) also noted the paucity of early aroid records. However, by early Tertiary, macrofossils are relatively widely dispersed (Daghlian 1981).

In the subfamily Aroideae we have an example of fossil intermediates that shed some light on generic evolution. Since 1821, the small, floating, aquatic lemnoid genera have usually been segregated as the family Lemnaceae. However, the cpDNA results of French *et al.* (1995) nested *Lemna* clearly within the subfamily Aroideae; in partial confirmation of these results are studies of *Limnobiophyllum expansum*, originally described by Kvaček (1995) from the Miocene of Europe. Stockey *et al.* (1997) recovered more than 200 specimens of *L. scutatum* (Dawson) Krassilov from the Paleocene of Alberta, specimens that include the first known flowering materials. In size the plants are transitional between modern lemnoids and *Pistia*. This study also described the first fossil aralean pollen, a monoporate, spheroidal grain with echinate spines. It is similar to grains previously described as *Pandaniidites* (Elsik 1968) and which resembles pollen of extant *Lemna* (see Hotton *et al.* 1994).

The cladistic analysis by Stockey *et al.* (1997), based on 34 vegetative and reproductive structures, nested *Limnobiophyllum* in a position between *Pistia* and the four lemnoid genera in a pectinally branched tree. This suggests an unexpected closeness of *Pistia* to the Lemnaceae. However, the cpDNA analysis separated *Pistia* from the lemnoid genera; they are paraphyletic rather than adjacent in the same clade. Further, if *Pistia* and the lemnoids are independently embedded within the Araceae, the family Lemnaceae cannot be recognized without making the Araceae paraphyletic. I would note that the morphological character list used by Stockey *et al.* (1997) is an excellent structural list and the character polarities are most probably interpreted correctly. However, these reduction trends are potentially homoplasious events of such character that homologies are very difficult or impossible to verify. All evidence at this time favours the position of including these reduced floating lemnoid genera within the Araceae.

The leaf record was reviewed by Daghlian (1981) and by Dilcher and Daghlian (1977). The family's record dating from the Paleocene is good. Specimens known as *Nityophyllites* from the Paleocene of Kazakhstan consist of fragments that seem araceous based on venation and cuticular features.

One of the interesting disjunctions among extant aroids is that of *Typhonodorum* of Madagascar and the apparently related *Peltandra* of eastern North America. This disjunction can only be understood if one or both genera were widely dispersed at one time, as now appears likely based on the following evidence. In 1977, Dilcher and Daghlian described a large well-preserved leaf from the Middle Eocene of Tennessee which they named *Philodendron limnestes* and placed in section *Meconostigma*, with which it seemed to agree in venation and cuticular features. Later, Mayo (1991) suggested that the specimen would be better placed near *Typhonodorum*, tribe Peltandreae. Hickey (1977: 114) described *Peltandra primaeva*, based on venation, from early Eocene rocks of North Dakota. Therefore, it appears quite likely that these two genera once had congeneric ranges. Wilde (1989) described cuticular features of three araceous leaves from the Middle Eocene of Germany.

Reproductive materials

Sahnipushpam shuklai, a possible spadix from Eocene Deccan Intertrappean of India, has uncertain affinities and may not be araceous. Berry (1930) described *Acorus heeri*, now the spadix form genus *Acorites* (Crepet 1978), occurring in the Middle Eocene of Tennessee. *Araceites hungaricus* Rusky is an inflorescence that can be included in the Araceae (Crepet 1978). Bogner's (1976) reexamination of *Acoropsis* Conwentz showed that *Acoropsis eximia* of the Eocene Baltic Amber is an infructescence that should be assigned to Monstereae.

Permineralized seeds and fruits of the Princeton chert (Middle Eocene) Allenby formation were assigned to Lasioideae by Cevallos-Ferriz and Stockey (1988) and Cevallos-Ferriz *et al.* (1991). Their specimen *Keratosperma allenbyensis* Cev. & Stoc. gen. et sp. nov. appears closely related to *Cyrtosperma* which constitutes evidence of subtropical elements in that flora. These are the oldest known fruit and seed remains of Lasioideae. The authors' table 2 provides a useful list of reproductive characters of Monsteroideae/Lasieae from several sources. They review reports on Oligocene seeds of Monstereae, Lasieae, Pothoideae, and Pistioideae, and a report by Bown *et al.* (1982) on fossil fruit from the Egyptian Oligocene.

Madison and Tiffney (1976) found fruits and seeds in the Oligocene representing Monstereae, Pothoideae and Lasioideae. These tribes are also recorded in Miocene and Pliocene sediments in a variety of Laurasian localities.

In reviewing records of fossil aroid pollen, Muller (1981) noted a report of a *Spathiphyllum* type from the Upper Miocene, and pollen similar to *Ambrosina, Amorphophallus, Arisarum, Holochlamys, Pseudodracontium* and *Thomsonia.* Graham (1976) reported a *Spathiphyllum* type in the Upper Miocene of Mexico as did Leopold (1969) from the Miocene of Palau (W. Pacific). At least four modern subfamilies were present by Oligocene times.

CHROMOSOMES

The family was first widely surveyed for chromosome counts by Marchant (1970, 1971a,b, 1972, 1973), who accounted for the existing literature and added many new counts. Petersen (1989) wrote a comprehensive review for the family and evaluated all published counts; new numbers were later added (Petersen 1993a,b). The most updated report with all dubious counts appropriately marked was provided in MBB with new information supplied by Petersen; counts are missing for only four genera. These counts are included in this volume.

Palaeopolyploidy is rampant in the family and the original basic number is rare or non-existent, perhaps occurring in *Ulearum* or *Filarum* ($n = 7$). The presumed basic (x) numbers as listed for the genera in MBB are 10–21 (continuous series), and 27. One or more genera with an extensive aneuploid or polyploid series occur in most subfamilies, making the cytological history of the family complex and not well understood.

BIOGEOGRAPHY

For a large, overwhelmingly tropical family, the nearly cosmopolitan presence of Araceae is unusual. The distribution of genera, subfamilies, and tribes has been treated by Willis (1949), Croat (1979), Grayum (1984, 1990), Bogner (1987) and MBB, and noted with generic treatments in this volume.

Most of the nine subfamilies are widely dispersed in the tropics and southern hemisphere including Pothoideae (including Monstereae), Lasioideae, Philodendroideae, Schismatoglottidoideae, and Aroideae. The Gymnostachydoideae (*Gymnostachys*) are eastern Australian, the Orontioideae (three genera) and Calloideae (*Calla*) are northern. The Lemnoideae are cosmopolitan aquatics.

At the generic level, the majority of tropical ranges are restricted to major floristic regions, rather than cosmopolitan. Figures given for tropical genera are Asian tropics: 44 genera; Neotropics: 36 genera; Africa: 20 genera. Neotropical genera have a higher percentage of endemic ranges and significantly more species.

The distribution and subfamilial structure of the family confirms its great age and the point of origin, or points of origin, of most subfamilies has not been solved. In their analysis of floristic origins, Raven and Axelrod (1974) considered the Araceae to be 'West Gondwanalandic-Laurasian'. However, the possibility of an east Gondwanan, or perhaps Laurasian origin has not been ruled out and will be discussed below.

An analysis attempting to explain the family's contemporary distribution, as well as the fossil record, has been provided by Grayum (1990) and will be briefly summarized here. The distribution of Engler's Lasieae and Monstereae strongly suggests that the family was well-established in west Gondwanaland before the separation of Africa and South America. However, Grayum pointed out that this inference does not establish west Gondwanaland as a point of origin for the family, and he proposed that the family originated in temperate or subtropical Laurasia. Later, it reached west Gondwanaland via Eurasia, perhaps in late Cretaceous whence it underwent a major radiation. In support of this concept, he noted that the four primitive (long-branch) taxa *Calla, Lysichiton, Orontium* and *Symplocarpus* are north temperate, and that there are no south temperate genera, unspecialized or not.

By this scenario, elements of this flora migrated secondarily into Eurasia and some perhaps reached North

America (*Typhonodorum, Peltandra*, and perhaps *Orontium*) via a north Atlantic route, this having been open prior to the Eocene (Raven and Axelrod 1974; Wolfe 1975). Confinement of *Peltandra* and *Orontium* to eastern North America is perhaps due to the barrier presented by the north–south epicontinental sea in the great plains region that lasted up to the Paleocene.

The final element of this proposal is that Tethyan distribution of certain Areae (Engler's Arinae) may reflect a Laurasian origin of the group. These taxa may have undergone a radiation on the northern margin of that sea extending from western Europe to Indo-Malaysia. Also included were the genera *Ambrosinia, Ariopsis, Arisaema, Arisarum, Pinellia, Pistia* and perhaps Thomsonieae. As evidence, Grayum (1990) put forward the poor representation of these groups in south-east Asia, Africa, and their absence in the Neotropics. He hypothesized that *Arisaema* only recently reached North America from Eurasia.

However, Hay (1992b) disagreed with the concept of a Laurasian origin for the Areae and he endorsed Riedl's (1980) suggestion of a Gondwanan origin for the tribe. Hay described *Lazarum*, new for the Areae, and considered it to be an autochthonous element of the Australian flora. Later, French *et al.* (1995) contributed an important piece of evidence supporting the idea of an east Gondwanan origin for the Areae. The six genera of Areae sampled for their cpDNA study are arranged topologically in a strict pectinal relationship, which, from base of the clade to distal-most are arranged as follows: *Typhonium* (south-east Asia, Australia), *Theriophonum* (India), *Sauromatum* (Africa, India), *Biarum* (Mediterranean), *Arum* (Mediterranean), and *Helicodiceros* (Mediterranean). *Lazarum* is considered most closely related to *Sauromatum*. This supports Hay's (1992b) idea that the tribe is southern in origin. He suggested that the progenitors of the northern genera, already seasonally adapted, were rafted northward on the Indian plate. The evidence certainly makes it less likely that the tribe is northern in origin. Li (1986) also argued for a south Asian (southern Laurasian) origin for the family.

CHEMISTRY

Chemical compounds in the family were surveyed by Hegnauer (1963, 1986) and briefly reviewed in Hegnauer (1987). The family is quite active in the production of secondary compounds. Among those widely detectable throughout the family are alkaloids and amines, saponins, cinnamic acids, and flavonoids. The latter are diverse enough to make some contribution to understanding classification in the family. Allergenes capable of causing contact dermatitis are present in some *Philodendron* species. Proanthocyanidins (condensed tannins and their precursors) are found in about half of the genera sampled. Essential oils, abundant in the outgroup, Acoraceae, are only known from two species of *Homalomena* within the Araceae. Cyanogenic glucosides, especially prussic acid, are found in many genera. They can produce enough HCN to be toxic. Finally, carbohydrates are abundant and are found as either mucilaginous glucomannans, or starch, or both.

Laticifer contents are chemically complex, as usual for this cell type, but they show systematically useful variation in the family that should be pursued (Fox and French 1988). New World 'colocasioids' *Xanthosoma* and *Syngonium* (two species each) have latex sterols which seem to vary systematically by genus. The Old World genera *Colocasia* and *Alocasia* (two species each), contain sterol esters, with no sterols detected (Fox and French 1988). The authors suggested that the antiherbivory hypothesis needs to be investigated regarding the role of sterols and related compounds.

A study of leaf flavonoids and anthocyanins of 61 genera of Araceae was made by Williams *et al.* (1981). Other studies of chloroplasts or their ontogeny include those of Guillot-Salomon *et al.* (1978), Chow *et al.* (1988), Stewart and Dermen (1979) and Buinova (1988). Bate-Smith (1968) found representation of a wide list of phenolic compounds showing no obvious relation to ecology or systematics in the family. Later, an analysis of the systematic distribution of tannins in angiosperm leaves was accomplished by Mole (1993). However, the systematic versus ecological significance of tannin presence has not been resolved.

The phenolic constituents of cell walls when studied by fluorescence microscopy (UV) form two major groups within the monocots (Harris and Hartley 1980), and the authors compared the results with Cronquist's (1968) classification. The Araceae (in group 2, three genera sampled) tested negative for the four wall extracts examined in this study, as did most of the Alismatidae. *Acorus* differed from the Araceae in testing positive for p-hydroxybenzoic acid. Groups reacting positively (group 1) included most of the Commelinidae and many Liliidae. Also, Harborne (1974), surveying flavonoids, found four monocot groupings. The one group having both flavonols and glycosyl-flavones comprised Araceae, Lemnaceae, and Commelinaceae.

Dring *et al.* (1995) surveyed chemistry in 55 genera from all aroid subfamilies. They stated that the family's chemistry is highly diverse and only a small part has been investigated; they studied alkyl resorcinols and polyhydroxy alkaloids. The various types reported showed some correlation with classification. In particular, the close relationship of *Anchomanes* to *Nephthytis* and *Pseudohydrosme* was strongly supported by the alkaloid evidence. Schmid *et al.* (1997) made a survey of seed lipids of 54 species representing 20 genera of the family. Using GC mass spectrometry, they reported finding 13-phenyltridecanoic acid (13-PTDA) in the seeds of seven genera of aroids, a first report from any plant family of a lipid with a substituted aromatic phenyl ring. All genera with this compound are in subfamily Aroideae (*sens. str.*).

Thermogenesis

A metabolic process that generates heat in inflorescences has been observed and studied for two centuries (Bay 1995). At least seven apparently unrelated thermogenic families are known but genera of Araceae provide the best studied examples. The process is complex, energy intensive, and apparently allows volatilization of odours that attract pollinators. The family shows at least four patterns of expression (Leick 1915). These vary from a generalized heating of the spadix for as long as two weeks at 35 °C above ambient in the cold season-blooming *Symplocarpus*, to a process in *Arum* where four distinct inflorescence zones volatilize distinct compounds during short but intense periods of heating. The increase in metabolism that accomplishes thermogenesis is triggered by salicylic acid (Raskin *et al.* 1989) and made possible by a very efficient electron transport process called the 'cyanide resistant pathway' (Meeuse 1975). The resulting high energy cost to the plant is thought to be justified on the basis of high pollinator specificity and pollination timing, yielding better genetic fitness of the seeds (see reviews by Bay 1995, and in MBB).

AROIDS AND HUMANS

In a review of aroid morphological variation, Bogner (1987) also surveyed economic uses. A chapter on the topic is included in MBB, and the family is noteworthy for its economic value. Many of the genera are important ornamentals, either as foliage plants, or based on their white or colourful inflorescences. Especially prominent ornamentals are species of *Aglaonema, Alocasia, Amorphophallus, Anthurium, Caladium, Colocasia, Dieffenbachia, Monstera, Philodendron, Spathiphyllum, Xanthosoma*, and *Zantedeschia*, but many others are also collected and grown for display throughout the world.

Equally important is the cultivation of several genera for their edible starchy tubers, corms or rhizomes. They are very valuable carbohydrate sources in many tropical nations and several taxa have been carried far from their original ranges. The two most important species, *Colocasia esculenta* (taro of south-east Asia), and *Xanthosoma sagittifolium* (South American yautia or cocoyam), are the source of ancient cultivars. They are well-established staple crops throughout the tropics. More locally, some species of *Alocasia, Amorphophallus paeoniifolius (campanulatus)* (elephant yam), and *Cyrtosperma chamissonis* (giant taro) are also cultivated for starchy tubers. *Monstera deliciosa* (ceriman), often grown as an ornamental, is prized in various places for its edible fruit, said to taste like pineapple. It is eaten only fully ripe to avoid the raphides and other irritants. The seeds of *Typhonodorum* (Madagascar) and *Montrichardia* (South America) also provide food.

Various lemnoid species are used as high-protein plant food for humans and domesticated animals, fish, and shrimp. They have a very favourable amino acid composition and high productivity, up to 40 t dry weight per hectare (Landolt 1998). In Thailand, *Wolffia globosa* is produced from pond culture as a vegetable for humans (Bhanthumnavin and McGarry (1971), and various species of other genera are being evaluated for larger scale production.

Purseglove (1972) and Coursey (1968) have discussed the edible aroids.

Regarding edibility, with very few exceptions, aroid parts must be cooked or heated before eating. Probably the large majority of species, including the 'edible' ones listed above, have a powerful irritant in virtually all plant parts. It is certainly associated with the ubiquitous raphide crystals, which penetrate soft epithelial surfaces of the gastro-intestinal tract. In the case of *Xanthosoma*, the barbed raphides will magnify mechanical damage (Sakai *et al.* 1972). However, Walter and Khanna (1972) argued that proteolytic enzymes (named 'dumbcain') work together with raphides. The combined mechanical and chemical damage probably activates kinin-releasing mechanisms in the body, triggering the inflammatory response. *Dieffenbachia* is commonly called 'dumb cane' since chewing on the plant parts may cause swelling of the throat sufficient to prevent speech and possibly cause death due to a blocked airway. The attractive seeds of *Arisaema* are also notorious in this respect, as are the leaves of *Pistia stratiotes*, 'water lettuce' (John MacDougal, personal communication). Raw araceous plant parts should be considered dangerous to humans in spite of the fact that uncooked starchy rhizomes are fed directly to livestock in various parts of the world.

Plowman (1969) and Bogner (1987) described folk uses of aroids, with examples as follows. In tropical Asia,

Typhonium blumei is used as a cure for diarrhoea. Schultes (1963) stated that a number of native tribes in north-western South America reported the use of parts of certain species of *Anthurium*, *Philodendron*, and *Urospatha* as female contraceptives. Other miscellaneous examples include basket-weaving materials (*Heteropsis spruceana* roots) and arrow poisons. See also the discussion in Bown (2000) and the listing in Duke and Vasquez (1994).

REVIEW OF VEGETATIVE ANATOMICAL LITERATURE

This review necessarily omits discussion of numerous contributions listed in the bibliography. The present intention is to provide an overview of diverse contributions and some other characteristics of the family.

General sources

Numerous anatomical details of the family are found throughout the monographic treatments of the family by Engler, often with Krause (1905, 1908, 1911, 1912, 1915, 1920a,b). Engler frequently used anatomical features in defining major subdivisions of the family. Solereder (1919) reviewed the anatomy of about 23 genera of aroids and Solereder and Meyer (1928) compiled observations on the general anatomy of about half the genera of Araceae, together with a compilation of early literature, although the treatment is not particularly systematically oriented.

The history of systematically oriented research on Araceae was reviewed by Croat (1998). In spite of his stated emphasis, Croat's comprehensive treatment covers history and influential workers, regional studies, as well as anatomy, cytology and other disciplines. One of the best general reviews of this or any family is MBB which provided a detailed account of essential literature and all aspects of classification, morphology, geography, and illustrated generic concepts. In an introductory chapter to MBB, French (1997) reviewed anatomical literature and gave a comprehensive, though not illustrated, summary of the anatomy of all organs and tissues. Grayum (1984) also reviewed anatomical literature in his thesis emphasizing palynology. Other recent surveys include Dahlgren and Clifford (1982), Dahlgren *et al.* (1985), and an abstract of *The genera of Araceae* by Mayo *et al.* in Kubitzki (1998).

The most prolific researcher in systematic anatomy of the family has been James French, whose work, often with collaborators, has covered many vegetative and reproductive tissues in a lengthy series of papers (1977–88). They are noteworthy for their wide generic sampling and are reviewed below.

Tissues

Collenchyma

The first account of collenchyma strands in the family was by Ambronn (1879–81). He reported that collenchyma development begins in the epidermal region in Engler's Aroideae and Amorphophallineae and is of subepidermal origin in *Colocasia esculenta* petioles with stranded collenchyma. He called attention to ring-forming (=banded) collenchyma in *Philodendron eximium*. Ittenbach and Boecker (1997) described the inflorescence axis anatomy of 13 African *Amorphophallus* species and commented on the support provided by collenchyma. However, the first surveys of this tissue and its trends in the family have only recently been published (Gonçalves *et al.* 2002; Keating 2000).

Sclerenchyma

Sclereids have long been observed in araceous tissues and were summarized first by Solereder and Meyer (1928) who listed the genera bearing trichosclereids. These cells, growing among ground tissue, can be branched, knobbed, or H-shaped, and are recorded at lengths of up to 1500 μm. They have only been reported from nine genera in the subfamily Pothoideae of Grayum (1990) and Keating (2002b; this volume) and have been described by Bloch (1946), Rao (1954, 1964, 1977, 1991), Nicolson (1960), Singh (1968) and Hinchee (1983); within the nine genera they vary from rare to numerous in one or all organs.

Later, Seubert (1997a) provided observations and a review of sclereid occurrence in the family and their distribution within organs. She reported them present in eleven genera and argued for acceptance of the subfamilies Monsteroideae and Pothoideae and perhaps raising *Spathiphyllum* to that level of recognition. However, note that the present study only accepts Pothoideae for the 16 genera (see discussion, p. 23). Brachysclereids occurring at the central cylinder boundary in *Pothoideum* were not mentioned in Seubert's review. Fibres in roots possessing isotropic layering were investigated by Mueller and Beckman (1979) (see roots, p. 16).

Xylem

Ko *et al.* (1990) looked at leaf sheaths, petioles, petiolules and roots of seven Korean spp of *Arisaema*, and found that all xylem is composed of tracheids. Tracheid walls are variously thickened and show no trend; examples from several genera, including *Pistia*, *Colocasia*, and

Scindapsus, are described in detail. A survey of xylem for the presence of vessels has been accomplished by Carlquist and Schneider (1997: Acoraceae; 1998: Colocasioideae: five genera) and Schneider and Carlquist (1998: Philodendroideae: five genera) as part of a series of studies of monocots. All genera studied have vessels in roots, as previously reported, and most genera have less specialized vessels in rhizomes (stems) as well. The absence of pit membranes, diagnostic for vessels, was confirmed by the authors using SEM. In the genera *Anthurium* and *Zantedeschia* (one species each studied), it could not be confirmed that stem vessels were present. No systematic meaning is clear but the Araceae have reached a higher level of physiological efficiency than if only tracheids were present.

Phloem

There are very few histological descriptions of phloem in the family. Shah and James (1971) described the ontogeny of sieve elements and companion cells of stolons in *Pistia*. Sieve mother cells produce sieve elements and companion cells. The latter continue to divide, producing up to 20 companion cells along a mature sieve element in up to two files. Sieve elements are short with oblique or transverse end walls having simple sieve plates, and containing protein bodies. In *Dieffenbachia* sp., Parthasarathy (1980) described phloem as being up to 4 years old and little or no P-protein was detected in mature sieve elements. It was noted that sieve tubes in monocot axes appear to function throughout organ life span. Behnke (1969a,b,c) described phloem contents in monocots and Behnke (1972) found that all sampled monocot species have exclusively P-type plastids in their sieve elements.

Laticifers

Early descriptions of laticifers in leaves were produced by Hanstein (1864), Trécul (1866, 1866–70), and Scott (1889). Tieghem (1872) described secretory tissue for seven genera. More recently, Pant and Kidwai (1966) described secretory canals and laticifers in the family. Articulated laticifers of the anastomosing type are said to begin as separate cells which then start lobing. Elongating tubular branches meet similar branches from adjacent cells and 'fuse by dissolution of the common walls to form articulate tubes'.

French (1988) surveyed about 75 genera of Araceae for laticifer presence. He noted that most genera have the non-anastomosing type. He concluded that the anastomosing type characterized the Colocasioideae (except *Ariopsis* which has non-anastomosing laticifers), which by implication were monophyletic. Later, the cpDNA study by French *et al.* (1995) implied independent (homoplasious) origins of Old World and New World anastomosing types. This begs some interesting evolutionary questions that have not yet been investigated.

Vágújfalvi (1971) recorded alkaloids as present in latex of five genera of Araceae and Fox and French (1988) investigated the contents of laticifers in a larger generic sample. Typical of laticifers in general, their contents included polyisoprene hydrocarbons among other compounds, and the authors found considerable variation across the family in latex chemistry, much of which could be systematically useful. Most groups have clear latex but milky latex characterizes *Dieffenbachia*. New World colocasioids possess sterols and sterol esters, which are also common in dicots.

Secretory tissue and glands

This topic subsumes a variety of different cells and tissues.

Tieghem (1872) described secretory tissue in seven aroid genera. He noted that the presence of laticifers and canals with secretory epithelium in the same organ is very uncommon among flowering plants. Belin-Depoux (1978) has described small circular glands near vascular bundles occurring on leaves of *Alocasia;* they consist of superficial columnar epidermal cells underlain by parenchyma. Later, Belin-Depoux (1989) described foliar hydathodes or foliar nectaries in *Alocasia*. Lindorf (1980) described secretory glandular cavities in the adaxial epidermis of *Anthurium bredemeyer*i that were 4–5 times the size of epidermal cells. Schnepf (1965a), in a TEM investigation, described the secretion of odorous eccrine type oil by the spadix apex of *Typhonium divaricatum*.

Cell inclusions

Starch

Reserve photosynthates in the family are quite variable. While of probable systematic significance, they have not been subject to any broad survey. Starch granules, certainly the most common reserve, were examined by Reichert (1913, 1919). He described physical, optical and chemical properties, with illustrations of starches of seven genera of Araceae. Observations of starch of seven araceous genera have been recorded from corms and rootstocks by Allen (1929), Winton and Winton (1935, 1945), and Fujimoto *et al.* (1990). Sakai and Hayashi (1973) recorded that nine aroid genera store carbohydrates as sugar rather than starch. The sugar-reserve genera are scattered throughout four subfamilies and Acoraceae.

Crystals

Nearly all of the known calcium oxalate crystal configurations, raphides, druses, prismatics, and crystal sand, have been reported in Araceae (Genua and Hillson 1985), and their presence was confirmed in this study. Among monocot families such diversity is only found in Araceae (Prychid and Rudall 2000). In the earliest anatomical study on the family, Turpin (1836) described a unique and specialized form of raphide cell, since then referred to as the biforine. This thick-walled, spindle-shaped cell has papillate ends capable of firing a stream of raphides out at either end as described by Middendorf (1968, 1982, 1983). Among systematic accounts of crystals are those of Buscalioni (1895–6: four genera), Solereder and Meyer (1928: 16 genera), Genua and Hillson (1985:14 genera), and Keating (2002a: family).

Development and chemical analysis of araceous crystals have been investigated by Pfeiffer (1925), Wakabayashi (1957a), Mollenhauer and Larson (1966), Pant and Kidwai (1966), Ledbetter and Porter (1970), Al-Rais *et al.* (1971), Rakovan *et al.* (1973), Sakai and Hanson (1974), Cody and Horner (1983), Genua and Hillson (1985), and Traquair (1987). Franceschi and Horner (1980) provided a thorough review of the topic. Where the chemistry has been examined, raphides have been found to be monoclinic calcium oxalate monohydrate (Whewellite) while druses are mostly calcium oxalate dihydrate (Weddelite). Sakai *et al.* (1972) and Cody and Horner (1983) described deep grooves in raphide surfaces of *Colocasia, Alocasia* and *Xanthosoma;* such crystals may also have backwardly oriented surface barbs which would certainly increase damage to mouth tissues when plants are grazed upon.

Investigation into what, in addition to raphides, causes the irritating effects experienced when chewing or swallowing aroid tissues was undertaken by Saha and Hussain (1983). They studied varieties of *Colocasia* and *Alocasia* and concluded that a non-volatile glycoside of 3, 4-dihydroxybenzaldehyde is the irritant. It is heat-labile; cooked or dried samples are considerably less acrid.

Silica

Netolitzky (1929) listed no Araceae as having siliceous bodies but Gertz (1916a,b) reported them as present in *Pothos ventricosus*. None was found in the present study.

Protrusions, pectic materials and warts, slime cells, mucilage

The chemistry and systematic importance of these materials is not well understood at present and they should be further investigated. Morassi Bonzi and Fabbri (1975) described chloroplast (stromatic) protrusions in *Arisarum, Arum, Dracunculus* and *Zantedeschia*. While there often appeared to be a close relationship between protrusions and groups of mitochondria, no function for them has been demonstrated. Such protrusions were frequently encountered in the present study on plastids and nuclei. Kisser (1928) recorded pectic warts within mesophyll cells of *Arum* and *Dieffenbachia*. Also, slime cells have been found suspended from neighbouring cells into ground tissue air cavities in *Dieffenbachia* (Kisser 1928; Lindorf 1980; Potgieter and Van Wyk 1992). A larger survey of intercellular pectic strands in leaf palisade parenchyma (Carr *et al.* 1980b) showed them to be present in *Arum sanctum*. Such strands apparently form the principal connection between palisade cells. As they occur at the limit of resolution of light microscopy, they were not confirmed in the present study. Carr *et al.* (1980a) also studied intercellular strands associated with stomata. Wakabayashi (1957b) reported on mucilage present in *Amorphophallus*.

Leaves

General compilations of aroid leaf anatomy were first made by Tieghem (1866), de la Rue (1866), and Dalitzsch (1886). The early work was summarized by Solereder (1919) who described the general leaf anatomy, stomata, crystals, and secretory tissue of 23 genera of aroids. Solereder and Meyer (1928) compiled observations on the general anatomy of about half the genera of Araceae, including all types of crystals; the treatment is not systematic in organization, nor accompanied by tables or index.

Ontogeny

Leaf ontogeny was described by Troll (1932b) for *Eminium, Helicodiceros*, and *Sauromatum*. He described shield-form leaf ontogeny (1932a) in *Alocasia, Ariopsis, Arisaema, Dracunculus, Orontium, Pothos, Remusatia*, and *Syngonium*. Troll and Meyer (1955) studied unifacial development in 11 genera of aroids. Marginal growth in *Philodendron oxycardium* leaves was examined by Pray (1957), who noted that superficial initials rarely if ever contribute to internal leaf tissue. Kaplan (1970, 1973a,b) made a series of studies of events occurring during shoot expansion of *Acorus*, providing observations more detailed than available for any genus of Araceae. Murata (1990a) investigated the morphological development of pedate leaves in seven genera of Araceae.

The phyllode theory

Arber (1918a) elaborated DeCandolle's phyllode theory of the monocot leaf and proposed that the lamina of bifacial leaves, such as found in the Araceae, represents the secondary expansion of the petiolar region. Kaplan (1973b) refined this theory by adding developmental observations of monocot dorsiventral leaves including *Zantedeschia aethiopica*. As in other monocots, leaves of Araceae present leaf primordium segmentation into upper and lower leaf zones, as opposed to the three zones of dicot primordia. However, in *Zantedeschia* and other aroid genera sampled by Kaplan, aroids appear to be unique among sampled monocots in having acroplastic maturation after the lower and upper leaf zones have been initiated. This makes them surprisingly more similar to the dicot maturation pattern than the other monocots sampled. Nevertheless, the dicot lamina always differentiates from the upper primordium zone, and in monocots from the lower primordium zone. The reticulate venation of aroid leaves has to be viewed as convergent with the dicot leaf pattern via a basically different ontogeny (Kaplan 1973b). Therefore, terms from Hickey's (1979) original leaf venation scheme, later modified to include the net-veined monocots (Leaf Architecture Working Group 1999), should be used only in a topographical sense and not suggestive of homologies.

Venation. The first comprehensive survey of leaf venation in the family was by Ertl (1932) who sampled 37 genera, and described the ontogeny for many. He concluded that higher orders of venation are always reticulate except in *Gymnostachys*. Ertl also described vascular bundle configuration in petiole transverse sections and related the patterns positionally to vascularization of the lamina. In general, the dorsal-most petiolar bundle becomes the central vein of the midrib. The lateral or ventral-most bundles diverge into the lamina closer to its base. However, the degree of bundle branching and crossing over (superpositioning) of other veins varies considerably among genera. Ertl held that most genera show basiplastic maturation (but see Kaplan 1973b) and connection of vascular bundles in the base of the lamina and petiole. Where pinnate venation occurs in monocots, as in Araceae, Kaplan (1973b) concluded that lamina construction is basically linear due to the presence of the multistranded 'midvein'. The wider blade is the result of differential expansion between the peripheral and central regions of the blade. Where pinnate secondary veins show an arcuate or diagonal course, they arrive at that orientation by developmental displacement of an originally longitudinal alignment (Ertl 1932; Troll 1939; Kaplan 1973b).

There are aroids with true reticulate higher order vasculature in addition to other typical dicot-like features including areoles, free vein endings and marginal venation. Kaplan (1973b) and Inamdar *et al.* (1983) offered reviews of leaf venation literature, describing in detail venation ontogeny of several genera including the parallelodromous *Pistia* (four to five secondaries along one side of a primary vascular bundle); the basally actinodromous *Colocasia esculenta* (peltate lamina, three primaries diverging from the base; six to eight secondaries along one side of the primary); and the campto-brochidodromous *Scindapsus aureus* (five to six secondaries along one side of the primary). Ertl's (1932) techniques should be revisited. They have the capacity to produce important trends, given the morphological variation in the family and our improved understanding of its classification.

A number of studies have outlined informative vascular patterns of smaller groups of taxa as shown by the following examples. Davie's (1917) early study of venation in *Philodendron selloum* concluded that leaf and pinna traces are related to the vascular system found in the stem in such a way as to be independent of the size of the leaf and its manner of development. Additional venation patterns were described by Knecht (1983) in African (Ivory Coast) *Amorphophallus, Anubias, Cercestis, Culcasia,* and *Xanthosoma*. Other genera treated include *Lemna minor* (Fletcher and Arnott 1963), and *Arum, Monstera, Philodendron* and *Pinellia* among observations on numerous families by Freundlich (1908).

Cuticle and blade surface. Barthlott and Wollenweber (1981) examined the taxonomic significance of epicuticular waxes. A comprehensive inventory of epicuticular waxes in monocots by Frölich and Barthlott (1988) revealed a wide variety of sculpturing, extrusions, rodlets, and irregular particles that have some systematic value. Arales (sixteen taxa examined) were found to have either no wax secretions, or unspecific patterns, or non-oriented platelets (*Alocasia*). Non-oriented types are widespread and not helpful for interfamilial comparisons. There is experimental evidence that cuticle development is influenced by the environment, which suggests that cuticle thickness should be interpreted cautiously (Sutter 1985).

Epidermis. Pfitzer (1872) noted the occurrence of a water-bearing epidermis in *Philodendron* and *Anthurium*, and Mittman (1888) described the spur or prick on the petiole of *Lasia*. Numerous observations of cell shape in surface view are available. Examples include the works of Linsbauer (1930) and Webber (1960) who also observed stomata. Pant and Kidwai 1966) noted that in the genera *Caladium, Schismatoglottis*, and *Zamioculas*, whose polygonal epidermal cells show straight anticlinal walls in surface view, the walls appear sinuous in a lower plane of focus; they described papillate and other forms of cells.

Pant and Kidwai (1966) recorded surface orientation of stomata as mostly irregular or random in all aroids, but longitudinal in *Gymnostachys* and Lemnoideae; stomatal frequency was found to be 18–159 per mm^2, with no mode.

Average guard cell length, stomatal index, and cuticle thickness was also calculated by Pant and Kidwai (1966) and all showed a broad range and mode. Shaw (1992) calculated mean stomatal frequencies for different parts of *Monstera deliciosa* leaves and found broad variances of frequencies depending on leaf form, position and size. In a given leaf, stomatal dimensions vary little, i.e., they belong to a single size class as is typical for monocot leaves (Dunn *et al.* 1965). Fryns-Claessens and Van Cotthem (1973) found Araceae to have no clear boundary between well-defined subsidiary cells and non-distinctive neighbouring cells. While most stomata in the family are brachyparacytic, there are percentages of non-typical subsidiary cell arrangement in most genera observed. Non-typical includes two polar subsidiary cells over guard cells at one end, or at both ends, or tricytic. Grau (1983) looked at cell orientation on upper and lower surfaces of 24 species in 15 genera, and Czaja (1962) brought polarization techniques to bear on guard cell and subsidiary cell orientation.

Developmentally, stomata in most Araceae sampled have been characterized as perigenous by Pant (1965) and Tomlinson (1974). Guard cell ontogeny was observed by Pant and Kidwai (1966) for eleven genera of Araceae. Guard cell initials divide once producing guard cells. An adjacent sister cell, usually unrecognizable from other neighbouring cells, becomes one of the subsidiary cells. However, these authors, as well as Fryns-Claessens and Van Cotthem (1973), found Lemnaceae and *Pistia* stomata to be anomocytic, since perigene neighbours remain undivided. The survey by Stebbins and Khush (1961) considered Araceae, Cannaceae, and Musaceae to have eumesogenous ontogeny. Nyawuane and Gill (1990) found the family to be mostly eumesogenous although eumeso-perigenous ontogeny was reported in *Dieffenbachia* and *Scindapsus*. They found no correlation between ontogenetic type and the presence of straight-walled polygonal or sinuous-walled mature epidermal cells.

Systematic use of stomatal types in subfamilies and tribes should be interpreted cautiously until more is learned of the contribution of ecological versus genetic factors. Numerous informative contributions include those of Porsch (1905), Mashima (1964: *Arisaema*), Grear (1973: *Orontium*), Roth and Merida de Bifano (1979), and Gill and Karatella (1983: *Colocasia*). Stomatal hydathodes have been reported in about 65 species of aroids by Ziegenspeck (1949).

Mesophyll. This tissue has been little studied systematically, although the current survey offers typology and shows some intriguing differences among tribes. The floating leaves of *Pistia* were reported by Sifton (1945) to have 713 parts/1000 air volume in its highly organized aerenchyma. Pant and Kidwai (1966) found that arm-cells are present in the spongy mesophyll in all aroids studied, except *Pistia* which has unlobed cells. Ogden's (1974) study of aquatic plant anatomy included observations on the leaves of four aroid genera; nearly all genera in the study have type 3 aerenchyma (classification of this volume). Evolutionary trends are not well understood regarding aerenchyma development.

Specializations, ecological anatomy. Miscellaneous studies have addressed ecological variation. An altitudinal gradient study of *Xanthosoma* by Korner *et al.* (1983) in New Guinea provided data on leaf size, stomatal frequency, cuticle thickness and cell-wall thickness. But the limits of variation are not well-understood for most systematic anatomical investigations. Lee and Richards (1991) noted substantial age-related variation when describing *Syngonium* and *Monstera* leaves, and Sutter (1985) found environmental-based variation in cuticle studies. Kasapligil (1961) investigated the seasonal geophytic *Arum orientale* of the Anatolian steppe flora and found that leaves showed very little anatomical variance when compared to normal mesophytic anatomy.

Sun/shade leaves

Lindorf (1980) examined shade leaf anatomy in *Anthurium*, *Dieffenbachia* and *Monstera*, and Inaba (1984) studied the effects of shading on the leaves of *Amorphophallus konjac*. Araus *et al.* (1986a), who included *Philodendron scandens* in a larger study of shade plants, noted that the internal mesophyll area in this C_3-photosynthesis type leaf is correlated with the light-saturated photosynthesis rate. Shade leaves were often found to have low palisade volume but higher spongy tissue volume, which caused the shade leaves to be thicker rather than thinner than sun leaves in this species. Chloroplast density was found to be lower but the chloroplasts larger. These findings imply that volumes and proportions of tissues, and air cavity space, would be difficult to use systematically. However, the typology and unusual cell shapes noted in the present study probably have some systematic utility.

Perforations (fenestration)

Trécul (1854) made the first descriptions of the ontogeny of leaf perforations in *Pothos* and *Monstera*. Later Melville and Wrigley (1969), studying *Monstera deliciosa*, determined that small necrotic spots occur in intercostal tissue when primordia are *c.* 14 mm in length. The spots expand toward the margin becoming sinuses or perforations depending on their original position. (See also Melville 1971.)

Variegation and colouration

Hara (1957) studied variegation in leaves of 55 species in 24 families and determined four types to exist. *Arisaema* was found to be of type II, having air spaces just beneath the lamina epidermis. Coloured leaf mottling in *Arum maculatum* is accompanied by a thinner lamina and more anthocyanin (Pethybridge 1903), a type IV variegation (Hara 1957). Hara's type I, having white streaks due to chlorophyll deficiency, has been found in *Zantedeschia* (Saxton 1913) and in six other genera by Bergdolt (1955). In white areas, the mesophyll is not differentiated and thinner than in adjacent green tissue, or may be normally developed. Araus *et al.* (1986b) have described the structural differences between green and white sectors of variegated leaves in *Scindapsus aureus*. Optical properties of the whitish segments of variegated leaves of *Dieffenbachia* sp. and *Spathiphyllum wallissii* were measured by Eller and Flach (1991); they found that absorption of energy took place in proportion to the amount of chlorophyll present. A number of studies reported investigations on the genetic basis of variegation including those of Zettler and El-Nil (1979) and Henny (1983a, 1986).

Foliar glands

In leaf tips (Vorläuferspitzen), Volkens (1883) described water excretion as a liquid in *Calla* and Gentner (1905) observed the structures in eleven genera of aroids. Most genera have a cylindrical hooded tip with the abaxial surface oriented outward on all sides. Guttation in *Anthurium* occurs through a large stoma with approximately seven subsidiary cells, not paracytically arranged. *Acorus* leaves have a basically different leaf tip with slime papillae (Gentner 1905).

Petiole

Ertl's (1932) ontogenetic study of araceous leaf venation remains the most thorough to date concerning the relationship between petiole and lamina venation patterns. He determined that the apparently scattered pattern of petiole venation, as seen on a transverse section, is related to lamina venation in a precise way. Early in development, in all genera examined, the young lamina shows a more typical monocotyledonous parallel venation. In some cases the medial to marginal relationship of petiole veins to the lamina is predictable and parallel (e.g. *Cryptocoryne, Orontium, Philodendron*). In others, the veins may cross over each other and branch more frequently (e.g. *Anthurium, Arum, Typhonodorum*) producing a more complex dorsiventral layered vascular development.

Niklas (1991) measured the elastic modulus of a *Spathiphyllum* petiole and found that values increase as the petiole dehydrates.

Stems

Investigations on stem anatomy constitute the largest body of work on aroid organ structure, much of which was accomplished by J. C. French, often with collaborators. Fisher and French (1978) investigated stem growth in 1–3 species of *Aglaonema, Anthurium, Pistia, Spathiphyllum*, and *Zantedeschia*. The two types of meristems found are uninterrupted meristems (UM) and intercalary meristems (IM). In the former, cell divisions are confined progressively to the distal region of a developing internode, causing acropetal tissue maturation. In the latter, localized regions of cell divisions occur in the lower (proximal) regions of developing internodes; maturation is basipetal. Frequently, species with sheathing leaf bases have IMs, while species with UMs have stems unprotected by leaf bases.

French and Tomlinson (1980) began studies in stem vasculature with a broad generic sample including a species each of *Culcasia, Epipremnum, Philodendron*, and *Syngonium*. Basing their approach on Tomlinson's earlier work on palm stem nodes and internodes, they characterized the vascular patterns and the resultant changes in vascular bundle histology as bundles change position in the stem and node. They concluded (see also French and Tomlinson 1986) that compound bundles are temporarily juxtaposed separate strands, and that aroid vasculature is flexible, often irregular, i.e., less predictable than the *Rhapis*-type of palms. This reflects the morphological diversity in the family (Tomlinson 1984).

Following this work, French and Tomlinson (1981a,b, c,d, 1983, 1984) studied the vascular patterns of a much larger sample of araceous stems, organized by subfamily, with the final paper dedicated to *Philodendron* alone. Another contribution (French and Tomlinson 1986) reviewed compound vascular bundles in monocotyledonous stems. They found bundles in aroids to be only superficially similar to those in the compared families Cyclanthaceae and Pandanaceae. Compound bundles were reported in *Cercestis, Dieffenbachia, Montrichardia, Philodendron, Rhodospatha, Stenospermation*, and *Zamioculcas*.

As reviewed by French (1997, in MBB, p. 14) a demarcation of the inner boundary of a stem cortex by an endodermis has been reported in *Amydrium, Anthurium, Orontium, Monstera, Pistia, Rhaphidophora, Scindapsus, Spathantheum, Symplocarpus*, and in the tribes Peltandreae and Schismatoglottideae. It was not encountered in most genera (French and Tomlinson 1980, 1981a–d, 1983, 1984). Van Fleet (1942) studied the endodermis and oxidase system in hundreds of monocot genera, including the centripetal rate of deposition of endodermal walls against certain variables. Casparian strips were found to originate as extracellular deposits first formed in intercellular space. Radial cell growth caused casparian strips to take a position between radial walls. Van Fleet (1942) observed four conditions in Araceae: (1) endodermis present in horizontal and erect underground stems: *Symplocarpus*, and also *Acorus* (Acoraceae); (2) endodermis in aerial portions of erect, prostrate or decumbent aerial stems and in culms and scapes as cylinder endodermis: *Scindapsus pictus*; (3) endodermis absent in sub-aerial stems which do not grow up to the ground line: *Arum* (3 spp.), *Colocasia esculenta*; (4) endodermis absent in aerial or sub-aerial stems, culms, scapes, or leaves: species of *Aglaonema, Amorphophallus, Arisaema, Peltandra, Philodendron*, and *Symplocarpus*. Wilson and Peterson (1983) made observations on the additions of lignin or suberin to the endodermis or other tissues.

Marlatt (1970) described wound cork formation from stem cuttings of *Dieffenbachia*, documenting eventual formation of a full periderm. Phellem tissue was shown by Keating (unpublished results) to bisect damaged *Epipremnum* stems isolating necrotic lesions. Additional literature on cicatrice formation, lenticels or periderm is found in Klebahn (1884), Muhldorf (1927), Bolli (1953), and El-Hadidi (1969).

Several miscellaneous studies exist on tuber and corm anatomy but no larger survey. *Pinellia cordata* tubers were found to possess a central ring of amphivasal vascular bundles surrounding a circular phloem strand (Higuchi and Okada 1980). The ground tissue contained enlarged circular cells each bearing a single raphide bundle, or multiple bundles in a clumped arrangement; the outer surface was protected by periderm. Ohtsuki (1960) investigated tuber reserve photosynthate, mostly of *Amorphophallus rivieri*. In that species the tuber ground tissue consisted of large cells each containing one large mucilage grain, 0.5–1.0 mm diameter, composed mostly of mannose and some d-glucose; small cells surrounding them were filled with small compound starch granules. It appears that tuber structure and its storage tissue can be quite specialized; it is sufficiently interesting to warrant a general survey. See additional contributions by Paliwal and Kavathekar (1972), Anandakumar *et al.* (1982), and Hather (1993).

Roots

All roots of araceous genera investigated are adventitious. They are usually divided among aerial or terrestrial, and Hinchee (1981) reported structural transitions as aerial roots enter soil in *Monstera*. The basic features of aerial and transitional roots are found to be quite similar and they are not reliably distinguished microscopically.

Aerial roots were first characterized for 20 species in the genera *Anthurium, Calla, Philodendron, Rhaphidophora* and *Scindapsus* by Olivier (1881). Goebel and Sandt (1930) made similar observations as did Went (1895) for seven genera of climbers and epiphytes. Sinnott and Bloch (1946) and Rakovan *et al.* (1973) recorded observations for *Monstera*, as did Napp-Zinn (1953) and Serobjan (1974: *Philodendron)*. Guttenberg (1940, 1968) reviewed available observations on aerial roots for over 30 genera.

Among terrestrial roots, the contractile type has been described for a few cormose genera. They are distinguished by unusual form and structure, particularly by their wrinkled surface after contraction as reported in detail by Scott (1908). Lamant and Heller (1967) described them for *Arum*. Contractile roots constitute a principal morphological adaptation of geophytes in the dormant season as they draw corms down to their appropriate 'physiological depth' (Pütz 1991). A 'channel effect' occurs where contractile roots pull the organ directly following it, thus clearing the way for the corm or offset (Pütz 1992). Contractile roots originate from the summit of the corm in *Sauromatum guttatum* and then exert downward pull against full organ resistance, hence having 0% channel effect. Contractile roots in *Arum italicum* exert a 20% channel effect.

Root zonation and histology

Structurally, aroid roots have many typical monocot features and the following zonation or layers have been reported for most of those studied: epidermis, exodermis, hypodermis, cortex, endodermis, and the stele, consisting of a pericycle surrounding equal numbers of xylem and phloem strands on alternating radii. In older roots, the stele centre may consist of lignified fibres or mature metaxylem. French (1987c), in his study of the sclerotic hypodermis (see below), described various anatomical features in roots of 280 species in 91 genera.

Ontogeny of young roots has been described by Berquam (1972) for *Syngonium* and the quiescent centre in the *Pistia* root primordium was described by Clowes (1984, 1985, 1992). Majumdar (1929) provided observations on the secondary root growth of *Amorphophallus*.

Root surface tissue varies according to age of the organ. In young roots, the epidermis is usually a single transient layer. The velamen, defined by Esau (1960) as a multiple epidermis, has been described in aerial roots of aroid epiphytes by Deshpande (1956), who also recorded one as occurring in *Spathiphyllum*, a terrestrial aroid. The velamen has 3–4 cell layers filled with granular matter. Root hairs were described by Pinkerton (1936) as arising from the hypodermis in *Philodendron cordatum*.

As reviewed by French (1987c) outer cortical tissues are divided among hypodermis, exodermis and cortical sclerenchyma. The hypodermis is highly localized and therefore potentially useful systematically. It forms the outer boundary of the cortex, distinguishing it from the epidermis. In most Araceae, the cortex consists entirely of thin-walled unlignified parenchyma. Some genera have peripheral collenchyma (e.g. *Anadendrum, Anthurium, Aridarum, Hottarum, Monstera, Pothoideum, Pothos*, and *Rhaphidophora*, as well as *Acorus* (Acoraceae)).

Trichosclereids were found by French (1987c) in some genera, where they develop in the intercellular ground space by intrusive growth (*Holochlamys, Monstera, Rhaphidophora, Rhodospatha, Scindapsus, Stenospermation*). Thick-walled pitted sclereids form narrow or wide bands adjacent and external to the endodermis in *Anthurium*, various Monstereae, and *Heteropsis*. Resin canals with sclerotic sheaths form in *Philodendron, Cercestis* and *Culcasia*. Systematic conclusions on the placement of eight genera are offered by French.

It appears that roots have a penetration or moisture barrier at the hypodermis as well as the endodermis. Endodermal casparian bands also appear in all species that have an exodermis or hypodermal casparian band. French (1987c) described the sclerotic hypodermis (SH) and other anatomical features in roots of 91 genera and 280 species. The SH has various forms but often has uniformly thickened elongated cells with lignified walls, 1–5 cells deep, adjacent to the exodermis. The SH is usually sharply delimited from subadjacent parenchyma. The exodermis is defined as having either differentially thickened periclinal walls, or lacking thickened walls.

Wilson and Peterson (1983) noted that the hypodermis of most roots has suberin and often lignin present. Perumalla *et al.* (1990) reported that most Araceae surveyed for this feature have a uniseriate hypodermis with casparian bands and an epidermis. In the Araceae, casparian bands may be suberized (S) or have lignin and suberin (LS). Aroids examined included species of *Anthurium* (S) or (LS), *Arisaema* (LS), *Dieffenbachia* (S) or (LS), *Philodendron* (S), and *Syngonium* (S). The results suggested that bands are systematically variable and not ecologically related. *Philodendron wendlandii* was found to have a multiseriate hypodermis and both layers have a lignosuberized casparian band (Peterson and Perumalla 1990). They found that the multiple hypodermis begins as one layer and matures centripetally, and may develop asynchronously. They followed Esau's (1977) definition that the hypodermis consists of a layer beneath the epidermis that is distinct from underlying tissues. They proposed that the exodermis be defined as a hypodermis containing a casparian band. That interpretation is followed in the present study but other definitions of exodermis appear in the literature.

Shishkoff (1987) surveyed 148 angiosperm families for the presence of a dimorphic hypodermis (DH), a structure that has suberized and non-suberized cells in an alternating regular pattern. More than half the genera sampled have the DH including one or more species of *Anthurium, Arum, Caladium, Dieffenbachia, Monstera, Philodendron* and *Remusatia (Gonatanthus)*.

Root periderm and related tissues have infrequently been investigated in the family. Weisse (1897) figured lenticels on aerial roots for several Araceae and reviewed earlier reports. Dutt (1956) described the lenticels of *Pothos* as being 'longitudinal' although Eames and MacDaniels (1947) had reported that lenticels are always transverse in roots. Causal factors in cork formation were studied in the family by Priestley and Woffenden (1922). In *Philodendron erubescens*, a cortical cork layer arises beneath the sclerenchymatous exodermal region in aerial roots and outside the incomplete secondary endodermis. In *Monstera deliciosa*, it was found that as aerial roots age they never produce a secondary endodermis. As the protoplasts of the primary endodermis become more permeable, cork appears beneath exodermis. Marlatt (1970) described details of periderm formation as wound cork in *Dieffenbachia* where suberin is deposited on wound surfaces up to 28 cells deep and starch granules disappear from neighbouring cells.

Tracheary elements in roots have long been known to be vessels in most monocots that have been investigated (Cheadle 1942; Wagner 1977). Using the SEM, Schneider and Carlquist (1998) and Carlquist and Schneider (1998) studied root samples of the subfamilies Philodendroideae and Colocasioideae, respectively. In all cases, vessels were present although they might be called incipient vessels. The absence or presence of pit membranes, a defining feature, cannot be distinguished by light microscopy. Perforation plates are not easily distinguished from wall pitting.

Tieghem (1870–1) supplied a list of *c.* 20 aroid genera that have secretory tissue. Resin canals were described by French (1987b) from a survey of roots of 91 genera in the family. They were found in all material examined of *Cercestis, Culcasia, Furtadoa, Homalomena* and *Philodendron*. Canal contents tested positively with Sudan IV and canals often had lignified or collenchymatous sheaths that may be diagnostic.

The ontogeny of root branching was observed in about 11 genera of Araceae by Tieghem and Duliot (1888). An observation of root branching in *Pistia* (Charlton 1983) revealed some regularity in lateral primordium spacing that began 100–350 μm behind the root cap junction.

Apogeotropic roots (roots that emerge from the ground) have been described by Sanford (1987). This phenomenon of nutrient-poor areas has been recorded in twelve species of five families. A *Philodendron* species from the Amazonian Tierra Firma and *Anthurium ellipticum* (Bruhn 1910) are given as examples but without observations on growth or anatomy. The total number of apogeotropic species and their mycorrhizal status are unknown.

Mycorrhizal symbiosis is believed to be widespread in both dicots and monocots but only limited information is available for aroids. Brundrett and Kendrick (1988, 1990) surveyed temperate plants with vescicular-arbuscular mycorrhizae (VAM) and described *Arisaema atrorubens* as being obligately mycorrhizal; fungal colonization occurred in 54% of roots sampled. In this species cortical air channels facilitated the spread of VAM fungi, but when roots became contractile, VAM colonies were lost. Nasim (1990) described roots containing VAM in *Colocasia*.

Only limited additional observations on cell-wall chemistry and structure are available for aroid roots. Wilson and Peterson (1983) found suberin and lignin in epidermal walls of *Dieffenbachia picta*, typical of half of flowering plant roots so tested, and walls also tested positive for cellulose. Central cortical walls were cellulosic and no differences were detected between geophytic and non-geophytic plants. In a study of isotropic layering of root fibre walls of several monocot families, Mueller and Beckman (1979) reported isotropic layers in the fibres of only *Monstera deliciosa* among Araceae examined; the several fibre layers were rich in pectin and lignin and corresponded to primary wall composition. Such layers were found only in mature root tissues, showing up infrequently in greenhouse-grown plants in the oldest roots.

ANATOMICAL CHARACTER STATE TRANSFORMATIONS AND THEIR POTENTIAL USE IN PHYLOGENETIC ANALYSIS

The order of genera presented in this volume follows the revised outline of Keating (2002b; Tables 1 and 5, this volume). It is based on a synthesis of data accumulated in this study combined with the topology produced by French *et al.* (1995), whose survey was based on cpDNA restriction site analysis. The morphology and geography found in MBB was also compared in detail. Anatomical variation was found here to match the French *et al.* topology with little ambiguity and most trends of specialization (listed below) can be interpreted parsimoniously. The family infrastructure presented by MBB is close to the new proposed outline as far as the first 31 genera, the bisexual-flowered Araceae. The placement by MBB of the remaining 74 unisexual-flowered genera within the subfamily Aroideae embeds a number of distinct groupings which are addressed in the revised classification presented here.

The bisexual (hermaphroditic) condition will be taken as plesiomorphic in common with all recent classifications of the family. This will begin the process of polarizing character states, many of which have not been analyzed across the family or reliably polarized before.

In spite of the great variety of habits and habitats in the family, most anatomical characters show gradual or subtle variation across subfamilies and tribes. The family seems to have settled into a characteristic anatomical Gestalt at an early stage in its radiation. There are very few characters or character suites that change abruptly across tribal or subfamilial lines. The greatest character richness occurs in the subfamily Pothoideae (including the monsteroids as a tribe) with a loss of, or simplification in, characters in more derived tribes. *Philodendron* encompasses the greatest anatomical variation of any single genus. As one moves across the Aroideae as defined here, there is often very little histological variation and histology alone is unreliable diagnostically. Trends in conspicuous and interesting tissues will be described as follows.

Collenchyma

This tissue, found in all subfamilies except Gymnostachydoideae, has emerged as the most interesting new character in this study (Keating 2000; Gonçalves *et al.* 2002). In general, collenchyma development is greatest where fibre development is least. In the Potheae, collenchyma does not coexist with fibre-ensheathed vascular bundles. In Anthurieae and throughout the remainder of that clade, collenchyma usually occurs as bands around the periphery of the petioles and midveins, and occasionally as bundle caps. It is present as bands or caps in the orontioids, except *Orontium*. Bands are present in most lasioids and absent in *Calla*.

Nearly all genera of philodendroids have banded collenchyma. Interrupted bands are also found in *Philodendron* and in *Zantedeschia*. The African and Asian genera of the Aglaonemateae (including five genera in the present outline) have interrupted bands and strands that

Table 1. Generic list of Araceae and comparisons of recent classifications

Genera	Keating 2002b	MBB 1997	Grayum 1990	Bogner & Nicolson 1991
Acorus	Acoraceae	Acoraceae	Acoraceae	Acoraceae
Gymnostachys	GY	GY	PO Gymnost	GY
Orontium	OR Oront	OR	LA Oront	LA Oront
Lysichiton	OR Symploc	OR	LA Symploc	LA Oront
Symplocarpus	OR Symploc	OR	LA Symploc	LA Oront
Pothos	PO Poth	PO Poth	PO Poth	PO
Pedicellarum	PO Poth	PO Poth	PO Poth	PO
Pothoidium	PO Poth	PO Poth	PO Poth	PO
Anthurium	PO Poth	PO Anthur	PO Anthur	LA Anthur
Holochlamys	PO Monster	MO Spathiph	PO Spathiph	MO Spathiph
Spathiphyllum	PO Monster	MO Spathiph	PO Spathiph	MO Spathiph
Rhodospatha	PO Monster	MO Monster	PO Monster	MO Monster
Stenospermation	PO Monster	MO Monster	PO Monster	MO Monster
Scindapsus	PO Monster	MO Monster	PO Monster	MO Monster
Rhaphidophora	PO Monster	MO Monster	PO Monster	MO Monster
Anadendrum	PO Monster	MO Anaden	PO Anaden	MO Anaden
Monstera	PO Monster	MO Monster	PO Monster	MO Monster
Alloschemone	PO Monster	MO Monster	PO Monster	MO Monster
Epipremnum	PO Monster	MO Monster	PO Monster	MO Monster
Amydrium	PO Monster	MO Monster	PO Monster	MO Monster
Heteropsis	PO Monster	MO Heterop	PO Monster	MO Heterop
Cyrtosperma	LA	LA	LA Las	LA Las
Lasimorpha	LA	LA		LA Las
Podolasia	LA	LA	LA Las	LA Las
Lasia	LA	LA	LA Las	LA Las
Anaphyllum	LA	LA	LA Las	LA Las
Urospatha	LA	LA	LA Las	LA Las
Anaphyllopsis	LA	LA		LA Las
Pycnospatha	LA	LA	LA Las	LA Las
Dracontium	LA	LA	LA Las	LA Las
Dracontioides	LA	LA	LA Las	LA Las
Calla	CA	CA	CA Calle	CA
Montrichardia	PH Philoden	AR Montrich	CA Montrich	LA Montrich
Anubias	PH Philoden	AR Anub	CA Anub	PH Anub
Furtadoa	PH Philoden	AR Homalom	CA Homalom	PH Philoden
Philodendron	PH Philoden	AR Philoden	CA Philoden	PH Philoden
Homalomena	PH Philoden	AR Homalom	CA Homalom	PH Philoden
Zantedeschia	PH Zanted	AR Zanded	CA Zanted	PH Zanted
Callopsis	PH Zanted	AR Callop	CA Callop	LA Callop
Stylochaeton	PH Stylochae	AR Stylochae	LA Stylochae	AR Stylochae
Gonatopus	PH Stylochae	AR Zamiocul	PO Zamiocul	LA Zamiocul
Zamioculcas	PH Stylochae	AR Zamiocul	PO Zamiocul	LA Zamiocul
Nephthytis	PH Aglaon	AR Nephthyt	CA Nephthyt	LA Nephthyt
Anchomanes	PH Aglaon	AR Nephthyt	CA Nephthyt	LA Nephthyt
Pseudohydrosme	PH Aglaon	AR Nephthyt	CA Nephthyt	LA Nephthyt
Aglaonema	PH Aglaon	AR Aglaon	CA Aglaon	PH Aglaon
Aglaodorum	PH Aglaon	AR Aglaon	CA Aglaon	PH Aglaon
Culcasia	PH Culcas	AR Culcas	CA Culcas	LA Culcas
Cercestis	PH Culcas	AR Culcas	CA Cerces	LA Nephthyt
Bognera	PH Spathicar	AR Dieffen	CA Bogner	PH Anubiad
Dieffenbachia	PH Spathicar	AR Dieffen	CA Dieffen	PH Dieffen
Spathantheum	PH Spathicar	AR Spathicar	CA Spathicar	AR Spathicar

Table 1. continued

Genera	Keating 2002b	MBB 1997	Grayum 1990	Bogner & Nicolson 1991
Gorgonidium	PH Spathicar	AR Spathicar	CA Spathicar	AR Spathicar
Synandrospadix	PH Spathicar	AR Spathicar	CA Spathicar	AR Spathicar
Gearum	PH Spathicar	AR Spathicar	CA Spathicar	AR Spathicar
Spathicarpa	PH Spathicar	AR Spathicar	CA Spathicar	AR Spathicar
Asterostigma	PH Spathicar	AR Spathicar	CA Spathicar	AR Spathicar
Mangonia	PH Spathicar	AR Spathicar	CA Spathicar	AR Spathicar
Taccarum	PH Spathicar	AR Spathicar	CA Spathicar	AR Spathicar
Phymatarum	SC Schismat	AR Schismat	CA Schismat	PH Philoden
Schismatoglottis	SC Schismat	AR Schismat	CA Schismat	PH Philoden
Aridarum	SC Schismat	AR Schismat	CA Schismat	PH Philoden
Piptospatha	SC Schismat	AR Schismat	CA Schismat	PH Philoden
Hottarum	SC Schismat	AR Schismat	CA Schismat	PH Philoden
Bucephalandra	SC Schismat	AR Schismat	CA Schismat	PH Philoden
Heteroaridarum	SC Schismat	AR Schismat	CA Schismat	PH Philoden
Cryptocoryne	SC Cryptocor	AR Cryptocor	AR Cryptocor	AR Areae
Lagenandra	SC Cryptocor	AR Cryptocor	AR Cryptocor	AR Areae
Spirodela	LE			
Lemna	LE			
Wolffia	LE			
Wolffiella	LE			
Amorphophallus	AR Thomson	AR Thomson	AR Thomson	AR Thomson
Pseudodracontium	AR Thomson	AR Thomson	AR Thomson	AR Thomson
Hapaline	AR Calad	AR Calad	CO Calad	CO Steudner
Syngonium	AR Calad	AR Calad	CO Calad	CO Syngon
Xanthosoma	AR Calad	AR Calad	CO Calad	CO Calad
Chlorospatha	AR Calad	AR Calad	CO Calad	CO Calad
Ulearum	AR Calad	AR Zomicar	CA Callop	AR Zomicar
Filarum	AR Calad	AR Zomicar	CA Callop	AR Zomicar
Zomicarpella	AR Calad	AR Zomicar	CA Callop	AR Zomicar
Caladium	AR Calad	AR Calad	CO Calad	CO Calad
Scaphispatha	AR Calad	AR Calad	CO Calad	CO Calad
Jasarum	AR Calad	AR Calad	CO Calad	CO Calad
Zomicarpa	AR Calad	AR Zomicar	CO Zomicar	AR Zomicar
Ambrosina	AR Arisar	AR Ambros	AR Ambros	AR Areae
Arisarum	AR Arisar	AR Arisar	AR Arisar	AR Areae
Peltandra	AR Peltand	AR Peltand	CA Peltand	PH Peltand
Typhonodorum	AR Peltand	AR Peltand	CA Peltand	PH Typhon
Colletogyne	AR Peltand	AR Arophyt	CA Arophyt	AR Arophyt
Carlephyton	AR Peltand	AR Arophyt	CA Arophyt	AR Arophyt
Arophyton	AR Peltand	AR Arophyt	CA Arophyt	AR Arophyt
Pistia	AR Pist	AR Pist	AR Pist	PI
Arisaema	AR Arisaem	AR Arisaem	AR Arisaem	AR Areae
Pinellia	AR Arisaem	AR Arisaem	AR Pinell	AR Areae
Typhonium	AR Areae	AR Areae	AR Areae	AR Areae
Theriophonum	AR Areae	AR Areae	AR Areae	AR areae
Sauromatum	AR Areae	AR Areae	AR Areae	AR Areae
Lazarum	AR Areae	AR Areae		
Biarum	AR Areae	AR Areae	AR Areae	AR Areae
Arum	AR Areae	AR Areae	AR Areae	AR Areae
Eminium	AR Areae	AR Areae	AR Areae	AR Areae
Dracunculus	AR Areae	AR Areae	AR Areae	AR Areae
Helicodiceros	AR Areae	AR Areae	AR Areae	AR Areae

Table 1. continued

Genera	Keating 2002b	MBB 1997	Grayum 1990	Bogner & Nicolson 1991
Ariopsis	AR Colocas	AR Colocas	AR Ariops	CO Ariops
Alocasia	AR Colocas	AR Colocas	CO Colocas	CO Colocas
Remusatia	AR Colocas	AR Colocas	CO Colocas	CO Steudner
Colocasia	AR Colocas	AR Colocas	CO Colocas	CO Colocas
Steudnera	AR Colocas	AR Colocas	CO Colocas	CO Steudner
Protarum	AR Colocas	AR Colocas	CO Colocas	CO Protar

Key to abbreviations: Subfamilies = pair of upper case letters; Tribes = more than two letters
Subfamilies: AR = Aroideae, CA = Calloideae, GY = Gymnostachydoideae, LA = Lasioideae, LE = Lemnoideae, MO = Monsteroideae, OR = Orontioideae, PH = Philodendroideae, PO = Pothoideae, SC = Schismatoglottoideae
Tribes: Aglaon = Aglaonemateae, Ambros = Ambrosineae, Anadend = Anadendreae, Anthur = Anthurieae, Anub = Anubiadeae, Areae = Areae, Ariops = Ariopsideae, Arisaem = Arisaemateae, Arisar = Arisareae, Arophyt = Arophyteae, Bogner = Bognereae, Calad = Caladieae, Calle = Calleae, Callop = Callopsideae, Cerces = Cercestideae, Colocas = Colocasieae, Cryptocor = Cryptocoryneae, Culcas = Culcasieae, Dieffen = Dieffenbachieae, Gymnost = Gymnostachydeae, Heterop = Heteropsideae, Homalom = Homalomeneae, Las = Lasieae, Monster = Monstereae, Montrich = Montrichardieae, Nephthyt = Nephthytideae, Oront = Orontieae, Peltand = Peltandreae, Philoden = Philodendreae, Pinell = Pinelleae, Pist = Pistieae; Poth = Potheae, Protar = Protareae, Schismat = Schismatoglottideae, Spathicar = Spathicarpeae, Spathiph = Spathiphylleae, Steudner = Steudnereae, Stylochae = Stylochaetoneae, Symploc = Symplocarpeae, Syngon = Syngonieae, Thomson = Thomsonieae, Typhon = Typhoneae, Zamiocul = Zamioculcadeae, Zanted = Zantedeschieae, Zomicar = Zomicarpeae.

are not aligned with vascular bundles (type Sb). This rare configuration contributes to their being placed in a single tribe.

The Aroideae are delimited by the presence of stranded (type Sv) collenchyma. At the base of the clade, the schismatoglottids show banded and banded-interrupted collenchyma that one could hypothesize as being transitional to the stranded type Sv.

However, there is another trend in collenchyma evolution that might also be operating. In a few genera below the Aroideae (*Lysichiton, Symplocarpus, Rhodospatha, Scindapsus, Monstera, Gonatopus*, and *Dieffenbachia)*, collenchymatous bundle caps develop as apparent replacement of fibre caps, suggesting a spatial homology (Patterson 1982), a different route to evolving type Sv strands. Normally, where fibre caps or sheaths occur, collenchyma, when present, is always banded. Stranded collenchyma, aligned with vascular bundles, normally occurs when fibres are absent, except in Caladieae where both may coexist.

Sclerenchyma

Fibres are best developed in the tribes Potheae and Monstereae where they frequently ensheath vascular bundles. The genus *Gymnostachys* shows a unique configuration for the family in having separate phloem and xylem caps that extend as transcurrent girders above and below the leaf bundles. Its phloem strands are circular and surrounded by deep fibre development, unlike any other aroid. Separate phloem and xylem caps remain frequent in the pothoids and monsteroids, the lasioids, or as phloem caps in *Orontium*. In the philodendroids, fibre sheaths are found only in the Culcasieae, *Asterostigma* (Spathicarpeae), and *Philodendron* (Philodendreae), not particularly closely allied. Fibre caps occur sporadically in nearly all clades of the subfamily.

Phloem fibre caps are common in schismatoglottids and the Caladieae. They are occasionally present in the distantly related *Lazarum* and *Colocasia*. In subfamily Aroideae phloem and xylem caps are less common.

The occurrence of trichosclereids in Araceae was reviewed by Seubert (1997a) who reported them present in one or several organs of eleven genera. She mentioned an additional report for *Montrichardia* by Solereder and Meyer (1928) which was not confirmed by Seubert or in the present study. Here, the trichosclereid-bearing genera are placed mostly in the Pothoideae: Monstereae. There is an occasional report for genera of Potheae (Nicolson 1960: *Pothos*, 2 species; and in the present study: *Anthurium pallens*, Croat 35265). The distribution of this unusual cell type also contributed to the decision to place all of these genera in one subfamily, rather than in two indistinctly bounded subfamilies. The capacity to express these cells is an apomorphy for this group that seems not to have been carried into more derived lines.

Crystals

The four types present all appear to be formed of calcium oxalate as determined by polarization behaviour, and by reaction to the usual microtechnical reagents (see anatomical literature review). Druses and crystal sand (clusters of fine prismatics) are common throughout the family although not detected in all genera. Both of these as recorded in descriptions show some architectural variation but information was insufficient to allow analysis of trends. Styloids, present in *Anadendrum* and *Amorphophallus*, are otherwise rarely found and are thus too little known for further comment.

The ubiquitous raphides can be characterized both by cell form and by configuration of crystalline contents. They often occur in at least two forms in ground tissue of most genera and there is no certain morphocline among these types. The simple, non-systematically useful form of raphides: thin-walled, rounded or oval cells with a single raphide bundle within, is nearly universal. The cells may be slightly or highly idioblastic, and variable in size with respect to neighbouring ground tissue, but in the majority of genera the simple type coexists with additional raphide types whose distribution is of considerable interest (Keating 2002a).

First, the subfamily Pothoideae commonly presents elongated raphide cells with ill-defined raphide bundles, that is, crystals occur in elongated, overlapping, sometimes helically oriented groups. Overlapping bundles occur sporadically elsewhere in Lasioideae, Culcasieae, Aglaonemateae, Zamioculcadeae, Cryptocoryneae, Schismatoglottideae, Caladieae, Arisaemateae, and in five genera of Areae.

Mostly elongated cells bearing multiple discrete bundles are common in bisexual-flowered subfamilies except Calloideae. They also occur sporadically in the unisexual Philodendroideae and Aroideae. Tubular cells, that extend greatly beyond the length of included raphide bundles, occur sporadically without any obvious trend. Rarely encountered articulated raphide tubes are distinctive apomorphies occurring in Philodendroideae: Spathicarpeae: *Taccarum* and *Asterostigma*, and in *Biarum* of Aroideae: Areae.

Probably the most specialized and unusual crystal phenomenon, described early by Turpin (1836), is the biforine cell. These specialized raphide cells are heavy-walled, spindle-shaped or cylindrical, usually with thin-walled and pointed (papillate) ends. When deformed, they are capable of shooting a flow of raphides into the mouth parts of grazing organisms (see Turpin 1836, Middendorf 1983). Similar cells, which may represent phylogenetic or ontogenetic stages, referred to as spindle-shaped or biforine-like in descriptions, have the same general idioblastic shape but lack thickened walls or terminal papillae. Spindle-shaped cells are found in *Symplocarpus, Cyrtosperma* and *Calla* of the hermaphroditic genera and are scattered through most tribes of the Philodendroideae and Aroideae.

True biforines are not encountered at all in the hermaphroditic aroids. They occur in only seven scattered genera of Philodendroideae, and 17 genera (seven tribes) of Aroideae. They are certainly the most specialized crystal cell form because of this exclusive distribution in the monoecious genera.

Many questions remain regarding the development of these crystals as, for instance, whether biforine-like cells are a phylogenetic or ontogenetic stage in the development of true biforines. Raphides of a given morphology may occur in various sizes and relationships to the surrounding tissue. In ontogeny, raphide cells can be identified early cytologically as they enlarge differentially just a few hundred micrometres proximal to the apical meristem or primordia surfaces. Beyond this little is known regarding the scope of crystal cell initiation timing and expansion in this family.

Summary of trends of specialization

For the above vegetative features and for those listed below, character state transformations (trends of specialization) are proposed reflecting anatomical and morphological evidence. In addition, the sequence and topology of genera as given by French *et al.* (1995) helped confirm the polarization of trends. The trends listed below appear those most likely or determinable from this study. In some cases, no trend is discernable, and in many cases the trends should be considered potentially homoplasious and not necessarily appropriate for family-wide cladistic analysis. Most trends will be useful for within-genus or within-tribe analysis.

Venation (leaf paradermal): Trend 1: venation pinnate in leaves or segments ⇒ acrodromous. Trend 2: secondary and tertiary veins reticulate ⇒ parallel or converging secondaries and tertiaries. Trend 3: brochidodromous loops ⇒ straightened ‘collectors’ (marginal or submarginal). Trend 4: brochidodromy ⇒ eucamptodromy.

Leaf structure: Trend: no differentiation ⇒ dorsiventral ⇒ isobilateral.

Trichomes: No trend. The several genera bearing trichomes seem not to be closely related on the basis of other characters. Therefore, trichomes are autapomorphies.

Epidermal cells (paradermal): Trend 1: cells straight-sided polygonal and nearly isohedral ⇒ cells elongated. Trend 2: cell walls straight ⇒ undulate ⇒ sinuous ⇒ interlocking.

Epidermis (TS): Trend: cells short ⇒ taller ⇒ columnar.

Stomata (PS): Trend 1: brachyparacytic (bpc) ⇒ anomocytic. Trend 2: brachyparacytic ⇒ brachypara-tetracytic ⇒ brachypara-hexacytic ⇒ amphibrachy-paracytic.

Subsidiary cells: Trend: neighbouring cells broad and irregular, little differentiated from epidermal cells ⇒ broad trapezoidal, angular ⇒ broad and rounded ⇒ narrow and rounded.

Stomata (TS): Trend 1: cells level with surface ⇒ cells on mounds. Trend 2: guard cells with double cuticular ledges ⇒ cells with single (outer) ledges.

Hypodermis: Trend: absent ⇒ present.

Palisade mesophyll: Trend 1: absent ⇒ columnar ⇒ squarish. Trend 2: columnar ⇒ T-shaped. Trend 3: columnar ⇒ armed. Trend 4: columnar ⇒ U-shaped.

Spongy mesophyll: Trend 1: type 1 ⇒ type 1a. Trend 2: type 1 ⇒ type 2 ⇒ type 3. Unknown: type 4 is apomorphic from an undetermined plesiomorphic state.

Air cavities: Trend: low substomatal cavities, mostly associated with types 1, 1a mesophyll ⇒ high substomatal cavities mostly associated with types 2 and 3 mesophyll.

Collenchyma: In general, collenchyma (in midribs and petioles) is least developed where fibre caps or sheaths are best developed. Trend 1: collenchyma absent ⇒ present. Trend 2: collenchyma absent ⇒ banded (type B) ⇒ bands interrupted (type Bi) ⇒ strands on alternate radii with vascular bundles (type Sb) ⇒ strands subtending peripheral vascular bundles (type Sv). Trend 3: phloem fibre caps ⇒ collenchyma cells replacing fibre strands (type Svc) ⇒ rounded strands subtending peripheral vascular bundles (type Sv).

Vascular bundles—pattern in TS midrib: Trend 1: vascular bundles appearing symmetrically (disposed) organized ⇒ irregularly scattered. Trend 2: vascular bundles numerous ⇒ few.

Vascular bundles—degree of development in TS midrib or petiole: Trend: type I ⇒ type II ⇒ type III.

Sclerenchyma (fibres): Trend: ensheathing vascular bundles ⇒ caps over phloem and xylem ⇒ caps over phloem ⇒ absent. (see collenchyma Trend 3.)

Sclerenchyma (trichosclereids): Trend 1 (in Pothoideae only): absent ⇒ present ⇒ absent. Trend 2: solitary ⇒ aggregated. (Variations in branching, if any, shape, degrees of aggregation have unknown polarities.)

Secretory ducts: Trend: absent ⇒ present (scattered in 14 genera of 4 subfamilies) ⇒ absent.

Laticifers: Trend: absent ⇒ present, articulated and non-anastomosing ⇒ articulated and anastomosing. (Anastomosing condition is an apparent homoplasious apomorphy in Old World and New World groups of Aroideae.)

Tannin cells: Widespread and scattered in the family; evolutionary patterns obscure.

Raphide cells (apomorphies for cells and crystals probably not due to a single morphocline). Trend 1: cells of type 1 (thin-walled and irregular, slightly larger than surrounding cells of ground tissue) ⇒ cells wider and circular. Trend 2: cells of type 1 ⇒ cells elongate ⇒ cells tubular. Trend 3: cells of type 1 ⇒ cells articulated, usually end to end. Trend 4: cells of type 1 ⇒ cells spindle-shaped ⇒ spindle cells with thick lignified walls (biforines).

Raphide crystals: Trend 1: single bundles ⇒ wide bundles. Trend 2: single bundles ⇒ crystals in oblique overlapping bundles. Trend 3: single bundles ⇒ multiple bundles per cell.

Stem structure: Trend: central cylinder demarcated from cortex ⇒ central cylinder not discernable.

Stem vascular bundles: Trend: collateral bundles ⇒ compound bundles ⇒ amphivasal bundles.

SCOPE OF ARACEAE AND COMPARISONS OF RECENT CLASSIFICATIONS

By any recent account, the family comprises over 100 genera placed in a variety of groupings. Four recent schemes are compared in Table 1 and listed in Tables 2–5. The generic sequence for the anatomical descriptions follows the proposed scheme by Keating (2002b, Table 5), which attempts to follow the topology of French *et al.* (1995, Table 6), and also most parsimoniously aligns the array of anatomical data.

The genus *Acorus*, traditionally treated as a member of the Araceae, is now considered on a variety of lines of evidence to be in its own family, Acoraceae (Grayum 1987; Buzgó 2001). Duvall *et al.* (1993), Davis (1995), Soltis *et al.* (1997) and Stevenson *et al.* (2000), using molecular evidence, concluded that *Acorus* is phylogenetically isolated and should be considered sister to all remaining monocots.

The planktonic lemnoids, *Landoltia*, *Lemna*, *Spirodela*, *Wolffia*, and *Wolffiella*, seem now to be clearly embedded within the Araceae. They are treated here as subfamily Lemnoideae and inserted before the Aroideae as indicated by French *et al.* (1995) and Maheshwari (1958). Landolt (1998) was more cautious. First, he noted Grayum's (1992) observations that the lemnoid uniporate pollen with the ulcerate aperture type is unknown in

Table 2. Classification of Grayum (1990) [*Acorus* excluded from Araceae]

Family ARACEAE Juss.

Subfamily Pothoideae Engl.
- Tribe Gymnostachydeae Nakai *Gymnostachys*
- Tribe Spathiphylleae Engl. *Spathiphyllum, Holochlamys*
- Tribe Anthurieae Engl. *Anthurium*
- Tribe Potheae Engl. *Pothos, Pedicellarum, Pothoidium*
- Tribe Anadendreae Bogner & J. French *Anadendrum*
- Tribe Monstereae Engl.
  - Subtribe Heteropsidinae Engl. *Heteropsis*
  - Subtribe Monsterinae Schott *Rhaphidophora, Monstera, Amydrium, Epipremnum, Scindapsus, Alloschemone, Stenospermation, Rhodospatha*
- Tribe Zamioculcadeae Engl. *Zamioculcas, Gonatopus*

Subfamily Calloideae Schott
- A. Calla Alliance
  - Tribe Calleae Schott *Calla*
- B. Nephthytis Alliance
  - Tribe Nephthytideae Engl. *Nephthytis, Anchomanes, Pseudohydrosme*
  - Tribe Callopsideae Engl. *Callopsis, Ulearum, Filarum, Zomicarpella*
  - Tribe Montrichardieae Engl. *Montrichardia*
- C. Aglaonema Alliance
  - Tribe Anubiadeae Engl. *Anubias*
  - Tribe Zantedeschieae Engl. *Zantedeschia*
  - Tribe Aglaonemateae Engl. *Aglaonema, Aglaodorum*
  - Tribe Spathicarpeae Schott *Mangonia, Asterostigma, Synandrospadix, Taccarum, Gorgonidium, Gearum, Spathantheum, Spathicarpa*
  - Tribe Dieffenbachieae Engl. *Dieffenbachia*
  - Tribe Bognereae Mayo & Nicolson *Bognera*
- D. Peltandra Alliance
  - Tribe Peltandreae Engl. *Peltandra, Typhonodorum*
  - Tribe Arophyteae Bogner *Arophyton, Carlephyton, Colletogyne*
  - Tribe Schismatoglottideae Nakai *Schismatoglottis, Piptospatha, Bucephalandra, Phymatarum, Aridarum, Heteroaridarum, Hottarum*
- E. Philodendron Alliance
  - Tribe Culcasieae Engl. *Culcasia*
  - Tribe Cercestideae Grayum *Cercestis*
  - Tribe Homalomeneae (Schott) M. Hotta *Furtadoa, Homalomena*
  - Tribe Philodendreae Schott *Philodendron*

Subfamily Colocasioideae Engl.
- Tribe Zomicarpeae Schott *Zomicarpa*
- Tribe Colocasieae Engl.
  - Subtribe Protarinae (Engl.) Grayum *Protarum*
  - Subtribe Steudnerinae Engl. & K.Krause *Steudnera*
  - Subtribe Remusatiinae Grayum *Remusatia, Gonatanthus*
  - Subtribe Colocasiinae Schott *Colocasia, Alocasia*
- Tribe Caladieae Schott
  - Subtribe Jasarinae Grayum *Jasarum*
  - Subtribe Scaphispathinae Grayum *Scaphispatha*
  - Subtribe Caladiinae Engl. & K. Krause *Caladium, Xanthosoma, Chlorospatha, Aphyllarum*
  - Subtribe Syngoniineae Schott *Syngonium*
  - Subtribe Hapalininae Engl. & K. Krause *Hapaline*

Subfamily Lasioideae Engl.
- Tribe Symplocarpeae Engl. *Symplocarpus, Lysichiton*
- Tribe Orontieae R. Br. ex Dumort. *Orontium*

Table 2. (*Cont.*)

Family ARACEAE Juss.
Tribe Lasieae Engl.
Subtribe Dracontiinae Schott *Cyrtosperma, Lasia, Anaphyllum, Podolasia, Urospatha, Dracontioides, Dracontium*
Subtribe Pycnospathinae Bogner *Pycnospatha*
Tribe Stylochaetoneae Schott *Stylochaeton*
Subfamily Aroideae Engl.
Tribe Thomsonieae Blume *Pseudodracontium, Amorphophallus*
Tribe Arisareae Dumort. *Arisarum*
Tribe Pinellieae Nakai *Pinellia*
Tribe Pistieae Blume *Pistia*
Tribe Cryptocoryneae Blume *Cryptocoryne, Lagenandra*
Tribe Ambrosineae Schott *Ambrosina*
Tribe Ariopsideae Engl. *Ariopsis*
Tribe Arisaemateae Nakai *Arisaema*
Tribe Areae Engl. *Arum, Dracunculus, Helicodiceros, Theriophonum, Typhonium, Sauromatum, Eminium, Biarum*

Araceae, and so is their utricular fruit type. However, we do know that pollen and fruit diversity is large in the Araceae, as is secondary chemistry. Also, extreme habit specialization among lemnoids makes direct structural comparisons difficult. As Landolt (1998) suggested, more detailed molecular genetic studies will be required among the putatively related aroids, lemnoids and adjacent Alismatiflorae in order to clarify their relationships.

The present total of 106 genera of Araceae includes *Gymnostachys* as the basal and most isolated long-branch genus (subfamily Gymnostachydoideae). The other bisexual aroid genera are placed among four subfamilies. The main difference with past practice is the recognition of a subfamily Pothoideae that includes two tribes, Potheae and Monstereae, as there seems insufficient evidence for their recognition at the subfamily level. None of the structural data allows for mutually exclusive subfamilial groupings nor does the topology of French *et al.* (1995). Genera of the subfamily Lasioideae are compact enough to need no additional infrastructure above the generic level.

In the MBB scheme, all remaining genera (over 2/3 of the total) are placed in the subfamily Aroideae. While this line is monophyletic as they suggest, there are branch-length differences (French *et al.* 1995) within that clade that readily support recognition of four subfamilies which are definitely supported by anatomical evidence. In the outline by Keating (2002b) the subfamily Philodendroideae includes six tribes based on the banded collenchyma pattern. The subfamily Schismatoglottidoideae includes the tribes Cryptocoryneae (two genera) and Schismatoglottideae (seven genera described here, now merged to five), followed by the subfamily Lemnoideae, recognized by several apomorphies. Finally, a restricted subfamily Aroideae is based primarily on stranded collenchyma (type Sv), holding eight tribes with 37 genera. The genus *Pistia*, often put at the end of the family because of several structural autapomorphies, is placed here as a monotypic tribe within the Aroideae according to the cpDNA evidence (French *et al.* 1995).

Taxonomic notes are added at the end of generic anatomical descriptions, and for many suprageneric groupings. Where none appears, no distinctive features or relationships were evident from the sample.

EVOLUTIONARY POSITION OF ARACEAE

Structural evidence

The earlier idea that Arales (Araceae and Lemnaceae) belong near the Arecales, Cyclanthaceae and Pandanales was illustrated by Dahlgren (1975) and Cronquist (1968, 1981: subclass Arecidae). At about this time, it was becoming clear on the basis of accumulating evidence that the nearest neighbours are elsewhere. Dahlgren (1980) placed the Arales (Aranae) near the base of the monocots, distal to the Alismatiflorae and Triuridiflorae and proximal to the Liliiflorae, probably the earliest such conclusion. On the basis of chemical characters, Dahlgren *et al.* (1981) reaffirmed that position. The 'Dahlgrenograms' of this period are compatible with the idea that the monocots arose as a clade from within the dicotyledons, as was stated by Dahlgren and Rasmussen (1983).

By 1992, Johri *et al.* had provided embryological evidence that palms and cyclanths cannot be close to Araceae.

Table 3. Classification of Bogner and Nicolson (1991) [*Acorus* excluded from family.]

ARACEAE

Subfamily Gymnostachydoideae Bogner & Nicolson *Gymnostachys*
Subfamily Pothoideae *Pothos, Pedicellarum, Pothoidium*
Subfamily Monsteroideae
 Tribe Anadendreae *Anadendrum*
 Tribe Monstereae *Amydrium, Rhaphidophora, Epipremnum, Scindapsus, Alloschemone, Stenospermation, Rhodospatha, Monstera*
 Tribe Heteropsideae *Heteropsis*
 Tribe Spathiphylleae *Spathiphyllum, Holochlamys*
Subfamily Calloideae *Calla*
Subfamily Lasioideae
 Tribe Orontieae *Lysichiton, Symplocarpus, Orontium*
 Tribe Anthurieae *Anthurium*
 Tribe Lasieae
  Subtribe Dracontiinae *Cyrtosperma, Lasimorpha, Lasia, Anaphyllum, Anaphyllopsis, Podolasia, Urospatha, Dracontioides, Dracontium*
  Subtribe Pycnospathinae *Pycnospatha*
 Tribe Zamioculcadeae *Zamioculcas, Gonatopus*
 Tribe Callopsideae *Callopsis*
 Tribe Nephthytideae *Pseudohydrosme, Anchomanes, Nephthytis, Cercestis*
 Tribe Culcasieae *Culcasia*
 Tribe Montrichardieae *Montrichardia*
Subfamily Philodendroideae
 Tribe Philodendreae
  Subtribe Homalomeninae *Furtadoa, Homalomena*
  Subtribe Schismatoglottidinae *Schismatoglottis, Piptospatha, Hottarum, Bucephalandra, Phymatarum, Aridarum, Heteroaridarum*
  Subtribe Philodendrinae *Philodendron*
 Tribe Anubiadeae *Anubias, Bognera*
 Tribe Aglaonemateae *Aglaonema, Aglaodorum*
 Tribe Dieffenbachieae *Dieffenbachia*
 Tribe Zantedeschieae *Zantedeschia*
 Tribe Typhonodoreae *Typhonodorum*
 Tribe Peltandreae *Peltandra*
Subfamily Colocasioideae
 Tribe Caladieae *Xanthosoma, Chlorospatha, Caladium, Scaphispatha, Jasarum*
 Tribe Steudnereae
  Subtribe Steudnerinae *Steudnera, Remusatia, Gonatanthus*
  Subtribe Hapalininae *Hapaline*
 Tribe Protareae *Protarum*
 Tribe Colocasieae *Colocasia, Alocasia*
 Tribe Syngonieae *Syngonium*
 Tribe Ariopsideae *Ariopsis*
Subfamily Aroideae
 Tribe Stylochaetoneae *Stylochaeton*
 Tribe Arophyteae *Carlephyton, Colletogyne, Arophyton*
 Tribe Spathicarpeae *Mangonia, Taccarum, Asterostigma, Gorgonidium, Synandrospadix, Gearum, Spathantheum, Spathicarpa*
 Tribe Zomicarpeae *Zomicarpa, Filarum, Zomicarpella, Ulearum*
 Tribe Thomsonieae *Amorphophallus, Pseudodracontium*
 Tribe Areae
  Subtribe Arinae *Arum, Dracunculus, Helicodiceros, Theriophonum, Typhonium, Sauromatum, Eminium, Biarum*
  Subtribe Arisarinae *Arisarum*
  Subtribe Arisaematinae *Arisaema*
  Subtribe Atherurinae *Pinellia*
  Subtribe Ambrosininae *Ambrosina*
  Subtribe Cryptocoryninae *Lagenandra, Cryptocoryne*
Subfamily Pistioideae *Pistia*

Table 4. Classification of Mayo, Bogner and Boyce (1997)

ACORACEAE C. Agardh *Acorus*
ARACEAE Juss.

Major Group *PROTO-ARACEAE*
Subfamily Gymnostachydoideae Bogner & Nicolson *Gymnostachys*
Subfamily Orontioideae Mayo, Bogner & P.C.Boyce *Orontium, Lysichiton, Symplocarpus*
Major Group *TRUE ARACEAE*
Subfamily Pothoideae Engl.
  Tribe Potheae Engl. *Pothos, Pedicellarum, Pothoidium*
  Tribe Anthurieae Engl. *Anthurium*
Subfamily Monsteroideae Engl.
  Tribe Spathiphylleae Engl. *Spathiphyllum, Holochlamys*
  Tribe Anadendreae Bogner & French *Anadendrum*
  Tribe Heteropsideae Engl. *Heteropsis*
  Tribe Monstereae Engl. *Amydrium, Rhaphidophora, Epipremnum, Scindapsus, Monstera, Alloschemone, Rhodospatha, Stenospermation*
Subfamily Lasioideae Engl. *Dracontium, Dracontioides, Anaphyllopsis, Pycnospatha, Anaphyllum, Cyrtosperma, Lasimorpha, Podolasia, Lasia, Urospatha*
Subfamily Calloideae Endl. *Calla*
Subfamily Aroideae
  PARAPHYLETIC GROUP: PERIGONIATE AROIDEAE
    Tribe Zamioculcadeae Engl. *Zamioculcas, Gonatopus*
    Tribe Stylochaetoneae Schott *Stylochaeton*
  MONOPHYLETIC GROUP: APERIOGONIATE AROIDEAE
    Dieffenbachia Alliance
      Tribe Dieffenbachieae Engl. *Dieffenbachia, Bognera*
      Tribe Spathicarpeae Schott *Mangonia Taccarum, Asterostigma, Gordonidium, Synandrospadix, Gearum, Spathantheum, Spathicarpa*
    Philodendron Alliance
      Tribe Philodendreae Schott *Philodendron*
      Tribe Homalomeneae (Schott) M.Hotta *Furtadoa, Homalomena*
      Tribe Anubiadeae Engl. *Anubias*
    Schismatoglottis Alliance
      Tribe Schismatoglottideae Nakai *Schismatoglottis, Piptospatha, Hottarum, Bucephalandra, Phymatarum, Aridarum, Heteroaridarum*
      Tribe Cryptocoryneae Blume *Lagenandra, Cryptocoryne*
    Caladium Alliance
      Tribe Zomicarpeae Schott *Zomicarpa, Zomicarpella, Ulearum, Filarum*
      Tribe Caladieae Schott *Scaphispatha, Caladium, Jasarum, Xanthosoma, Chlorospatha, Syngonium, Hapaline*
    No Alliance
      Tribe Nephthytideae Engl. *Nephthytis, Anchomanes, Pseudohydrosme*
      Tribe Aglaonemateae Engl. *Aglaonema, Aglaodorum*
      Tribe Culcasieae Engl. *Culcasia, Cercestis*
      Tribe Montrichardieae Engl. *Montrichardia*
      Tribe Zantedeschieae Engl. Zantedeschia
      Tribe Callopsideae Engl. *Callopsis*
      Tribe Thomsonieae Blume *Amorphophallus, Pseudodracontium*
      Tribe Arophyteae Bogner *Arophyton, Carlephyton, Colletogyne*
      Tribe Peltandreae Engl. *Peltandra, Typhonodorum*
      Tribe Arisareae Dumort. *Arisarum*
      Tribe Ambrosineae Schott *Ambrosina*
      Tribe Areae L. *Arum, Eminium, Dracunculus, Helicodiceros, Theriophonum, Typhonium, Sauromatum, Lazarum, Biarum*
      Tribe Arisaemateae Nakai *Pinellia, Arisaema*
      Tribe Colocasieae Engl. *Ariopsis, Protarum, Steudnera, Remusatia, Colocasia, Alocasia*
      Tribe Pistieae Blume *Pistia*

Table 5. Classification of Araceae and notes on certain distinguishing leaf and petiole characters (Keating, this volume)

Family ARACEAE
Series I Bisexual-flowered aroids
Subfamily GYMNOSTACHYDOIDEAE Collenchyma absent; vascular bundles unique, type I; fibres unique: vascular bundle sheaths and girders. *Gymnostachys*
Subfamily ORONTIOIDEAE
Tribe Orontieae Collenchyma absent; fibres as phloem caps; laticifers non-anastomosing. *Orontium*
Tribe Symplocarpeae Collenchyma types Svc, B; fibres absent; laticifers absent. *Lysichiton, Symplocarpus*
Subfamily POTHOIDEAE
Tribe Potheae Collenchyma absent (type B in *Anthurium*); vascular bundle type I; fibres ensheathing bundles; laticifers absent; prismatic crystals. *Pothos, Pedicellarum, Pothoidium, Anthurium*
Tribe Monstereae Collenchyma absent or type B; vascular bundle types I, II; fibres as phloem or xylem caps or ensheathing bundles; trichosclereids usually present; secretory ducts in most genera; laticifers non-anastomosing, occasional. *Holochlamys, Spathiphyllum, Rhodospatha, Stenospermation, Scindapsus, Rhaphidophora, Anadendrum, Monstera, Alloschemone, Epipremnum, Amydrium, Heteropsis*
Subfamily LASIOIDEAE
Collenchyma absent or type B; vascular bundle types I, II; fibres: caps or ensheathing bundles; midrib or petiole ground tissue type 4. *Cyrtosperma, Lasimorpha, Podolasia, Lasia, Anaphyllum, Urospatha, Anaphyllopsis, Pycnospatha, Dracontium, Dracontioides*
Subfamily CALLOIDEAE
Collenchyma absent; vascular bundle types I, II; laticifers non-anastomosing. *Calla*
Series II Monoecious aroids
Subfamily PHILODENDROIDEAE
Collenchyma type B; vascular bundle types I, II; fibres as bundle caps; secretory ducts present; laticifers non-anastomosing
Tribe Philodendreae Collenchyma type B; vascular bundle types I, II; fibres ensheathing or as bundle caps; laticifers non-anastomosing. *Montrichardia, Anubias, Furtadoa, Philodendron, Homalomena*
Tribe Zantedeschieae Collenchyma type Bi; vascular bundle types I, II, III. *Zantedeschia, Callopsis*
Tribe Stylochaetoneae Collenchyma type B; vascular bundles type II; fibres absent; laticifers absent. *Stylochaeton*
Tribe Zamioculcadeae Collenchyma type B; vascular bundles type I, II; fibres absent or occasional bundle caps; secretory ducts present or absent; laticifers absent. *Gonatopus, Zamioculcas*
Tribe Aglaonemateae Collenchyma types B, Bi, Sb; fibres absent, or bundle caps, or ensheathing bundles; laticifers non-anastomosing. *Nephthytis, Anchomanes, Pseudohydrosme, Aglaonema, Aglaodorum*
Tribe Culcasieae Collenchyma type B; vascular bundle type I; fibres absent, ensheathing, or xylem and phloem caps; secretory ducts present; laticifers non-anastomosing (*Cercestis*). *Culcasia, Cercestis*
Tribe Spathicarpeae Collenchyma types B, Bi; vascular bundle types I, II, III; fibres ensheathing or bundle caps; laticifers non-anastomosing or absent. *Bognera, Dieffenbachia, Spathantheum, Gorgonidium, Syandrospadix, Gearum, Spathicarpa, Asterostigma, Mangonia, Taccarum*
Subfamily SCHISMATOGLOTTIDOIDEAE
Vascular bundle types II, III.
Tribe Cryptocoryneae Collenchyma type Sv; laticifers non-anastomosing or absent. *Cryptocoryne, Lagenandra*
Tribe Schismatoglottideae Stomata mostly brachypara-hexacytic or brachypara-octocytic; collenchyma types B, Bi; laticifers non-anastomosing. *Phymatarum, Schismatoglottis, Aridarum, Piptospatha, Hottarum**, *Bucephalandra, Heteroaridarum**
Subfamily LEMNOIDEAE
Collenchyma and fibres absent; vascular bundle type II; laticifers absent. *Spirodela, Landoltia, Lemna, Wolffia, Wolffiella*
Subfamily AROIDEAE
Collenchyma type Sv; vascular bundle types II, III; laticifers: 2 types occur.
Tribe Thomsonieae Laticifers non-anastomosing. *Amorphophallus, Pseudodracontium*
Tribe Caladieae Laticifers anastomosing. *Hapaline, Syngonium, Xanthosoma, Chlorospatha, Ulearum, Filarum, Zomicarpella, Caladium, Scaphispatha, Jasarum, Zomicarpa*
Tribe Arisareae Laticifers non-anastomosing. *Ambrosina, Arisarum*
Tribe Peltandreae Laticifers non-anastomosing *Peltandra, Typhonodorum, Colletogyne, Carlephyton, Arophyton*
Tribe Pistieae Collenchyma types Svc, Sv; laticifers absent. *Pistia*
Tribe Arisaemateae Laticifers non-anastomosing. *Arisaema, Pinellia*
Tribe Areae Laticifers non-anastomosing. *Typhonium, Theriophonum, Sauromatum**, *Lazarum**, *Biarum, Arum, Eminium, Dracunculus, Helicodiceros*
Tribe Colocasieae Laticifers anastomosing or non-anastomosing. *Ariopsis, Alocasia, Remusatia, Colocasia, Steudnera, Protarum*

*Genera recently reduced to synonymy but treated separately in this volume

The apparent isolation of the Arales among monocots was expressed in Takhtajan (1997) where the families were placed within the derived subclass, Aridae, that included orders: Arales, Acorales, Cyclanthales, Pandanales, Typhales, but this placement contradicted his discussion in which he stated that the Arales are closer to the alismatids and triurids.

The detailed reviews of monocots by Dahlgren and Clifford (1982), Dahlgren and Rasmussen (1983), and Dahlgren *et al.* (1985) evaluated the potential of structural characters in understanding the relationships of the monocot groups. These compendia present valuable arrays of structural and chemical character states, many of which can be polarized within families.

The distinctive and specialized morphology of several monocot groups including orchids and grasses suggests that monocot evolution should be subject to decipherment. But among the lower groups, the main phyletic lines remain obscure. Whole suites of characters, certainly useful diagnostically, and for which morphoclines can be uncovered within families, are of essentially no use in inferring the plesiomorphic condition of lower monocots, or understanding interfamilial relationships. Parallel trends are very common within these orders or superorders. Examples of interesting characters which exhibit homoplasious development or loss across families include: growth habits, ranging from vines to bulbs; variations in atactostele structure; leaves varying from simple phyllomes to large compound structures; supervolute ptyxis in Araceae and many other families.

In histology, the occurrence of root-hair short cells is widespread; vessels are widespread in roots where stems are vesselless; laticifers are scattered among several otherwise unrelated monocot families; sieve-element plastids of the PIIc type (Behnke 1972, 1995) virtually define the monocots as a synapomorphy for all; oxalate raphides are common among alismatids, triurids, commelinids but are present in all other superorders. They are probably common in monocot ancestors; perhaps a synapomorphy for the monocots and lost in some groups. Stomatal subsidiary cells are virtually unusable as a taxonomic character at this time. Rasmussen (1983) has shown that there is little correspondence between the appearance of a mature stomatal complex and its ontogenetic type in the Liliiflorae. Therefore, one cannot use the mature stomatal complex as a basis for phylogenetic comparisons and we do not know the plesiomorphic state of development or mature appearance in the monocots. Intravaginal squamules (scales in the axils of leaves) are present in Araceae and probably all alismatids. Epicuticular waxes are absent in alismatids and Ariflorae but scattered in many orders. The cladogram offered by Dahlgren and Rasmussen (1983) shows the Ariflorae to be sister to the Triuridiflorae and Alismatiflorae. They also noted that the lemnoids probably arose within ancestral Araceae.

Molecular phylogenetic studies

Monocot–dicot divergence has been placed by Li *et al.* (1993) at 200 mya, a conclusion supported by analysis of nuclear genes encoding 26S and 18S ribosomal RNAs (Wolfe *et al.* 1989). Studies by Soltis *et al.* (1997) using 18S rDNA, Chase *et al.* (1993) and Duvall *et al.* (1993), using the plastid gene encoding for the large subunit of ribulose, *rbc*L, concluded that the monocots are a well-supported monophyletic group and that *Acorus* is the 'primal extant monocot'. The monocots are probably derived from within the monosulcate Magnoliidae with the paleoherbs as the immediate sister group. With *Saururus* as the dicot outgroup, *Acorus* emerged as the sister group to all other monocots, and therefore certainly belongs in its own family. The Alismataceae are sister to the Araceae. The genus *Pleea* (Melantheaceae) is basal to the adjacent remaining monocot lines. See also reviews by Chase *et al.* (2000) and Stevenson *et al.* (2000).

In a wider survey of the Magnoliidae, also using the *rbc*L gene, Qiu *et al.* (1993) agreed that the monocots are monophyletic and nested among ranalian orders (see also Källersjö *et al.* 1998). As concluded by Burger (1977) on the basis of morphology and anatomy, Piperales were found by Qiu *et al.* (1993) to be a sister group to the Laurales, monocots, Magnoliales and Nymphaeales. The sequence of the six monocot genera sampled, beginning with the long-branch genera, is *Pleea, Gymnostachys, Spathiphyllum, Pistia, Sagittaria* and *Alisma*.

The Angiosperm Phylogeny Group (APG 1998) produced an ordinal classification of flowering plant families. It was based on a compilation of cladistic analyses using structural and genetic sequence data, and resulted in recognition of 462 families in 40 putatively monophyletic orders, and 'monophyletic informal higher groups'. In the monophyletic monocots, they began with the monogeneric Acorales, followed by Alismatales consisting of the 13 families: Alismataceae, Aponogetonaceae, Araceae (including lemnoids), Butomaceae, Cymodoceaceae, Hydrocharitaceae, Juncaginaceae, Limnocharitaceae, Posidoniaceae, Ruppiaceae, Scheuchzeriaceae, Tofieldiaceae, Zosteraceae. This order is followed by the Asparagales (29 families), Dioscoreales (five families), Liliales (nine families), Pandanales (four families), Commelinids (4+ orders including Arecales). At this time the identification of the nearest dicot relative of these monocot groups remains unsettled. It awaits a combined

Table 6. Strict consensus tree of French *et al.* (1995), with the classification of Keating (2001b) shown on the right. Reproduced by permission of J. C. French *et al.*

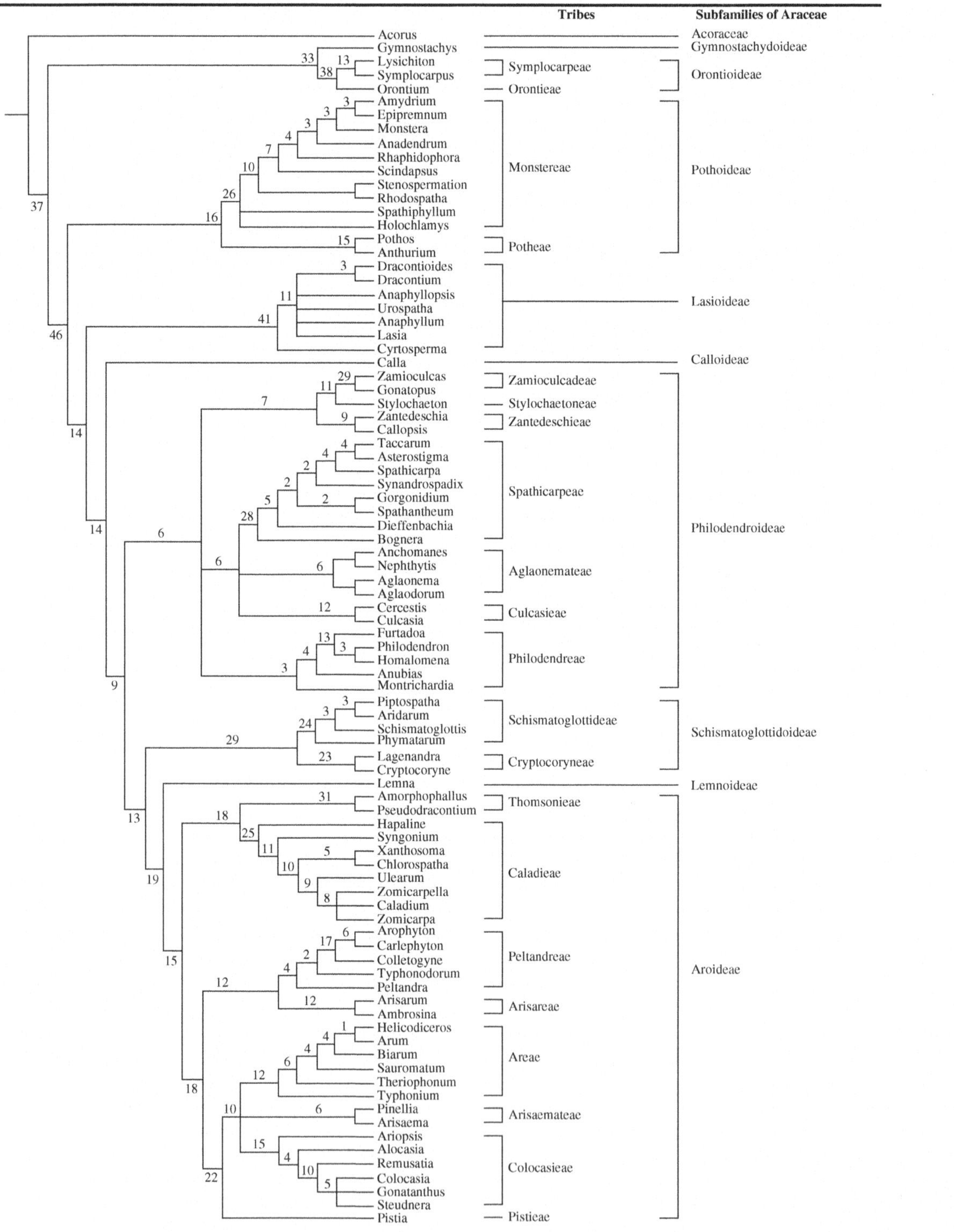

multigenomic and morphological character analysis (Duvall 2000).

The consensus, then, of these lines of evidence is that: (1) the monocots are a monophyletic line embedded within the lower dicots; (2) *Acorus* is sister to all monocots; (3) the Araceae are close enough to the alismatids for them to be placed together within a large order; (4) the Araceae are distant from the Arecaceae, Pandanaceae and Cyclanthaceae.

Structural relationships with the Alismatidae

The thorough structural study of the Helobiae (Alismatidae) by Tomlinson (1982) makes possible detailed comparisons. In spite of their being largely aquatic, the alismatid families show many similarities among themselves and with the Araceae. In leaf tissue, the cuticle is simple and thin, stomata when present are often paracytic, usually with narrow subsidiary cells. Leaf mesophyll shows bifacial development, and palisade cells may have lateral projecting lobes making them H- or U-shaped, a condition found in some genera of Araceae. The aerenchyma shows a generally lacunose ground tissue, common in emergent and floating leaves. Using the aerenchyma classification of the present study, the alismatids are mostly type 3 with some development of type 4, both well represented in terrestrial Araceae. Intravaginal squamules are widespread in the alismatids and present in a few aroids. Vascular bundle architecture can be very different within monocot groups but the Alismatidae types seem to be homologous in architecture with types I, II, and III of the Araceae, the last type being especially common among alismatids. Tannins are common or uncommon. Oxalate crystals are present as rhomboids, styloids and occasionally druses in the alismatids. 'Laticifers' appear to be present in some families although they are described as having 4–6 sheath cells, i.e. they are secretory ducts rather than true laticifer cells. Tieghem (1872) recorded circles of secretory canals bordered by 4–10 small cells in *Alisma* and fewer in *Sagittaria;* his description resembled secretory canal structure found in *Philodendron* or *Homalomena*. True raphides of any type appear to be missing among alismatids.

The alismatids are highly adapted aquatics in terms of habit, structure and ecology (Tomlinson 1982) and very heterobathmic (Takhtajan 1997: 563). While there are many similarities between the alismatids and aroids, the most defining feature in this comparison is the presence and elaboration of raphide crystals in the aroids.

MATERIALS AND METHODS

Approximately 380 specimens representing 105 out of 106 currently accepted genera of Araceae plus *Acorus* were available for anatomical study. While most were prepared from liquid-preserved materials, a few were only available dry and had to be restored.

Early in this study, vegetative tissue samples were prepared as leaf transsections (TS), through the midrib and margin, and oriented to cut secondary or tertiary veins in TS. Petioles were cut in TS proximally and distally, and often in longisection (LS). Stems and roots were cut in TS and occasionally in LS. In this phase, section preparation involved conventional paraffin microtechnique, including Safranin-O and Fast Green FCF staining (Berlyn and Miksche 1976; O'Brien and McCully 1981). The most recent third of the specimens were prepared by hand sectioning and wet mounting (Herr 1992; Keating 1996) and stained using mostly three dyes: iodine-potassium iodide (IKI), Toluidine blue-O, and cresyl violet acetate (CVA). The sections were then mounted in 30% aqueous calcium chloride, which renders the slides virtually permanent if they are stored horizontally.

A note regarding calcium chloride: this salt is hygroscopic and its final concentration and volume as a liquid slide mountant is dependent on equilibration with atmospheric humidity. The salt solution is heavy, highly ionic, and has the refractive index of glass. Because of its ability to weaken hydrogen bonds, it stimulates a marked metachromatic (polychromic) effect when using CVA and a number of other commonly used aniline dyes, especially of the quinone-imine group (Herr 1992; Keating 1996). In addition to indicating the presence and position of starch granules, IKI is polychromic on cellulosic and lignified cell walls. In an excess of iodine cellulose becomes magenta in a few days, a modified zinc-chlor-iodide test (Artschwager 1921). Lignin and cutin become yellow-orange. These preparations are usually just as photogenic as thinner paraffin sections and are often more informative.

Dried material was restored to its original hydrated dimensions using a dilute surfactant such as Kodak Photoflo®. If more drastic treatment was necessary, 1–5% sodium hydroxide worked well with the covered specimen dish placed on a warming tray for 8–24 h and removed by inspection.

Photomicrographic images were recorded on 35 mm Kodak Technical Pan® film.

HISTOLOGY AND CHARACTER STATE DEFINITIONS FOR VEGETATIVE SHOOT AND ROOT ANATOMY OF ARACEAE

This survey of histology of the Araceae includes the most reliable anatomical characters that can be used for diagnostic, evolutionary, or other studies. It includes characters that can be routinely determined for most specimens but whose trends of specialization are in some cases poorly understood. Conspicuously useful characters have been reviewed in the Introduction in the sections on anatomical character state transformations and on the evolutionary position of the Araceae; they will also be discussed later in the taxonomic notes sections at the end of tribal and generic descriptions.

For all cell and tissue types, two additional pieces of data are recorded in generic descriptions as appropriate. *Position*, the location among other tissues and organs, whether restricted, associated with other tissues, or found randomly throughout an organ. Also *frequency*, a subjective assessment as to whether a feature is rare, uncommon, common, very common or abundant.

This systematic survey is broad rather than deep for most genera. In the absence of statistically valid population samples, observations on presence, position and frequency should be taken as applying to specimens rather than species. The position and frequency of idioblastic cells containing crystals, tannin, or other materials, for example, can vary between samples and locationally within samples, due to unknown factors.

'Lists of genera in which certain diagnostic features occur' have been compiled and begin on p. 290.

CHARACTERS RECORDED

Leaf structure

Leaves are dorsiventral (Fig. 1A) or rarely isobilateral (Fig. 1B,C). Most leaves are dorsiventral and hence this is rarely discriminating in the Araceae. Differentiation of mesophyll tissues may vary from undifferentiated (Fig. 1B), weakly differentiated (Fig. 1D) to strongly differentiated (Fig. 1A).

Venation

Configurations of vascular bundles are described as they appear in surface view in clearings and paradermal sections of fresh or dried leaves. Venation in most aroid leaves is characterized by a distinct midrib (or primary). The term 'midvein' is also used to refer to one or more vascular bundles in close proximity (*sensu* Esau 1965), but the term midrib is preferred here due to the highly variable number and spacing of vascular bundles within the rib. There is a clear lack of homology with the midrib of dicots. Secondary veins (also frequently with several vascular bundles per vein) are pinnately arranged, at least above the base.

Sagittate, cordate, or pedate leaves or tetrahedral leaves may be acrodromous, having multiple primaries arising from the apex of the petiole. In the case of compound leaves, the venation of segments, not the whole leaf, is described. Secondaries, and the tertiaries arising from them, may become aligned or parallel in some genera. In others, they and higher order veins may be reticulate and closely resemble the condition found in dicot leaves. Only a few basic terms from Hickey's (1979) system of venation patterns are used here because of the lack of homology with dicot patterns. The terms brochidodromous and eucamptodromous are adequately applied (with no implication of homology) to describe the pattern or relationships of secondary veins near the margins. Brochidodromous loops can become straightened or intramarginal veins termed collectors.

There are several interesting, probably diagnostic variations in the nature of vascularization of the lamina from the midrib. For example, primary, secondary and tertiary veins in some genera produce branches by monopodial branching of individual bundles: one branch remains with the larger vein, and one enters a smaller vein. In other genera, most secondaries and tertiaries are formed as one or more bundles of the larger vein bend into the branch, thereby diminishing the number of bundles in the larger vein remaining distal to the point of branching. Also, the level of order or randomness in the pattern of lateral vein connection from base to apex of the midrib varies significantly. However, due to the lack of sufficient comparative material, these patterns are not pursued further in this study.

Midrib shape

In TS, the midrib shape and size show systematically useful variation in a few genera. Adaxial and abaxial

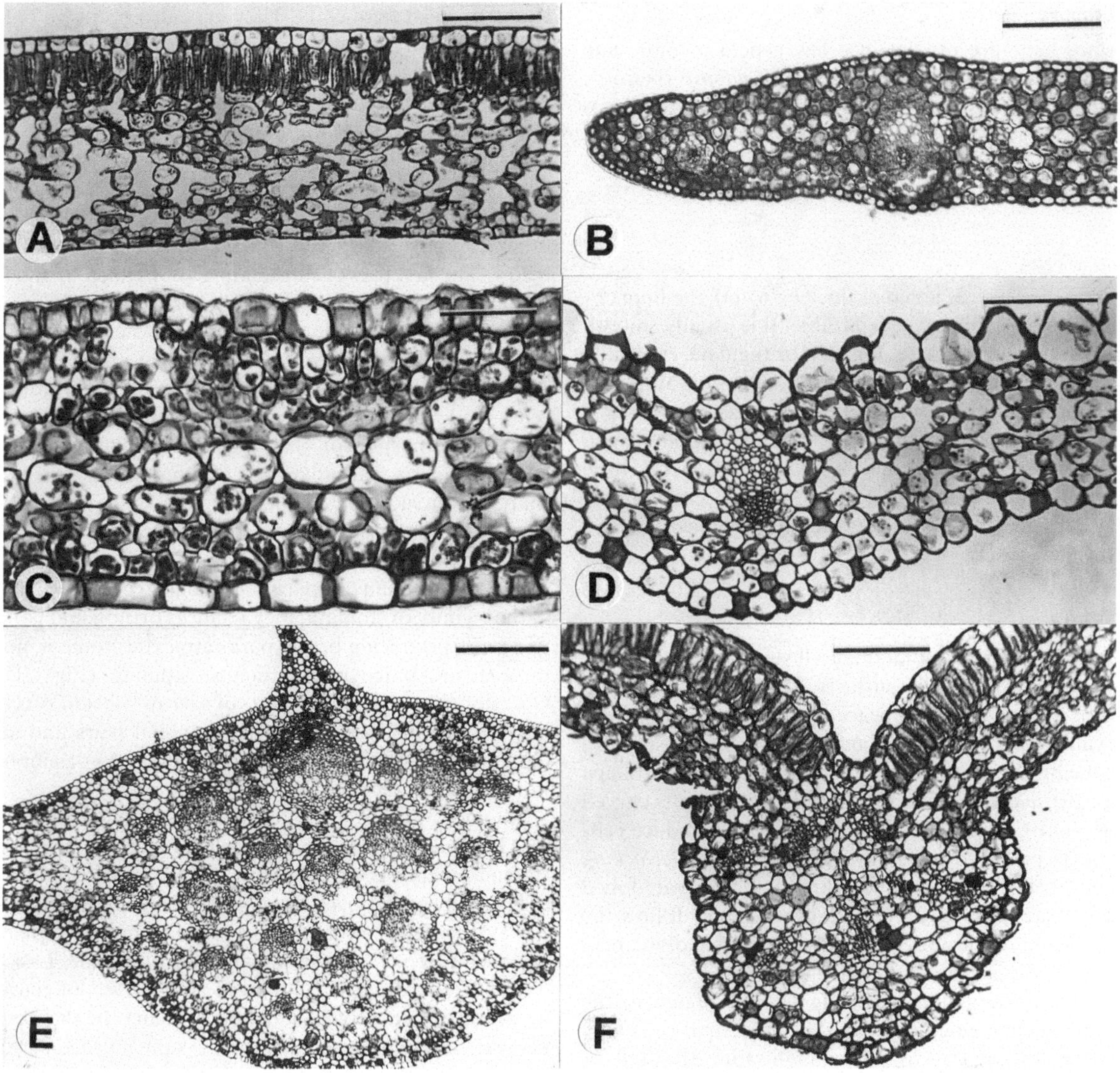

Fig. 1. Leaf sections. (A) *Rhaphidophora decursiva*, TS dorsiventral leaf showing marked differentiation of mesophyll into columnar palisade cells and type 1 spongy cells. Stoma and substomatal cavity also present on adaxial surface; (B) *Gymnostachys anceps*, TS leaf showing mesophyll with no differentiation; (C) *Pothoidium lobbianum*, TS isobilateral leaf with short palisade cells beneath each epidermal layer; (D) *Culcasia rotundifolia*, TS dorsiventral leaf showing weak mesophyll differentiation and papillate adaxial epidermis; (E) *Anthurium veitchii*, TS midrib showing broadly convex adaxial profile with single well-defined narrow angular ridge, and broadly and deeply rounded-convex abaxial surface; (F) *Amorphophallus hildebrandtii*, leaf segment TS midrib, narrowly grooved adaxially and irregularly semicircular abaxially. Scale bars: A,B,F = 200 μm; C = 50 μm; D = 150 μm; E = 300 μm.

midrib surfaces are described as convex, rounded or ridged, flat, or concave. Three examples are shown here (Figs 1E,F, 2A) and characterized in the figure legends. In this study, midrib samples are taken from the centre of the leaf and it should be understood that midribs vary in robustness within species and within plants. The ontogenetic status of the leaves in this study is occasionally not known even though mature leaves were preferred. Therefore, the midrib shape should be noted but interpreted cautiously.

Trichomes

Surface hairs are present in a few genera in uniseriate form. They may be unicellular, e.g. in *Callopsis* (petiole), *Dieffenbachia, Asterostigma* (Fig. 2A,B) and *Schismatoglottis*, 2–3 celled in *Lasia*, or multicellular in *Philodendron* (petiole, rarely), *Homalomena peltata* (rarely), and *Pistia* (Fig. 2C).

Cuticle

This surface layer is noted as thick ($\geq 6\,\mu m$), medium (2–5 μm), thin ($\leq 1\,\mu m$) or indetectable. It is usually smooth but may be rough, striate (Fig. 2D) or rugulate, especially at the midrib or leaf margins. In TS, intrusion of flanges between epidermal cells is noted in a few genera. In *Schismatoglottis*, the cuticle may produce cuticular papillae in the centre of flat epidermal cell walls. In general, little systematic use is inferred from this layer since the sample size is inadequate for assessing environmental and ontogenetic variation.

Epidermis

In paradermal view, epidermal cell shape is recorded for both adaxial and abaxial surfaces. The length/width ratio (*l*/*w*) is estimated and appears reliable for species. Cells are called *polygonal* if anticlinal walls are mostly straight, producing the geometric straight-sided polygon shape (Fig. 2E); *undulate* when anticlinal walls are gently curved (Fig. 2D,F); *sinuous* when anticlinal walls produce cells with a deeply lobed appearance (Fig. 2G) and *interlocking* when anticlinal walls produce a knobbed or clavate-lobed cell appearance (Fig. 2H). Sinuous walls may follow the entire depth of the anticlinal cell walls or the lobing may occur only near the surface (Fig. 3A).

In TS, epidermal tissue is always interpreted in the family as being single-layered. The epidermal cells may be short or square (25–55 μm high, Fig. 3B), tall or columnar (28–112 μm high, Fig. 3C), or they may be horizontally elongate (flattened) (15–30 μm high, Fig. 3D). Rather than flat, most cells are somewhat rounded internally and/or externally. They may be papillate on the external surface (Fig. 3E) and rounded or flat on the internal surface.

Stomata

In surface or paradermal view, stomata are defined to include pairs of guard cells together with their slit-like aperture (*sensu* Fahn, 1990). In Araceae, they usually appear randomly oriented and ± uniformly dispersed in the leaf epidermis (not clustered) and occasionally on petiole or stem surfaces. On leaves, stomata occur abaxially most frequently. They may also occur on both surfaces equally, be more frequently or exclusively found on either surface, or be absent (in submerged leaves of *Jasarum*). They are usually level with the surface, but may occasionally be raised on mounds above it.

Stomata of the family are basically brachyparacytic, where a pair of subsidiary cells are adjacent and parallel to the guard cells and do not come in contact with each other over the guard cell poles, with exceptions noted below. The first pair of subsidiary cells, parallel to and in contact with the guard cells, can be broad (Figs 2G, 3G) or narrow (Fig. 4A), rounded or angular, where their outer edges are more sharply squared as they abut adjacent epidermal cells (Fig. 3G). In the Lasioideae the subsidiary cells tend toward trapezoidal or semicircular shape (Fig. 2H). In spite of their positional relationship to guard cell pairs, subsidiary cell differentiation (size and shape) from adjacent epidermal cells may vary from marked to non-existent, in which case they are referred to as neighbouring cells.

When more than one pair of subsidiary cells is modified in the vicinity of stomata, they relate as orthogonal pairs, two pairs producing brachypara-tetracytic stomata, and three pairs, brachypara-hexacytic stomata (Fig. 3H). Occasionally a total of 4–5 pairs of narrow subsidiary cells may be present as concentric orthogonal pairs and are designated as brachypara-octocytic (Fig. 4A) or amphi-brachypara-hexacytic (Fig. 74C).

Anomocytic (Fig. 3F, tendency) and unipolar (Fig. 2F,G) types are occasionally encountered. In unipolar stomata, members of the first pair of parallel subsidiary cells are in contact over one pole of a pair of guard cells. Always in small percentages where found, unipolar stomata occur in 12 genera of the subfamilies Pothoideae, Lasioideae, Philodendroideae and Aroideae (see Lists of genera in which certain diagnostic features occur). In the Lemnoideae, in addition to anomocytic stomata, actinocytic stomata are encountered having 4–9 neighbouring cells.

As seen in TS, guard cells usually show double cuticular ledges (Fig. 4B), but the guard cell profile at the stomatal opening can be variable. This feature was not surveyed in detail in this account.

Hypodermis

This tissue, uncommon in the family, is usually one layer deep in the adaxial lamina (Fig. 4C) or two layers in one specimen (Fig. 4D). It may occur in the vicinity of the midrib (Fig. 3C) or other large veins, as well as in otherwise chlorenchymatous outer petiole tissue. In leaf mesophyll, it is distinguished by lack of plastids, not necessarily by shape differentiation.

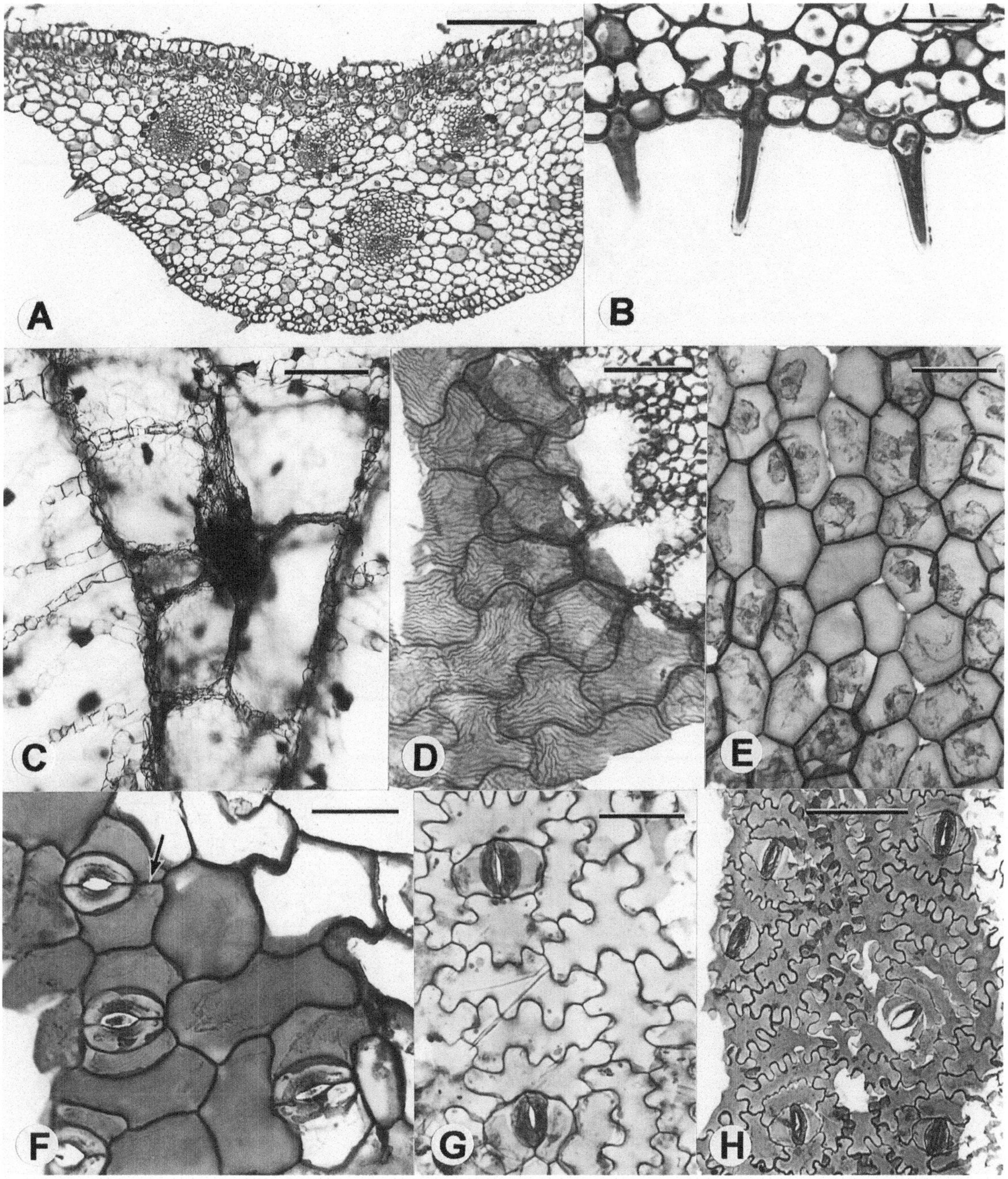

Fig. 2. Leaf and petiole sections. (A) *Asterostigma riedelianum*, TS midrib with three short trichomes visible abaxially; (B) *Callopsis volkensii*, TS distal petiole, lateral portion with three short unicellular trichomes; (C) *Pistia stratiotes*, TS lamina, abaxial rib showing numerous multicellular, uniseriate trichomes; (D) *Amorphophallus (Thomsonia) sumawongii*, PS adaxial epidermis showing undulate anticlinal walls with finely striate cuticle; (E) *Scaphispatha gracilis*, PS adaxial surface with polygonal epidermal cells; (F) *Ambrosina bassii*, PS adaxial epidermis with brachyparacytic stomata and unipolar stoma where subsidiary cells extend to contact each other at one pole (arrow); (G) *Podolasia stipitata*, PS lamina showing sinuous abaxial epidermal cells; note unipolar stoma at bottom; (H) *Nephthytis afzelii*, PS lamina showing interlocking abaxial epidermal cell lobes. Scale bars: A = 200 μm; B,D–G = 50 μm; C = 300 μm; H = 150 μm.

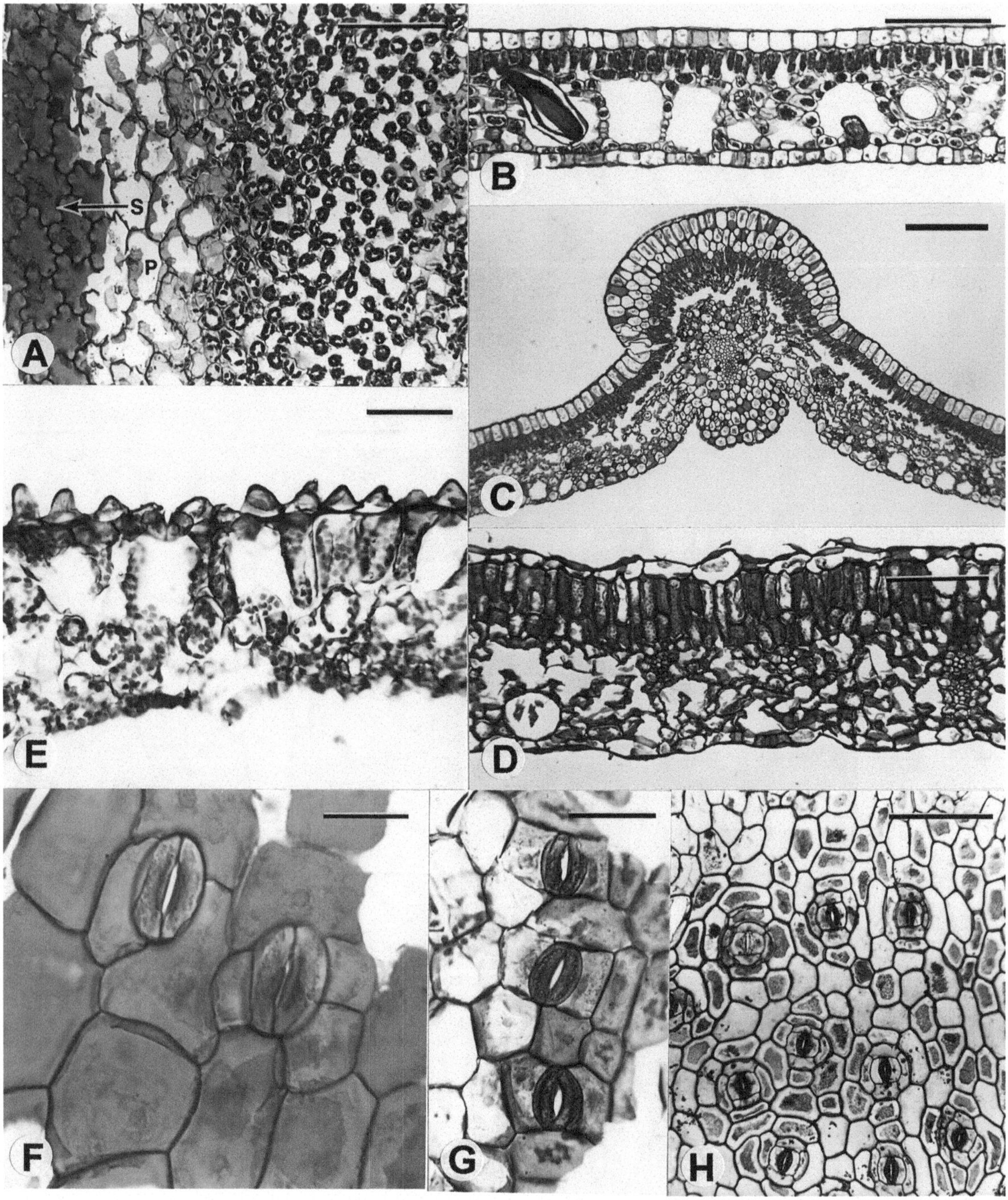

Fig. 3. Leaf sections. (A) *Aglaonema costatum*, PS adaxial epidermis showing sinuous-edged epidermal cells at surface (arrows), polygonal shape (p) near cell base, and mesophyll; (B) *Homalomena lindenii*, TS lamina with adaxial epidermal cells appearing as tall as wide, refered to in text as 'square'; (C) *Bucephalandra motleyana*, TS lamina and midrib with adaxial epidermal cells tall and columnar; (D) *Arum maculatum*, TS lamina with epidermal cells short and broad (flattened) on both surfaces; (E) *Orontium aquaticum*, TS lamina with adaxial epidermal cells papillate; (F) *Biarum tenuifolium*, PS abaxial epidermis with tendency to anomocytic stomata, rare in Araceae; (G) *Peltandra sagittifolia*, PS abaxial surface with polygonal epidermal cells and brachyparacytic stomata. Subsidiary cells have narrow or broad angular corners; (H) *Monstera oreophila*, PS abaxial epidermis with undulate epidermal cells and brachypara-hexacytic stomata. Scale bars: A,B,D,H = 150 μm; C,F = 300 μm; E,G = 50 μm.

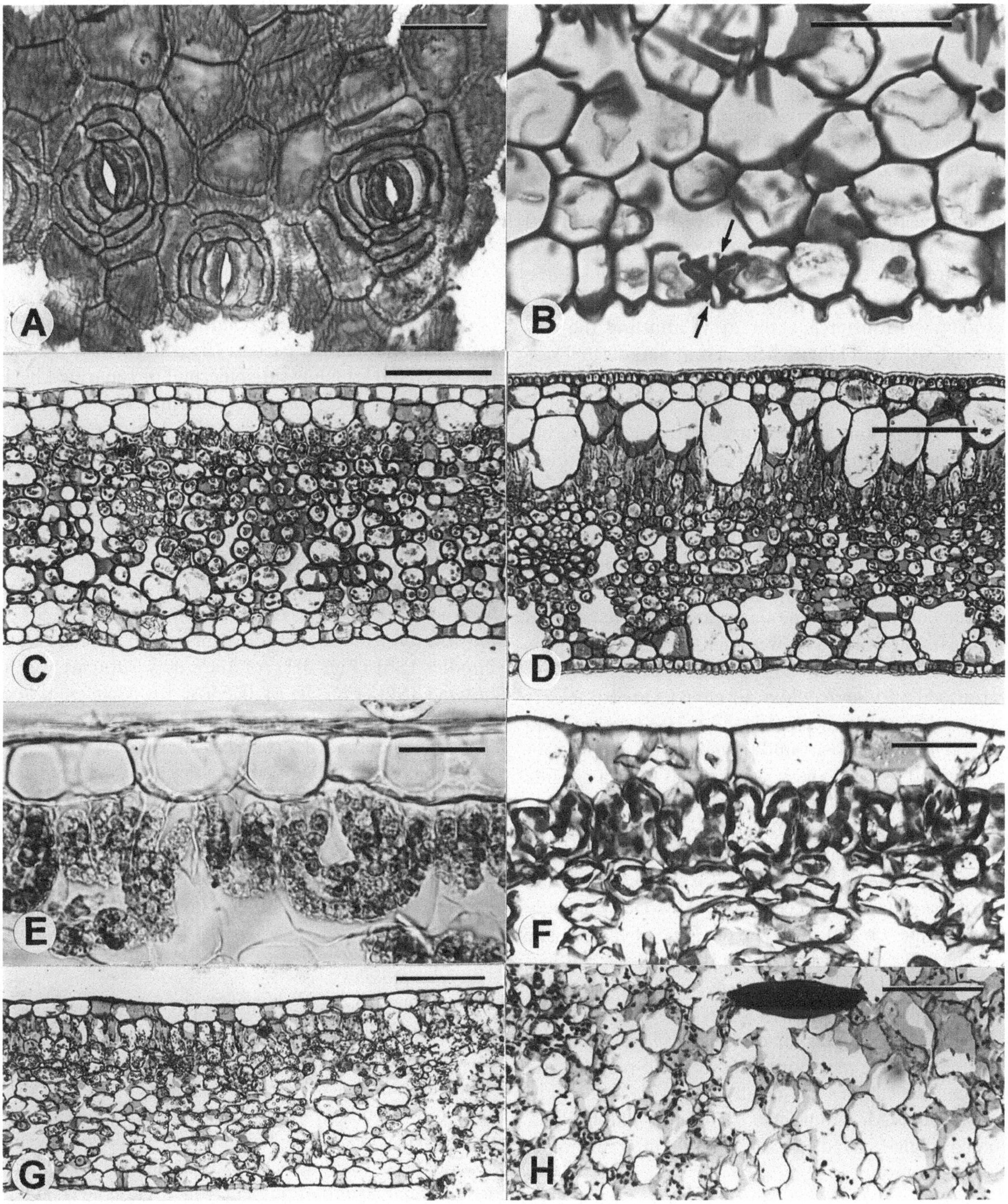

Fig. 4. Leaf sections. (A) *Bucephalandra motleyana*, PS abaxial epidermis with polygonal epidermal cells and a striate-rugulate cuticle. Brachypara-octocytic stomata with up to 4 pairs of orthogonally arranged subsidiary cells; (B) *Spathiphyllum cochlearispathum*, TS petiole margin. Stomatal guard cells level with surface and showing inner and outer cuticular ledges (arrows); (C) *Anthurium schlechtendalii*, TS lamina showing single-layered adaxial hypodermis. Hypodermis also discontinuous abaxially; (D) *Scindapsus scortechinii*, TS lamina showing small-celled outer hypodermal layer and deep irregular inner hypodermal layer; (E) *Pseudohydrosme gabunensis*, TS lamina adaxial surface. U-shaped 'palisade' cells with 6–8 lobes when seen in paradermal section; (F) *Holochlamys guineensis*, TS lamina adaxial surface. One palisade layer of multi-armed lobed cells; (G) *Stenospermation sessile*, TS lamina showing type 1 spongy mesophyll. Cells rounded with no distinct arms and irregular air cavities; (H) *Anaphyllopsis americana*, PS lamina showing type 1a spongy mesophyll. Air cavities small between strong intercellular connections. Scale bars: A,B,E,F = 50 μm; C,D,H = 150 μm; G = 200 μm.

Mesophyll

In most but not all cases, the mesophyll is differentiated into an adaxial palisade and abaxial spongy layers. The systematic utility of these tissues proved to be unexpectedly high. Palisade cells may be columnar (Fig. 3D), squarish (Fig. 1C), rounded or T-shaped, with the broad top of the 'T' facing the adaxial epidermis. In rare cases, the cells are large and multi-armed forming a U-shape (Fig. 4E), H-shape or irregular W-shapes (Fig. 4F) as in *Holochlamys* (Monstereae), the arms abutting the epidermis and with 6–8 lobes when seen in paradermal section. Cell length/width ratio (*l*/*w*) is occasionally included.

Leaf spongy tissue or other ground tissues of petiole or stem show variations in structure that can be reliably scored: **Type 1**: simple irregular cells with or without definable, round-ended arms (Fig. 4G). The air cavities (intercellular spaces) are the same size as, or smaller in diameter than the cell body. **Type 1a**: only subtly different from type 1, cells have five or more discrete, flat-ended arms that form distinctive bridges to the next cell due to arm alignment (Fig. 4H). The air spaces appear to be small and circular and situated between cell bodies. This type is usually restricted to leaf mesophyll. **Type 2**: cells are longer-armed and intercellular air cavities are up to 2–3 times the cell body diameter (Fig. 5A). This type is found in mesophyll or midrib ground tissue and occasionally in petiole ground tissue. **Type 3**: spongy cells form a net-like aerenchyma of uniseriate partitions radiating from corners (Fig. 5B), with one to several *partition cells* between the *corner cells*. This type, and type 2, often produce columnar air cavities oriented as substomatal wells extending the height of the spongy tissue (Fig. 5C). Air space is usually estimated to be 95% of this type of tissue. Type 3 tissue may occur in petiole or stem ground tissue as well as leaf mesophyll. **Type 4**: a ground tissue that is restricted to midribs, petioles and stems. In TS, this type of aerenchyma usually has large circular air cavities that are separated by two-multiseriate partitions (Fig. 5D). In petioles and stems, these air cavities are usually very wide and longitudinally elongated. Transverse or oblique diaphragms, normally single-layered, occur at long intervals of *c.* 10–30 mm or more.

Air cavities

In mesophyll tissue, intercellular air-cavity volume is subjectively estimated as the percentage of air space to cell volume on a section surface.

Collenchyma

This tissue is easily distinguished from fibres by staining reaction, even when wall thickening pattern is similar. For instance, in wet mounts collenchymatous thickenings are sharply distinguished from lignified fibres when using cresyl violet acetate-stained sections mounted in calcium chloride. A study of this previously neglected tissue demonstrates major trends of specialization within the family (Keating 2000). Collenchyma is occasionally absent but, when present, the cells are axially elongate (Fig. 5E) and most often angularly thickened in TS (Fig. 5F). The pattern in TS midribs, petioles, and stems is distinctive and reliably characterizes tribes and subfamilies. The banded pattern (B) is illustrated by a continuous band which is subepidermal or separated by 1–2 ground tissue layers from the epidermis. The band or layer is two to many cells thick (Fig. 5G). In a given species the band is usually uniformly thick around the periphery of an organ. Interrupted bands (Bi) are found in some tissues where ribs of chlorenchymatous tissue or ordinary ground tissue may radiate through gaps in a basically banded pattern (Fig. 5F). Stranded (S) collenchyma comprises irregular or rounded, discrete strands usually occurring around the peripheries of leaf midribs or petioles. Strands may occur on alternate radii of an organ between vascular bundles (Sb) (Fig. 5H), or they may subtend vascular bundles (Sv) (Fig. 5I) or be close to xylem or phloem, resembling fibre caps (Svc).

Collenchyma tends to be best developed where sclerenchyma is least developed. In a few genera, collenchyma and sclerenchyma fibres occupy identical loci and grade from one to the other, depending on position in the organ. In these cases they pass the congruence test for homology but it is not known whether these two tissues are always homologous in origin.

Vascular bundles

Vascular bundles vary from numerous to few or one per midrib; they may appear random or scattered, or disposed in distinct bilateral symmetry. As they appear in TS, vascular bundles within leaves and petioles are always simple and collateral and may be practically classified into three types. **Type I**: numerous metaxylem cells in several files, forming a broad boundary with a wide, semicircular strand of phloem (Fig. 6A). **Type II**: xylem is composed of a small or large protoxylem lacuna and two to several narrow metaxylem cells (Fig. 6B), forming a narrow boundary with a small circular or laterally narrowed strand of phloem. **Type III**: a wide functional late protoxylem cell or lacuna with 0 or only 1–2 very narrow metaxylem cells (Fig. 6C); the xylem opposes a small and laterally narrowed phloem strand.

The three above types can occur in stem tissue as well. In addition, central cylinders of stems may have

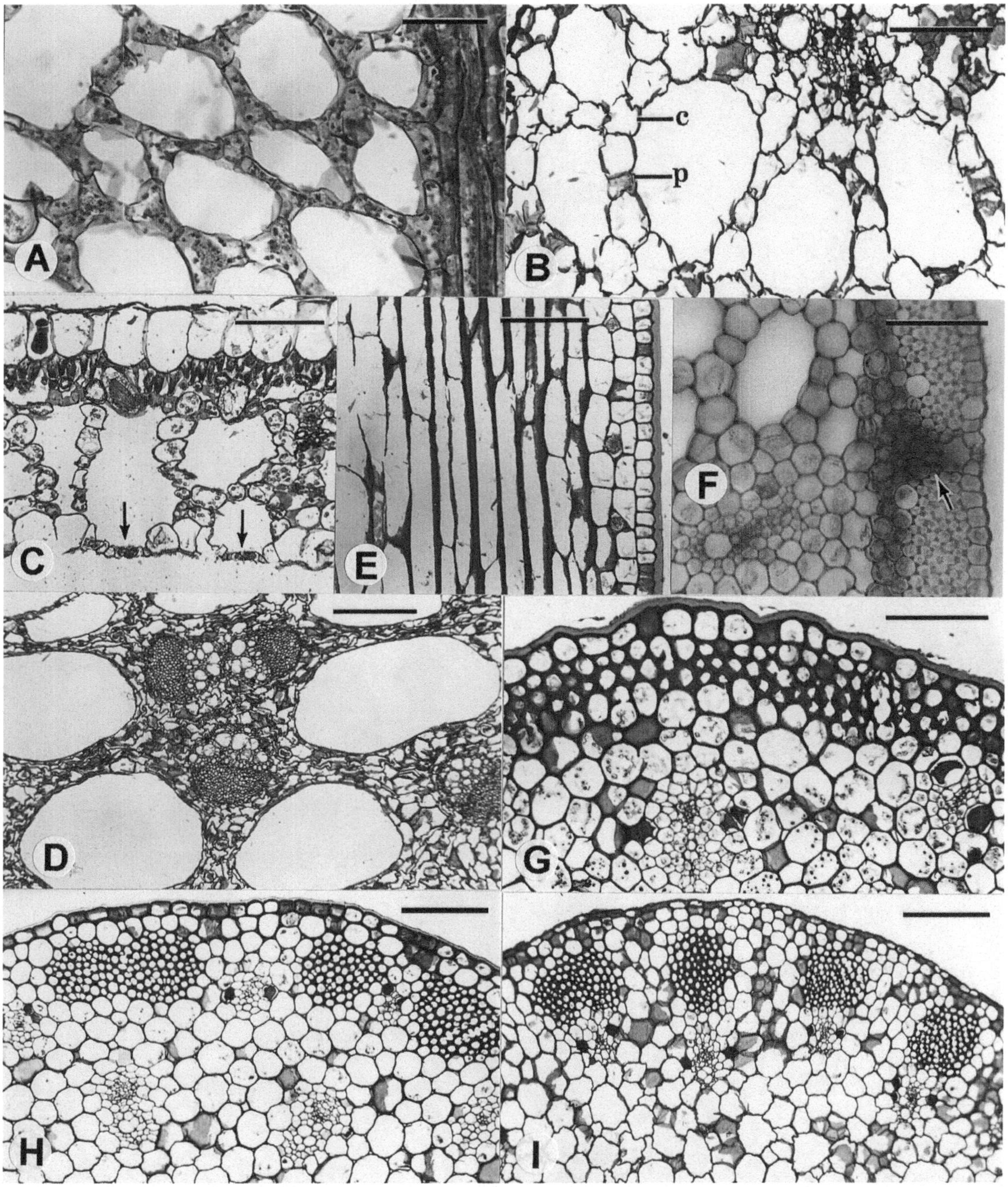

Fig. 5. Leaf, petiole and stem sections. (A) *Steudnera colocasiifolia*, PS lamina showing type 2 spongy aerenchyma. Mesophyll cells long-armed and all partitions between air cavities are extensions of corner cells; (B) *Jasarum steyermarkii*, TS proximal petiole showing type III vascular bundle and type 3 aerenchyma. This tissue contains partition cells (p) separating air cavities attached to corner cells (c); (C) *Aridarum nicolsonii*, TS lamina showing adaxial epidermal cells columnar or square and larger than abaxial cells. Abaxial stomata (arrows) subtended by high substomatal air cavities; (D) *Cyrtosperma johnstonii*, TS distal petiole, central tissue with vascular bundles dispersed among type 4 air cavities; (E) *Dieffenbachia lentii*, LS stem showing elongated collenchyma cells having 8–10:1 (length/width ratio); (F) *Zantedeschia albomaculata*, TS petiole showing subepidermal collenchyma band (type Bi) interrupted by chlorophyll-bearing cortical cells (arrow); (G) *Hottarum lucens*, TS distal petiole abaxial side showing subepidermal collenchyma band (type B), 3–5 cell layers deep; (H) *Aglaonema brevispathum*, TS distal petiole showing strands of collenchyma (type Sb), not subtending vascular bundles but rather on alternate radii with vascular bundles; (I) *Syngonium auritum*, TS distal petiole showing collenchyma in independent ±circular strands (type Sv), each subtending small vascular bundle. Scale bars: A = 50 μm; B,F,G = 150 μm; C,E,H,I = 200 μm; D = 300 μm.

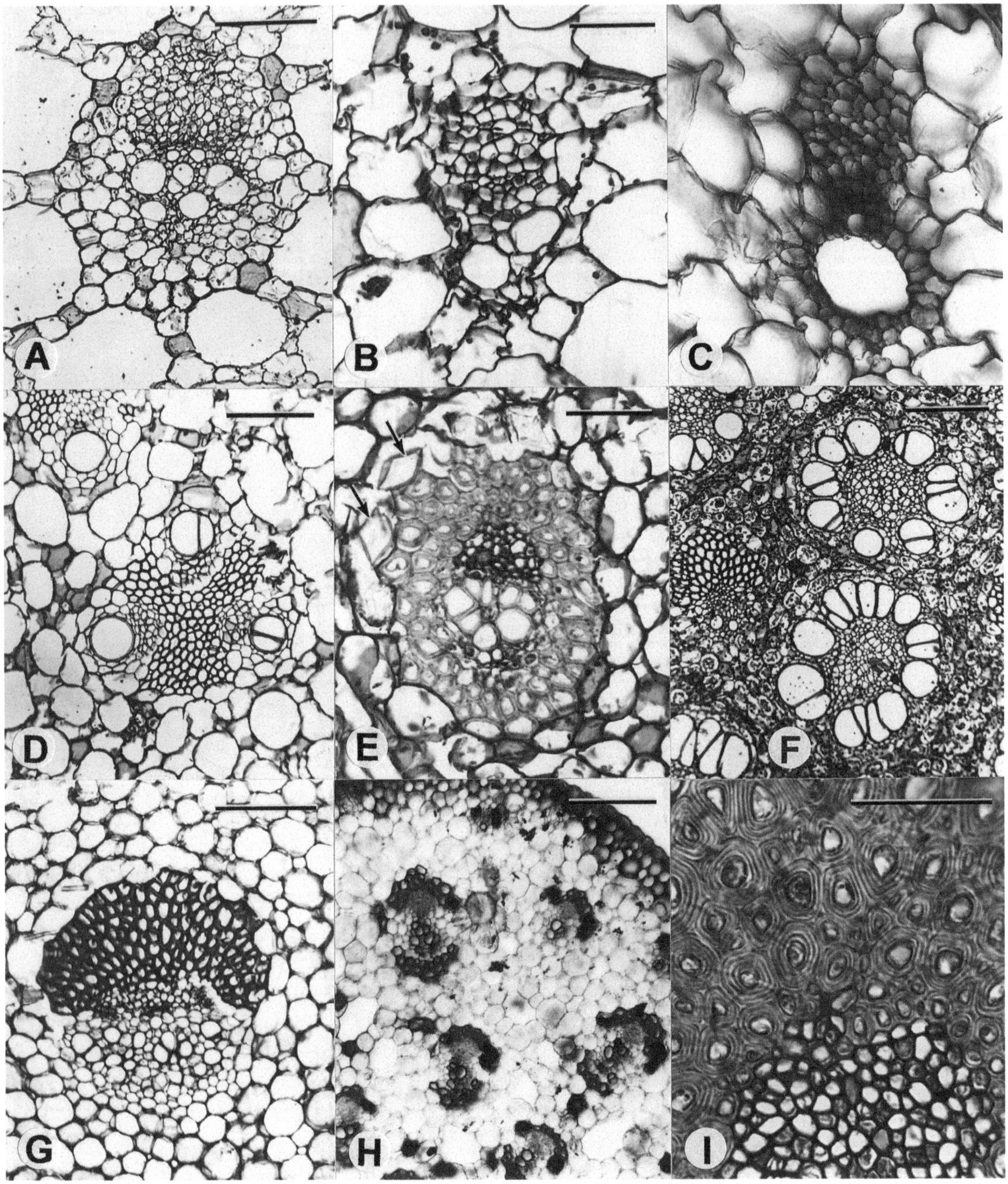

Fig. 6. Vascular bundles in petiole and stem sections. (A) *Orontium aquaticum*, TS distal petiole, central tissue showing type I vascular bundle in type 3 aerenchyma; (B) *Spathicarpa sagittifolia*, TS distal petiole central tissue. Type II vascular bundle with few metaxylem cells and a circular strand of phloem; (C) *Sauromatum guttatum*, TS petiole central tissue. Type III vascular bundle with large persistent protoxylem element and laterally narrowed phloem strand; (D) *Montrichardia arborescens*, TS stem central tissue. Compound bundle shown with three xylem poles, phloem strands and a central triangular fibre strand; (E) *Pothos scandens*, TS petiole base with fibres ensheathing type I vascular bundle. Note rhomboidal crystals next to sheath; (F) *Syngonium hoffmannii*, TS stem near cortex-central cylinder boundary showing amphivasal vascular bundles; (G) *Monstera siltepecana*, TS stem cortical ground tissue with small type I vascular bundle and thick phloem fibre cap; (H) *Cercestis ivorensis*, TS petiole abaxial side with central cylinder absent. Separate fibre strands over xylem and phloem; (I) *Heteropsis spruceana*, TS distal petiole showing thick sheathing layer of fibres surrounding phloem. Fibre walls bearing distinct isotropic wall layering. Scale bars: A,C,G = 150 μm; B,E,I = 50 μm; D,H = 200 μm; F = 300 μm.

compound bundles and amphivasal bundles. *Compound bundles* appear to be 2–6 fused bundles with central phloem and discrete xylem strands at the periphery. This type may have a single central circular or fluted phloem strand or separate phloem strands, separated by a fused central fibre cap of various shapes (Fig. 6D). *Amphivasal bundles* are circular rings of tracheary elements surrounding a central, usually circular phloem strand (Fig. 6F).

Sclerenchyma

Two forms occur. *Fibres*, when present, are always associated with vascular bundles where they may ensheath bundles entirely (Fig. 6E), or form caps over the phloem or xylem, or both (Fig. 6G,H). In some cases they are sparsely present as several individual fibres associated with vascular bundles. Thick fibre walls may show distinct concentric layering in TS (Fig. 6I) in some genera. These layers, also called isotropic layers by Mueller and Beckman (1979), are the result of banded deposition of cellulose, pectins, hemicelluloses and lignin such that isotropic and anisotropic components produce visible alternating wall layers. The layers can be seen using polarised light, bright field (usually), and electron microscopy. The significance of visible layering is not known but it appears characteristic of certain tribal groups, e.g. Monstereae.

Sclerenchyma may also occur as *trichosclereids* that are straight, branched, or H-shaped (Fig. 7A). In some branched types, an enlarged central body may occur (Fig. 7B). The cells may be solitary (Fig. 7C) or aggregated (Fig. 7D,E) and occur in leaf blades and petioles. In the lamina they are usually aligned parallel to certain veins. In axial organs, they are parallel with the main vascular bundles.

Secretory ducts are elongated cylindrical cavities surrounded by one or more layers of secretory epithelial cells that are distinctly smaller than surrounding ground tissue cells (Fig. 7F). The ducts vary in diameter with two distinct sizes appearing in some species. They are easily distinguished from type 4 air cavities by their epithelium and usually mucilaginous contents. They tend to run parallel to vascular bundles but are never attached to them.

Laticifers

While absent in many tribes, laticifers are the articulated type in all genera where they appear. In TS, they occur laterally to vascular bundles, often in pairs, and usually near the phloem. Morphologically, they are non-anastomosing (Fig. 8B) or anastomosing (Fig. 8C). Anastomosing laticifers are noteworthy for branching away from vasculature and ramifying among the cells of neighbouring ground tissue (Fig. 8D). They vary considerably in diameter and in stainability of contents. Studies by Fox and French (1988) indicated that laticifers in colocasioids contain sterols and sterol esters. In two tribes, Spathicarpeae and Areae, laticifers often contain clear contents although pigmentation and stainability vary widely throughout the family.

Tannin cells

These idioblastic cells are distinguished from neighbours by the presence of dense, granular, amorphous, often dark-staining contents. The cells are not usually differentiated by size and shape from surrounding ground tissue (Fig. 8A). Using the caffeine fixation test, the contents often test positive for tannins or phlobaphenes, but in many cases the contents are unknown. The usual staining reaction with metachromatic dyes is the same as for lignin, e.g. blue/violet in the case of cresyl violet acetate, or blue when using Toluidine blue-O. On the basis of appearance and staining reactions, there appears to be only one type of such idioblastic cells.

Starch granules

These vary considerably within specimens, probably in terms of ontogenetic state and nutrition of the sampled plant. In midribs and stems, starch granules are most often concentrated in the vicinity of vascular bundles. Granule size and shape are distinct in a few genera and a systematic survey of granules beyond the observations of this study would be valuable. Except in those few genera where they are distinct, starch granules are not considered diagnostically useful in these sample sizes.

Crystal-bearing cells

Whenever crystals of Araceae have been tested, their appearance as well as chemical and optical behaviour is typical for calcium oxalate. The crystals of *Xanthosoma sagittifolia* were so diagnosed by Cody and Horner (1983) as were those of *Xanthosoma*, *Colocasia* and *Alocasia* by Sakai and Hanson (1974). Raphides, druses, crystal sand, prismatic crystals and styloids (rarely) are found, often in the same organ.

Raphide-bearing cells are nearly universally present in the family and they show highly variable morphology. The various forms are reliably characterized within genera and two or three types may be present in a single specimen. The needle-like crystals are arranged in bundles of closely stacked, aligned crystals surrounded by a membrane within a cell. The parent cell can be:

(a) thin-walled, short and irregular, wide (Fig. 8E), rounded or oval-shaped (Fig. 8F), varying from smaller to larger than surrounding ground tissue cells. The crystals are usually arranged in single bundles in these cells;

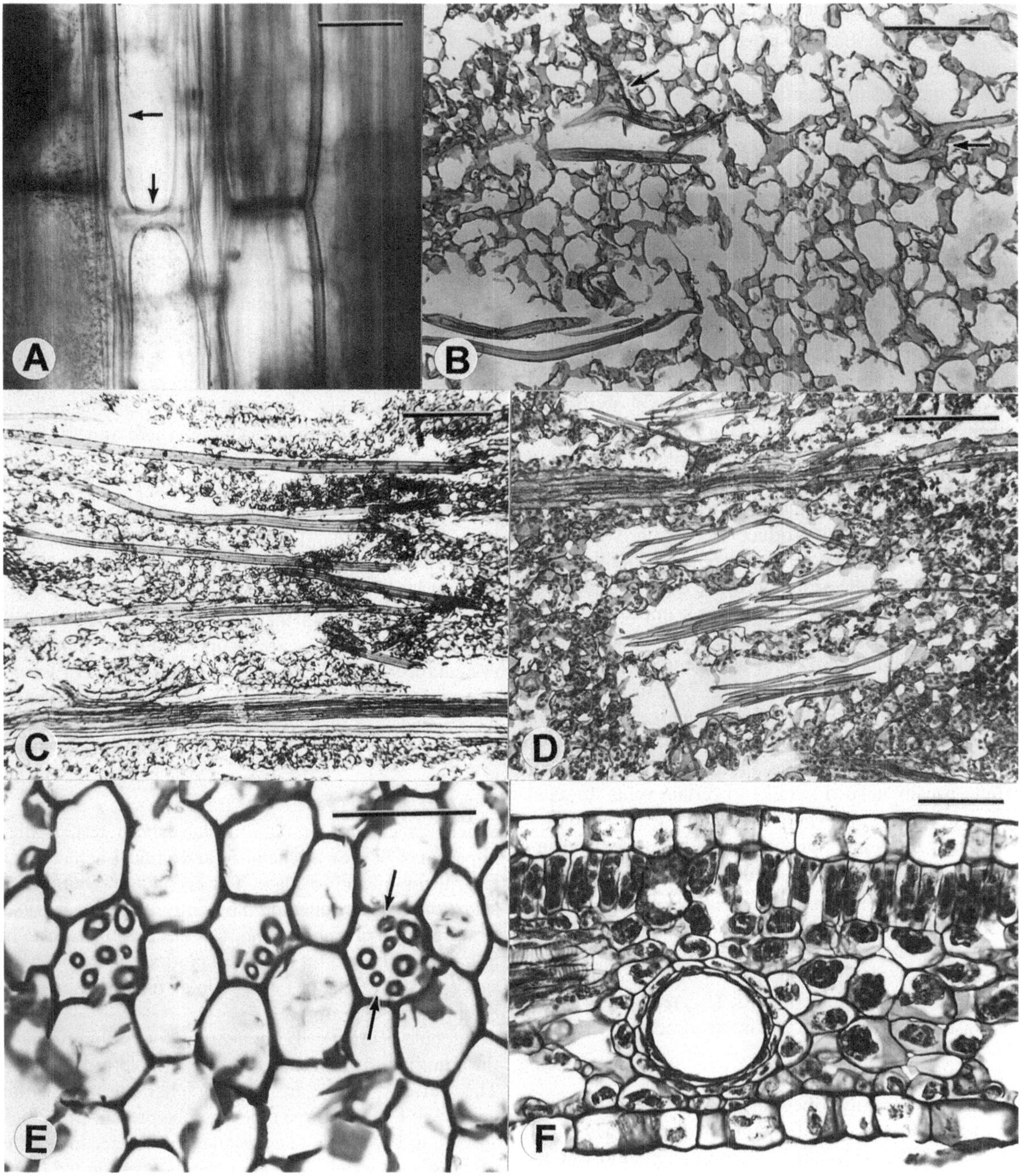

Fig. 7. Leaf and petiole sections. (A) *Alloschemone occidentalis*, LS petiole section showing centre of H-shaped trichosclereid (arrows); (B) *Rhaphidophora africana*, PS lamina showing spongy mesophyll. Branched sclereids have enlarged central bodies (arrows); (C) *Monstera oreophila*, PS lamina showing isolated trichosclereids running parallel in mesophyll; (D) *Spathiphyllum* sp., PS lamina showing oblique view of parallel groups of trichosclereids traversing mesophyll; (E) *Spathiphyllum cannaefolium*, TS petiole marginal tissue showing intercellular cavities containing trichosclereids (arrows); (F) *Homalomena lindenii*, TS lamina with large secretory duct having two epithelial layers. Scale bars: A,E,F = 50 μm; B,D = 150 μm; C = 200 μm.

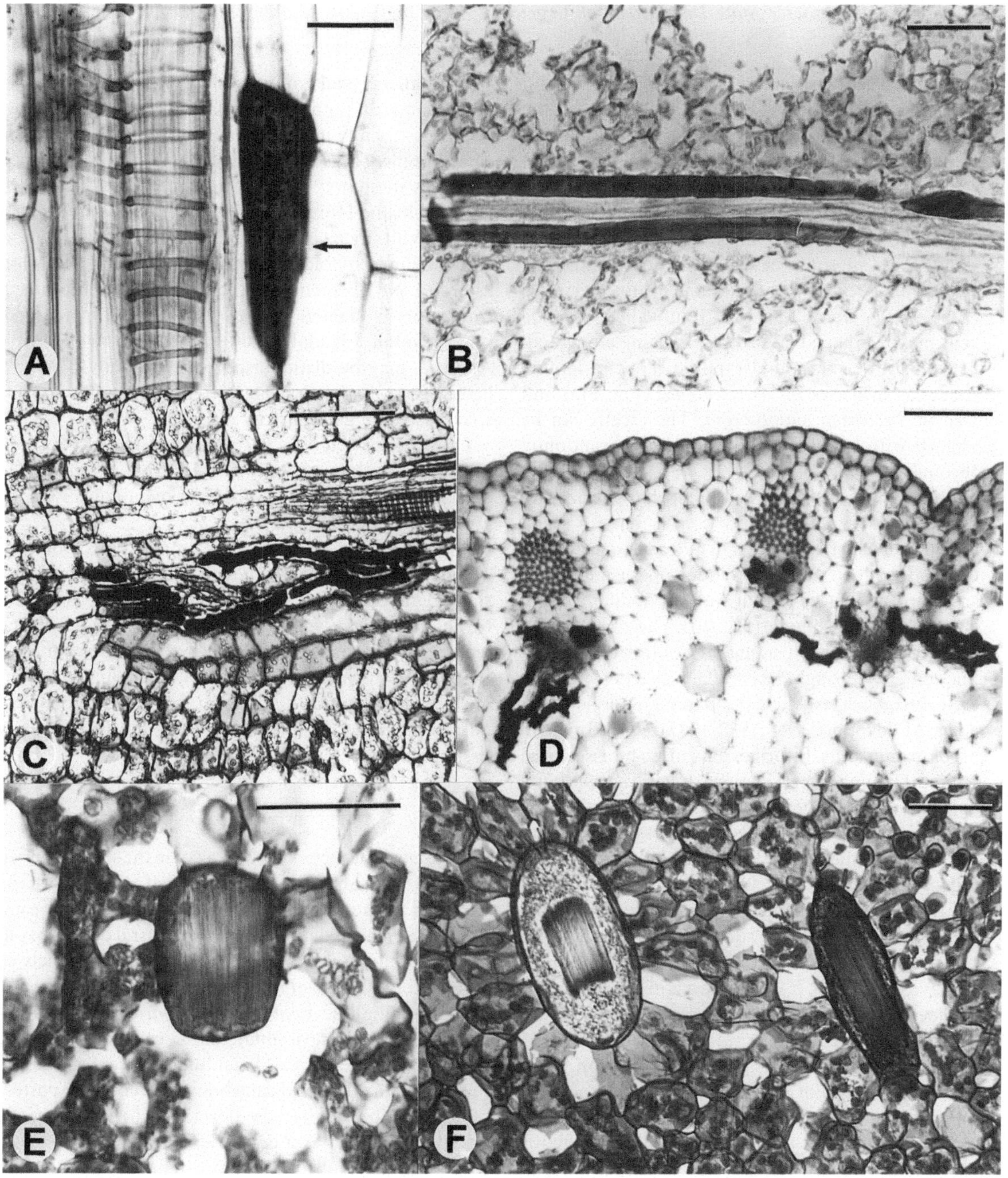

Fig. 8. Leaf, petiole and stem sections. (A) *Pycnospatha arietina*, LS petiole with tannin-bearing cell (arrow) near protoxylem elements. Cell distinguished only by content from neighbouring ground tissue; (B) *Pseudodracontium siamense*, PS lamina showing non-anastomosing pair of laticifers lateral to vascular bundle; (C) *Syngonium hoffmannii*, LS stem showing anastomosing laticifers near vascular bundle; (D) *Zomicarpella amazonica*, TS petiole margin showing anastomosing laticifers ramifying into ground tissue in vicinity of vascular bundles; (E) *Chlorospatha longipoda*, PS lamina showing wide raphide bundle in broad cell suspended within spongy mesophyll; (F) *Anubias barteri* var. *nana*, PS lamina with type 1 spongy mesophyll. Thin-walled oval raphide cell and thin-walled narrow raphide cell. It is not known whether these are developmental stages of the same crystal cell type. Scale bars: A,B,E,F = 50 μm; C = 150 μm; D = 200 μm.

(b) thin-walled cells becoming elongate in some genera, extending considerably beyond the ends of the crystal bundles (Fig. 9A). Such cells are referred to here as tubular, and when quite long, have been called 'raphide vessels' by various authors. The crystals within may be in single bundles, multiple discrete bundles (Fig. 9A), multiple overlapping bundles, or in some cases, elongated bundles of overlapping helically or obliquely arranged crystals (Fig. 9B,C). Rarely, short, thin-walled raphide cells may be aligned end to end forming elongated, articulated or compound raphide tubes in lamina tissue, as in leaves of *Asterostigma* (Fig. 9D) and also in *Taccarum* and *Biarum*. These cells can be aligned with vascular bundles or course randomly through mesophyll tissue;

(c) **biforines**: torpedo-shaped cells, first characterized by Turpin (1836). These unusual cells have a thick lateral wall which is lignified in at least some genera. Cell ends (or one end) are rounded and usually have small protuberances or thin-walled papillae. Within is a single raphide bundle (Fig. 9E). These specialized cells are capable of extruding or shooting the crystalline contents through the papillae when deformed or disturbed, such as when the parent tissue is being eaten. Middendorf (1968, 1982, 1983) referred to them as 'blow-gun' cells. Biforines may often be associated with leaf lamina or petiole tissue having a type 3 or 4 aerenchyma. They can be wholly within the aerenchyma partition, or be oriented across (perpendicular to) a partition with both ends protruding into adjacent air spaces (Fig. 9E). A thin-walled variant of the biforine (biforine-like, or spindle-shaped, in this volume) is attached by one end to an aerenchyma partition, and has the free end suspended within an air cavity (Fig. 9F); it is usually distinguished by the absence of a constricted papilla on the free end. The variations in form and position, presence of intergrading forms, as well as systematic position, leave open the question of their homology and evolution.

Druses are clusters of small calcium oxalate prismatic crystals of various sizes (Fig. 9G), usually found in rounded cells. While common in dicot families they are rarely found in monocots except in *Acorus*, Araceae and *Tofieldia* (Prychid and Rudall 2000). The cells may or may not be differentiated from surrounding ground tissue. There are differences between druses in terms of orientation, texture, size and the density of their prismatic components, but these are mostly not reliably characterized by light microscopy and no precise typology was developed in this study.

Prismatic crystals and **styloids**, large crystalline parallelograms or pointed flat plates, are found in a few genera, especially in ground tissue cells ensheathing vascular bundles (Figs 6E, 9C) or at the periphery of central cylinders in stems. These types are defined here as by Franceschi and Horner (1980), Fahn (1990) and Dickison (2000). While their sizes and specific shapes vary, Dickison (2000) defined styloids as being at least four times as long as broad. No distinction of the smaller prismatic crystals from more elongated styloids was drawn for monocots by Prychid and Rudall (2000). Both types, while rare in Araceae, may be distinguished in the family. Rhombohedral crystals are always shorter and usually not conspicuously elongated in any direction.

Crystal sand refers to varying quantities of very small prismatic crystals of various sizes and textures. They occupy a small portion of otherwise metabolically typical ground tissue cells.

All crystalliferous cells vary in the detectability of crystal membranes, and in the presence or absence of dark-staining mucilaginous sheaths. Birefringence, as detected by polarized light, varies in terms of whether the crystals are covered or mixed with other mucilaginous materials. At present, such variation cannot be evaluated systematically. Much visible variation could be due to age, nutritional state, or other environmental history of the tissue examined.

Stem

Histological characters as they occur in the stem are mostly included in the above discussions of leaf and petiole tissue. In terms of stem architecture, vasculature and ground tissue are separated into cortex and central cylinder zones in some genera. There are usually differences between vascular bundles mostly assignable to their being leaf trace or axial bundles. Leaf traces, always collateral and having protoxylem, originate in the central cylinder and pass through the cortex. Axial bundles are distinguished by their metaxylem and can show more complex patterns of aggregation. In its best developed form, the central cylinder boundary is a cylindrical concentration of vascular bundles embedded in a confluent lignified fibre barrier separating central ground tissue from cortical tissue (Fig. 10A). The presence of an endodermal cylinder surrounding the central cylinder is identified by the presence of casparian strips deposited in their anticlinal walls. It is recorded here in the genera *Acorus* and *Epipremnum*, and as surrounding individual meristeles in *Schismatoglottis* and *Bucephalandra*. In addition it is recorded in literature as surrounding the central cylinder in *Amydrium*, *Anthurium*, *Monstera*, *Orontium*,

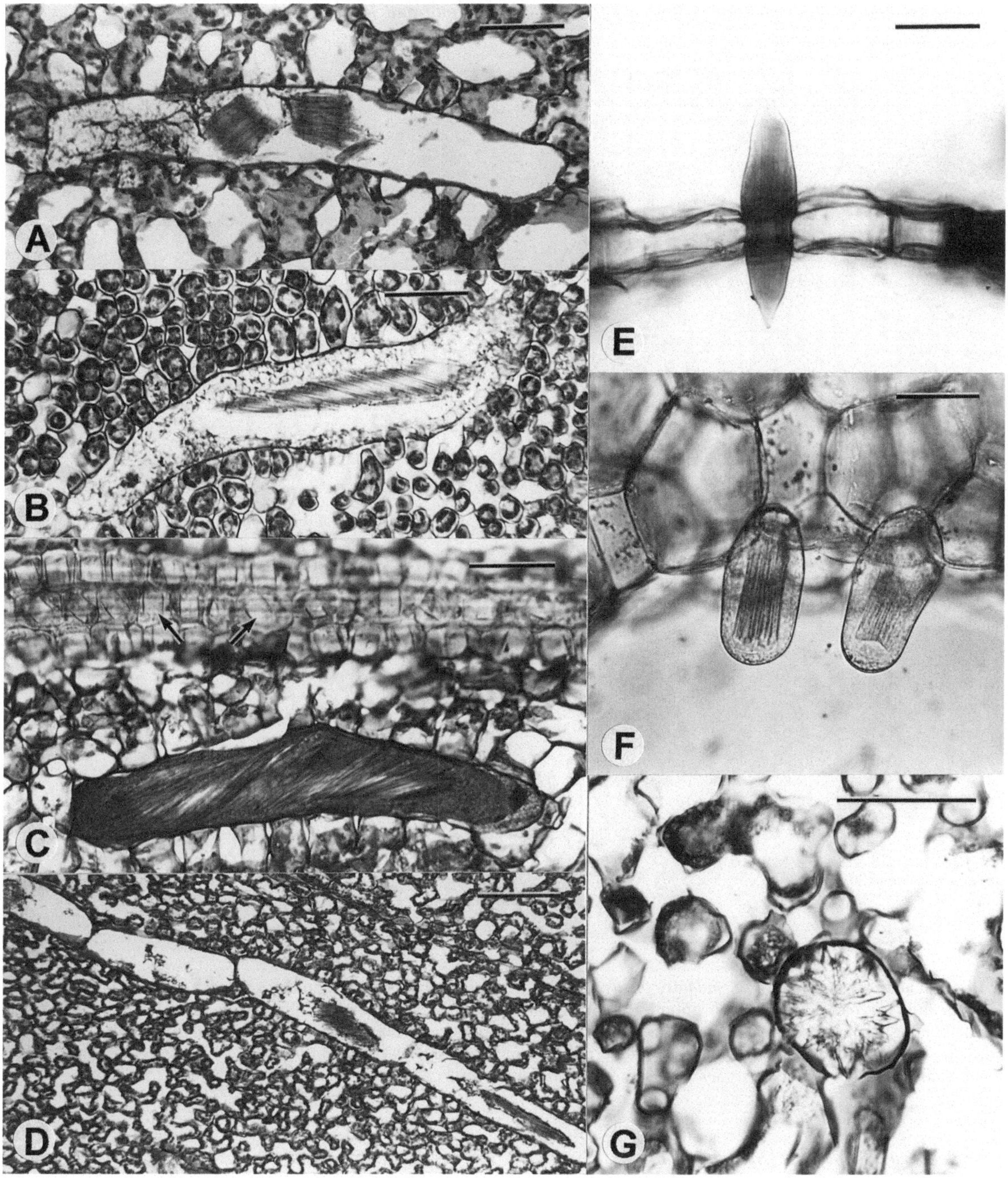

Fig. 9. Leaf and petiole sections. (A) *Spathantheum orbignyanum*, PS lamina showing raphide tube with double bundles suspended within spongy mesophyll; (B) *Anthurium polyschistum*, PS lamina through palisade mesophyll. Raphide tube bearing elongated oblique raphide bundle; (C) *Pedicellarum paiei*, PS lamina with raphide tube bearing elongated oblique crystal bundle. Bundle sheath bearing prismatic crystals (arrows); (D) *Asterostigma riedelianum*, PS lamina with raphide cells within mesophyll connected to form articulated raphide tube; (E) *Pistia stratiotes*, TS leaf base showing type 3 aerenchyma partition bearing biforine raphide cell protruding orthogonally into two adjacent cavities; (F) *Alocasia cucullata*, TS petiole showing type 4 aerenchyma partition and adjacent cavity. Raphide cells, anchored by one end, suspended in cavity; (G) *Amydrium medium*, PS lamina with enlarged druse cell in spongy mesophyll. Scale bars: A–C,E–G = 50 µm; D = 300 µm.

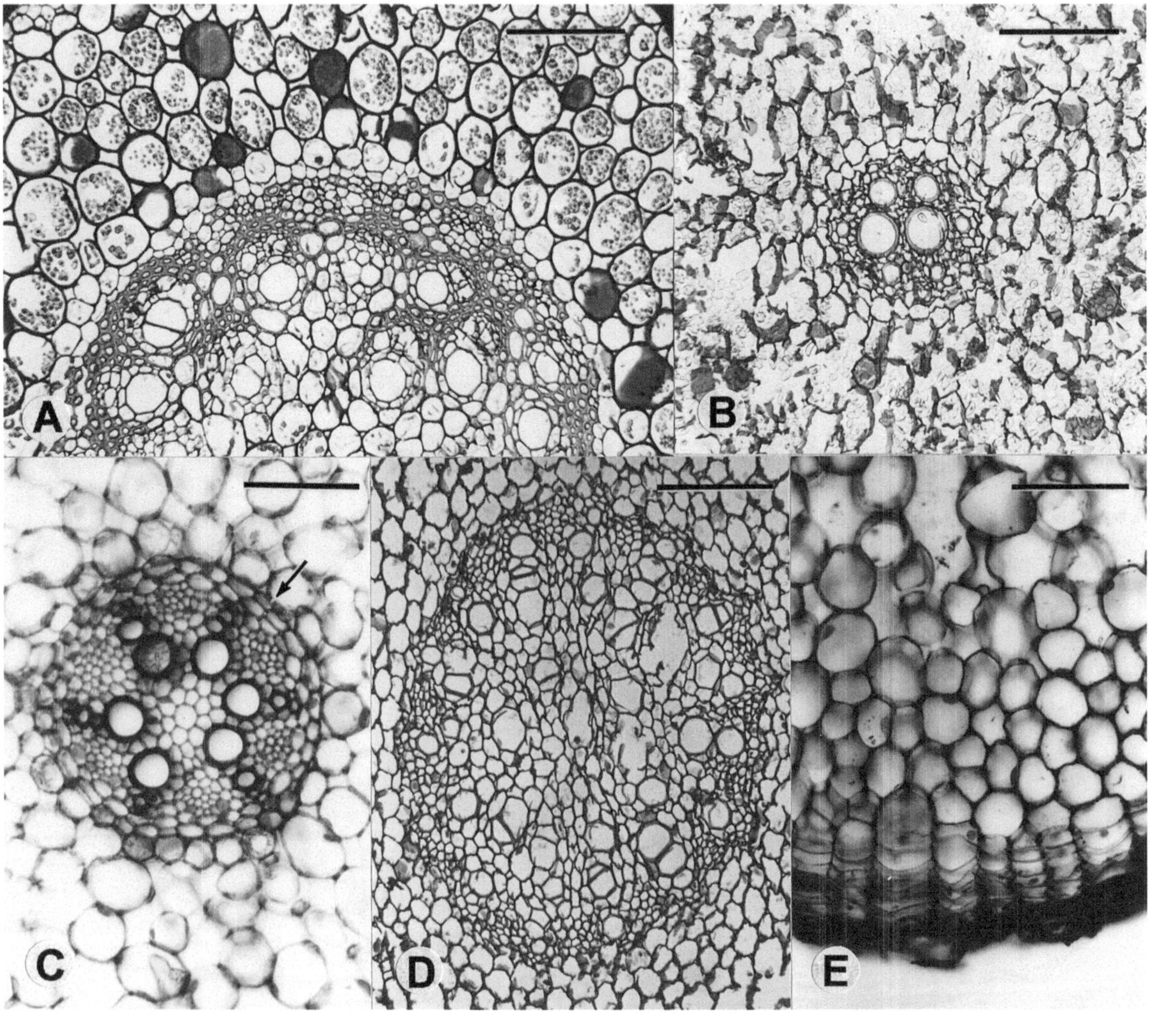

Fig. 10. Stem and root sections. (A) *Scindapsus pictus*, TS stem showing cortical parenchyma loosely packed with numerous corner air spaces and cells packed with starch granules. Central cylinder clearly demarcated by fibre band surrounding closely spaced vascular bundles; (B) *Cryptocoryne spiralis*, TS root centre showing inner cortex and stele with 6 protoxylem poles; (C) *Gearum brasiliense*, TS root centre with clearly defined endodermis (arrow); (D) *Zamioculcas zamiifolia*, TS root showing stele and less well-defined endodermis; (E) *Gearum brasiliense*, TS root showing outer cortex and periderm. Scale bars: A–E = 150 μm.

Rhaphidophora, Scindapsus, Spathantheum and *Symplocarpus*, and as surrounding meristeles in *Hottarum* and *Typhonodorum* (French 1997; French and Tomlinson 1981a–d, 1983). Van Fleet (1942) observed patterns of endodermal occurrence or its absence in stems of the family. Its presence in the relatively small stem sample of this study is noted but not analysed in detail. Central cylinders may lack fibres and be otherwise only vaguely delimited. In stems where compound (Fig. 6D) and amphivasal (Fig. 6F) vascular bundles are present, they always occur within the central cylinder, however weakly it is otherwise defined. Corms usually do not have central cylinders but there are various ring-like or solid cylinder patterns formed by the dense deposition of starch granules in the ground tissue.

Root

The least studied organ in this survey, general features of roots are described for some genera. Roots are adventitious and may be subterranean (where they are often contractile), aerial, or transitional (Hinchee 1981). There are anatomical differences among these types but they are often subtle and more a question of developmental timing. Therefore, it is not often possible to identify the type of root microscopically.

The root cap has a columella containing starch granules. The young surface of the root is an epidermis which may have root hairs in the soil feeder roots. Root hairs are occasionally abundantly formed on one side where an aerial root is in close contact with a substrate. Beneath the epidermis is an exodermis which develops lignified thickenings in a characteristic pattern. In some cases there is a periderm layer and no exodermis (Fig. 10E). In *Philodendron* roots, beneath the epidermis there is a multiple hypodermis with a ligno-suberized casparian band (Peterson and Perumalla 1990). A multi-layered velamen may occur on the surface of aerial roots of some genera. The patterns formed by epidermis, exodermis and hypodermis, as well as the presence of periderm or velamen, should be surveyed in detail for systematic purposes. French (1987c), for instance, noted the presence of a multi-layered sclerotic hypodermis in seven genera.

Cells of the outer cortex are usually randomly oriented, often grading into orderly radial files nearer the stele. Collenchyma may occur rarely in the outer cortex as reported in *Monstera* (Hinchee 1981) and *Philodendron* (Porsch 1911). In mature contractile and aerial roots, the cortex is often dissected by radial schizogenous cavities. The inner cortical boundary is an endodermis (Fig. 10C, arrow) which has casparian strips in all cases. The outer stele is usually a single-layered pericycle.

The conducting tissue is polyarch, varying from a low of 5–6 protoxylem poles (Fig. 10B,C) up to 30 poles in some species of *Monstera*. The number of protoxylem poles may vary from low in the young root to higher in sections of the same older root (Hinchee 1981) and hence is not reliably diagnostic. Phloem strands are irregularly circular and on alternate radii between the xylem poles (Fig. 10 C,D). In mature roots the stele centre is a varying mixture of large diameter metaxylem cells, ground tissue, thin- or thick-walled lignified fibres, or some combination of ground tissue admixed with fibres or brachysclereids. Trichosclereids occur in cortical tissue of some members of the Monstereae. Resin canals in root cortical ground tissue have been reported by French (1987b) to occur only in *Cercestis, Culcasia, Furtadoa, Homalomena* and *Philodendron*. Laticifers have been reported in the root cortex of *Philodendron*, some Caladieae and Colocasieae (Weiss 1866; Porsch 1911). Raphides and druses are frequent in most species.

ACORACEAE C. Agardh

The family comprises the single genus *Acorus*.

Acorus L.

This genus of two species is very widely distributed. The species range from Siberia, across northern Europe as far north as 63° north latitude, and across northern North America. In Asia, they extend southward to subtropical India and from Japan, south to Formosa, south-west China, Thailand and Vietnam, Celebes and south of the equator in Java. *Acorus calamus* is found over most of the range. *Acorus gramineus* is native to Asia. The plants are rhizomatous herbs inhabiting wet meadows, open habitats, and along or in bodies of water. The species are distinct from aroids in having triangular, distichous leaf bases and unifacial blades. (See illustration in Rost 1979b: 298.) Both species were examined. Chromosomes in *A. gramineus*: 2n = 22; in *A. calamus*: 2n = 18, 22, 24, 36, 44, 48.

Leaf surface

Cuticle smooth or occasionally striate. **Epidermis**: cells tending to be elongate parallel to longitudinal axis of leaf. Cells with end walls in one row alternating with lateral walls in the next row. **Stomata** on both surfaces, brachyparacytic with one pair of subsidiary cells (Fig. 11D). Guard cells oriented parallel to longitudinal axis of leaf. Distal to groove, both leaf surfaces morphologically abaxial. **Venation**: parallel major and minor veins within chlorenchyma. Small cross-connecting veins, common or uncommon, running obliquely, sometimes across several larger veins.

Leaf TS

Lamina sharply laterally folded, equitant and ensiform. **Structure** isobilateral, having a narrow peripheral band of chlorenchyma and central aerenchyma (Fig. 11A,B). **Size**: 12–20 mm wide. **Midrib** symmetrically enlarged with deeply rounded-convex or deltoid-convex expansion on both surfaces. Mesophyll continuous beneath both surfaces. One exceptionally large vascular bundle at apex of each ridge (Fig. 11E). **Cuticle** thin. **Epidermis**: small square cells on both surfaces. Idioblastic oil cells common, with deltoid extensions into chlorenchyma (Fig. 11G); cells containing essential oils (Rost 1978, 1979). **Stomata** with large guard cells and inner and outer cuticular ledges. **Mesophyll**: chlorenchyma cells forming 4–6 layers of compact spongy tissue at periphery; cells irregularly rounded and short-armed, arranged in files in LS. Substomatal air cavities few and small. **Aerenchyma**: 75% of leaf, varying from small circular spaces between multi-armed spongy cells near chlorenchyma (type 1a), to very large cavities in centre of lamina and midrib (type 3) (Fig. 11B). Air cavity partitions uniseriate and 1–4 cells long between triangular 'corner' cells where three partitions join. Aerenchyma less strongly developed in *A. gramineus* (Fig. 11C).

Vascular bundles: large and small bundles around periphery of lamina, with one to several metaxylem cells per bundle (type I). Larger bundles ± evenly spaced. Phloem always oriented facing surface. Xylem: tracheids with reticulate or scalariform wall pitting and very elongate end walls with an overlap of up to 62 μm. Phloem strands often in discrete pairs in vascular bundles of *A. gramineus* (Fig. 11F), but as single semicircular strand at broad and straight or curved xylem boundary in *A. calamus* (Fig. 11E). Sieve elements having oblique end walls. **Sclerenchyma**: fibres as vein sheaths or phloem caps with *c.* 8 layers of fibres, caps always oriented toward exterior. **Laticifers**: idioblastic laticifer (-like) cells, hollow with dark amorphous lining, common throughout. Occasional large rounded idioblastic cells with smooth dark contents present in chlorenchyma. **Secretory cells**: epidermal oil cells (Fig. 11G) present, having a ring of dense material and square openings to surface. **Tannin** cells scattered, parenchyma-sized and common. **Starch**: Sakai and Hayashi (1973) reported that *A. gramineus* stores more sugar than starch in leaf chloroplasts. **Crystals**: druses common (in some specimens) in mesophyll and in subepidermal cells. Squarish prismatics and irregular crystals, sometimes in clusters, in aerenchyma partitions and in rows of small cells on surface of fibrous vein sheaths.

Rhizome TS

Outline circular. **Size**: diameter *c.* 1.8 cm. **Cuticle** medium-thick and smooth. **Epidermis**: cells small and pointed or rounded-papillate on surface, or cells as flattened remnants in vicinity of periderm. **Cork**: 3–6 layers of empty cells arising subepidermally. **Ground tissue**: subepidermal chlorenchyma in 8–12 layers composed of rounded or short-armed cells. Cortex mostly (90%) type 3 aerenchyma with 1–3 partition cells. Central cylinder an aerenchyma with smaller air cavities. **Air cavities**: 20% of chlorenchyma, 95% of inner cortex. **Central cylinder** delimited

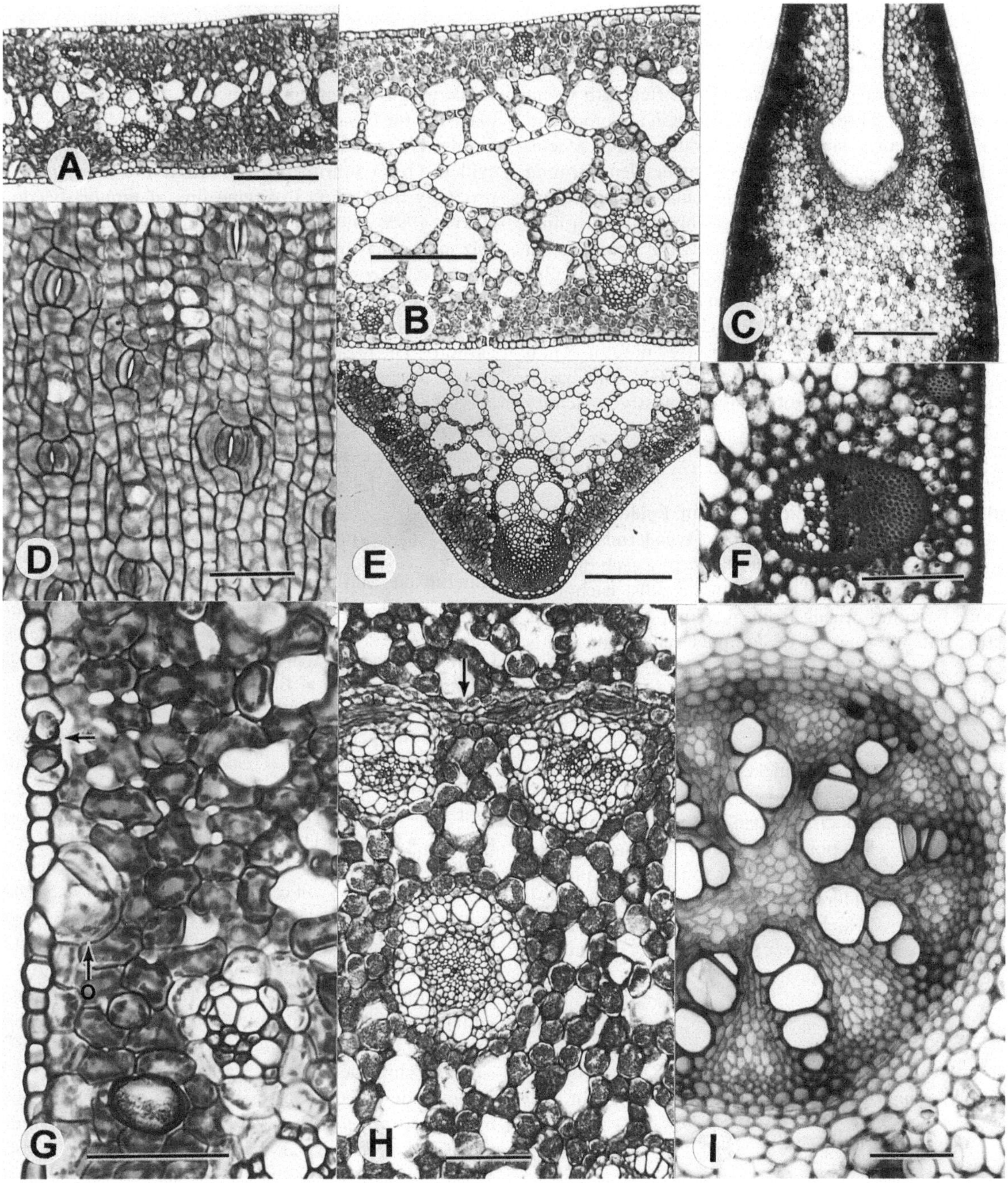

Fig. 11. *Acorus* leaf, rhizome and root sections. (A,B,D,E,G,H) *A. calamus;* (C,F,I) *A. gramineus;* (A) Lamina TS distal; (B) Lamina TS basal, above fusion of basal groove; (C) Lamina TS basal groove. Most chlorenchyma restricted to darkened peripheral portions of leaf; (D) Lamina PS showing epidermis and brachyparacytic stomata; (E) Midrib TS with largest vascular bundle; (F) Lamina TS lateral margin and vascular bundle; (G) Lamina showing oil cell (arrow o), stoma (arrow), and transition from peripheral chlorenchyma to aerenchyma; (H) Rhizome TS showing amphivasal bundles of central cylinder and endodermis (arrow); (I) Root TS showing stele. Scale bars: A,B,F = 150 μm; C = 300 μm; D,G–I = 50 μm; E = 200 μm.

by endodermis with narrow (*A. calamus*) or broad (*A. gramineus*) casparian thickenings. **Vascular bundles**: collateral and few in cortex; amphivasal, numerous and scattered in central cylinder (Fig. 11H), often with xylem clustered into 6–8 groups suggesting origin as compound bundles; phloem a large circular strand within xylem. Cortical bundles with several files of metaxylem forming straight broad boundary with semicircular phloem strand (type I). **Sclerenchyma**: fibres occasionally ensheathing larger cortical bundles, or amphivasal bundles in *A. gramineus*. **Secretory cells**: oil cells large and circular among aerenchyma partitions. **Tannin** cells uncommon among aerenchyma partitions. **Starch** granules abundant in ground tissue, especially in centre (see note on Sakai and Hayashi 1973, cited above). **Crystals**: prismatics occasional and small in ground tissue, mostly in vicinity of vascular bundles.

Root TS

Outline circular. **Size**: diameter 2–3 mm. **Epidermis**: when present, of flattened tile-like cells. **Hypodermis** of two layers of thin-walled palisade cells abutting each other at zig-zag margin. **Trichoblasts**: elongate cells, each with bulbous base and arising from outer hypodermal layer. **Cortex**: outer 2–6 layers of compact cells, with evenly thickened pectic wall layers. Inner cortex, *c*. 90% of total, of type 3 aerenchyma with one-celled partitions in *A. calamus* or of rounded cells with numerous cell-sized air cavities in *A. gramineus*.

Endodermis: casparian thickenings wide and thick or thin in anticlinal walls. **Pericycle**: single parenchymatous layer beneath endodermis giving rise to lateral roots opposite protoxylem poles. **Vasculature**: xylem with nine poles alternating with nine circular to triangular phloem strands (Fig. 11I). **Fibres**: occasional isolated cells or cell pairs in centre of stele. **Oil cells**: occasional enlarged circular cells within cortex. **Tannin** cells occasional in cortex, same size as cortical cells.

Material examined

Acorus calamus L. cult. M; Bouford 18710, Missouri (MO); *A. gramineus* L. 75p215; 772042; Bogner s.n., 12/1995.

Taxonomic notes

The literature (Kaplan 1970; Rost 1978, 1979a, b) and these observations point to considerable variation in the anatomy of these wide-ranging plants. The leaves are referred to as unifacial by Grayum (1987) but it should be noted that the leaves have equitant grooved bases where they show normal bifacial development. Only in the fused upper portions are they structurally isobilateral.

Grayum (1987) has reviewed a broad array of evidence concerning the question of whether *Acorus*, traditionally placed as the basal genus of Araceae, really belongs in the family. He concluded that it is not an aroid. Molecular evidence has confirmed this isolation and that *Acorus* is probably the earliest diverging line of monocotyledons (Davis 1995; Duvall *et al.* 1993). Studies of root and rhizome vessels by Carlquist and Schneider (1997) show those cells to represent a plesiomorphic state. They possess incompletely lysed pit membranes, probably adequate for plants of small leaf area and that live in wet habitats. The present study agrees that *Acorus* has several unique vegetative features that isolate it. Its unifacial epetiolate leaves, presence of ethereal oil cells, and the lack of raphides, make it stand apart from all genera of Araceae. The evidence presented by this study supports recognition of a separate family for the genus.

Species reported on in the literature

Amelunxen and Gronau (1968, 1969): *A. calamus*: rhizome, oil cells. • Arber (1918a, b): *A. calamus*: leaf. • Arber (1921): *A. calamus*: 'spathe'. • Baranetzky (1897): *A. calamus*: shoot apex. • Barlund (1939): *A. calamus*: rhizome. • Bhat *et al.* (1992): *A. calamus*: leaf. • Biermann (1898): *A. calamus*: oil cells. • Blanc (1978): ontogeny. • Buchholz (1921): *A. gramineus*: shoot. • Bugnon (1981): *A. calamus*: leaf. • Carlquist and Schneider (1997): *A. calamus, A. gramineus*: stem and root, vessels. • Chauveaud (1897): *A. gramineus*: root, phloem. • Chrysler (1904): *A. calamus*: stem. • Dalitzsch (1886): *A. calamus, A. gramineus*: leaf. • Denissova (1975): *A. calamus*: oil cells. • Dey and Das (1982, 1988) *A. calamus*: rhizome. • Duval-Jouve (1873): leaf. • Engler (1905): *A. calamus, A. gramineus*: root, leaf. • Ertl (1932): *A. calamus*: leaf. • French (1987c): *A. calamus, A. gramineus*: root [sclerotic hypodermis absent]. • French and Tomlinson (1981c): *A. calamus, A. gramineus*: stem. • Gentner (1905): *A. calamus*: leaf. • Guillaud (1878): *A. calamus*: rhizome. • Gulliver (1865): *A. calamus*: crystals [no raphides]. • Guttenberg (1968): *A. calamus*, A. gramineus: root. • Hather (1988, 1993): rhizome. • Higley (1880): *A. calamus*: crystals. • Holfert (1889): root. • Holm (1911b): *A. calamus*: root, rhizome, leaf. • Hotta (1971): *A gramineus*: leaf, root. • Jackson and Snowdon (1968, 1990): *A. calamus*: rhizome. • Kaplan (1968, 1970, 1971, 1973a, 1975): *A. calamus*: leaf. • Kattein (1897): *A. calamus*: root. • Kimura *et al.* (1963): *A. calamus*: rhizome. • Kimura and Nagamachi (1935): *A. gramineus*: root. • Kirchner *et al.* (1934): *A. calamus*: rhizome, root, leaf. • Konstanty (1926): *A. calamus*: rhizome, root. • Kroemer (1903): root. • Laux (1887): probably *A. calamus*: rhizome. • Lierau (1888): *A. calamus, A. gramineus*: root. • Linsbauer (1930): epidermis. • Luhan (1957): *A. calamus*: rhizome. • Maillefer (1916–17, 1920): *A. calamus*: root. • Mangin (1880): *A. calamus*: stem, leaf. • Mangin (1882): *A. calamus, A. gramineus*: root. • Mbaya (1976): *A. calamus*: leaf. • Meischner and Schröder (1990) *A. calamus*: root. • Möbius (1887): *A. calamus*: rhizome. • Moll and Janssonius (1923): *A. calamus*: rhizome. • Ogden (1974): *A. calamus*: stem. • Parmelee and Savile (1954): *A. calamus*: leaf (rusts). • Raunkiaer (1895–9): anatomy. • Rodrigue (1900): *A. calamus*: leaf. • Roth (1949): *A. calamus*: leaf. • Rywosch (1909): *A. calamus*: root. • Sakai and Hayashi (1973): *A. gramineus*: leaf starch/sugar. • Schantz (1958): *A. calamus*: rhizome, root, leaf. • Schwendener (1882): root, rhizome. • Selvaraj and Subramanian (1987): *A. calamus*: root, leaf. • Shishkoff (1987): *A. calamus, A. gramineus*: root. • Solereder and Meyer (1928): *A. calamus, A. gramineus*: all organs. • Stewart and Dermen (1979): *A. gramineus*: leaf ontogeny. • Studer and Brandle (1984): *A. calamus*: rhizome. • Thielke (1948, 1960): *A. calamus, A. gramineus*: leaf. • Tieghem (1866): *A. calamus, A. gramineus*: all organs. • Tieghem (1870–1): *A. gramineus*: root, secretory tissue. • Tieghem (1891): stem. • Tieghem (1905): *A gramineus*: rhizome. • Tieghem and Douliot (1888): *A. calamus*: branch root. • Tschirch (1894): secretory tissue. • Tschirch (1913): *A. calamus*: rhizome. • Tschirch (1909–25): root, rhizome. • Turpin (1836): *A. calamus, A. gramineus*: leaf, crystals. • Van Fleet (1942): *A. calamus, A. gramineus*: stem, endodermis. • Wattendorff (1974): rhizome. • Wichtl (1989): rhizome. • Wollenweber (1897): *A. calamus*: leaf. • Zacharias (1879): *A. calamus, A. gramineus*: secretory tissue. • Zinke (1924): *A. calamus, A. gramineus*: leaf.

Additional literature

Batygina and Yakovlev 1990; Bochenska and Kozlowski 1974; Bogner and Nicolson 1991; Buzgó and Endress 2000; Clausen 1927; Daghlian 1981; Dahlgren 1989; Davis 1966; Domin 1911; Duvall 2001; Duvall *et al.* 1993; Erdtman 1952; Eyde *et al.* 1967; Franck 1975; French 1985a, 1986c, 1987a; Grayum 1987, 1990; Hegnauer 1986; Huang 1972; Irmisch 1874; Marchant 1973; McClure 1970; Motley 1994; Parmelee and Savile 1954; Pelt 1966; Ray 1988; Rost and Bos 1979; Sattler 1973; Saville 1971; Scaramella Petri and Spadoni 1968; Schultes 1970a; Schultes and Farnsworth 1980; Schultes and Hofmann 1973; Sharma 1967; Smolenski *et al.* 1973, 1974; Thanikaimoni 1969; Tieghem 1907; Tillich 1985; Williams *et al.* 1981; Wulff 1954.

ARACEAE Juss.

SUBFAMILY GYMNOSTACHYDOIDEAE BOGNER & NICOLSON

A monotypic subfamily bearing bisexual perigoniate flowers.

Gymnostachys R.Br.

This genus is native to eastern Australia. Plants are terrestrial evergreen herbs occurring in open shrublands, or in sheltered gullies or rain forests. The leaves are ensiform and linear with grooved, equitant bases. Chromosomes: 2n = 48. One specimen was available for study.

Leaf surface

Cuticle smooth. **Epidermis**: cells linear, elongate-rounded (1–3:1 *l*/*w*). **Stomata** on both surfaces, parallel, aligned with epidermal cells and brachyparacytic (Fig. 12D), usually with 2 narrow pairs of subsidiary cells. **Venation**: larger veins parallel, smaller ones having occasional oblique cross connections.

Leaf TS

Structure: lamina isobilateral (Fig. 12A,B). **Leaf margin** finely sinuate (Fig. 12G). **Size**: 42 cm × 9 mm (*l*/*w*). **Midrib** thinner than remainder of lamina; adaxially shallowly V-grooved, abaxially convex-rounded V. **Cuticle** medium-thick. **Epidermis**: cells longitudinally elongated, slightly rounded toward interior and exterior. **Stomata** level with surface and subtended by very shallow substomatal air cavities (Fig. 12B). **Mesophyll** not differentiated, of rounded cells with several short arms that contact other cells (type 1a) (Fig. 1B). One row of tall bulliform cells at midrib adaxial groove and adjacent surface. **Air cavities**: *c*. 10% of mesophyll space.

Vascular bundles: small circular bundle in midrib, ensheathed by fibres, most layers over phloem, but not transcurrent. Widely spaced, large, well-differentiated bundles in lamina, transcurrent with fibre sheaths (nearest type I but unique for Araceae); small bundles between larger ones. Xylem: 20–30 metaxylem cells in 3–5 irregular files. Phloem: 1–3 small circular strands separated from xylem by wide fibre boundary (Fig. 12E,F). **Sclerenchyma**: thick-walled fibres, with no visible wall layering, forming massive girder-like sheaths surrounding vascular bundles in midrib and lamina, up to 20 cells deep over xylem or phloem. Smaller bundles with smaller xylem and phloem caps. Phloem completely surrounded and isolated from xylem. **Crystals**: raphides common in large, often elongated cells with 1–4 bundles, randomly distributed in mesophyll (Fig. 12C,G). Druses common, distributed subepidermally in mesophyll. Crystal sand common, 1–2 per cell, in numerous cells.

Material examined

Gymnostachys anceps R.Br., Rodd 2326, Australia, cult. M.

Taxonomic notes

In addition to its unusual reproductive features, the genus is isolated by its linear leaves that are ensiform with grooved equitant bases. Anatomically it is the only genus bearing an undifferentiated lamina, longitudinally elongated epidermal cells, bulliform cells at the midrib adaxial groove, and having unusually robust, widely spaced transcurrent laminar bundles. Buzgó (2001) found that flower development of *Gymnostachys* more closely resembles that of *Lysichiton* and *Symplocarpus* than that of other aroids.

Species reported on in the literature

French (1987c): root. • French and Tomlinson (1981c): stem vasculature. • Gulliver (1865): crystals. • Hotta (1971): leaf.

Additional literature

Domin 1911; Eyde *et al.* 1967; French 1985a, 1986c, 1987a; Grayum 1987, 1990; Hegnauer 1986; Ray 1988; Thanikaimoni 1969; Williams *et al.* 1971.

SUBFAMILY ORONTIOIDEAE MAYO, BOGNER & BOYCE

A group of herbs bearing bisexual, perigoniate flowers. They are divided among the tribe Orontieae (*Orontium*) and the tribe Symplocarpeae.

Tribe Orontieae

Orontium L.

This monotypic genus is restricted to Atlantic North America. Plants are aquatic herbs rooted on the bottom

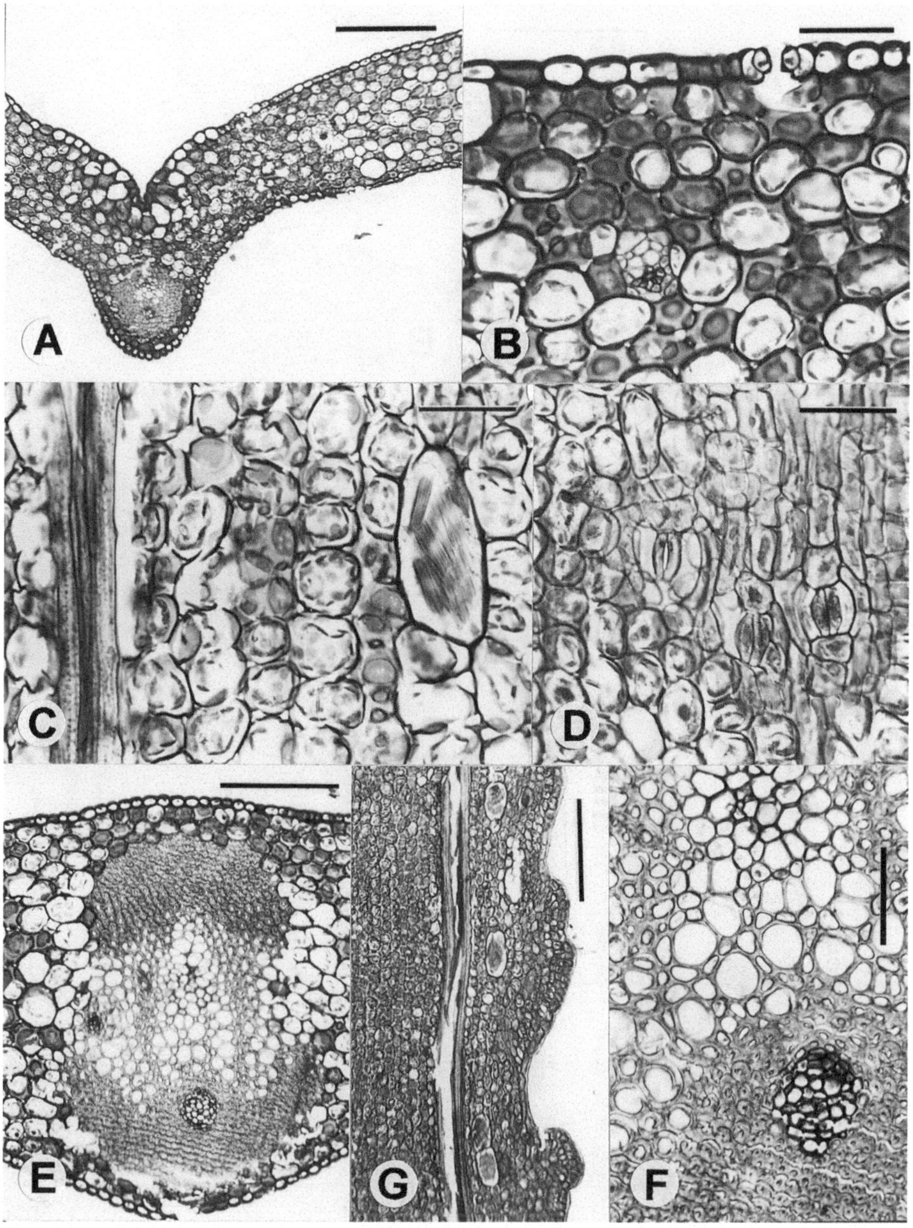

Fig. 12. ***Gymnostachys anceps*** **leaf sections**. (A) Midrib TS; (B) Lamina TS showing shallow substomatal cavity and compact undifferentiated mesophyll; (C) Leaf mesophyll PS with raphide cell; (D) Leaf mesophyll PS showing brachyparacytic stomata; (E) Mid-lamina large vascular bundle showing enlarged bundle caps; (F) Detail of (E) showing circular phloem strand embedded in fibre cap; (G) Leaf PS showing detail of leaf margin and raphide cells. Scale bars: A,G = 200 µm; B–D,F = 50 µm; E = 150 µm.

in shallow fresh water with emergent, broad elliptic leaves. They have a wide ecological amplitude occurring in cool or warm climates. Chromosomes: 2n = 26. One specimen was available for study.

Leaf surface

Cuticle striate adaxially and striate or smooth abaxially. **Epidermis**: adaxial cells polygonal and isohedral (1–1.5:1 *l*/*w*). Abaxial cells polygonal (3–4:1 *l*/*w*) (Fig. 13F).

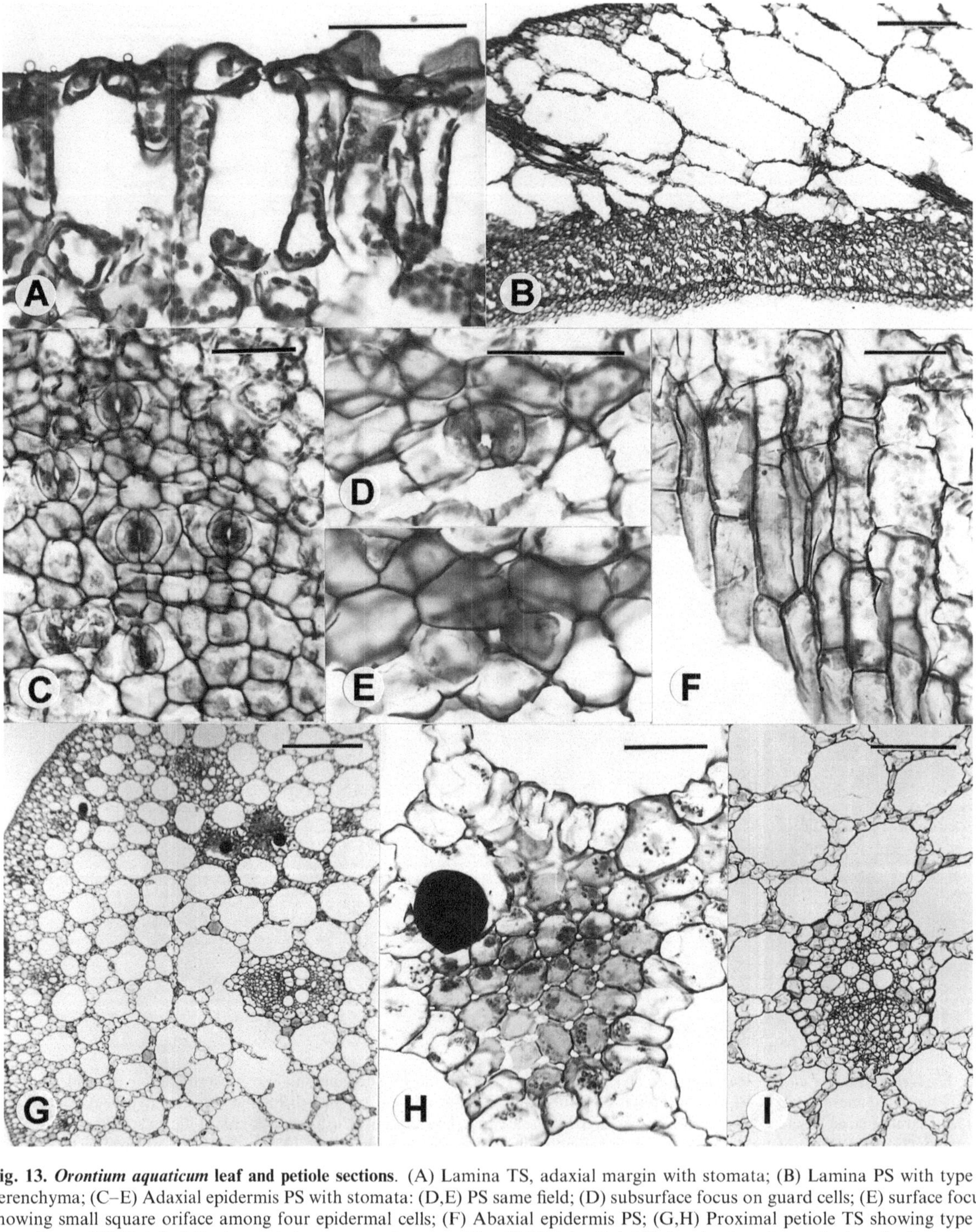

Fig. 13. *Orontium aquaticum* leaf and petiole sections. (A) Lamina TS, adaxial margin with stomata; (B) Lamina PS with type 3 aerenchyma; (C–E) Adaxial epidermis PS with stomata: (D,E) PS same field; (D) subsurface focus on guard cells; (E) surface focus showing small square oriface among four epidermal cells; (F) Abaxial epidermis PS; (G,H) Proximal petiole TS showing type 3 aerenchyma; (H) Detail of diaphragm between aerenchyma cavities; (I) Distal petiole TS showing type I vascular bundle. Scale bars: A,C–F,H = 50 μm; B,G = 300 μm; I = 200 μm.

Stomata: adaxial and brachyparacytic or anomocytic (Fig. 13C). Subsidiary cells broad, angular and irregular, or undifferentiated from neighbouring epidermal cells (Fig. 13D,E). **Venation**: midrib prominent and wide at base with numerous vascular bundles, diminishing markedly in width distally as secondaries diverge laterally and ascend acutely. Distal lamina vascularized by parallel secondaries and some parallel tertiaries. Quaternaries and higher orders forming orthogonal cross connections. Some higher orders reticulate. Veins merging at margin in straightened eucamptodromous pattern.

Leaf TS

Size: 58 × 48 mm (*l*/*w*). **Midrib** flat adaxially, shallowly and broadly convex abaxially. **Cuticle** very thin. **Epidermis**: cells short abaxially and distinctly papillate adaxially (Fig. 3E). **Stomata** subtended by well-developed substomatal cavities. **Mesophyll**: one palisade layer (Fig. 13A) underlain by 2–3 spongy layers (type 1), followed by type 3 aerenchyma with high, wide air cavities with uniseriate partitions up to five cells long (Fig. 13B). **Air cavities** often elongate parallel with direction of secondary and tertiary veins, up to 98% of aerenchyma space.

Vascular bundles: widely-spaced single row of small or large vascular bundles in midrib. Xylem: 3–5 files of well-developed metaxylem (10–15 cells). Phloem: semicircular strand (type I). **Sclerenchyma**: fibres forming small, 1–2-layered phloem caps. **Laticifers** occasional, lateral to vascular bundles. **Crystals**: raphides as single bundles in small cells, or as multiple bundles in elongate cells, common throughout mesophyll. In type 3 aerenchyma, raphide cells oriented parallel and within partitions. Druses in circular cells beneath adaxial epidermis and near large vascular bundles, common or uncommon.

Petiole TS

Outline: crescent-shaped proximally, subterete distally and adaxially flattened. **Size**: diameter 4–5 mm. **Cuticle** thick and smooth, grooved externally and flanged internally at cell boundaries. **Epidermis**: cells small and taller than wide or square. **Chlorenchyma**: plastids in outer compact ground tissue. **Ground tissue**: outer 3–5 layers of compact small parenchyma cells surrounding central type 3 aerenchyma (Figs 6A, 13G). Partitions 2–4 cells long. **Air cavities** large, separated longitudinally by occasional, oblique, 1-layered diaphragms with small corner intercellular air spaces (Fig. 13H). Total space up to 98% of aerenchyma.

Vascular bundles scattered symmetrically; *c.* 5 large and *c.* 10 small. Xylem: one medium-sized protoxylem tracheid, 6–12 metaxylem tracheids of various sizes (Fig. 13I). Phloem: large strand, semicircular to elliptic (type I). **Laticifers**: large rounded cells with dark homogeneous contents, occasional in ground tissue or near vascular bundles. Rosendahl (1911) reported 'latex ducts' near phloem but not verified in this material. French (1988) reported non-anastomosing laticifers present. **Tannin** cells occasional throughout. **Crystals**: raphides in compact ground tissue, or in aerenchyma partitions in large thin-walled cells, common throughout. Druses uncommon in ground tissue near vascular bundles.

Material examined

Orontium aquaticum L. Boufford 18027, North Carolina, (MO).

Taxonomic notes

The three genera of Orontioideae, all north temperate herbs, appear similar in 'grade' but anatomical characters and the cpDNA restriction site study by French *et al.* (1995) require that they be divided among two tribes. Differentiating anatomical characters include position of stomata, raphide variation, and presence or position of collenchyma, sclerenchyma, and laticifers. In comparison to *Lysichiton* and *Symplocarpus*, *Orontium* is distinguished by the presence of sclerenchyma fibres, the absence of collenchyma, and the presence of laticifers.

Species reported on in the literature

Ertl (1932): leaf venation. • French (1987c): root, hypodermis. • French (1988): laticifers. • French and Tomlinson (1981a): stem, vasculature. • Grear (1973): stem. • Gulliver (1865): crystals. • Krause (1908): root, stem, leaf. • Ogden (1974): petiole. • Rosendahl (1911): laticifers. • Solereder and Meyer (1928): all organs. • Troll (1932a): leaf ontogeny. • Troll and Meyer (1955): leaf ontogeny. • Turpin (1836): *O. japonicum*: leaf, crystals.

Additional literature

Barabé and Forget 1987; Casadoro *et al.* 1982; Eyde *et al.* 1967; French 1985b, 1986c, 1987a; Grayum 1990; Guéguen 1900, 1901; Hegnauer 1986; McClure 1970; Ray 1988; Richter 1929; Thanikaimoni 1969; Tieghem 1907; Tillich 1985; Williams *et al.* 1981.

Tribe Symplocarpeae

This tribe comprises the small genera *Lysichiton* and *Symplocarpus*, two north temperate herbs.

Anatomical summary

Leaf surface

Cuticle smooth or finely striate. **Epidermis**: cells straight-sided polygonal. **Stomata** in both surfaces and brachyparacytic. Subsidiary cells broad and irregular, little differentiated from neighbouring cells. **Venation**: midrib prominent, secondaries basally acrodromous, distally pinnate and curved-ascending, eucamptodromous at margins or with straightened brochidromous loops.

Leaf TS

Structure dorsiventral. **Midrib** shallowly concave adaxially, broadly convex abaxially. **Cuticle** thin, or not seen. **Epidermis**: cells slightly papillate or rounded on both surfaces, or flattened. **Mesophyll**: palisade 1–2-layered, cells compact. Spongy cells: type 3 aerenchyma. **Collenchyma**: thin-walled cells forming xylem and phloem caps at largest midrib bundles in *Lysichiton*. In *Symplocarpus*, collenchyma subepidermally banded (type B) in abaxial midrib. **Vascular bundles** scattered symmetrically in midrib, type I. **Tannin** cells common in ground tissue in *Lysichiton*. **Reserve photosynthate**: reported to be sugar in *Lysichiton*. **Crystals**: raphides as single bundles in small, wide or slightly elongated thin-walled cells. Midrib raphide cells in biforine form in *Symplocarpus*. Druses common near vascular bundles in *Lysichiton*. Prismatics rare and subepidermal in *Symplocarpus*.

Petiole TS

Outline shallowly concave proximally, subcircular distally in *Lysichiton*, bilaterally compressed in *Symplocarpus*. **Cuticle** thin and smooth. **Epidermis**: cells rounded or rounded-papillate, flattened in *Symplocarpus*. **Collenchyma**: phloem caps at larger bundles in both genera, also as xylem caps and banded (type B) at periphery in *Symplocarpus*. **Ground tissue**: outer cells compact, becoming type 3 aerenchyma at centre. **Vascular bundles** scattered and numerous, type I. **Tannin** cells occasional throughout in *Lysichiton*. **Crystals**: raphides as single bundles in thin-walled cells in ground tissue. Druses occasional, near vascular bundles.

Taxonomic notes

In spite of their extratropical ranges, these genera possess some very unspecialized characters including well-developed collenchyma caps at vascular bundles. These two genera show some diagnostic differences between them but are closer to each other than they are to *Orontium*. The last genus possesses laticifers and sclerenchyma fibres as phloem caps.

Lysichiton Schott

This genus of two rhizomatous species is found in Pacific North America, eastern Siberia and Japan. Plants are terrestrial rhizomatous herbs occurring in swampy hardwood forests or broad swampy meadows and creek edges. Leaves are erect, short-petiolate, elliptic, appearing succulent, and having a broad midrib. Chromosomes for both species: $2n = 28$. Both species were available for study.

Leaf surface

Cuticle smooth. **Epidermis**: all cells polygonal with straight anticlinal walls and mostly isodiametric (1–1.2:1 *l*/*w*). **Stomata** on both surfaces, brachyparacytic (Fig. 14B). Subsidiary cells mostly broad and angular, differing little from adjacent epidermal cells. **Venation**: secondaries arising acutely from midrib and curving toward eucamptodromous marginal connections. Tertiaries and higher orders orthogonal-reticulate.

Leaf TS

Size: 21×8 cm (*l*/*w*). **Midrib** broadly and shallowly concave adaxially and broadly convex abaxially. **Cuticle** very thin or not seen. **Epidermis**: cells somewhat papillate adaxially or rounded on both surfaces (Fig. 14A). Adaxial cells somewhat taller than or same height as abaxial cells. **Mesophyll**: two palisade layers of short, tightly packed cells. Spongy tissue: well developed aerenchyma, more than 10 cells deep, with large air cavities. Cavity partitions 1–2 cells wide (type 3). **Air cavities** in midrib with occasional oblique, single-layered diaphragms in LS. Circular air passages occurring at diaphragm cell corners. **Collenchyma**: thin-walled cells forming xylem and phloem caps at largest midrib vascular bundles.

Vascular bundles: in midrib about four symmetrically disposed larger bundles and four smaller bundles, or widely spaced and numerous scattered bundles; arranged in islands of ground tissue with radiating uniseriate partitions of aerenchyma (Fig. 14E). Bundles well-defined (type I) (Fig. 14E). Xylem: 5–10 metaxylem strands opposing a broad semicircular phloem strand. **Tannin** cells same shape as ground tissue cells, with dark granular contents, subepidermal and randomly distributed in lamina and midrib. **Reserve photosynthate**: sugar rather than starch stored in ground tissue (Sakai and Hayashi 1973). **Crystals**: raphides as single bundles, common in aerenchyma partitions and in palisade. Cells wide or slightly elongate in adaxial mesophyll (Fig. 14C). Druses common and small near vascular bundles (Fig. 14D).

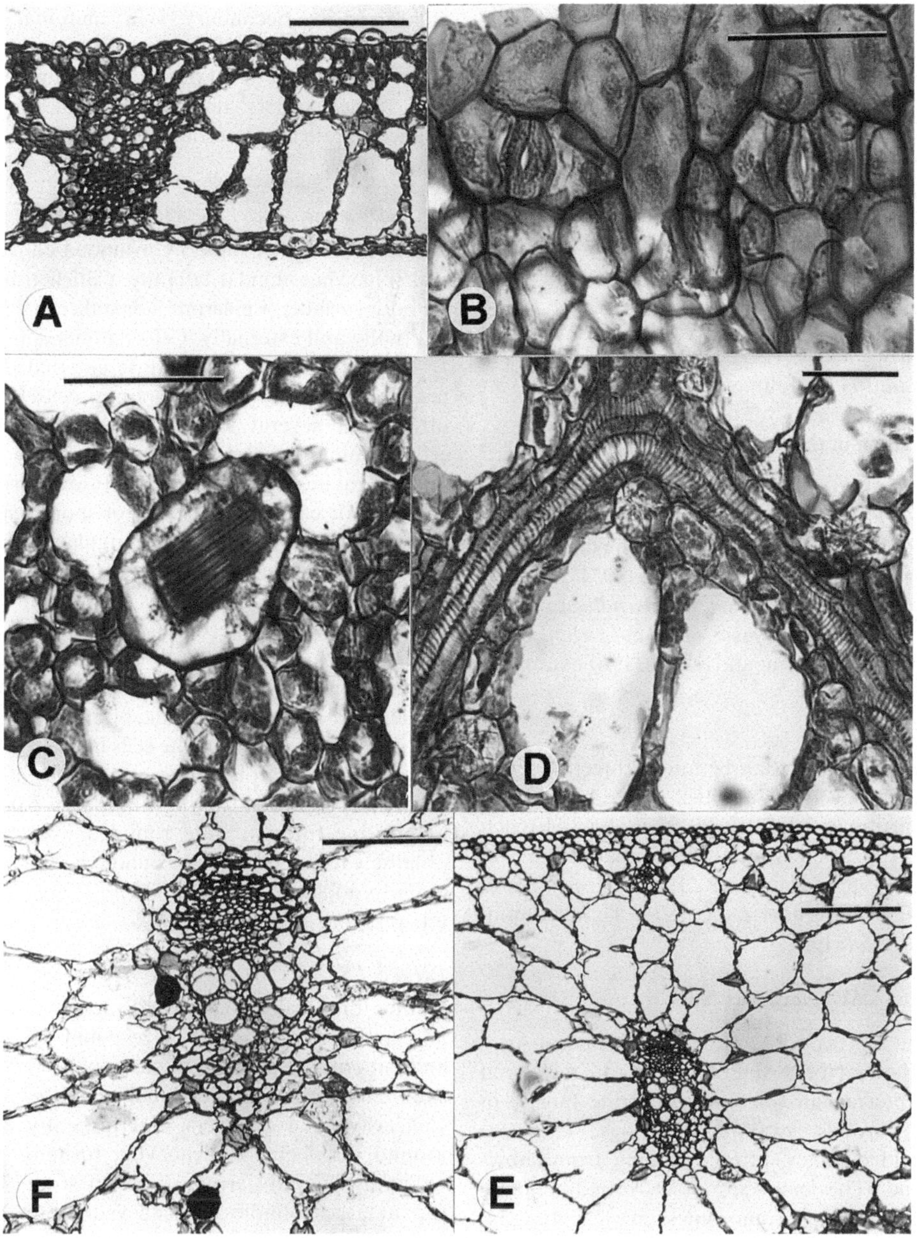

Fig. 14. *Lysichiton camtschatcense* leaf and petiole sections. (A) Lamina TS; (B) Leaf epidermis PS showing brachyparacytic stomata; (C) Leaf mesophyll PS with wide raphide cell; (D) Leaf mesophyll PS with vein in type 3 mesophyll. Druses common in vein sheath; (E) Midrib TS showing type I vascular bundle in type 3 aerenchyma (abaxial surface at top); (F) Proximal petiole lateral TS showing type I vascular bundle and type 3 aerenchyma (lateral surface toward top). Scale bars: A,F = 150 μm; B–D = 50 μm; E = 200 μm.

Petiole TS

Outline shallowly concave adaxially, broadly and deeply convex-rounded abaxially. **Size**: width 1.5 cm. **Cuticle** very thin and smooth. **Epidermis**: cells rounded adaxially, rounded-papillate abaxially. **Collenchyma**: phloem caps at larger vascular bundles. **Ground tissue**: two outer layers of small cells. Most tissue type 3 aerenchyma with 1–4-celled partitions. Periodically, 1-layered diaphragms with corner openings delimiting air cavities in LS. **Air cavities** up to 98% of aerenchyma.

Vascular bundles scattered and numerous, type I (Fig. 14F). Xylem: 3–8 metaxylem cells. Phloem: broad semicircular strand. **Tannin** cells occasional throughout. **Crystals**: raphides in single bundles in wide or rounded cells throughout; common near vascular bundles or in partitions. Druses occasional in partition cells near vascular bundles.

Material examined

Lysichiton camtschatcense (L.) Schott, cult. M; *L. americanus* Hultén and H.St. John, cult. M.

Species reported on in the literature

Ertl (1932): *L. camtschatcense*: leaf. • French (1987c): *L. americanus, L. camtschatcense*: root. • French (1988): *L. americanus*: laticifers. • French and Tomlinson (1981a): *L. americanus*: stem vasculature. • Hotta (1971): *L. camtschatcense*: leaf. • Krause (1908): *L. camtschatcense*: root, stem, leaf. • Sakai and Hayashi (1973): *L. camtschatcense*: chloroplasts. • Stebbins and Khush (1961): stomata.

Additional literature

Barabé and Forget 1987; Barabé and Labrecque 1984; Barabé *et al.* 1986, 1987; Campbell 1899, 1900, 1902; Davis 1966; Erdtman 1952; Eyde *et al.* 1967; French 1985b, 1986c, 1987a; Grayum 1990; Guinaudeau *et al.* 1975; Halevy 1989; Hegnauer 1986; Hesse 1980; Rosendahl 1911; Smolenski *et al.* 1974; Thanikaimoni 1969; Williams *et al.* 1981.

***Symplocarpus* Nutt.**

This genus contains two species found in north-east Asia, Japan, and Atlantic North America. Plants are terrestrial herbs with subterranean rhizomes occurring mostly in wet meadows, woods, swamps and bogs. They are adapted to cold climates, often emerging from snow-covered ground. The leaves are erect, broad-petiolate with an elliptic lamina. Chromosomes: 2n = 28, 30, *c.* 54, 60. One species was available for study.

Leaf surface

Cuticle finely striate. **Epidermis**: cells polygonal with straight anticlinal walls (1–2:1 *l/w*) on both surfaces. **Stomata** brachyparacytic (Fig. 15C), on both surfaces. Subsidiary cells broad and variously shaped. **Venation**: midrib prominent. Secondaries basally acrodromous, distally pinnate. Partially straightened submarginal eucamptodromous connections from tertiaries arising from midrib. Secondary vein marginal connections eucamptodromous or with straightened brochidodromous loops in distal portion of lamina. Quaternaries and higher orders reticulate. Areoles elongate and aligned parallel with tertiaries.

Leaf TS

Size 162 × 110 mm (*l/w*). **Midrib** broadly and shallowly concave adaxially, broadly rounded-convex abaxially and somewhat angular laterally. **Cuticle** thin or not seen on both surfaces. **Epidermis**: adaxial cells flat to rounded internally and externally. Cells flat abaxially. **Mesophyll**: 1–2 layers of short, poorly differentiated palisade cells, resembling tightly packed spongy cells. Palisade cells often with several short, pointed processes. Spongy cells forming type 3 aerenchyma with very large, polygonal air cavities with 1-seriate partitions forming net-like pattern. **Air cavities** up to 99% of spongy space.

Collenchyma: 2–4 layers of angularly thickened subepidermal cells on abaxial side of midrib (type B). Collenchyma caps 2–5 cell layers over xylem and phloem. **Vascular bundles**: fewer than 10 bundles, in ± symmetrical pattern; bundles well-formed (Fig. 15A) with numerous metaxylem cells (type I). **Crystals**: raphides in single bundles, narrow or wide, often with gum sheaths in thin-walled short or elongate cells in lower palisade layer. Cells occasionally projecting at one end into air cavities but mostly embedded among other cells. Midrib raphides in biforine form in wide rounded oval cells attached obliquely to larger vascular bundle margins and projecting into adjacent air cavities (Fig. 15D). Prismatic crystals rare and subepidermal.

Petiole TS

Outline bilaterally compressed, deeply V-shaped adaxially and keeled abaxially. **Size**: 8 mm wide. **Cuticle** thin and smooth. **Epidermis**: cells small and mostly flat externally. **Stomata** occasional and level with surface. **Collenchyma**: 1–2 layers, intermittently subepidermal around periphery. Collenchyma forming phloem caps wrapping laterally around phloem strands, also xylem caps at largest bundles; better developed in distal sections.

Chlorenchyma: outer 4–5 layers of ground tissue. **Ground tissue** mostly type 3 aerenchyma with air cavities up to 90% of tissue (Fig. 15B). **Vascular bundles** scattered, type I. Xylem: 12–14 metaxylem cells opposing a large circular or semicircular phloem strand. **Crystals**: raphides occasional in thin-walled rounded cells.

Stem TS

Outline terete. **Size**: diameter 9 mm. **Cuticle** thin and smooth. **Epidermis**: cells small, rounded internally and

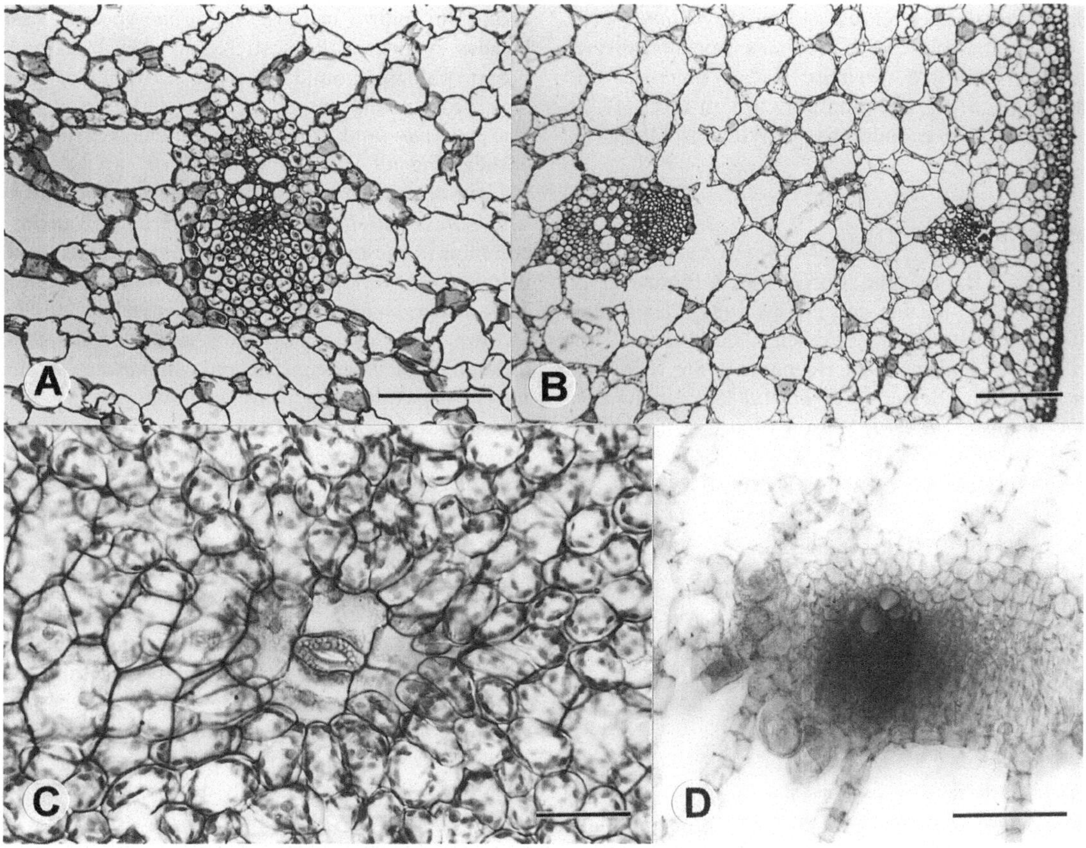

Fig. 15. *Symplocarpus foetidus* leaf and petiole sections. (A) Midrib TS central vascular bundle suspended in aerenchyma; (B) Petiole TS with vascular bundles; (C) Lamina PS showing brachyparacytic stoma from within substomatal cavity; (D) Hand section of midrib TS showing raphide cells at margin of vascular bundle and projecting into air cavities. Scale bars: A,D = 150 μm; B = 300 μm; C = 50 μm.

externally. **Collenchyma**: 1–2-layered continuous subepidermal band. Collenchyma caps over phloem and xylem strands of larger bundles. **Ground tissue**: type 3 aerenchyma; 1–2 partition cells between corners. In LS, horizontal or oblique uniseriate diaphragms spaced 1.2–3.5 mm apart.

Vascular bundles scattered, collateral, type I. Xylem: 12–14 metaxylem and some protoxylem cells forming broad boundary with large semicircular phloem strand. **Crystals**: raphides common or occasional at edge of air cavities, mostly embedded within partitions. Druses small, common near vascular bundles.

Material examined

Symplocarpus foetidus (L.) Nutt., Boufford 18176, North Carolina, USA; Keating 2275, Illinois, USA.

Taxonomic notes

The biforines noted here are highly idioblastic, wider than others in the family, and without the thin-walled ends that characterize the most highly evolved of this cell type. Also they do not occur embedded in other tissue. But it is questionable whether this cell type is homologous throughout the family.

Species reported on in the literature

Chrysler (1904): *S. foetidus*: stem vasculature. • Ertl (1932): *S. foetidus*: leaf venation. • French (1987c): *S. foetidus*: root. • French (1988): *S. foetidus, S. nipponicus*: laticifers. • French and Tomlinson (1981a): *S. foetidus*: stem vasculature. • Gentner (1905): leaf tip. Genua and Hillson (1985): *S. foetidus*: crystals. • Higley (1880): *S. foetidus*: crystals. • Hotta (1971): *S. latifolius,*

S. nipponensis: leaf. • Krause (1908): *S. foetidus*: stem, leaf. • Netolitzky (1929): *S. foetidus*: crystals: review. • Solereder (1919): *S. foetidus*: leaf. • Solereder and Meyer (1928): *S. foetidus*: anatomy. • Van Fleet (1942): *S. foetidus*: stem, endodermis. • Williams (1919): *S. foetidus*: raphides.

Additional literature

Barabé 1982; Barabé and Forget 1987; Barabé *et al.* 1986b, 1987b; Camazine and Niklas 1984; Davis 1966; Domin 1911; Erdtman 1952; French 1985b, 1986c, 1987a; Grayum 1990; Halevy 1989; Hegnauer 1986; Ray 1988; Rosendahl 1911; Seymour *et al.* 1983; Thanikaimoni 1969; Williams *et al.* 1981; Wilson 1960.

SUBFAMILY POTHOIDEAE

Tribe Potheae

This tribe comprises the Asian tropical genera *Pothos, Pedicellarum*, and *Pothoidium*, and the large New World tropical genus *Anthurium*.

Anatomical summary

Leaf surface

Cuticle smooth, or striate in *Anthurium*. **Epidermis**: cells polygonal with straight anticlinal walls or undulate, occasionally sinuous in *Anthurium*. Secretory glands present in some *Anthurium* species. **Stomata** abaxial and brachyparacytic to brachypara-hexacytic; amphibrachyparacytic occasionally in *Anthurium*. Subsidiary cells variable, often broad and angular, often little varying from neighbouring cells. **Venation** pinnate or acrodromous with multiple parallel primaries, both conditions present in *Anthurium*. Secondaries brochidodromous or eucamptodromous at lamina margin.

Leaf TS

Structure dorsiventral or weakly so, isobilateral in *Pothoidium*. **Midrib** broadly concave adaxially, broadly convex abaxially. Distinctive with high deltoid adaxial ridge in some *Anthurium* species. **Cuticle** thick or thin. **Epidermis**: many cells bearing druse crystals in *Pothos;* cells often tall in *Anthurium* or cells short near glands or stomata. **Hypodermis**: adaxial and frequently present in *Anthurium*. **Mesophyll**: palisade tissue often in 1–2 compact layers. Spongy tissue compact, types 1 or 1a aerenchyma or with longer arms (type 2) in *Pothos* and *Anthurium*. Palisade tissue absent in *Pedicellarum* and some *Anthurium* species. **Collenchyma** band (type B) abaxial in midrib in most *Anthurium* species. **Vascular bundles** symmetrically scattered in midrib, large, with broad phloem boundary (type I). Smaller bundles scattered around large central bundle in midrib of *Pedicellarum* and *Pothoidium*. **Sclerenchyma**: fibres ensheathing all vascular bundles, or cap at xylem or phloem of some *Anthurium* species. Isotropic wall layers usual in *Pedicellarum* and *Anthurium*. **Tannin** cells common in mesophyll of *Anthurium*. **Crystals**: raphides in multiple overlapping bundles in short or elongated cells, also as single bundles in smaller cells in *Anthurium*. Druses in epidermis of *Pothos* and *Pothoidium*, or in mesophyll of *Pedicellarum* and *Anthurium*. Prismatics may be present in vein sheaths of Asian genera or ground tissue of *Anthurium*. Crystal sand in epidermal cells of *Pothoidium* and *Anthurium*.

Petiole TS

Outline concave adaxially, subcircular abaxially; irregular or squarish in *Anthurium*. Shape similar to lamina in *Pothoidium*. **Cuticle** thick, striate-rugulate, or smooth in *Anthurium*. **Epidermis**: cells small. **Collenchyma**: entire cortical layer in *Anthurium*. **Ground tissue** compact, small near periphery, larger at centre; medium-sized throughout in *Pedicellarum* and *Pothoidium*. Type 3 aerenchyma at centre in *Anthurium* and one *Pothos* species. **Vascular bundles** few and symmetrically disposed, type I. Bundles numerous, in concentric circles in *Pedicellarum*, or scattered in *Anthurium*. **Sclerenchyma**: fibres ensheathing vascular bundles; wall layers usually isotropic. Trichosclereids ungrouped, throughout ground tissue, seen in *Pothos rumphii* only. **Tannin** cells in *Anthurium* ground tissue. **Crystals**: raphides as single bundles in thin-walled cells in ground tissue. Raphides in tubular cells of *Pedicellarum*. Druses common in epidermal layers of *Pothos*, or ground tissue of *Pedicellarum* and *Anthurium*. Prismatics common in bundle sheaths of *Pedicellarum*.

Stem TS

Outline circular, or square-ridged in *Pothoidium*. **Cuticle** thick and smooth, or thin in *Anthurium*. **Epidermis**: cells small and square. **Cork** in *Anthurium*, of hypodermal origin and thin-walled and columnar. **Collenchyma** absent or banded in *Anthurium*, in outer cortex. **Ground tissue** compact, smaller cells in cortex, larger in central cylinder or uniform. **Central cylinder** abruptly bounded by heavily lignified fibrous cylinder in Asian genera. In *Anthurium*, central cylinder defined by circle of vascular bundles and 1–3 fibre layers. **Vascular bundles** variable in size and shape, types I, II, III. Mostly type I in *Pedicellarum* and *Anthurium*. **Sclerenchyma**:

fibres ensheathing all vascular bundles or as phloem caps, the cell walls mostly with isotropic layering. Brachysclereids numerous at central cylinder boundary in *Pothoidium*. **Tannin** cells in ground tissue of *Anthurium*. **Crystals**: raphides mostly in single bundles in thin-walled cells. Druses common near vascular bundles and near central cylinder boundary, in central cylinder in *Pothoidium*, and in ground tissue in *Anthurium*. Prismatics abundant near central cylinder boundary in *Pedicellarum* and *Pothoidium*.

Taxonomic notes

All four genera are appropriately placed in the same tribe. They show an array of characters that are expressed in a mosaic such that no genus stands out as more isolated. *Anthurium* because of its huge size expresses more character states but this still does not obscure its obvious relationship to the Asian genera. Each genus shows diagnostic autapomorphies.

Pothos L.

This genus of about 50 species is centred in Indomalaya. Most species are confined to the tropics and subtropics of south and south-east Asia: India to the Himalayas, Sri Lanka, to south China, Taiwan, Philippines, the Moluccas and New Guinea. One species is found in tropical Australia, and one in Madagascar and Comoro Islands. Plants occur in primary and secondary forests. They are mostly hemi-epiphytic root climbers with non-heteroblastic, distichous leaves, but some are terrestrial on steep banks. The shoot is a shrubby branched monopodium. Leaves are ovate to elliptic and articulated with a broad laminar petiole. The petiole summit at the laminar junction is variable and probably diagnostic. Chromosomes are small (4 counts): 2n = 24, 36. Four species were available for study.

Leaf surface

Cuticle roughly striate on both surfaces. **Epidermis**: cells polygonal to isodiametric with straight anticlinal walls on both surfaces (1–1.5:1 *l*/*w*) (Fig. 16C). Many cells appearing small and rounded and containing a druse crystal; neighbouring epidermal cells radiating from each crystal cell. **Stomata** brachypara-hexacytic, except brachypara-tetracytic in *P. scandens* (Fig. 16C). Subsidiary cells broad, angular, often varying little except by position from epidermal cells. Tomlinson (1970) noted that mature stomata of *Pothos* have four subsidiary cells and that four initials are involved in stomatal development.

Venation: In *P. scandens*: midrib a weakly defined cluster of primary vascular bundles, ascending and spreading from apex of petiole. Basal secondaries steeply ascending, exmedially acute, admedially obtuse from primaries. Veins merging eucamptodromously at margin.

In *P. rumphii*: midrib prominent, secondaries acute, mostly straight, ending abruptly in mid-lamina at ± straightened brochidodromous collector. Tertiaries branching acutely from collector, merging at submarginal collector; smaller collectors present at margin. Tertiaries parallel to secondaries or reticulate. Higher orders reticulate.

In *P. remotiflorus*: prominent midrib with pinnate, irregularly acute secondaries, merging at straightened submarginal collector. Tertiaries and higher orders reticulate.

In *P. macrocephalus*: midrib prominent. Secondaries basal and suprabasal ascending. Tertiaries abmedially broadly acute from secondaries, and obtuse admedially. Tertiaries parallel to each other, not to secondaries. Margin eucamptodromous. Higher orders reticulate.

Leaf TS

Structure dorsiventral. **Size**: *P. macrocephalus* 58 × 31 mm, *P. remotiflorus* 88 × 14 mm, *P. rumphii* 162 × 52 mm, *P. scandens*: 74 × 29 mm (*l*/*w*). **Midrib**: broadly and shallowly concave adaxially, broadly rounded convex abaxially (Fig. 16A). **Cuticle** medium-thick adaxially, thin abaxially. **Epidermis**: cells square and large adaxially and somewhat shorter abaxially. Cells flat on both surfaces and rounded internally. Adaxial epidermal cells sclerotic in *P. rumphii*. **Stomata** level with surface. **Mesophyll** well differentiated. Palisade: 1-2 layers of squarish cells. Spongy tissue: up to four layers of rounded cells (up to eight layers in *P. scandens*) with short tuberculate lobes (type 1a) (Fig. 16B), or more usually long-armed (type 2). **Air cavities** up to 25% of spongy layer.

Vascular bundles scattered, well defined and symmetrically disposed in midrib; four smaller bundles flanking one large median bundle. In *P. rumphii*, large midrib with large median bundle and 10 additional large and small bundles symmetrically placed in three tiers. In *P. scandens*, midrib narrow with three symmetrically placed large bundles. Vascular bundles type I. Xylem: 1–2 narrow protoxylem elements and numerous metaxylem elements. **Sclerenchyma**: fibres ensheathing all vascular bundles. **Crystals**: raphides uncommon in mid-mesophyll, multiple overlapping bundles in short or very long tubular cells (Fig. 16D). Druses in either epidermis with large points oriented toward surface. In surface view, druse

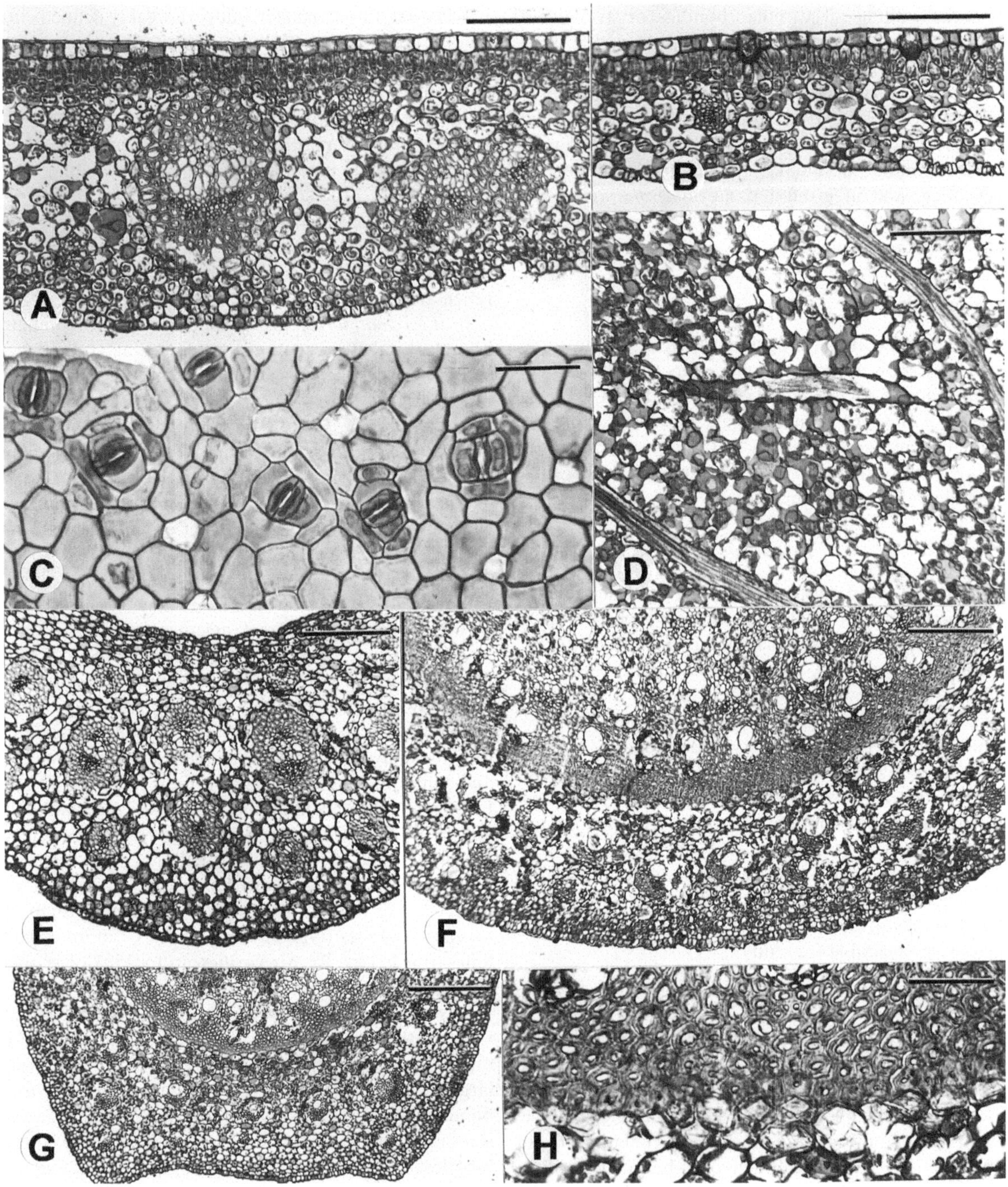

Fig. 16. *Pothos* leaf, petiole and stem sections. (A–C,E,F,H) *P. scandens*; (D,G) *P. macrocephalus*; (A) Midrib TS showing type I vascular bundle ensheathed by fibres; (B) Lamina TS showing enlarged druse cells in adaxial epidermis; (C) Leaf abaxial epidermis PS showing brachypara-tetracytic stomata; (D) Mesophyll showing type 1 aerenchyma and elongate raphide cell; (E) Petiole base TS; (F) Stem TS showing deep fibrous boundary at central cylinder; (G) Stem TS; (H) Stem TS showing prismatic crystals in inner cortical layer. Scale bars: A,B,D = 150 μm; C,H = 50 μm; E,F = 200 μm; G = 300 μm.

cells small, circular, with 7–8 adjacent epidermal cells radiating from them. Druses common in mid-mesophyll and unmodified in *P. remotiflorus*. Prismatics uncommon in vein sheaths or absent.

Petiole TS

Outline winged with no midrib demarcated. Broadly concave adaxially, deeply and broadly convex abaxially. **Size**: diameter 1 × 5 mm. **Cuticle**: thick and broadly striate-rugulate, often with flanges between outer epidermal anticlinal walls. **Epidermis**: cells small and rounded. **Stomata** occasional, level with abaxial surface, with shallow substomatal cavity. **Ground tissue**: cells small near epidermis grading to large at centre. **Air cavities** at corners of cells, up to 3% of space. Type 3 aerenchyma with 1-celled partitions in *P. rumphii*.

Vascular bundles: large central bundle, several smaller bundles symmetrically disposed around it (type I) (Fig. 16E). Xylem: 5–10 metaxylem elements in 2–4 files. Phloem: semicircular strand. **Sclerenchyma**: fibres ensheathing vascular bundles up to several layers deep (Fig. 6E); cell walls in several discrete isotropic layers. Trichosclereids: large diameter cells throughout ground tissue, ungrouped, seen in *P. rumphii* only. **Starch** granules abundant and small, 2–5 μm diameter. **Crystals**: raphides common in ground tissue in thin-walled cells. Druses common in epidermal layers, often with enlarged, sharp, chisel-shaped facet pointing toward surface. Druses abundant in abaxial ground tissue in *P. rumphii*. Prismatics found in vein sheaths of *P. scandens* (Fig. 6E), otherwise not seen. Crystal sand common in outer walls of adaxial epidermal cells in *P. macrocephalus*.

Stem TS

Outline terete. **Size**: diameter 2.5 mm. **Cuticle** thick and smooth, flanged between outer anticlinal epidermal walls. **Epidermis**: cells small and square, rounded externally. **Chlorenchyma**: outer 8–10 layers of ground tissue. **Ground tissue**: cells small near epidermis, grading to large in inner cortex and central cylinder. **Air cavities** at cell corners, up to 5% of ground tissue. **Cortex**: well defined and narrow or wide. **Central cylinder**: bounded abruptly by heavily lignified outer sclerenchyma, grading toward centre to thinner-walled sclerenchyma 12–15 layers deep (Fig. 16F,G).

Vascular bundles: 1–3 rows of small bundles in cortex; central cylinder bundles collateral; types I, II or III. Xylem: 1–8 medium-width metaxylem cells or one very wide metaxylem element. Phloem strand semicircular at broad boundary with larger xylem strands or irregular and smaller at small xylem strands. Hotta (1971) noted advanced vessels with 2–20-barred perforation plates in *P. atropurpurascens*. **Sclerenchyma**: fibre sheaths surrounding all vascular bundles, the cell walls with distinct isotropic bands. **Starch** granules abundant in most ground tissue, *c.* 10–12 μm diameter. **Crystals**: raphides common in thin-walled cells in outer cortical ground tissue. Druses variably-sized and common near vascular bundles in *P. scandens*, or in epidermis and uncommon at central cylinder boundary in *P. macrocephalus*. Prismatics abundant at central cylinder boundary outside sclerenchyma sheath (Fig. 16H); very common in ground tissue adjacent to most vascular bundles.

Node TS (French and Tomlinson 1981; *P. scandens* only)

Nodal vascular bundles course: leaf trace system an arc of *c.* 10 wide bundles, each continuing into central cylinder. Central petiole bundle a continuous median trace. Narrow leaf traces numerous, entering and remaining confined to cortex. Branch trace penetrating deeply in broad arc into stem central cylinder. Branch trace elements mostly undifferentiated vascular bundles; proximal extensions ± recognizable as discrete traces. Cortical bundles often anastomosing. Adventitious root traces, 6–8, immediately below node, each inserted into opposite lateral leaf traces.

Material examined

Pothos macrocephalus Scort., Bogner 1550, cult. M; *P. remotiflorus* Hook., Jayasuriya and Dassanayake 1016, Ceylon, K; *P. rumphii* (Presl) Schott, Forman 240, Celebes; *P. scandens* L., Bogner 156, Madagascar, cult. M.

Taxonomic notes

The species of *Pothos* form a coherent group but there is much diagnostic variation among them. The genus would repay a detailed study of its morphology and anatomy. For instance, Hotta (1971) reported that many *Pothos* species have no trichosclereids but he reported them present in *P. repens* and *P.* sp. (Hotta 14227). They were found in the present study in *P. rumphii*. Hotta reported that other species have fibrous tracheids in stems similar to trichosclereids.

Species reported on in the literature

Arzt (1937): *P. aureus*, *P. celatocaulis*: cuticle. • Beck and Mannagetta (1912): crystals. • Buscalioni (1895–96): *P. platicaule*: crystals. • De la Rue (1869):

P. crassinervis: petiole crystals. • Dutt (1956): root lenticels. • El-Gazzar *et al.* (1977): *P. aureus*: stomata. • Engler (1905): root. • Ertl (1932): *P. loureirii, P. macrophyllus*: leaf venation. • Francini (1946): root cap. • French (1987c): several spp.: root. • French and Tomlinson (1981c): *P. scandens*: stem vasculature. • Gertz (1916a): *P. celatocaulis*: tyloses. • Goebel (1922): root. • Gulliver (1865): *P. acaulis*: crystals, starch. • Guttenberg (1968): *P. aureus, P. celatocaulis*: root. • Hotta (1971): several spp.: leaf, stem. • Janse (1897): root endophytes. • Lierau (1888): *P. celatocaulis, P. scandens*: root. • Linsbauer (1930): *P. scandens*: epidermis. • Müller (1919): *P. aureus, P. celatocaulis, P. gracilis*: hydathodes, leaf tip. • Netolitzky (1929): crystals. • Nicolson (1960): *P. repens*: stem, leaf, petiole. • Paliwal (1969): stomata. • Pant and Kidwai (1966): *P. aureus*: leaf. • Richter (1901): *P. celatocaulis*: root. • Rosanoff (1867): *P. argyrea*: petiole crystals. • Roth (1949): *P. celatocaulis, P. macrophyllus*: leaf ontogeny. • Schwendener (1882): shoot. Solereder (1919): *P. celatocaulis, P. rumphii, P. scandens*. leaf. • Solereder and Meyer (1928): several spp.: several organs. • Tieghem (1866): several spp.: all organs. • Tomlinson (1970): *P.* sp.: stomata. • Trécul (1854): *P. repens*: leaf. • Troll (1932a): *P. macrophyllus*: leaf ontogeny. • Turpin (1836): *P. foetida, P. violacea*: leaf, crystals. • Wakker (1888): *P. scandens*: crystals. • Webber (1960): *P. hermaphroditus, P. jambea* (illus.): epidermis, stomata. • Went (1895): several spp.: root.

Additional literature

Blanc 1978; Davis 1966; Domin 1911; Erdtman 1952; Eyde *et al.* 1967; Glück 1901; French 1985a, 1986c, 1987a; Gertz 1916a; Grayum 1990; Huang 1972; Kapoor *et al.* 1975; Meusel 1951; Mookerjea 1955; Sharma 1967; Sharma *et al.* 1985; Shirley and Lambert 1918; Thanikaimoni 1969; Williams *et al.* 1981.

Pedicellarum M. Hotta

This monotypic genus is found in moist forests in Borneo. Leaves are small, elliptic and short-petiolate on spreading shoots. No chromosomes counts are available. One dry specimen was available for study.

Leaf surface

Cuticle smooth. **Epidermis**: cells polygonal with straight to undulate anticlinal walls adaxially (1–1.2:1 *l*/*w*) and polygonal, undulate, or irregularly elongate abaxially (2–4:1 *l*/*w*). **Stomata** abaxial and brachyparacytic (Fig. 17D), or brachypara-tetracytic. Subsidiary cells narrow or broad, usually rounded and smaller than surrounding epidermal cells. **Venation**: pinnate with ± straightened brochidodromous collector. Tertiaries and higher orders reticulate.

Leaf TS

Structure weakly dorsiventral. **Size**: 75 × 18 mm (*l*/*w*). **Midrib** broadly concave adaxially and broadly rounded-fluted abaxially (Fig. 17A). **Cuticle** thick on both surfaces. **Epidermis**: cells tall abaxially and horizontally elongate adaxially. Abaxial midrib outer cell walls sclerotic beneath thick cuticle. **Stomata** level with surface. **Mesophyll** little differentiated; upper adaxial layer horizontally elongated, not palisade. Spongy tissue: 3–4 layers, cells with 1–2 lobes (type 1). **Air cavities**: *c.* 10% of spongy layer.

Vascular bundles: smaller vascular bundles scattered symmetrically around one very large central midrib bundle. Bundles broadly oval (type I). Xylem: numerous metaxylem elements. Phloem: a large semicircular strand. **Sclerenchyma**: fibres ensheathing bundles (Fig. 17A,B) in *c.* six layers over phloem and fewer elsewhere; isotropic wall layers prominent, up to four. **Crystals**: raphides in large tubular cells in multiple overlapping bundles surrounded by mucilage (Figs 9C, 17C), common near or between vascular bundles. Druses common and randomly distributed in mesophyll, occasional in adaxial epidermis. Prismatics large, abundant in cells forming lines along vascular bundles.

Petiole TS

Outline: subterete (Fig. 17E) and shallowly flattened adaxially. **Size**: diameter 1 mm. **Epidermis**: cells small and square. **Chlorenchyma**: outer 1–2 ground tissue layers, adaxially. **Ground tissue**: medium-sized compact parenchyma throughout. **Air space** minimal, but 2–3 large schizogenous cavities.

Vascular bundles: 16–18 symmetrical bundles, in three concentric circles; bundles ensheathed (type I). Xylem: 5–10 medium-sized metaxylem elements in 2–5 tiers. Phloem: broad semicircular strand. **Sclerenchyma**: fibres numerous and thick-walled, with discrete wall-layering, ensheathing all vascular bundles, deepest layers over phloem. Outer rows of bundles connected by confluent fibre sheaths. **Crystals**: raphides in tubular cells, rare in central ground tissue. Druses coarse, common in all ground tissue. Prismatics large and very common outside all bundle sheaths.

Stem TS

Outline circular. **Size**: diameter 2 mm. **Cuticle** thick and smooth. **Epidermis**: cells small, outer surface flat. **Ground tissue** cells uniform-sized throughout. **Cortex** very narrow,

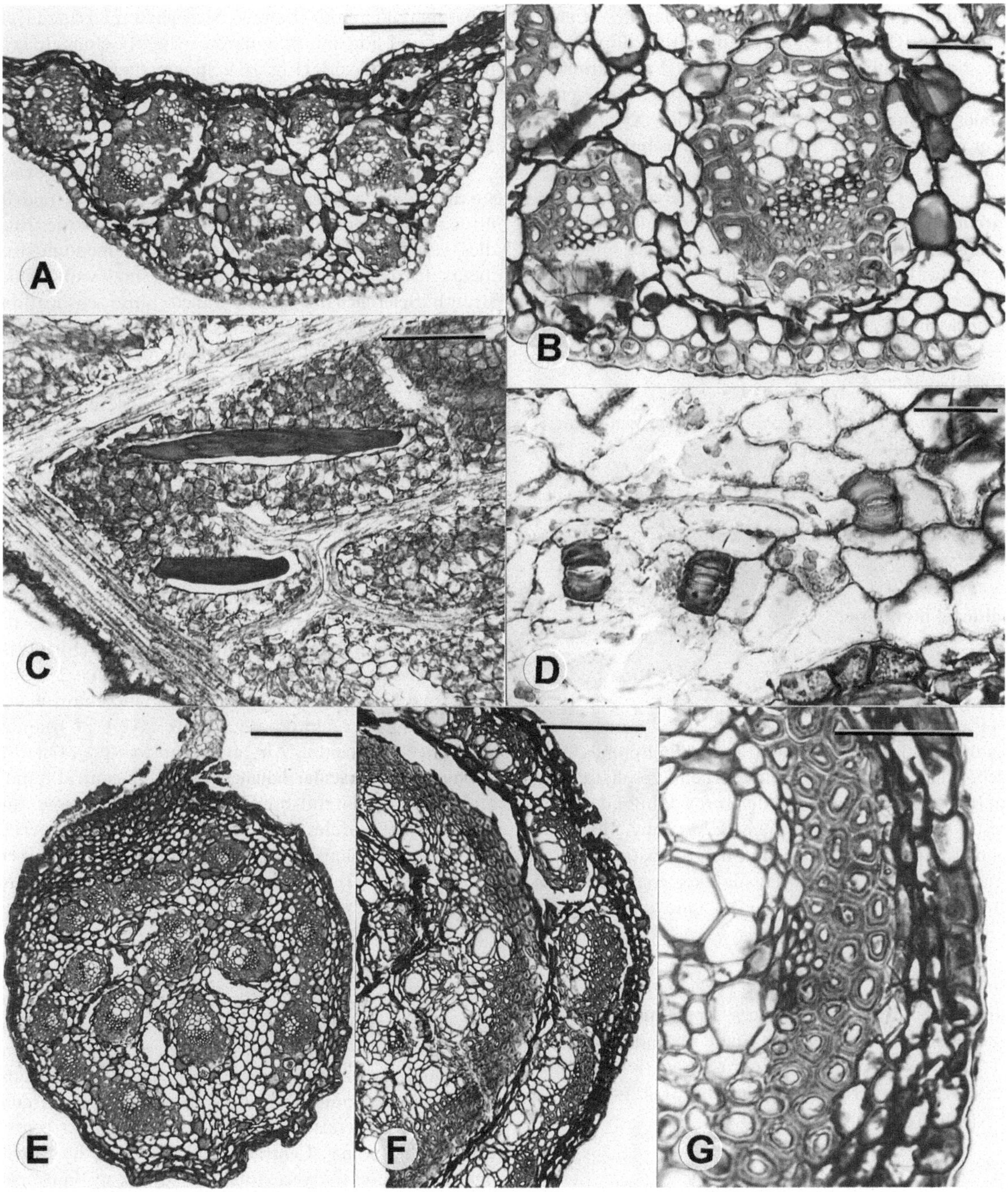

Fig. 17. *Pedicellarum paiei* leaf, petiole and stem sections. (A,B) Midrib TS; (B) Abaxial surface; (C) Mesophyll PS showing raphide cells; (D) Epidermis PS showing brachyparacytic stomata; (E) Petiole TS; (F,G) Stem TS: (F) sclerotic central cylinder boundary and narrow cortical layer; (G) Epidermis and cortical bundle cap. Small phloem strand in centre. Scale bars: A,C,F = 150 μm; B,D,G = 50 μm; E = 200 μm.

width equal to one row of vascular bundles. **Central cylinder** large (Fig. 17F), sharply demarcated by continuous ring of fibres 3–6 cells deep. **Vascularbundles** numerous, type I, closely packed in cortex and central cylinder leaving little ground tissue space. Xylem: metaxylem cells wide and numerous. Fewer but very wide metaxylem cells in central cylinder bundles. Phloem small diameter strands (Fig. 17G). **Sclerenchyma**: fibres encircling vascular bundles; cells thick-walled with 4–5 isotropic layers. **Crystals**: raphides in single bundles in inner cortical tissue. Rhombohedral crystals common in outer cortical tissue (Fig. 17G) and near outer phloem sclerenchyma sheaths.

Material examined

Pedicellarum paiei M.Hotta, Paie 16354, Sarawak, cult. K.

Species reported on in the literature

French and Tomlinson (1981c): *P. paiei*: stem vasculature.

Additional literature

French 1985a; Grayum 1990.

Pothoidium Schott

This monotypic genus is found in the Philippines, Java, Celebes and Moluccas. Plants are hemi-epiphytic scandent herbs occurring in the understorey of humid primary forests. The shoot is a branched, shrubby, distichous monopodium. Blades of the non-heteroblastic leaves are elliptic and articulated onto a blade-like, elongated petiole segment. Chromosomes: 2n = 24. One specimen was available for study.

Leaf surface

Cuticle smooth on both surfaces. **Epidermis**: cells polygonal with anticlinal walls straight or undulate (1–2:1 *l*/*w*). **Stomata**: abaxial, brachypara-hexacytic (Fig. 18D). Subsidiary cells narrow. **Venation**: multiple parallel primaries and reticulate higher orders.

Leaf TS

Structure weakly dorsiventral (Fig. 18C) to isobilateral. **Size**: 4.7 × 0.7 cm (*l*/*w*). **Midrib** slightly thickened abaxially near base. **Cuticle** medium-thick adaxially and thinner abaxially. **Epidermis**: cells tall and square, flat or externally rounded adaxially and flat abaxially; rounded internally on both layers. **Stomata** level with surface; substomatal cavities shallow. **Mesophyll**: palisade layers abaxial and adaxial, of square or slightly elongate cells packed with plastids (Fig. 1C). Spongy tissue: five central layers of larger parenchyma cells with very short lobes (type 1a aerenchyma). **Air cavities** at cell corners, up to 20% of spongy layer.

Vascular bundles: one large midvein with several smaller veins symmetrically placed in centre of midrib; no marked boundary with adjacent lamina tissue. Bundles type I. Xylem: numerous small to medium-sized metaxylem cells in several files. Phloem: small semicircular strand. **Sclerenchyma**: thick-walled fibres ensheathing vascular bundles. Lignin stains varying in hue on inner and outer groups of fibres. **Crystals**: raphides common in multiple overlapping bundles in elongated large cells, in mid-mesophyll (Fig. 18E). Druses mostly in epidermal layers, often with elongated points facing surface. Prismatics: large parallelograms, abundant near vascular bundle sheaths. Crystal sand coarse and very common in epidermal cells.

Petiole TS

Outline: structure very similar to leaf lamina, only differences described below. **Midrib** more distinct, flat adaxially, slightly convex-rounded abaxially (Fig. 18B). **Ground tissue**: little differentiation (Fig. 18A); 2–3 subepidermal layers of small cells packed with chloroplasts, on both surfaces. Internally, *c.* five layers of rounded spongy cells in centre. **Air cavities** up to 10% of central spongy cells. **Vascular bundles**: one large central bundle and two large lateral bundles in centre of petiole, and several small bundles closer to abaxial surface; numerous metaxylem elements and a large semicircular phloem strand (type I). **Sclerenchyma**: thick-walled fibres ensheathing all vascular bundles.

Stem TS

Outline: subterete, four broad, shallow ridges forming squarish corners (Fig. 18F). **Size**: diameter 2 mm. **Cuticle**: thick and smooth, flanges often between outer anticlinal epidermal walls. **Epidermis**: cells small and square. **Chlorenchyma**: outer 4–5 ground tissue layers. **Ground tissue**: outer cortical cells small, cells larger at central cylinder boundary. **Central cylinder** marked by heavily lignified sclereid layer, followed by deep fibre ring (Fig. 18G,H).

Vascular bundles: 1–2 rows of bundles in cortex, bundles scattered and numerous in central cylinder; bundles collateral (types II and III). Xylem: 1–3 large late protoxylem cells and up to 10 medium-sized metaxylem cells. Phloem: small, dorsiventrally narrowed,

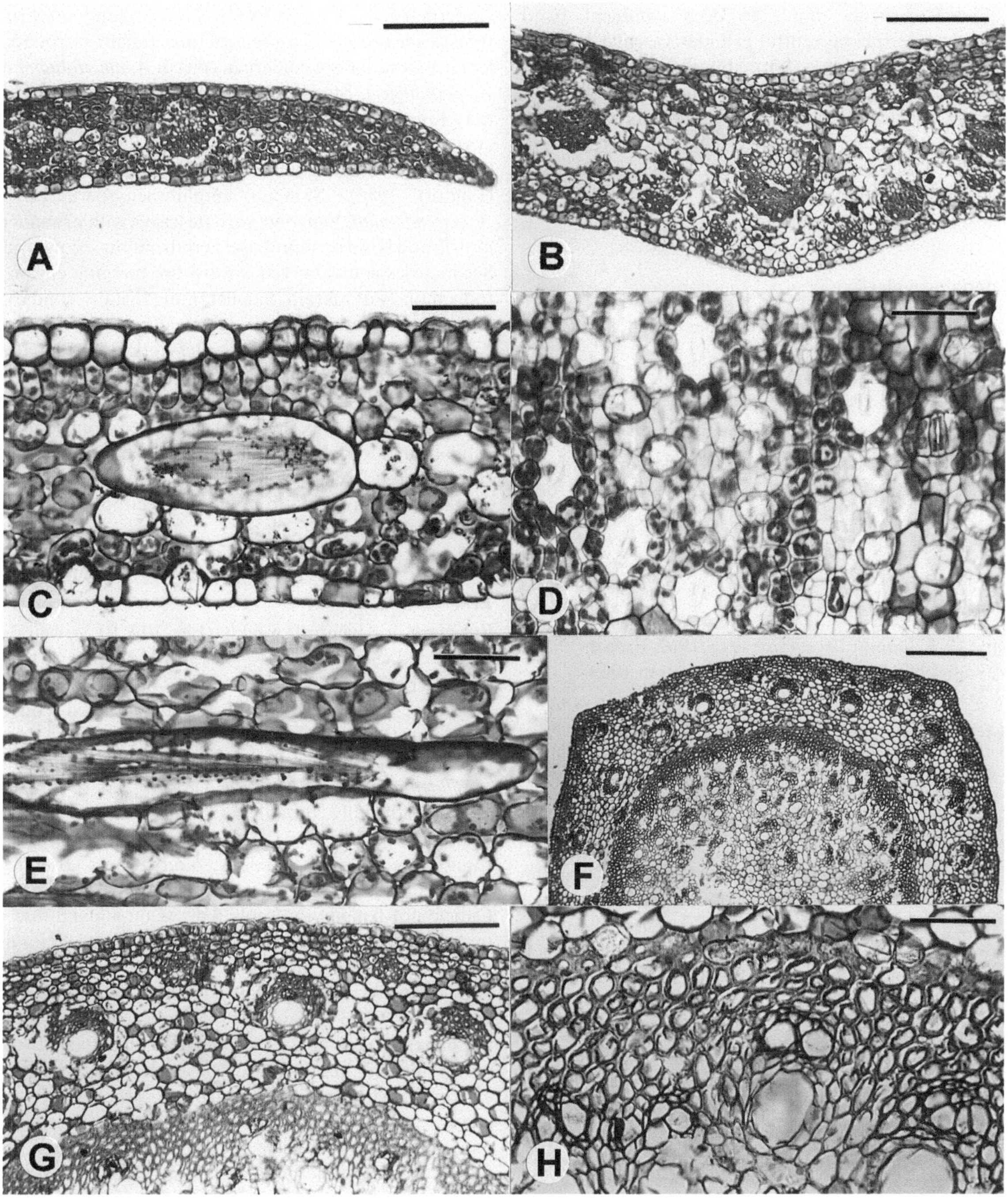

Fig. 18. *Pothoidium lobbianum* leaf, petiole and stem sections. (A) Proximal petiole margin TS; (B) Proximal petiole midrib TS; (C) Lamina TS showing raphide cell; (D) Leaf mesophyll PS showing substomatal air cavities and brachypara-tetracytic stomata; (E) Leaf mesophyll PS showing elongated raphide cell; (F) Stem TS showing well-delimited central cylinder; (G) Stem TS showing cortical bundles; (H) Stem TS showing bundles inside central cylinder boundary. Scale bars A,B,G = 150 μm; C–E,H = 50 μm; F = 300 μm.

oval. **Sclerenchyma**: fibres in deep confluent band embedding peripheral central cylinder vascular bundles and small phloem caps in central bundles. Fibres forming deep C-shaped caps over phloem in cortical bundles (Fig. 18G). Sclereids heavily lignified at central cylinder boundary. **Crystals**: raphides in thin-walled cells common in cortex. Druses large and common in central cylinder. Prismatics large and very common in inner and outer cortex.

Material examined

Pothoidium lobbianum Schott, Bogner 1272, cult. M.

Species reported on in the literature

(Species are *P. lobbianum*, except as noted.)

French (1987c): root. • Hotta (1965): *P. atropurpurescens, P. leptostachys, P. repens*: leaf, stem, root. • French and Tomlinson (1981c): stem vasculature. • Rosanoff (1867): *P. argyrea*: crystals.

Additional literature

Blanc 1978; Domin 1911; Eyde *et al.* 1967; French 1985a; Grayum 1990; Meusel 1951; Thanikaimoni 1969.

Anthurium Schott

The approximately 1000 species of this genus are placed in up to 19 sections based on morphology and cytology. They are distributed in the American tropics from Mexico to Argentina and Paraguay, and the West Indies. The plants are mostly epiphytic appressed climbers in the understorey of primary, mostly humid to wet forests; less frequently in higher branches. Occasionally, plants are terrestrial, epilithic, hemi-epiphytic, or rheophytes (such as *A. amnicola* Dressler (Croat 1988)). They are common at lower and middle elevations and in cloud forests (Croat 1988). Plants are of varied habit with large, often erect petiolate leaves varying from elliptic to cordate to hastate. Chromosomes: 2n = 30 (most species), 20, 24, 28, 40, 48, 56, 60, 90, 124. Thirty-five species were available for study, too few to divide the descriptions into sections with confidence.

Leaf surface

Cuticle smooth to slightly striate. **Epidermis** of polygonal cells with straight anticlinal walls (1–3:1 *l/w*), or undulate or sinuous walls in *A. michelii*. Circular secretory glands in epidermis of *A. bredemeyeri*, 4–5 × size of typical epidermal cells (Lindorf 1980). Sunken glandular structures ('punctations') with amorphous contents surrounded by narrow radiating epidermal cells in *A. lancifolium* and *A. wendlingeri*. **Stomata** brachyparacytic (Fig. 19C) but next four adjacent cells often loosely arranged in hexacytic pattern, or brachypara-hexacytic in *A. pariense*. Subsidiary cells medium to wide with square edges (Lindorf 1980). Stomata amphibrachy-paracytic in *A. oerstedianum*. **Venation**: cordate leaves with prominent midrib and basal or suprabasal acrodromous secondaries. Secondaries acute, curved toward tip, merging eucamptodromously at margin. Smaller, more distal secondaries broadly acute ending at straightened brochidodromous collectors. Tertiary and higher orders reticulate. Pinnately veined species: prominent midrib with irregular secondaries forming irregular intercostal areas. Secondaries acute or widely acute, slightly straightened, connecting in submarginal brochidodromous loops, or straightened collector vein well within margin. Tertiaries also with straightened brochidodromous connections. Tertiaries and higher orders reticulate.

Leaf TS

Structure dorsiventral, weakly (Fig. 19A,B) or well differentiated. **Size**: 15–40 × 7–15 cm *l/w*. **Midrib** variable, characteristically distinctive with a high deltoid ridge adaxially, and broadly and deeply rounded convex abaxially (Fig. 1E). Some species with symmetrical rounded deltoid ridges extending from both surfaces. In *A. clidemioides* and *A. flexile*, midrib flat or broadly concave adaxially, broadly and deeply convex-rounded or fluted abaxially. In *A. corrugatum*, midrib with small rounded convexity adaxially, and deeply rounded-convex abaxially. **Cuticle** thick to thin or not detected. Cuticle not flanged to deeply flanged on adaxial side of midrib, and occasionally flanged over portions of lamina; often best developed at margin and lignified. Cuticle deep and bilayered, deeply flanged on both surfaces in *A. erskinei*.

Epidermis: cells tall or medium-sized, flat adaxially and more rounded abaxially. Cells flat or deeply rounded externally with cuticular flanges in grooves between adaxial walls, and adaxial and abaxial walls in *A. clidemioides*. Sunken glandular structures with dark amorphous contents at surface in *A. lancifolium* and *A. wendlingeri*. Epidermal cells shorter in vicinity of stomata which are often raised above surface by mound of small subsidiary cells. **Stomata**: guard cells with symmetrical double ledges, or with one ledge, or two ledges at outer opening. **Hypodermis**: 1-layered, present in five species beneath adaxial or both surfaces (Fig. 4C). **Mesophyll**: palisade tissue absent

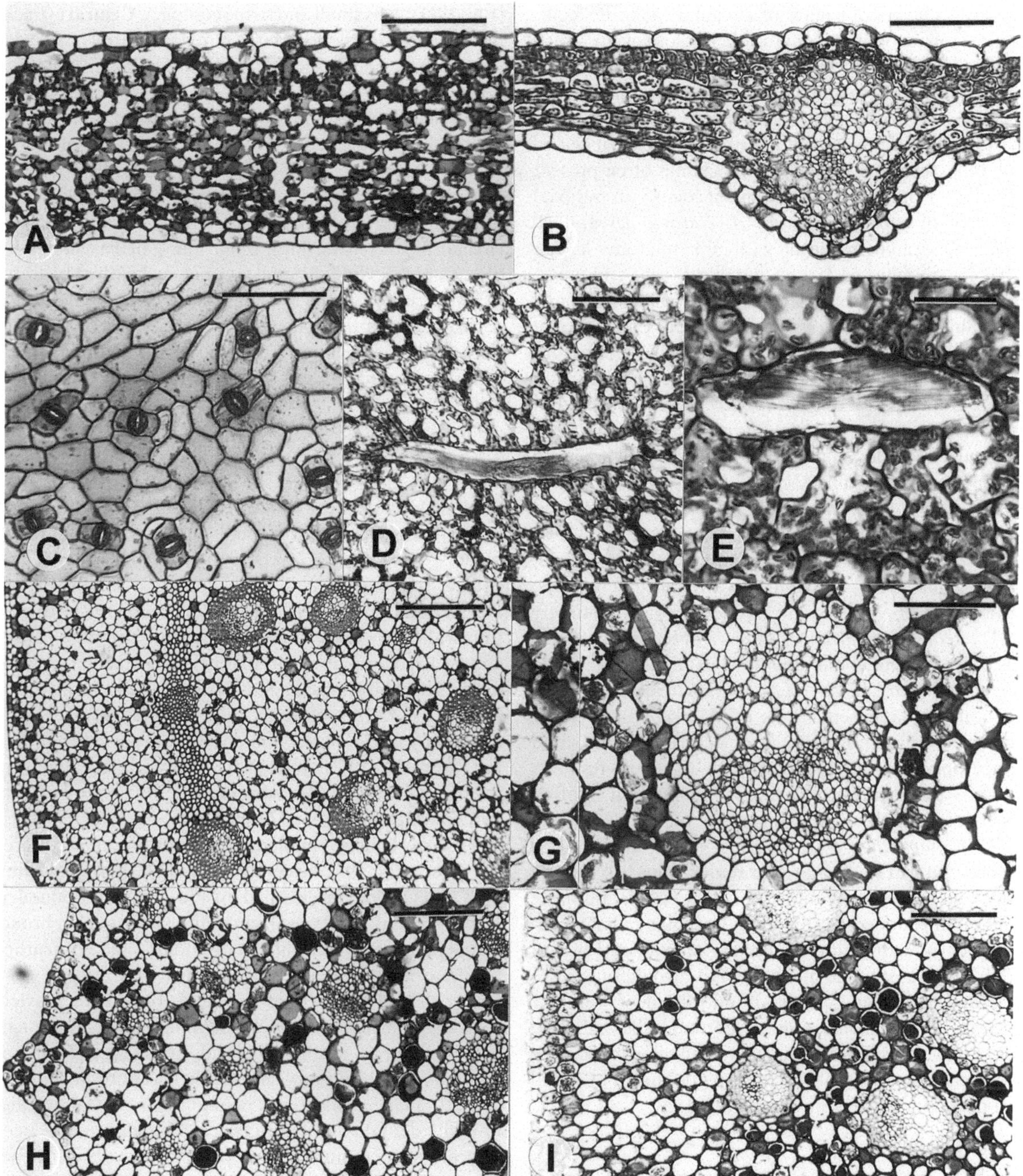

Fig. 19. *Anthurium* leaf, petiole and stem sections. (A) *A. veitchii*; (B,C) *A. salvadorense*; (D) *A. hoffmannii*; (E) *A. schlechtendalii*; (F) *A. watermaliense*; (G) *A. ravenii*; (H) *A. polyschistum*; (I) *A. wendlingeri*. (A) Lamina TS showing adaxial hypodermis and weakly differentiated mesophyll; (B) Lamina TS with secondary vascular bundle; (C) Abaxial epidermis PS with stomata; (D) Lamina PS showing spongy mesophyll and elongated raphide cell; (E) Lamina PS showing type 1a mesophyll and raphide cell; (F) Petiole TS showing mechanical tissue band confluent with outer vascular bundles; (G) Petiole TS, ground tissue and type I vascular bundle; (H) Petiole TS; (I) Stem TS. Scale bars: A–C,G = 150 μm; D,H,I = 200 μm; E = 50 μm; F = 300 μm.

or with 1–3 layers of square or irregular 'palisade' cells. Spongy cells rounded, multi-lobed, somewhat horizontally elongate, up to 25 layers (types 1, 1a or 2). Spongy type 1 or 2 with tall air cavities above stomata in *A. willdenowii*. Spongy type 2 in species studied by Lindorf (1980). Plastids evenly distributed, well packed in some cells, several layers of chlorenchyma often present adaxially at midrib. Plastids most dense in adaxial mesophyll of some species. In *A. clidemioides* and *A. flexile*, mesophyll reduced with up to four layers: 1-layered columnar or T-shaped palisade cells and type 1 spongy tisssue. **Air cavities** irregular, 25–60% of spongy tissue. Substomatal cavities small or large. **Collenchyma**: weakly formed band of angularly thickened cells, subepidermal on abaxial side of midrib in most species; band varying from 2-layered in *A. huixtlense* to very broad in *A. hookeri*.

Vascular bundles: numerous scattered bundles in midrib or bundles symmetrical in *A pariense, A. michelii*, and *A. subcordatum* (type I). Bundles broadly rounded with up to four protoxylem elements and up to 12 metaxylem elements and with semicircular phloem strand. **Sclerenchyma**: fibres ensheathing midrib vascular bundles, 1–5 layers deep, thickest over phloem; walls distinctly 4–5-layered. In *A. pariense*, fibres as phloem or xylem caps only. Phloem containing nests of fibres in *A. cerro-campanense and A. cinereopetiolatum*. **Surface glands** common in adaxial epidermis of *A. wendlingeri*; semicircular cup of glandular palisade cells leading to small surface opening with cuticle absent and dark resinous materials occluding pore. **Tannin** cells in mesophyll, often numerous, some elongate, with amorphous dark contents. In *A. pentaphyllum* var. *bombacifolium* tannin cells numerous at leaf base. **Starch** granules in midrib ground tissue, and occasionally in mesophyll. **Crystals**: raphides common in rounded or elongated irregular cells, irregularly oriented with multiple overlapping bundles (Figs 9B, 19D,E). Druses small to medium-sized throughout mesophyll and in ground tissue of midrib, near vascular bundles. In *A. concinnatum*, epidermal cells radiating around few druse-containing cells. Irregular loose prismatics uncommon in cells of lamina and midrib. Crystal sand occasionally present in mesophyll.

Petiole TS

Outline irregular or squarish, adaxially grooved. **Size**: diameter 8–10 mm. **Cuticle** medium-thick and smooth. **Epidermis**: cells small, horizontally elongate and level with surface. **Collenchyma** comprising entire cortex, wall layers thin. **Chlorenchyma**: plastids common in cortical cells. **Ground tissue**: cortex bounded by 8–9 layers of mechanical tissue (Fig. 19F), or not so bounded (Fig. 19H); cells small and tissue compact. **Central cylinder** mostly compact tissue of large parenchymatous cells with air spaces at cell corners, or sharply delimited type 2, 3 aerenchyma.

Vascular bundles numerous and scattered. Outer circle of bundles connected by interfascicular ring of fibres; central bundles larger than outer bundles; type I (Fig. 19F,G). Xylem: 6–10 wide metaxylem cells in 3–4 irregular files. Phloem: broad elliptic strand. **Sclerenchyma**: fibres ensheathing outer circle of vascular bundles. Central cylinder bundles either ensheathed by fibres or with fibre strands capping phloem. **Tannin** cells occasional, scattered in ground tissue. **Starch** granules uncommon or common, scattered or throughout central cylinder ground tissue. **Crystals**: raphides in thin-walled cells, scattered in ground tissue. Druses infrequent or common, scattered in ground tissue.

Stem TS

Outline: terete. **Size**: diameter 6–35 mm. **Cuticle** thin and smooth. **Epidermis**: cells small and squarish. **Cork**: in older specimens 4–6 layers of hypodermal origin; cells empty, thin-walled and often columnar. **Collenchyma**: absent or as a band of 6–8 outer cortical layers, with cells thin-walled. **Ground tissue**: cortex with compact collenchyma tissue, or up to 40 parenchyma layers; cells uniform throughout, except smaller near epidermis. Cortex from $\frac{1}{2}$–$\frac{1}{10}$ radius of stem (Fig. 19I). **Central cylinder** defined by circle of medium and small vascular bundles, not confluent, or by circle of 1–3 fibre layers and confluent bundle caps. **Air cavities** none or up to 2–5% at corners of ground tissue cells.

Vascular bundles small, numerous or few and scattered in cortex (Fig. 19I); larger and scattered in central cylinder, especially concentrated inside cylinder boundary. Bundles collateral (type I), circular or oblate, with broad, usually curved xylem/phloem boundary, and ensheathed by uniseriate circle of small parenchyma cells. Xylem: large or small protoxylem lacuna, 4–14 wide metaxylem cells in 3–5 irregular files. Phloem: large, broad semicircular or elliptic strand at broad, straight, irregular or curved xylem boundary. **Sclerenchyma**: fibres as phloem caps, 2–3 layers thick over bundles near central cylinder boundary and over all cortical bundles; caps deep crescent-shaped surrounding phloem, or very deep, not crescent-shaped. Fibres may form confluent caps between bundles at central cylinder boundary. Fibre walls with 4–8 clearly distinct layers in most specimens. **Tannin** cells absent or common, scattered in ground tissue and phloem, especially concentrated near periphery of cortex. **Starch** granules packed in central cylinder, fewer in inner cortex or only sparsely present. **Crystals**: raphides in discrete bundles common in axially aligned cells scattered in ground tissue; cells thin-walled, enlarged or not. Druses common to abundant in all ground tissue and cortex, or especially in cortex; cells usually not enlarged.

Root TS (seen in eight species)

Outline circular, diameter 3–8 mm. **Trichoblasts** absent or common in epidermis of *A. micheliii*. **Velamen** often present, 4–10 irregular layers. **Exodermis** beneath velamen or epidermis; cells square to columnar, thin-walled. Lignified thickenings heavy on inner periclinal walls in *A. michelii*; lignified outer periclinal walls in *A. ravenii*. **Cortex** uniform, compact parenchyma, not arranged in files. Regular radiating schizogenous openings in *A. polyschistum*. **Endodermis** uniformly lignified with passage cells opposite xylem poles. **Pericycle** 1–2-layered. **Stele**: 9–14 protoxylem and phloem poles. Vessels observed in roots of *A. scherzerianum* (Schneider and Carlquist 1998). **Sclerenchyma** absent or fibres occupying centre of stele. Partitions usually radiating out from lignified centre isolating rounded phloem strands from adjacent xylem files and surrounding phloem distally. **Tannin** cells common throughout cortex. **Starch** granules occasional in cortex. **Crystals**: raphides in small thin-walled cells in outer cortex. Druses: infrequent coarse aggregations or loose prismatic aggregations in inner cortex near endodermis.

Material examined

Anthurium cerrocampanense Croat, Croat 37365, Panama, MO; *A. cinereopetiolatum* Croat, Croat 35868, Panama, MO; *A. circinatum* Croat, Croat 33102, Panama, MO ; *A. clidemioides* Standl., cult. M 729/76; *A. colonicum* Krause, Croat 33519, Panama, MO; *A. concinnatum* Schott, Croat 34992, Panama, MO; *A. corrugatum* Sodiro, cult. M; *A. erskinei* Mayo, Erskine 34, Bahia, Brazil; *A. flexile* Schott, Boutin 5153, Belize; Croat 42654, Honduras, MO; *A. gargaranum* Standl., Croat 34112, Panama, MO; *A. hoffmannii* Schott, Croat 35266, Costa Rica, MO; *A. hookeri* Kunth, Metcalfe s.n., 1959, K; *A. huixtlense* Mutuda, Croat 42673, Honduras, MO; *A. lancifolium* Schott, Croat 33255, Panama, MO; *A. michelii* Guill., Croat 33142, Panama, MO; *A. nervatum* Croat, Croat 33145, Panama, MO; *A. obtusilobum* Schott, Croat 43356, Costa Rica, MO; *A. ochranthum* Schott, Croat 36881, Panama; *A. oerstedianum* Schott, Croat 35267, Costa Rica, MO; *A. pallens* Schott, Croat 35265, Costa Rica, MO; *A. pariense* Bunting, Venezuela, cult. M; *A. pentaphyllum* (Aubl.) G.Don var. *bombacifolium* Schott, Croat 42637, Honduras, MO; *A. polyschistum* Schultes & Idobro, cult. M; *A. ptarianum* Steyerm., Bogner 980, Venezuela, cult. M; *A. ravenii* Croat & Baker, Croat 35264, Costa Rica, MO; *A. rhodostachyum* Sodiro, Bogner 870, Ecuador; *A. salvadorense* Croat, Croat 42169, El Salvador, MO; *A. schlechtendalii* Kunth, Croat 39414, Mexico, MO; *A. schottianum* Croat & Baker, Croat 43247, Costa Rica, MO; *A. standleyi* Croat & Baker, Croat 43439, Costa Rica, MO; *A. subcordatum* Schott, Croat 42396, El Salvador, MO; *A. veitchii* Mast., cult. M; *A. watermaliense* Hort. ex Bailey & Nash, Croat 33265, Panama, MO; *A. wendlingeri* Barroso, cult. M; *A. willdenowii* Kunth, Metcalfe s.n., St Vincent, WI, K.

Taxonomic notes

The ground plan of *Anthurium* as a whole is distinctive yet the variation is sufficient that definite trends of specialization could be determined if a larger sample were available. It is very likely that at least some of the sections should be recognizable anatomically.

Species reported on in the literature

Arzt (1937): *A. regale, A. roeseanum*: leaf. • Bachmann (1879–81): *A. longifolium, A. scherzerianum*: leaf. • Bergdolt (1955): *A. crystallinum*: leaf. • Blanc (1977b): *A. scandens*: buds. • Blanc (1978): ontogeny. • Bloch (1926): *A. andreanum, A. longifolium*: root. • Boyd (1932): *A. scherzerianum*: seedling. • Bretfeld (1879–81): *A. variabile*: leaf. • Bruhn (1910): *A. ellipticum*: root. • Brunswik (1921): *A. binotii*: dermal tissue. • Brunzema (1928): *A. andreanum, A. cordatum, A. scherzerianum*: crystals. • Chauveaud (1897): *A. crystallinum*: root, phloem. • Chrysler (1904): *A. acaule*: stem. • Costantin (1885): several spp.: root. • Costerus (1875): root, lenticels. • Dalitzsch (1886): several spp.: leaf. • DeCapite and Milletti (1967): *A. andreanum*: root. • El-Gazzar *et al.* (1977): *A. andreanum*: stomata. • Elisei (1941): *A. brevipes*: casparian strips. • Engler (1905): several spp.: leaf, shoot, root. • Ertl (1932): several spp.: leaf. • Fisher and French (1978): *A. andreanum, A. salviniae, A.* sp.: stem. • French (1987c): several spp.: root. • French and Tomlinson (1981c): *A. polyschistum*: stem. • Fricke (1926): *A. grusonii*: leaf. • Funke (1931): *A. variabile, A.* sp.: leaf. Gentner (1905): *A. andreanum*: leaf. • Genua and Hillson (1985): *A. andreanum*: crystals. • Giovannozzi (1911): *A. angustinum*: root. • Goebel (1878): *A. longifolium*: root. • Goebel (1922): *A. ellipticum*: root. • Gravis *et al.* (1943): *A. hoffmannii*: seedling. • Gulliver (1865): *A. coriaceum, A. harrisii*: crystals. • Guttenberg

(1940): several spp.: root. Guttenberg (1968): *A. ellipticum*: root. • Hanstein (1864): *A. wagnerianum*: laticifers, sclereids. • Hassack (1886): *A. crystallinum, A. harrisii, A. magnificum*: leaf. • Higaki *et al.* (1984): *A. andreanum*. • Howard (1969): *A. dominicense*: stomata. • Karagyozova (1968–9): *A. scherzerianum*: crystals. • Kattein (1897): *A. dominicense*: root. • Kroemer (1903): *A. andreanum*: root. • Leavitt (1904): several spp.: root. • Lierau (1888): several spp.: root. • Lindorf (1980): *A. bredemeyeri, A. longissimum*: leaf. • Linsbauer (1930): *A. binotii, A. cannaefolium, A.* sp.: epidermis. • Mangin (1882): root. • Melville and Wrigley (1969): *A. acaule*: leaf. • Meyer (1925): *A. wagnerianum*: root. • Meyer (1959): *A. andreanum, A. scheuchzerianum*: leaf. • Meyer (1962): *A. andreanum, A. scheuchzerianum*: leaf. • Möbius (1899): petiole. • Müller (1919): several spp.: leaf tip. • Netolitzky (1929): crystals. • Olivier (1881): several spp.: root. • Pant and Kidwai (1966): *A. signatum*: stomata. • Pfitzer (1872): *A. membranuliferum*: leaf. • Pigneur (1947): *A. scherzerianum*: phloem. • Poulsen (1875): leaf. • Rada and Jaimez (1992): *A. bredermeyeri*: leaf, root. • Reinhardt (1885): *A. angustianum, A. digitatum, A. longifolium*: root. • Richter (1901): *A. digitatum*: root. • Rosanoff (1867): *A. rubricaulis, A. selloum*: petiole. • Roth (1949, 1980, 1990): leaf. • Roth (1952): *A. veitchii*: leaf. • Roth and Fast (1954): *A. scherzerianum*: leaf. • Scherer (1904): *A. angustinum, A. undatum*: vasculature. • Schimper (1884, 1888): several spp.: root. • Schwendener (1882): root. • Shishkoff (1987): several spp.: root. • Solereder (1919): several spp.: leaf. • Solereder and Meyer (1928): several spp.: all organs. • Stahl (1896): *A. crystallinum*: leaf. • Staudermann (1924): *A. magnificum, A. leuconeurum* • Tieghem (1866): several spp.: all organs. • Tieghem (1870–1): several spp.: root, secretory tissue. • Tieghem (1872): *A. crassinervum*: secretory tissue. • Tieghem and Douliot (1888): *A. leuconeurum*: root. • Traquair (1987): *A. scherzerianum*: leaf. • Trécul (1866): vasculature. • Troll (1932a): *A. crystallinum, A. digitatum, A. podophyllum:* leaf ontogeny. • Troll and Meyer (1955) *A. andreanum, A. hookeri*: leaf ontogeny. • Uphof *et al.* (1962): hairs. • Voronkina (1975): root. • Wakker (1888): *A. hookeri*: crystals. • Webber (1960): *A. crystallinum, A. scandens, A. undatum*: leaf, stomata. • Weisse (1897): several spp.: root, lenticel. • Went (1895): *A. digitatum, A. pseudopodophyllum, A. tripartitum*: root. • Zinke (1924): *A. acaule*: leaf.

Additional literature

Barahona Carvajal 1977; Batygina and Yakovlev 1990; Belin-Depoux *et al.* 1987; Bhattacharya 1976; Blanc 1977a; Brandza 1891; Campbell 1900, 1905; Clausen 1927; Croat 1980, 1983, 1985, 1991; Daumann 1930, 1970; Domin 1911; Erdtman 1952; French 1985a, b, 1986c, 1987a; Gaiser 1923; Gertz 1916a; Grayum 1990; Guéguen 1900, 1901; Hay 1986; Hegnauer 1986; Higaki *et al.* 1980a,b; Irmisch 1874; Jayalakshmi and Ramachandran 1993; Juel 1915; Leitgeb 1864; Lesica and Antibus 1990; Lindinger 1908; Madison 1979; Maercker 1965; Manya-Chernej 1978; Meusel 1951; Mookerjea 1955; Nahrstedt 1975; Ray 1988, 1992a; Samuels 1913; Sargant 1903; Savchenko and Mania 1970; Smolenski *et al.* 1974; Szendel and Weryszko 1973; Thanikaimoni 1969; Tillich 1985; Vij *et al.* 1982; Wannakrairoj and Kamemoto 1990; Williams *et al.* 1981.

Tribe Monstereae Engl.

The tribe is composed of 12 genera distributed throughout the tropics: *Holochlamys, Spathiphyllum, Rhodospatha, Stenospermation, Scindapsus, Rhaphidophora, Anadendrum, Monstera, Alloschemone, Epipremnum, Amydrium*, and *Heteropsis*.

Anatomical summary

Leaf surface

Cuticle smooth or striate, rugulate, occasionally flanged. **Epidermis**: cells polygonal, undulate or sinuous. **Stomata** adaxial, abaxial, or on both surfaces, ranging from brachyparacytic to brachypara-hexacytic. Subsidiary cells narrow or broad, rounded or angular. Second pair of paracytic subsidiary cells distinctive in *Anadendrum*. **Venation** acrodromous or with prominent midrib. Secondaries eucamptodromous or brochidodromous and straightened collector vein may be present.

Leaf TS

Structure mostly dorsiventral, but isobilateral in *Scindapsus scortecchinii*. **Midrib** broad and rounded abaxially or V-rounded in *Holochlamys* and *Scindapsus*. **Cuticle** thick or thin. **Epidermis**: cells tall or short on either surface. **Hypodermis**: 1–2 layers deep beneath adaxial epidermis in *Scindapsus* only. **Stomata** mostly level with surface, but raised in *Amydrium* and slightly raised in *Spathiphyllum*. **Mesophyll** weakly or strongly differentiated. Palisade 1–3 layers deep; cells U-shaped, H-shaped or otherwise lobed in *Holochlamys* (Fig. 4F) and *Spathiphyllum*. Spongy cells in 2–8 layers, tissue types 1, 1a, 2 or 3. **Collenchyma** in peripheral band in abaxial midrib (type B), when present. **Vascular bundles**: mostly well-formed, type I, occasionally type II, often symmetrically arranged in midrib. **Sclerenchyma**: fibres commonly ensheathing bundles or as xylem or phloem caps. Walls often isotropic-layered. Trichosclereids absent in *Amydrium, Heteropsis* and two *Scindapsus* species; present in other genera, mostly as ungrouped simple or H-shaped cells, but in groups in *Holochlamys, Spathiphyllum* and *Alloschemone*. **Laticifers**: non-anastomosing, occasionally present. **Secretory ducts** and cavities with 1-layered epithelium present in most genera. **Tannin** cells occasional in ground tissue of some genera. **Crystals**: raphides common in multiple overlapping bundles in elongate or tubular cells or as single bundles in short cells. Cells with double bundles in *Stenospermation*. Druses and crystal sand common.

Petiole TS

Outline crescent-shaped proximally and terete distally in most genera. **Cuticle** thin or thick, smooth, striate, papillate or rough. **Epidermis**: cells short or tall or papillate; cells with pectic striae in *Spathiphyllum*. **Collenchyma** banded (type B), also present as strands or vein sheaths or phloem caps in *Rhaphidophora* and *Anadendrum*. **Ground tissue** divided into cortex and central cylinder in some genera; uniform in *Heteropsis*. Cells compact or type 2 aerenchyma; type 2 or 4 in *Monstera;* type 3 in *Stenospermation* and *Scindapsus;* type 3 or 4 in *Spathiphyllum*. **Vascular bundles** mostly numerous, type I or occasionally type II.

Sclerenchyma: fibres ensheathing vascular bundles, or present as phloem caps. Fibres replaced distally by collenchyma in *Scindapsus*, some *Rhaphidophora* species, *Epipremnum* and *Amydrium*. Fibres 10–15 layers below epidermis in *Monstera* and delimiting central cylinder in *Alloschemone*. Trichosclereids branched or unbranched, and grouped or ungrouped in ground tissue; absent in *Stenospermation, Anadendrum* and *Heteropsis*. Brachysclereids present in *Rhaphidophora*. **Laticifers** mostly absent; non-anastomosing in *Rhodospatha, Stenospermation* and *Rhaphidophora*. **Secretory ducts** occasionally present in ground tissue. **Starch** widely distributed; formed in large compound crystalloids packed with many small granules in *Amydrium*. **Crystals**: raphides in multiple overlapping bundles or commonly in small cells in single bundles. Other crystal types widely dispersed: crystal sand, aggregated prismatics (*Alloschemone*), large prismatics (*Rhaphidophora*), coarsely jagged druses (*Rhodospatha*), ordinary druses (*Stenospermation, Rhaphidophora, Amydrium*).

Stem TS

Outline terete or laterally compressed or ridged. **Cuticle** smooth, except striate in *Alloschemone*, and thin or thick. **Epidermis** of small or larger cells, smooth or undulate-ridged; with thick, lignified walls in *Epipremnum*. **Hypodermis**: 1-layered in *Heteropsis*. **Periderm** present, of epidermal origin in *Spathiphyllum*, of subhypodermal origin in *Rhodospatha, Monstera* and *Epipremnum*. **Collenchyma**: thin subepidermal band in *Anadendrum*. **Ground tissue** compact, cells grading from smaller to larger in centre. **Central cylinder** delimited by ring of fibres in most genera. **Vascular bundles** scattered throughout, larger in central cylinder; collateral, bicollateral or compound in central cylinder; types I, II, or occasionally III. **Sclerenchyma**: fibres may form central cylinder boundary, or as phloem caps, the fibres often having isotropic wall layering; ensheathing vascular bundles in *Alloschemone* and often in *Monstera*, and cortical bundles in *Stenospermation*. Trichosclereids present in cortex of several genera. **Secretory ducts** in cortex of *Rhaphidophora*. **Starch** granules usually common in ground tissue. **Crystals**: raphides as single bundles in small cells, or in large cells in *Spathiphyllum*, or in multiple overlapping bundles; absent from *Monstera*. Druses common in ground tissue. Prismatics common in *Rhaphidophora*.

Root TS (Scindapsus, Monstera, Alloschemone, Epipremnum)

Epidermis: cells medium-sized and ± square, or cells collapsed. **Exodermis** 1–2 layers beneath epidermis in *Scindapsus*. **Periderm** occasionally present on outer surface of cortex. **Cortex**: outer cells with or without regular files separated by radial cavities; inner ground tissue usually in radial files near stele. **Endodermis** with narrow anticlinal casparian strips in *Scindapsus*, or forming sclerified inner cortical boundary in *Monstera* and *Alloschemone*. **Sclerenchyma**: trichosclereids may be present in cortex. Central tissue usually lignified as small fibre strands or brachysclereids. **Vascular tissue**: protoxylem poles varying from nine in *Scindapsus* up to 25 in *Epipremnum*. **Crystals**: raphides in multiple overlapping bundles or as single bundles. Druses and crystal sand present.

Taxonomic notes

The included genera have a complex taxonomic history but available evidence provides insufficient breaks between them to justify separation. The genera of this subfamily collectively express robust differentiation of lignified cells and a series of other characters, more than in all the other more derived tribes. *Anadendrum* is the most distinctive with its unique stomatal subsidiary cell pattern and the presence of collenchyma that ensheaths or caps vascular bundles. *Spathiphyllum* and *Holochlamys* show a relationship on the basis of lobed palisade cells and other characters.

The genera of both tribes of the subfamily (Potheae and Monstereae) are clearly more closely related to each other than to other generic groupings. The genera of Philodendroideae: Philodendreae show the closest similarity of character state expression.

Holochlamys Engl.

This genus of three species is found in New Guinea. Plants are terrestrial herbs with short creeping fusiform tubers occurring in primary forest understorey and on stream banks. The leaves are long-petiolate and erect with elliptic or narrow elliptic blades. Chromosomes: 2n = 60. Two species were available for study.

Leaf surface

Cuticle smooth adaxially, striate abaxially and markedly striate on midrib. **Epidermis**: cells deeply undulate to isodiametric or elongate on either surface (1.5–4:1*l*/*w*). **Stomata** on both surfaces, brachyparacytic to brachparahexacytic (Fig. 20F) and unipolar. Subsidiary cells mostly narrow with angular edges. **Venation**: midrib prominent, secondaries narrow, acutely ascending, merging eucamptodromously at margin. Tertiaries parallel with secondaries. Quaternaries irregular or orthogonal with tertiaries.

Leaf TS

Structure dorsiventral and poorly or well differentiated. **Size**: 91 × 26 mm (*l*/*w*). **Midrib** broadly rounded-concave adaxially, and broadly rounded-convex (Fig. 20C) or broadly V-shaped abaxially (Fig. 20A). **Cuticle** thin. **Epidermis**: cells square on both surfaces or square to rectangular abaxially (Fig. 20B,D), flat to convex externally and irregularly rounded internally. **Stomata**: guard cells level with surface and with double ledges. **Mesophyll**: one layer of palisade tissue, slightly differentiated (Fig. 20B,D) from *c.* eight layers of armed spongy cells (type 2 or small type 3). Palisade cells often H-shaped or lobed (Fig. 4F). **Air cavities** substomatal, up to 50% of spongy tissue. **Collenchyma**: thin band, subepidermal in abaxial midrib.

Vascular bundles numerous and scattered in midrib or comprising *c.* three rows of small type II bundles. Xylem: 2–4 narrow metaxylem elements opposing small circular phloem strand. **Sclerenchyma**: trichosclereids in aggregates of *c.* 6, closely spaced in rows parallel to vascular bundles throughout midrib and abaxial mesophyll (Fig. 20E); some branching present. Weakly-formed fibres occasionally ensheathing secondary vascular bundles. **Crystals**: raphides uncommon and large, often in elongate mesophyll cells in multiple overlapping bundles (Fig. 20G). Raphide cells tending to be parallel to secondaries, mostly near centre of leaf. Druses mostly in palisade layer, common to uncommon in adaxial mesophyll. Crystal sand common in all mesophyll cells.

Petiole TS

Outline subterete with small adaxial groove distally, semicircular or crescent-shaped proximally (Fig. 20H). **Size**: diameter 6 × 5 or 3 × 2 mm. **Cuticle** thin to medium-thick and striate. **Epidermis**: cells tall and rounded. **Collenchyma**: subhypodermal band 14–20 cells deep distally, or a discontinuous band 1–2 cells deep proximally. **Chlorenchyma** 2–3 cells deep subepidermally in *H. guineensis*. **Ground tissue**: cells large, with corner air cavities, up to 5% air space.

Vascular bundles small (Fig. 20H), scattered and numerous, type II. Xylem: one large protoxylem cell and cluster of 2–8 small metaxylem cells. Phloem: circular strand. **Sclerenchyma**: fibres forming phloem caps near periphery in *H. beccarii*. Trichosclereids appearing ungrouped or grouped throughout ground tissue in *H. beccarii*, or mostly in abaxial outer ground tissue in *H. guineensis*. **Starch** granules small and uncommon near phloem in ground tissue. **Crystals**: raphides uncommon in ground tissue, near periphery. Druses uncommon and subepidermal. Crystal sand sparse in all ground tissue.

Material examined

Holochlamys beccarii Engl., Bogner 1269, New Guinea, cult. M; *H. guineensis* Engl. & Krause, MBG 772040, cult. MO.

Species reported on in the literature

French (1987c): *H. beccarii*: root. • French and Tomlinson (1981d): *H. beccarii*: stem vasculature. • Nicolson (1960): *H. beccarii, H. guineensis*: leaf, stem, petiole. • Solereder and Meyer (1928): sclereids.

Additional literature

Eyde *et al.* 1967; French 1985a, 1986c, 1987a; Grayum 1990; Hotta 1971; Lugardon *et al.* 1987–88; Thanikaimoni 1969; Williams *et al.* 1981.

Spathiphyllum Schott

This genus of 60 species is found mostly in the American tropics, with a few species in Asia and the Pacific islands. The plants are understorey, rhizomatous, acaulescent herbs, often deeply rooted, and found in primary forests or regrowth, usually in wet places. Some species are rheophytes. Others are occasionally hemi-epiphytes. The

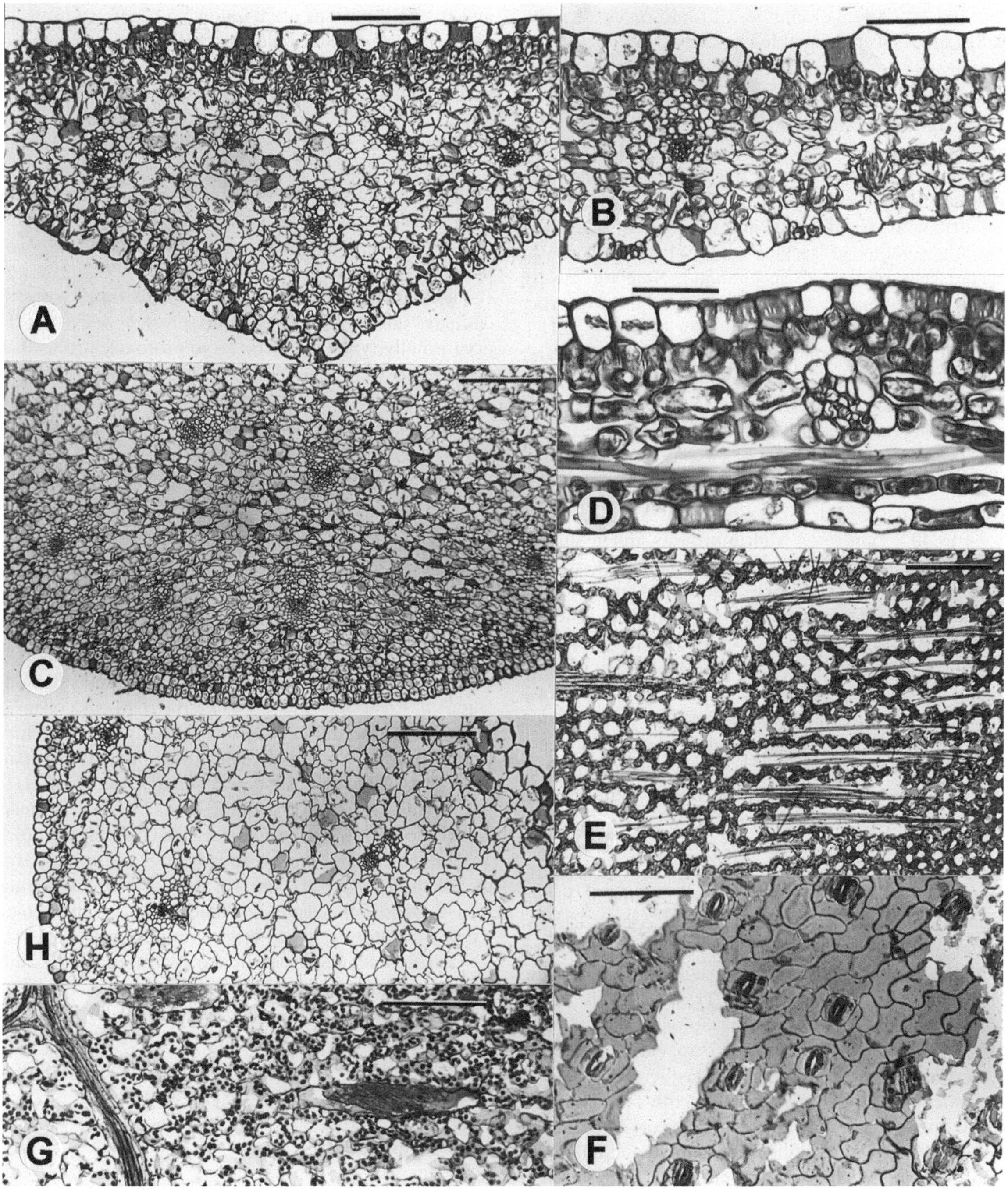

Fig. 20. *Holochlamys* leaf and petiole sections. (A,B,G,H) *H. guineensis*; (C,D,E,F) *H. beccarii*; (A,C) Midrib TS; (B,D) Lamina TS. Note grouped trichosclereids in (D) and (E); (E) Mesophyll PS showing grouped trichosclereids in spongy mesophyll; (F) Leaf epidermis PS showing epidermis and stomata; (G) Mesophyll PS showing raphide cell; (H) Proximal petiole TS showing type II vascular bundles. Scale bars: A,E,H = 200 μm; B,F,G = 150 μm; C = 300 μm; D = 50 μm.

leaves are lanceolate or elliptic. Chromosomes: 2n = 30, 60. Six species were available for study.

Leaf surface

Cuticle smooth adaxially, striate-rugulate abaxially over midrib and lamina in *S. cannaefolium*. In other species, striate surface due to striate outer (probably pectate) epidermal walls. **Epidermis**: adaxial cells polygonal, isodiametric with straight anticlinal walls (1–1.5:1 *l*/*w*), or adaxially sinuous at surface and polygonal beneath. Abaxial cells polygonal or undulate. **Stomata** mostly abaxial and unipolar, brachyparatetra- or -hexa-cytic, with differentiated narrow subsidiary cells (Fig. 21E). **Venation**: pinnate, secondaries acute, closely spaced and regular. Tertiaries parallel with secondaries. Quaternaries orthogonal to tertiaries. Areoles irregular. Trichosclereid groups easily mistaken for vascular bundle patterns.

Leaf TS

Structure dorsiventral. **Size**: specimens ranging from 18 × 4 to 40 × 9 cm (*l*/*w*). **Midrib** broadly and shallowly concave adaxially, broadly convex-rounded V-shaped abaxially (Fig. 21A). **Cuticle** thick over midrib and lamina. **Epidermis**: cells tall on both surfaces, shorter abaxially. Cells flat externally and rounded internally. **Stomata** level with surface or slightly raised; may be adaxial over midrib and occasional over lamina. **Mesophyll**: one palisade layer often weakly differentiated (Fig. 21B); cells often U-shaped in *S.* sp. Spongy layers *c.* 8, short-armed, type 1 or 1a. **Air cavities** irregular, 30–80% of spongy tissue. **Collenchyma**: multi-layered subepidermal band at abaxial midrib.

Vascular bundles: well-formed midrib bundles, numerous and scattered (Fig. 21A), type I. Xylem: many large and small metaxylem elements. Phloem: broad elliptic strand at curved boundary with xylem. **Sclerenchyma**: fibres ensheathing vascular bundles, thickest at xylem and phloem poles, thinner laterally, or capping xylem and phloem (Fig. 21F). Trichosclereids unaggregated and common in mid-mesophyll, or aggregated in clusters (Figs 7D, 21D) of 15–25, appearing in cleared preparations to resemble closely spaced tertiary veins. In *S. cannaefolium*, trichosclereids frequently with enlarged central body forming H-shape or other curved branched idioblast. **Crystals**: raphides in obliquely overlapping bundles in short or elongated cells (Fig. 21C), common in upper mesophyll. Druses uncommon in mesophyll.

Petiole TS

Outline subterete and shallowly grooved adaxially or crescent-shaped. **Size**: diameter 8–15 mm. **Cuticle** thin and striate, striations often following pectic striations on outer epidermal walls. **Epidermis**: cells small and rounded. **Stomata** occasional; level with surface (Fig. 4B). **Collenchyma** banded; slight corner thickenings on epidermal and 3–10 layers of subepidermal cells, around periphery. **Chlorenchyma**: plastids few in outer 3–4 layers. **Ground tissue**: cells small at periphery and larger in centre. **Air cavities**: central ground tissue mostly type 3 aerenchyma with 1–2 partition cells in *S. friedrichstahlii*. Cavities mostly small and numerous in *S. cochlearispathum*. Cavities 4–5, type 4 in *S.* sp.

Vascular bundles numerous and scattered, mostly circular, unsheathed (Fig. 21G) or sheathed (type II, occasionally type I). Xylem: large protoxylem cells 0–1; large metaxylem cells 3–6. Phloem: circular strand at broad or narrow boundary with xylem. **Sclerenchyma** absent, or fibres forming deep phloem caps, or phloem and xylem caps, or bundles ensheathed. Trichosclereids abundant in ground tissue, especially near periphery, in clusters of 2–8, present in all air cavities (Fig. 7E); cells showing several-fold variability in diameter and wall thickness within specimens; cells mostly linear, occasionally with enlarged central body giving rise to H-shape or more branches. **Starch** granules common in ground tissue.

Stem TS

Outline terete. **Size**: diameter 2.5 cm. **Cuticle** thin and smooth. **Epidermis** mostly obliterated in specimen of *S.* sp. **Periderm**: well-developed; originally from periclinal divisions of epidermis. **Ground tissue**: mostly compact tissue with large cells, or type 3 aerenchyma (Fig. 21H). **Air cavities**: 1–2% of ground tissue. **Central cylinder** weakly defined by presence of starch and vascular bundles.

Vascular bundles numerous in centre; in cortex sparse, circular and collateral (type I). Xylem: semicircular strand of 5–8 metaxylem elements. Phloem strand semicircular. Central cylinder bundles larger, amphivasal (Fig. 21I), 4–6 xylem poles and large circular central phloem strand. **Starch** granules abundant in central cylinder ground tissue. **Crystals**: raphides very common as large bundles in large circular cells surrounded by radiating ground tissue cells. Druses common in ground tissue; large, very sharp-pointed crystals in large circular cells.

Material examined

Spathiphyllum brevirostre (Liebm.) Schott, Croat 39885, Mexico, MO; *S. cannaefolium* (Dryand.) Schott, Metcalfe s.n., West Indies; *S. cochlearispathum* Engl., cult. M; *S. friedrichstahlii* Schott, Croat 38312, Canal Zone, MO; *S. laeve* Engl., Croat 38059, Panama, MO; *S.* sp., cult. MO.

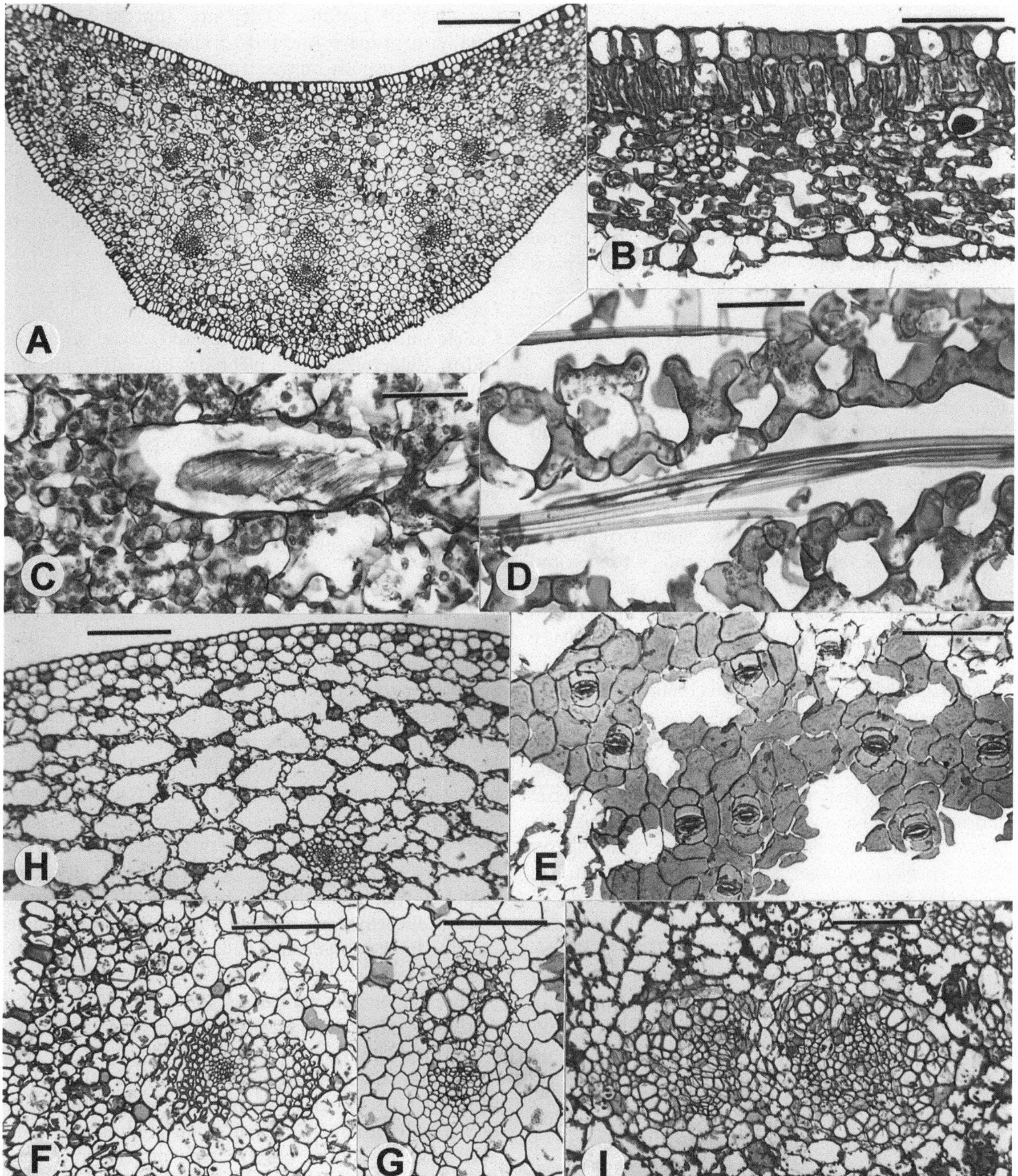

Fig. 21. *Spathiphyllum* leaf, petiole and stem sections. (A,C–F) *S*. sp.; (B,G) *S. cannaefolium*; (H,I) *S. friedrichstahlii*. (A) Midrib TS; (B) Lamina TS; (C) Lamina mesophyll PS showing type 1 aerenchyma and raphide cell; (D) Lamina spongy tissue PS showing clustered trichosclereids; (E) Leaf epidermis PS showing brachypara-tetracytic stomata; (F) Midrib abaxial TS showing tall epidermal cells and vascular bundles; (G) Petiole central tissue TS showing vascular bundle; (H) Stem TS showing cortical type 1 aerenchyma; (I) Stem central cylinder TS showing compound vascular bundles. Scale bars: A,H = 300 μm; B,E–G = 150 μm; C,D = 50 μm; I = 200 μm.

Taxonomic notes

All species examined have similar anatomy. They share trichosclereid structure, a special type of cuticle striation originating as collenchymatous striations on epidermal outer surfaces, vascular bundle architecture, type of collenchymatous thickenings, and texture of ground tissue. Dalitzsch (1886) illustrated the adaxial epidermal cells of *S. blandum* as having a bevelled edge and a depressed centre. This surface feature was not encountered in these specimens and could have been an artifact of specimen preparation.

Species reported on in the literature

Beck and Mannagetta (1912): crystals. • Dalitzsch (1886): *S. blandum, S. cochlearispathum, S. lanceifolium*: leaf. • Deshpande (1956): *S. patinii*: root, velamen. • Eller and Flach (1991): *S. wallisii*: leaf. • Engler (1908): several spp.: leaf. • Ertl (1932): *S. cannaefolium, S. cochlearispathum*: leaf venation. • Fisher and French (1978): *S. cochlearispathum, S. floribundum, S. kochii*: stem vasculature. • French (1987c): *S. fulvovirens, S.* sp.: root. • French and Tomlinson (1981d): *S. cannaefolium, S. wallisii*: stem venation. • Gentner (1905): leaf tip. • Genua and Hillson (1985): *S. commutatum*: crystals. • Guttenberg (1940): root. • Lierau (1888): several spp.: root. • Meyer (1962): *S. cochlearispathum*: leaf. • Nicolson (1960): several spp.: stem, leaf, petiole. • Niklas (1991): *S. 'clevlandii'*: petiole. • Palandri (1967): *S. wallisii*: leaf. • Pant and Kidwai (1966): *S. dechardii*: leaf, stomata. • Rajagopal *et al.* (1987): *S. wallisii*: leaf. • Rambour (1965): *S. wallisii*: raphides. • Solereder (1919): *S. cochlearispathum, S. kochii*: leaf, sclereids. • Solereder and Meyer (1928): *S. cochlearispathum, S. kochii, S. patini*: anatomy. • Sutter (1985): *S. wallisii*: cuticle. • Tieghem (1866): *S. lancaefolium*: all organs. • Tieghem (1870–1): *S. lancaefolium*: root, secretory tissue. • Webber (1960): *S. cannaefolium, S. floribundum*: epidermis, stomata. • Weidner *et al.* (1985): leaf. • Zinke (1924): *S. blandum*: leaf.

Additional literature

Barabé and Chrétien 1986; Bunting 1960; Davis 1966; Erdtman 1952; Eyde *et al.* 1967; French 1985a, 1986c, 1987a; Grayum 1990; Guéguen 1901; Huynh 1974; Kanerva *et al.* 2000; Lugardon *et al.* 1987–88; Manya-Chernej 1978; Meusel 1951; Ray 1988; Thanikaimoni 1969; Ueno 1980; Vogel 1962, 1990; Williams *et al.* 1981.

Rhodospatha Poepp.

This New World genus of 67 species ranges from Mexico to Brazil. Plants are hemi-epiphytic and occasionally terrestrial or epilithic. They are appressed climbers, usually in primary lowland to mid-elevation humid to wet forests, usually on tree trunks and less frequently on steep banks. Plants have non-heteroblastic juvenile and adult forms that flower at more than 2.5 m above the ground. Terrestrial species may have reclining stems extending more than a metre over the forest floor. The leaf is long-petiolate with an elliptic lamina. Chromosomes: 2n = 28, 56. Three species were available for study.

Leaf surface

Cuticle smooth on both surfaces and striate on adaxial midrib. **Epidermis**: cells polygonal with straight anticlinal walls (1–1.5 *l/w*) on both surfaces. **Stomata** mostly abaxial and brachyparacytic or brachyparatetra- (or -hexa-) cytic (Fig. 22D). Subsidiary cells broad and square-cornered. **Venation**: midrib prominent. Secondaries and intersecondaries broadly acute. Tertiaries closely spaced and parallel with secondaries and merging eucamptodromously. Higher orders reticulate.

Leaf TS

Structure dorsiventral. **Size**: lamina length 11–30 cm. **Midrib** broadly curved-concave adaxially, and broadly and deeply rounded-convex abaxially (Fig. 22A). **Cuticle** medium-thick adaxially and thin abaxially; thick on adaxial midrib. **Epidermis**: cells large adaxially and large or smaller abaxially; flat externally and rounded internally on both surfaces. **Stomata** level with surface. **Mesophyll**: 1–2 palisade layers of short, well-differentiated cells (Fig. 22B) and 10–12 layers of short-armed spongy cells (types 1, 1a), or longer-armed cells in *R. wendlandii* (type 2). **Air cavities** small and up to 40% of spongy layer. **Collenchyma**: a band 6–8 layers deep abaxially in midrib. In *R. wendlandii*, collenchyma replacing sclerenchyma as fibre caps and also in hypodermal band.

Vascular bundles numerous and scattered in midrib, robust (type I), including small parenchyma zone. Xylem: 3–4 wide metaxylem elements opposing a broad semicircular phloem strand. **Sclerenchyma**: fibres ensheathing small bundles, or as phloem caps in midrib bundles, or occasionally absent. Trichosclereids common in midrib and lamina, solitary, parallel, and mostly unbranched (Fig. 22B). Absent in lamina of *R. bogneri*. **Tannin** cells common in mesophyll, numerous; lobed and articulated idioblasts structurally similar to neighbours, bearing dark contents (Fig. 22C). **Laticifers** non-anastomosing, common by vascular bundles. **Secretory ducts** large and few, without epithelial cells, common in midrib of *R. moritziana*, containing amorphous material.

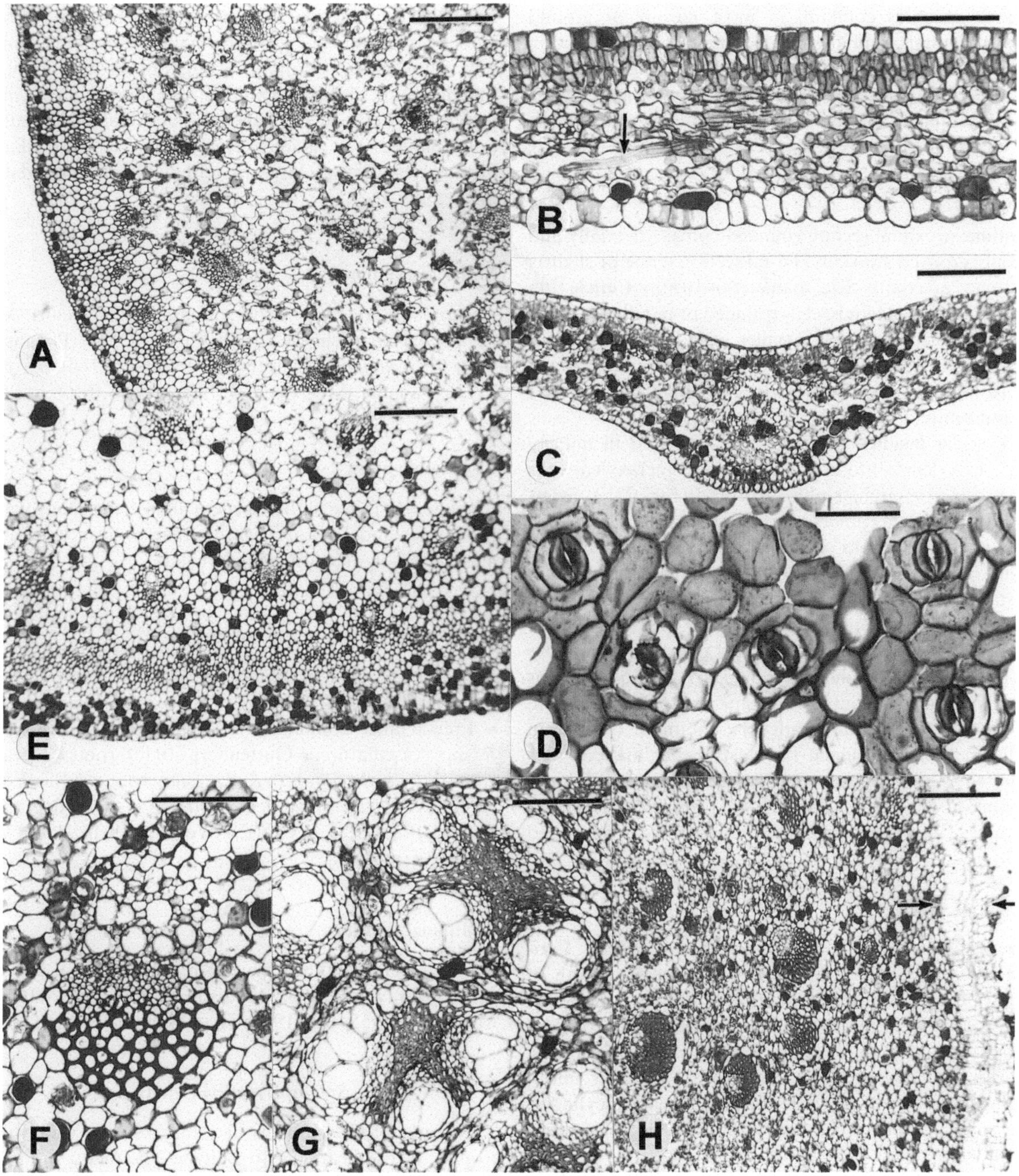

Fig. 22. *Rhodospatha* leaf, petiole and stem sections. (A) *R. wendlandii* var. *folsomii*; (B,D) *R. moritziana;* (C,E–H) *R. wendlandii*. (A) Midrib TS side; (B) Lamina TS, with solitary sclereid in LS at arrow ; (C) Lamina TS with vascular bundle; (D) Abaxial epidermis PS with stomata; (E) Petiole TS; (F) Petiole TS with type I vascular bundle; (G,H) Stem TS: (G) central cylinder, compound vascular bundles; (H) cortical bundles and periderm (arrows). Scale bars: A,E,H = 300 μm; B,F = 150 μm; C,G = 200 μm; D = 50 μm.

Starch granules common in most mesophyll ground tissue. **Crystals**: raphides in wide single bundles common in mid-mesophyll, or in multiple overlapping bundles in elongated cells. Druses in mesophyll, coarse and jagged, most common in palisade layer and midrib ground tissue in large well-defined circular cells.

Petiole TS

Outline: proximal section rounded-convex abaxially and deeply grooved adaxially. Distal section terete or slightly flattened adaxially. **Size**: diameter 6–10 mm. **Cuticle** thin and smooth. **Epidermis**: cells rounded or papillate externally. **Ground tissue**: cells smaller near periphery, larger toward centre, loosely packed. Chloroplasts in outermost 8–10 layers. Type 2 aerenchyma in small regions near organ centre.

Vascular bundles numerous and scattered in midrib, type I. Xylem a broad strand of 6–10 metaxylem cells opposing semicircular phloem strand (Fig. 22F). **Sclerenchyma**: fibres as phloem caps, vascular bundle sheaths, and as discontinuous sheath of cells just within chloroplast layer (Fig. 22E). Fibres replaced by collenchyma distally in *R. bogneri*, *R. moritziana* and *R. wendlandii*. Trichosclereids common as solitary cells, scattered throughout ground tissue, oriented parallel to long axis of petiole. **Secretory ducts** few and large, with amorphous contents and without epithelium, scattered in ground tissue in *R. moritziana*. **Laticifers**: common, 1–3 within phloem strands. **Tannin**: idioblasts with dark contents common throughout ground tissue. **Crystals**: raphides common or uncommon, in wide or elongate cells. Druses: large and jagged in circular cells concentrated near petiole periphery.

Stem TS and LS

Outline: mature stem terete. **Size**: 3.5 cm diameter. **Periderm**: multi-layered; inner layer palisade-shaped, a sclerotic layer toward periphery followed by 3–5 files of empty tile-like cells (Fig. 22H). **Ground tissue**: compact polyhedral cells, uniform throughout with few air spaces. **Cortex** distinct, bounded internally by continuous, undulating double row of fibres defining central cylinder. **Vascular bundles**: scattered and crowded in central cylinder and less frequent in cortex. Central cylinder bundles uniformly compound with 2, 3, 4, or 5 xylem/phloem units, phloem oriented toward centre (Fig. 22G); phloem partially surrounded by fibre caps, fused into single ramifying (fluted) fibre strand in bundle centre. Some compound bundles in cortex, but most cortical bundles simple, collateral (type I), surrounded by deeply enveloping fibre cap or sheath, deepest over phloem (Fig. 22H); broad row of metaxylem cells bounded by semicircle of phloem except in smallest bundles. **Tannin** cells idioblastic, common throughout ground tissue. **Starch** granules large and abundant in most parenchyma except outer cortex. **Crystals**: raphides in single or multiple bundles, in axially elongated cells, abundant throughout. Druses large, in circular cells, abundant throughout.

Material examined

Rhodospatha bogneri Croat, Bogner 1281, cult. M; *R. moritziana* (Schott) Croat, Croat 33540, Panama, MO; *R. wendlandii* Schott var. *folsomii* Croat, Croat 33726, Panama, MO; Croat 43792, Mexico, MO.

Taxonomic notes

Rhodospatha is one of three genera in the tribe, along with *Scindapsus* and *Stenospermation*, that possess secretory ducts and non-anastomosing laticifers.

Species reported on in the literature

Engler (1908): *R. heliconiifolia, R. latifolia, R. oblongata*: stem, leaf, root. • French (1987c): several spp.: root. • French and Tomlinson (1981d, 1986): *R. latifolia*, *R.* spp.: vasculature. • Guttenberg (1940): root. • Lierau (1888): *R. heliconiifolia, R. latifolia*: root. • Nicolson (1960): *R. picta*: stem, leaf, petiole. • Solereder and Meyer (1928): leaf, crystals. • Webber (1960): *R. picta*: epidermis, stomata.

Additional literature

Cevallos-Ferriz and Stockey 1988; French 1985a, 1986c, 1987a; Grayum 1990; Madison and Tiffney 1976; Mookerjea 1955; Ray 1988; Thanikaimoni 1969.

Stenospermation Schott

This genus of 60 species is distributed in the American tropics, from Nicaragua to Bolivia. The plants are epiphytes in primary cloud forests at mid to high elevations, usually on tree trunks and lower branches. Occasional species are terrestrial or epilithic. They have appressed growth, usually short internodes, and are rarely scandent or terrestrial on steep banks. Leaves are long-petiolate, oblanceolate, elliptic or ovate. Chromosomes medium-sized: 2n = 28. Four species were available for study.

Leaf surface

Cuticle smooth. **Epidermis**: cells polygonal with straight anticlinal walls (1–2:1 *l*/*w*) on both surfaces. **Stomata** abaxial, brachyparacytic (Fig. 23D) to brachyparahexacytic. Subsidiary cells mostly narrow and angular. **Venation**: midrib prominent, dissipating distally by loss of steeply ascending secondaries. Tertiaries closely spaced and parallel or irregularly parallel with secondaries, merging eucamptodromously. Quaternaries oblique, orthogonal or reticulate.

Leaf TS

Structure dorsiventral. **Size** 15 × 6 cm *l*/*w*. **Midrib** slightly concave or flat adaxially, broadly and shallowly convex abaxially (Fig. 23A). **Cuticle** thin on both surfaces, slightly flanged adaxially, and patchy abaxially. **Epidermis**: cells tall on both surfaces, flat externally and rounded internally (Fig. 23B). **Stomata** level with surface. **Mesophyll**: one poorly differentiated palisade layer and *c*. 12 distinctive spongy layers of round-bodied cells bearing several short lobes (type 1) (Fig. 4G). **Air cavities** irregular, about 20% of spongy tissue.

Vascular bundles numerous and scattered in midrib, well-formed and small (types I, II). Several metaxylem cells opposing broad semicircular phloem strand. **Sclerenchyma**: fibres ensheathing vascular bundles with 1–2 cell layers. Trichosclereids common in mesophyll between secondary and tertiary vascular bundles, large or small (length 800–1500 μm), ungrouped (Fig. 23C), straight or branched, or having ramified ends. **Starch** granules common in mesophyll. **Crystals**: raphides common in mesophyll (Fig. 23E), in single or double bundles, with no visible sheath; elongated cells common. Large druses common in palisade layer. Crystal sand common in mesophyll.

Petiole TS

Outline semicircular with crescentiform adaxial extensions. **Size**: width 9 mm. **Cuticle** medium-thick and smooth. **Epidermis**: cells small and irregular. **Collenchyma** subepidermal in proximal sections, in peripheral band (Fig. 23F), *c*. 12 layers of angularly thickened cells (type B). Collenchyma absent in distal sections. **Ground tissue**: parenchyma ± uniform-sized and compact in outer tissue. **Air cavities** numerous and cell-sized in central ground tissue. Type 3 aerenchyma on adaxial side, with one-celled partitions.

Vascular bundles numerous and scattered (type II) (Fig. 23G). Xylem: 1–2 wide protoxylem elements and 1–2 narrower metaxylem elements. Phloem: laterally narrowed or rounded strand separated by short parenchyma strand from xylem. **Sclerenchyma**: fibres in crescent-shaped cap over phloem, 1–7 cells deep. Young petiole with no lignified fibre caps. **Secretory ducts** medium-sized and common in outer cortex with 2 epithelial layers. **Laticifers**: non-anastomosing, 2–3 per vascular bundle lateral to phloem. **Tannin** cells common throughout ground tissue and collenchyma. **Starch** granules rounded or elongate, common in all ground tissue. **Crystals**: raphides common in collenchyma in single bundles, cells not enlarged. Druses common in subepidermal zone, uncommon throughout central ground tissue.

Stem TS and LS

Outline terete. **Size**: diameter 15–18 mm. **Cuticle** thin and smooth. **Epidermis**: cells vestigial and collapsed. **Hypodermis**: outer 2–3 layers heavily lignified and fibrous. **Ground tissue**: cortical ground tissue with numerous cell-sized air spaces. Inner cortex *c*. 20 cells deep (type 2 aerenchyma). **Central cylinder**: outer boundary sharply differentiated; ± continuous lignified band, two cells deep except deeper in vicinity of confluent vascular bundle caps.

Vascular bundles numerous and crowded in central cylinder, fewer in cortex. Central cylinder bundles collateral or mostly compound with 2–6 collateral units (Fig. 23H). Xylem with 3–8 wide metaxylem cells and semicircular phloem strand (type I). Cortical bundles collateral, xylem and phloem very small strands, together circular. **Sclerenchyma**: fibres ensheathing cortical bundles. Central cylinder bundles with thick fluted strand in centre comprising fused phloem caps; phloem strands occupying grooves in fibre strand. In LS, fibres having oblique end-walls and occasional horizontal septa; pits unbordered. **Secretory ducts** occasional in cortex; epithelial tissue absent. **Tannin** cells common or infrequent, scattered in ground tissue. **Starch** granules rounded, scattered in all ground tissue. **Crystals**: raphides in single oblique bundles in axially elongate short tubes, infrequent in ground tissue (Fig. 23I).

Material examined

Stenospermation angustifolium Hemsl., Croat 35255, Costa Rica, MO; *S. longipetiolatum* Engl., Ecuador, cult. SEL; *S. sessile* Engl., Croat 33268, Panama, MO; *S.* sp., Croat 33216, Panama.

Taxonomic notes

Raphide crystals in the leaves are not of the overlapping type seen in all other members of this tribe.

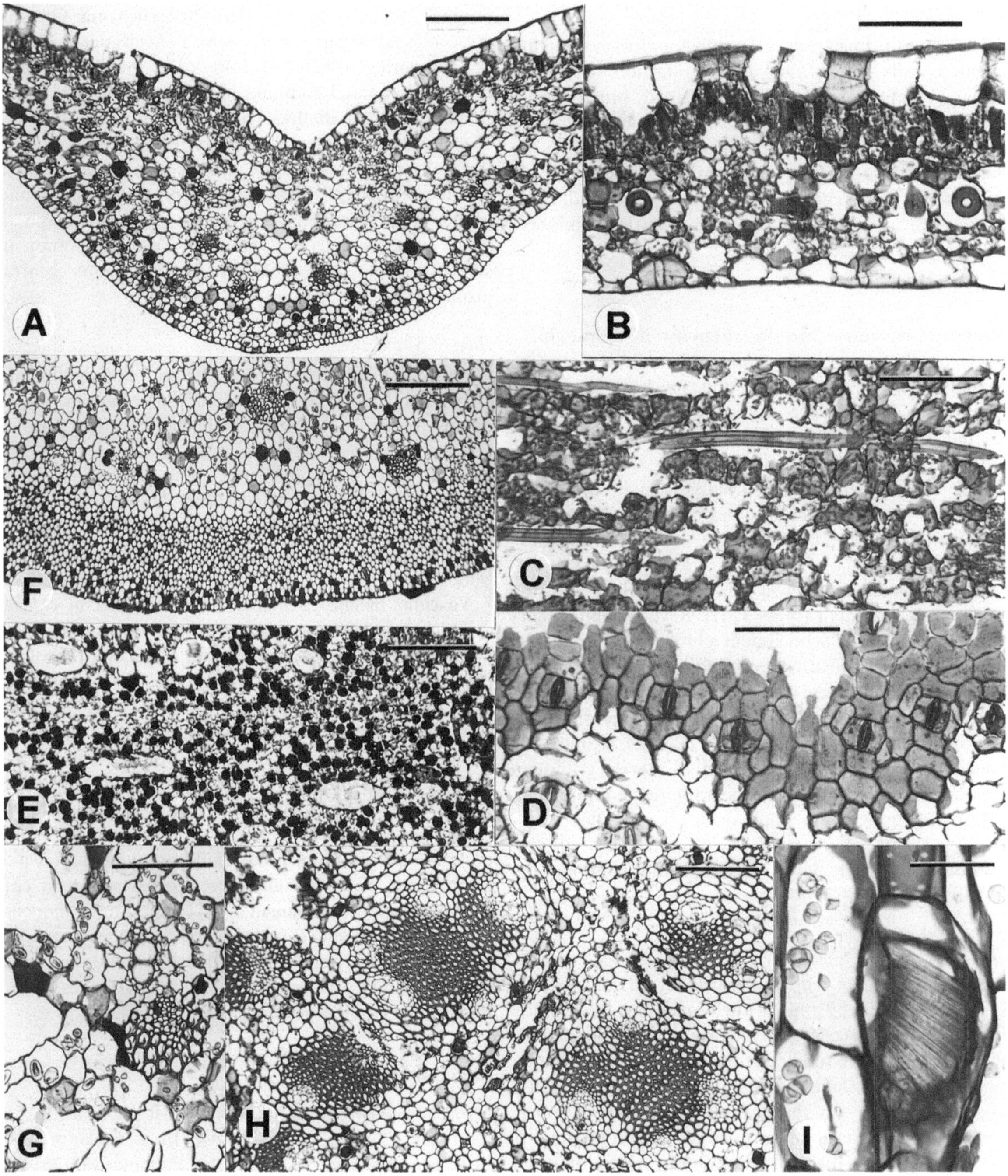

Fig. 23. *Stenospermation* leaf, petiole, stem sections. (A,B) *S. longipetiolatum*; (C,D,F,G) *S. sessile*; (E,H,I) *S.* sp. (A) Midrib TS; (B) Lamina TS showing 2 trichosclereids in TS; (C) Mesophyll PS. Note solitary trichosclereids; (D) Leaf PS epidermis with brachyparacytic stomata. (E) Leaf mesophyll PS showing small raphide cells; (F) Petiole abaxial TS showing broad subepidermal collenchyma band; (G) Petiole, adaxial central type II vascular bundle; (H) Stem TS centre showing compound vascular bundles; (I) Stem LS ground tissue and broad raphide bundle. Scale bars: A,F,H = 300 μm; B–D,G = 150 μm; E = 200 μm; I = 50 μm.

Species reported on in the literature

Engler (1908): root. • Ertl (1932): leaf venation. • French (1987c): several spp.: root. • French and Tomlinson (1981d, 1986): several spp.: stem vasculature. • Lierau (1888): *L. pompayense*: root. • Müller (1919): *S. popayanense*: leaf tip. • Nicolson (1960): *S. marantifolium, S. popayanense*: stem, leaf, petiole. • Solereder and Meyer (1928): crystals.

Additional literature

Domin 1911; French 1985a, 1986c, 1987a; Grayum 1990; Madison and Tiffney 1976; Ray 1988; Thanikaimoni 1969; Williams *et al.* 1981.

Scindapsus Schott

This genus of 25 species is distributed from India and Sikkim to south-east Asia, and to the Solomon Islands. One species said to occur in Brazil is treated separately in this work as the monotypic genus *Alloschemone*. Plants are hemi-epiphytic appressed climbers occurring mostly in the understoreys of wet primary forests to dry deciduous forests. They also occur as terrestrial or epilithic forms. The plants are heteroblastic showing alternating adult and juvenile growth phases. The stem is a two-ranked liane with a long-petiolate leaf and an ovate, cordate, lanceolate or elliptic lamina. Chromosomes: 2n = 42, 56, 58, 60, 64, 70, 112. Three species were available for study.

Leaf surface

Cuticle smooth or striate-rugulate abaxially in *S. scortechinii* and striate in *S. pictus*. **Epidermis**: cells polygonal with straight or curved anticlinal walls (1–3:1 *l/w*) on both surfaces. In *S. pictus*, adaxial cells lobed and deeply sinuous; abaxial cells irregularly polygonal. In *S. scortechinii*, abaxial cells polygonal-elongate (2–6:1 *l/w*) and arranged in parallel groups of 2–6 due to abundant stomata. **Stomata** more common abaxially, brachyparacytic to brachypara-hexacytic (Fig. 24D–F), sometimes unipolar. Subsidiary cells narrow with rounded edges. Stebbins and Khush (1961) illustrated *S. aureus* with eight subsidiary cells called haphazard but actually bipolar. **Venation**: midrib prominent. Secondaries and intersecondaries steeply acute, apically curved, and submarginally eucamptodromous. Tertiaries and higher orders reticulate. In *S. scortechinii*, larger veins parallel with orthogonally branched or reticulate higher-order veins.

Leaf TS

Structure dorsiventral or with isobilateral features in *S. scortechinii*. **Size**: *S. lucens* 85 × 51 mm (*l/w*), *S. pictus*: 60 × 35 mm (*l/w*), *S. scortechinii* 168 × 90 mm (*l/w*). **Midrib**: little deeper than lamina, slightly broadly convex on both surfaces, or flat adaxially, a broadly rounded V abaxially in *S. scortechinii*. **Cuticle** thin or thick, flanged adaxially; medium-thick abaxially in *S. scortechinii*. **Epidermis**: cells widely varying, square and large or smaller, flat or rounded externally and internally. Cells of two sizes in *S. pictus* var. *argyraeus*: tall and subcircular, and very short as stomatal subsidiary cells (Fig. 24C). Abaxial epidermal cells highly rounded-papillate in *S. pictus*. **Stomata** level or slightly sunken; abundant in *S. scortechinii*. **Hypodermis**: 1–2 layers beneath adaxial epidermis, abaxial in places, in *S. scortechinii* (Fig. 4D); cells irregular, cells of innermost layer large and deeply rounded internally. **Mesophyll**: may be weakly differentiated. Palisade layers 1–3 adaxially and abaxially in places, of armed cells in *S. pictus*. Spongy cells 2–15 layers, small to large and lobed (types 1 or 1a). **Air cavities**: large substomatal cavities, and smaller cavities, up to 40% of spongy tissue; up to 90% in *S. pictus*. **Collenchyma**: up to eight layers of thin-walled cells, forming an abaxial subepidermal band in midrib of *S. lucens* and *S. scortechinii*.

Vascular bundles: midrib with one large central bundle and two smaller lateral bundles, or numerous scattered bundles in *S. scortechinii*. Bundles mostly type I with 2–6 xylem elements bordering a broad semicircular strand of phloem, except in *S. scortechinii*, where bundles type II, well-formed, but with small cross-sectional conducting area (Fig. 24A); xylem elements 4–5, opposing small circular phloem strand. **Sclerenchyma**: fibre bundle sheaths or xylem and phloem caps at larger bundles, absent at smaller bundles. Trichosclereids present as isolated cells in lamina or in midrib ground tissue on abaxial side. Absent in *S. lucens*. In *S. pictus* var. *argyraeus* cells large (diameter 30–50 μm; length 3500–4000 μm), aligned with secondaries in mid or lower mesophyll. Rao (1977) reported lengths averaging 1000–1500 μm for nine species. **Secretory ducts**: dark mucilaginous material produced in central air cavities in midrib in *S. scortechinii*. **Laticifers**: non-anastomosing, on both sides of some vascular bundles; absent in *S. scortechinii* and *S. lucens*. **Starch** granules common in midrib ground tissue and in spongy tissue. **Crystals**: raphides common or uncommon and randomly oriented in mid-mesophyll. Raphides in single bundles in short cells, or overlapping bundles in elongated cells up to 300–400 μm long. Double-ended biforines suspended across air cavities in *S. pictus*. Druses common, small or large, in mid-mesophyll.

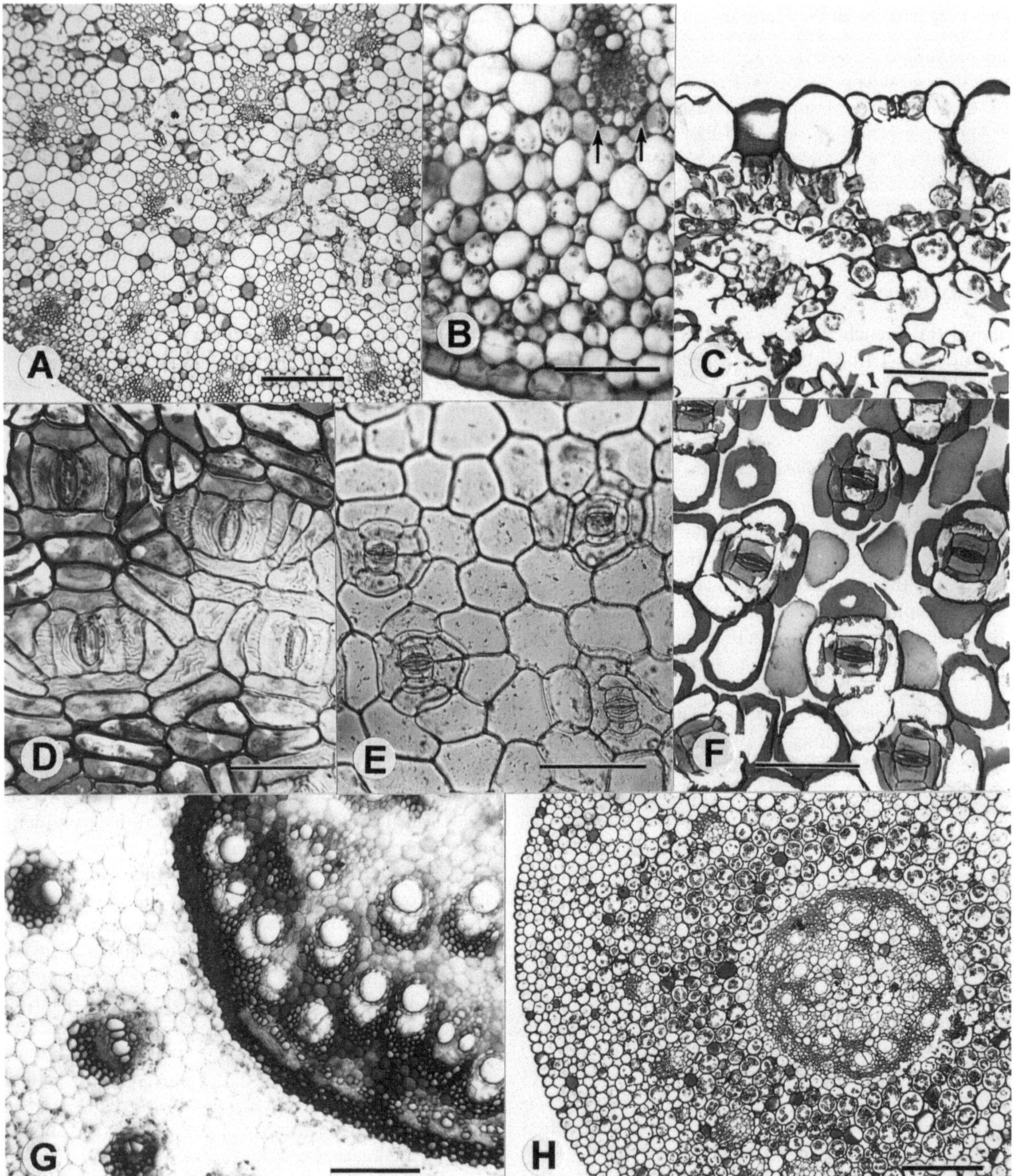

Fig. 24. *Scindapsus* leaf, petiole and stem sections. (A,D) *S. scortechinii*; (B,E,G) *S. lucens* (hand sections, calcium chloride mounts); (C,F,H) *S. pictus* var. *argyraeus*; (A) Midrib TS abaxial side; (B) Petiole TS, abaxial side with vascular bundle. Note thin collenchymatous thickenings (arrows); (C) Lamina TS, tall adaxial epidermal cells and stoma with small subsidiary and guard cells; (D) Lamina PS, stomata and elongated abaxial epidermal cells; (E) Lamina PS, stomata and polygonal abaxial epidermal cells; (F) Lamina PS, stomata and rounded adaxial epidermal cells; (G) Stem TS showing fibre boundary of central cylinder and phloem fibre caps (TBO stained); (H) Stem TS showing small central cylinder; starch abundant in inner cortex. Scale bars: A,H = 300 μm; B,C,E,F = 150 μm; D = 50 μm; G = 200 μm.

Crystal sand common in all mesophyll cells, or in lower mesophyll.

Petiole TS

Outline: proximally crescent-shaped. Petiole terete distally, somewhat flattened adaxially, but elliptic with adaxial groove in *S. scortechinii*. **Size**: diameter 3 mm basally, 2 mm distally. **Cuticle** thin and smooth, or thick and rugulate in *S. scortechinii*. **Epidermis**: cells square in TS and axially elongate; cells rounded externally, flat internally in *S. scortechinii*. **Stomata** occasional and level with epidermis. **Collenchyma** in distal petiole: band of light pectic thickenings in cells of outer abaxial ground tissue. Collenchyma with laminar and angular thickenings in phloem cap cells distally; proximally, few collenchyma cells among sclerenchyma fibre caps. In *S. lucens* (distal petiole), collenchyma present as phloem caps only (Fig. 24B). **Chlorenchyma**: deep layer of outer ground tissue; plastids uncommon. **Ground tissue**: cells large, compact, with corner air cavities, up to 5% of ground tissue in *S. lucens* (Fig. 24B) and *S. pictus*; central ground tissue type 3 aerenchyma, vascular bundles at corners of large rounded air cavities. In *S. scortechinii*, cells smaller near periphery, larger in centre.

Vascular bundles: 15–25 bundles; median abaxial bundle and right and left rows of bundles arranged symmetrically, or randomly scattered in *S. scortechinii*. Bundles type I. Xylem: 2–8 elements bordering a broad semicircular or rounded strand of phloem. In *S. scortechinii*, 1–3 wide late protoxylem or early metaxylem elements and 0–3 narrow metaxylem elements (type II); phloem a small circular strand. **Sclerenchyma**: phloem fibre caps proximally mixed with a few collenchyma cells; all fibres replaced by collenchyma distally. In *S. pictus*, fibres absent at vascular bundles. In *S. scortechinii*, fibre band 3–4 cells deep and 6–8 cells beneath epidermis, following abaxial surface; adaxial bundles ensheathed by fibres; abaxially, fibres fewer, becoming irregular phloem caps, finally absent near abaxial margin; isolated trichosclereids very common at outer edge of fibre band. **Laticifers** absent, or rare near vascular bundles in *S. pictus*. **Starch** granules throughout ground tissue proximally, concentrated in central zone distally. **Crystals**: raphides in small bundles, common in ground tissue proximally. Distally, druses becoming very common in outer ground tissue and around vascular bundles.

Stem TS and LS (S. lucens; S. pictus var. *argyraeus)*

Outline terete. **Size**: diameter 3 mm. **Cuticle** medium-thick and smooth. **Epidermis**: cells medium-sized and square in section, axially elongate. Stem divided into cortex and central cylinder. **Cortex**: compact ground tissue, cells large, chloroplasts common in outer layers; some air cavities at cell corners (Fig. 10A). **Central cylinder** delimited by 3–5-layered ring of lignified fibres (Fig. 24G,H).

Vascular bundles scattered, concentrated within central cylinder; a single ring of widely spaced bundles in cortex (type II) (Fig. 24H). Outer bundles of central cylinder with confluent phloem fibre caps that may merge with central cylinder ring. Cortical bundles with 1–5 metaxylem elements bordering a small elliptical or semicircular strand of phloem. Bundles of central cylinder (types II, III) usually with one very wide metaxylem element bordering strand of phloem with *c.* 6 large sieve elements and their companion cells. Some bundles with up to six xylem elements and a strand of smaller phloem cells, more resembling cortical bundles. **Sclerenchyma**: fibre sheath cells delimiting central cylinder having visible (isotropic) wall layers. Fibres also forming semicircular phloem caps (Fig. 24G). **Starch** granules packed solidly in central cylinder ground tissue; granules less densely present in inner cortical ground tissue. **Crystals**: raphides occasional in slender single bundles in ground tissue; much more common as axially very elongated multiple overlapping bundles up to 700–1100 μm. Druses common near vascular bundles.

Root TS (S. pictus var. *argyraeus)*

Outline terete. **Size**: 1.5 mm diameter. **Exodermis**: 1–2 lignified cell layers beneath collapsed epidermis. **Cortex**: uniform ground tissue, in radial files near stele. **Endodermis**: narrow casparian strip present in anticlinal walls. **Stele**: nine xylem and phloem poles. **Sclerenchyma**: fibres in centre of stele in small strand. Trichosclereids isolated and uncommon in cortex. **Crystals**: raphides common in outer cortex in elongated raphide tubes with multiple overlapping bundles. Druses and crystal sand common throughout cortical ground tissue.

Material examined

Scindapsus lucens Bogner & Boyce, cult. M; *S. pictus* Hassk. var. *argyraeus* Engl., cult. M; *S. scortechinii* Hook. f., Koerper s.n., Borneo, cult. M.

Taxonomic notes

Each of the specimens examined is quite distinctive anatomically. *Scindapsus lucens* is one of few aroid plants observed showing a sclerenchyma to collenchyma transition occurring in the petiole, thus establishing positional and transitional homologies in this genus (see Donoghue and Sanderson 1992: p. 397). Basal sections have lignified fibres as phloem caps, while distal sections show lamellar and angularly thickened collenchyma caps in the position occupied by fibres at lower levels. Successively higher sections show a cell by cell replacement of cell types. Note: Dalitzsch (1886) illustrated the adaxial epidermis of *S. argyraeus* showing cells with pointed tips (not quite papillae) with spongy or palisade tissue near surface. See treatment of *Alloschemone* for comments on its level of similarity to *Scindapsus*.

Species reported on in the literature

Araus *et al.* (1986a): *S. aureus*: variegated leaf. • Bergdolt (1955): *S. aureus, S. pictus*: leaf. • Bretfeld (1879–81): *S. glaucus*: leaf. • Czaja (1962): *S. pictus*: leaf epidermis. • Dalitzsch (1886): *S. argyraeus*: leaf. • Elisei (1941): *S. pictus*: casparian strips. • Engler (1908): *S. aureus, S. pictus*: stem, root. • Fisher and French (1976): *S. aureus*: stem meristem. • French (1987c) several spp.: root. • French and Tomlinson (1981d): several spp.: stem vasculature. • Funke (1929): *S. aureus*: root. • Funke (1931): *S. manilliferus*: leaf, petiole. • Genua and Hillson (1985): *S. aureus*: crystals. • Gertz (1916a b): *S. pinnatifidus*: tyloses. • Guttenberg (1968): *S. lingulatus*: root. • Hassack (1886): *S. argyraeus*: leaf. • Hotta (1971): *S.* sp.: stem; *S. aureus*: leaf venation. • Inamdar *et al.* (1983): *S. aureus*: leaf venation. • Lierau (1888): several spp.: root. • Linsbauer (1930): *S. pinnatus*: epidermis. • Majumdar and Pal (1961): *S. officinalis*: leaf sheath. • Mangin (1882): *S. multijugus, S. pinnatus, S.* sp.: root. • Napp-Zinn (1953): *S. aureus*: root. • Neger (1922): *S. decursivus, S. pinnatus*: root. • Nicolson (1960): several spp.: stem, leaf, petiole. • Nyawuane and Gill (1990): *S. pictus*: stomata. • Olivier (1880, 1881): *S. pertusus*: root. • Petersen (1893): *S. pinnatifidus*: meristems. • Rao (1954): *S. officinalis, S. pictus*: leaf sclereids. • Rao (1964, 1977): several spp.: leaf sclereids. • Rao and Bhattacharya (1978): leaf sclereids. • Reinhardt (1885): *S. pinnatifidus*: root. • Schwendener (1882): shoot. • Singh (1968): several spp.: leaf sclereids. • Solereder and Meyer (1928): *S. pictus*: all organs. • Stebbins and Khush (1961): *S. aureus*: stomata. • Sueur (1866–67): *S. pertusus*: sclereids. • Swamy and Sivaramakrishna (1972): *S. aureus*: stem. • Swamy and Sivaramakrishna (1976): root, apical meristem. • Thaler (1955–56): *S. aureus*: epidermis. • Tieghem (1866): *S. pictus*: all organs. • Tieghem (1870–71): *S. pictus*: root, secretory tissue. • Trécul (1866): vasculature. • Turlier (1968): *S. argyraeus*: leaf. • Van Fleet (1942): *S. pictus*: stem, endodermis. • Voronkina (1975): *S. aureus*: root apex. • Webber (1960): *S. argyraeus, S. aureus, S. pictus*: epidermis, stomata. • Went (1895): several spp.: root. • Zinke (1924): *S. argyraea*: leaf.

Additional literature

Blanc 1977a; Domin 1911; Erdtman 1952; French 1985a, 1986c, 1987a; Gentner 1905; Grayum 1990; Hegnauer 1986; Madison and Tiffney 1976; Majumdar and Pal 1961; Meusel 1951; Thanikaimoni 1969; Vágújfalvi 1971; Williams *et al.* 1981.

***Rhaphidophora* Hassk.**

This genus of about 60 species ranges from west Africa, India, Ceylon, Nepal to south-east Asia and Samoa. Plants are mostly hemi-epiphytic appressed climbers inhabiting a range of wet evergreen to deciduous primary forests. Some species are terrestrial or epilithic, and two are recorded as being rheophytic. Stems are liane-like with 2-ranked leaves showing heteroblastic development and alternating adult-juvenile growth phases. The leaves are entire, pinnatifid, or pinnate and sometimes perforate. Lamina shapes are ovate or elliptic or with bases unequally cordate. Petioles are sheathing and grooved. Chromosomes medium-sized: 2n = 42, 54, 56, 60, *c.*120. Three species were available for study.

Leaf surface

Cuticle smooth on both surfaces, striate on *R. sulcata*. **Epidermis**: cells polyhedral with undulate (Fig. 25D) or straight-sided anticlinal walls, isodiametric (Fig. 25G) to elongate (1–3:1 *l*/*w*) on both surfaces. **Stomata** mostly abaxial and brachyparacytic (Fig. 25E) to brachyparahexacytic. Subsidiary cells broad or narrow, rounded, square, or irregular. **Venation**: midrib prominent. Secondaries broadly or narrowly acute, often closely spaced, curved-ascending, merging at straightened eucamptodromous marginal collector. Tertiaries parallel with secondaries or reticulate. Quaternaries having irregular cross connections or reticulate. Higher orders reticulate.

Leaf TS

Structure dorsiventral. **Size**: 136 × 43 mm (*l*/*w*) or smaller. **Midrib** large, shallowly rounded-concave or flat adaxially, broadly rounded-hemispheric abaxially (Fig. 25A). **Cuticle** thick adaxially and thin abaxially; flanged or not on both surfaces and appearing granular when viewed in polarized light. **Epidermis**: cells tall (Fig. 25A), or even palisade-like adaxially (Fig. 1A), flat on surface and rounded internally. Cells tall or usually short abaxially, flat externally and internally. **Stomata** with distinct subsidiary cells, level with surface. **Mesophyll**: one well-differentiated palisade layer (8–10:1 *l*/*w*) (Fig. 25B) and *c.* 9 spongy layers, horizontally elongated. Paradermally, spongy cells thick-walled and lobed with irregular or

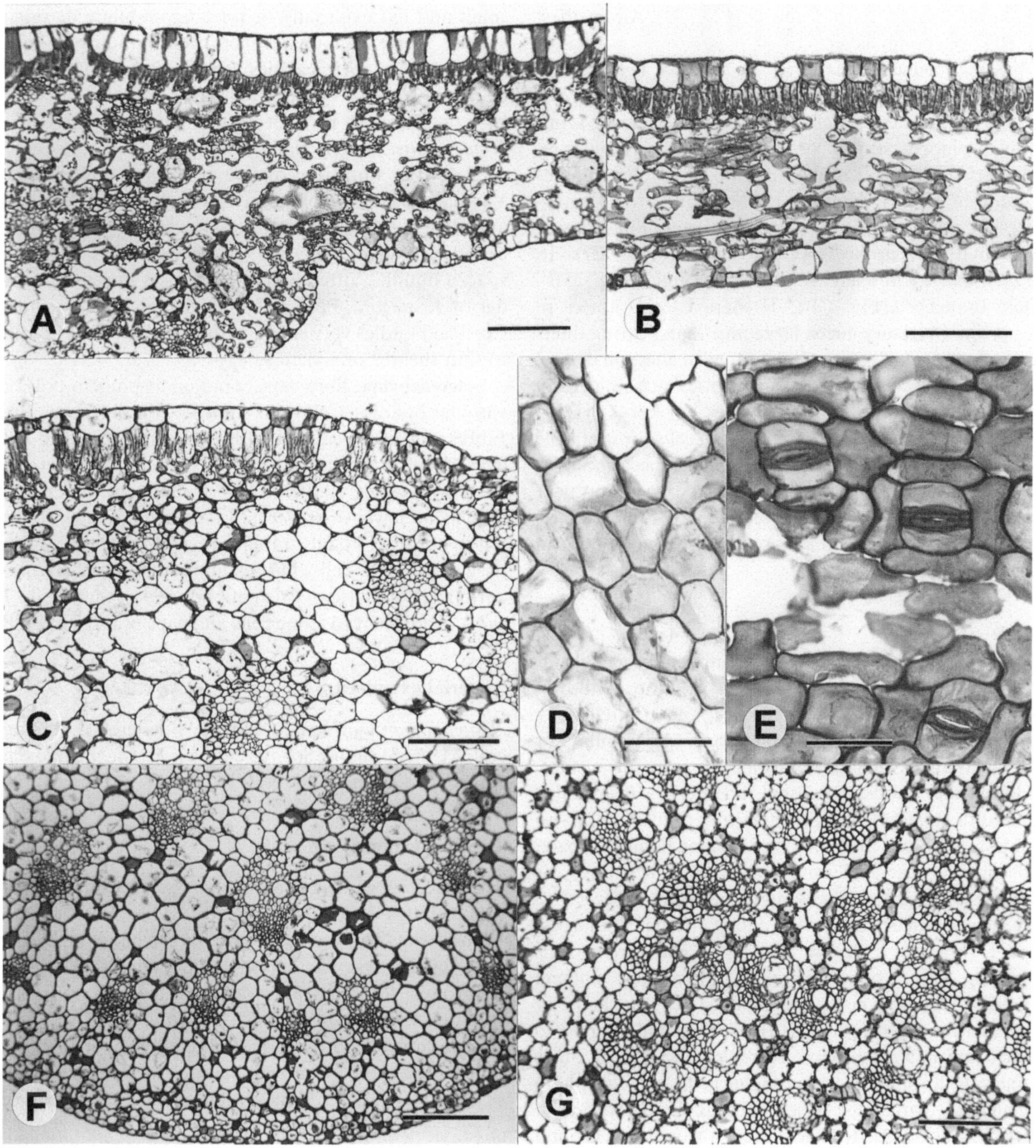

Fig 25. *Rhaphidophora* leaf, petiole and stem sections. (A) *R. sulcata;* (B,D) *R. africana;* (C,E–G) *R. decursiva.* (A) Lamina and midrib TS. Note frequent secretory ducts; (B) Lamina TS with trichosclereid in spongy tissue; (C) Midrib TS with adaxial groove at upper right. Note palisade tissue and adaxial stomata; (D) Adaxial epidermis PS; (E) Abaxial epidermis PS and stomata; (F) Petiole TS abaxial side. Note type I and II vascular bundles; (G) Stem TS showing central cylinder vascular bundles. Scale bars: A,F, G = 300 μm; B = 150 μm; C = 200 μm; D,E = 50 μm.

regular air spaces (types 1 and 2) (Fig. 7B). **Air cavities**: 30–80% of spongy layer.

Vascular bundles scattered in midrib (Fig. 25C), arrangement sometimes bilaterally symmetrical; type I. Xylem: 2–6 metaxylem elements opposing a large semicircular phloem strand, or narrower type II bundles. **Sclerenchyma**: fibres ensheathing most vascular bundles or as fibre caps. Trichosclereids and long-armed astrosclereids, not parallel, ungrouped and common in lower mesophyll and up to 1500 µm long (Rao 1977). In *R. africana*, trichosclereids common, single (Fig. 25B), often branched (Fig. 7B), H-shaped; not noted in *R. sulcata*. **Secretory ducts**: large mucilage-bearing ducts or short irregular cavities numerous in mesophyll and midrib of *R. sulcata* (Fig. 25A); cavities surrounded by single row of mesophyll cells. **Laticifers**: not seen. **Crystals**: raphides as single bundles in small cells or multiple overlapping bundles in elongated cells, occasional in mesophyll. Druses occasional in palisade layer and near abaxial epidermis, often with very sharp pointed edges. Crystal sand common in spongy layer, less common in palisade.

Petiole TS and LS

Outline: proximal section rounded abaxially; deeply or shallowly grooved adaxially. Distal section shallowly grooved adaxially. **Size**: diameter 6–15 mm. **Cuticle** thick and smooth. **Epidermis**: cells small or large, papillate or flat externally. **Stomata** occasional, level with surface. **Collenchyma**: bands occasional in outer 8–10 layers of ground tissue. Phloem caps often present in *R. decursiva* and less developed in *R. africana*. **Ground tissue** compact, cells smaller near epidermis and larger near centre. **Air cavities**: few and scattered among parenchyma.

Vascular bundles numerous and scattered, the largest type I (Fig. 25F). Xylem: 4–8 wide metaxylem cells opposing broad boundary with semicircular phloem strand. **Sclerenchyma**: trichosclereids branched or unbranched, ungrouped, scattered throughout ground tissue; rare in *R. decursiva*. **Secretory ducts**: large ducts numerous throughout ground tissue in *R. sulcata* with no specialized epithelium; ducts absent in *R. africana*. **Laticifers** not seen. **Starch** present throughout ground tissue, granules 12–20 µm diameter. **Crystals**: raphides few and randomly found in ground tissue, probably in axially elongated cells. Druses common throughout, especially abundant near periphery.

Stem TS

Outline terete or laterally compressed. **Size**: 9–25 mm diameter. **Cuticle** thick and smooth. **Epidermis**: cells small and flat externally or forming undulate or ridged surface. **Cortex**: small peripheral cells, grading to larger cells in centre. **Air cavities** few in cortex and central cylinder. **Central cylinder** delimited, usually unevenly, by concentrated ring of larger vascular bundles in *R. decursiva*, or partially confluent fibre caps in *R. decursiva*, or fibre ring of two cell layers in *R. sulcata*. **Vascular bundles** collateral throughout (types I, II and III) (Fig. 25G) or occasionally compound in central cylinder. Xylem: very wide metaxylem cells in closely spaced bundles, often filled with tyloses in central cylinder of *R. sulcata*. Phloem forming semicircular strands. Cortical bundles very small, with large fibre caps grading to thin sheaths on xylem side.

Sclerenchyma: fibre caps common at phloem poles of vascular bundles. Cell walls showing concentric layering. Fibres externally capped with row of thick-walled brachysclereids with ramified pits in *R. sulcata* only. **Secretory ducts** with single epithelial layers abundant in cortex of *R. sulcata*, fewer in *R. decursiva*. **Starch** granules abundant in cortex. **Crystals**: raphides few throughout ground tissue in narrow cells. Druses numerous, especially near periphery of cortex. Prismatics large, common in *R. sulcata*, adjacent to stone cells and caps of large bundles.

Material examined

Rhaphidophora africana N.E.Br., Bogner 602, Gabon, cult. M; *R. decursiva* Schott, cult. M; *R. sulcata* Gagnep., Bogner 1313, Thailand, cult. M.

Taxonomic notes

The three specimens examined form a coherent group that belongs clearly among the trichosclereid-bearing Monstereae. Their anatomy is character-rich and varies in striking ways. The species do not divide neatly into subgroups. *Rhaphidophora sulcata* and *R. decursiva* have gum ducts. *R. decursiva* and *R. africana* have trichosclereids. Scrutiny of the list of characters produces a taxonomic plexus and points out the need for a detailed study of the genus in order to ascertain trends of specialization.

Species reported on in the literature

Dalitzsch (1886): *R. pertusa*: leaf. • Engler (1908): *R. decursiva, R. korthalsii, R. pertusa*: root, stem, leaf. • Ertl (1932): *R. celatocaulis, R. decursiva*: leaf venation. • French (1987c): several spp.: sclerotic hypodermis absent. • French and Tomlinson (1981d): *R. celatocaulis, R. decursiva, R. sulcata*: stem vasculature. • Funke (1931): *R. pallidivenia*: leaf, petiole. • Guttenberg (1940): several spp.: root. • Hotta (1971): *R. latevaginata*, *R.* sp.: leaf, stem, root. • Leblois (1887): *R. pinnata*: secretory tissue. • Lierau (1888): *R. decursiva*: root. • Mangin (1882): root. • Müller (1919): *R. decursiva*: leaf tip. • Netolitzky (1929): *R. pinnata*: crystals. • Nicolson (1960): *R. celatocaulis, R. decursiva, R. reineckei*: stem, leaf, petiole. • Olivier (1880, 1881): *R. angustifolia, R. pinnata, R.* sp.: root. • Pant and Kidwai (1966): *R. calophyllum, R. celatocaulis*: leaf. • Rao (1977): several spp.: leaf, sclereids. • Rao and Bhattacharya (1978): leaf, sclereids. • Reinhardt (1885): *R. peepla*: root. • Schwendener (1882): root. • Singh (1968): several spp.: sclereids. • Solereder and Meyer (1928): several spp.: all organs. • Spratt (1920): *R. decursiva*: root. • Tieghem (1866): *R. angustifolia, R. pinnata*: all organs. • Tieghem (1970–71): *R. angustifolia, R. pinnata:* root, secretory tissue. • Tieghem (1872): *R. pinnata*: secretory tissue. • Webber (1960): *R. celatocaulis*: epidermis, stomata.

Additional literature

Blanc 1977a, 1978; Domin 1911; Erdtman 1952; Eyde *et al.* 1967; French 1985a, 1986c, 1987a; Grayum 1990; Hegnauer 1986; Hotta 1965; Huang 1972; Knecht 1983; Madison and Tiffney 1976; Meusel 1951; Mookerjea 1955; Ray 1988; Sharma 1967; Thanikaimoni 1969; Williams *et al.* 1981.

Anadendrum Schott

The nine species in this genus are distributed in southeast Asia and Malaysia. Plants are hemi-epiphytic and scandent, rooting on trees, shrubs, or on rocks in moist to wet forests, or along streams (Croat 1988). Leaves are long-petiolate, broad ovate or cordate with an acuminate tip. Chromosomes: $2n = 60$. One species was examined.

Leaf surface

Cuticle smooth adaxially, with marked randomly oriented striations on both laminar surfaces and midvein (Fig. 26D). **Epidermis**: cells polygonal with straight anticlinal walls adaxially (2–3:1 *l*/*w*), and polygonal with straight, undulate, or irregularly elongate walls abaxially (2–4:1 *l*/*w*). **Stomata** mostly abaxial, variable and distinctive. Mostly brachypara-hexacytic with second pair of parallel subsidiary cells smaller than paracytic subsidiary cells (Fig. 26D). Orthogonal terminal cells wrapping around guard cell/subsidiary cell pairs, ending at second pair of parallel subsidiary cells. Subsidiary cells occasionally occurring in 5–6 sets, each set orthogonal to layers internal to it. Occasional stomatal complexes have apparently radiating subsidiaries. **Venation**: midrib prominent with pinnate secondaries. Secondaries broadly acute and curved-ascending, merging eucamptodromously at margin. Tertiaries and higher orders reticulate and irregular.

Leaf TS

Structure dorsiventral. **Size** 57 × 44 mm (*l*/*w*). **Midrib** slightly thicker than lamina, concave adaxially, broadly rounded abaxially (Fig. 26A). **Cuticle** thick. **Epidermis**: cells square to paradermally elongate, abaxial cells often largest. **Stomata** level with epidermis. **Mesophyll**: *c.* 8 layers including one weakly differentiated palisade layer (Fig. 26B). Spongy cells irregular or armed with air spaces at corners (type 1). **Air cavities** up to 40% of spongy layer. **Collenchyma**: poorly defined band abaxially in midrib (Fig. 26A); cells comparatively thin-walled.

Vascular bundles symmetrically arranged in midvein, three large bundles, one medial, two lateral, plus two smaller bundles on each side in upper and lower regions of midrib (Fig. 26A); type I. Xylem: 1 to several metaxylem cells forming broad boundary with semicircular phloem strand. **Sclerenchyma**: fibres ensheathing larger veins of lamina and midrib (Fig. 26F). **Crystals**: raphides medium-coarse, large, in overlapping bundles in elongate cells (Fig. 26E), tending to run parallel to vascular bundles. Druses large, throughout mesophyll and randomly oriented. In one specimen, druses abundant near leaf tip, uncommon elsewhere.

Petiole TS

Outline: base of petiole broadly sheathing and smoothly concave–convex (Fig. 26C). Distally petiole laterally compressed, irregularly rounded abaxially and smoothly concave adaxially. **Size**: diameter 2–4 mm. **Cuticle**: medium-thick, smooth to deeply striate; thin on adaxial surface of proximal petiole. **Epidermis**: cells small, rounded internally and externally. **Collenchyma**: forming vein sheaths or phloem caps, best developed near lamina insertion; 4–5 subepidermal layers abaxially. **Ground tissue**: uniform rounded parenchyma with air cavities at corners.

Vascular bundles: about 12, type I. Xylem: metaxylem of fewer than 12 cells forming boundary with semicircular or elliptical phloem strand. **Crystals**: raphide cells uncommon, in 4–5 subepidermal layers. Druses very common in proximal petiole, especially in adaxial subepidermal cells.

Stem TS

Outline terete. **Size**: diameter 5 mm. **Cuticle** thin and mostly smooth. **Epidermis**: cells small, rounded to

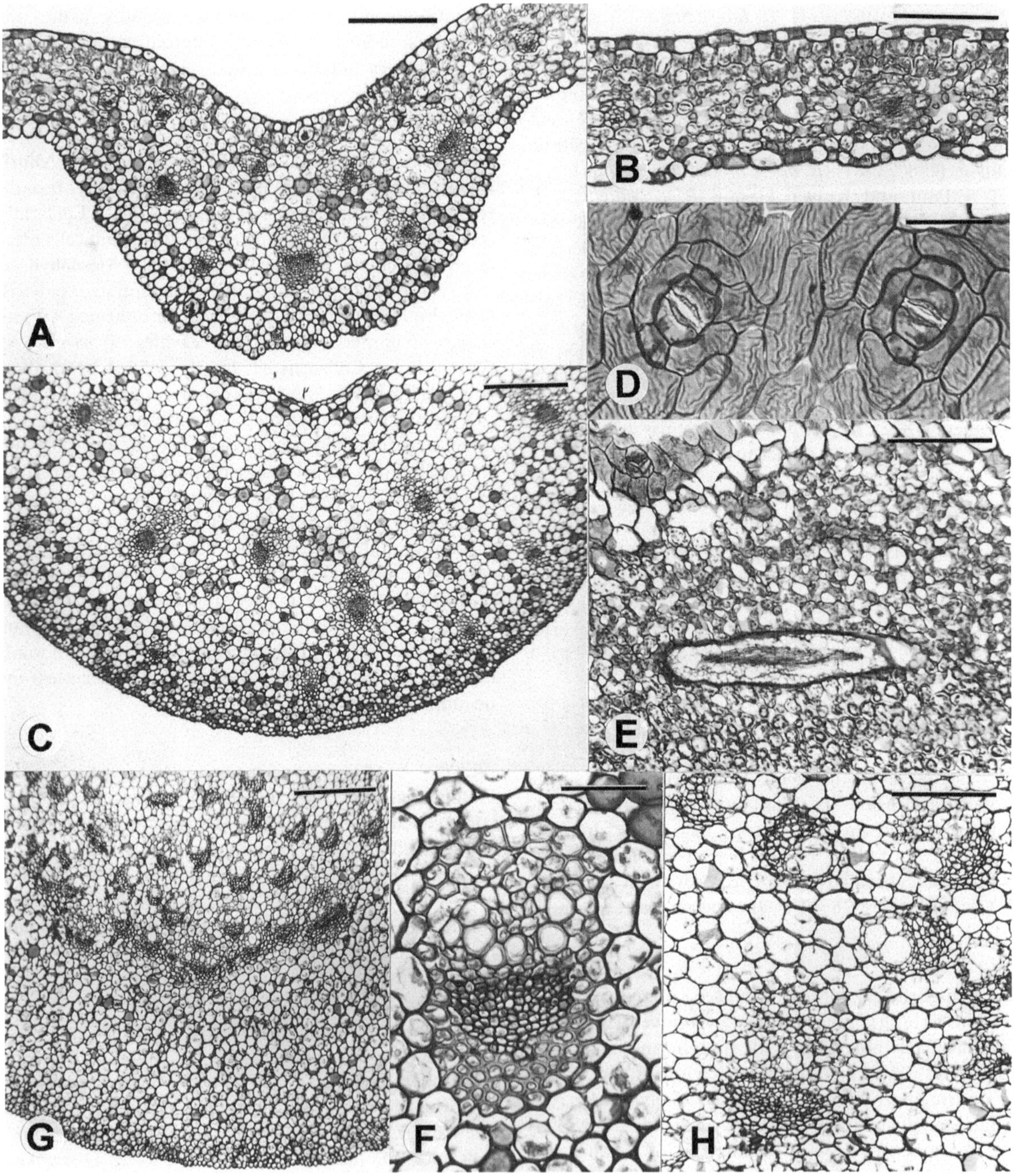

Fig. 26. *Anadendrum microstachyum* leaf, petiole and stem sections. (A) Midrib TS with symmetrical arrangement of vascular bundles; (B) Lamina TS; (C) Petiole TS; (D) Leaf abaxial epidermis PS showing brachypara-hexacytic stomata; (E) Leaf mesophyll PS showing raphide cell; (F) Midrib TS, central type I vascular bundle; (G) Stem TS showing cortex/central cylinder boundary; (H) Stem central cylinder TS. Scale bars: A = 200 μm; B,E,H = 150 μm; C,G = 300 μm; D,F = 50 μm.

papillate externally. **Collenchyma**: irregular band in 2–3 poorly defined subepidermal layers. **Ground tissue**: cells smaller near periphery, large and ± uniform internally. Cortex weakly delimited (in young specimen) from central cylinder by irregular band of vascular bundles with ± confluent sclerenchyma on one side (Fig. 26G). **Air cavities** common at corners of ground tissue.

Vascular bundles: central cylinder containing largest bundles and most of vasculature (type I) (Fig. 26H). Xylem: several wide metaxylem cells bounded by semicircular strand of phloem. Bundles less well-defined in cortex. **Sclerenchyma**: phloem caps or poorly differentiated sheaths in a few bundles. **Crystals**: raphides common throughout, in cells not otherwise distinguished from ground tissue. Druses common throughout ground tissue. Crystal sand in most ground tissue cells.

Material examined

Anadendrum microstachyum (DeVr. & Miq.) Back & Alderw., Bogner 1314, Bogor Botanical Garden, cult. M.

Taxonomic notes

Two diagnostic characters of *Anadendrum* stand out in this tribe. First, the stomatal subsidiary cells form a unique pattern for the family. Second, there is a tendency in this genus for collenchyma tissue to ensheath or cap vascular bundles instead of the usual band-forming zone that normally occurs in this subfamily.

Species reported on in the literature

French (1987c): *A. microstachyum*: root. • French and Tomlinson (1981c): *A. microstachyum*: stem vasculature. • Hotta (1971): leaf.

Additional literature

Domin 1911; Eyde *et al.* 1967; French 1985a, 1986c, 1987a; Grayum 1990; Hotta 1965; Ray 1988; Thanikaimoni 1969.

Monstera Adans.

This genus of about 60 species ranges from Mexico to Brazil and the Lesser Antilles. Plants are mostly epiphytic or hemi-epiphytic, with occasional terrestrial or epilithic species. They are found in primary forests, occasionally in regrowth, and are abundant in open areas along forest edges. They are appressed climbers with short internodes, having juvenile and adult growth forms that may develop pendent flowering branches. Lianes have 2-ranked leaves. Juvenile leaves are ovate-orbicular or cordate. Adult leaves are broadly ovate-cordate, perforated and/or pinnately lobed. Chromosomes medium-sized: $2n = 24$, 48, 56, 58, 60, 70. Seven species were available for study.

Leaf surface

Cuticle striate or rugulate on both surfaces. **Epidermis**: cells isodiametric with straight to undulate anticlinal walls on both surfaces (1.5:1 *l/w*). **Stomata** more frequent abaxially, brachyparatetra-(hexa-)cytic and unipolar (Figs 3H, 27D). Subsidiary cells broad or narrow and irregular. **Venation**: midrib prominent. Secondaries pinnate and broadly acute ending at straightened marginal collector. Tertiaries parallel with secondaries or tertiaries. Higher orders reticulate.

Leaf TS

Structure dorsiventral. **Size**: liquid-preserved fragments of large leaves measured in decimetres. **Midrib** flat adaxially; broadly and shallowly rounded-convex abaxially (Fig. 27A). **Cuticle**: medium-thick on both surfaces. **Epidermis**: cells large, outer walls flat, internal walls rounded on both leaf surfaces. **Stomata** level with surface. **Mesophyll**: palisade layers (1–2), irregular, packed with chloroplasts (Fig. 27B). Spongy layers *c.* 12, cells with several short arms (type 1a); illustrated as type 2 by Lindorf (1980). **Air cavities** fairly regular, up to 40% of spongy layers. Substomatal chambers shallow. **Vascular bundles**: midrib with numerous scattered bundles, arrangement may show bilateral symmetry. Bundles robust (type II). Xylem: eight or more narrow or wide metaxylem elements, irregularly opposing large circular phloem strand.

Sclerenchyma: fibres ensheathing vascular bundles, about five layers thick over phloem and xylem, thinner laterally. Confluent fibre band frequent at outer bundle row of abaxial midrib (Fig. 27E). Trichosclereids occasional to common in mid-mesophyll (Fig. 7C), solitary and branched, irregular or H-shaped, called zosterosclereids by Rao (1977) and not aligned. **Mucilage cells** occasional with dark contents in mid-mesophyll. **Crystals**: raphides common in elongate cells in overlapping bundles (Fig. 27C), or as single bundles in short cells (Fig. 27F); bundles tending to run parallel with secondary veins. Druses mostly in upper mesophyll and near vascular bundles. Crystal sand aggregated, very common in mesophyll.

Petiole TS

Outline: proximally semicircular abaxially, grooved and winged adaxially; distally terete. **Size**: diameter 5–12 mm.

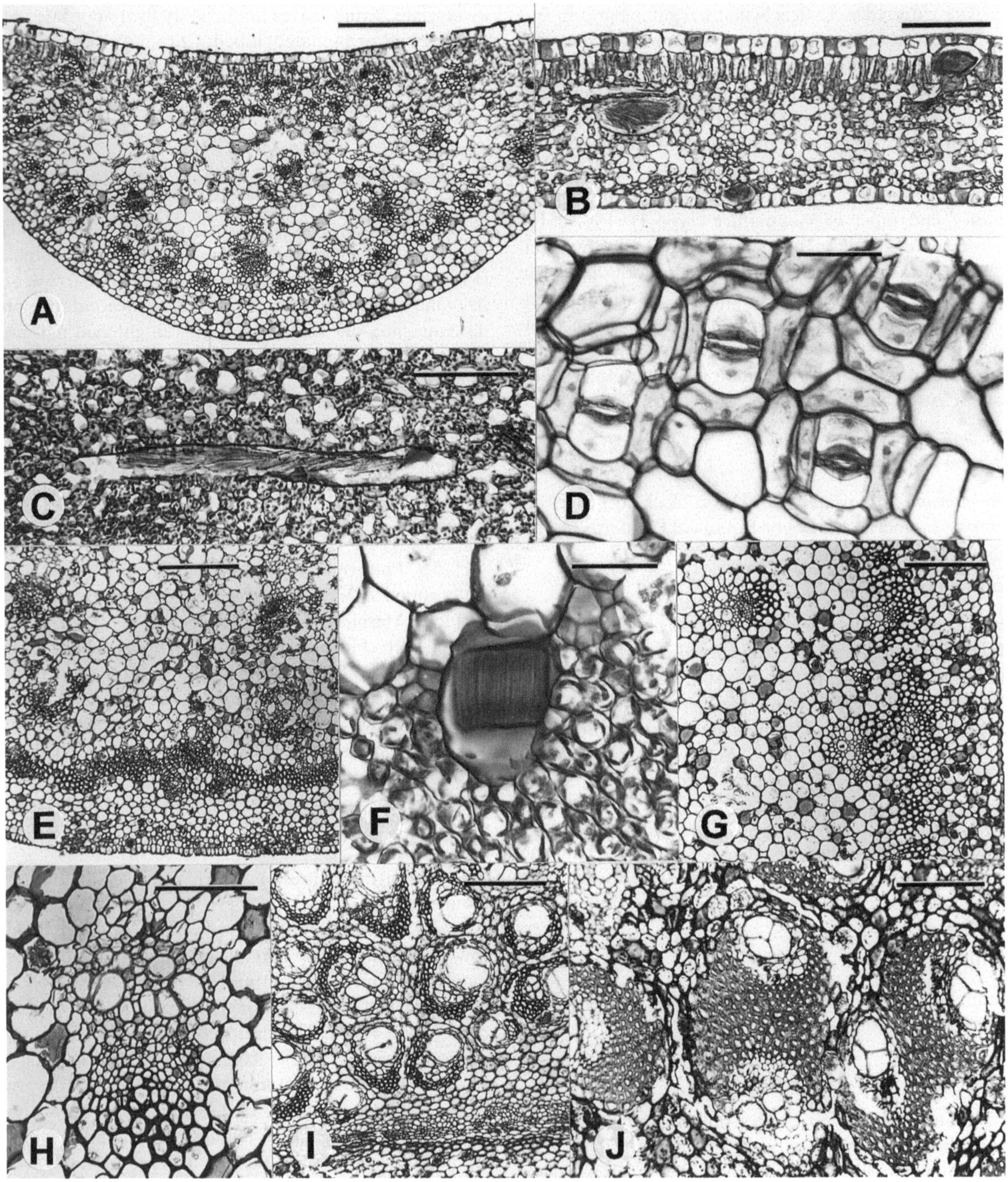

Fig. 27. *Monstera* leaf, petiole and stem sections. (A) *M. pittieri*; (B,E,F) *M. acuminata*; (C,I) *M. siltepecana*; (D) *M. dilacerata;* (G,H,J) *M. oreophila;* (A) Midrib TS; (B) Lamina TS; (C) Mesophyll PS showing elongated raphide cell and multiple overlapping bundles; (D) Abaxial epidermis PS showing brachypara-hexacytic stomata; (E) Midrib TS; (F) Mesophyll PS showing wide raphide bundle; (G) Lateral petiole TS; (H) Petiole TS, central vascular bundle, type III; (I) Stem TS showing cortical boundary (below) and central cylinder vascular bundles; (J) Stem central cylinder TS showing compound vascular bundles. Scale bars: A,E,G,I = 300 μm; B,J = 200 μm; C,H = 150 μm; D,F = 50 μm.

Cuticle thick and smooth, or papillate abaxially in *M. acuminata*. **Epidermis**: cells flat externally and internally, except rounded externally and very unevenly aligned in *M. acuminata*, and markedly papillate in *M. dilacerata* with thick striate cuticular caps on papillae. **Collenchyma**: thin-walled layers in outer ground tissue in *M. siltepecana*, or as phloem caps in *M. acuminata*. Collenchyma as phloem caps in absence of fibres in distal petiole of *M. pittieri*. **Chlorenchyma**: outer 4–5 layers of ground tissue. **Ground tissue**: cells becoming larger toward centre. **Air cavities** numerous, aerenchyma types 2 and 4.

Vascular bundles numerous and scattered. Outer bundle row, within 10–12 cortical ground tissue layers, aligned with discontinuous bands of sclerenchyma (Fig. 27G). Bundles in inner cortex may align allowing fibre caps to form confluent boundary between cortical and central ground tissue. Xylem: 1–2 wide metaxylem cells and occasionally 2–3 narrower metaxylem cells bounding circular or semicircular phloem strand (type II) (Fig. 27G, H). **Sclerenchyma**: abaxial fibre band, continuous, 10–15 layers beneath epidermis, 2–3 cells deep, often merging with vascular bundle caps in vicinity, absent in one specimen of *M. acuminata* and discontinuous in *M. pittieri*. Fibres common as phloem caps or sheaths, thinner over xylem. Trichosclereids very common, ungrouped, mostly toward periphery of organ; also in centre in *M. dilacerata*. **Starch** granules thinly distributed in ground tissue or concentrated near vascular bundles. **Crystals**: raphides common, especially in outer ground tissue; cells axially elongate, with crystals surrounded by mucilage in their bundles. Coarse druses common, especially in ground tissue near vascular bundles.

Stem TS and LS

Outline terete or subterete. **Size**: diameters 1.7 mm (*M. acuminata*), 1.2 × 1.8 mm (*M. oreophila*), 8.5 mm (*M. pittieri*), 10 × 19 mm (*M. siltepecana*). **Cuticle** thin or thick and smooth. **Epidermis**: cells irregular externally and internally, or flat or rounded. **Cork**: broad discontinuous bands, in three zones: two thin-walled cell layers, 2–6 lignified cell layers, not elongate, and 3–5 outer, thin-walled, empty cell layers; absent in *M. pittieri*. **Ground tissue**: ± uniform cell size range. **Cortex** broad. **Air spaces** small and infrequent. **Central cylinder**: (Fig. 27I) sharply delimited by 2–5 layers of brachysclereids, often with ramified pits, and inner layer of heavily lignified fibres.

Vascular bundles scattered in cortex, abundant and closely spaced in central cylinder (Fig. 27I, J). Cortical bundles: broad strand of 5–10 medium-sized metaxylem cells bounded by broad phloem strand (type I). Central cylinder bundles: xylem with 1–2 very wide metaxylem cells, occasionally additional 2–3 narrower metaxylem cells opposing broad irregular phloem strand (type II); bundles randomly oriented. **Sclerenchyma**: thin-walled fibres capping phloem in cortex (Fig. 6G). In central cylinder, fibre caps larger, often ensheathing bundles, tending to be confluent between bundles, especially at periphery of central cylinder (Fig. 27J). Concentric wall layering often present in thick fibre cell walls within central cylinder. Inner fibre walls tending to be gelatinous in *M. siltepecana*. Trichosclereids common in cortex, random and solitary. **Starch** granules common in cortical ground tissue, common or abundant in central cylinder ground tissue, or absent. **Crystals**: raphides absent. Druses large and coarse, very common near vascular bundles, common randomly in ground tissue. Prismatic crystals in every stone cell at central cylinder boundary, or absent.

Material examined

Monstera acuminata C.Koch, Croat 42933, Nicaragua, MO; Croat 42530, Honduras, MO; *M. dilacerata* C.Koch, Croat 42742, Honduras, MO; *M. membranacea* Madison, Croat 35142, Costa Rica, MO; *M. oreophila* Madison, Croat 33267, Panama, MO; *M. pittieri* Engl., Croat 33775, Panama, MO; *M. punctulata* (Schott) Engl., Croat 33510, Panama, MO; *M. siltepecana* Matuda, Croat 43049, Nicaragua, MO; Croat 42930, Nicaragua, MO.

Taxonomic notes

Monstera is one of the nine trichosclereid-bearing genera of the tribe Monstereae. The species examined demonstrate few structurally distinguishing differences, except for the stem. Unique for this genus the stem central cylinder is delimited by brachysclereids with ramified pits and associated prismatic crystals, an absence of raphides, and numerous large, coarse druses. Unique for the subfamily the petiole develops type 4 air cavities. However, considered together, most leaf and petiole characters of this genus embed it well within its subfamily neighbours.

Species reported on in the literature

(Species called *Tornelia* or *Philodendron pertusum*, considered here as *Monstera*.)

Al-Rais *et al*. (1971): *M. deliciosa*: crystals. • Areschoug (1879): '*Philodendron pertusum*': leaf. • Arzt (1937): *M. deliciosa*: leaf cuticle. • Becker and Ziegenspeck (1931): *M. deliciosa*: raphides. • Blanc (1977a): *M. pertusa*: branch buds. • Bloch (1926): *M. deliciosa*: root, wounding. • Bloch (1944): *M. deliciosa*: meristems. • Bloch (1946): *M. deliciosa*: air root trichosclereids. • Chauveaud (1897): *M. deliciosa*: root, phloem. • Chrysler (1904): *M. deliciosa*: stem central cylinder. • Cody and Horner (1983): crystals • Costantin (1885): *M. repens, T. fragrans*: root. • Costerus (1875): *T. fragrans*: aerial root. • Dalitzsch (1886): *M. deliciosa* [as *T. fragrans*]: leaf. • DeCapite and Milletti (1968): *M. deliciosa*: aerial root. • De la Rue (1869): '*P. pertusum*': crystals. • El-Gazzar *et al.* (1977): *M. deliciosa*: stomata. • Elisei (1941): *M. deliciosa*: casparian strips. • Engler (1908): several spp.: root, stem, leaf. • Ertl (1932): *M. acuminata, M. deliciosa*: leaf venation. • Eschrich (1983): *M. deliciosa*: root phloem. • Franceschi and Horner (1980): *M. deliciosa*: crystals. • Francini (1946): *M. deliciosa*: root cap. • French (1987c): *M. acuminata, M. adansonii*: root. • French and Tomlinson (1981d): several spp.: stem vasculature. • Freundlich (1908): *M. pertusa*: leaf vasculature. • Genua and Hillson (1985): *M. deliciosa*: leaf crystals. • Gertz (1916b): *M. deliciosa*: tyloses. • Giovannozzi (1911): *M. deliciosa*: periderm. • Greger (1920): *M. deliciosa*: root. • Gulliver (1865): '*P. pertusum*', *M. deliciosa*: raphides. • Guttenberg (1968): several spp.: root. • Hinchee (1981): *M. deliciosa*: root. • Hinchee (1983): *M. deliciosa*: sclereids and crystals. • Hotta (1971): *M. deliciosa, M.* sp.: leaf. • Kaplan (1984): several spp.: leaf, petiole. • Klein (1923): *M. deliciosa*: root, tyloses. • Kovacs and Rakovan (1975): *M. deliciosa*: root, crystals. • Kroemer (1903): *M. deliciosa*: root. • Langlet (1927): *M. deliciosa*: root meristem. • Leavitt (1904): *M. deliciosa*: root. • Lee and Richards (1991): leaf. • Lierau (1888): several spp.: root. • Lindorf (1980): *M. henri-pittieri*: leaf. • Linsbauer (1924): leaf. • Linsbauer (1930): *M. fragrans, M.* sp. (and as *Tornelia*): epidermis. • Mangin (1882): several spp.: root. • Melville (1971): *M. pertusa*: leaf venation. • Melville and Wrigley (1969): *M. deliciosa, M. pertusa*: leaf. • Meyer (1925): *M. deliciosa, M. ovata*: root. • Mollenhauer and Larson (1966): *M. deliciosa*: crystals. • Mueller and Beckman (1979): *M. deliciosa*: root fibres. • Müller (1919): *M. deliciosa*, '*P. pertusum*': leaf tip. • Neger (1922): '*P. pertusum*': root, lenticel. • Nicolson (1960): seven spp.: stem, leaf, petiole. • Olivier (1880, 1881): several spp.: root. • Pant and Kidwai (1966): *M. deliciosa*: leaf. • Philipp (1923): *M. deliciosa*: abscission, periderm. • Priestley and Woffenden (1922): *M. deliciosa*: root cork. • Rakovan *et al.* (1973): *M. deliciosa*: root crystals. • Rao (1977): *M. deliciosa*: leaf sclereids. • Rao and Bhattacharya (1978): leaf, sclereids. • Reinhardt (1885): *M. lennea*: root. • Richter (1901): *M. acuminata*: root. • Roth (1980, 1990): leaf. • Rywosch (1909): *M. deliciosa*: root. • Sakai *et al.* (1984): crystals. • Scherer (1904): *M. deliciosa, M. dilacerata*: vasculature. • Schlotmann (1933): *M. deliciosa*: trichosclereids. • Schoute (1902): meristem. • Schwarz (1878): '*P. pertusum*': leaf. • Schwendener (1882): root, rhizome. • Serobjan (1974): '*P. pertusum*': aerial root. • Shaw (1992): *M. deliciosa*: stomata. • Shishkoff (1987): root hypodermis. • Sinnott and Bloch (1946): *M. deliciosa*: aerial root. • Solereder (1919): several spp.: leaf. • Solereder and Meyer (1928): several spp.: all organs. • Spratt (1920): *M. deliciosa*: root. • Szujko-Lacza *et al.* (1974): *M. deliciosa*: aerial root. • Tieghem (1866): several spp.: all organs. • Tieghem (1870–1): several spp.: root, secretory tissue. • Tieghem (1872): *M. surinamensis*: secretory tissue. • Tieghem (1891): stem. • Tieghem and Douliot (1888): *M. repens, M. (T.) fragrans*: root. • Trécul (1854): *M. adansonii*: leaf. • Trécul (1866): vasculature. • Troll and Meyer (1955): *M. deliciosa*: leaf ontogeny. • Ullisch (1930): *M. deliciosa*: leaf. • Voronkina (1975): *M. deliciosa*: root apex. • Webber (1960): *M. acuminata, M. deliciosa, M. friedrichsthalii*: epidermis, stomata. • Weisse (1897): *M. deliciosa, M. dilacerata, M. pertusa*: root, lenticels. • Went (1895): *M. (T.) dilacerata*: root. • Wiesner (1875): '*P. pertusum*': leaf. • Wittlin (1896): '*P. pertusum*': crystal formation.

Additional literature

Barabé and Chrétien 1985b; Blanc 1978, 1980, 1986; Domin 1911; French 1985a, 1986b, 1987a; Gertz 1916a; Grayum 1990; Haydon and Shaw 1991; Hegnauer 1986; Hofmann 1930; King 1894; Lesica and Antibus 1990; Lopez-Portillo *et al.* 2000; Madison 1977; Madison and Tiffney 1976; Meusel 1951; Mühldorf 1925; Preda 1910; Ray 1988, 1990, 1992a; Sharma 1967; Shirley and Lambert 1918; Smolenski *et al.* 1972; Strong and Ray 1975; Thanikaimoni 1969; Tillich 1985; Williams *et al.* 1981.

Alloschemone Schott

This monotypic genus from Brazil is considered by some to be the New World species of the otherwise Asian genus *Scindapsus*. It occurs in primary, humid to wet forests. The plants are hemi-epiphytic appressed-climbers on understorey trees (Croat 1988). Leaves show heteroblastic development as the plants pass through juvenile to adult form. Blades are ovate-pinnatifid. Chromosomes: 2n = 84. One specimen was available for study.

Leaf surface

Cuticle smooth. **Epidermis**: cells undulate adaxially (1.5–2:1 l/w) (Fig. 28C); cells sinuous at surface but polygonal with straight anticlinal walls at cell interior abaxially (1–3:1 l/w) (Fig. 28D). **Stomata** abaxial and brachyparacytic

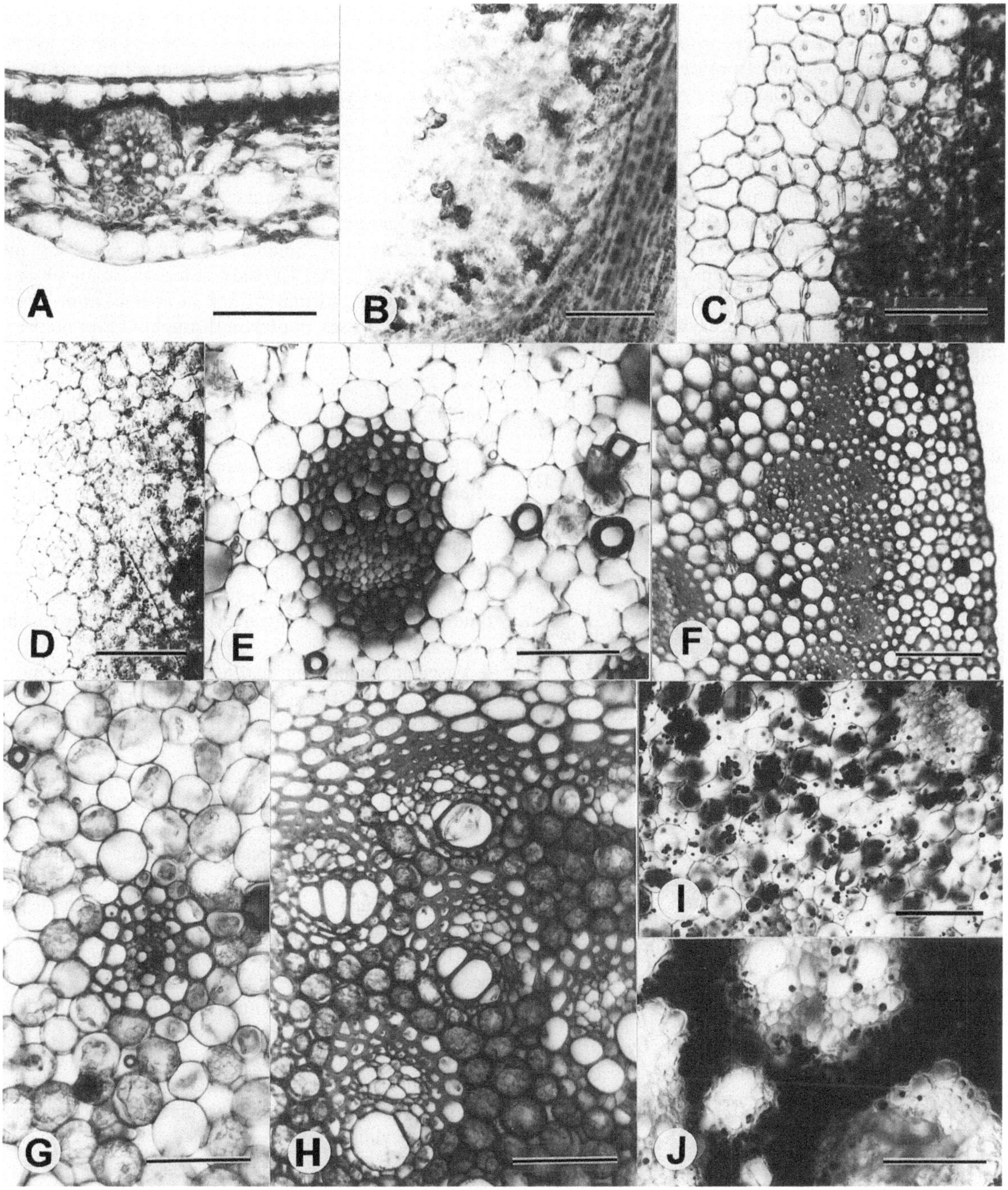

Fig. 28. *Alloschemone occidentalis* hand sections of leaf, petiole and stem. (A) Lamina TS showing secondary vascular bundle; (B) Lamina PS showing idioblastic tannin cells in mesophyll; (C) Lamina PS showing adaxial epidermis; (D) Lamina PS showing abaxial epidermis; (E) Petiole TS showing vascular bundle and three trichosclereids at right; (F) Petiole TS showing outer row of vascular bundles with confluent fibre sheaths; (G) Petiole TS showing cortical vascular bundle; (H) Stem TS, central cylinder vasculature and ground tissue; (I,J) Stem TS, iodine starch test: (I) cortical region showing irregular granule deposition pattern; (J) central cylinder showing starch deposition in all ground tissue. Scale bars: A,C,E,G,H,J = 150 μm; B,D,F,I = 200 μm.

to brachypara-hexacytic. Subsidiary cells narrow and rounded. **Venation**: pinnate with broadly-acute secondaries. Tertiaries narrowly acute from secondaries, merging at straightened marginal collectors.

Leaf TS

Structure dorsiventral. **Size** 65 × 27 mm (*l*/*w*). **Midrib** shallowly concave adaxially and broadly convex-rounded abaxially. **Cuticle** thin. **Epidermis**: cells large, square adaxially, somewhat thinner abaxially. **Stomata** level with surface. **Mesophyll**: palisade cells short, compact (2–4:1 (*l*/*w*). Spongy layers 6–8, type 1 near abaxial epidermis, type 2 in centre of tissue. **Air cavities**: up to 60% of spongy layer. Substomatal cavities shallow. Small cavities at palisade/spongy boundary and larger cavities regularly spaced within spongy layer. **Collenchyma** weakly developed abaxially at midrib; thin wall layering on epidermal and subepidermal walls.

Vascular bundles: 1–2 bundles in midrib (type II). Xylem: 1 small protoxylem cell, 2–5 medium-sized metaxylem cells. Phloem a small irregular strand. **Sclerenchyma**: fibres ensheathing larger vascular bundles, without detectable isotropic wall layering (Fig. 28A). Trichosclereids in small clusters in air cavities just beneath palisade layer. Larger diameter trichosclereids (up to 40–50 μm) in larger air cavities lower in spongy layer. Sclereids 1–5 per cluster, appearing in parallel groups in paradermal section. **Tannin** cells scattered, with dark contents, very common in mesophyll (Fig. 28B). **Crystals**: raphide cells large and thin-walled, often elongate, filled with multiple overlapping, often obliquely tiered crystals. Druses in small cells in palisade layer, very common.

Petiole TS and LS

Outline: terete distally, deeply grooved proximally. **Diameter** 8–10 mm. **Cuticle** medium-thick, rough or striate-ridged. **Epidermis**: cells of medium height, rounded internally and externally. **Chlorenchyma**: chloroplasts sparse in outer ground tissue. **Ground tissue** divided into cortex and central cylinder. Cortex of 8–10 parenchyma layers. Central cylinder delimited sharply by nearly continuous band of fibres. **Air cavities** numerous, cell-sized (Fig. 28G) and axially contiguous cavities throughout ground tissue. **Vascular bundles** scattered and numerous in central cylinder (type I) (Fig. 28E); xylem with 6–8 metaxylem elements forming broad line with semicircular or elliptic phloem strand. Cortical bundles scattered and frequent, circular, fibre-ensheathed; xylem: 1–3 narrow metaxylem cells; phloem: small ± circular strand (Fig. 28G).

Sclerenchyma: fibres ensheathing all vascular bundles and forming more or less continuous circle delimiting central cylinder boundary (Fig. 28F). Boundary sclerenchyma heavily ensheathing a circle of very small vascular bundles. Trichosclereids numerous in all ground tissue, divided roughly among large and small, varying in diameter by a factor of 6; most unaggregated (Fig. 28E), the smallest cells occasionally aggregated. Sclereids occupying intercellular spaces of ground tissue, with lateral processes or irregular plates ramifying between ground tissue cell boundaries, perpendicular to direction of axis elongation. Smaller cells often ramifying into H-shapes (Fig. 7A). **Crystals**: unusual, small or large aggregations of prismatics, fine or coarse, none resembling typical druses; abundant throughout ground tissue.

Stem TS and LS

Outline terete, flattened on side distal to bud initiation. **Size**: diameter 9–11 mm. **Cuticle** thin and irregular or striate. **Epidermis**: cells square with rounded surfaces internally and externally. **Ground tissue** sharply divided by fibres into cortex and central cylinder. Cortical ground tissue with chloroplasts in about third to tenth layers. Parenchyma cells circular in TS, elongate in LS, loosely packed with numerous cell-sized axially contiguous air spaces. Central cylinder of similar ground tissue.

Vascular bundles numerous and scattered, types I (largest), II and III in central cylinder (Fig. 28H); collateral, occasionally bicollateral or irregularly fused with other bundles through branching and anastomosing. Bundles small and collateral in cortex. Xylem of largest bundles with 2–4 wide metaxylem elements and narrower metaxylem elements. Phloem: a broad elliptic strand. **Sclerenchyma**: fibres ensheathing vascular bundles. Trichosclereids common in cortex, especially outer cortex; large and robust or small and thin-walled; lateral processes common, often forming H-shaped ramified cells or short plates between adjacent parenchyma cells, perpendicular to axis. **Starch**: large circular granules packed in all parenchyma of central cylinder (Fig. 28J) and innermost half of cortical parenchyma, less frequent in outer cortex (Fig. 28I). **Crystals**: raphides in single bundles in elongate cells, rare in central cylinder. Druses, small and irregular, often weakly clustered in cortex, rare in central cylinder.

Root TS

Outline terete. **Size**: diameter 2 mm. **Periderm** over cortex. **Ground tissue**: cells not in radial files, irregular, circular with numerous intercellular corner spaces. **Endodermal** layer forming a sclerified boundary with cortex. **Stele** with *c.* 12 xylem and phloem poles. Central tissue lignified. **Sclerenchyma**: trichosclereids common throughout cortex; smaller diameter cells near centre, larger toward periphery; cells often ramified. **Tannin** cells occasional. **Crystals**: raphides in single bundles uncommon in cortex.

Material examined

Alloschemone occidentalis (Poepp.) Engl. & Krause, Brazil, Sakurgui 1050, cult. M.

Taxonomic notes

There has been a suggestion that the genus be moved to *Scindapsus* but it continues to be recognized. *Alloschemone* is anatomically similar to *Scindapsus* and the argument for its recognition seems to be mostly geographical. *Scindapsus* shows much diagnostic variation which, with comprehensive study, should prove to be taxonomically significant. More material and a more detailed study would be necessary to identify the closest affinities of *Alloschemone*.

Species reported on in the literature

Engler (1908): *A. occidentalis*: leaf. • French and Tomlinson (1981d): *A. occidentalis:* stem vasculature.

Additional literature

Grayum 1990.

Epipremnum Schott

This genus of about 15 species ranges from Burma and south-east Asia, Formosa, Japan, the Philippines to Samoa, the Marshall Islands, and Australia. Plants are hemi-epiphytic appressed-climbing vines, or may occur in terrestrial or epilithic forms. They occur mostly in humid primary forests or in regrowth on trees or rocks. Leaf development is heteroblastic with small leaves in the juvenile form, becoming larger and more robust in the canopy of the host tree. Leaves broadly ovate to shallowly cordate to pinnatisect and rarely perforate. Chromosomes medium-sized: $2n = 50, 56, 60, 70, 74, 84$. Three species were available for study.

Leaf surface

Cuticle striate with coarse wrinkles. **Epidermis**: cells isodiametric with straight anticlinal walls adaxially (1–1.5:1 *l*/*w*), and polygonal abaxially (2–3:1 *l*/*w*). **Stomata** abaxial only or on both surfaces, brachyparacytic with 1–3 pairs of subsidiary cells (Fig. 29D). Subsidiary cells mostly narrow or trapezoidal. **Venation**: midrib prominent. Secondaries basal or suprabasal acrodromous, pinnate distally. Secondaries ending at straightened submarginal collector, merging brochidodromously or eucamptodromously. Tertiaries irregular or parallel with secondaries, connecting orthogonally with collector. Higher orders reticulate.

Leaf TS

Structure dorsiventral. **Size**: 56 × 42 mm, 123 × 34 mm (*l*/*w*). **Midrib** prominent, flat or slightly concave adaxially and rounded abaxially. **Cuticle** thick adaxially, thinner abaxially and very thin in some specimens. **Epidermis** composed of square cells adaxially, flat externally and rounded internally, abaxial cells elongated and flat, or both epidermal layers thick, of columnar cells, flat externally and rounded internally. **Stomata**: guard cells level with epidermis. **Mesophyll**: one palisade layer (2–6:1 *l*/*w*), strongly or weakly differentiated from 6–9 spongy layers (Fig. 29C). Spongy cells horizontally elongated or very irregular, forming tall air cavities over stomata; spongy cells appearing tightly fitted against each other among numerous circular substomatal air spaces (types 1, 2). **Air cavities** 50–75% of spongy layer.

Vascular bundles numerous in midrib. Bundles broad and circular with wide, straight xylem/phloem boundary (type I) (Fig. 29A). Metaxylem cells more than 12 in largest bundles. **Sclerenchyma**: fibres ensheathing vascular bundles uniformly in 2–5 layers. Discontinuous fibre band connecting most outer bundles laterally (Fig. 29B). Trichosclereids uncommon throughout leaf, especially midrib. **Crystals**: raphides abundant in overlapping bundles in elongated cells (Fig. 29E), or in wide bundles in shorter cells. Raphide cells tending to run parallel with secondaries and most concentrated at base and distal centre of lamina. Druses common in palisade layer, rare in spongy layer and occasional near vascular bundles. Crystal sand abundant in lamina and midrib.

Petiole TS

Outline crescent-shaped at base. **Size**: width 1.4 cm. **Cuticle** thin or medium-thick and smooth. **Epidermis**: cells small and square. **Stomata** occasional, with shallow substomatal air cavities. **Collenchyma**: phloem caps, deep and well-developed in mature tissue (Fig. 29F); band in outer ground tissue joining phloem caps of outer row of vascular bundles. **Chlorenchyma**: outer few ground tissue layers. **Ground tissue** compact, cells small near periphery, large in centre with few corner air cavities.

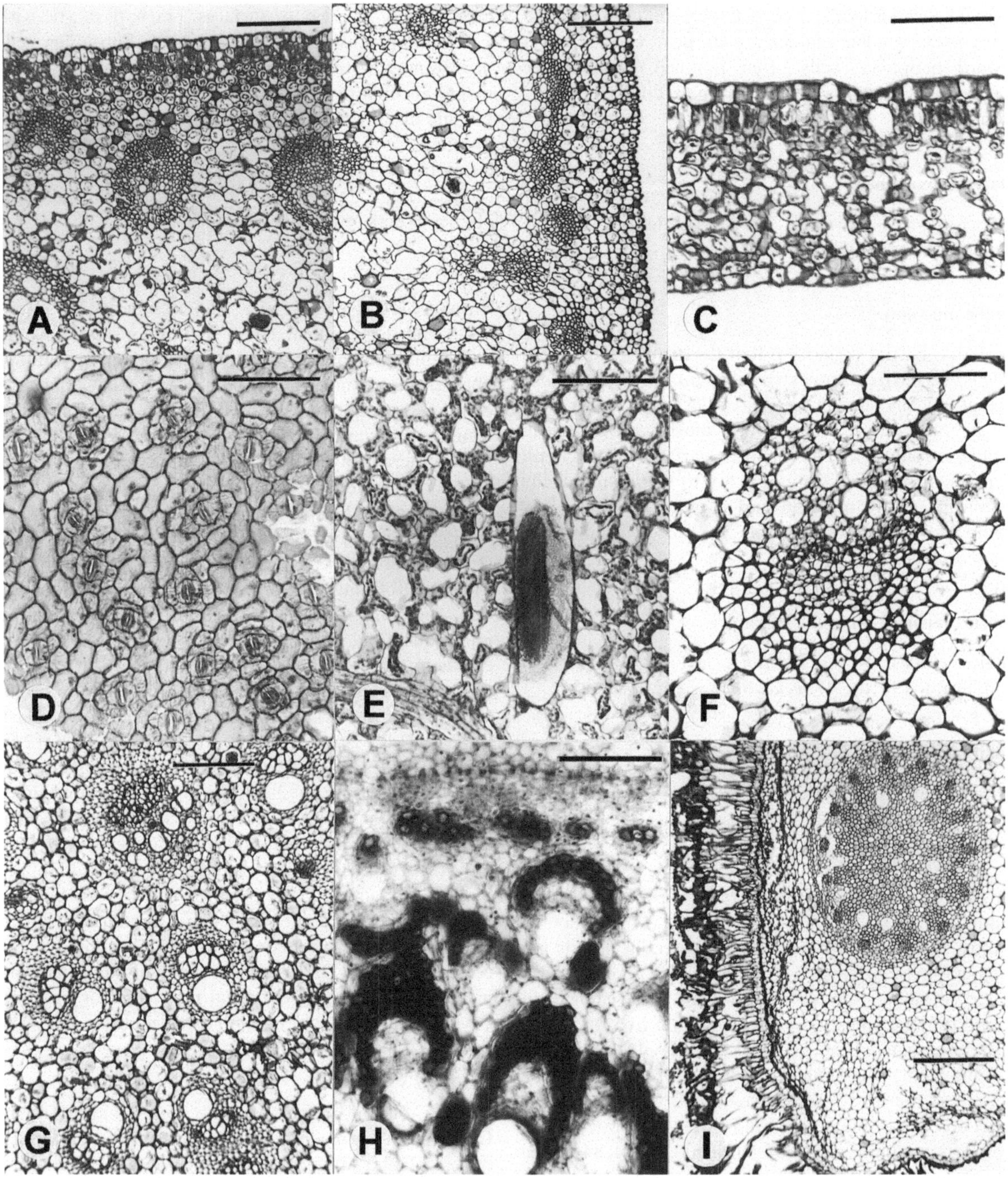

Fig. 29. *Epipremnum aureum* leaf, petiole, stem and root sections. (A) Midrib TS, adaxial surface; (B) Midrib TS, lateral surface with confluent subcortical fibres; (C) Lamina TS; (D) Abaxial epidermis PS, stomata; (E) Lamina PS, raphide cell; (F) Petiole centre TS showing type I vascular bundle; (G) Stem TS centre showing types I, III vascular bundles; (H) Stem centre TS, hand section in TBO/calcium chloride showing fibre caps, strands and endodermis; (I) Root TS showing central cylinder, cortex, and root hairs at adherent margin. Scale bars: A,B,G,I = 300 μm; C–F,H = 150 μm.

Vascular bundles numerous and scattered, type I (Fig. 29F). Xylem: 3–10 wide metaxylem cells opposing broad semicircular or oval phloem strand with large sieve tube elements. **Sclerenchyma**: single trichosclereids common throughout ground tissue. **Crystals**: raphides in single, often oblique bundles in large, thin-walled cells, often axially oriented. Druses occasional, expecially near vascular bundles.

Stem TS

Outline terete with 3–5 ridges; stem scrambling with long internodes. **Epidermis**: outer and subepidermal layers with thick, apparently lignified walls in older stems; cells polygonal or deltoid, elongated longitudinally. **Periderm** beginning as subhypodermal periclinal divisions. **Chlorenchyma**: outer 10–12 cortical layers. **Ground tissue**: cells smaller in chlorenchyma zone but composed of large cells in both inner cortex and central cylinder. **Central cylinder** bounded by endodermis with simple casparian strips showing some lignification on anticlinal walls (Fig. 29H).

Vascular bundles of cortex types I, II or III; bundles of central cylinder large. Xylem of variable extent: may have simple protoxylem lacuna, 1 wide metaxylem cell or up to 12 medium-sized metaxylem elements (Fig. 29G). Phloem tissue variable in shape and extent but with organized grid pattern of sieve tube elements and companion cells. Sieve tube elements enlarged in central cylinder, much smaller in phloem of cortical bundles. **Sclerenchyma**: fibres as bundle caps over phloem, 6–8 cells deep (Fig. 29H); fibre strands present distal to phloem only, or wrapping deeply around in contact with metaxylem cells. Trichosclereids common in cortex as elongate idioblastic fibres, occasionally bearing short branch stubs. **Crystals**: raphides uncommon throughout ground tissue.

Root TS

Outline circular. **Size**: diameter 5–8 mm. **Epidermis**: cells medium-sized and ± square. **Root hairs** very common, abundant on those root surfaces appressed against substrate (Fig. 29I). **Exodermis**: epidermis dying back against maturing exodermis surface; outer walls accumulating lignified thickening. **Ground tissue**: cells uniform and compact in young tissue. Older cortical tissue developing radially elongate schizogenous cavities isolating remaining ground tissue into alternating radial strips. **Endodermis**: single layer with wide casparian bands on radial walls. **Pericycle**: one layer of large irregularly-shaped cells. **Vascular tissue** of 14–25 protoxylem poles alternating with equal number of circular or irregular phloem strands. Stele centre mostly ground tissue with irregular groups of brachysclereids or fibres. Secretory duct-like openings in outer stele, probably immature metaxylem cells. **Crystals**: raphides occasional in ground tissue near root periphery.

Material examined

Epipremnum aureum (Linden & André) Bunting, cult. M; cult. SIUC; *E. pinnatum* Engl., cult. MO; *E.* sp. cult. MO (75c1562).

Species reported on in the literature

Arber (1922b): *E. mirabile*: leaf. • Baranetzky (1897): *E. mirabile*: leaf. • DiBenedetto and Garcia (1992): leaf. • Engler (1911): *E. pinnatum, E.* sp.: stem, leaf. • French (1987c): *E. falicifolium, E. pinnatum*: root [sclerotic hypodermis absent]. • French and Tomlinson (1980): *E. pinnatum*: stem vasculature. • French and Tomlinson (1981d): *E. falcifolium*: stem. • Funke (1931): *E. pinnatum*: leaf. • Guttenberg (1940): *E. pinnatum*: root. • Hotta (1971): *E.* sp.: leaf; *E. pinnatum*: stem. • Lierau (1888): *E. mirabile*: root. • Müller (1919): *E. mirabile*: leaf tip. • Nicolson (1960): *E. pinnatum*: leaf, petiole, stem. • Solereder and Meyer (1928): *E. pinnatum, E.* sp.: root, other organs. • Webber (1960): *E. pinnatum*: epidermis, stomata. • Went (1895): *E. mirabile*: root.

Additional literature

Blanc 1977a; Cevallos-Ferriz and Stockey 1988; French 1985a, 1986c, 1987a; Grayum 1990; Huang 1972; Madison and Tiffney 1976; Mai and Gregor 1982; Meusel 1951; Smolenski *et al.* 1972; Thanikaimoni 1969; Williams *et al.* 1981.

Amydrium Schott

This genus of 4–8 species is found in Malaysia. The plants are mostly hemi-epiphytic appressed low climbers on trees, or terrestrial on steep banks. They occur mostly in primary forests or late secondary regrowth. Leaves are long-petiolate, ovate-cordate to trilobed or pinnatisect, and sometimes perforate and show heteroblastic development with alternating juvenile-adult growth phases. Chromosomes: $2n = 60$. One species was examined.

Leaf surface

Cuticle smooth. **Epidermis**: cells polygonal with straight anticlinal walls on both surfaces, 1–3:1 l/w. **Stomata** in both surfaces brachyparacytic to brachypara-hexacytic

(Fig. 30C). Subsidiary cells narrow. Second pair of subsidiary cells parallel to first pair and always longer. **Venation**: basally acrodromous with strong midrib and curved ascending secondaries merging at straightened collector vein. Higher orders reticulate.

Leaf TS

Structure dorsiventral. **Size**: lamina length 22 cm. **Midrib** flat adaxially and broadly rounded abaxially (Fig. 30A). **Cuticle** thin, or not seen on all surfaces. **Epidermis**: cells flat externally on both surfaces. **Stomata** level to slightly raised with obvious subsidiary cells. **Mesophyll** moderately differentiated (Fig. 30B). Palisade cells in 1–2 poorly defined layers. Spongy tissue forming 5–6 layers of irregularly armed cells among large air cavities (type 2). **Air cavities** irregular, occupying about 60% of mesophyll. **Collenchyma**: abaxial band in midrib, slightly thickened cell walls in a narrow hypodermal zone, not well defined.

Vascular bundles: symmetrical pattern in midrib with one large median bundle, one large lateral bundle on each side where lamina joins midrib, two small bundles medial, adaxial, and two sets of small bundles lateral abaxially (Fig. 30A). Bundles type II. Large bundles with two rows of metaxylem with *c*. 8 cells. Phloem in an irregular strand; sieve cells 80–400 μm long, with end walls horizontal to moderately oblique. **Sclerenchyma**: at larger bundles, fibres forming two bundle caps, one thicker at phloem pole, the other wider at xylem pole. **Tannin cells** occasional in mid to lower mesophyll, packed with dark granular contents. **Crystals**: fine raphides in multiple overlapping bundles in elongate cells (Fig. 30D,E), most common at palisade-spongy boundary. Druses common in mid to upper mesophyll in circular idioblastic cells (Fig. 9G). Crystal sand abundant, especially on abaxial side of cells.

Petiole TS

Outline: proximal end broadly rounded abaxially, shallowly concave adaxially. Distal outline becoming laterally compressed and half-sized but rounded abaxially and shallowly-grooved adaxially. **Size**: *c*. 8 mm wide. **Cuticle** thin and smooth. **Epidermis**: cells flat on surface and rounded internally, bearing occasional stomata. **Collenchyma**: vaguely defined band, in 2–3 hypodermal layers beneath all surfaces (Fig. 30F). **Ground tissue** compact, with no chloroplasts; divided into cortex and central zone. **Vascular bundles** numerous and scattered, broad, type I or II. Xylem with 2 rows of metaxylem, up to 8–10 cells. Phloem an elliptical strand with collenchyma phloem cap (Fig. 30G).

Cellular inclusions: central zone ground tissue with large round starch 'crystalloids', *c*. 20 μm diameter, 3–12 per cell; each containing numerous small spheroids 1–3 μm diameter, generally within crystalloid membrane; each spheroid forming visible cross pattern in polarized light. **Crystals**: raphides uncommon in ground tissue, in multiple overlapping bundles filling elongated cells, not near veins. Druses occasional in adaxial mesophyll. Crystal sand abundant in all mesophyll and bundle sheath cells; sand abundant outside crystalloid cell zone but also present within.

Material examined

Amydrium medium (Zoll. & Morr.) Nicolson, Bogner 433, Thailand, cult. M.

Taxonomic notes

Hotta (1971) noted that Rosendahl (1911) found trichosclereids. Seubert (1997a) also reported trichosclereids in leaf sheath and petiole of *A. humile*. Hotta found none in five species and none was found in this study.

Species reported on in the literature

Blanc (1978): development. • French (1987c): *A. medium*: root. • French and Tomlinson (1981d): *A. medium*: stem vasculature. • Hotta (1971): *A. medium*: leaf venation. • Webber (1960): *Epipremnopsis huegeliana*: epidermis, stomata.

Additional literature

Blanc 1977a; Eyde *et al*. 1967; French 1985a, 1986b, 1987a; Grayum 1990; Hay 1986; Hotta 1965; Madison and Tiffney 1976; Thanikaimoni 1969; Williams *et al*. 1981.

Heteropsis Kunth

This genus of 13 species is distributed in the American tropics, mostly in South America. They are hemiepiphytic, occasionally epilithic, scandent plants on rocks or trees. They usually occupy the lower canopy or tree trunks in primary forests. Plants may produce slender roots hanging toward the ground. The shoot is somewhat branched, monopodial and shrubby. Petioles are usually short, bearing oblong to lanceolate blades. Chromosomes: $2n = 26–28$. Two species were available for study.

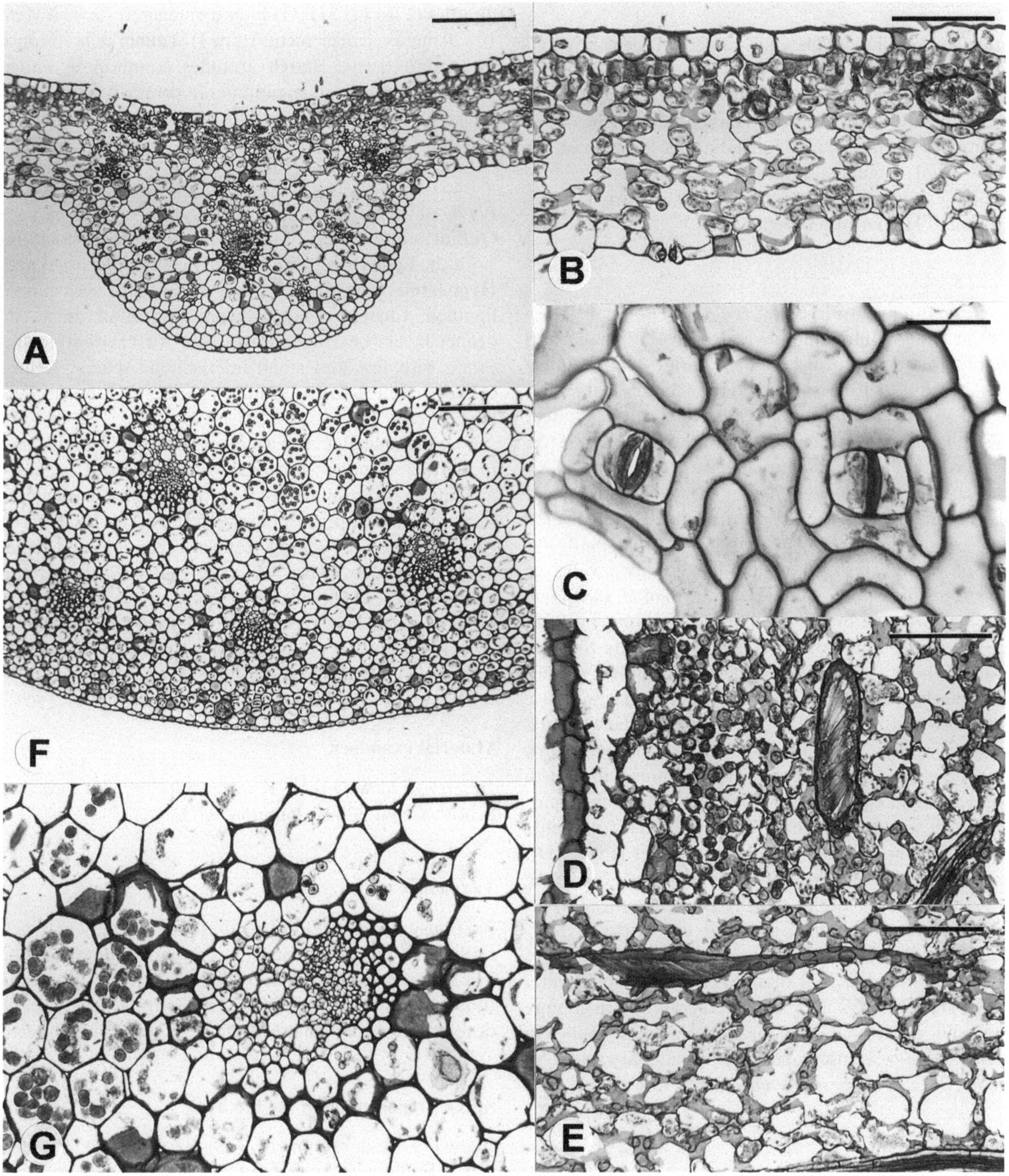

Fig. 30. *Amydrium medium* leaf and petiole sections. (A) Midrib TS; (B) Lamina TS with raphide cell TS; (C) Leaf epidermis PS with brachypara-tetracytic stomata; (D) Leaf mesophyll PS with raphide cell having oblique overlapping crystal bundles; (E) Leaf mesophyll PS, raphide tube with one oblique raphide bundle; (F) Proximal lateral petiole TS; (G) Distal petiole TS central tissue with type I vascular bundle and collenchyma phloem cap. Scale bars: A,F = 300 μm; B,D,E,G = 150 μm; C = 50 μm.

Leaf surface

Cuticle smooth. **Epidermis**: cells polygonal with anticlinal walls straight or slightly undulate (1–1.5:1 *l/w* in *H. spruceana*, 1–4:1 *l/w* in *H. oblongifolia*) on both surfaces. **Stomata**: abaxial and brachyparacytic (Fig. 31E) to brachypara-hexacytic. Subsidiary cells broad or narrow and angular. **Venation**: pinnate with acute-ascending secondaries merging with submarginal collector vein. Higher orders reticulate.

Leaf TS

Structure dorsiventral. **Size**: *H. oblongifolia* 40–47 × 5–7 mm (*l/w*). **Midrib** shallowly concave adaxially and broadly rounded-convex abaxially (Fig. 31C). **Cuticle** thick or thin. **Epidermis**: adaxial external walls flat, abaxial walls flat or rounded; internal walls rounded. Cells tall on both surfaces. **Stomata** level with surface. **Mesophyll**: 2–3 layers of short palisade cells (Fig. 31A). (One layer in *H. oblongifolia*, to 30% of lamina, Fig. 31B). Spongy cells: 2–6 layers of rounded short-armed type 1 cells, type 2 in *H. spruceana*. **Air cavities** small, 20–50% of spongy layer. **Vascular bundles**: two rows of symmetrically disposed bundles in midrib of *H. spruceana* (Fig. 31C), or 1–2 bundles in midrib of *H. oblongifolia*; type I. Xylem: numerous metaxylem elements forming broad boundary with broad semicircular phloem strand.

Sclerenchyma: fibres ensheathing all vascular bundles. Fibre wall with 2–4 clearly visible secondary layers. **Tannin** cells: same shape as ground tissue cells, with dark amorphous contents, common or uncommon in midrib and mesophyll. **Crystals**: raphides very common in mid-mesophyll, usually in tubular cells with dark contents and continuous overlapping crystal bundles (Fig. 31D), the largest 500–640 μm long. Raphide tubes occurring between vascular bundles, mostly parallel with midvein or secondaries. Raphides in small cells rare in mid-mesophyll. Druses common or uncommon throughout mesophyll; abundant and small in circular subepidermal cells in *H. oblongifolia*. Crystal sand common in spongy layer and occasional in palisade cells; not found in *H. oblongifolia*. Druses and crystal sand abundant in mesophyll cells of *H. spruceana*.

Petiole TS

Outline: adaxial side shallowly concave; abaxial side broadly convex-curved; winged. **Size**: diameter *c*. 1 mm, length 2 mm. **Ground tissue**: cells uniform-sized, chloroplasts common. **Vasculature**: *H. oblongifolia*: 1–2 vascular bundles, ensheathed by unlignified cells. Xylem of 1–3 narrow elements opposing a broad oval phloem strand. In *H. spruceana* 24 fibre-ensheathed vascular bundles (Figs 6I, 31F,G) in two staggered rows. Xylem: 10–20 metaxylem elements (type I). **Tannin** cells common in ground tissue. **Starch** granules common in midrib ground tissue. **Crystals**: raphides in rounded cells (in TS); cells elongated parallel to leaf traces, and subepidermal abaxially.

Stem TS

Outline terete. **Size**: *c*. 1 mm diameter. **Cuticle** thick and smooth. **Epidermis**: rounded cells of medium thickness. **Hypodermis**: one layer of thick-walled cells, occasionally lignified. **Cortex**: *c*. three layers of ground tissue, the deeper layers bearing chloroplasts. **Air cavities**: ground tissue with few and small intercellular spaces. **Central cylinder**: well defined by fibre cylinder 3–5 layers deep (Fig. 31H).

Vascular bundles: two to several cortical bundles and numerous bundles scattered within central cylinder. Xylem: 1–2 large elements or up to four smaller elements, bordering an elliptical phloem strand (type II). **Sclerenchyma**: fibres bounding central cylinder and small irregular phloem caps; two to several well-defined wall layers for each cell. **Starch** granules rare, occasional in phloem and ground tissue of *H. oblongifolia*; abundant in ground tissue of *H. spruceana*. **Crystals**: raphides in multiple bundles in raphide tubes in cortex (Fig. 31I).

Material examined

Heteropsis oblongifolia Kunth, cult. M; *H. spruceana* Schott, Croat 33808, Panama, MO.

Taxonomic notes

No unusual characters or combinations suggest sufficient reason for placement of *Heteropsis* in a separate tribe. The genus is distinguished by the character of the raphide tubes. Banded collenchyma, normally at least weakly developed in other genera of this tribe, is not encountered.

Species reported on in the literature

Blanc (1978): development. • Dalitzsch (1886): *H. ovata*: leaf. • Engler (1905): leaf. • French (1987c): several spp.: root. • French and Tomlinson (1981c): *H. jenmanii*: stem vasculature. • Potiguara and Nascimento (1994): *H. jenmanii*: all organs. • Solereder (1919): *H. ovata*: leaf. • Tieghem (1866): *H. ovata*: all organs. • Tieghem (1870–1): *H. ovata*: root, secretory tissue. • Trécul (1866): vasculature.

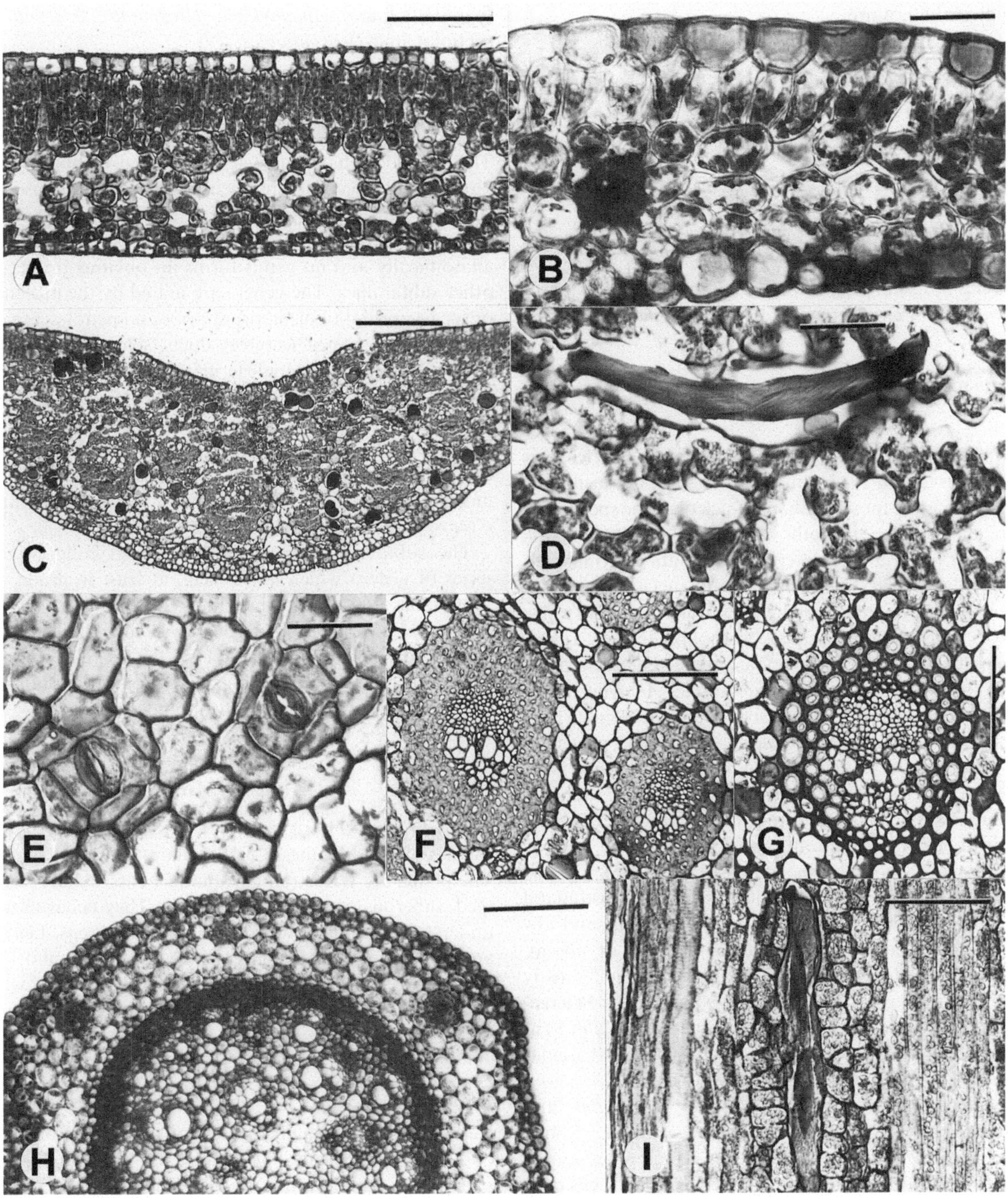

Fig. 31. *Heteropsis* leaf, petiole and stem sections. (A,C–G,I) *H. spruceana*; (B,H) *H. oblongifolia*. (A,B) Lamina TS; (C) Midrib TS; (D) Leaf mesophyll PS with raphide cell bearing two overlapping crystal bundles; (E) Leaf epidermis PS showing stomata; (F) Distal petiole TS showing type I fibre-ensheathed vascular bundles; (G) Petiole TS, type I vascular bundle; (H) Stem TS showing well-delimited central cylinder. (I) Stem LS showing elongated raphide cell with multiple bundles. Scale bars: A,F–I = 150 μm; B,D, E = 50 μm; C = 300 μm.

Additional literature

Domin 1911; Erdtman 1952; French 1985a, 1986c, 1987a; Grayum 1990; Meusel 1951; Ray 1988; Thanikaimoni 1969; Williams *et al.* 1981.

SUBFAMILY LASIOIDEAE ENGL.

This subfamily is not further subdivided above the generic level. It includes *Cyrtosperma, Lasimorpha, Podolasia, Lasia, Anaphyllum, Urospatha, Anaphyllopsis, Pycnospatha, Dracontium* and *Dracontioides*.

Anatomical summary

Leaf

Structure strongly or weakly dorsiventral. **Trichomes** absent, except in *Lasia*, but petiolar spines occasionally present. **Cuticle** mostly thin, but thick on some specimens; cuticular surface smooth or striate. In *Dracontium*, cuticular material may interpenetrate subepidermal cells of abaxial midrib. **Epidermal cells**: in surface view polygonal, with straight, undulate or sinuous anticlinal walls. In TS cells flattened, square, or papillate (in *Lasia*). **Stomata** mostly brachyparacytic but may be brachypara-tetra- or -hexa-cytic in *Cyrtosperma, Anaphyllum* or *Dracontioides*; mostly or exclusively abaxial. Stomata level with surface in all genera. Wide subsidiary cells present in all genera but cells may also be narrow in some genera; cell outlines irregular, trapezoidal, angular or rounded. **Venation** pinnate or acrodromous. Secondaries brochidodromous or eucamptodromous near margin.

Mesophyll: spongy tissue of types 1, 1a, 2, or 3. **Ground tissue** in midrib and petiole mostly type 4. Substomatal cavities high with types 2 and 3 ground tissue, and low with types 1 and 1a. **Collenchyma** banded or absent. **Vascular bundles** in midrib TS mostly few and mostly symmetrical; in leaf and petiole types I and II. **Sclerenchyma**: fibres may be absent but most commonly present as phloem and xylem caps, or ensheathing vascular bundles. Sclereids absent.

Secretory tissue absent, except in *Urospatha*, and **laticifers** absent. **Tannin** cells mostly present. **Starch** observed in most genera. **Crystals**: raphides present as single or multiple bundles in rounded, elliptical cells or in short or elongate tubular cells. Round-ended biforine-like cells observed in *Cyrtosperma*, noted occasionally in *Lasimorpha* and rarely in type 4 partitions in petiole of *Dracontium*. Druse cells, when present, of various sizes and druses fine-textured. Styloids observed only in *Pycnospatha*. Crystal sand present in *Cyrtosperma* (petiole), *Lasia, Anaphyllum, Urospatha, Dracontium* (petiole), and *Dracontioides*.

Taxonomic notes

The Lasioideae have an ancient tropical distribution (see biogeography section) and their possession of corms and rhizomes suggests long isolation and adaptation to dry season dormancy. The genera form a distinctive group anatomically and no genus forms an obvious bridge to other subfamilies. The genera are united by the presence of type 4 ground tissue in the midribs and petioles as well as by the absence of secretory ducts, sclereids, and laticifers. Within the Lasioideae many of the genera show distinctive apomorphies, for example *Anaphyllum* and *Dracontioides* share specialized stomata, *Pycnospatha* has distinctive epidermal cells and crystals, *Anaphyllum* has an unusual mesophyll, and *Lasia* possesses an unusual epidermis and trichomes. *Dracontium* and *Cyrtosperma* both have specialized raphides.

The subfamily belongs to the banded collenchyma series of genera which allies them, at least in grade, to subfamilies Pothoideae and Philodendroideae.

The presence of types I and II vascular bundles and ensheathing or capping fibres places the Lasioideae close to the proto-Araceae but their stem modifications (corms and rhizomes) are distinctively apomorphic.

Cyrtosperma Griff.

This genus of 11–12 species is distributed widely in Malaysia and Melanesia to Solomon and Fiji Islands. The plants are terrestrial or aquatic, coarse, short-stemmed, tuberous or rhizomatous herbs. They occur in wet understoreys of open woods or in marshy areas. Leaves are long-petiolate with spiny petioles and lower midribs. Blades are deeply cordate to irregularly hastate with elongate basal lobes. Chromosomes: $2n=26$. Four specimens (three species) were available for study.

Leaf surface

Cuticle smooth adaxially and smooth to striate abaxially on veins. **Epidermis**: adaxial cells polygonal with straight or undulate to sinuous anticlinal walls (1–4:1 *l/w*), and abaxially undulate (Fig. 32E) to sinuous (2–3:1 *l/w*). **Stomata** mostly abaxial and brachyparacytic (Fig. 32E), and tending toward brachypara-tetracytic. Subsidiary cells broad, rounded or angular. **Venation** acrodromous with basal recurved primaries; strongly tetrahedral primaries in hastate leaves. Secondaries from

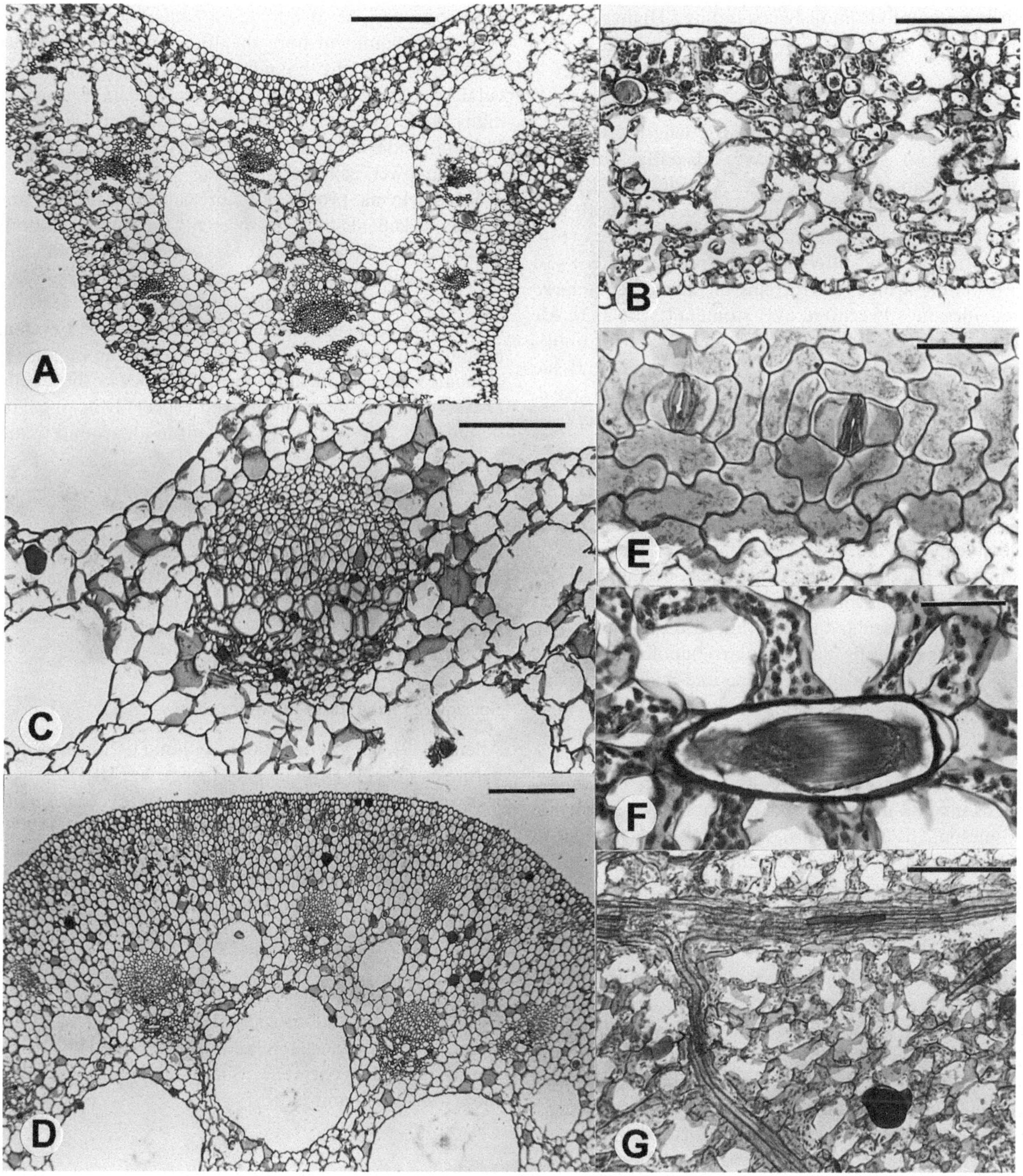

Fig. 32. ***Cyrtosperma johnstonii*** **leaf sections and** ***Lasimorpha senegalensis*** **leaf and petiole sections**. (A,B,E,F) *C. johnstonii*; (C,D,G) *L. senegalensis*; (A) Midrib TS with type 4 air cavities; (B) Lamina TS; (C) Proximal petiole TS with type I vascular bundle; (D) Distal petiole TS with type 4 air cavities; (E) Leaf abaxial epidermis PS with brachyparacytic stomata; (F) Leaf spongy mesophyll PS and raphide cell; (G) Leaf mesophyll PS with type 2 spongy aerenchyma and tannin cell. Scale bars: A,D = 300 μm; B,C,G = 150 μm; E,F = 50 μm.

midrib eucamptodromously ascending. Higher orders reticulate.

Leaf TS

Structure weakly differentiated. **Size**: length 25–30 cm. **Midrib** flat to rounded-concave adaxially, rounded-semicircular abaxially (Fig. 32A). **Cuticle** thin. **Epidermis**: cells mostly square and medium-sized on both surfaces. **Mesophyll**: 1–2 short-celled palisade layers (Fig. 32B), often weakly differentiated. Spongy tissue: 8–10 layers of long-armed cells forming medium-sized to large air cavities, mostly centred over stomata (types 2, 3). **Air cavities**: midrib aerenchyma of 1–8 cavities with multiseriate partitions (type 4) (Fig. 32A). Spongy tissue air cavities well developed with 30–80% air space. **Collenchyma**: subepidermal band of 1–2 layers of thin-walled cells in abaxial midrib.

Vascular bundles: *c.* 8 well-formed, scattered bundles in midrib; bundles broad (type I). Xylem: 10–18 tracheids. Phloem: a broad semicircular strand. **Sclerenchyma**: phloem cap fibres in 5–6 layers; a smaller xylem cap, or continuous bundle sheath. **Tannin** cells irregular, rounded, with dark amorphous contents, randomly dispersed in midrib, in vascular bundles, and abaxial mesophyll; small elongate cells in vascular bundles. **Crystals**: raphides common in spongy tissue and midrib with one wide bundle in small, round-ended cells, similar to biforines but with non-lignified walls and terminal papillae absent (Fig. 32F). Some raphide cells elongate, bearing multiple crystal bundles, common in lamina. Druses small to medium-sized, common or occasional in mid-mesophyll and midrib, randomly distributed in mesophyll. Large druses present in enlarged adaxial epidermal cells.

Petiole TS

Outline: proximally deeply grooved and winged adaxially, and broadly rounded abaxially. Distally, rounded-elliptic and dorsiventrally slightly flattened. **Size**: diameter 10 mm proximally, 5 mm distally. **Cuticle** thin and smooth. **Epidermis**: cells small and isohedral, rounded externally. **Collenchyma**: peripheral band up to eight layers deep. **Chlorenchyma**: outer cortex 3–5-layered, bounded internally by ring of peripheral vascular bundles. **Ground tissue**: cells small at periphery and larger in centre; cells axially elongate. **Air cavities** axially elongate, up to 3 mm long between horizontal or oblique, 1-layered diaphragms; mostly type 4, occasionally type 3. **Vascular bundles** broad (type I) and scattered in centre, interspersed among large air cavities in ground tissue with type 4 multiseriate partitions (Fig. 5D). In distal section, numerous peripheral bundles in outer cortex forming band with confluent fibre sheaths. Xylem: semicircular strands at straight boundary with semicircular, circular, or irregular phloem strands. Xylem formed of many late protoxylem and few metaxylem elements. Phloem: wide diameter sieve elements in central bundles, narrower near periphery.

Sclerenchyma: proximally, fibre caps over phloem and usually xylem, distally not present; or bundles entirely sheathed by fibres in *C. beccarianum*. **Tannin** cells common and randomly oriented; individual cells 200–800 μm long within phloem strands. **Starch** granules common in cells surrounding vascular bundles. **Crystals**: raphides uncommon, in multiple overlapping bundles in elongated cells, or in single broad bundles in short cells. Druses abundant in ground tissue, especially in air cavity diaphragms. Crystal sand abundant in all ground tissue.

Material examined

Cyrtosperma beccarianum A.Hay, Papua New Guinea, cult. M; *C. ferox* N.E.Br., Bogner 1510, Malaysia, cult. M; *C. johnstonii* N.E.Br., Bogner 2341, cult. M; NYBG intro., cult. M.

Species reported on in the literature

Allen (1929): *C. merkusii*: starch. • Blanc (1978): ontogeny. • Engler (1911): *C. giganteum*: stem. • Ertl (1932): *C. johnstonii*: leaf venation. • French (1987c): *C. cuspidispathum, C. johnstonii, C. merkusii*: root. • French (1988): several spp.: laticifers (absent). • French and Tomlinson (1981a): *C. johnstonii*: stem. • Hay (1986): *C. griffithii*: stem. • Webber (1960): *C. johnstonii*: epidermis, stomata.

Additional literature

Cevallos-Ferriz and Stockey 1988; Eyde *et al.* 1967; French 1985b, 1986c, 1987a; Grayum 1990; Hay 1992a; Hotta 1971; Knecht 1983; Simmonds 1976; Thanikaimoni 1969; Williams *et al.* 1981.

Lasimorpha Schott

A monotypic genus native to west tropical Africa. *Lasimorpha senegalensis* Schott is a species of large terrestrial rhizomatous herbs or emergent aquatics found at forest edges and in swamps and ravines of savannas. It can produce underground stolons, many metres long, that propagate vegetatively producing large colonies.

Petioles are very long and spiny and bear sagittate leaves. Chromosomes: 2n = 26. One species was available for study.

Leaf surface

Cuticle smooth adaxially, striate abaxially. **Epidermis**: cell walls straight or undulate. **Stomata** mostly abaxial. Subsidiary cells broad or narrow, rounded and trapezoidal. **Venation**: tetrahedral primaries, two entering basal lobes. Looped secondaries ascending eucamptodromously.

Leaf TS

Structure dorsiventral, well differentiated. **Size**: 30 × 8.8 cm (*l*/*w*). **Midrib**: flat to rounded-concave adaxially, rounded-convex abaxially. **Cuticle** thin. **Epidermis**: cells small, square, rounded externally. **Mesophyll**: two palisade layers. Spongy tissue: long-armed cells forming mostly type 2 aerenchyma (Fig. 32G). **Air cavities**: midrib aerenchyma formed of eight or more large type 4 cavities with multiseriate partitions. Spongy tissue air cavities high below stomata, forming *c*. 70% space. **Collenchyma**: subepidermal band of *c*. 3 layers of thin-walled cells. **Vascular bundles** broad, type I; *c*. 6 bundles symmetrically scattered in midrib. Xylem: 4–6 files of wide metaxylem cells. Phloem: a broad elliptic strand. **Sclerenchyma**: fibres as phloem caps, well-developed in midrib. **Tannin** cells common near vascular bundles (Fig. 32G) and in midrib ground tissue. **Crystals**: raphides in single bundles; cells short or elongate with oblique overlapping bundles, or wide oblique bundles. Occasional biforine-like cells, occasionally lignified with short oblique bundles. Druses occasional in ground tissue or mesophyll.

Petiole TS

Outline: rounded, slightly rounded adaxial keel. **Size**: diameter 11 mm proximally, 5 mm distally. **Cuticle**: proximally medium-thick and smooth; distally thin and striate. **Epidermis**: cells small and square. **Collenchyma**: 1–2-layered subepidermal band, cells thin-walled. **Chlorenchyma**: 3–5 layers in outer ground tissue. **Ground tissue**: cells compact. **Air cavities**: large, scattered type 4 circular cavities (Fig. 32D). **Vascular bundles** numerous and scattered; broad, type I (Fig. 32C). Xylem: 3–7 files of wide metaxylem cells. Phloem: broad semicircular strand. **Sclerenchyma**: fibres forming several-layered phloem caps proximally, absent distally. **Tannin** cells in ground tissue and xylem, frequent proximally, occasional distally. **Starch** granules occasional in bundle sheath parenchyma. **Crystals**: raphide cells simple, few in outer cortex. Druses occasional, scattered in ground tissue, some cells projecting into air cavities.

Material examined

Lasimorpha senegalensis Schott, Bogner 691, W. Africa, cult. M.

Taxonomic notes

This genus is nearly identical anatomically with *Cyrtosperma*.

Species reported on in the literature

French (1988): *L. senegalensis*: laticifers [absent].

Additional literature

Hay 1986, 1992a.

Podolasia N.E.Br.

This monotypic genus is found in the Malay archipelago. Plants are aquatic herbs occurring in swampy places along streams. They have a short erect caudex or a rhizome. The leaves have long spiny petioles and a sagittate lamina with elongate basal lobes. Chromosomes: 2n = 26. One specimen was available for study.

Leaf surface

Cuticle smooth. **Epidermis**: adaxial cells polygonal (Fig. 33E) with straight or slightly undulate anticlinal walls (1–3:1 *l*/*w*); abaxial cell walls undulate-sinuous with 3–6 or more shallow lobes per cell (1–4:1 *l*/*w*) (Fig. 2G). **Stomata** mostly abaxial and simply brachyparacytic. Occasional unipolar stomata present (Fig. 2G). Subsidiary cells wide or narrow with rounded or mostly angular corners. **Venation**: midrib prominent with pedate basal lobe primaries and basal acrodromous curved-ascending secondaries. Basal secondaries forming mid-laminar collector, straight with abrupt joints and merging with tertiaries. Midrib tertiaries irregular-reticulate. Higher orders reticulate and widely spaced. Outer laminar veins merging brochidodromously at margin.

Leaf TS

Structure poorly differentiated. **Size**: at midrib 80 × 50 mm (*l*/*w*); basal lobe length 60 mm. **Midrib** large and flat adaxially, a semicircle or 3/4 circle abaxially and sharply indented at lamina insertion (Fig. 33A). **Cuticle** thin on both surfaces. **Epidermis**: cells flat or rounded externally and internally on both surfaces. Abaxial epidermal cells same size as or smaller than adaxial. **Mesophyll**:

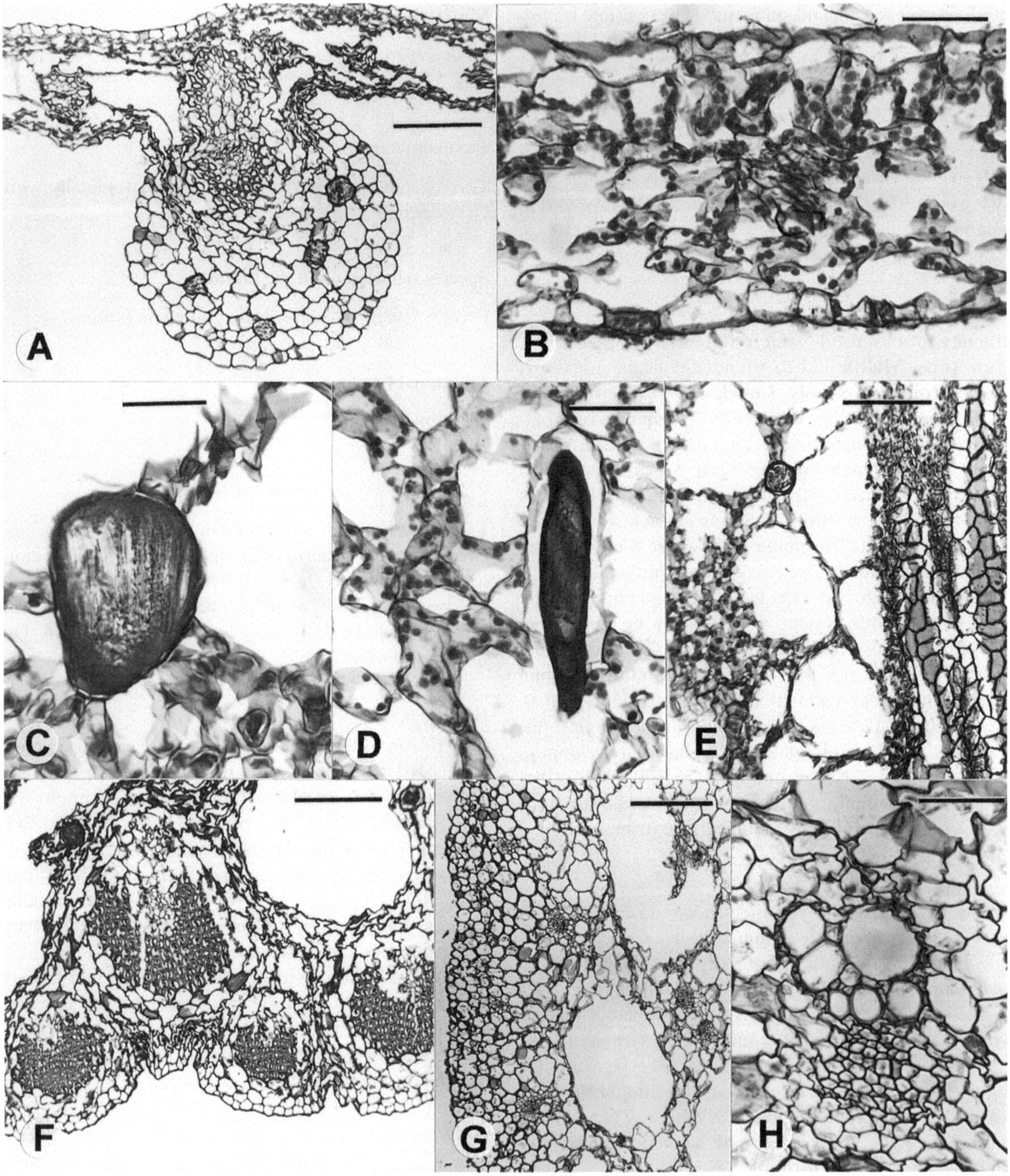

Fig. 33. *Podolasia stipitata* leaf and petiole sections. (A) Midrib TS with single vascular bundle; (B) Lamina TS showing weakly-differentiated mesophyll; (C) Mesophyll PS, wide raphide cell in type 3 aerenchyma; (D) Mesophyll PS showing narrow raphide cell; (E) Leaf PS showing polygonal adaxial epidermal cells, type 3 aerenchyma and druse cell; (F) Distal petiole TS showing fibre caps over phloem; (G) Distal petiole TS showing type 4 aerenchyma; (H) Distal petiole TS showing type II vascular bundle. Scale bars: A,E,F = 200 μm; B–D,H = 50 μm; G = 300 μm.

1–2 palisade layers, occupying up to 20% of lamina thickness, not well differentiated from 7–12 spongy layers (Fig. 33B). Spongy cells forming large, high, type 3 air cavities in central lamina. Partitions 1–4 cells long having small and discrete rounded air spaces at cell corners. Spongy cells of type 1 near abaxial surface, cells of various sizes. **Air cavities** up to 95% of spongy tissue. One or two large air cavities often occupying most midrib space (type 4). Other midrib ground tissue compact.

Vascular bundles: 1–3 symmetrically placed bundles in midrib; type I. Xylem: medium-sized protoxylem element and numerous metaxylem elements in several files. Phloem: large oval or semicircular strand. **Sclerenchyma**: fibres forming large C-shaped caps surrounding xylem and phloem, or ensheathing entire bundle (Fig. 33A). **Tannin** cells idioblastic, often larger than mesophyll cells and common throughout mesophyll. **Crystals**: raphides in narrow (Fig. 33D) or wide, single (Fig. 33C) or double bundles in large rounded cells, infrequent in mesophyll. Raphides common in corner cells and in aerenchyma partitions. Druses large, uncommon, in large rounded cells with mesophyll cells radiating from periphery of crystalliferous cell (Fig. 33E).

Petiole TS

Outline terete with rounded adaxial groove proximally. **Surface spines**: falcate, sharp tissue, composed of proliferation of epidermal and subepidermal tissues; cell walls slightly lignified. **Cuticle** thin and smooth. **Epidermis**: cells flat or rounded internally and externally. **Chlorenchyma**: *c.* 6 outer cortical layers bearing few chloroplasts. **Ground tissue**: cells mostly of uniform size, slightly larger in centre. **Air cavities**: large and rounded, *c.* 5, separated by multiseriate type 4 partitions (Fig. 33F,G). Numerous smaller cavities near periphery, separated by type 3 uniseriate partitions.

Vascular bundles: numerous, scattered, type I, smaller near periphery. Larger bundles with 1–2 protoxylem elements and 1–8 metaxylem elements. Phloem: oval or semicircular strand (Fig. 33H). **Sclerenchyma**: fibres forming large semicircular phloem caps (Fig. 33F) and smaller shallow xylem caps. Peripheral bundles more deeply capped or ensheathed. **Tannin** cells frequent among ground tissue, some forming pairs axially. **Starch** granules occasional in ground tissue near some vascular bundles. **Crystals**: raphides common in rounded or elongate cells, especially bordering air cavities; bundles mostly single and wide, occasionally multiple in elongate cells.

Stem TS

Outline terete. **Size**: diameter 1.8 cm. **Epidermis**: cells small and square. **Ground tissue**: uniform storage tissue with cells in one specimen densely filled with large spherical starch granules (*c.* 16 μm diameter). Ground tissue not demarcated into central cylinder and cortex. **Air cavities** small, parenchyma cells loosely packed.

Vascular bundles scattered, those in centre compound, of various forms, some amphivasal, others with a massive, central, shallowly fluted fibre strand surrounded by 3–5 separate phloem/xylem units; xylem cells mostly metaxylem. Outer vascular bundles collateral, small with deep rounded fibre caps. **Tannin** cells idioblastic, common throughout ground tissue. **Crystals**: occasional druses in ground tissue.

Material examined

Podolasia stipitata N.E.Br., Anderson 26757, Sarawak, K.

Species reported on in the literature

(Both *P. stipitata*) • French (1987c): root. • French (1988): laticifers (absent).

Additional literature

French 1985b, 1986c, 1987a; Grayum 1990; Hay 1986, 1992a; Hotta 1971; Thanikaimoni 1969.

Lasia Lour.

This genus of two species is distributed in India and Sri Lanka, south-east Asia, and Malaysia. Plants are aquatic or terrestrial, short-stemmed spiny herbs with underground rhizomes. They occur usually in wet forests, open marshes or wetlands, or in permanently standing water. The leaves are long-petiolate with spiny petioles. Blades are sagittate, hastate, or pinnately divided with hastate lobes. Chromosomes: 2n = 26. One species and two forms were available for study.

Leaf surface

Cuticle smooth or finely striate adaxially, not observed abaxially. **Epidermis**: cells polygonal with anticlinal walls straight to undulate adaxially (1–3:1 *l*/*w*) and undulate-sinuous with 3–8 lobes per cell abaxially (Fig. 34C). **Trichomes** occasional abaxially near large veins, 2–3 cells long, uniseriate, slightly hooked and lignified. **Stomata** brachyparacytic and mostly abaxial. Subsidiary cells broad and trapezoidal (Fig. 34C), or narrow. **Venation**: lamina and lobes appearing pinnately veined. Basal secondaries acrodromous. Medial segment with irregular intercostal areas, curved-ascending secondaries,

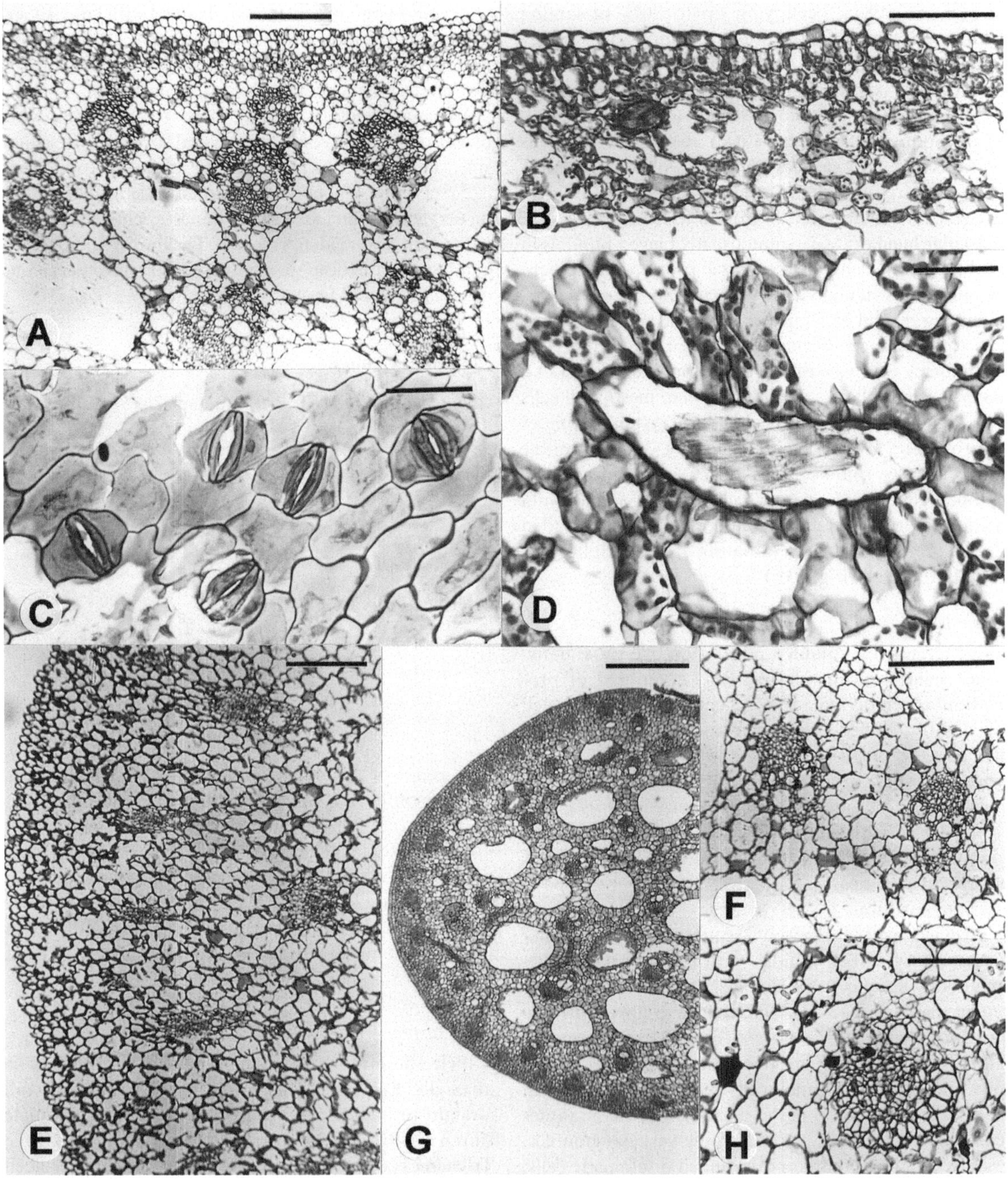

Fig. 34. *Lasia spinosa* leaf, petiole and stem sections. (A) Midrib TS, adaxial surface; (B) Lamina TS; (C) Abaxial epidermis PS with brachyparacytic stomata; (D) Lamina PS with raphide cell; (E) Proximal petiole margin TS and small type III vascular bundles; (F) Proximal petiole centre TS: type I vascular bundles in type 4 aerenchyma; (G) Young stem TS with type 4 aerenchyma; (H) Mature stem TS showing amphivasal vascular bundle. Scale bars: A,E,G = 300 μm; B,F = 150 μm; C,D = 50 μm; H = 200 μm.

eucamptodromous basally and forming collector vein distally. Higher orders reticulate.

Leaf TS

Structure well differentiated. **Size**: 28 × 8 cm (*l*/*w*). **Midrib** slightly concave adaxially (Fig. 34A), large and broadly semicircular abaxially. **Cuticle** thin, patchy. **Epidermis** somewhat papillate adaxially. Abaxial epidermis somewhat thinner with cells somewhat rounded but mostly flat externally. **Mesophyll**: palisade 1–2-layered, of rounded cells (Fig. 34B). Spongy tissue *c*. 10-layered, of irregular long-armed cells often arranged in vertical partitions between large substomatal air cavities (type 2). **Air space** *c*. 90% of spongy layer. Large type 4 air cavities in midrib (Fig. 34A). **Collenchyma**: weakly-defined band, subepidermal on abaxial side of midrib.

Vascular bundles scattered in midrib and well developed (type I). Xylem: 10–15 metaxylem cells forming a broad strand of cells. Phloem: a broad flattened or semicircular strand. **Sclerenchyma**: fibres forming well-developed caps over phloem and xylem of larger vascular bundles. **Crystals**: raphides common in mid-mesophyll, in elongate tubes containing multiple bundles (Fig. 34D). Druses common in adaxial mesophyll. Crystal sand common throughout.

Petiole TS

Outline terete. **Size**: diameter 8 mm. **Cuticle** thick and smooth. **Epidermis**: cells small and rounded externally. **Collenchyma**: slight development in peripheral band, 3–5 layers deep (Fig. 34E). **Chlorenchyma**: plastids in outer 8–10 cell layers. **Ground tissue**: cells small subepidermally, grading to large in centre. **Air cavities**: medium-sized to very large and circular type 4 openings (Fig. 34F), comprising up to 70% air space.

Vascular bundles scattered, wide, type I, rounded and well developed toward centre or smaller near periphery (Fig. 34E). Xylem: 1–2 wide protoxylem cells, 8–12 wide metaxylem cells in wide band. Phloem: wide semicircular strand. **Sclerenchyma**: fibres as phloem caps over outer vascular bundles. **Tannin** cells occasional in ground tissue. **Starch** granules cylindrical, up to >30 μm long, common in ground tissue. **Crystals**: raphides in single bundles common in outer ground tissue, in thin-walled cells, occasional in air cavity partitions. Druses occasional, especially in central ground tissue. Crystal sand common in ground tissue.

Stem TS

Outline elliptic. **Size**: diameter 9 × 16 mm. **Cuticle** thin and smooth, with flanges between epidermal cell margins. **Epidermis**: cells small, flattened. **Cork**: patchy subepidermal development. **Ground tissue**: mostly uniform parenchyma, loosely packed in centre, or type 4 cavities (Fig. 34G), up to 50% air space.

Vascular bundles irregularly scattered, slightly concentrated at centre; bundles large and amphivasal in centre of mature stem (Fig. 34H), the smaller bundles collateral. Xylem: numerous metaxylem elements in circular or wide strand. Phloem: variously-shaped, rounded to broad semicircular strand. **Tannin** cells common, isolated in ground tissue. **Starch** granules common, cylindrical, up to >30 μm long. **Crystals**: raphides in multiple bundles in large cells, common throughout ground tissue. Druses small, occasional in ground tissue. Crystal sand and tiny druses scattered in ground tissue.

Material examined

Lasia spinosa (L.) Thw. f. *simplex* Thw., Bogner 428, Thailand, cult. M; *L. spinosa* f. *spinosa* Thw., cult. M.

Species reported on in the literature

Dalitzsch (1886): *L. aculeata, L. ferox*: leaf. • Engler (1911): *L. spinosa*: root, leaf, petiole. • Ertl (1932): *L. aculeata*: leaf. • French (1987c): *L. spinosa*: root. • French (1988): *L. spinosa*: laticifers (absent). • French and Tomlinson (1981a): *L. spinosa*: stem vasculature. • Guttenberg (1968): *L. spinosa*: root. • Hay (1986): leaf. • Hotta (1971): leaf. • Lierau (1888): *L. spinosa*: root. • Mittmann (1888): *L. spinosa*: petiole. • Müller (1919): *L. aculeata*: leaf tip. • Solereder and Meyer (1928): *L. spinosa* : all organs. • Tieghem (1866): *L. aculeata, L. ferox*: all organs. • Tieghem (1870–71): *L. ferox*: root, secretory tissue. • Trécul (1866–70): vasculature. • Uphof *et al.* (1962): hairs. • Voronkina (1975): *L. aculeata*: root.

Additional literature

Eyde *et al.* 1967; French 1985b, 1986c, 1987a; Grayum 1990; Hay 1992a; Hegnauer 1986; Hotta 1965; Manya-Chernej 1978; Nahrstedt 1975; Thanikaimoni 1969.

Anaphyllum Schott

This genus of two species is restricted to south India. Plants are rhizomatous terrestrial understorey herbs found in evergreen to semi-evergreen forests, and occasionally wet places. The leaves are long-petiolate with a dominant midrib, lateral lobing, sagittate bases, and coarsely crenate margin. Chromosomes: 2n = 26. One species was available for study.

Leaf surface

Cuticle smooth on abaxial midrib. **Epidermis**: anticlinal cell walls sinuous on both surfaces (1–4:1 *l*/*w*) (Fig. 35D). **Stomata** abaxial, brachypara-tetra- or -hexa-cytic (Fig. 35D). Subsidiary cells broad or narrow and rounded. **Venation**: prominent midrib and basal lobe primaries. Secondaries pinnate, ± opposite on midrib. Acutely ascending secondaries ending in submarginal brochidodromous collector. Small marginal collectors merging eucamptodromously. Tertiaries and higher orders reticulate.

Leaf TS

Structure weakly differentiated. **Size**: specimen 96 × 90 mm (*l*/*w*); lobe 80 × 34 mm (*l*/*w*). **Midrib** slightly concave adaxially, convex, semicircular and grooved abaxially (Fig. 35A). **Cuticle** thin or not detected on lamina, thick on abaxial midrib. **Epidermis**: cells square to elongate and tall on both surfaces; smooth-surfaced. **Mesophyll** poorly differentiated, all five layers composed of spongy tissue. Cells nearest adaxial surface more tightly packed, sometimes U-shaped (Fig. 35B,E) or with fewer arms than type 1 spongy layers beneath (Fig. 35B). **Air cavities** near and within midrib (type 4). Laminar air space about 50%.

Vascular bundles in midrib robust, about 5–6 symmetrically arranged; type II (Fig. 35A). Xylem mostly of protoxylem elements, variable, up to several cells. Phloem strand rounded, convex toward xylem. Sieve elements with scattered sieve areas on side walls. **Crystals**: raphides occasional in mid-mesophyll, single bundles in centre of occasionally articulated, elongate cells (Fig. 35B,C); crystal cells randomly oriented with respect to veins. Crystal sand throughout, including epidermis.

Petiole TS

Outline broadly elliptic. **Size**: diameter 3 mm proximally and 2 mm distally. **Cuticle** thick and smooth, grooved at cell margins. **Epidermis**: cells rounded to squarish. **Stomata** abaxial, occasional and level with surface. **Collenchyma**: 1–2-layered peripheral subepidermal band distally; less collenchyma development proximally. **Ground tissue**: compact parenchyma except for cavities. **Air cavities**: ring of 5–6 large cavities (type 4), evenly disposed between epidermis and centre (Fig. 35G). **Vascular bundles** scattered, with phloem oriented outward, type I. Xylem: several metaxylem cells in 2–6 files (Fig. 35F,G). **Sclerenchyma**: proximally, outermost bundles with well-developed fibre caps around phloem; fibres less well developed around central bundles. Distally, bundle caps with little lignification. **Crystals**: raphides few in subepidermal layers. Crystal sand common throughout.

Material examined

Anaphyllum beddomei Engl., Blasco s.n., 1/5/74, S. India, cult. M.

Species reported on in the literature

French (1987c): *A. wightii*: root. • French (1988): *A. wightii*: laticifers (absent). • Hay (1986): stem architecture.

Additional literature

French 1985b, 1986c, 1987a; Grayum 1990; Hay 1992a; Thanikaimoni 1969.

Urospatha Schott

This genus of about 10 species is distributed from Central America through lowland Amazonia. Plants are aquatic rhizomatous herbs growing in coastal swamps and only rarely in drier areas. Petioles are erect, elongate, and bear sagittate to hastate leaves. Chromosomes: $2n = 52$. Two species were available for study.

Leaf surface

Cuticle smooth or finely striate on both surfaces. **Epidermis**: adaxial cells polygonal with straight anticlinal walls; abaxial cells polygonal with straight to sinuous walls (1–2:1 *l*/*w* for both surfaces). **Stomata** on both surfaces and brachyparacytic (Fig. 36D). Subsidiary cells broad or narrow and irregular. **Venation**: strong midrib and basal lobe primaries. Secondaries acutely-curved and parallel, merging eucamptodromously near margin or in outer lamina. Tertiaries and higher order veins reticulate.

Leaf TS

Structure well differentiated. **Size**: centre lobe fragments 4 × 8 cm. **Midrib** shallowly concave-rounded adaxially and broadly convex-rounded abaxially (Fig. 36A). **Cuticle** thin. **Epidermis**: cells medium-sized on both surfaces, flat to rounded externally and internally. **Mesophyll** well differentiated; 1–2 palisade layers, up to 20% of lamina. Spongy tissue of *c*. 15 layers; 3–5 layers of type 1 on both sides of well-formed type 3 aerenchyma (Fig. 36B). **Air cavities**: large squarish air spaces in central spongy layer separated by 1-seriate partitions (type 3), occupying up to 99% of aerenchyma space (Fig. 36A,B,C). Large substomatal air cavities present. **Collenchyma**: abaxial midrib band occurring when fibres absent.

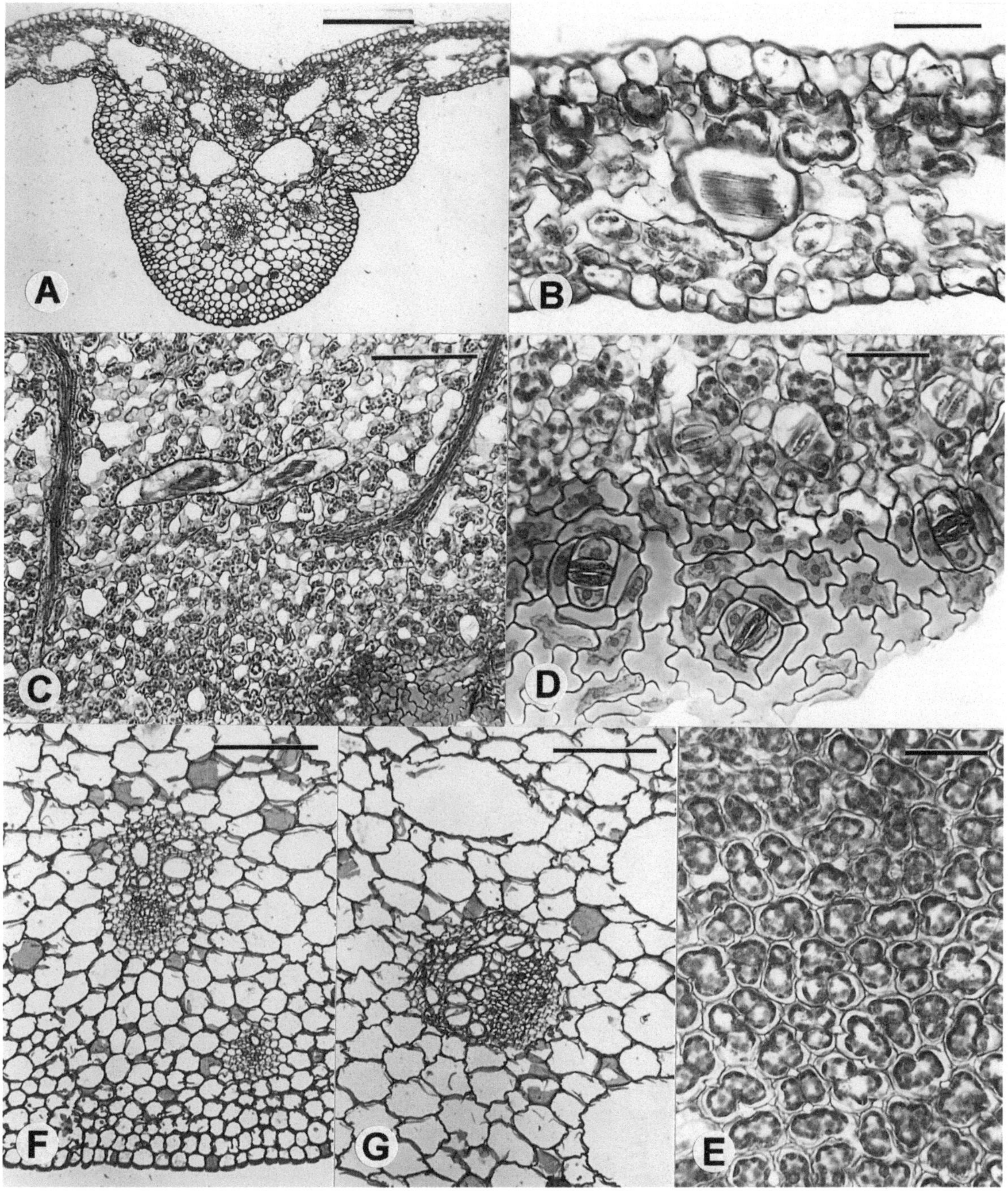

Fig. 35. *Anaphyllum beddomei* leaf and petiole sections. (A) Midrib TS, note type 4 aerenchyma; (B) Lamina TS with lobed palisade cells and central raphide cell; (C) Mesophyll PS showing pair of articulated raphide cells with oblique crystal bundles, in type 1 spongy tissue; (D) Abaxial epidermis PS showing brachypara-tetracytic and -hexacytic stomata; (E) Mesophyll PS showing lobed palisade cells; (F,G) Distal petiole TS showing type I vascular bundles: (F) margin; (G) centre, also showing type 4 aerenchyma. Scale bars: A = 300 μm; B,D,E = 50 μm; C,F,G = 150 μm.

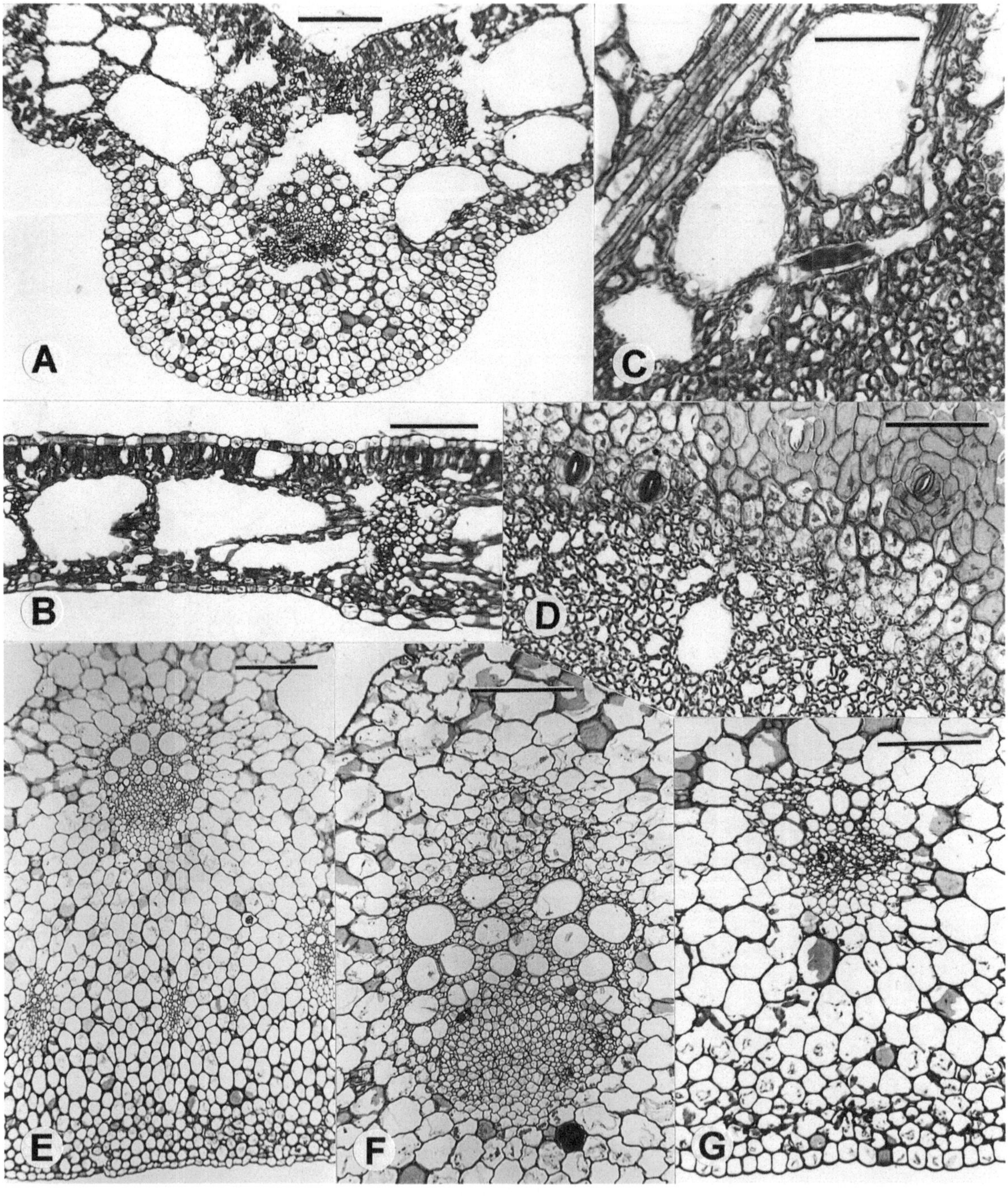

Fig. 36. *Urospatha grandis* leaf and petiole sections. (A) Midrib TS showing type 3 aerenchyma; (B) Lamina TS with type 3 spongy aerenchyma; (C) Mesophyll PS showing elongated raphide cell; (D) Mesophyll PS and adaxial epidermis showing stomata and substomatal air cavity in palisade tissue; (E) Petiole TS margin; (F) Petiole TS centre showing large type I vascular bundle; (G) Distal petiole TS, lateral margin. Scale bars: A,E = 300 μm; B = 200 μm; C,D,F,G = 150 μm.

Vascular bundles numerous in midrib, or one large vascular bundle flanked by one smaller bundle on either side (Fig. 36A); bundles well developed and wide (type I). Numerous metaxylem cells in *c.* 5 files forming concave boundary opposing broad semicircular phloem strand. **Sclerenchyma**: fibres forming broad caps over xylem and phloem, or absent. **Secretory tissue**: large gland at surface with large slit opening surrounded by cluster of stomata; vascular bundles converging toward opening. **Crystals**: raphides common in midrib and lamina. In lamina, raphides mostly in single bundles in elongated cells (Fig. 36C), common in adaxial or abaxial mesophyll; also in 1–2 bundles in large rounded cells in aerenchyma partitions. Druses small, occasional in palisade.

Petiole TS

Outline: subterete or laterally compressed with shallow adaxial groove and laminar wings. **Size**: 4–5 mm to 1–1.7 cm diameter. **Cuticle** thick or thin and smooth or striate. **Epidermis**: cells small, rounded externally and internally. **Stomata** few, with large substomatal cavities. **Collenchyma**: thin-walled band in outer 5–15 layers of ground tissue (Fig. 36E). **Ground tissue**: cells small near periphery, grading to larger near centre (Fig. 36E,G). **Air cavities**: *c.* 10 very large cavities and numerous smaller ones, all type 4 aerenchyma (Fig. 36E). Infrequent 2–3-layered diaphragms separating cavities longitudinally.

Vascular bundles numerous, scattered symmetrically, and very large in centre (type I) (Fig. 36F,G); peripheral bundles small (Fig. 36E). Xylem: 4–8 files, 16–35 metaxylem cells forming broad concave boundary, wrapping halfway around phloem in largest bundles. Phloem: a broad semicircular or broad elliptical strand. **Sclerenchyma**: proximally: large vascular bundle caps at phloem near periphery; smaller narrow phloem caps 3–4 layers deep near centre. Distally absent. **Tannin**: large and small cells with dark contents, near vascular bundles or randomly distributed in ground tissue. **Starch** granules uncommon near vascular bundles and abundant in diaphragms. **Crystals**: raphides small, in single or gradually overlapping bundles in axially aligned ground tissue cells. Crystals rare in distal sections, frequent proximally.

Material examined

Urospatha grandis Schott, Croat 35764, Costa Rica, MO; *U. tonduzii* Engl., Bogner 1115, Costa Rica, cult. M.

Species reported on in the literature

Engler (1911): petiole. • French (1987c): *U. sagittifolia*: root. • French (1988): *U. sagittifolia*: laticifers. • French and Tomlinson (1981a): *U. sagittifolia*: stem vasculature. • Solereder and Meyer (1928): secretory tissue.

Additional literature

Cevallos-Ferriz and Stockey 1988; Erdtman 1952; French 1985b, 1986c, 1987a; Grayum 1990; Hay 1992a; Ray 1988; Schultes 1970b; Thanikaimoni 1969.

Anaphyllopsis A.Hay

This genus of three species has a neotropical range. Plants are terrestrial or aquatic, rhizomatous understorey herbs of primary forests, especially swampy, sandy areas along streams. The short erect stems are at least partly underwater for part of the year (Croat 1988). Leaves are heteroblastic. Mature petioles are erect, bearing pinnatifid to pinnatisect, or perforate blades. Chromosomes: 2n = 26. One specimen was available for study.

Leaf surface

Cuticle smooth on both surfaces. **Epidermis**: cells with anticlinal walls undulate on both surfaces (1–4:1 *l*/*w*). **Stomata** brachyparacytic (Fig. 37G), very common abaxially, rare adaxially. Subsidiary cells irregular or trapezoidal; cells broad or narrow with rounded margins. **Venation**: from available basal segment, one small eccentric midrib, secondaries steeply ascending, merging submarginally, brochidodromously or eucamptodromously. Higher order veins reticulate.

Leaf TS

Structure: lamina well differentiated. **Size** of triangular segment 50 × 15 mm *l*/*w*. **Midrib** flat adaxially, broadly and shallowly protruding abaxially (Fig. 37A). **Cuticle** thin. **Epidermis**: cells on both surfaces flat externally and slightly rounded internally. **Mesophyll**: palisade cells 1-layered, compact (5:1 *l*/*w*) (Fig. 37B,F). Spongy cells 5–7-layered, type 1a (Fig. 37B,C,E). Chloroplasts slightly more numerous in palisade cells. **Air cavities** 20–40% of spongy layer. Substomatal air chambers very shallow (Fig. 37C).

Vascular bundles: 1–2 small bundles in midrib; all bundles small, type I. Xylem of largest bundles with 1–2 narrow protoxylem cells and several short files of 6–10 metaxylem cells. Phloem: a semicircular strand. **Tannin** cells common, idioblastic, similar in shape to other mesophyll cells (Fig. 37D), especially common near margin; also as elongated, non-interconnecting cells within vascular bundles. **Starch** granules uncommon, in isolated cells near phloem of larger bundles. **Crystals**:

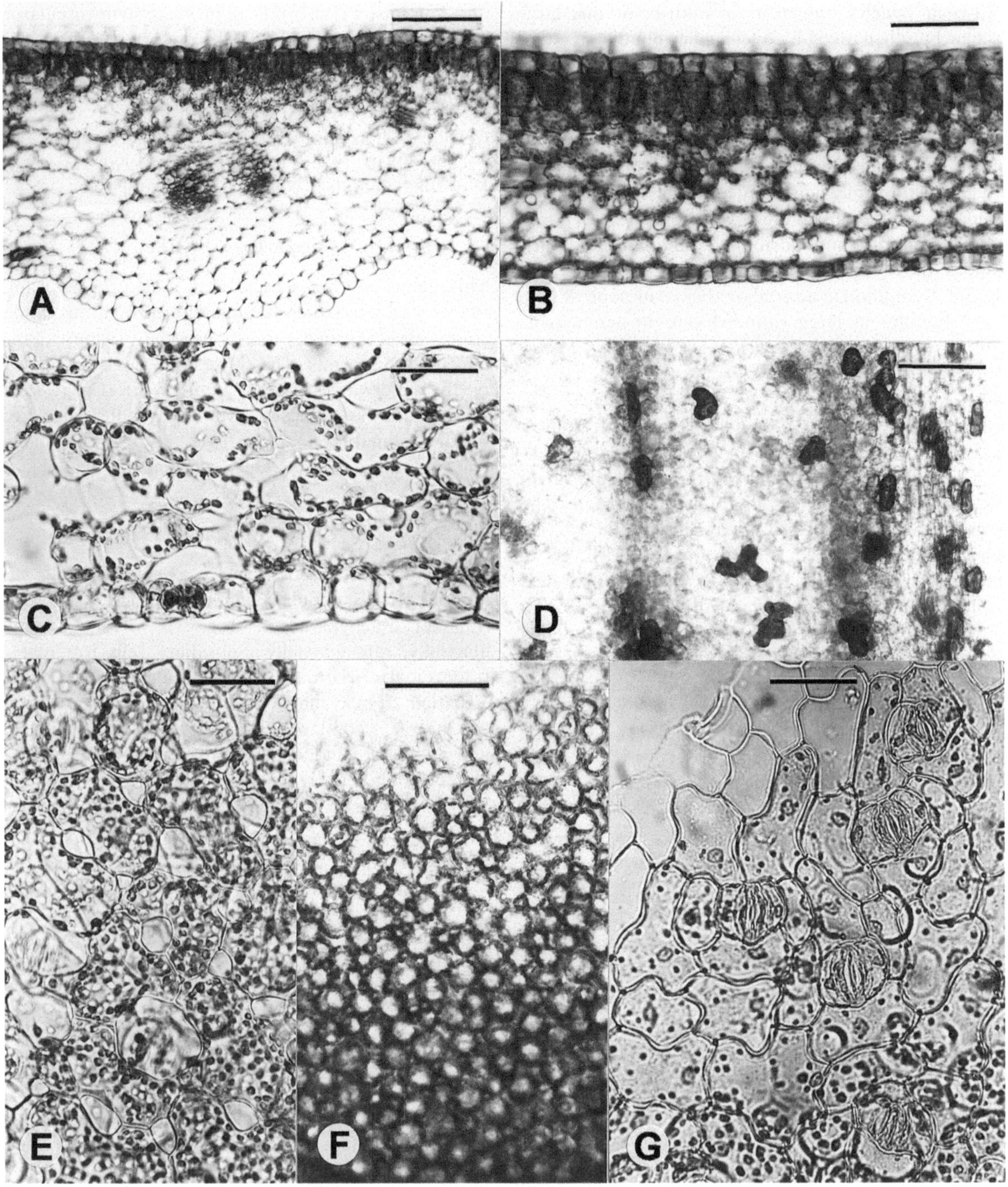

Fig. 37. *Anaphyllopsis americana* leaf sections. Wet mounts: iodine in calcium chloride. (A) Midrib TS; (B) Lamina TS; (C) Lamina TS showing abaxial stoma and shallow substomatal air cavity; (D) Lamina PS showing tannin cells in mesophyll; (E) Mesophyll PS showing type 1a spongy aerenchyma; (F) Mesophyll PS, palisade tissue; (G) Abaxial epidermis PS showing brachyparacytic stomata. Scale bars: A,B,D = 200 μm; C,E,G = 50 μm; F = 150 μm.

raphide cells elongated, randomly oriented, at palisade-spongy boundary; crystals in elongated overlapping masses, rather than discrete parallel bundles.

Material examined

Anaphyllopsis americana (Engler) A.Hay; A.Hay 2811, French Guiana, cult. M.

Species reported on in the literature

French (1987c): *A. americana*: root. • French (1988): *A. americana*: laticifers (absent). • Hay (1986): stem.

Additional literature

Hay 1992a.

Pycnospatha Thorel ex Gagnep.

This genus of two species is distributed in Thailand, Laos and Vietnam. The plants are terrestrial herbs with deep tubers, occurring in shady places. Leaves are long-petiolate with occasional petiolar spines. Lamina development is heteroblastic and varies from cordate, panduriform, to digitately decompound with acute tips. Chromosomes: 2n = 26. One specimen was available for study.

Leaf surface

Cuticle: smooth, but striate in vicinity of larger veins. **Epidermis**: cells elongate; with anticlinal walls undulate adaxially and sinuous abaxially (2–4:1 *l/w*). **Stomata** more frequent abaxially and brachyparacytic (Fig. 38E), rarely unipolar. Subsidiary cells wide-elongate and variably-shaped, rounded-edged or angular. **Venation**: suprabasal acrodromous primaries and prominent midrib in each segment. Secondaries widely spaced and irregular, narrowly acute basally, widely acute distally, forming irregular intercostal areas. Secondaries merging at marginal brochidodromous collector following marginal indentations of lamina contours. Tertiaries and higher orders reticulate (Fig. 38D).

Leaf TS

Structure: mesophyll weakly differentiated. **Size** (irregularly elliptic segment): 80 × 32 mm (*l/w*). **Midrib** small, outline level with lamina adaxially and deeply convex-rounded abaxially (Fig. 38A). **Cuticle**: medium thickness. **Epidermis**: cells large, slightly taller adaxially, rounded internally and externally. **Mesophyll**: one palisade layer of weakly-differentiated cells (Fig. 38B), sometimes lobed or H-shaped. Spongy tissue of *c*. 5 layers, up to 80% of mesophyll; cells long-armed, type 2. **Air cavities** up to 40% of spongy tissue. Large elongate cavities (type 4) in midrib (Fig. 38A), separated by multicellular partitions and divided longitudinally by occasional 1-layered diaphragms. **Collenchyma**: continuous, abaxial sub-epidermal midrib band, two cells deep (Fig. 38C).

Vascular bundles: six midrib bundles (type I) symmetrically disposed among type 4 air cavities. Xylem: 4–8 irregularly grouped proto- and meta-xylem cells, proximal to broadly elliptic phloem strand (Fig. 38C). **Sclerenchyma**: fibres forming broad caps 1–3 cells deep over phloem strands, less well developed or absent over xylem. **Starch** granules uncommon; well-formed small granules in guard cells. **Crystals**: raphide cells very common throughout mesophyll; raphides variable: some in large wide cells as wide single bundles, or in multiple overlapping bundles in broad or elongate cells. Styloids occasional and clustered, larger than usual druses and oriented beneath abaxial stomata.

Petiole TS

Outline circular. **Size**: basal end diameter 6 mm, distal end diameter 4 mm. **Cuticle**: medium-thick and smooth. **Epidermis**: cells appearing square in TS, elongate in LS; cells in TS rounded internally and externally. **Chlorenchyma**: plastids sparse in outer ground tissue. **Ground tissue**: smaller cells near periphery, grading to larger cells in centre. **Air cavities**: 12–14 large circular air cavities, axially elongate, separated by multiseriate partitions (type 4) (Fig. 38F), with occasional 1-layered diaphragms seen in LS. Diaphragms with organized triangular cell openings at cell corners.

Vascular bundles scattered in central ground tissue, and a well defined outer ring near periphery; 1–4 narrow or wide, mostly protoxylem elements (Figs 8A, 38F) proximal to circular phloem strand (type II). **Sclerenchyma**: fibre strands forming multi-layered and semicircular phloem caps near periphery. Caps over xylem single-layered and irregular. **Tannin** cells rare in ground tissue (Fig. 8A). **Starch** granules rare and scattered in ground tissue. **Crystals**: raphides in single or double bundles in expanded cells in diaphragms, or in cells bordering air cavities. Raphides in multiple or overlapping bundles in axially elongate cells in peripheral ground tissue.

Stem TS

Outline circular. **Size**: diameter 4–5 mm. **Cork**: several files of thin-walled empty cells. **Ground tissue**: parenchyma uniform throughout organ. **Vascular bundles** scattered, often running obliquely and appearing to be leaf traces; small, collateral and weakly formed (type II). Xylem: 1–4 narrow metaxylem cells. Phloem: very small irregular

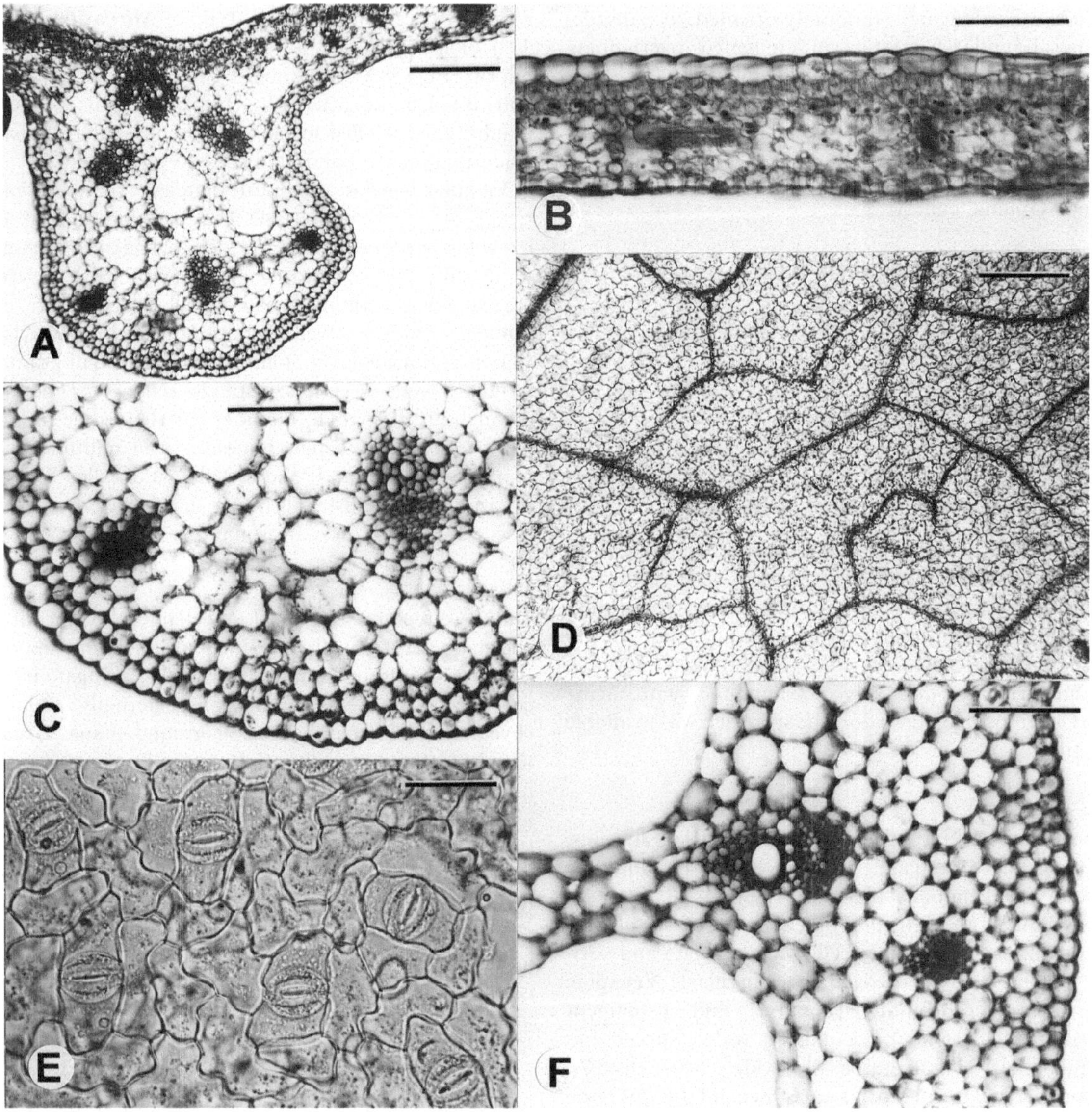

Fig. 38. *Pycnospatha arietina* leaf and petiole sections. Wet mounts: iodine in calcium chloride. (A) Midrib TS showing type 4 aerenchyma; (B) Lamina TS; (C) Midrib TS showing abaxial collenchyma band and type I vascular bundle; (D) Lamina clearing showing fine venation; (E) Abaxial epidermis PS showing brachyparacytic stomata; (F) Mid-petiole TS showing type 4 aerenchyma and vascular bundles. Scale bars: A,D = 300 μm; B,C,F = 150 μm; E = 50 μm.

strand. **Starch** granules densely present in all ground tissue. **Crystals**: druses and raphides in single bundles, abundant throughout ground tissue.

Contractile root TS

Epidermis present, cells axially elongate and irregularly shaped. **Ground tissue** unorganized near periphery, in radial files near stele. **Stele** with six xylem and six phloem poles. Stele centre not sclerified. **Starch** granules scattered in ground tissue.

Material examined

Pycnospatha arietina Gagnep., Bogner 395, Thailand, cult. M.

Taxonomic notes

The presence of subepidermal styloids in the leaves is unique in the subfamily.

Species reported on in the literature

French (1988): *P. arietina*: secretory tissue. • Hotta (1971): leaf.

Additional literature

French 1985b, 1986c, 1987a; Grayum 1990; Hay 1986, 1992a; Thanikaimoni 1969.

Dracontium L.

This genus of 18–23 species is distributed in tropical America. Plants are terrestrial tuberous herbs of seasonally dry tropical moist and tropical dry forests, savannas, and less commonly in wet areas. Following production of the inflorescence, plants produce one erect, elongate, robust petiole bearing a large, trisect, irregularly-cut compound foliage leaf. Chromosomes: $2n = 26$. Five species were available for study.

Leaf surface

Cuticle smooth to slightly rough, striate, or forming reticulate ridges on lamina; strongly striate on abaxial midrib. **Epidermis**: cells with anticlinal walls undulate to sinuous on both surfaces (Fig. 39C). **Stomata** brachyparacytic (Fig. 39C). Subsidiary cells mostly broad and rounded or trapezoidal, or irregular, occasionally not distinguished from undulate epidermal cells; subsidiary cells occasionally unipolar. **Venation**: midrib prominent, forking in mid-lamina in some segments. Secondaries acute, forming irregular intercostal areas and submarginal brochidodromous loops. Some eucamptodromous connections present at margin. Intersecondaries parallel with secondaries. Tertiaries and higher orders reticulate.

Leaf TS

Structure well or poorly differentiated. **Size** (elliptic segments): 85 × 36 mm (*l*/*w*) or smaller. **Midrib** adaxially grooved, abaxially large rounded-convex (Fig. 39A), or midrib nearly embedded in lamina, slightly and broadly convex abaxially. **Cuticle** thin on both surfaces. In some specimens, cuticular material interpenetrating abaxial midrib ground tissue intercellular spaces as deep as two cell layers. **Epidermis**: cells medium-sized, slightly larger adaxially and somewhat elongate or square; cells rounded or flat externally and internally. Occasionally, abaxial midrib epidermis with elongated papillae. **Stomata**: guard cells often with only outer stomatal ledges or inner ones much reduced. **Mesophyll**: 1–2 palisade layers of tightly packed cells (8:1 *l*/*w*), up to 30% of lamina thickness. Palisade cells often either cup-shaped or U-shaped and 2–multi-armed and often bearing large peripheral chloroplasts. Spongy tissue with 7–8 layers of irregular cells with short lobes (Fig. 39B), about three cells surrounding more or less circular air cavities, or spongy cells with up to six longer arms surrounding abundant larger air spaces (types 1, 2). **Air cavities**: 50–70% of spongy layer. High air cavities often not aligned beneath stomata. Larger diameter midribs with large type 4 aerenchyma cavities scattered among vascular bundles (Fig. 39A).

Collenchyma: a single-layered abaxial band in midrib, or absent. **Vascular bundles**: *c*. 6 symmetrically disposed bundles in midrib; types I–II. Xylem often with one wide protoxylem lacuna and 9–10 narrower metaxylem elements. Phloem: a large semicircular strand. **Sclerenchyma** absent, or present as few fibres on adaxial or abaxial sides of vascular bundles, or well-developed forming deep phloem caps. **Crystals**: raphides, single or double bundles of fine crystals in wide rounded cells or elongate cells, common in mid-mesophyll and abaxially in midrib. Raphides occasionally in very long cells with a long row of crystal bundles per cell (Fig. 39D); tending to be parallel with secondary veins.

Petiole TS

Outline circular. **Size**: diameter 8 mm to 4 cm. **Cuticle** of medium thickness and smooth or finely striate. **Epidermis**: cells of variable sizes and shapes, usually rounded externally and internally. **Chlorenchyma**: 1–3 subepidermal ground tissue layers. **Ground tissue**: *c*. 12 outer parenchyma layers encircling central aerenchyma of eight very large circular cavities separated by 3-seriate partitions (type 4) (Fig. 39E,F). Cavities axially elongate and separated by horizontal or slightly oblique 3-seriate diaphragms. Parenchyma cells axially elongated (1.5–2:1 *l*/*w*). Outer cortical ground tissue typically thin-walled but lignified in a band extending 1/3 around organ. **Vascular bundles** numerous and scattered, concentrated near margin (types I or II). Xylem usually a very wide protoxylem lacuna and 3–8 narrower metaxylem tracheids. Phloem: a broad oval strand.

Sclerenchyma: fibres forming deep phloem caps at outermost bundles (Fig. 39E,F), broad phloem caps and scattered fibres below xylem at larger bundles deeper in cortex, and as small strands of 3–5 cells above phloem in bundles in aerenchyma where partitions join. **Tannin**:

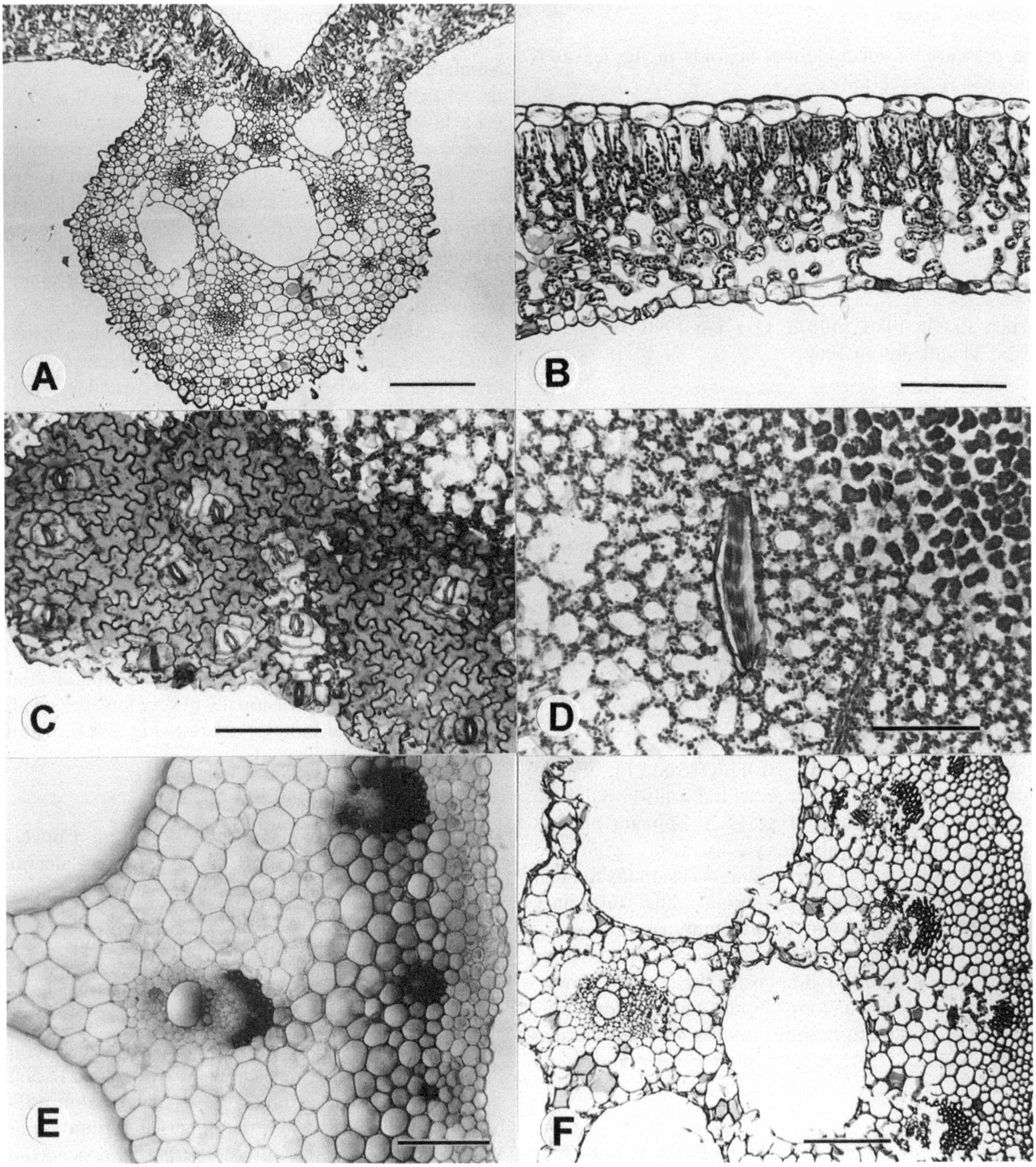

Fig. 39. *Dracontium* leaf and petiole sections. (A,B) *D. soconuscum*; (C,D,F) *D. spruceanum*; (E) *D. gigas*; (A) Midrib TS showing type 4 aerenchyma; (B) Lamina TS; (C) Lamina PS showing sinuous abaxial epidermal cells and brachyparacytic stomata; (D) Lamina PS showing elongated raphide cell; (E) Hand section of petiole TS showing fibre caps over phloem and type 4 aerenchyma cavities; (F) Petiole TS, lateral margin and type 4 aerenchyma cavities. Scale bars: A,F = 300 µm; B,C,D = 150 µm; E = 200 µm.

cells with dark amorphous or granular contents beneath chlorenchyma, seen in LS as individual or rows of two elongated cells. **Starch** granules in 1–2 axial rows of cells lateral to about 1/3 of vascular bundles. **Crystals**: multiple raphide bundles in axially elongated cells up to 2–3 mm long, subepidermal and in ground tissue. Raphides in LS as single or double bundles in individual, short-elliptic biforine-like cells attached by one end to horizontal diaphragms between cavities. Crystal sand uncommon, mostly in ground tissue near vascular bundles.

Root TS

Stele: protoxylem poles 13 in all cases.

Material examined

Dracontium gigas (Seem.) Engl., Bogner 1267, Costa Rica, cult. M; *D. pittieri* Engl., Croat 35104, Costa Rica, MO; *D. prancei* G.Zhu, Bogner 1132, cult. M; *D. soconuscum* Matuda, cult. SEL, K(132/78); *D. spruceanum* (Schott) G.Zhu, Croat 34334, Panama, MO.

Taxonomic notes

Cytological note: unique in this genus, nuclei in leaf and petiole cells are usually surrounded by conspicuous clusters of adherent spherical organelles or vesicles, 2–3 µm in diameter. Among other members of the tribe, *Dracontium* is unique in having armed palisade cells.

The specimens observed suggest a structurally coherent genus but with numerous, possibly systematically useful variations. Trends of specialization would probably emerge from a more detailed study. While the genus is clearly close to other lasioid genera, no closer alliance is discernible with these data.

Species reported on in the literature

Blanc (1978): ontogeny. • Engler (1911): *D. asperum*, *D. foecundum*, *D. polyphyllum*: stem, leaf. • French (1987c): several spp.: root. • French (1988): several spp.: laticifers (absent). • Hay (1986): leaf. • Solereder and Meyer (1928): secretory tissue.

Additional literature

Erdtman 1952; Eyde *et al.* 1967; French 1985b, 1986c, 1987a; Grayum 1990; Hay 1992a; Thanikaimoni 1969; Williams *et al.* 1981.

Dracontioides Engl.

This monotypic genus is found on the Bahia coast of Brazil. The plants are emergent aquatic herbs having spongy rhizomes and occurring in standing water. The several leaves are erect, long-petiolate and linear-elliptic with a cordate or deeply sagittate base. Lamina perforations occasional. Chromosomes: 2n = 26. One specimen was available for study.

Leaf surface

Cuticle thin and smooth. **Epidermis**: cells polygonal with anticlinal walls straight to undulate on both surfaces (1–2:1 *l*/*w*). **Stomata**: brachypara- (-hexa-) cytic (Fig. 40C), more frequent abaxially. Subsidiary cells: 1–3 pairs, inner cells mostly broad and rounded-edged. **Venation**: midrib and basal lobe primaries diverging at petiole summit. Secondaries variously acute and merging marginally or deeply submarginally at brochidodromous collectors. Higher orders reticulate.

Leaf TS

Structure weakly differentiated. **Size**: 8 × 1.5 cm (*l*/*w*). **Midrib**: small, flat adaxially and rounded-convex abaxially (Fig. 40A). **Cuticle** very thin or present only as flanges on both surfaces. **Epidermis**: cells flat externally and rounded internally. **Mesophyll**: one weakly differentiated palisade layer (cells 2:1 *l*/*w*) (Fig. 40B). Spongy cells in *c.* 7 layers, horizontally elongated in irregular columns (type 1). **Air cavities** substomatal, small and numerous.

Vascular bundles: one large bundle in midrib with 6–10 medium or wide metaxylem elements (type I) (Fig. 40A). Phloem: a semicircular strand. **Sclerenchyma**: vascular bundle caps over xylem and phloem, in midrib and secondary laminar bundles. **Tannin** cells numerous and same shape as mesophyll cells or more elongate; with dark amorphous or granular contents (Fig. 40E), throughout mesophyll and next to vascular bundles. **Crystals**: raphides in elongate cells, 1–2 bundles per cell, subepidermal adaxially in mesophyll and uncommon (Fig. 40D). Druses common in palisade tissue and subepidermal on both surfaces. Crystal sand uncommon throughout, 1–few crystals per mesophyll cell.

Petiole TS

Outline: subterete (Fig. 40G) with small adaxial rounded groove. **Size**: diameter 3 mm. **Cuticle** thick and smooth. **Epidermis**: cells small and rounded. **Chloroplasts**: few in outer 2–3 layers of ground tissue. **Ground tissue**: cells small near periphery grading to large in centre. **Air cavities**: very large circular cavities, type 4 (Fig. 40G).

Vascular bundles large, scattered. Xylem: 1–3 wide late protoxylem cells (Fig. 40F,H), 2–5 narrow metaxylem cells in irregular file (type II). Phloem: small, circular or irregular strand. **Sclerenchyma**: fibre caps, 3–4 cells deep over phloem (Fig. 40F). **Tannin** cells common and scattered in ground tissue. **Starch** granules small and sparse

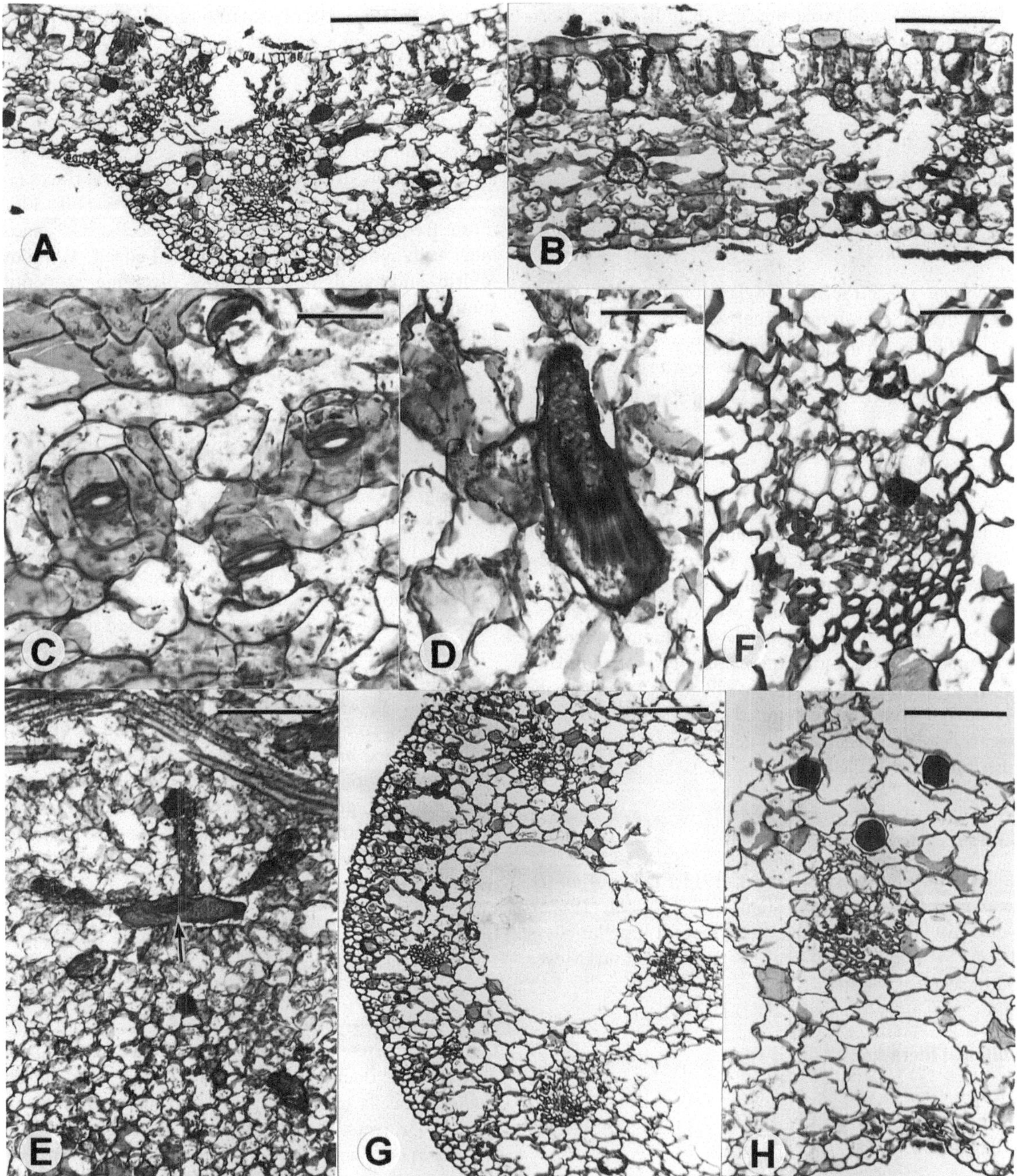

Fig. 40. *Dracontioides desciscens* leaf and petiole sections. (A) Midrib TS; (B) Lamina TS; (C) Leaf abaxial epidermis PS showing brachyparacytic stomata; (D) Mesophyll PS showing raphide cell; (E) Mesophyll PS with unusual tannin cell (arrow); (F) Petiole TS, abaxial side showing type II vascular bundle; (G) Distal petiole TS showing type 4 aerenchyma; (H) Petiole TS, abaxial side showing type II vascular bundle with fibre caps. Scale bars: A,G = 200 μm; B,E,H = 150 μm; C,D,F = 50 μm.

in ground tissue. **Crystals**: raphide cells few, crystals in single bundles in thin-walled ground tissue cells. Druses few, large, scattered in ground tissue.

Material examined

Dracontioides desciscens Engl., Storr 075, cult. M.

Species reported on in the literature

(All *D. desciscens*)

Engler (1911): stem, leaf. • French (1987c): root. • French (1988): laticifers (absent). • French and Tomlinson (1981a): stem vasculature. • Hay (1986): leaf. • Hotta (1971): leaf.

Additional literature

French 1985b, 1986c, 1987a; Grayum 1990; Hay 1992a; Thanikaimoni 1969.

SUBFAMILY CALLOIDEAE ENDL.

A monotypic subfamily having bisexual flowers.

Calla L.

The genus is circumboreal, ranging from Japan to Formosa, and North America as far south as New Jersey, Indiana and Pennsylvania. Its preferred habitat is swampy grasslands, cold water bogs and pond margins. The plants are terrestrial or emergent aquatics having long creeping rhizomes. Leaves are 2-ranked and cordate. Chromosomes: small, 2n = 36, 63, 69, 70, 72. One specimen was available for study.

Leaf surface

Cuticle striate. **Epidermis**: cells polygonal with straight anticlinal walls on both surfaces (1–2:1 *l/w*). **Stomata** brachyparacytic, a few anomocytic and mostly abaxial (Fig. 41D). Subsidiary cells mostly broad and irregular. **Venation**: midrib prominent at base, dissipating to a narrow vascular bundle by distal third of leaf. Secondaries obtuse at base, acutely-ascending distally, diverging and not branching from midrib cluster. Secondaries converging with marginal vasculature and near leaf tip in anastomosing plexus. Tertiaries irregularly parallel with secondaries. Higher orders reticulate.

Leaf TS

Structure dorsiventral. **Size**: lamina 44 mm along midrib, 54 mm tip to basal lobe × 50 mm (*l/w*). **Midrib** broad and shallow, flat adaxially and broadly curved convex abaxially. Shallow abaxial grooves common. **Cuticle** very thin or not seen. **Epidermis**: cells thin and flat on both surfaces. **Mesophyll** with one, occasionally two palisade layers (Fig. 41A,B). Spongy layer a type 3 aerenchyma composed entirely of widely spaced uniseriate partitions (Fig. 41A). Partition length 5–8 cells.

Vascular bundles: 3–4 bundles in midrib; types I, II. Xylem: up to seven elements of various sizes (Fig. 41F). Phloem: a small, somewhat irregular strand. **Laticifers** non-anastomosing with short lateral processes, dark contents, lateral to phloem of all vascular bundles (Fig. 41E). **Tannin** cells uncommon, scattered in epidermal layers. **Starch**: Sakai and Hayashi (1973) noted absence of starch storage in plastids. **Crystals**: raphides in single bundles; cells thin-walled, narrow, biforine-like but with rounded, non-papillate ends (Fig. 41C) and randomly oriented in spongy aerenchyma partitions. Druses occasional in palisade layer. Crystal sand occasional in mesophyll cells.

Petiole TS

Outline semicircular, flattened and slightly concave adaxially. **Size** 5 × 6 mm diameter. **Cuticle** thin and smooth. **Epidermis**: cells large and square or columnar. **Chlorenchyma**: plastids present in outer 2–3 ground tissue layers. **Ground tissue**: type 3 aerenchyma with partitions of 5–7 cells surrounding wide, axially elongated air cavities (Fig. 41G). Cavities separated along organ axis by infrequent 1-layered diaphragms (Fig. 41H). **Air space**: 97% of aerenchyma.

Vascular bundles scattered, uniformly spaced; large and well developed, type I (Fig. 41H). Xylem: 4–6 mostly very wide tracheids and two narrow protoxylem lacunae. Phloem: large circular strand. **Laticifers** non-anastomosing, lateral to most phloem strands. **Starch** granules scattered in ground tissue near vascular bundles. **Crystals**: raphides few in thin-walled cells in outer ground tissue. Crystal sand common in ground tissue.

Material examined

Calla palustris L., Boufford 18801, Ontario, Canada, MO.

Taxonomic notes

The anastomosing plexus of secondaries in the distal portion of the lamina is unlike that of other aroids and does not resemble any named type. This genus is similar in grade to Orontioideae but is not particularly close overall. Placement in its own subfamily is justified.

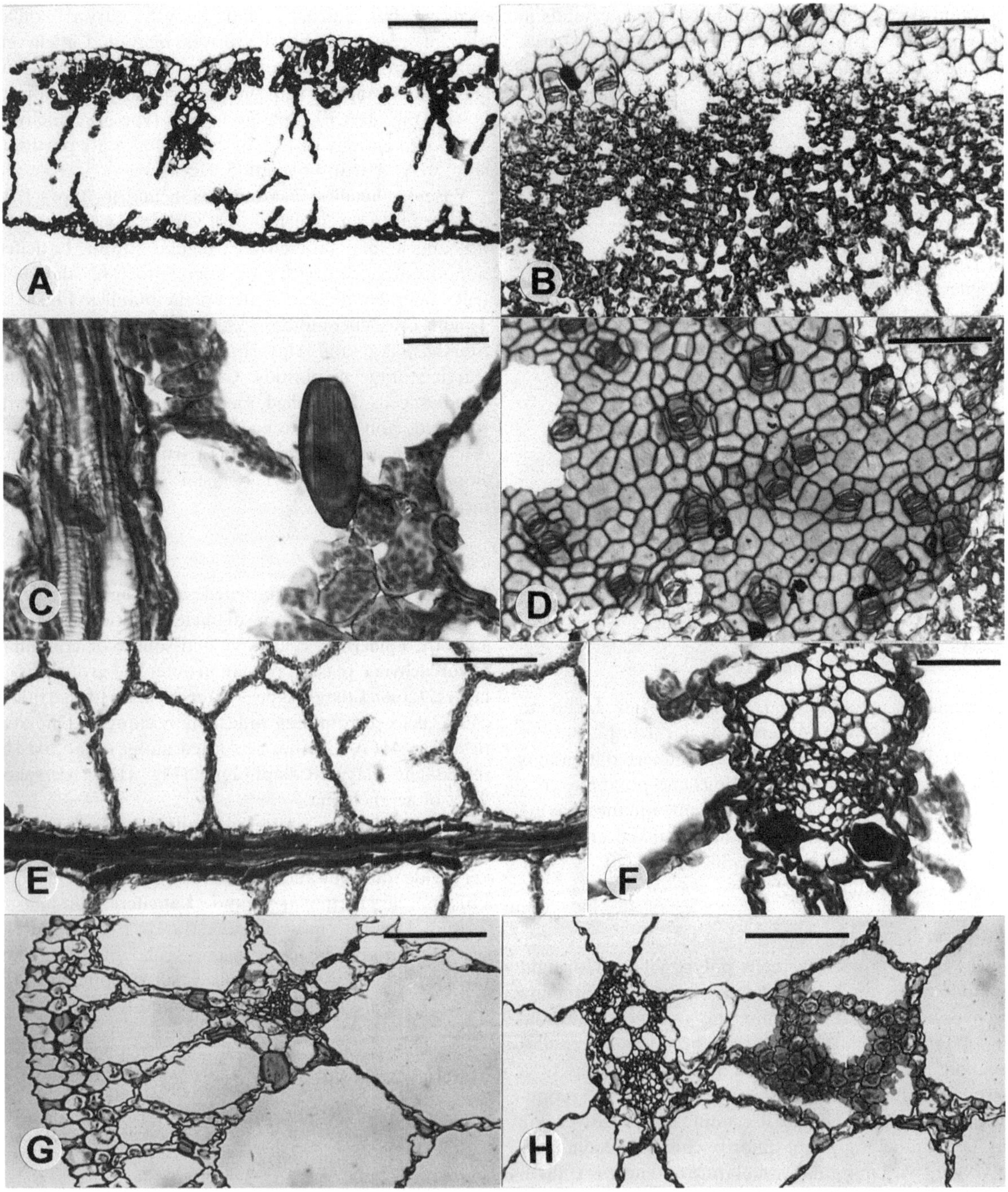

Fig. 41. *Calla palustris* leaf and petiole sections. (A) Lamina TS showing spongy layer entirely type 3 aerenchyma; (B) Lamina PS showing adaxial epidermis (top) and palisade cells. Gaps in palisade tissue are substomatal cavities; (C) Lamina PS showing type 3 aerenchyma and raphide cell; (D) Abaxial epidermis showing stomata; (E) Lamina PS showing type 3 aerenchyma and non-anastomosing laticifers at vascular bundle; (F) Midrib TS, type I vascular bundle; (G) Petiole TS margin; (H) Petiole centre showing vascular bundle and partial section of diaphragm. Scale bars: A = 200 µm; B,D,E,G,H = 150 µm; C,F = 50 µm.

Species reported on in the literature

(All species are *C. palustris;* listed as *C. aethiopica* in some references.)

Arber (1922a): rhizome, cambium. • Arber (1922b): leaf. • Buinova (1988): leaf. • Chauveaud (1897): root, phloem. • Chrysler (1904): stem. • Dalitzsch (1886): leaf. • De La Rue (1866): leaf. • Dudley (1937): rhizome, root. • Ertl (1932): leaf. • Flahault (1878): root. • French (1987c): root. • French (1988): laticifers. • French and Tomlinson (1981a): stem vasculature. • Gentner (1905): leaf. • Giovannozzi (1911): tuber. • Gravis *et al.* (1943) seedling. • Gulliver (1865): crystals. • Guttenberg (1968): root. • Hotta (1971): leaf. • Krause (1908): leaf, stem, root. • Kroemer (1903): root. • Lierau (1888): root. • Linsbauer (1930): epidermis. • Luhan (1957): rhizome. • Mangin (1882): root. • Metsävainio (1931): root. • Müller (1919): leaf tip. • Netolitzky (1929): crystals. • Ogden (1974): stem. • Olivier (1881): root. • Porsch (1905): leaf. • Roth (1949): leaf • Sakai and Hayashi (1973): leaf. • Schwendener (1882): shoot. • Siedler (1892): root. • Solereder and Meyer (1928): all organs. • Tieghem (1866): all organs. • Tieghem (1891): stem. • Tieghem and Douliot (1888): root. • Troll and Meyer (1955): leaf ontogeny. • Volkens (1883): leaf tip. • Voronkina (1975): root.

Additional literature

Antony 1898; Barabé and Forget 1987; Barabé and Labrecque 1983; Batygina and Yakovlev 1990; Bochenska and Kozlowski 1974; Davis 1966; Domin 1911; Erdtman 1952; French 1985b, 1986c, 1987a; Grayum 1990; Ibrahim 1965; Kirchner *et al.* 1934; Lehmann and Sattler 1992; Lipayeva 1989; McClure 1970; Meusel 1951; Ohlendorf 1907; Ravn 1894; Ray 1988; Scribailo and Tomlinson 1992; Thanikaimoni 1969; Tillich 1985; Williams *et al.* 1981.

SUBFAMILY PHILODENDROIDEAE

Tribe Philodendreae Schott

A tribe composed of *Philodendron*, second largest genus in the family, and *Montrichardia, Anubias, Furtadoa*, and *Homalomena.* Species vary in habit and range from the American tropics to south-east Asia and west Africa.

Anatomical summary

Leaf surface

Cuticle smooth to finely striate; rough or with raised ridges in *Furtadoa.* **Epidermis**: cells polygonal with straight or undulate anticlinal walls. **Trichomes** only in *Homalomena*, multicellular and uniseriate, noted rarely. **Stomata** on both surfaces or abaxial, and usually brachyparacytic. Stomata occasionally brachypara-hexacytic or -octocytic in *Furtadoa* and brachypara-tetracytic or -hexacytic in *Homalomena* and *Philodendron.* Subsidiary cells broad and rounded or angular; narrow in *Anubias.* **Venation**: midrib prominent, secondaries, intersecondaries and tertiaries often parallel. Secondaries merging bluntly at straightened brochidodromous collectors, or merging eucamptodromously.

Leaf TS

Structure dorsiventral, or tending toward isobilateral in *Montrichardia.* **Midrib** flat adaxially and rounded-convex abaxially or strongly protruding in *Anubias* and *Furtadoa*, or nearly embedded in *Homalomena*; variable in *Philodendron.* **Hypodermis** absent or present adaxially in midrib in *Philodendron.* **Mesophyll**: 1–4 palisade layers adaxially, often also lesser development abaxially in *Montrichardia.* Spongy layers: 5–10 (up to 15 in *Philodendron*), type 1 in *Furtadoa*, types 2 or 3 aerenchyma often present in *Philodendron.* **Collenchyma**: narrow or broad subepidermal band in abaxial midrib. **Vascular bundles**: few to numerous and scattered in midrib, type I or II in *Montrichardia*, type I in *Anubias*, type II in *Homalomena* and *Philodendron*, few and types I and II in *Furtadoa.* **Sclerenchyma**: fibre strands as phloem caps at larger bundles in *Montrichardia* and *Philodendron*, or ensheathing bundles in *Anubias* where isotropic wall layering often present, and in *Philodendron* subgenus *Pteromischum.* Strands showing variable cap development, including xylem and phloem caps in *Furtadoa* and *Homalomena.* **Secretory ducts** in mesophyll of *Furtadoa, Homalomena* and *Philodendron.* **Laticifers** non-anastomosing and 1–4 lateral to vascular bundles. **Tannin** cells common in *Montrichardia* ground tissue and *Philodendron* subgeuns *Meconostigma*, absent in *Furtadoa* and other *Philodendron* subgenera. **Crystals**: raphides in single bundles in small cells in *Montrichardia* and *Philodendron* or in large cells in *Furtadoa;* these plus single bundles in biforine cells in *Anubias* and *Philodendron*, except subgenus *Pteromischum.* Raphides as single or multiple bundles in large cells, and biforine-like in *Homalomena* and sometimes in *Philodendron.* Druses small or large in mesophyll.

Petiole TS

Outline irregularly terete distally, crescent-shaped or grooved proximally, size and outline highly variable in *Philodendron.* **Cuticle** smooth or striate; flanged in *Anubias.* **Epidermis**: cells small or square in most genera, often papillate or sub-trichome size in *Philodendron*. **Hypodermis** of several layers in *Philodendron* (except subgenus *Meconostigma*). **Ground tissue** compact near periphery, grading to type 3 or 4 aerenchyma at centre in *Montrichardia, Homalomena* and *Philodendron*. Cells all compact in *Anubias, Furtadoa* and some *Homalomena* species. **Collenchyma** strands replacing fibres distally in *Anubias*. In *Furtadoa, Homalomena* and *Philodendron*, collenchyma in an abaxial peripheral band. **Vascular bundles** usually types II or III, rarely type I in *Philodendron*. Bundles numerous and scattered, or arranged in concentric circles in *Anubias*, and occasionally *Homalomena*. **Sclerenchyma** fibre strands as semicircular phloem caps, absent in *Furtadoa*, or ensheathing vascular bundles in *Philodendron* subgenus *Pteromischum*. **Secretory ducts** present near organ periphery in *Furtadoa, Homalomena* and *Philodendron*. **Laticifers** non-anastomosing, lateral to vascular bundles in *Furtadoa, Philodendron* subgenera *Philodendron* and *Pteromischum, Homalomena* and *Anubias*. **Tannin** cells scattered in ground tissue. **Crystals**: raphides in single bundles in small cells in ground tissue, and biforine-like in most *Philodendron* species, true biforines in subgenus *Meconostigma*. Druses sometimes present in ground tissue.

Stem TS (Montrichardia and *Philodendron)*

Outline terete. **Cuticle** often thick and rough-surfaced in *Philodendron*. **Epidermis**: outer two cell layers thickly lignified without discrete cuticle in *Montrichardia*; small cells in *Philodendron* as well as some hypodermis development. Some papillate cell development in *Philodendron* subgenus *Pteromischum*. **Cork** cells occasional on surface of older *Philodendron* stems. **Collenchyma**: a peripheral band in *Philodendron*. **Ground tissue**: compact outer layers grading to type 2 aerenchyma at centre. **Vascular bundles** scattered, showing slight tendency to separation of cortex and central cylinder. In *Philodendron* subgenus *Philodendron*, central cylinder well-demarcated by circle of vascular bundles, often with fused boundary strands of xylem or phloem. Bundles mostly type III. Compound bundles of pairs and threes occasional in *Montrichardia*, numerous in *Philodendron*. **Sclerenchyma**: fibre strands as phloem caps of varying development. Fibre strands in centre of compound bundles, variable in *Philodendron* subgenus *Pteromischum*. **Secretory ducts** numerous in outer cortex of *Philodendron*. **Laticifers** present lateral to vascular bundles in *Philodendron*. **Tannin** cells scattered in ground tissue. **Crystals**: raphides in single bundles in small cells. Druses small, grading to crystal sand, except large in central cylinder in *Philodendron* subgenus *Pteromischum*.

Taxonomic notes

This group of genera near the base of the Philodendroideae expresses more characters than any other tribe in the family. This is mostly due to the large radiation of species in *Philodendron* and its anatomical complexity. Diagnostic differences are present and expected in this tribe but the reticulate nature of their character expression makes it a coherent group.

French (1987c) found the five genera now placed in this tribe to be united in having a sclerotic hypodermis in the root. This character is otherwise only known in the family in *Cercestis* and *Culcasia* (tribe Culcasieae), also at present placed in this subfamily. At the time of French's work, the genera bearing this character were commonly distributed among three subfamilies.

Montrichardia Cruger

This genus of two species is found in tropical America. Plants are rooted aquatics occurring in warm, standing or slowly moving fresh water at low elevations, and occasionally in tidal zones. The shrub-like herbs have an erect spongy stem. Leaves are long-petiolate, cordate, sagittate, or hastate. Chromosomes: $2n = 48$. One species was available for study.

Leaf surface

Cuticle smooth adaxially and finely striate abaxially. **Epidermis**: cells polygonal with straight anticlinal walls adaxially and undulate abaxially. Cells slightly smaller adaxially (1–2:1 *l/w*). **Stomata** on both surfaces and brachyparacytic (Fig. 42F). Subsidiary cells broad and rounded or angular. **Venation**: midrib prominent, secondaries and tertiaries parallel. Secondaries merging bluntly at straightened brochidodromous collector. Some tertiaries merging obliquely with other tertiaries in intercostal areas. Quaternaries orthogonal with tertiaries or reticulate.

Leaf TS

Structure dorsiventral and tending toward isobilateral. **Midrib** mostly flat adaxially and rounded-convex abaxially (Fig. 42A). **Cuticle** thin. **Epidermis**: cells somewhat larger adaxially. Cells flat externally and rounded internally. **Stomata** level with surface. **Mesophyll**: 1–2 palisade layers of tightly packed cells adaxially (Fig. 42C,D) and one layer often present abaxially. Spongy layers 5–8, in central third of lamina, arranged mostly in vertical partitions (type 2, occasionally type 3) with elongate-armed cells. **Air cavities** often partitioned as vertical sub-stomatal wells, up to 80% of spongy layer. **Collenchyma**:

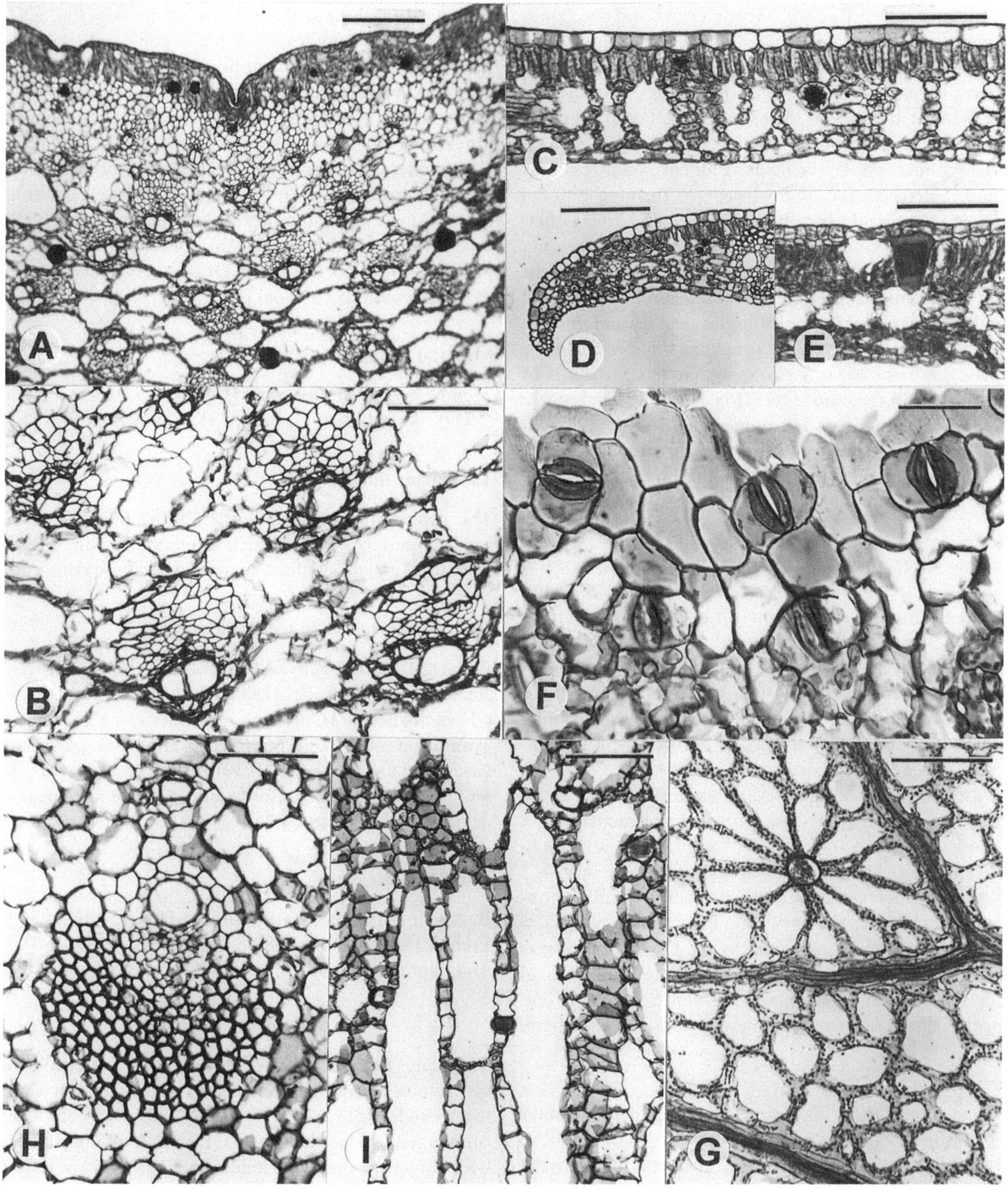

Fig. 42. *Montrichardia arborescens* leaf and stem sections. (A,B) Adaxial midrib TS showing type I or II vascular bundles in type 3 aerenchyma; (C,D,E) Lamina TS, note druse cell in (C), margin in (D) and raphide cell in palisade tissue in (E); (F) Leaf abaxial epidermis PS showing brachyparacytic stomata; (G) Mesophyll PS showing crystal cell centred in type 2 aerenchyma; (H) Stem TS near cortex-central cylinder boundary. Note type III vascular bundle with large fibre cap over phloem; (I) Stem LS showing elongate type 3 aerenchyma cavities and transverse diaphragm. Scale bars: A = 300 μm; B,C,E,G,H = 150 μm; D,I = 200 μm; F = 50 μm.

broad band, 6–8 cells deep, subepidermal and abaxial in midrib (type B).

Vascular bundles numerous and scattered in midrib at corners of partitions between large, type 3 air cavities (Fig. 42A,B), bundles large or small (type I or II). Xylem: 1–2 wide late protoxylem cells or lacunae with several narrower metaxylem tracheids. Phloem: small circular strand. **Sclerenchyma**: fibres absent or forming phloem caps, less frequently forming broad xylem caps. **Laticifers** lateral to vascular bundles and non-anastomosing, occasionally with dark granular contents. **Tannin** cells common in midrib ground tissue and lamina. **Crystals**: raphides occasional in single bundles, embedded in cells, often with dark amorphous contents. Cells rounded or short-cylindrical, at corners of aerenchyma partitions or vertically within palisade layer (Fig. 42E). Druses: coarse and large in upper or lower mesophyll (Fig. 42C), in circular cells at corners of radiating partitions of aerenchyma (Fig. 42G). Crystal sand common in mesophyll.

Petiole TS

Outline terete distally with flattened adaxial surface. Crescent-shaped proximally. **Size**: diameter 5–8 mm. **Cuticle** medium-thick and smooth or striate. **Epidermis**: cells small and square, rounded externally. **Stomata** occasional, level with surface and with small substomatal cavities. **Chlorenchyma**: plastids common in outer 6–8 layers. **Ground tissue**: cells small near periphery grading to large in central aerenchyma. **Air cavities**: type 3 aerenchyma in centre with partitions 1–3 cells long.

Vascular bundles numerous and scattered (usually type III). Xylem: very wide late protoxylem tracheid, 0–2 narrow metaxylem tracheids. Phloem: a circular strand. **Sclerenchyma**: fibres as semicircular phloem caps, 4–5 layers deep, best developed on peripheral vascular bundles and some larger central bundles. **Tannin** cells scattered and occasional in ground tissue. **Crystals**: raphides in single bundles in thin-walled irregular cells, occasional in peripheral ground tissue. Crystal sand common in epidermis, peripheral ground tissue and scattered elsewhere.

Stem TS and LS

Outline terete. **Size**: diameter 15 mm. **Epidermis**: outer two cell layers thickly lignified without discrete cuticle. **Ground tissue**: outer 6–10 layers of compact parenchyma cells grading to well-differentiated type 2 central aerenchyma. Longitudinally, elongate air cavities divided by occasional oblique, single-layered diaphragms with small circular corner air passages (Fig. 42I).

Vascular bundles scattered, slight tendency for uneven concentration of bundles dividing cortex from central cylinder. Bundles mostly collateral (type III) (Fig. 42H), few compound bundles, in elliptical pairs and triangular threes attached by confluent fibre caps (Fig. 6D). Xylem: one very wide early metaxylem tracheid per bundle. Phloem: small or large circular strand. **Sclerenchyma**: fibres in 5–10 thick-walled, crescent-shaped cell layers over phloem strands (Fig. 42H). **Tannin** cells common and scattered throughout. **Crystals**: raphides occasional to common, single bundles in rounded or short cylindrical, thin-walled cells; cells at corners of aerenchyma partitions or in cortical area. Druses very small, grading to crystal sand, occasional in ground tissue.

Material examined

Montrichardia arborescens (L.) Schott, Metcalfe s.n. (1959), West Indies; Croat 38331, Canal Zone, MO.

Taxonomic notes

The genus fits well within the Philodendroideae anatomical pattern. No distinctive characters are present and tribal isolation is not suggested by these data.

Species reported on in the literature

Blanc (1978): ontogeny. • Engler (1911): *M. arborescens*: stem, leaf. • French (1987c): *M. arborescens*: root. • French (1988): *M. arborescens*: laticifers. • French and Tomlinson (1981a, 1986): *M. arborescens*: stem and vasculature. • Solereder (1919): *M. arborescens, M. linifera*: leaf. • Solereder and Meyer (1928): *M. arborescens, M. linifera*: all organs. • Tobler (1922): *M. linifera*: stem.

Additional literature

Barabé and Lacroix 2001a; Domin 1911; Erdtman 1952; French 1985b, 1986c, 1987a; Grayum 1990; Mayo 1986; Thanikaimoni 1969.

Anubias Schott

This genus of eight species is found in west Africa. Most species are terrestrial rhizomatous herbs in wet shady forest understoreys. Three species may live completely submerged as true rheophytes. Leaves are long-petiolate with blades ranging from lanceolate to triangular, hastate to trifid. Chromosomes: $2n = 48$ for most species; one species, $2n = 72$. Five species were available for study.

Leaf surface

Cuticle smooth to slightly striate. **Epidermis**: adaxial cells polygonal with straight anticlinal walls (1–2:1 *l/w*).

Abaxial cells polygonal to curved, same size or larger than adaxial (1–4:1 *l/w*). **Stomata** on both surfaces or abaxial only, brachyparacytic, or with 2–3 pairs of narrow subsidiary cells parallel to guard cells (Fig. 43E).

Venation: in *A. heterophylla*, midrib prominent with pinnate, closely spaced, acute secondaries and inter-secondaries merging eucamptodromously at margin. Tertiaries crossing orthogonally or obliquely between secondaries. Some tertiaries and quaternaries reticulate. In *A. afzelii*, secondaries steeply acute, merging sub-marginally at a straightened brochidodromous collector. Tertiaries orthogonal to midrib, sometimes forming areoles with quaternaries. Quaternaries tending to be parallel with tertiaries. In *A. minima*, secondaries closely spaced, acute, curved-ascending, merging brochidodromously or eucamptodromously at margin, with straightened collector. Tertiaries and higher orders reticulate.

Leaf TS

Structure strongly or weakly dorsiventral. **Size**: 110 × 25 mm (*l/w*). **Midrib** flat or slightly concave adaxially, slightly convex to strongly protruding-rounded abaxially (Fig. 43A). **Cuticle** thicker adaxially. Cuticle flanged along anticlinal epidermal walls in one specimen. **Epidermis**: cells large, larger adaxially or same size on both surfaces. Outer cell walls flat; inner walls rounded. **Stomata**: guard cells level with surface, with double cuticular ledges. **Mesophyll** with 1–3 layers of palisade, up to 30% of lamina. Spongy tissue: 5–7 layers of loose, irregular cells, or long-armed cells forming a regular aerenchyma (Fig. 43B,C) (occasionally type 1, mostly type 2, rarely type 3). **Air cavities** as substomatal spaces, occupying 60–90% of spongy layers. Well-structured aerenchyma with cavities running between abaxial epidermis and palisade layer in *A. gilletii* and *A. heterophylla*. **Collenchyma**: 1–2-layered subepidermal band of thin-walled cells abaxially in midrib in most specimens.

Vascular bundles well-formed in midrib and lamina; midrib with up to 15 scattered bundles (type I) (Fig. 43A,F). Xylem with several helically-thickened, elongate protoxylem cells and 2–8 metaxylem cells. Phloem: 1–3 layers of medium-length sieve elements with horizontal sieve plates forming a strand with a generally broad boundary with xylem. **Sclerenchyma**: fibre sheaths encircling vascular bundles; up to three discrete wall layers visible in fibres. **Laticifers**: 2–3 non-anastomosing, occasionally branched strands along vascular bundles, lateral in fibre zone at juncture of xylem and phloem, common in all species. **Tannin** cells with dark contents scattered in mesophyll. **Crystals**: raphides common, single bundles in oval biforine cells in upper layers of type 2 or 3 spongy tissue; walls sclerotic, some with thin papillae on tips. Larger oval raphide cells with thin walls also common (Figs 8F, 43D). Druses small or large, common in sub-epidermal cells or upper mesophyll. Crystal sand very common in mesophyll. Frequency and position of crystals varying with species.

Petiole TS

Outline subterete distally, crescent-shaped or subterete and shallowly grooved adaxially. **Size**: diameter 1.5–4 mm. **Cuticle** thick on surface. Deep flanges may occur along anticlinal epidermal walls, or with flanges also penetrating ground tissue cell walls beneath epidermis. **Epidermis**: cells small, rounded. **Collenchyma**: strands forming phloem caps in distal sections; occasional weakly thickened subepidermal band 2–4 layers deep. **Ground tissue**: cells small near epidermis, grading to large in centre. **Air cavities** at cell corners in most species, or commonly as type 3 aerenchyma in *A. afzelii*.

Vascular bundles numerous, scattered, or in three concentric circles. Bundles small, type III in *A. heterophylla*, type II in other species. Xylem: one wide late protoxylem tracheid and 1–2 narrow metaxylem cells in *A. heterophylla*. Other species with up to two protoxylem cells and 1–6 narrow or medium-width metaxylem cells. Phloem: broad semicircular strand in *A. afzelii*, or small irregular circle in other species. **Sclerenchyma**: deep fibre caps, with thick-walled cells nearly encircling phloem at peripheral bundles (Fig. 43G). Fibre caps thinner and thin-walled in central bundles. In *A. gilletii*, outer ground tissue a *c.* 4-layered subepidermal sclerotic layer, confluent with outer phloem caps. Fibres often absent distally or replaced by thin-walled collenchyma thickenings. **Laticifers** usually two, lateral to phloem. **Tannin** cells common, scattered in ground tissue. **Starch** granules common or uncommon in all ground tissue. **Crystals**: raphides common in all ground tissue, in thin-walled cells. Basal tissue (proximal sections) with numerous raphides in *A. gilletii*, in large mucilaginous cells. Druses common in central ground tissue, in large circular cells.

Material examined

Anubias afzelii Schott, cult. K 9/62; *A. barteri* Schott var. *nana* (Engl.) Crusio, Bogner 1301, WAG, cult. M; *A. gilletii* Wildem. & Th.Dur., Bogner 1296, cult. M; *A. heterophylla* Engl., Bogner 1300, Cameroon; *A. minima* A.Chev., Knecht 11/53, Côte d'Ivoire.

Taxonomic notes

The species sampled here are anatomically diverse suggesting a need for a more thorough anatomical study.

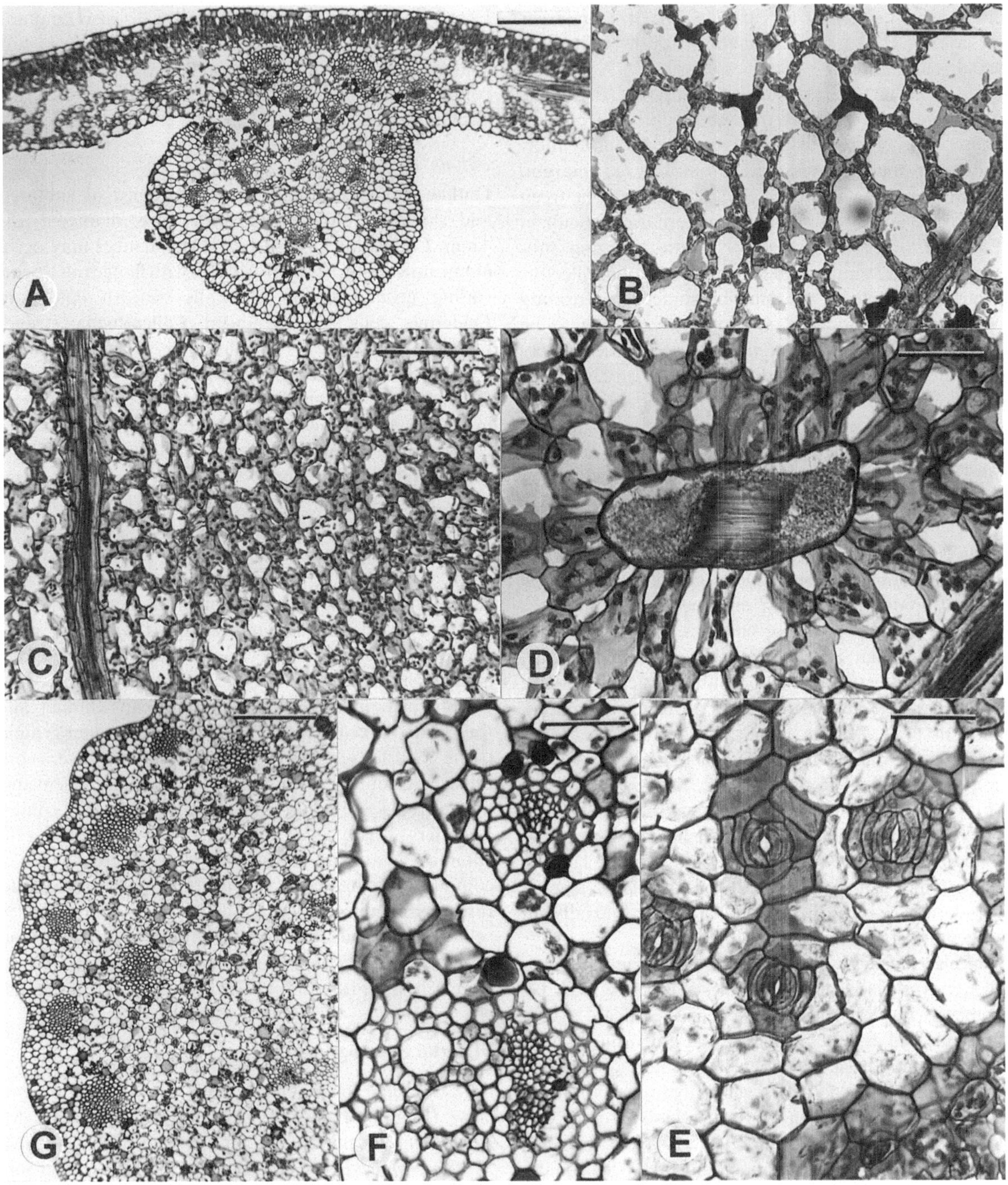

Fig. 43. *Anubias* leaf and petiole sections. (A,F,G) *A. heterophylla*; (B) *A. gilletii*; (C) *A. afzelii*; (D,E) *A. barteri* var. *nana*. (A) Midrib and lamina TS; (B) Mesophyll PS showing type 3 aerenchyma; (C) Mesophyll PS showing type 2 aerenchyma; (D) Mesophyll PS with raphide cell; (E) Abaxial epidermis PS showing brachypara-hexacytic stomata; (F) Midrib TS showing vascular bundles; (G) Petiole TS showing fibre caps over phloem. Scale bars: A,G = 300 μm; B,C = 150 μm; D–F = 50 μm.

Species reported on in the literature

Engler and Krause (1912): leaf. • French (1987c): several spp.: root hypodermis. • French (1988): *A. afzelii, A. barteri, A. gigantea*: laticifers. • French and Tomlinson (1981b): *A. barteri, A. heterophylla*: stem vasculature. • Knecht (1983): *A. barteri, A. gigantea*: leaf venation. • Solereder and Meyer (1928): *A. barteri, A.* sp.: leaf, crystals, secretory tissue.

Additional literature

Engler 1915; French 1985b, 1986c, 1987a; Grayum 1990; Mayo 1986; Meusel 1951; Ray 1988; Williams *et al.* 1981.

Furtadoa M. Hotta

This monotypic genus of epilithic rheophytes with rhizomatous stems grows on wet rocks along streams in Sumatra. The leaves are long-petiolate with an elliptic lamina. Chromosomes: 2n = 40. One specimen was available for study.

Leaf surface

Cuticle smooth to rough, usually with raised ridges forming coarse anastomosing pattern abaxially, or finer reticulate pattern adaxially. **Epidermis**: adaxial cells polygonal with straight or slightly curved anticlinal walls (1–3:1 *l/w*). Cells similar abaxially (1–4:1 *l/w*). **Stomata**: mostly abaxial, brachyparacytic, a few (10%) with up to 4–5 narrow, angular sets of subsidiary cells (brachypara-hexacytic or octocytic) (Fig. 44B). **Venation**: midrib prominent. Secondaries pinnate, narrowly acute, closely spaced, parallel and ascending. Secondaries merging eucamptodromously at margin. Tertiaries parallel with secondaries. Quaternaries mostly orthogonal to secondaries and tertiaries, and reticulate.

Leaf TS

Structure dorsiventral. **Size**: 44 × 14 mm (*l/w*) and 108 × 44 mm (*l/w*). **Midrib** flat adaxially. Abaxial side with strongly protruding rounded rib subtending one large vein (Fig. 44A). **Cuticle** medium-thick with irregular thick protuberances and thin spots. **Epidermis**: cells large adaxially, of variable size abaxially. Cells flat externally and flat or rounded internally on both surfaces. **Stomata** level with surface. **Mesophyll**: one palisade layer, up to 15% of lamina thickness. Spongy tissue: *c.* 10 layers of type 1 aerenchyma. **Air cavities** irregular and small, *c.* 25% of spongy tissue. **Collenchyma**: very slight tendency toward collenchymatous thickenings in subepidermal band in abaxial midrib.

Vascular bundles: three large bundles in midrib. Bundles mostly type II, largest type I. Xylem: largest bundles with 3–6 metaxylem elements opposing broad semicircular phloem strand, or small irregular or circular phloem strand. **Sclerenchyma**: fibres present as xylem caps, phloem caps, xylem and phloem caps, or absent. **Secretory ducts** common in mid-mesophyll, with single epithelial layer. **Laticifers** non-anastomosing, 2–3 lateral to all vascular bundles. **Starch** granules commonly stored in ground tissue near phloem. **Crystals**: raphides occasional in broad single bundles in large circular cells in mid-mesophyll. Crystal bundles tending to run parallel with secondaries.

Petiole TS and LS

Outline terete distally, adaxially grooved at base. Basally, rounded ridges separated by shallow grooves (Fig. 44C). **Cuticle** thick and striate-ridged. **Epidermis**: cells tall and square. **Collenchyma**: band of angularly thickened cells following periphery of organ, beginning three layers within epidermis and *c.* 10 layers thick (Fig. 44C,D); absent in adaxial groove of proximal petiole zone. **Chlorenchyma**: some chloroplasts infrequent in cells internal to collenchyma layer. **Ground tissue** uniform, mostly with no corner air spaces. **Air cavities**: cell-sized air channels common, especially near vascular bundles.

Vascular bundles randomly scattered, small (type II) (Fig. 44C). Xylem: one protoxylem lacuna and 1–3 metaxylem elements opposing large circular phloem strand (Fig. 44F). **Secretoryducts** present in collenchyma layer or interior to it. Ducts narrow with one epithelial layer surrounded by one cell layer of modified smaller ground tissue. **Laticifers**: non-anastomosing; 2–3 large laticifers, lateral and adjacent to phloem of all vascular bundles (Fig. 44D,E,F). **Tannin** cells common in ground tissue. **Starch** granules throughout ground tissue, especially abundant in cells adjacent to phloem (Fig. 44E). **Crystals**: raphides occasional in ground tissue. Cells small with single bundles, somewhat axially elongate (2:1 *l/w*).

Root TS

Outline terete. **Size**: diameter 1 mm. **Epidermis**: narrow or collapsed cells, dimpled on surface. **Exodermis**: 1–2 layers of enlarged cells with unevenly thickened walls. Thickenings mostly on internal periclinal walls and some interior anticlinal walls. **Hypodermis**: 1–2 lignified fibrous layers with uniformly thickened walls separated from exodermis by one thin-walled parenchyma layer. **Cortex**: thin-walled ground tissue, outer cells large and irregular; inner cells arranged in radial files, ending at endodermis

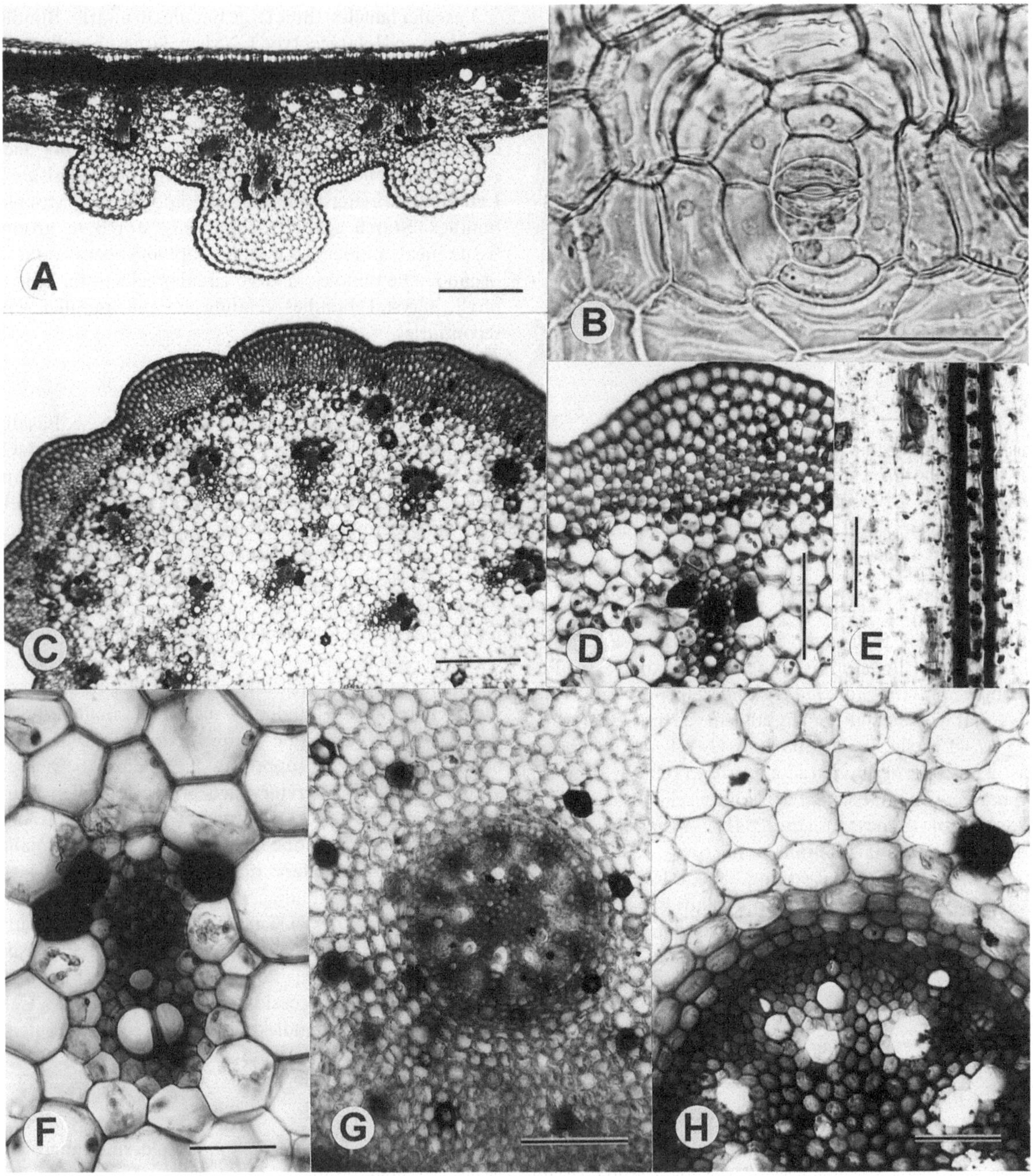

Fig. 44. *Furtadoa sumatrensis* leaf, petiole and root hand sections. (A) Lamina TS midrib cut distal to origin of secondaries. Note concentration of chloroplasts in palisade layer; (B) Lamina PS, abaxial epidermis and stoma; (C,D) Petiole TS, lateral side showing collenchyma band: (D) detail of vascular bundle and two laticifers; (E) Petiole LS showing two non-articulated laticifers and row of starch cells between them; (F) Petiole TS, type II vascular bundle; (G) Root TS, iodine-stained. Note stele and cortical tannin cells; (H) Root TS, trypan blue-stained showing sclerified stele centre. Scale bars: A,C = 300 µm; B,F,H = 50 µm; D,G = 150 µm; E = 200 µm.

(Fig. 44G,H). **Air cavities**: 20% air spaces among loosely packed cortical parenchyma. **Endodermis**: cylinder of cells bearing broad, thin casparian bands in anticlinal walls. **Pericycle**: single layer within endodermis. **Stele**: 10 protoxylem poles alternating with 10 circular phloem strands (Fig. 44G). **Sclerenchyma**: fibres in centre of stele (Fig. 44H). **Tannin cells**: scattered singly in cortex and stele (Fig. 44G).

Material examined

Furtadoa sumatrensis M.Hotta, cult. Bogor; M.

Species reported on in the literature

French (1987b, c): root. • French (1988): leaf.

Additional literature

French 1986c, 1987a; Grayum 1990.

Philodendron Schott

This genus consists of about 700 species of various habits and diverse ecology, which are widely distributed in the American tropics. Plants are primarily hemi-epiphytes as appressed climbers on trees. Some hemi-epiphytes become epiphytes by eventually losing connection with the ground. Internodes are long during the juvenile phase, becoming shorter at sexual maturity. Some are true epiphytes, germinating in trees and producing roots that later reach the ground. The leaves are highly variable and can be elliptic or deeply pinnately-lobed cordate, sagittate, pinnatifid-lobed, lanceolate-sagittate, pinnately-parted, or falcate-lobed. Some species are deeply rooted terrestrial, epilithic, or occasionally aquatic plants. They may grow in primary forests from low to middle elevations, and may occur as weedy species of disturbed sites. Some species are pachycauls growing in open, usually dry areas. Chromosomes small: $2n = 30, 32, 34, 36, 48$. The subgenera *Philodendron* (31 species examined), *Pteromischum* (four species examined), and *Meconostigma* (three species examined) are treated separately.

Subgenus *Philodendron*

Leaf surface

Cuticle smooth on both surfaces or striate abaxially. **Epidermis**: adaxial cells straight-sided polygonal (1–2:1 *l*/*w*). Abaxial cells similar to or larger than adaxial cells or more elongate (2–4:1 *l*/*w*). **Stomata**: abaxial, randomly oriented and brachyparacytic to brachypara-hexacytic (Fig. 45G). Subsidiary cells narrow or broad, curved, angular or irregular. **Venation**: midrib prominent or weak. Secondaries obtuse, widely or narrowly acute (proximal to distal) and parallel, curving toward margins or apically ascending, often converging at tip. Secondaries merging eucamptodromously at margin. Tertiaries orthogonal or quaternaries orthogonal or reticulate. Higher orders reticulate.

Leaf TS

Structure dorsiventral and usually well-differentiated. **Size**: 45 × 66 mm (*l*/*w*) to 95 × 62 mm (*l*/*w*). Leaves becoming very large. **Midrib** little thicker than lamina or shallowly concave adaxially; shallowly rounded-convex (Fig. 45D) or rounded V-shaped abaxially. Lamina may be grooved over larger secondary veins. **Cuticle**: medium-thick and sometimes deeply flanged between epidermal cells (Fig. 45C). Abaxial cuticle may be thinner. **Epidermis**: cells large or small, square to columnar or tile-like adaxially. Cells larger or same size abaxially, outer and inner walls flat. **Stomata** level with surface; conspicuous double ledges often present. **Hypodermis**: absent or 1–3 layers adaxially in midrib; or isodiametric cells beneath both epidermal layers; cells may contain dark material. In *P. crassispathum*, hypodermis of 1–4 layers adaxially and 1–3 layers abaxially (Fig. 45B). **Mesophyll**: 1–4 palisade layers of elongated or rounded cells, usually packed with plastids, occupying 10–30% of lamina (Fig. 45A,B). Spongy layers up to 15 cells deep (type 2). **Air spaces**: large substomatal cylindrical cavities perpendicular to surface, 30–90% of space in spongy layer, or air cavities irregular in some species. Paradermally, spongy tissue with one layer of unarmed cells below abaxial epidermis, then small cells or short-armed cells surrounding large air cavities extending below stomata to palisade cells; 3–5 spongy cells per uniseriate partition (type 3 aerenchyma). Large cavities in midrib separated by uniseriate partitions of ground tissue, partitions 3–4 cells long between junctions. **Collenchyma**: band of 4–5 discrete subepidermal layers of small cells on abaxial side of midrib, often extending to vascular bundles.

Vascular bundles: one to several large bundles in midrib, often at corners of net-like aerenchyma. Bundles well-delimited (type II) (Fig. 45B,D). Xylem: 1–5 wide metaxylem cells; xylem cells often ringed with single layer of small parenchyma cells. Phloem: small, circular or elliptical strand. **Sclerenchyma**: fibres, usually as phloem caps 1–3 layers deep and less commonly capping xylem; fibres ensheathing some laminar bundles; cell walls with up to five discrete wall layers. **Secretory ducts**: occasional in mid or upper lamina or midrib, with 1–2 epithelial

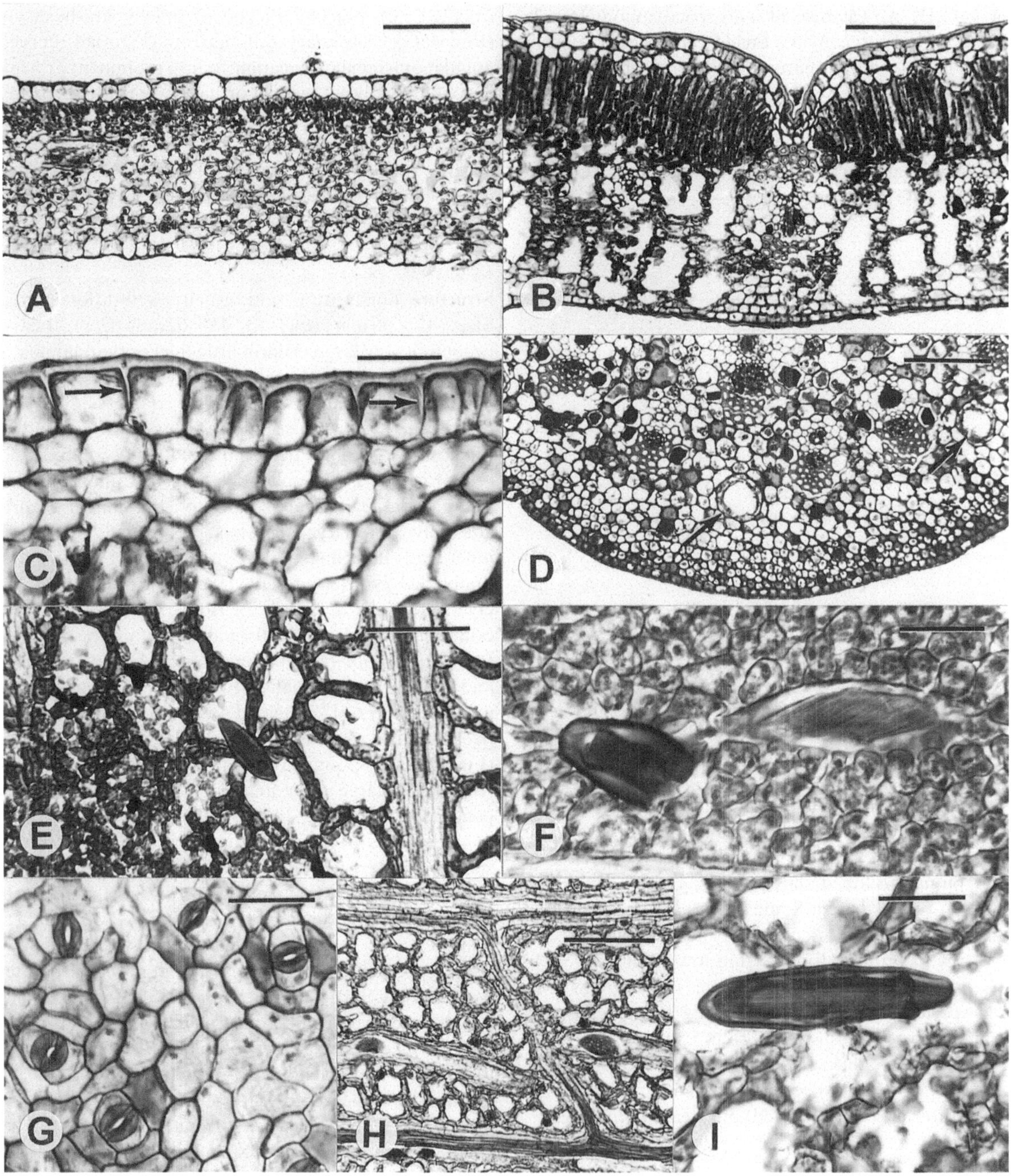

Fig. 45. *Philodendron* subgenus *Philodendron* leaf sections. (A,H,I) *P. pseudauriculatum*; (B,C) *P. crassispathum*; (D) *P. fragrantissimum*; (E) *P. smithii*; (F) *P. schottii*; (G) *P. brenesii*. (A,B) Lamina TS: (B) Note adaxial hypodermis and adjacent secretory ducts; (C) Lamina TS, adaxial surface showing cuticular flanges (arrows); (D) Midrib TS showing abaxial surface and secretory ducts (arrows); (E) Lamina PS showing type 3 aerenchyma and biforine raphide cell; (F) Lamina PS showing raphide cell and portion of another. Note oblique overlapping crystals; (G) Abaxial epidermal PS showing stomata; (H) Lamina PS showing parallel veins and portions of parallel secretory ducts; (I) Lamina PS showing biforine raphide cell. Scale bars: A,B,D,H = 200 µm; E = 150 µm; C,F,G,I = 50 µm.

layers (Fig. 45B,D,H). **Laticifers**: non-anastomosing and lateral to vascular bundles, 2–3 per bundle, outside fibres, often with short processes protruding between adjacent mesophyll cells; with dark amorphous contents. **Starch** granules abundant in midrib and not seen in lamina. **Crystals**: raphides common or uncommon in single bundles in short (Fig. 45F) or elongate cells, often biforine type (Fig. 45E,I), ends protruding across partitions into both adjacent air spaces (Fig. 45E), or in large rounded cells of upper or lower mesophyll. Raphide bundles occasionally surrounded by dark sheaths. In *P. crassispathum*, crystal cells elongated parallel with vascular bundles. Druses rare or common, often coarse and sharp-pointed, in rounded cells in palisade and spongy tissue partitions. Crystal sand absent or common in most mesophyll cells.

Petiole TS

Outline broadly oval with short wings facing adaxially; or somewhat laterally compressed and shallowly grooved adaxially. **Size**: diameter 0.8–2.5 cm. **Cuticle** smooth and thin. **Epidermis**: cells small, flat or rounded on surface and rounded to angular internally. Papillae round-tipped, unicellular and thin-walled, on abaxial surface, becoming 2–3 times height of ordinary epidermal cells in *P. verrucosum*, especially common on midrib and larger secondary veins. **Collenchyma**: continuous band of 4–10 cell layers (Fig. 46C,D) irregularly separated from epidermis by 1–4 ground tissue layers. Cells angularly thickened and longitudinally elongated (8–10:1 *l*/*w*). **Chlorenchyma**: up to five rows of cells inside collenchyma. In *P. jodavisianum*, chlorenchyma forming occasional perpendicular partitions (anticlinal to surface) dividing collenchyma band, reaching subepidermal layers. **Ground tissue**: cells of various sizes among numerous small to medium-sized air cavities separated by multi- or uniseriate partitions up to three cells long (types 3,4) (Fig. 46B). Cells often longitudinally elongated (3–6:1 *l*/*w*). **Vascular bundles** numerous and scattered, parallel in LS with some higher-order angular connections. Phloem mostly facing periphery (types II, III, or type I in *P. schottii*). Xylem: 1–3 protoxylem cells per bundle, with spiral to annular thickenings; occasionally one metaxylem element. Phloem: circular strands with sieve cells and companion cells forming organized grid. In LS, sieve elements with horizontal to 45° oblique end walls. **Sclerenchyma**: fibres forming phloem caps, 2–4 cells deep, usually not fully surrounding phloem laterally. In some species, fibre caps surrounding xylem and phloem, or ensheathing bundle.

Secretory ducts surrounded by epithelium and 1–2 layers of additional small cells, very common in collenchyma and common in ground tissue (Fig. 46D). **Laticifers**: non-anastomosing, 1–3, on one or both flanks of vascular bundles. **Starch** granules common throughout ground tissue, rare in collenchyma. **Crystals**: raphides common or occasional, cells narrow and biforine or biforine-like, oriented across aerenchyma partitions with ends pointing into air cavities, or cells randomly oriented. True biforines rarely seen. Druses present subepidermally and in cells bordering air cavities. Prismatics small, in small groups common in ground tissue.

Stem TS

Outline circular. **Size**: diameter 1–3 cm. **Cuticle**: medium-thick and smooth to rough-surfaced. **Epidermis**: cells small with rounded outer walls. **Cork** cells occasionally present subepidermally. **Hypodermis**: one layer of large cells external to 1–2 layers of thick-walled fibres, followed by thin-walled periderm cells. **Collenchyma**: band 3–10 layers deep, beneath periderm; cell walls angularly thickened. **Cortex**: loosely packed parenchyma. **Air cavities**: equivalent in size to adjacent cells or smaller, throughout ground tissue. **Central cylinder** weakly delimited by a circle of vascular bundles, some of which tangentially fused in pairs or threes by phloem fibres and confluent phloem strands. Cortical bundles randomly scattered. **Vascular bundles** collateral, highly variable in form and orientation. Compound bundles numerous in central cylinder (Fig. 46F,H) with irregular fibre strands in centre of groups of 2–8 bundle units. Small phloem strands outside fibres facing groups of 1–2 metaxylem elements on outer side. **Sclerenchyma**: simple collateral bundles may have large wide fibre strand surrounding phloem (Fig. 46G), a very small irregularly shaped strand of fibres, or no fibres. Fibre cell walls with up to seven discrete layers visible.

Secretory ducts small and numerous in collenchyma and outer cortex, less common elsewhere; duct cavities surrounded by two modified layers of small cells. Contents of cavities testing positively for lipids. **Laticifers**: absent or present as 1–3 cells lateral to many vascular bundles, occasional in phloem. **Tannin** cells occasional in ground tissue. **Crystals**: raphides as single, often oblique, bundles in elongate cells (Fig. 46E), most common in central cylinder. Druses of all sizes common in small cells in cortex.

Subgenus *Pteromischum*

Leaf surface

Cuticle smooth on both surfaces. **Stomata** brachyparacytic and abaxial. **Epidermis**: cells polygonal with straight anticlinal walls.

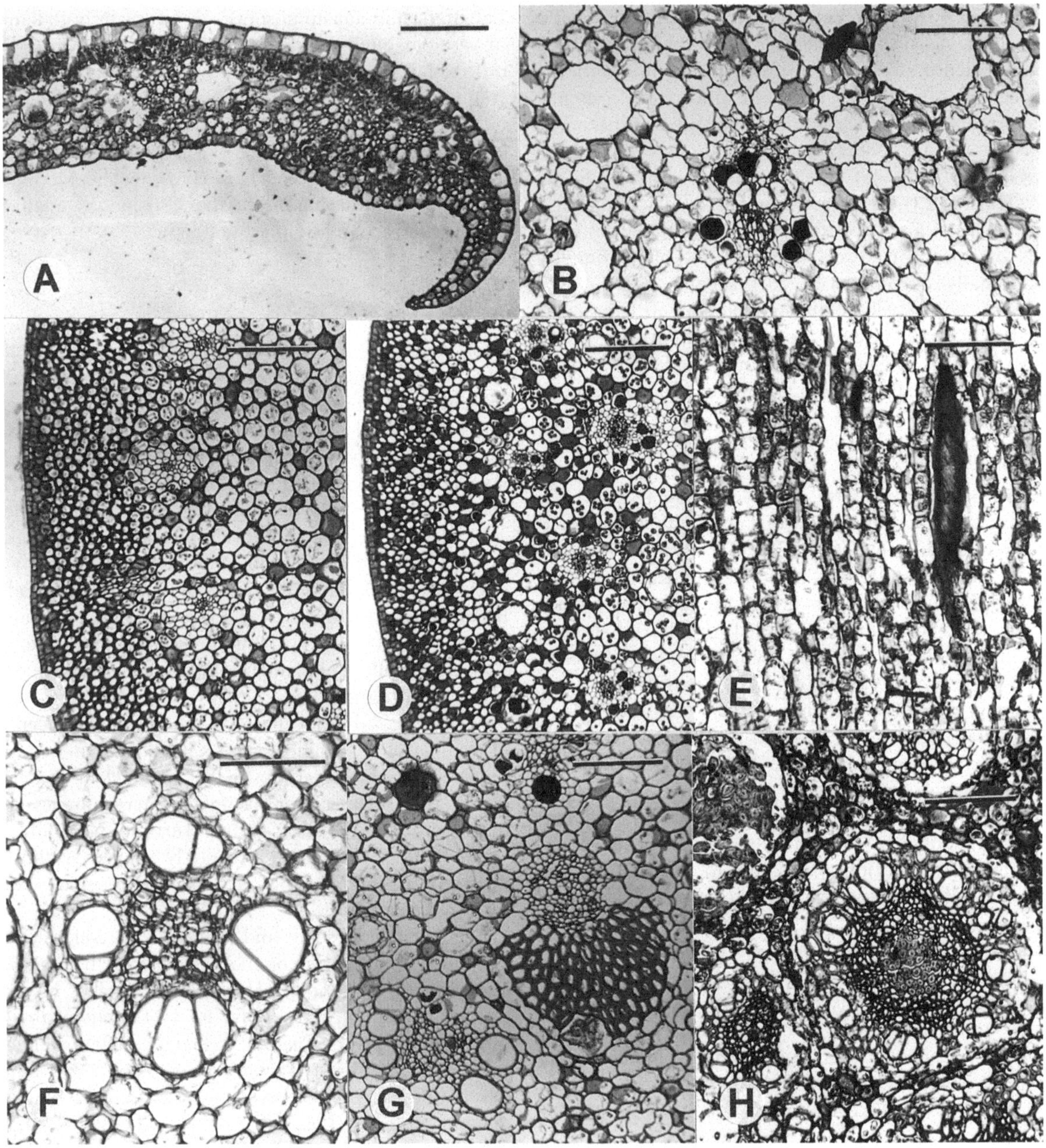

Fig. 46. *Philodendron* subgenus *Philodendron* petiole and stem sections and subgenus *Pteromischum* leaf section. (A) subgenus *Pteromischum*; (B–H) subgenus *Philodendron*; (A) *P. rhodoaxis*; (B) *P. brenesii*; (C) *P. acuminatissimum*; (D) *P. fragrantissimum*; (E) *P. microstictum*; (F) *P. warszewiczii*; (G) *P. bakeri*; (H) *P. scalarinerve*. (A) Lamina TS at margin; (B) Petiole TS, centre showing type II vascular bundle and type 4 air cavities; (C) Petiole TS, lateral side showing deep subhypodermal collenchyma band; (D) Proximal petiole TS, lateral side showing row of secretory ducts interior to collenchyma band; (E) Stem LS showing elongated raphide cell with overlapping crystals; (F) Stem TS showing compound vascular bundle; (G) Stem TS showing compound bundles and collateral leaf trace with enlarged fibre strand adjacent to phloem; (H) Stem TS, central cylinder showing compound, nearly amphivasal bundle with circular fibre strand in centre. Scale bars: A–C,E,G,H = 200 μm; D = 300 μm; F = 150 μm.

Leaf TS

Structure dorsiventral. **Size**: 10 × 6 cm. **Midrib** prominent, concave adaxially and semicircular abaxially. **Cuticle** thin or thick on both surfaces. **Epidermis**: cells large and square to columnar adaxially, somewhat smaller to half as large and square abaxially (Fig. 46A). **Stomata** level with surface. **Mesophyll**: *c*. 7 poorly differentiated layers. Palisade of 1–3 layers of squarish cells (Fig. 46A). Spongy cells in about six layers, large, irregularly shaped, and somewhat horizontally elongate (type 2). Chloroplasts concentrated mostly in adaxial half of mesophyll, or only slightly more abundant adaxially. **Air cavities** small and irregular, up to 10–25% of spongy layer. **Collenchyma**: abaxial subepidermal band of 2–6 angularly thickened layers in midrib.

Vascular bundles numerous and scattered in midrib, with phloem oriented abaxially throughout or phloem oriented randomly on adaxial side; type II. Xylem: 1–5 metaxylem cells opposing circular or elliptic phloem strand. **Sclerenchyma**: fibres ensheathing larger bundles and often separating xylem and phloem; or only as phloem caps in smaller bundles in lamina. Larger laminar bundles may also have tall transcurrent caps over xylem and phloem. **Secretory ducts** of two sizes. Small ducts in midrib forming subepidermal abaxial row, with second row beneath peripheral row of vascular bundles, and some scattered in centre. Secretory ducts 4–5, each the size of 4–6 vascular bundles, surrounding centre of midrib. Laminar ducts common between spongy and palisade layers, or in abaxial mesophyll evenly spaced between vascular bundles. **Laticifers** non-anastomosing, 1–2 per vascular bundle in lamina. **Starch** granules in all ground tissue but densest in centre of midrib. **Crystals**: raphide cells uncommon subepidermally. Druses large and coarse in both epidermal layers, every 3–4 cells, in cells partially displaced from surface; each druse with one especially long, sharp prismatic point oriented toward organ surface. Large druses may be common throughout mesophyll.

Petiole TS

Outline crescent-shaped. **Size**: 4–9 mm diameter. **Cuticle** medium-thick and smooth. **Epidermis**: cells small and rounded, occasionally pointed or papillate on outer surface. **Periderm**: occasional regions of periclinal divisions beneath epidermis, or absent. **Collenchyma**: 4–10 layers beneath epidermis, occasionally interrupted by radial chlorenchyma partitions. **Chlorenchyma** in outer ground tissue, cells small. **Ground tissue** gradually becoming larger at centre, cells closely packed with occasional cell-sized air cavities or small type 3 aerenchyma in adaxial central region of *P. ensifolium*.

Vascular bundles numerous and scattered; types II and III. Xylem: 2–3 wide metaxylem cells at broad boundary with broad phloem strand. **Sclerenchyma**: fibres ensheathing vascular bundles and dividing xylem and phloem. **Secretory** ducts occasional or common, in collenchyma and central ground tissue; small or larger with 1–2 epithelial and modified surrounding layers. **Laticifers** non-anastomosing, on each side of phloem strand. **Crystals**: raphides occasional in single bundles. Druses frequent in subepidermal cells, bearing enlarged points facing epidermis in *P. ensifolium*.

Stem TS

Outline terete or flattened at axillary bud. **Size**: diameter 9 mm. **Cuticle** smooth and thin. **Epidermis**: cells regularly or irregularly papillate. **Collenchyma**: band of 5–12 layers of angularly thickened cells beneath epidermis. Chloroplasts common in collenchyma and in up to eight layers of ground tissue beneath outer cortex. **Cortex**: parenchyma cells rounded and loosely packed with numerous nearly cell-sized intercellular spaces. **Central cylinder** forming irregular or sharp boundary with cortex but without endodermis. Parenchyma similar to that of cortex, with numerous cell-sized air cavities and conspicuous intercellular spaces at corners, or air spaces absent. **Vascular bundles**: cortex with simple, collateral bundles (types II, III), phloem normally oriented outward. Outer central cylinder with loosely scattered bundles; bundles collateral or mostly compound in central cylinder. Xylem: usually 1–3 wide xylem cells. Phloem: circular strand. Central compound bundles with irregular strands of phloem fibres irregularly confluent among groups of 3–6 vascular bundles linked at phloem end. In central part of central cylinder of *P. alliodorum*, 2–8 phloem/xylem units clustered into one compound vascular bundle, with xylem oriented outward.

Sclerenchyma: fibres in large fluted strands in centre of compound bundles. Phloem in grooves of fibre strand; 4–10 layers of cells in simple or collective phloem caps; walls thin with large lumina. French and Tomlinson's (1984) typology cannot readily be applied to these specimens. **Secretory ducts**: small, with 1–2 layers of epithelial cells, common in cortex and collenchyma area, occasionally with dark contents. In *P. inaequilaterum*, very large ducts (2–4 times the size of vascular bundles) occurring with 1-layered epithelium of 8–10 cells and up to two additional modified surrounding rows of cells; these ducts forming a conspicuous single circle in inner cortex, producing copious quantities of mucilage when cut. **Laticifers** laterally attached to vascular bundles, 2–4 per bundle, and also more commonly found in cortex. **Starch** granules common in areas of cortical and central cylinder ground tissue. **Crystals**: raphides in single bundles, common in rounded cells in cortex. Druses large and common in central cylinder; smaller and fewer in cortex, except abundant in discontinuous subepidermal groups; fewer elsewhere, occasionally up to 4 druse cells around veins. Druses may be sharp-pointed and abundant in hypodermal area.

Subgenus *Meconostigma*

Leaf surface

Cuticle smooth. **Epidermis**: cells polygonal (1.5–2:1 l/w) on both surfaces and of equivalent size. **Stomata**: abaxial only, brachypara-tetracytic or occasionally brachypara-hexacytic. Subsidiary cells mostly narrow. **Venation**: midrib prominent. Secondaries and tertiaries acute, straight, close and parallel, collected eucamptodromously at margin. Quaternary bridge veins crossing between tertiaries.

Leaf TS

Structure dorsiventral. **Size**: length >30 cm. **Midrib** flat or slightly convex adaxially and irregularly rounded or more usually prominent and semicircular abaxially. **Cuticle** thick adaxially and thin abaxially. **Epidermis**: cells flattened or large and square. **Stomata** slightly sunken, cuticular ledges present or absent. **Mesophyll** well differentiated; 1–3 palisade layers, cells of irregular height in narrow band (*c.* 10% of mesophyll), and packed with plastids, or cells taller, in a more prominent band (Fig. 47A). Spongy tissue of 7–8 layers, formed nearly entirely of uniseriate partitions surrounding large circular substomatal air cavities occupying up to 90% of spongy layer (type 3); cells horizontally elongate and not lobed. **Collenchyma**: abaxial and adaxial bands, 2–15 subepidermal layers. Band usually continuous and thicker abaxially, thinner and more frequently interrupted by chlorenchyma adaxially (Fig. 47B).

Vascular bundles scattered and not numerous in midrib (Fig. 47B); type II. Xylem: in midrib bundles up to 3–6 protoxylem and metaxylem elements opposing small elliptic phloem strand. **Sclerenchyma**: fibres present in lamina as irregular sheath to secondary or higher order bundles, or as two strands capping phloem and xylem poles. **Secretory ducts** with single epithelial layer and dark contents, present in spongy layer parallel to and between secondary vascular bundles (Fig. 47A). **Tannin**: idioblastic cells with dark contents common and randomly found in mesophyll. **Laticifers**: 2–4 lateral to most vascular bundles. **Crystals**: raphides in single bundles very common in large rounded cells, mostly among palisade cells, occasionally in spongy layer. Cells often biforine-like, anchored by one end and suspended in aerenchyma spaces (Fig. 47C). Raphide bundles often surrounded in cells with dark amorphous material. Druses small, common in mesophyll.

Petiole TS

Outline: slightly dorsiventrally compressed, mostly convex-rounded abaxially with concave groove edged by two low sharp ridges adaxially. **Size**: up to 2 cm diameter. **Cuticle** thick and smooth. **Epidermis**: cells small, outer walls flat, inner walls variably shaped. **Collenchyma**: an irregular subepidermal band of 10–50 layers of small angularly thickened cells (Fig. 47F,H). **Chlorenchyma** forming isolated radial ribs through the collenchyma, often flared at outer or inner end, or both. **Ground tissue**: central cells ± larger than peripheral cells. **Air cavities** small in outer cortical area, becoming very large in centre; cavities separated by 2–3-seriate partitions (type 4) (Fig. 47E). **Vascularbundles** scattered throughout (types I and II) (Fig. 47E–G). Xylem: up to 10 cells in several files opposing rounded phloem strand. **Sclerenchyma**: fibres forming up to 10-layered caps over phloem (Fig. 47H), and 1–2 loosely packed cell layers over xylem parenchyma.

Secretory ducts small, size of parenchyma cells, surrounded by epithelium plus 1-2 modified cell layers; ducts scattered throughout ground tissue and collenchyma (Fig. 47F). **Starch** granules present in one layer of ground tissue surrounding phloem bundle caps, mostly of outer vascular bundles. **Crystals**: biforine raphides abundant in specialized cylindrical cells with pointed cells, attached by one end to air cavity partitions, and suspended in cavities. Cells readily extruding raphides through papillae (Fig. 47D) at thin-walled ends when disturbed or deformed. Druses common in small rounded cells lining air cavities and generally in ground tissue.

Material examined

Subgenus*Philodendron*

P. acuminatissimum Engl. ssp. *doratophyllum* 27-74-30, cult. SEL; *P. bakeri* Croat, Croat 33710, Panama, MO; *P. brenesii* Standl., Croat 35519, Costa Rica, MO; *P. crassispathum* Croat, Croat 33150, Panama, MO; *P. cretosum* Croat & Grayum, Croat 37344, Panama, MO; *P. dolichophyllum* Croat, Croat 77008, Panama, MO; *P. dominicalense* Croat, Croat 35268, Costa Rica, MO; *P. fragrantissimum* (Hook.) Kunth, Croat 76548, Panama, MO; Croat 37574, Panama, MO; *P. grazielae* Bunting, Bogner 1286, cult. M; *P. hederaceum* (Jacq.) Schott, Croat 37049, Panama, MO; *P. helleniae* Croat, Croat 37576, Panama, MO; *P. jodavisianum* Bunting, Croat 77011, Panama, MO; Croat 34771, Panama, MO;

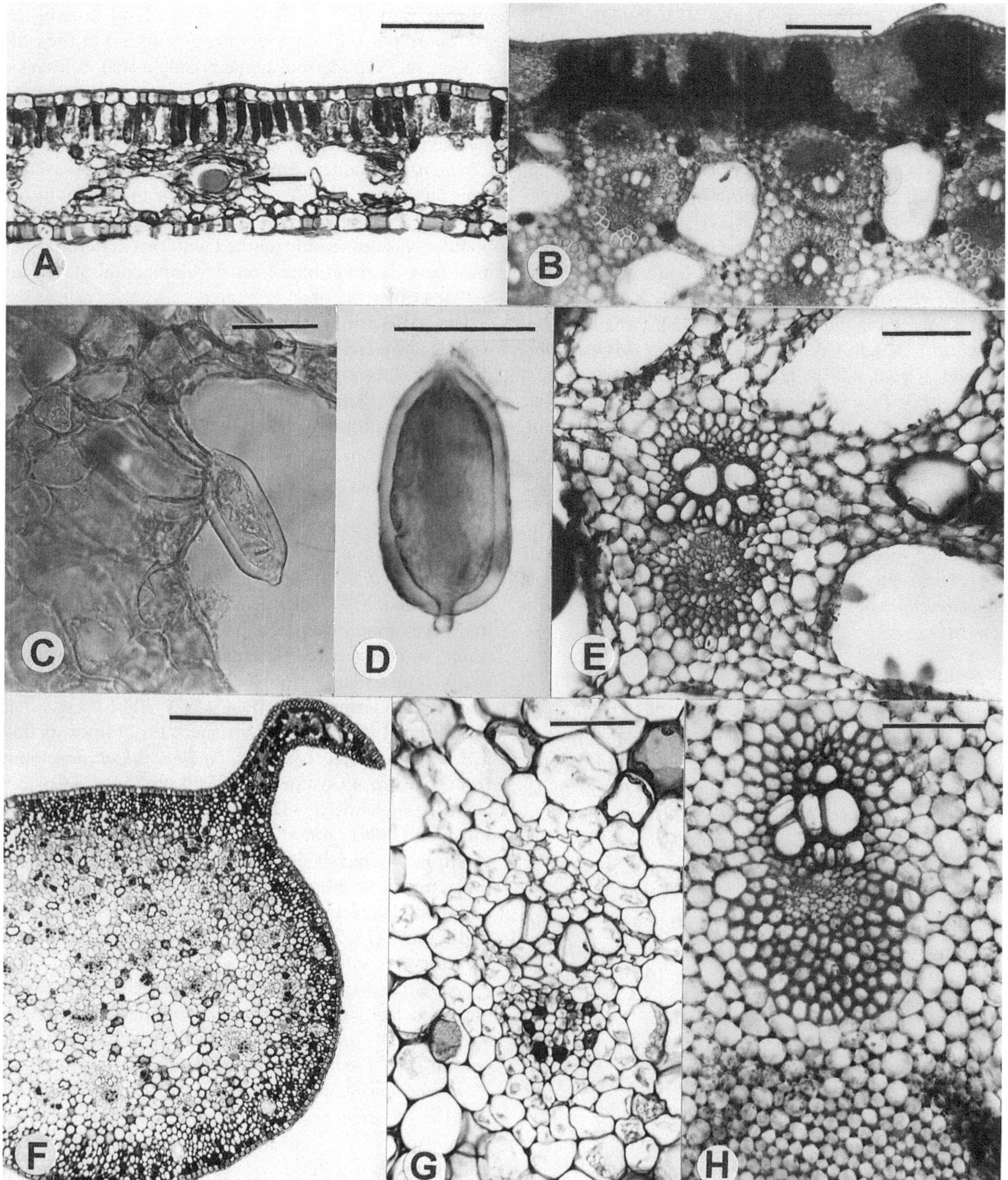

Fig. 47. *Philodendron* subgenus *Meconostigma* leaf and petiole sections. (A) *P. goeldii*; (B–E,H) *P. bipinnatifidum*; (F,G) *P. leal-costae*. (A) Lamina TS showing type 3 spongy aerenchyma and secretory duct (arrow); (B) Midrib TS showing adaxial epidermis and zones of chlorenchyma separated by subepidermal strands of collenchyma; (C) Midrib TS showing air cavity with thick-walled biforine attached to wall; (D) Petiole centre TS, air cavity with detached biforine, crystal bundle barely visible in cell centre; (E) Petiole TS centre, with type II vascular bundle and type 4 aerenchyma; (F–H) Petiole TS: (F) centre of organ with type 4 aerenchyma; (G) type II vascular bundle; (H) type II vascular bundle with fibre caps and adjacent collenchyma band. Note wedge of chlorenchyma in band at lower right corner. Scale bars: A,H = 150 μm; B,E = 200 μm; C,D,G = 50 μm; F = 300 μm.

P. lentii Croat & Grayum, Croat 37481, Panama, MO; *P. microstictum* Standl. & L.O.Williams, Croat 35189, Costa Rica, MO; *P. pedatum* Kunth, Croat 77249; *P. pseudauriculatum* Croat, Croat 76554, Panama, MO; *P. purpureoviride* Engl., Croat 35348, Costa Rica, MO; *P. rothschuhianum* (Engl.) Croat & Grayum, Croat 35657, Costa Rica, MO; *P. rugosum* Bogner & Bunting, Bogner 1522, Ecuador; *P. sagittifolium* Liebm., Croat 77012, Panama, MO; Croat 35254, Costa Rica, MO; Croat 35343, Costa Rica, MO; Croat 40197, Mexico, MO; Croat 39415, Mexico, MO; *P. scalarinerve* Croat, Croat 37569, Panama, MO; *P. scandens* K.Koch & Sello, Croat 22909, Panama, MO; Croat 34454, Panama, MO; *P. schottii* K.Koch, Croat 33684, Panama, MO; *P. smithii* Engl., Croat 42675, Honduras, MO; Croat 40223, Mexico, MO; *P. straminicaule* Croat & Grayum, Croat 33146, Panama, MO; *P. sulcicaule* Croat & Grayum, Croat 33734, Panama, MO; *P. tenue* C.Koch & Augustin, Croat 35691, Costa Rica, MO; *P. verapazense* Croat, Croat 41656, Guatemala, MO; *P. verrucosum* Mathieu ex Schott; Croat 33991, Panama, MO; Croat 76558, Panama; *P. warszewiczii* C.Koch & Bouché, Croat 42891, Nicaragua, MO; *P. wilburii* Croat & Grayum var. *longipedunculatum* Croat & Grayum, Croat 35249, Costa Rica, MO.

Subgenus ***Pteromischum***

P. alliodorum Croat & Grayum, Croat 76557, Panama, MO; Croat 77025, Panama, MO; *P. ensifolium* ssp. *ensifolium* Croat & Grayum, Croat 76558, Panama, MO; Croat 77010; Croat 34783, Panama, MO; *P. inaequilaterum* Liebm., Croat 77009, Panama, MO; *P. rhodoaxis* Bunting ssp. *lewisii* Croat & Grayum, Croat 33761.

Subgenus ***Meconostigma***

P. bipinnatifidum Schott, cult. SIUC; *P. goeldii* G.M.Barroso, Plummer 136, Brazil, cult. Burle-Marx; *P. leal-costae* Mayo & Barroso, Harley 19428, Brazil.

Subgenus undetermined

P. auyantepuiense Bunting, Bogner s.n., Venezuela; *P. martianum* Engl. (Sect. Paursia), cult. M; Croat 33526, Panama, MO; *P. squamiferum* Poepp. & Endl., Croat 77248.

Taxonomic notes

Within the Araceae, the genus *Philodendron* has the widest array of character states for any genus. The characters discovered in this study support a unified concept of the genus although there are subtle differences among the subgenera. For example, large, thick-walled biforine raphides are the only type found in subgenus *Meconostigma* while a wider array is found in the other subgenera. Hypodermal tissue is only noted in leaves of subgenus *Philodendron*. Only type 4 spongy aerenchyma is found in petioles of subgenus *Meconostigma*. Very few identifying characters are noted in subgenus *Pteromischum;* the smaller sample of species shows less variation than subgenus *Philodendron*. French and Tomlinson's (1981b) proposed typology for compound vascular bundles was not readily applied with this material. There may be variations based on developmental stages and perhaps other factors.

The MBB tribes Anubiadeae and Homalomeneae fit well near *Philodendron* but are distinguished by having the more elaborate type I vascular bundles in the leaves and petioles. Those genera fit well in the tribe Philodendreae as proposed here (Table 5).

Species reported on in the literature

Ambronn (1879–81): *P. eximium*: collenchyma. • Araus *et al.* (1986b): *P. scandens*: leaf. • Baranetzky (1897): *P. macrophyllum, P. pinnatifidum, P. tripartitum*: shoot apex. • Blanc (1977a, b): *P. fenzlii, P. speciosum*: axillary bud development. • Bloch (1926): *P. glaziovii, P. populneum*: root tissue redifferentiation. • Bloch (1944): *P. glaziovii*: meristem, wound redifferentiation. • Bolli (1953): wound periderm. • Bretfeld (1879–81): *P.* sp.: leaf. • Bunting (1968): *P. scandens*, etc. : shoot apex, petiole, leaf. • Chauveaud (1897): *P. bipinnatifidum*: developing roots, phloem. • Costantin (1885): 4 spp.: root. • Costerus (1875): root, lenticels. • Dalitzsch (1886): several spp.: leaf. • Daniel (1899): grafts. • Davie (1917): *P. selloum*: leaf venation. • Dilcher and Daghlian (1977): *P. limnestes*: (palaeo) leaf venation, cuticle. • Engler (1911): *P. bipinnatifidum, P. crassinervium, P. pinnatifidum*: root. • Engler and Krause (1912): Philodendroideae: several spp.: root, secretory tissue, leaf, crystals. • Ertl (1932): *P. oxycardium, P. pinnatifidum, P. triumphans*: leaf venation. • Fisher and French (1976): *P. oxycardium*: intercalary meristems. • French (1987b,c): several spp.: root. • French (1988): several spp.: laticifers. • French and Tomlinson (1980): *P. mexicanum:* stem vasculature. • French and Tomlinson (1981b): *P. seguine:* stem vasculature. • French and Tomlinson (1984): >40 spp.: stem vasculature. • French and Tomlinson (1986): *P. selloum, P.* sp.: compound vascular bundles. • Freundlich (1908): *P. erubescens*: leaf vascular bundles. • Gentner (1905): *P. panduraeforme*: leaf tip. • Genua and Hillson (1985): *P. fenzlii*: leaf crystals. • Giovannozzi (1911): *P. corsinianum, P. eximium, P. selloum*: root suberin. • Goebel and Sandt (1930): 4 spp.: aerial roots. • Grayum (1990): anatomical review. • Gulliver (1865): *P. giganteum*: crystals. • Guttenberg (1940): several spp.: root. • Guttenberg (1968): *P. selloum*: root. • Hotta (1971): *P. bipinnatifidum:* leaf. • Kattein (1897): *P. ceratocaule*: root stele, phloem. • Keating (1997): several spp.: leaf, petiole, stem. • Klebahn (1884): lenticels. • Leblois (1887): secretory canals, laticifers. • Lierau (1888): several spp.: root. • Lindinger (1905–6): *P. selloum*: root. • Linsbauer (1930): root, leaf. • Lohse (1924): *P. erubescens, P. eximium, P. imbe*: tyloses. • Mangin (1882): root, stem. • Mayo (1991): *P.* subg. *Meconostigma* spp.: leaf venation, root. • Meyer (1925): *P. bipinnatum, P. speciosum*: root. • Meyer (1959, 1962): *P. bipinnatifidum*: leaf function, assimilation. • Möbius (1885): several spp.: secretory tissue. • Molisch (1899): *P. cannaefolium, P. tanyphyllum, P. vetteranum*: general. • Morel (1910): *P. crassinervium*: root. • Müller (1919): several spp.: leaf tip. • Netolitzky (1929): *P .melanochrysum, P. oxycardium*: crystals. • Olivier (1880, 1881): several spp.: root. • Pant and Kidwai (1966): *P. lacerum:* leaf, stomata. • Perumalla *et al.* (1990): *P. bipennifolium, P. scandens* ssp. *scandens*: root hypodermis. • Peterson and Perumalla (1990): *P. wendlandii*: root hypodermis. • Petit (1889): *P. lacerum*: petiole. • Pfitzer (1872): *P. cannaefolium*: leaf epidermis. • Pinkerton (1936): *P. cordatum:* root, root hairs. • Politis (1914): *P. oxycardium, P. melanochrysum*: starch, crystal cells. • Porsch (1911): *P. selloum*: root. • Potgieter and van Wyk (1992): intercellular pectin. • Pray (1957): *P. oxycardium*: leaf ontogeny. • Priestley and Woffenden (1922): *P. erubescens*: root endodermis, periderm. • Ritterbusch (1971): *P. houlletianum*: stem vasculature. • Rosanoff (1867): *P. sellowianum*: petiole crystals. • Roth (1949): leaf ontogeny. • Roth (1990): leaf. • Roth and Fast (1954): leaf development. • Rywosch (1909): branch root development. • Scherer (1904): *P. grandifolium, P. selloum, P. melanchrysum*: vasculature. • Schimper (1884, 1888): epiphyte structure, roots. • Schwendener (1882): endodermis. • Shishkoff (1987): root hypodermis. • Siedler (1892): root. • Solereder (1919): *P. selloum:* leaf. • Solereder and Meyer (1928): several spp.: anatomy. • Stahl (1896): *P. lindeni, P. verrucosum*: leaf. • Tieghem (1866): several spp.: all organs. • Tieghem (1870–1): *P. crinipes, P. lacerum, P. micans*: root, secretory tissue. • Tieghem (1872): several spp.: secretory ducts. • Trécul (1865, 1866): several spp.: vasculature. • Van Fleet (1942): *P. cordatum, P. hastatum*: stem, endodermis. • Voronkina (1975): root development. • Webber (1960): several spp.: leaf epidermis, stomata. • Weisse (1897): several spp.: roots, lenticels. • Went (1895): several spp.: root. • Wittlin (1896): *P. argyrea*: druse crystals and formation. • Zinke (1924): *P. longilaminatum, P. warszewiczii*: leaf mesophyll.

Additional literature

Antony 1898; Barabé and Lacroix 2001b; Barabé *et al.* 2000; Barahona Carvajal 1977; Bate-Smith 1968; Belin-Depoux 1987; Campbell 1900; Croat 1997; Daghlian 1981; Domin 1911; Erdtman 1952; Eyde *et al.* 1967; French 1985b, 1986b, 1987a; Gottsberger and Amaral 1984; Grayum 1996; Guéguen 1901; Hay 1986; Hegnauer 1986; Irmisch 1874; Kraus 1896; Krause 1913; Lindinger 1908; Madison 1979; Mayo 1986, 1989, 1991; Meeuse 1975; Meusel 1951; Mookerjea 1955; Murashige 1974; Pacini and Franchi 1982; Plowman 1969; Pohl 1931a, b; Preda 1910; Ray 1986, 1988, 1990, 1992; Reffstrup and Boll 1985; Roth 1949; Roth and Merida de Bifano 1979; Sanford 1987; Seymour *et al.* 1983; Thanikaimoni 1969; Tieghem 1907; Vágújfalvi 1971; Van Fleet 1942;Williams *et al.* 1981; Zimmermann 1932.

Homalomena Schott

This genus of 130 species occurs mostly in tropical Asia, but includes 17 species found in South America. Plants are terrestrial, occasionally rheophytic, usually understorey herbs with short erect, sometimes creeping stems. The leaves are elliptic, ovate, or cordate to sagittate. They may occur in primary forests, in partial shade, or in wet areas, sometimes tolerating submerged conditions. Chromosomes small: $2n = 38$, 40, 42, 56, 80. Ten specimens, including seven species, were available for study.

Leaf surface

Cuticle smooth on both surfaces, or striate-rugulate in *H. wallisii*. **Epidermis**: cells polygonal with straight to undulate anticlinal walls on both surfaces (1–4:1 *l*/*w*). **Trichomes**: noted only in *H. peltata;* multicellular, uniseriate, on abaxial surface. **Stomata** abaxial, or mostly abaxial in *H. engleri*. Mostly brachypara-hexacytic (Fig. 48D) but often brachyparacytic. Subsidiary cells narrow and rounded. **Venation**: midrib prominent. Secondaries obtuse to broadly acute, to narrowly acute distally. Intercostal areas ± regular. Tertiaries closely spaced and becoming parallel with secondaries. Vascular bundles merging eucamptodromously at margin. Higher-order veins parallel with tertiaries, or orthogonal with them, or reticulate.

Leaf TS

Structure dorsiventral. **Size**: 32 × 20 mm to 150 × 120 mm (*l*/*w*). **Midrib**: nearly embedded in lamina, slightly concave adaxially, slightly convex-rounded abaxially in *H. engleri*. In other species, midrib large, flat or shallowly concave adaxially, deeply convex-rounded abaxially (Fig. 48A).

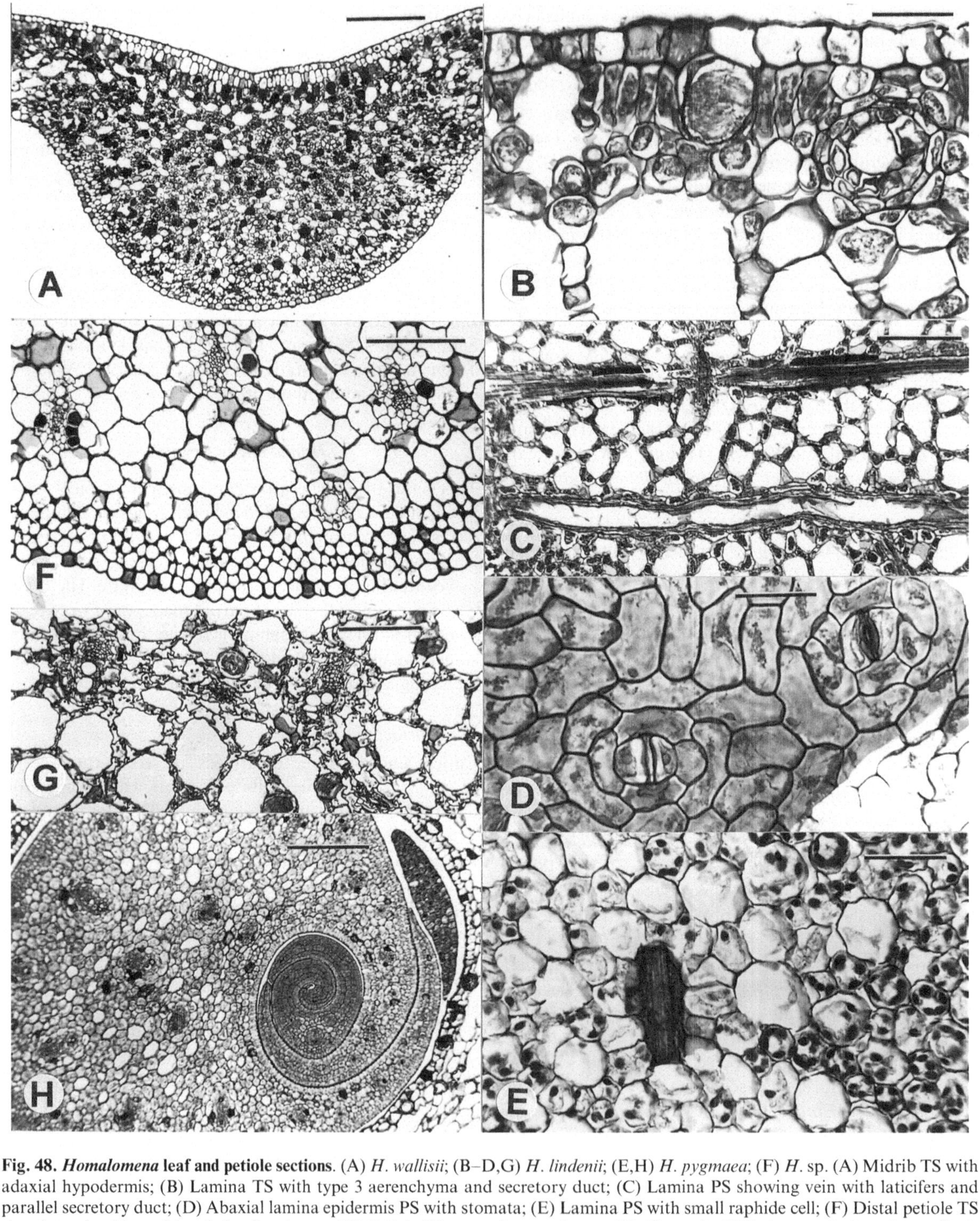

Fig. 48. *Homalomena* leaf and petiole sections. (A) *H. wallisii*; (B–D,G) *H. lindenii*; (E,H) *H. pygmaea*; (F) *H.* sp. (A) Midrib TS with adaxial hypodermis; (B) Lamina TS with type 3 aerenchyma and secretory duct; (C) Lamina PS showing vein with laticifers and parallel secretory duct; (D) Abaxial lamina epidermis PS with stomata; (E) Lamina PS with small raphide cell; (F) Distal petiole TS showing subepidermal banded collenchyma; (G) Petiole TS, central aerenchyma; (H) Young leaf TS above apical meristem. Scale bars: A = 300 μm; B,D,E = 50 μm; C,G,H = 200 μm; F = 150 μm.

Cuticle thick adaxially, medium-thick to thick at margins; thin or not seen in *H. engleri*. **Epidermis**: cells medium-sized, the same on both surfaces; flat externally and internally (Fig. 3B). Short flattened cells on both surfaces in *H. engleri*. **Stomata** level with surface. **Hypodermis**: one layer of columnar cells in adaxial midrib (Fig. 48A). **Mesophyll**: one palisade layer of loosely packed, elongate cells. Spongy tissue: *c.* 5 layers forming well-developed, organized aerenchyma; corner cells triangular with vertically elongate partition cells (type 3) (Fig. 48B). Spongy aerenchyma type 1–2 in *H. wallisii*. **Air cavities** very large and net-like, 95% of spongy layer. **Collenchyma**: 3–5-layered, subepidermal, abaxial band in midrib.

Vascular bundles: few to many scattered bundles in midrib; well developed (type II). Xylem: several proto and metaxylem cells of various sizes opposing a circular phloem strand. **Sclerenchyma**: fibres in 1–2-layered xylem and phloem caps in midrib and larger laminar bundles. **Secretory ducts** common, parallel with secondaries and tertiaries, in spongy layer and with 1–2 layers of epithelium (Figs 7F, 48B,C); ducts not noted in *H. engleri*. **Laticifers**: non-anastomosing, arranged 3–4 per bundle lateral to vascular bundles (Fig. 48C). **Crystals**: raphides in single (Figs 3B, 48B,E) or multiple bundles in large rounded or oval cells, subepidermal on both surfaces. Cells occasionally elongate when containing multiple bundles. Some raphides in *H. wallisii* in cylindrical, thick-walled cells, i.e. biforine-like, but without distinct terminal papillae. In *H. engleri*, raphide cells often suspended across aerenchyma partitions, ends pointing into adjacent air spaces. Crystal sand common in mesophyll and epidermal cells, as large clusters. In midrib, many crystals intermediate between raphides and crystal sand. In *H. engleri*, druses common, mostly in palisade zone and midrib.

Petiole TS

Outline distally terete and shallowly grooved adaxially; proximally, crescent-shaped. **Size**: diameter 3–5 mm. **Cuticle**: medium-thick to thick and striate. **Epidermis**: cells large and flat or rounded. **Collenchyma**: well-developed peripheral band, subepidermal or separated from epidermis by 1–2 layers of ground tissue. Band up to 8–10-layered, of angularly thickened cells (Fig. 48F). **Chlorenchyma**: in outer 1–2 collenchyma layers and subepidermal ground tissue, when present. **Ground tissue**: cells large and uniform with corner air cavities, up to 5% of air space; or type 3 aerenchyma with long or short partitions and up to 95% of air space (Fig. 48G, see Fig. 48H for young stage).

Vascular bundles numerous and scattered, or arranged in series of concentric circles in *H. wallisii*. Bundles unsheathed (type II) (Fig. 48F). Xylem: one wide late protoxylem cell and 2–6 medium-sized metaxylem cells in irregular strand. Phloem: circular strand. **Sclerenchyma**: fibres as small 2–3-layered phloem caps at most vascular bundles, especially in proximal sections. **Secretory ducts**: narrow or wide, with 1-layered epithelium, in outer ground tissue near or within collenchyma band (Fig. 48F); rare in *H. engleri*. **Laticifers**: large diameter, and non-anastomosing, 2–4 lateral to each vascular bundle. **Tannin** cells common and scattered in most species. **Starch** granules very common in ground tissue, circular, 6–8 μm diameter in *H. wallisii*, elongate, to 15 μm long in *H. peltata*. **Crystals**: raphides common in ground tissue in irregular, usually thin-walled cells. Common in wide oval cells in partitions of type 3 aerenchyma in *H. lindenii*. Druses common near epidermis and in outer collenchyma in *H. wallisii*; common throughout ground tissue in *H. peltata*.

Material examined

Homalomena engleri Bogner [*Diandriella nova-guineensis* Engl.], Pulle 303, New Guinea, K; *H. lindenii* (Rod.) Ridley, Bogner 464, cult. Thailand, cult. M; *H. peltata* (Poepp.) Mast., 760520-2, cult. MO; Croat 38285, Canal Zone; Croat 32979, Canal Zone; *H. pygmaea* (Hassk.) Engl. var. *purpurescens* (Schott) Engl., Bogner 1114, cult. M; *H. speariae* Bogner & Moffler, E. Spear s.n., Columbia, cult. M; *H. wallisii* Regel, Bogner 618, cult. M; *H.* sp., 75p1669, 772036.

Taxonomic notes

Homalomena specimens examined show a number of anatomical variations that are potentially diagnostic at the species level. This genus would repay a careful comparative study with an expanded species and organ representation. *Homalomena engleri* [*Diandriella*] shows some distinctive variations such as the absence of secretory ducts in the petiole, and a distinctive position of raphides in leaf. While it is uncertain what a larger sample of *Homalomena* would show, anatomical arguments can be made from this study that would support recognition of *Diandriella*.

Species reported on in the literature

Blanc (1978): ontogeny. • Dalitzsch (1886): *H. coerulescens*: leaf. • Engler and Krause (1912): several spp.: root, secretory tissue, crystals. • French (1987b, c): several spp.: root. • French (1988): several spp.: laticifers.

• French and Tomlinson (1981b): several spp.: stem vasculature. • Guttenberg (1968): *H. coerulescens, H. rubescens*: root. • Hay (1986): leaf. • Hotta (1971): *H.* 'aff.' *lancifolia*: leaf; *H. paucinerva*: root. • Leblois (1887): *H. rubescens*: secretory tissue. • Lierau (1888): several spp.: root. • Marcatili (1887): *H. rubra*: laticifers. • Philipp (1923): *H. rubra*: periderm. • Pirotta and Marcatili (1885): *H. rubra*: laticifers. • Solereder and Meyer (1928): several spp.: all organs. • Tieghem (1866): *H. rubescens, H. wendlandii*: all organs. • Tieghem (1870–71, 1872): *H. rubescens*: root, secretory tissue. • Trécul (1865, 1966): *H. porteana, H. rubescens*: vasculature. • Uphof *et al.* (1962): hairs. • Webber (1960): *H. pygmaea*: epidermis, stomata.

Additional literature

Eyde *et al.* 1967; French 1985b, 1986c, 1987a; Gertz 1916a; Grayum 1990; Hegnauer 1986; Mayo 1986; Moffler 1983; Mookerjea 1955; Ray 1988; Schultes 1970a; Schultes and Hofmann 1973; Sharma 1967; Thanikaimoni 1969; Williams *et al.* 1981.

Tribe Zantedeschieae

A tribe composed of the small African genera *Zantedeschia* and *Callopsis*.

Anatomical summary

Leaf surface

Cuticle striate or smooth. **Epidermis**: cells polygonal with straight anticlinal walls, or undulate to sinuous. **Stomata** brachyparacytic to brachypara-tetracytic on both surfaces. **Venation**: midrib prominent with secondaries at least distally acute, curved ascending, merging eucamptodromously.

Leaf TS

Structure dorsiventral or weakly so. **Midrib** broadly rounded-concave adaxially, broadly rounded-convex abaxially. **Mesophyll**: adaxial palisade in 1–3 layers, some development on abaxial surface in *Zantedeschia*. Spongy cells in 5–9 layers, mostly type 3 aerenchyma in *Zantedeschia*, type 1a in *Callopsis*. **Collenchyma**: discontinuous subepidermal band within abaxial midrib (type Bi) in *Zantedeschia*. **Vascular bundles** scattered in midrib, at corners of aerenchyma partitions, or single bundle in *Callopsis*. Bundles small, mostly type III in *Zantedeschia*, type I or II in *Callopsis*. **Sclerenchyma**: scanty development of fibre caps over larger bundles in *Callopsis*. **Laticifers** non-anastomosing, 2–4 large cells lateral to phloem. **Tannin** cells common in palisade layer of *Callopsis*. **Crystals**: raphides in single bundles in narrow cells, variously oriented.

Petiole TS

Grooved proximally, nearly rounded distally or slightly flattened adaxially in *Callopsis*. **Cuticle** medium-thick and striate or smooth. **Trichomes** in *Callopsis* only. **Ground tissue**: mostly type 3 aerenchyma or large rounded air spaces in *Zantedeschia*, cells small near periphery in *Callopsis* grading to larger in centre. **Collenchyma**: banded at periphery, type Bi, or type B in *Callopsis*. **Vascular bundles** scattered in *Zantedeschia*, at corners of type 3 aerenchyma; bundles weakly formed (type III). In *Callopsis*, 10–12 bundles symmetrically disposed in ± uniform ground tissue, bundles ensheathed by small cells (type I). **Sclerenchyma**: fibres forming semicircular layers over phloem or entire bundle in *Callopsis* only. **Laticifers** non-anastomosing, lateral to phloem strands. **Tannin** cells scattered in ground tissue in *Callopsis*. **Crystals**: single bundles in thin-walled cells variously oriented; biforines rarely present in *Zantedeschia* only.

Taxonomic notes

The genera have diverged from a common ground plan and show diagnostic differences particularly in development of ground tissue, collenchyma, sclerenchyma and vascular bundles. *Callopsis* appears to be the less specialized with its type I vascular bundles, presence of fibre caps or sheaths, and lack of type 3 aerenchyma. *Zantedeschia* shows specializations that may be related to its occupation of habitats of more marked seasonality.

Zantedeschia Spreng.

This genus of six species is distributed in South Africa. Plants are terrestrial, well-rooted rhizomatous or tuberous herbs. Their habitat requirements vary from dry to swampy substrates. A pair of long-petiolate leaves are produced with a cordate base and attenuate tip in *Z. aethiopica*. The leaves are elliptic in *Z. albomaculata*, or cordate, sagittate or hastate with short basal lobes directed orthogonally to the midrib. Chromosomes (35 counts, 5 spp.): $2n = 32$. Two species were available for study.

Leaf surface

Cuticle striate or smooth. **Epidermis**: cells polygonal with straight anticlinal walls (1–4:1) adaxially, and more elongate-polygonal (1–6:1 *l/w*) abaxially. **Stomata** on both surfaces and brachyparacytic (Fig. 49D). Subsidiary cells mostly narrow and angular. **Venation**: midrib prominent. Secondaries acute, curved-ascending and merging eucamptodromously at margin. Tertiaries parallel with secondaries. Quaternaries with orthogonal or oblique connections, or reticulate. Higher orders reticulate.

Leaf TS

Structure weakly dorsiventral. **Size**: 152 × 64 mm (*l*/*w*); basal lobe 67 × 31 mm (*l*/*w*). **Midrib** broadly rounded-concave adaxially and broadly rounded-convex abaxially. **Cuticle** thin. **Epidermis**: cells large on both surfaces, or smaller abaxially; becoming smaller in vicinity of guard cells. Cells rounded or flat externally, more rounded internally. **Stomata**: guard cells with double cuticular ledges. **Mesophyll**: adaxial palisade of 1–2 or occasionally three layers; patchy development of palisade occasional on abaxial surface. About 5–9 type 1 spongy layers appearing in places (Fig. 49B) but mostly type 3 aerenchyma with uniseriate partitions in *Z. aethiopica*.

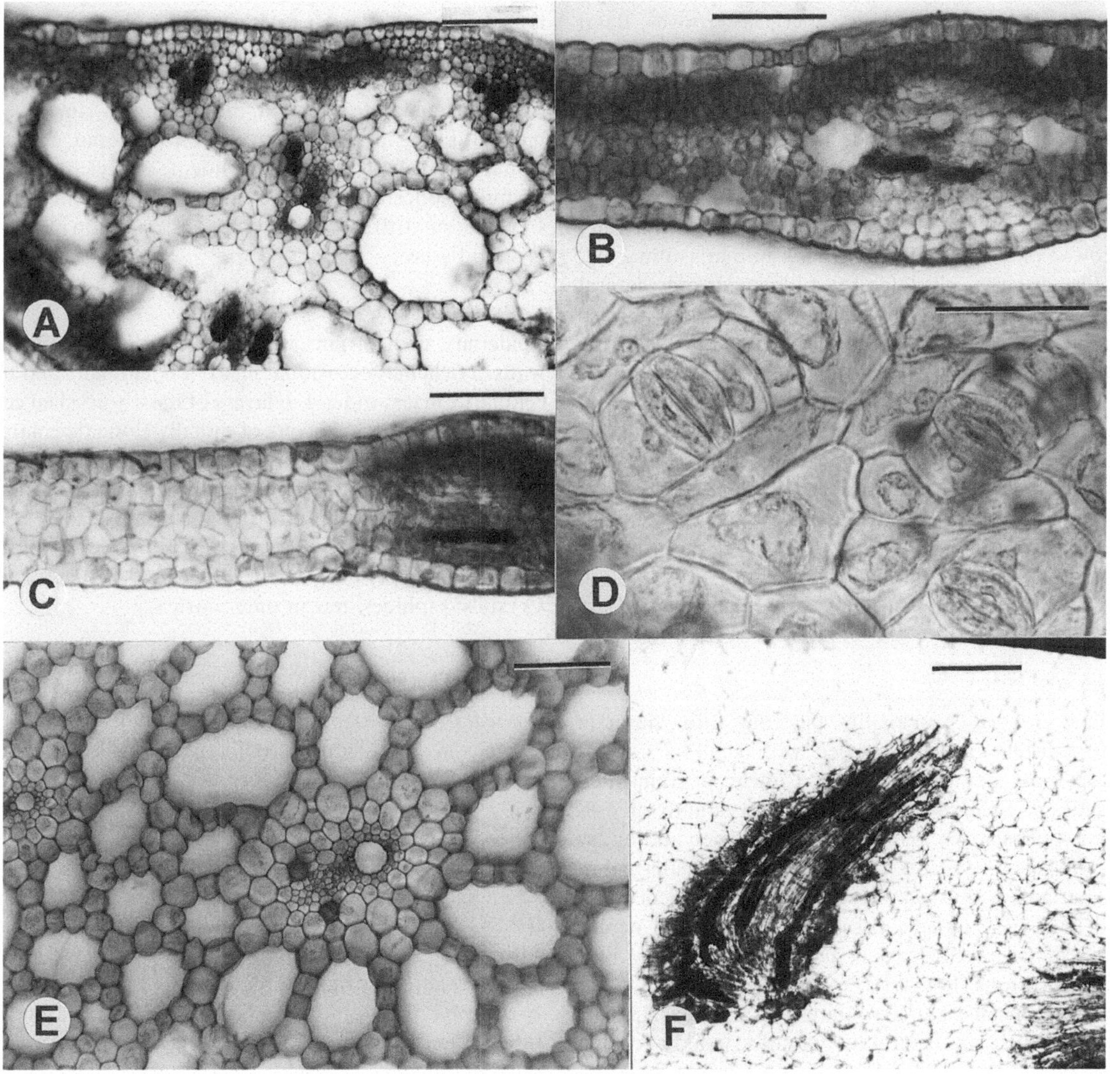

Fig. 49. *Zantedeschia albomaculata* leaf, petiole and stem hand sections. (A) Midrib TS; (B) Lamina TS; (C) Lamina TS showing undeveloped mesophyll in unpigmented zone; (D) Adaxial epidermis PS with stomata; (E) Petiole TS showing type 3 aerenchyma; (F) Stem TS showing vascular bundle and laticifers. Scale bars: A,E = 200 μm; B,C = 150 μm; D = 50 μm; F = 300 μm.

In *Z. albomaculata*, as also noted by Dalitzsch (1886) and Saxton (1913), white patches on leaf due to undifferentiated and unpigmented islands of thinner lamina (3–5 parenchyma layers) compared to neighbouring normal bifacial tissues (Fig. 49C). **Air cavities**: small substomatal cavities, otherwise small and irregular, up to 30% of air space in spongy tissue of *Z. albomaculata*. Air cavities large and circular in much of lamina, up to nearly 80% of space in *Z. aethiopica*, and in nearly all of midrib. **Collenchyma**: discontinuous subepidermal band, 4–8 cells deep on abaxial side of midrib (type Bi); band interrupted by small patches of chlorenchyma, meeting surface at stomata in shallow superficial grooves. Band units subtending but not directly adjacent to vascular bundles.

Vascular bundles scattered in midrib (Fig. 49A); also in centre of ground tissue islands at corners of aerenchyma partitions; bundles small (mostly type III). Xylem: one wide functional protoxylem element and 0–4 narrow metaxylem elements bounding a circular or laterally narrowed phloem strand. **Laticifers**: non-anastomosing, 2–4 files of large cells lateral to phloem strands; lateral cell walls bearing short processes. **Crystals**: raphides in narrow straight or curved cells, and with rounded or somewhat pointed ends (biforine-like). Cells oriented obliquely across aerenchyma partitions, or embedded within them, or attached by one end with small constricted rounded tip. Biforines noted by Gulliver (1865) in *Z. aethiopica* were not confirmed in either of these species. Raphides in lamina in single bundles or oblique overlapping bundles in elongated cells. Crystal sand or coarse prismatics abundant in central and lower spongy mesophyll and in guard cells. Small or large druses occasional in *Z. albomaculata*.

Petiole TS

Outline grooved proximally, tapering into laminate sheath; distally nearly rounded with shallow adaxial groove. **Size**: diameter 7–8 mm. **Cuticle** medium-thick and striate. **Epidermis**: cells small and square or columnar. **Collenchyma**: band up to 6 cells deep and angularly thickened (type Bi). **Chlorenchyma** subepidermal in small strands within gaps in discontinuous collenchyma band (Fig. 5F). **Ground tissue** mostly a type 3 aerenchyma of large circular air spaces, separated by uniseriate partitions (Fig. 49E).

Vascular bundles scattered, in ground tissue islands of type 3 aerenchyma (Fig. 49E); bundles weakly formed (type III). Xylem: single wide functional protoxylem element, 0–3 narrow metaxylem elements and a small circular phloem strand (type III). **Laticifers** non-anastomosing, uncommon near vascular bundles, or 1–3, common and lateral to phloem (Fig. 49E). **Starch** granules common in ground tissue near vascular bundles. **Crystals**: raphides near collenchyma and in aerenchyma, cells often vertically aligned with partitions, or oriented across partitions. Biforines, noted as present by Solereder and Meyer (1928), rarely seen, biforine-like and only in *Z. albomaculata*; cell ends smooth or with small warty processes. Coarse crystal sand occasional in partitions.

Stem TS

Outline circular. **Size**: diameter 2.7 cm. **Periderm**: 12–15 layers of thin-walled phellem. **Groundtissue**: uniform, large compact cells. **Air cavities** mostly absent. **Vascular bundles**: *c.* 25–30, scattered throughout ground tissue; circular, with irregular orientation (Fig. 49F). Xylem: 4–6 strands of metaxylem in discontinuous amphivasal pattern. Phloem: single circular central strand, or occasional circular strands independent of xylem. **Starch** granules concentrated in central half of ground tissue. **Crystals**: small druses and crystal sand common in central ground tissue.

Contractile root TS

Epidermis with striate wall thickenings on anticlinal walls. **Trichomes** occasional in section, elongate and unicellular. **Cortex**: outer 4–6 layers of mostly crushed cells. Central cortex of 3–5 layers of radially elongate enlarged cells. Inner cortex of gradually smaller cells occurring in tile-like files, extending to endodermis. **Endodermis** with conspicuous casparian strips in anticlinal walls. **Xylem** arranged in 10 radii each with one protoxylem pole and one larger metaxylem element. **Phloem** not differentiated. **Crystals**: raphides, few in outer cortex.

Material examined

Zantedeschia aethiopica (L.) Spreng., cult. MO 8/3/78. *Z. albomaculata* (Hook.f.) Baill., 31/70, cult. M.

Species reported on in the literature

Zantedeschia (syn. *Richardia* Kunth) • Arzt (1937): *Z. aethiopica*: cuticle. • Beck and Mannagetta (1912): crystals. • Borzi (1887–88): *Z. africana*: root ontogeny. • Boyd (1932): *Z. elliottiana*: seedling. • Catalano (1915): *Z. africana*: contractile root. • Chauveaud (1897): *Z. africana*: root, phloem. • Chrysler (1904): *Z. albomaculata, Z. aethiopica*: stem. • Czaja (1962): *Z. aethiopica*: leaf epidermis. • Dalitzsch (1886): *Z. africana Z. albomaculata*: leaf. • Deinega (1898): *Z. aethiopica*: leaf. • El-Gazzar *et al.* (1977): *Z. africana*: stomata. • Engler and Krause (1912): root, leaf, crystals. • Ertl (1932):

Z. aethiopica: leaf venation. • Fisher and French (1978): *Z. aethiopica*: stem ontogeny. • Fleischer (1885): *Z. aethiopica*: leaf. • French (1987c): *Z. aethiopica*: root. • French (1988): *Z. aethiopica*: laticifers. • French and Tomlinson (1981b): *Z. albomaculata*: stem vasculature. • Gentner (1905): *Z. aethiopica, Z. africana*: leaf tip. • Genua and Hillson (1985): *Z. aethiopica*: crystals. • Golden (1894): petiole. • Gravis *et al.* (1943): *Z. albomaculata*: seedling. • Gronegress (1974): *Z. elliottiana*: plastids. • Gulliver (1865): *Z. aethiopica*: crystals. • Hassack (1886): *Z. albomaculata*: leaf. • Kaplan (1973b): *Z. aethiopica*: leaf. • Knoll (1948): *Z. aethiopica*: leaf ontogeny. • Leavitt(1904): *Z. aethiopica*: root. • Lierau (1888): *Z. aethiopica, Z. albomaculata*: root. • Molisch (1899): *Z. aethiopica*: tissues. • Morassi Bonzi and Fabbri (1975): *Z. aethiopica*: leaf. • Müller (1919): leaf tip. • Netolitzky (1929): crystals. • Reichert (1919): several spp.: starch. • Saxton (1913): *Z. albomaculata*: leaf. • Solereder (1919): leaf. • Solereder and Meyer (1928): *Z. aethiopica, Z. albomaculata*: anatomy. • Tieghem (1866): *Z. africana*: all organs. • Tieghem (1870–1): *Z. africana*: root, secretory tissue. • Tieghem and Douliot (1888): *Z. africana*: branch root ontogeny. • Trécul (1865): *Z. africana*: vasculature. • Treub (1876): *Z. africana*: root meristem. • Witsch (1939): *Z. aethiopica*: root.

Additional literature

Buchholz 1921; Clarke *et al.* 1978; Davis 1966; Domin 1911; Erdtman 1952; French 1985b, 1986c, 1987a; Grayum 1990; Halevy 1989; Hegnauer 1986; Mayo 1986; Pais 1981; Richter 1929; Sharma 1967; Smolenski *et al.* 1972; Thanikaimoni 1969; Tieghem 1907; Tillich 1985; Wheaton and Stewart 1970; Williams *et al.* 1981.

Callopsis Engl.

This monotypic genus is distributed in east Africa (Kenya and Zanzibar). Plants are terrestrial, small rhizomatous herbs of the understorey of lowland evergreen forests. The leaves are long-petiolate and cordate. Chromosomes are small: 2n = 36. One specimen was available for study.

Leaf surface

Cuticle smooth to striate on both surfaces. **Epidermis**: at surface, anticlinal walls of adaxial cells undulate to sinuous (Fig. 50E); abaxial cells isodiametric to undulate (2:1 *l/w*, on both surfaces). Anticlinal walls of cells of both layers forming straight-sided polygons near internal surface. **Stomata** brachyparacytic (Fig. 50C) or brachypara-tetracytic, more common abaxially. Subsidiary cells broad or narrow, mostly rounded laterally. **Venation**: midrib prominent. Secondaries obtuse near base, arising acutely distally. Tertiaries tending to be orthogonal to midrib. Higher orders reticulate.

Leaf TS

Structure dorsiventral. **Size**: lamina 9.5 × 4 cm (*l/w*). **Midrib** concave rounded-grooved adaxially, broadly rounded-convex abaxially (Fig. 50A). **Cuticle** thin to absent on both surfaces. **Epidermis**: cells flat externally and internally. **Stomata** on slight mounds above epidermal surface. **Mesophyll**: one poorly differentiated palisade layer (1–3:1 *l/w*) (Fig. 50B). Spongy tissue in 7–8 loosely packed layers; cells very short-armed with small circular air spaces at corners (type 1a). **Air cavities**: up to 75% of substomatal spongy tissue.

Vascular bundles: one large bundle in midrib (Fig. 50A); bundles types I and II. Xylem of 1–2 protoxylem cells and 4–6 metaxylem cells opposing a circular to semicircular strand of phloem. **Sclerenchyma**: few fibres as phloem caps over larger bundles. **Laticifers** non-anastomosing, 2–4 large cells lateral to phloem. **Tannin** cells common in palisade layer. **Crystals**: raphides in single bundles of fine crystals in elongated, thin-walled cells in mid-mesophyll and midrib (Fig. 50D). Druses large, common in mesophyll cells. Crystal sand widespread in mesophyll, attached to, or within, cellular membranes.

Petiole TS

Outline subterete (Fig. 50F), adaxially slightly flattened. **Size**: diameter 2 × 3 mm. **Cuticle** thick and smooth. **Trichomes** sparse, short and unicellular, extending from single epidermal cells around periphery (Fig. 2B). **Epidermis**: cells small and rounded. **Collenchyma**: continuous subepidermal band, 6–10 cells deep (Fig. 50F). **Chlorenchyma**: plastids common in outer 7–8 cell layers. **Ground tissue** grading from small cells beneath epidermis to large in centre. **Air cavities** at cell corners only.

Vascular bundles: 10–12 bundles symmetrically disposed; bundle outline circular, well differentiated and ensheathed by smaller ground tissue cells (type I) (Fig. 50F). Xylem: 1–2 protoxylem cells, 10–14 metaxylem cells forming straight to concave boundary with phloem (Fig. 50G). Phloem: large circular strand. **Sclerenchyma**: fibres forming 1–3 semicircular layers around phloem (Fig. 50G), completing lignified ring of xylem. Larger bundles with fibres encircling xylem, but separated from xylem by ground tissue. **Laticifers** non-anastomosing, 1–3 lateral to phloem. **Tannin** cells uncommon and scattered in ground tissue. **Starch** granules few, scattered in ground tissue. **Crystals**: raphides in thin-walled cells common in outer ground tissue.

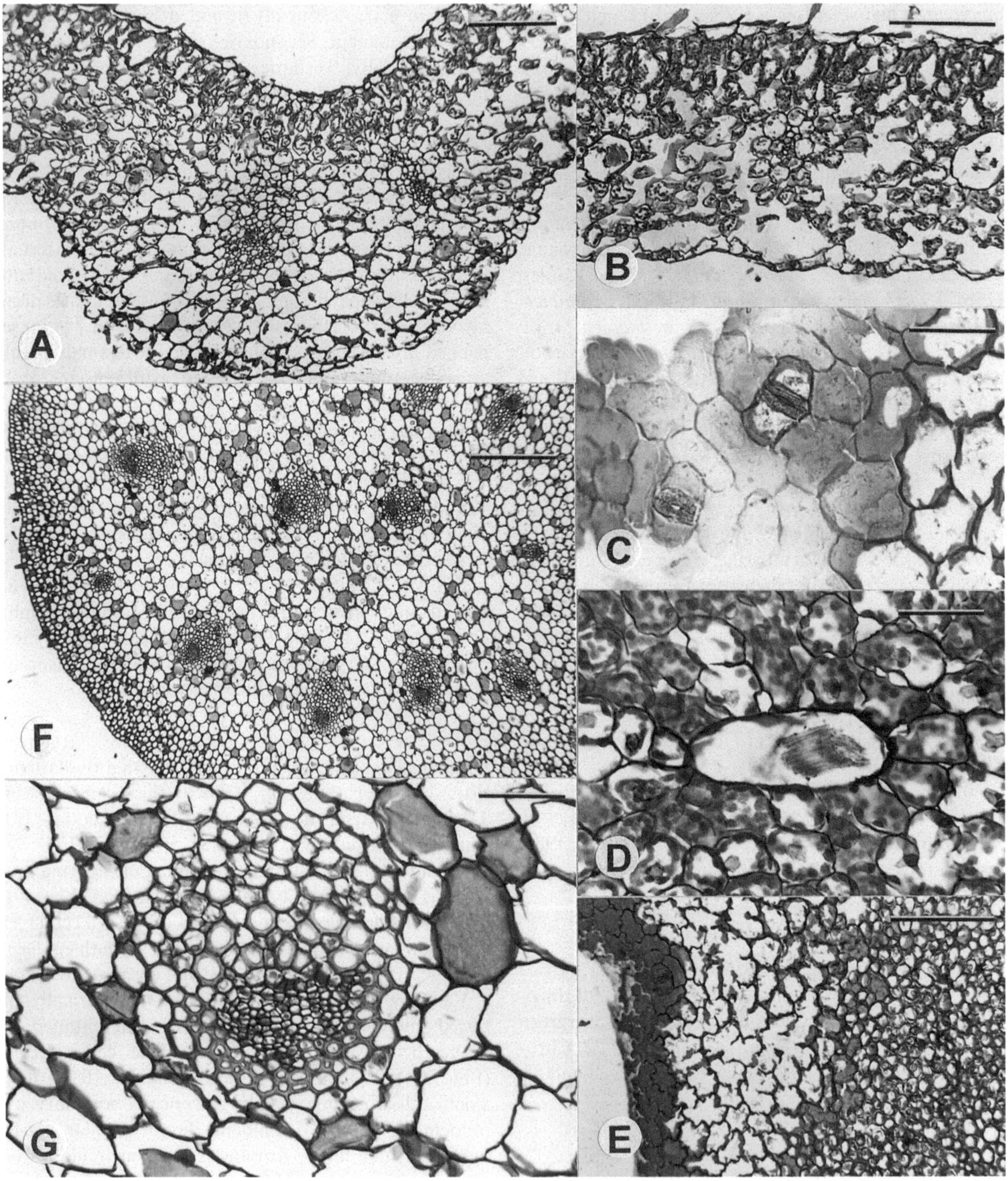

Fig. 50. *Callopsis volkensii* leaf and petiole sections. (A) Midrib TS; (B) Lamina TS; (C–E) Lamina PS: (C) Abaxial epidermis showing brachyparacytic stomata; (D) Mesophyll showing raphide cell; (E) Transition from palisade to adaxial epidermal tissue. Epidermal cells polygonal internally and sinuous at surface; (F) Proximal petiole TS, showing peripheral collenchyma band; (G) Distal petiole TS, showing type I vascular bundle. Scale bars: A = 200 μm, B,E = 150 μm, C,D,G = 50 μm, F = 300 μm.

Material examined

Callopsis volkensii Engl., Bogner 218, Tanzania, cult. M.

Species reported on in the literature

French (1987c): root. • French (1988): laticifers. • French and Tomlinson (1983): stem.

Additional literature

French 1986a, c, 1987a; Grayum 1990; Hegnauer 1986; Ray 1988; Thanikaimoni 1969; Williams *et al.* 1981.

Tribe Stylochaetoneae Schott

Consisting of one genus.

Stylochaeton Lepr.

This genus of 21 rhizomatous species is distributed in central and south-east tropical Africa. The plants are terrestrial, seasonally dormant rhizomatous herbs with stoloniferous stems and succulent underground roots. Leaves are hastate or have shallow auriculate basal lobes. Engler (1920) reported species to have rhizomatous tubers. Chromosomes: 2n = 28, 56. Five species were available for study of leaf and two for petiole.

Leaf surface

Cuticle smooth on both surfaces and striate on abaxial midrib. **Epidermis**: cells polygonal with straight anticlinal walls (1–2:1 *l*/*w*). **Stomata**: abaxial, brachyparahexacytic, except brachyparacytic (Fig. 51F) in *S. barteri*. Subsidiary cells narrow and angular or rounded. **Venation**: primaries basal and suprabasal acrodromous. Medial primary with pinnate, acute, steep-ascending secondaries becoming parallel with midrib. Secondaries merging eucamptodromously at margin. Tertiaries parallel with secondaries or oblique to reticulate. Higher orders reticulate.

Leaf TS

Structure dorsiventral. **Size** 108 × 68 mm (*l*/*w*) or smaller. **Midrib** shallowly rounded-convex adaxially and broadly or narrowly rounded-convex abaxially. **Cuticle** very thin to absent. **Epidermis** tall adaxially, cells flat externally and unevenly rounded internally. Abaxial cells flat and short. **Stomata** very common and level with surface. **Mesophyll**: 2–5 palisade layers of elongated and well-packed cells (Fig. 51D), or small, rounded cells in *S. zenkeri* (Fig. 51B). Spongy tissue of *c.* 5 cell layers forming an aerenchyma with 1–2-seriate partitions, 8–10 cells long (type 3), delimiting large regular air cavities (Fig. 51B,E). Type 1 spongy cells near epidermis and palisade boundary. Spongy tissue only type 1 in *S. barteri* (Fig. 51C,D). **Air cavities** up to 99% of spongy tissue, or up to 60% in *S. barteri*. **Collenchyma** in midrib, angularly thickened cells in abaxial subepidermal band.

Vascular bundles: well-formed scattered bundles in midrib (type II). Numerous metaxylem elements opposing a large circular phloem strand. Mesophyll vascular bundles smaller. Vascular bundles with elongated wide parenchymatous cells forming bundle sheaths, especially in *S. barteri*. **Tannin** cells with dark granular contents common in mid-mesophyll, not usually associated with vascular bundles. **Crystals**: raphides in single wide bundles, rare or common in rounded cells in palisade tissue. In *S. barteri*, parallel-walled biforines common in mid-mesophyll (Fig. 51C). Druses small and very common in palisade layer. Crystal sand abundant in all spongy and midrib cells in large clusters.

Petiole TS

Outline crescent-shaped (Fig. 51A). **Size**: width 7 mm. **Cuticle** thin and smooth or striate. **Epidermis**: cells small, rounded externally. **Stomata** occasional adaxially with well-developed outer and inner ledges in *S. barteri*. **Collenchyma**: band continuous subepidermally on abaxial side, 2–5 layers thick or 1–2 layers in *S. barteri* (Fig. 51G). **Ground tissue**: uniform compact cells in *S. zenkeri*, type 2 or 3 aerenchyma throughout in *S. barteri*. **Vascular bundles** scattered and numerous; type I in *S. zenkeri*, more irregular type II in *S. barteri* (Fig. 51H). Xylem strand wide, with 10–15 metaxylem cells in *S. zenkeri*, with 1–6 metaxylem cells in *S. barteri*. Phloem strand semicircular with broad boundary with xylem in *S. zenkeri*, a small laterally narrowed strand in *S. barteri*. **Tannin** cells occasional throughout ground tissue. **Starch** common in ground tissue, granules 6–8 μm diameter. **Crystals**: druses common in ground tissue, occasionally in pairs in adjacent cells in *S. zenkeri*. Raphides common in outer ground tissue in thin-walled cells in single bundles and occasionally in partition cells in *S. barteri*.

Material examined

Stylochaeton barteri N.E.Br., K-Jodrell spirit coll.; *S. crassispathus* Bogner, Bogner 143, Tanzania, cult. M; *S. maximus* Engl., Bogner 216, Tanzania, cult. M; *S. obliquinervis* Peter, 416/71, Transvaal, cult. M; *S. zenkeri* Engl., Bogner 631, Gabon, cult. M.

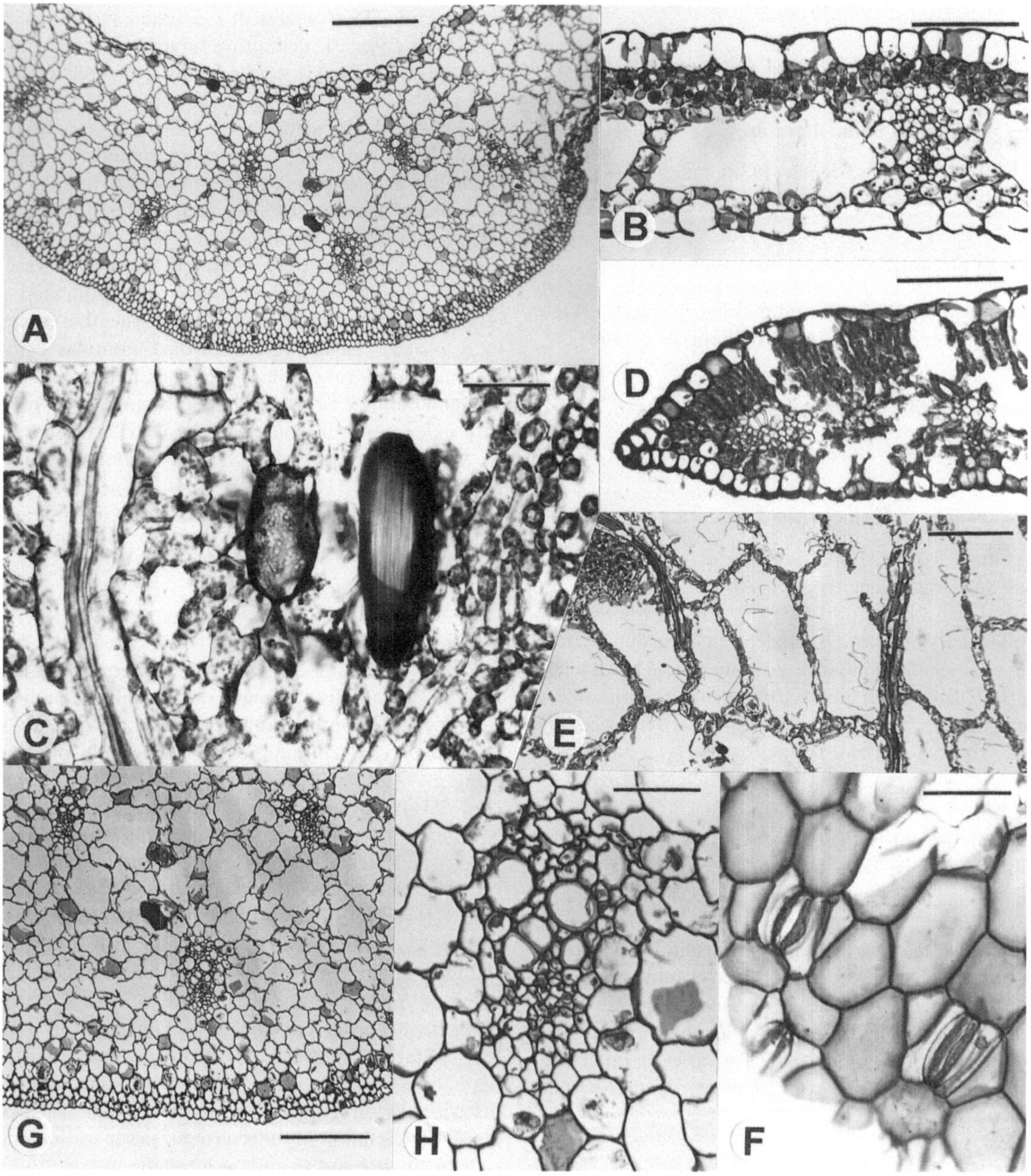

Fig. 51. *Stylochaeton* leaf and petiole sections. (A,C,D,F–H) *S. barteri*; (B,E) *S. zenkeri*. (A) Petiole TS; (B) Lamina TS; (C) Lamina PS showing elongated sheath cells along small vein and raphide cells; (D) Lamina TS margin; (E) Lamina PS showing type 3 spongy mesophyll; (F) Lamina abaxial surface PS showing brachyparacytic stomata; (G) Petiole TS showing type 3 central aerenchyma; (H) Petiole TS, central type II bundle. Scale bars: A,E = 300 μm; B,D = 150 μm; C,F,H = 50 μm; G = 200 μm.

Taxonomic notes

Stylochaeton and the two genera of Zamioculcadeae are the only perigoniate genera in this subfamily and the only ones apparently lacking non-anastomosing laticifers. The rhizomatous stems and assignment to the banded collenchyma series place all three genera within this subfamily. The specimens of *Stylochaeton* are distinguished from Zamioculcadeae in having only straight-sided polygonal epidermal cells, having up to brachypara-hexacytic stomata, numerous palisade layers, and development of type 3 spongy aerenchyma.

Species reported on in the literature

Engler (1920c): *S. natalensis, S. lancifolius*: stem. • French (1987c): *S. bogneri, S. zenkeri*: root. • French (1988): *S. bogneri, S. zenkeri*: laticifers. • French and Tomlinson (1983) *S. maximus, S. puberulus, S. zenkeri*: stem vasculature. • Solereder and Meyer (1928): *S. warneckii*: anatomy.

Additional literature

French 1986a, c, 1987a; Grayum 1990; Hesse *et al.* 2001; Knecht 1983; Marchant 1970; Thanikaimoni 1969; Williams *et al.* 1981.

Tribe Zamioculcadeae Engl.

This tribe comprises the genera *Gonatopus* and *Zamioculcas*.

Anatomical summary

Leaf

Leaves pinnately compound and dorsiventral. **Cuticle** medium-thick to thin and smooth or striate. **Epidermis**: cell surface shape ranging from sinuous or undulate to polygonal. Epidermal cells in TS tall and square or mixed rounded and flattened. **Stomata** mostly abaxial and brachyparacytic. Subsidiary cells variable in shape. **Mesophyll**: palisade cells armed in *Zamioculcas* or lobed in *Gonatopus* and well-differentiated. Spongy cell types 1, 1a. **Collenchyma**: few banded layers. **Venation** pinnate with eucamptodromous or brochidodromous connections at straightened marginal collector. **Midrib** with 1–3 vascular bundles, types I, II. Fibres absent. **Sclereids** and **laticifers** absent. **Tannin** cells numerous and common in ground tissue and mesophyll. **Starch** granules common in ground tissue. **Crystals**: raphides common in circular, elliptical or larger cells, in one or more bundles or multiple overlapping bundles. Druses present. Crystal sand absent or present as tiny prismatics in anticlinal walls of leaf epidermis in *Gonatopus boivinii*.

Petiole

Outline subterete or semicircular and flattened adaxially. **Size**: diameter 3–6 mm. **Cuticle** medium-thick and striate or smooth. **Epidermis**: cells small and square or broad. **Collenchyma** absent or in peripheral band of up to six layers; cells thin-walled. **Ground tissue**: cells small near periphery, larger in centre. **Chlorenchyma**: outer five layers of ground tissue. **Air cavities** small, at cell corners. **Vascular bundles** scattered and random (type II), may be larger in centre. Xylem: metaxylem cells 6–12, bounding small circular or irregular phloem strand. **Sclerenchyma**: fibres capping some peripheral vascular bundles in *Gonatopus*. **Starch** granules in ground tissue of *Zamioculcus*. **Crystals**: raphides in large cells in *Gonatopus*; raphide and druse cells uncommon in *Zamioculcas*.

Stem

Central cylinder sharply but irregularly delimited by starch layer. **Vascular bundles** collateral. **Starch** granules large truncated cylinders or cones. **Crystals**: raphides in multiple bundles, often in large mucilage cells.

Taxonomic notes

These data, together with chromosome number and geographic range, place the genera of Zamioculcadeae closer to each other than either is to the Stylochaetoneae. *Gonatopus* and *Zamioculcas* have only type 1 spongy tissue and wide raphide cells. *Zamioculcas* is distinguished by its unusual armed palisade cells, sinuous adaxial epidermal cells, and presence of type I vascular bundles.

Gonatopus Engl.

This genus of five species is distributed from east tropical Africa to South Africa. Plants are terrestrial, tuberous or rhizomatous understorey herbs in dune forests, shady rocky pockets, or at the edges of primary forests. The corm produces a single erect leaf, digitately divided at the petiole summit and having pinnately compound branches. Chromosomes: $2n = 34$, *c.* 68. Two species were available for study.

Leaf surface

Cuticle smooth on lamina, striate on midrib. **Epidermis**: cells polygonal with straight anticlinal walls to undulate adaxially and undulate to sinuous abaxially (1–4:1 *l*/*w*). **Stomata** mostly abaxial and brachyparacytic (Fig. 52D), some unipolar. Subsidiary cells broad and rounded or square-cornered. **Venation**: midrib prominent. Secondaries steeply ascending, especially at base, merging at brochidodromous submarginal collector. Secondaries broadly acute distally in *G. boivinii*. Tertiaries and higher orders reticulate but tending to be parallel with secondaries.

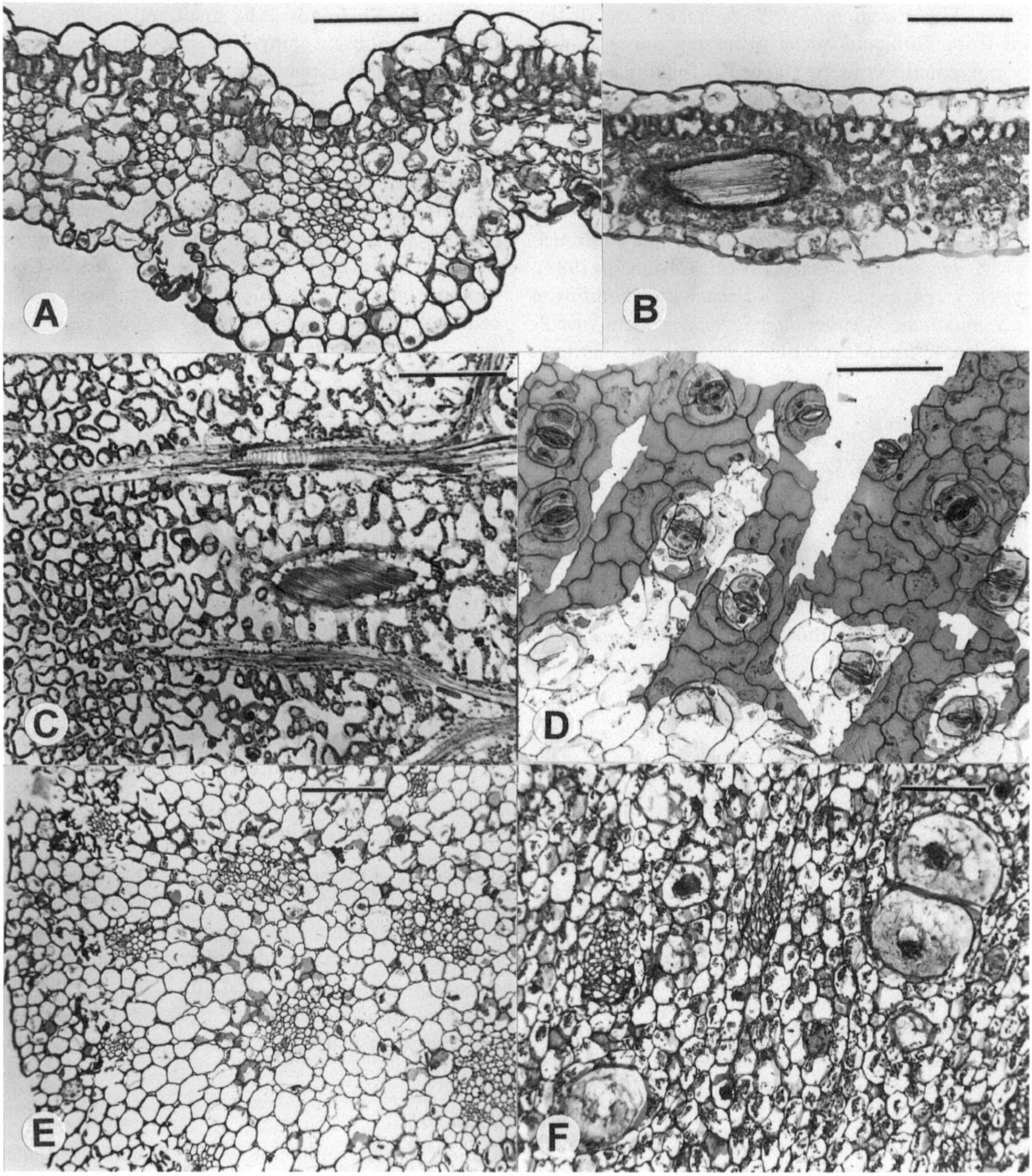

Fig. 52. *Gonatopus* leaf, petiole and stem sections. (A,C,E,F) *G. rhizomatosus*; (B,D) *G. boivinii*. (A) Midrib TS with single vascular bundle; (B) Lamina TS showing raphide cell and lobed palisade cells; (C) Mesophyll PS showing raphide cell and lobed palisade cells; (D) Leaf epidermis showing undulate epidermal cells and stomata, some with unipolar subsidiary cells (three adjacent subsidiary cells); (E) Petiole TS; (F) Stem TS showing raphide cells. Scale bars: A–D = 150 μm; E,F = 300 μm.

Leaf TS

Structure dorsiventral and well-differentiated. **Size** (terminal segment): 80 × 35 mm; 59 × 12 mm (*l*/*w*). **Midrib** concave and rounded-grooved adaxially, broadly convex abaxially, little thicker than lamina (Fig. 52A). **Cuticle** thin on both lamina surfaces, thicker on midrib. **Epidermis** on both surfaces: mixed flattened cells interspersed with larger rounded cells in *G. rhizomatosus*; large tall or squarish cells in *G. boivinii*. **Stomata** level with surface and guard cells with double cuticular ledges (cf. Fig. 4B). **Mesophyll**: one palisade layer, cells lobed, up to 35% of lamina (Fig. 52B,C). Spongy layers 4–5, irregular (types 1, 1a). Chloroplasts numerous in all mesophyll cells; nuclei large, up to 17 μm diameter. **Air cavities** up to 50% of spongy layer.

Vascular bundles: 1–2 small bundles in midrib. Bundles in lamina very small (type II). Xylem: 2–10 tracheary elements opposing a small irregular phloem strand. **Tannin** cells rounded and with dark granular contents, common in epidermis and mesophyll. **Starch** granules common throughout mesophyll. **Crystals**: raphides, coarse or fine, common in large circular to elongated mesophyll cells (Fig. 52B) or in multiple overlapping bundles in elongated cells (8–10:1 *l/w*). Numerous spongy cells adherent to periphery of enlarged raphide cells. Raphide cells tending to be parallel to adjacent vascular bundles (Fig. 52C). Crystal sand: small individual prisms inserted within anticlinal epidermal walls in *G. boivinii*.

Petiole TS

Outline subterete. **Size**: diameter 6 mm. **Cuticle** medium-thick and striate. **Epidermis**: cells small and square. **Collenchyma**: mostly absent or in faint peripheral band. Strands forming caps over outer circle of vascular bundles. **Ground tissue**: cells small near periphery, large in centre. **Chlorenchyma**: outer five layers of ground tissue. **Air cavities** at cell corners, up to 5% of ground tissue.

Vascular bundles scattered and random, unsheathed (type II) (Fig. 52E). Xylem: 6–12 small metaxylem cells bounding small irregular phloem strand. **Sclerenchyma**: fibres capping phloem of some peripheral vascular bundles. **Crystals**: raphides in large cells, common throughout ground tissue and concentrated near periphery.

Stem TS (G. rhizomatosus)

Outline terete. **Size**: diameter 14 mm. **Cuticle** thin and smooth. **Epidermis**: cells small and tile-like. **Cork**: well developed over most of surface, 6–8 layers of thin-walled cells. **Ground tissue**: cells large throughout, divided into cortex and central cylinder by chloroplast-starch granule boundary. **Chlorenchyma**: outer 10 layers of ground tissue.

Vascular bundles: few scattered, well-organized bundles, simple (type II), collateral and unsheathed or unusual. Xylem in larger bundles with 3–5 metaxylem strand clusters nearly surrounding small irregular phloem strand. **Secretory cells** or ducts: numerous, large, mucilage-filled cells with no epithelium. **Starch**: long cylindrical granules up to 24 μm long, common in all ground tissue internal to outer chloroplast zone. **Crystals**: raphide bundles often in large mucilage cells; very wide single or multiple bundles (Fig. 52F).

Material examined

Gonatopus boivinii (Decne.) Engl., cult. M; *G. rhizomatosus* Bogner & Oberm., 315/68, cult. M.

Species reported on in the literature

Engler (1905): *G. boivinii*: root. • Ertl (1932): *G. boivinii*: leaf. • French (1987c): *G. angustus, G. boivinii*: root. • French and Tomlinson (1981c): *G. marattioides*: stem. • Hay (1986): leaf. • Lierau (1888): *G. boivinii*: root. • Meyer (1962): *G. boivinii*: leaf. • Netolitzky (1929): *G. boivinii*: crystals. • Shishkoff (1987): root. • Solereder (1919): *G. boivinii*: leaf. • Solereder and Meyer (1928): *G. boivinii*: all organs.

Additional literature

French 1985a, 1986c, 1987a; Grayum 1990; Hesse *et al.* 2001; Sakai *et al.* 1984; Thanikaimoni 1969; Williams *et al.* 1981.

Zamioculcas Schott

This monotypic genus is native to east tropical Africa. Plants are seasonally dormant herbs with a rhizomatous tuber. They occur in coastal forest understoreys and are adapted to open areas or shade. The leaves are pinnately compound to tripinnate with swellings some distance below the pinnae (Bown 2000). Chromosomes are large (four counts): 2n = 34. One specimen was available for study.

Leaf surface

Cuticle smooth on both surfaces. **Epidermis**: adaxial cells finely sinuous with 9–12 lobes at the surface; cells below surface polygonal with straight anticlinal walls. Abaxial cell walls undulate at surface and polygonal internally. **Stomata** abaxial and brachyparacytic (Fig. 53E). Subsidiary cells broad or narrow and irregular. **Venation**: midrib prominent. Secondaries irregularly acute, merging eucamptodromously at proximal margin, and merging distally in irregularly sized brochidodromous loops or straight collector vein. Tertiaries and higher orders reticulate.

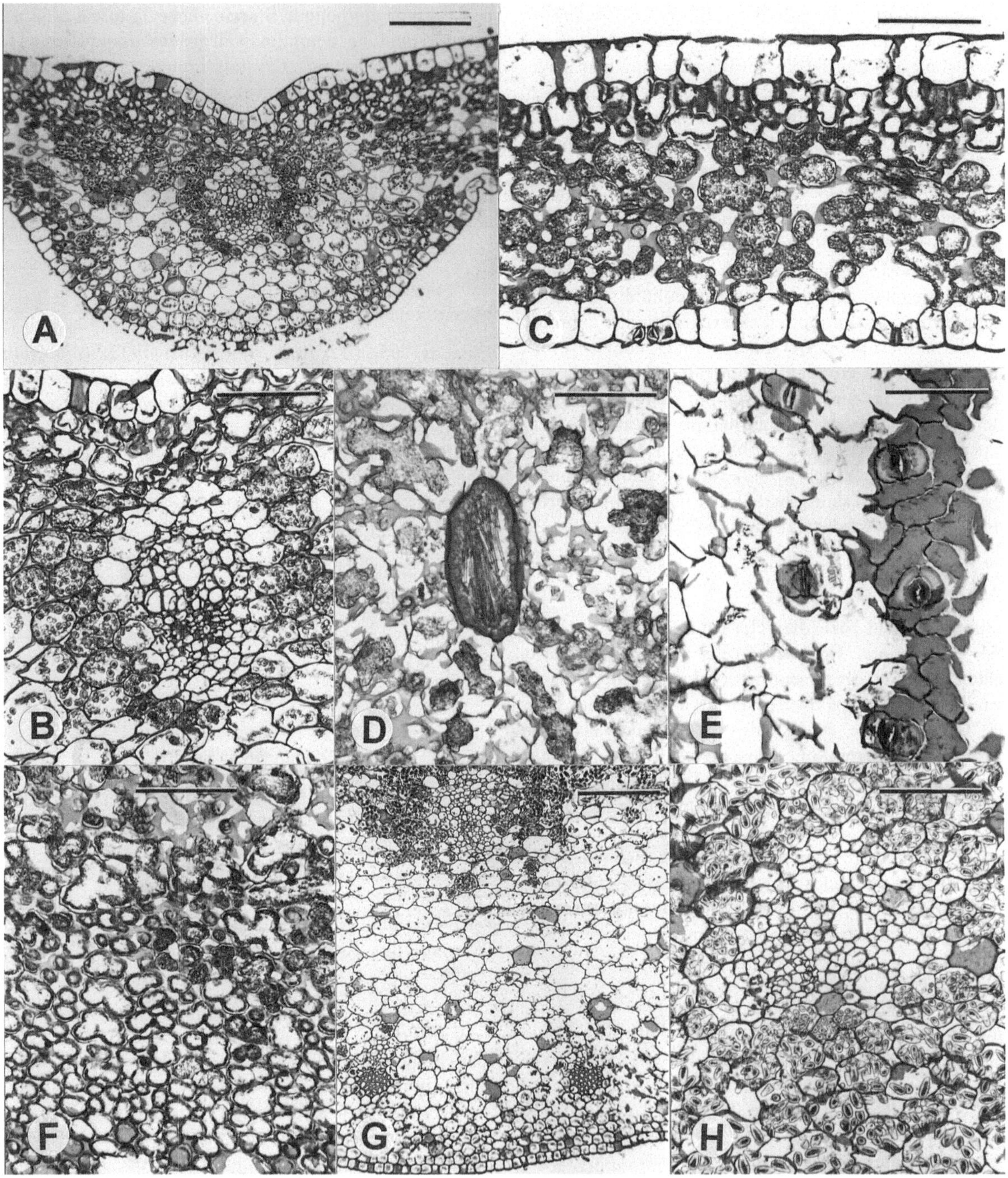

Fig. 53. *Zamioculcas zamiifolia* leaf and stem sections. (A,B) Midrib TS showing large central type I vascular bundle with starch in ground tissue surrounding bundle sheath; (C) Lamina TS showing shallow substomatal cavities and lobed palisade cells; (D) Mesophyll PS showing raphide cell in spongy tissue; (E) Leaf abaxial epidermis showing brachparacytic stomata and one unipolar stoma; (F) Mesophyll PS showing spongy tissue and lobed palisade cells; (G) Stem TS showing cortex and boundary of central cylinder; (H) Stem TS showing collateral vascular bundle surrounded by starch-bearing ground tissue within central cylinder. Scale bars: A,G = 300 μm; B–F,H = 150 μm.

Leaf TS

Structure dorsiventral. **Specimen (segment) sizes**: 80 × 30; 87 × 35 mm (*l/w*). **Midrib** shallowly grooved adaxially and broadly rounded abaxially (Fig. 53A). **Cuticle** medium-thick on both surfaces. **Epidermis**: cells tall and square on both surfaces, flat externally and rounded internally. **Stomata** level with surface. **Mesophyll**: 1 unusual palisade layer, cells variously columnar (1:5 *l/w*), or armed to produce U-shaped or H-shaped connected pairs (Fig. 53C,F). Spongy layers *c*. 5, lobed or short-armed (type 1). **Air cavities** small, ramified and irregular, up to 25% of spongy layer.

Collenchyma: subepidermal 2-layered band abaxial in midrib. **Vascular bundles**: three symmetrically disposed robust bundles in midrib (type I). Xylem with numerous metaxylem elements opposed to a broad semicircular phloem strand. Vascular bundles ensheathed by one layer of large parenchyma cells (Fig. 53B). **Tannin**: numerous branched mesophyll cells with dark granular contents, up to 40–50% of mesophyll. **Starch**: abundant granules in spongy parenchyma. **Crystals**: raphides coarse, 1–2 bundles in large circular or elliptical cells common in mesophyll (Fig. 53D). Coarse druses uncommon in large circular mesophyll cells.

Petiole TS

Outline semicircular and flattened adaxially. **Size**: diameter 3 × 5 mm. **Cuticle**: medium-thick and smooth. **Epidermis**: cells small and square or broad. **Collenchyma** band, 5–6 subepidermal layers, thin-walled or absent. **Ground tissue** cells becoming larger toward centre. **Air cavities** at corners between cells.

Vascular bundles scattered, 4–5 large bundles at centre, smaller and scattered bundles near periphery. Bundles unsheathed (type II). Xylem: *c*. 12 metaxylem elements in 3–4 files in largest bundles. Phloem: a circular or irregular strand. **Starch** granules common in ground tissue. **Crystals**: raphides and druses uncommon, in circular cells near periphery.

Stem TS

Outline subterete. **Size**: diameter 7 × 9 mm. **Cuticle**: thick and smooth. **Epidermis**: cells square and rounded externally. **Ground tissue**: cortex of small chlorenchyma cells in outer five layers. Cells becoming larger internally. **Air cavities** at corners of parenchyma cells. **Central cylinder** sharply but irregularly demarcated by starch boundary (Fig. 53G).

Vascular bundles: scattered large bundles in central cylinder; smaller bundles in irregular circle in cortex. Bundles collateral (type I) (Fig. 53H). Xylem: within central cylinder: numerous metaxylem elements in large bundles. Phloem: an irregular or semicircular strand. In cortex: 2–3 metaxylem cells and small irregular phloem strand (type II). **Sclerenchyma**: fibres forming thick-walled, circular or semicircular caps over phloem of cortical bundles. **Starch** granules elongate, truncated conical shapes, very large, up to 30 µm long, nearly confined to, and abundant in, central cylinder ground tissue. **Crystals**: raphides in multiple bundles in rounded cells surrounded by copious mucilage, rare in ground tissue. Crystal sand: scattered in ground tissue.

Root TS

Outline subcircular. **Size**: diameter 3 × 4 mm. **Epidermis**: few collapsed cells present. **Hypodermis**: 4–6 layers of thin-walled irregularly-shaped cells with all walls thinly suberized or lignified. **Exodermis**: (following definition of Peterson and Perumalla 1990) a single layer separating hypodermis from outer cortex having thickened anticlinal walls only, yielding casparian-strip stain reaction. **Cortex**: mostly large randomly oriented parenchymatous cells. Near vascular cylinder, 6–8 layers of cortical cells arranged in radial files. **Air cavities** few, at cell corners. **Endodermis**: unusually large irregularly-shaped cells forming undulate cylinder, with anticlinal walls recognizable only by small ± circular casparian thickenings. **Pericycle**: single layer within endodermis. **Vascular cylinder** large, with ground tissue at centre (Fig. 10D). Xylem: protoxylem poles 15, each adjacent to 2–4 wider metaxylem cells. Phloem: 15 small irregularly shaped strands. **Crystals**: raphides: single bundles in small cells, few in outer cortex.

Material examined

Zamioculcas zamiifolia Engl., Bogner 126, Tanzania, cult. M.

Species reported on in the literature

Czaja (1962): leaf. • Engler (1905): leaf, root. • Ertl (1932): leaf venation. • French (1987c): root. • French and Tomlinson (1981c, 1986): stem vasculature. • Hotta (1971): leaf. • Kaplan (1984): leaf, petiole. • Lierau (1888): root. • Linsbauer (1930): epidermis. • Netolitzky (1929): crystals. • Pant and Kidwai (1966): leaf, stomata. Solereder (1919): leaf. • Solereder and Meyer (1928): anatomy. • Troll and Meyer (1955): leaf development.

Additional literature

Barabé and Forget 1988a; French 1985a, 1986c, 1987a; Grayum 1990; Hay 1986; Hesse *et al.* 2001; Thanikaimoni 1969; Williams *et al.* 1981.

Tribe Aglaonemateae

The tribe is composed of three tropical African genera, *Nephthytis*, which also occurs in Sarawak, *Anchomanes*, and *Pseudohydrosme*, and the two south-east Asian genera *Aglaonema* and *Aglaodorum*.

Anatomical summary

Leaf surface

Cuticle smooth to striate and rough. **Epidermis**: cells polygonal, undulate or deeply sinuous in the African genera, to fusiform in *Aglaodorum*. **Stomata** abaxial and brachyparacytic, brachypara-tetracytic; also brachypara-hexacytic in *Aglaodorum*. Subsidiary cells narrow or broad and rounded. **Venation** acrodromous in the African genera, often with prominent midrib in a segment. Venation strictly pinnate and secondaries closely parallel in *Aglaonema* and *Aglaodorum*. Secondary veins at margins eucamptodromous, brochidodromous or looped.

Leaf TS

Structure dorsiventral or weakly so. **Cuticle** usually thin. **Midrib** often flat or slightly convex adaxially, mostly convex-rounded or deeply rounded abaxially. **Mesophyll**: palisade 1-layered, cells tending to be lobed in the African genera. Spongy layers 2–5, or up to 14 in *Aglaodorum*, forming types 1 or 2 aerenchyma, or type 3 aerenchyma occasionally in *Aglaonema*. **Collenchyma** forming bands, but mostly interrupted bands or strands not aligned with vascular bundles (types B, Bi, Sb). **Vascular bundles** in midrib few or numerous, and scattered or symmetrically placed, and weakly developed, type II. **Fibres** mostly absent, but forming phloem and xylem caps in *Nephthytis*. **Laticifers** non-anastomosing, lateral to phloem. **Tannin** cells common in *Aglaonema* and *Aglaodorum*. **Crystals**: raphides as biforines, biforine-like cells or as single bundles in thin-walled cells; also present in multiple bundles in elongate cells. Druses and crystal sand also present.

Petiole TS (*Pseudohydrosme* not included)

Outline oval or subcircular distally. **Cuticle** smooth, varying in thickness. **Epidermis**: cells small or irregular in size and shape. **Ground tissue** compact, cells small at periphery, larger in centre, type 3 aerenchyma in centre in *Aglaonema* and *Aglaodorum*. **Collenchyma**: type Sb near periphery; more deeply situated in *Aglaonema*. **Fibres** may ensheath vascular bundles or cap xylem or phloem. **Vascular bundles** scattered and numerous, types II or III. **Laticifers** non-anastomosing, variable number of strands near phloem. **Tannin** cells may be present. **Crystals**: raphides in single bundles in thin-walled cells. Biforines traversing aerenchyma partitions in *Aglaonema* and *Aglaodorum*.

Taxonomic notes

The five genera make a coherent group on the basis of the rare collenchyma types Bi and Sb. They are isolated from the other 'No alliance' genera of MBB. The collenchyma type is probably related to the break up of banded collenchyma and not related to the type Sv of the Aroideae.

The three African genera *Anchomanes, Nephthytis* and *Pseudohydrosme* show a few characteristics that make them more closely related to each other. These include long-petiolate, hastate or divided leaves, sinuous epidermal cells and lobed palisade cells. The five are placed in the same tribe here partly due to the evidence of French *et al.* (1995). Nearest neighbours of this tribe are probably the Culcasieae.

Nephthytis Schott

This genus of five species is distributed in tropical Africa. Plants are terrestrial understorey herbs with creeping rhizomes. They are found in dense shade in lowland forests, on a variety of substrates. The leaf is deeply hastate and long-petiolate. Chromosomes are large: 2n = 40, *c*.50, 60. Four species were available for study.

Leaf surface

Cuticle smooth. **Epidermis**: anticlinal walls deeply sinuous on both surfaces, cells having 9–12 lobes interlocking with neighbouring cells (Fig. 2H). **Stomata** abaxial and brachyparacytic. Subsidiary cells narrow and rounded. **Venation**: basal acrodromy with prominent midrib. Secondaries broadly acute. Basal primaries and midrib secondaries connecting in submarginal irregular pattern with looping beyond main arches. Composite intersecondaries ± parallel with secondaries. Tertiaries and higher orders reticulate with orthogonal branching.

Leaf TS

Structure dorsiventral with large abaxially protruding veins. **Size**: 70 × 120 mm (*l*/*w*); basal lobes 75 × 47 mm (*l*/*w*). **Midrib** rounded-convex adaxially and large, protruding 3/4 circle abaxially (Fig. 54A). **Cuticle** thin on both surfaces. **Epidermis**: cells large, flat externally and rounded internally. **Stomata** level with surface. **Mesophyll**: one palisade layer, cells isodiametric or irregularly expanded, often lobed, lobes oriented toward adaxial epidermis (Fig. 54B). Spongy tissue: *c*. 5 layers of long-armed cells

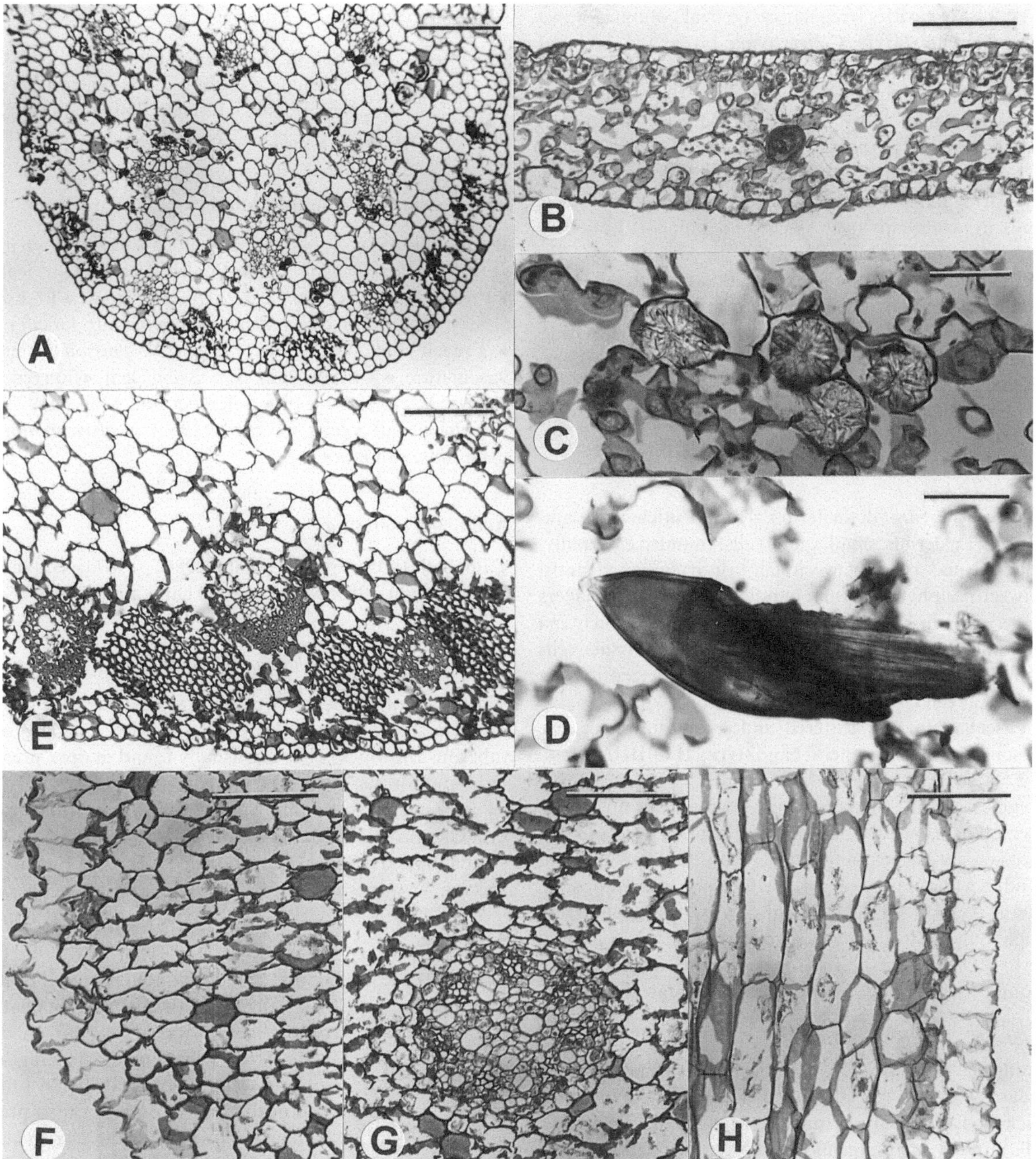

Fig. 54. ***Nephthytis afzelii*** **leaf, petiole and root sections.** (A) Midrib TS, abaxial; (B) Lamina TS; (C) Mesophyll PS, druse cells; (D) Mesophyll PS, broken biforine cell; (E) Proximal petiole margin. Note vascular bundles with fibre caps or sheaths and collenchyma strands; (F) Root TS, outer cortex; (G) Root TS, stele with seven protoxylem poles; (H) Root cortex LS. Scale bars: A = 300 μm; B,F–H = 150 μm; C,D = 50 μm; E = 200 μm.

(type 2). **Air cavities** irregular or vertical columns, up to 80% of spongy layers. **Collenchyma**: broad adaxial band in midrib, separated from epidermis by a layer of ground tissue. Discontinuous in places abaxially (types Bi or Sb).

Vascular bundles scattered in midrib and weakly formed (Fig. 54A); type II. Xylem: large protoxylem strand and 2–7 small or large metaxylem strands. Phloem: semicircular, irregular or crescent-shaped strand. **Sclerenchyma**: fibres in small caps over xylem and phloem. **Laticifers** common, non-anastomosing and with dark contents, on both sides of vascular bundles. **Crystals**: raphides in single bundles, often in lignified short cylindrical cells similar to biforines (Fig. 54D), common in mid-mesophyll. Other raphides of coarse texture in multiple bundles in elongated cells. Some tendency for raphide cells to run parallel along primary vascular bundles. Druses in circular cells common in mid-mesophyll (Fig. 54C).

Petiole TS

Outline oval. **Size**: diameter 6 × 4 mm. **Cuticle** thick and smooth. **Epidermis**: small square cells, rounded externally. **Collenchyma**: prominent circular or irregular strands, between peripheral vascular bundles (Fig. 54E), two layers below epidermis (type Sb) (cells staining like collenchyma but walls thinly layered like fibres). **Ground tissue**: cells small subepidermally, grading to large in centre. **Air cavities** at cell corners, up to 5% space.

Vascular bundles scattered and numerous in centre, most in irregular peripheral band (types II or III). Xylem: 1–2 large early metaxylem strands. Phloem: small irregular strand. **Sclerenchyma**: semicircular fibre caps over phloem, mostly on peripheral bundles (Fig. 54E), occasionally ensheathing smaller bundles. Most central bundles without fibre caps. **Laticifers**: non-anastomosing, lateral to most vascular bundles. **Crystals**: raphides in single bundles in thin-walled cells, occasional in ground tissue. Druses large, in circular cells, occasional in ground tissue. Crystal sand occasional in ground tissue.

Root TS

Outline circular, size 1.5–2.0 mm. **Exodermis**: cells large, square and thin-walled (Fig. 54F,H) or with lignified thickened anticlinal and inner periclinal walls. **Epidermis**: cells present in fragments. **Cortex**: uniform polyhedral cells mostly without air cavities (Fig. 54F,H). **Endodermis**: single layer of large cells with narrow anticlinal casparian strips. **Pericycle**: one irregular layer. **Stele**: seven protoxylem poles (Fig. 54G). Phloem: seven irregular, few-celled strands. Centre: ground tissue and immature xylem. **Starch** granules abundant in ground tissue, 3–4 μm diameter. **Crystals**: raphides in single bundles in thin-walled cells, common in ground tissue.

Material examined

Nephthytis afzelii Schott, Knecht 23/198, Côte d'Ivoire; *N.* aff. *gravenreuthii* Engl., Knecht 25/243, Côte d'Ivoire; *N. hallaei* (Bogner) Bogner, Bogner 750, Gabon; *N. poissonii* N.E.Br., Bogner s.n., Nigeria.

Species reported on in the literature

Bergdolt (1955): *N. picturata*: leaf. • El-Gazzar *et al.* (1977): *N. afzelii*: stomata. • Engler (1911): stem, leaf. • French (1987c): *N. hallaei, N. poissonii*: root. • French (1988): *N. afzelii, N. hallaei, N. poissonii:* laticifers. • French and Tomlinson (1981a): *N. poissonii*: stem vasculature. • Meyer (1962): *N. poissonii*: leaf. • Solereder (1919): *N. poissonii*: leaf. • Solereder and Meyer (1928): *N. poissonii*: all organs. • Webber (1960): *N. gravenreuthii*: epidermis, stomata.

Additional literature

Campbell 1905; French 1985b, 1986c, 1987a; Grayum 1990; Knecht 1983; Mayo 1986; Thanikaimoni 1969.

Anchomanes Schott

This genus of 10 species occurs in tropical Africa. Plants are terrestrial, sparsely-leaved herbs with creeping or tuberous rhizomes. They are usually found in open areas of humid forests, at forest edges, or savanna forests. The leaves have a single tall stipe with three tetrahedral branches producing pinnately divided blades, truncate or double-tipped distally. Chromosomes: 2n = 40 for three species. Two species were available for study.

Leaf surface

Cuticle striate. **Epidermis**: cells with undulate, sinuous to interlocking anticlinal walls on both surfaces. **Stomata** abaxial and brachyparacytic. Subsidiary cells narrow or broad and rounded (Fig. 55C). **Venation**: midrib straight or sinuous, with ± pinnate secondaries. Midrib may fork distally. Suprabasal acrodromous secondaries also present. In midrib, secondaries broadly acute, irregular, apically curved. Tertiary and higher order veins reticulate.

Leaf TS

Structure dorsiventral, poorly differentiated. **Size**: leaflet 73 × 29 mm (*l*/*w*). **Midrib** small, flat adaxially, convex-rounded abaxially (Fig. 55A). **Cuticle** thin at margin and midrib, not noted elsewhere. **Epidermis**: cells medium-tall on both surfaces, cells rounded externally and internally.

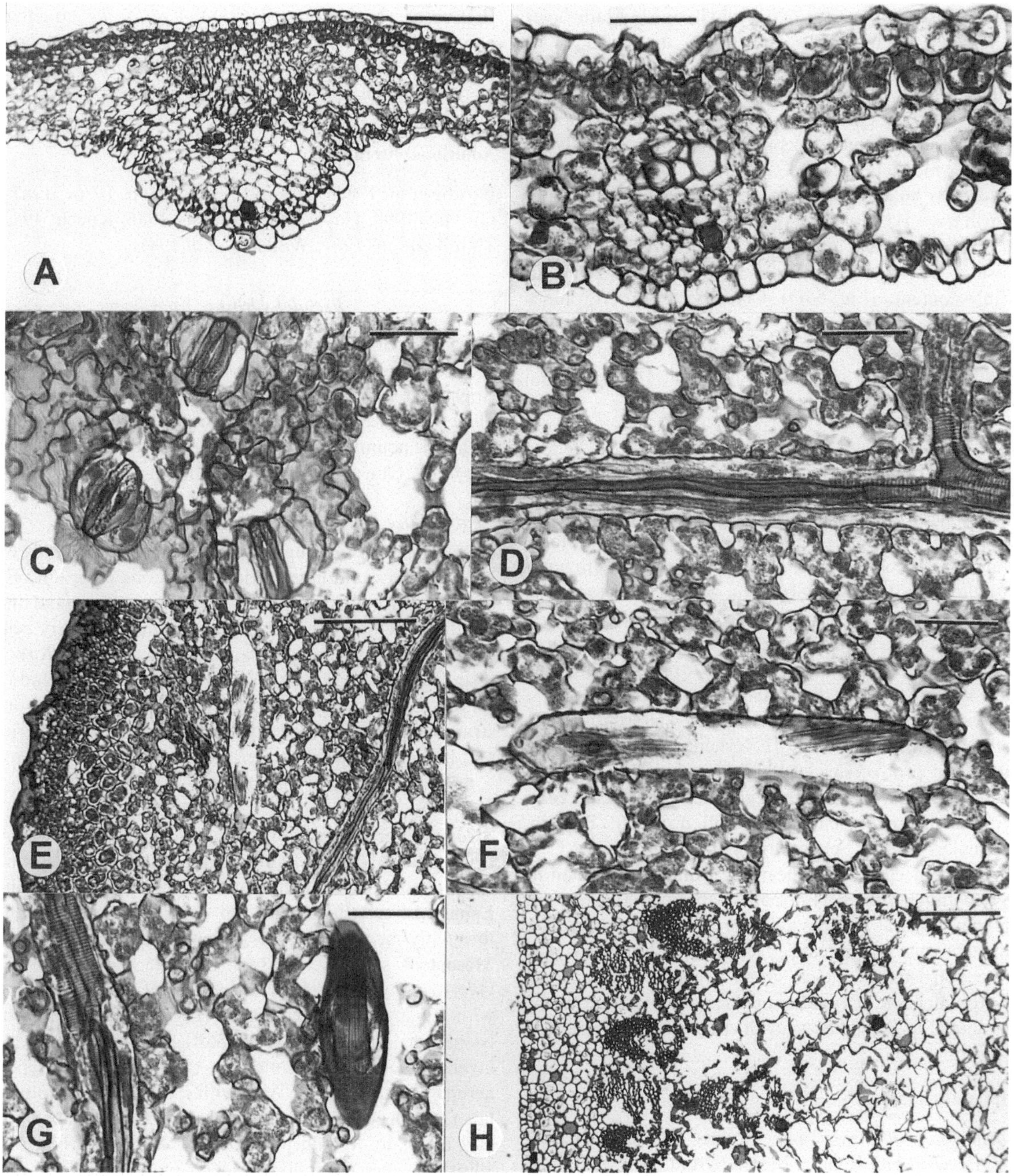

Fig. 55. *Anchomanes difformis* leaf and petiole sections. (A) Midrib TS; (B) Lamina TS; (C) Abaxial leaf epidermis PS with brachyparacytic stomata; (D) Leaf mesophyll PS with vein and lateral non-anastomosing laticifers; (E,F) Leaf mesophyll PS showing elongate raphide cells; (G) Leaf mesophyll PS with biforine-like raphide cell; (H) Petiole margin TS. Note vascular bundles with fibre caps, on alternate radii with collenchyma strands. Scale bars: A = 200 μm; B–D,F,G = 50 μm; E = 150 μm; H = 300 μm.

Stomata level with surface. **Mesophyll**: one palisade layer of lobed, U-shaped cells (Fig. 55B), up to 30 % of mesophyll. Spongy layers 4–5, cells armed (type 1–2) (Fig. 55B). **Air cavities**: clearly formed substomatal cavities, irregular deeper in mesophyll, up to 50% of spongy layers. **Collenchyma**: subepidermal band abaxially in midrib (type B) (Fig. 55A).

Vascular bundles: one large and two small bundles in midrib (type II). Xylem: several metaxylem elements and an irregular strand of phloem. **Laticifers** non-anastomosing, lateral to vascular bundles and often without visible contents (Fig. 55D). **Crystals**: biforine raphides common in mesophyll (Fig. 55G). Raphides also common in thin-walled short cells and in multiple overlapping bundles in elongate cells (Fig. 55E,F). Crystal sand few in each mesophyll cell.

Petiole TS

Outline subterete near summit. **Size**: 9 × 13 mm diameter. **Cuticle** very thin and smooth. **Epidermis**: cells small and rounded externally. **Collenchyma**: large elliptical strands, situated 8–10 cells below epidermis, alternating with outer vascular bundles (type Sb) (Fig. 55H). **Chlorenchyma**: irregular groups of cells, 3–5 cells deep in cortical region. **Ground tissue**: cells small near epidermis, grading to large in centre. **Air cavities** large, few in central ground tissue, type 4. Many cell-sized air cavities.

Vascular bundles: scattered, outer row alternating with collenchyma strands; type II. Xylem: usually 1–3 wide functional protoxylem lacunae or metaxylem cells surrounded by row of very small ground tissue sheath cells; very narrow metaxylem elements often present. Phloem: a circular strand. **Sclerenchyma**: fibres forming xylem and phloem caps or ensheathing bundles, 1–3 bundle rows within periphery (Fig. 55H); caps much thicker over phloem. Central bundles at corners of aerenchyma with sclerenchyma much reduced or absent. **Laticifers**: 1–2 present on flanks of phloem strands. **Tannin** cells occasional in ground tissue. **Crystals**: raphides common, randomly distributed in ground tissue with cells axially aligned.

Material examined

Anchomanes difformis Engl., Bogner 727, Gabon, cult. M; *A. welwitschii* Rendle, Magill (Croat 69773), Gabon, cult. MO.

Species reported on in the literature

Engler (1911): *A. difformis*: root, stem leaf. • Ertl (1932): *A. difformis*: leaf venation. • French (1987c) *A. difformis*: root. • French (1988): *A. difformis* : laticifers. • Guttenberg (1968): root. • Hotta (1971): leaf. • Lierau (1888): *A. dubius*: root. • Solereder and Meyer (1928): secretory tissue.

Additional literature

Barabé et al. 1986a, 1987a; French 1985b, 1986c, 1987a; Grayum 1990; Hay 1986; Hegnauer 1986; Knecht 1983; Thanikaimoni 1969; Williams *et al* 1981.

***Pseudohydrosme* Engl.**

This genus of two species is found in west Africa. Plants are rare terrestrial and tuberous understorey herbs. The single leaf has an elongate, erect petiole with a trisect blade. The leaf tips are dracontioid, blunt and square, with two acuminate tips at the corners. Chromosomes: 2n = *c*.40. One species was available for study.

Leaf surface

Cuticle smooth or rough. **Epidermis**: anticlinal walls on both surfaces undulate to deeply sinuous (Fig. 56D,F), with up to 10–12 lobes (3–6:1 *l*/*w*). **Stomata**: abaxial and brachyparacytic with wide, semi-elliptic subsidiary cells (Fig. 56D). **Venation**: midrib prominent and often forked distally. Secondaries basal and suprabasal acrodromous. Basal secondaries forming ascending brochidodromous straightened loops collecting distal secondaries. Tertiaries and higher orders reticulate.

Leaf TS

Structure dorsiventral. **Size** (ovate segment): 70 × 44 mm (*l*/*w*). **Midrib** shallowly rounded-convex adaxially and deeply rounded-semicircular abaxially. **Cuticle** thin. **Epidermis**: cells large and flat externally and rounded internally on both surfaces. **Stomata** level with surface. **Mesophyll**: structure unusual. Palisade zone of one cell layer occupying 30–50% of mesophyll. Cells U-shaped with 2–6 arms facing epidermis (Figs 4E, 56A), the wide basal portion internal (Fig. 56B). Spongy tissue: 2–4 layers of elongate and multi-armed cells, forming a type 1 aerenchyma (Fig. 56C). **Air cavities**: up to 20% of spongy layer. **Collenchyma** discontinuously banded abaxially and adaxially at midrib (type Sb). Strands broad and elliptical subepidermally. Vascular bundles and chlorenchyma on same radii as the several gaps in the discontinuous collenchyma. Small wide strand at adaxial midrib ridge.

Vascular bundles scattered, *c*. 6 per midrib, more or less symmetrically disposed; bundles not well demarcated or formed (type II). Xylem: 4–8 metaxylem cells at irregular

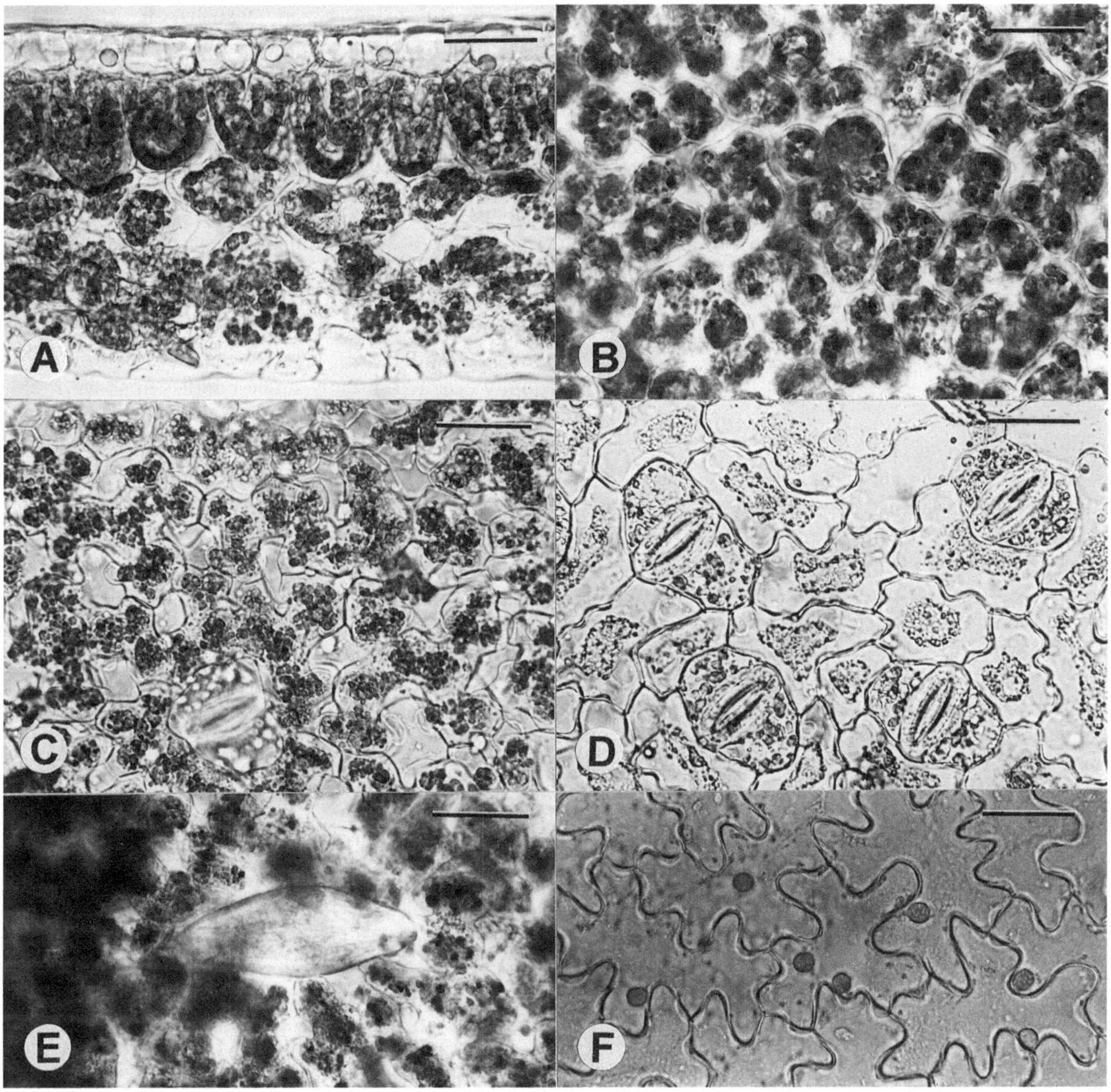

Fig. 56. *Pseudohydrosme gabunensis* leaf hand sections in iodine/calcium chloride mounts. (A) Lamina TS showing U-shaped palisade cells; (B) Lamina PS showing palisade cells; (C) Lamina PS showing type 1a spongy aerenchyma; (D) Lamina abaxial surface PS showing undulate epidermal cells and brachyparacytic stomata with trapezoidal subsidiary cells; (E) Mesophyll PS showing biforine-like cell; (F) Lamina adaxial surface PS showing sinuous epidermal cells. Scale bars = 50 μm.

boundary with irregular phloem strand. **Laticifers** not detected. **Crystals**: raphides common, in wide bundles in large circular cells. Occasional biforines present in central mesophyll, variously oriented (Fig. 56E). Occasional elongate raphide cells bearing multiple overlapping bundles. Crystal sand abundant in mesophyll and midrib ground tissue.

Material examined

Pseudohydrosme gabunensis Engl., Breteler s.n., Gabon, cult. Leiden.

Additional literature

French 1985b, 1986c; Grayum 1990; Hay 1986.

Aglaonema Schott

This genus of 21 species is distributed from southeastern Asia to northeastern India, through southern China, Indonesia and New Guinea (Nicolson 1969). Plants are terrestrial herbs with erect stems and creeping rhizomes growing in the understorey of humid evergreen or deciduous forests. They have erect or decumbent epigeal stems that are mostly unbranched, and generally have distinct internodes and continuously growing leaves. Leaves are long-petiolate, broadly ovate, ovate, or broadly elliptic, and may have acuminate tips. Chromosomes: 2n = 16, 40, 42, 56, 60, 80, 110, 118, 120. Five species have been examined.

Leaf surface

Cuticle smooth. **Epidermis**: cells polygonal with straight, undulate or sinuous anticlinal walls on both surfaces (2–4:1 *l*/*w*) (Fig. 3A). **Stomata** abaxial and randomly oriented, brachyparacytic to brachypara-tetracytic (Fig. 57D) with one, occasionally two, pairs of subsidiary cells. Subsidiary cells broad or narrow and rounded. **Venation**: midrib prominent with acute, ascending secondaries ending eucamptodromously at margins. Intersecondaries and tertiaries parallel with secondaries. Secondaries and primary intersecondaries often curving outward near their ends, attaching orthogonally to collector vein. In *A. rotundum*, tertiaries arising from secondaries regularly exmedially acute and admedially obtuse. Higher-order veins orthogonal between larger veins, or reticulate. Ultimate veinlets often describing square or rectangular areoles.

Leaf TS

Structure dorsiventral. **Size**: up to 93 × 70 mm (*l*/*w*). **Midrib** large, flat to slightly convex-rounded adaxially, semicircular abaxially (Fig. 57A). **Cuticle** thin adaxially and not flanged; not detected abaxially, except thick on both surfaces in *A. simplex*. **Epidermis**: cells square to rectangular. Adaxial cells larger than abaxial, or the same size in *A. simplex*. Outer walls flat, inner walls rounded. **Stomata** level with epidermis. **Mesophyll** weakly or well differentiated with 1-layered palisade (Fig. 57B) occupying 10–20% of lamina; chloroplasts concentrated in palisade. Spongy cells large, irregular, mostly horizontally elongate, type 1 grading to type 2, and with few plastids. Aerenchyma type 3 in midrib of *A. simplex*. **Air cavities**: 20–60% of spongy parenchyma. **Collenchyma** in discrete strands, broadly-elliptic or circular, up to 10 cells deep on abaxial side of midrib (type Sb) and 1–2 layers within epidermis (Fig. 57A); also adaxial in *A. simplex*. Cell walls angularly thickened. **Vascular bundles**: few to numerous weakly developed bundles in midrib (type II). Xylem: 1–5 wide metaxylem cells. Phloem: small strand without sheath. Bundles well developed in *A. simplex* with 3–5 wide metaxylem strands and broad phloem strand encircled by distinct parenchymatous sheath.

Laticifers: non-anastomosing, 2–4 wide cells around phloem of vascular bundles (Fig. 57C), containing amorphous droplets; found anywhere around bundles in *A. simplex*. **Tannin**: idioblastic mesophyll cells present in *A. simplex*, containing plastids and other dark contents. **Starch** granules common in cells of bundle sheath and other ground tissue. **Crystals**: raphides common, randomly oriented, usually single bundles in short or elongate, rounded or narrow cells. Raphides also in biforine cells with lignified walls and papillate tips (Fig. 57E). Druses in small rounded cells common throughout mesophyll. Crystal sand may be common in palisade cells.

Petiole TS

Outline crescent-shaped, becoming elliptic distally, and laterally narrowed. **Size**: diameter 8 mm. **Cuticle** thin to medium-thick and ± smooth. **Epidermis** of short flat cells. **Collenchyma** subepidermal in discrete circular strands, on abaxial side only (type Sb) (Fig. 5H). **Ground tissue**: small rounded compact cells at periphery. Tissue becoming type 3 aerenchyma in centre with 2–4 rounded, uniseriate partition cells extending between unmodified corner cells.

Vascular bundles: *c*. 12, scattered, nearly symmetrically disposed (types II and III). Xylem: 1–3 wide metaxylem cells in radial file. Phloem: small laterally narrowed strand. **Laticifers**: large cells, usually two lateral to phloem. **Crystals**: cells with single raphide bundles found anywhere in ground tissue. Biforines commonly transversing aerenchyma partitions or adjacent to them; cell ends rounded and smooth or warty, or with short papillae. Druses small and common in rounded cells, anywhere in ground tissue.

Stem TS

Outline circular. **Size**: diameter 2.4 cm. **Cuticle** thin and smooth. **Epidermis**: cells small and tile-like. **Periderm** patchy, of subepidermal origin. **Collenchyma**: slight development; few subepidermal layers with thin collenchymatous walls. **Ground tissue**: cells uniform in centre and small near periphery. **Air cavities** few, as small corner spaces.

Vascular bundles numerous and scattered; small, irregularly shaped, often compound and weak (Fig. 57F,G), usually type II when collateral. Xylem: 1–4

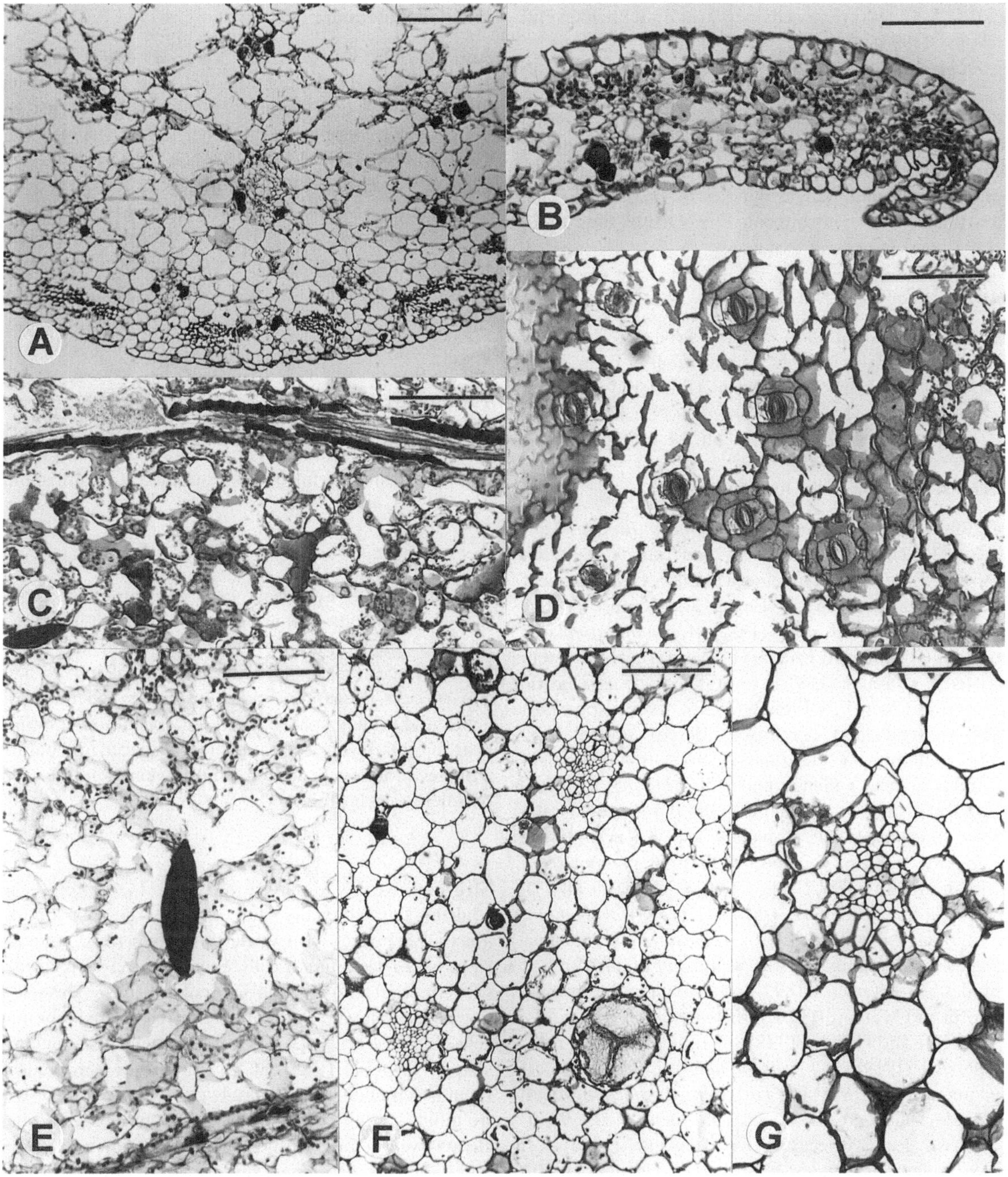

Fig 57. *Aglaonema* leaf and stem sections. (A,B,E–G) *A. costatum*; (C,D) *A. brevispathum*; (A) Midrib TS, abaxial region showing type Sb collenchyma; (B) Lamina TS margin; (C) Lamina PS showing non-anastomosing laticifers along vein; (D) Lamina abaxial epidermis showing stomata and epidermal cells, sinuous near surface (at left); (E) Lamina PS with biforine raphide cell in spongy mesophyll; (F,G) Stems TS: (F) with secretory duct; (G) detail of bicollateral (small compound) bundle. Scale bars: A = 300 µm; B–E,G = 150 µm; F = 200 µm.

strands, 2–4 metaxylem cells per strand. Schneider and Carlquist (1998) recorded vessel elements in stems of *A. pictum*. Phloem: one or more central strands. **Secretory ducts** large, with one epithelial layer, scattered in ground tissue (Fig. 57F). **Tannin** cells common, isolated and scattered in ground tissue. **Starch** granules irregular, common in ground tissue, 6–15 µm diameter. **Crystals**: raphides occasional, single bundles in small oval cells; often thick-walled and biforine-like. Druses few, small and scattered in ground tissue.

Material examined

Aglaonema brevispathum (Engl.) Engl. f. *hospitum* (Williams) Nicolson, cult. M; *A. costatum* N.E.Br. 1954, Kew spirit collection; *A. crispum* (Pitcher & Manda) Nicolson, cult. M, s. loc.; *A. rotundum* N.E.Br., Bogner 303, cult. M; *A. simplex* Blume, cult. M 115/76.

Species reported on in the literature

Bergdolt (1955): *A. roebelinii, A. marantifolium*: leaf. • Czaja (1962): *A. roebellinii*: leaf epidermis. • Dalitzsch (1886): *A. simplex*: leaf. • Deinega (1898): *A. simplex*: leaf developmental anatomy. • Engler and Krause (1912): several spp.: root, leaf. • Ertl (1932): *A. commutatum*: leaf venation. • Fisher and French (1978): *A. roebelinii*: stem meristems. • French (1987c): several spp.: root [sclerotic hypodermis absent]. • French (1988): several spp.: laticifers. • French and Tomlinson (1981b): several spp.: stem vasculature. • Gardiner (1889): *A. mannii*: leaf glandular organs. • Genua and Hillson (1985): *A. modestum*: leaf crystals. • Guttenberg (1940): root. • Hassack (1886): *A. commutatum, A. pictum*: leaf. • Knoll (1948): *A. simplex*: leaf tip. • Leavitt (1904): *A. pictum*: root. • Lierau (1888): several spp.: root. • Marcatili (1887): *A. commutatum*: laticifers. • Molisch (1899): *A. commutatum*: tissues. • Müller (1919): *A. costatum, A. oblongifolium, marantifolium*: leaf tip. • Noro *et al.* (1992): leaf raphides. • Petersen (1893): *A. simplex*: meristems. • Pirotta and Marcatili (1885): *A. commutatum*: laticifers. • Schneider and Carlquist (1998): *A. pictum*: stem, root. • Solereder (1919): *A. marantifolium*: raphides. • Solereder and Meyer (1928): *A. costatum, A. marantifolium, A. simplex*: all organs. • Stewart and Dermen (1979): *A. modestum*: leaf variegation. • Tieghem (1866): *A. marantifolium, A. simplex*: all organs. • Tieghem (1870–1): *A. simplex*: root, secretory tissue. • Tieghem (1872): *A. marantifolium*: secretory tissue. • Trécul (1865): *A. simplex*: vasculature. • Van Fleet (1942): *A. simplex*: stem, endodermis. • Voronkina (1975): *A. modestum*: root apex. • Webber (1960): *A. costatum, A. marantifolium, A. pictum*: epidermis, stomata.

Additional literature

Antony 1898; Batygina and Yakovlev 1990; Campbell 1903, 1912; Daumann 1930, 1970; Davis 1966; DiBenedetto and Cogliatti 1990; DiBenedetto and Garcia 1992; Domin 1911; French 1985b, 1986b, 1987a; Grayum 1990; Halevy 1989; Hassack 1886; Henny 1983a, 1986, 1992; Hotta 1965; Knoth 1981; Kulkarni *et al.* 1990; Lindinger 1908; Mayo 1986; Meusel 1951; Nicolson 1969; Pfitzer 1962; Ray 1988; Savchenko and Mania 1970; Thanikaimoni 1969; Tieghem 1907; Troll 1928; Williams *et al.* 1981.

Aglaodorum Schott

This monotypic genus is found in Sumatra and Borneo. Plants are aquatic or terrestrial, stout rhizomatous herbs growing in tidal swamps. The leaves are elliptic and long-petiolate. One specimen was examined. Chromosomes: 2n = 34, 36, 40.

Leaf surface

Cuticle smooth. **Epidermis**: cells elongate, fusiform or polygonal with straight anticlinal walls on both surfaces. **Stomata** brachyparacytic, -tetracytic or -hexacytic, on both surfaces (Fig. 58D) with 2–3 sets of narrow subsidiary cells. **Venation**: midrib prominent. Secondaries acute and widely spaced, merging eucamptodromously near margin. Intersecondaries closely spaced and parallel with secondaries. Scattered smaller veins tending to be parallel to each other.

Leaf TS

Structure weakly dorsiventral. **Size**: 14.5 × 9 cm (*l*/*w*). **Midrib**: flat adaxially and broadly convex-rounded abaxially (Fig. 58A). **Cuticle** thin. **Epidermis**: rounded, squarish or rectangular cells. **Stomata** level with surface, with internal and external cuticular ledges. **Mesophyll**: one palisade layer of moderately differentiated, compact cells. Spongy cells: 12–14 layers, larger and irregularly shaped (type 1) (Fig. 58C). **Air cavities** conspicuous, regular circular substomatal cavities, together with irregular spaces providing *c.* 50% air space. **Collenchyma**: well-defined strands, abaxial in midrib (Fig. 58B) and *c.* 2 layers inside epidermis; strands on alternate radii with peripheral vascular bundles (type Sb).

Vascular bundles scattered in midrib at corners of type 3 aerenchyma partitions (Fig. 58A); bundles small (type II). Xylem: 1–5 metaxylem elements. Phloem: circular or irregular strand at irregular boundary with xylem. **Laticifers**: non-anastomosing and lateral to phloem

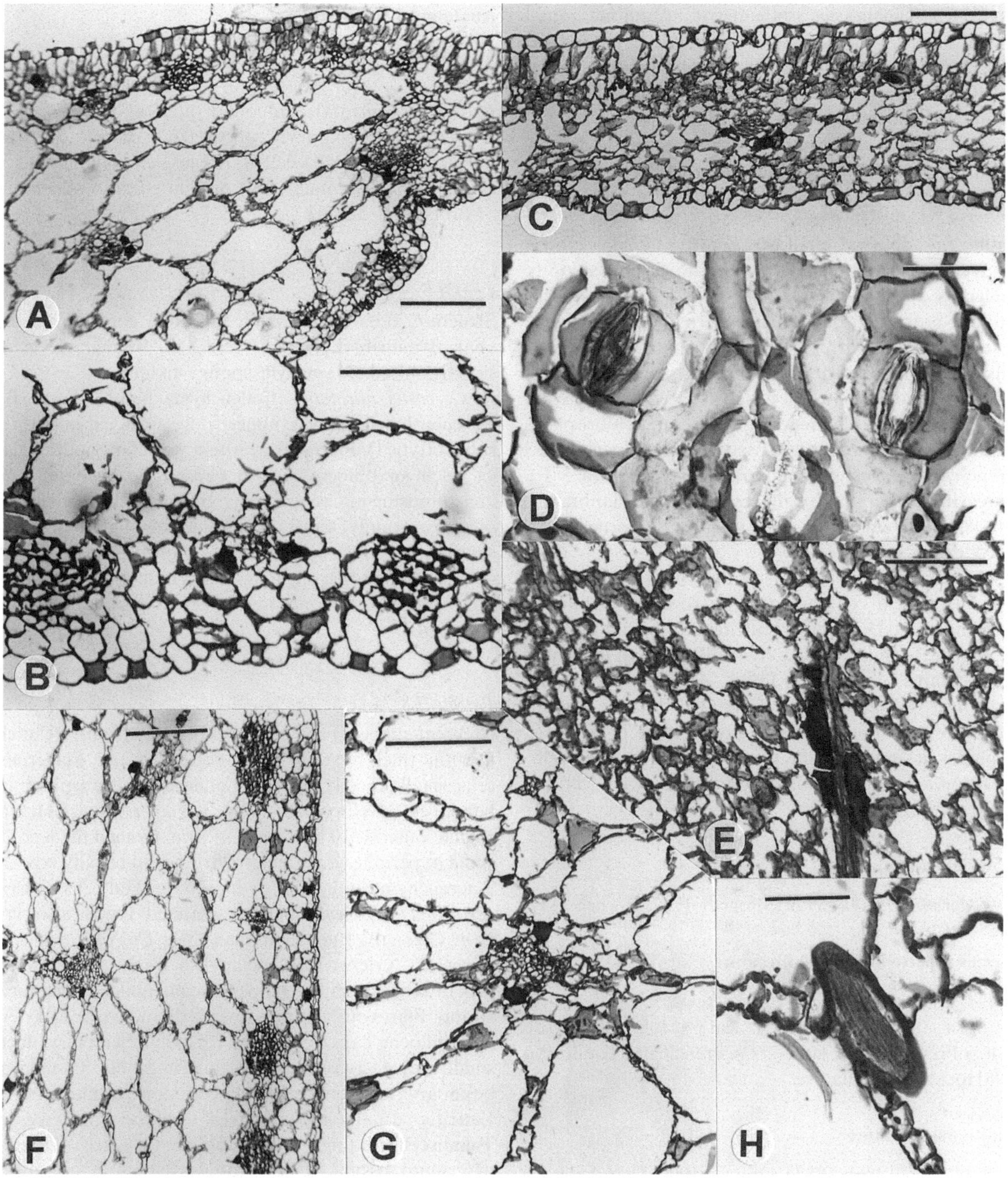

Fig. 58. ***Aglaodorum griffithii*** **leaf and petiole sections.** (A) Midrib TS showing type 3 aerenchyma, also seen in (B,F–H); (B) Midrib TS abaxial region showing type Sb collenchyma strands, often not adjacent to vascular bundles; (C) Lamina TS; (D) Abaxial epidermis PS with brachyparacytic or -tetracytic stomata; (E) Mesophyll PS with small vein and non-anastomosing laticifers; (F) Petiole TS showing type Sb collenchyma (at right); (G) Petiole TS, central tissue with type II vascular bundle and two adjacent laticifers; (H) Petiole TS, biforine suspended across type 3 aerenchyma partition. Scale bars: A,F = 300 μm; C = 200 μm; B,E, G = 150 μm; D,H = 50 μm.

(Fig. 58E). **Tannin** cells isodiametric in mid-mesophyll with granular dark contents. Occasionally as bundle sheath cells, 2–4:1 *l/w* ratio. **Crystals**: biforine raphides occasional, arranged without relation to veins in upper lamina and subepidermal in midrib; raphides also occasional in single bundles in cells. Druses common, mostly in palisade tissue. Crystal sand common throughout.

Petiole TS

Outline broadly oval; shallowly concave adaxially. **Size**: diameter 6 × 9 mm. **Cuticle** medium-thick and smooth. **Epidermis**: cells irregular in size and shape, some cells rounded-protruding or papillate externally. **Collenchyma**: strands rounded or irregular, 2–4 subepidermal layers of cells, not subtending vascular bundles; one or two small vascular bundles between widely spaced collenchyma strands (type Sb). **Chlorenchyma**: scattered chloroplasts in outer 4–5 layers of ground tissue. **Ground tissue**: type 3 aerenchyma, 3–5 cells per uniseriate partition (Fig. 58F). **Air cavities** separated axially by 1-layered diaphragms with small circular air passages at cell corners. Air space 95%.

Vascular bundles scattered, small (type II). Xylem: 1–3 wide metaxylem cells and 0–3 narrow ones. Phloem: irregular strand (Fig. 58G). **Laticifers**: 1–3 strands lateral to phloem. **Tannin** cells scattered and common in ground tissue. **Starch**: scattered granules, especially in ground tissue near vascular bundles. **Crystals**: biforines, suspended across partitions in aerenchyma, rare (Fig. 58H). Raphides in single bundles in rounded thin-walled cells, uncommon in aerenchyma. Crystal sand: a few crystals in most ground tissue.

Material examined

Aglaodorum griffithii Schott, Bogner 1767, Borneo (M).

Species reported on in the literature

(All *A. griffithii*)

Engler and Krause (1912): leaf. • French (1987c): root. • French (1988): laticifers. • French and Tomlinson (1981b): stem vasculature.

Additional literature

French 1985b, 1986b, 1987a; Grayum 1990; Thanikaimoni 1969.

Tribe Culcasieae Engl.

A tribe composed of the African genera *Culcasia* and *Cercestis*.

Anatomical summary

Leaf surface

Cuticle smooth or striate, thin to thick. **Epidermis**: cells polygonal (*Culcasia*) to sinuous or interlocking (*Cercestis*). **Stomata** brachyparacytic to brachypara-tetracytic or -hexacytic. Subsidiary cells variable. **Venation** pinnate or acrodromous, secondaries at margin eucamptodromous or brochidodromous.

Leaf TS

Structure dorsiventral with broadly convex-rounded abaxial midrib. **Epidermis**: cells tall. **Stomata** level or slightly raised. **Mesophyll**: spongy tissue type 2 (type 1 in *Cercestis mirabilis*). **Collenchyma** banded (type B). **Vascular bundles** few, symmetrically arranged in midrib, robust (type I). **Fibres** ensheathing vascular bundles, also as xylem or phloem caps in *Cercestis*. **Laticifers**: non-anastomosing in *Cercestis* only. **Secretory ducts** in midrib of *Culcasia* only. **Crystals**: raphides in single bundles in small cells in both genera, and in biforine-like cells; they may also occur in elongated cells in multiple overlapping bundles (*Culcasia*), or as multiple bundles in tubular cells (*Cercestis*).

Petiole TS

Crescent-shaped proximally, subterete distally. **Cuticle** medium-thick to thick and smooth or striate. **Epidermis**: cells small, mostly square. **Collenchyma**: subepidermal band, best developed abaxially, 3–8 layers deep. **Chlorenchyma**: outer 8–10 layers in *Culcasia*. **Ground tissue**: cells small at periphery, grading to larger, and usually type 2–3 aerenchyma in centre, or loosely packed cells in *Cercestis afzelii*. **Vascular bundles** scattered, type I, also type II in *Cercestis*. Bundles numerous in *Culcasia*, fewer in *Cercestis*. Xylem: 3–12 metaxylem tracheids in several files in larger bundles. Phloem: semicircular or elliptical strand. **Fibres** ensheathing vascular bundles, or as xylem and phloem caps, or absent distally. **Secretory ducts** small, with 1–2-layered epithelium in ground tissue near boundary with collenchyma, common or uncommon. **Laticifers** usually in pairs lateral to phloem in *Cercestis*. **Tannin** cells frequent or occasional in *Cercestis*, scattered in ground tissue. **Starch** granules common in central ground tissue. **Crystals**: raphides biforine-like in central ground tissue in *Culcasia*, in air cavity partitions in *Cercestis*. Crystal bundles also in thin-walled irregular cells in outer ground tissue, common or uncommon. Druses in large or small cells common subepidermally (*Cercestis*) and in other ground tissue.

Taxonomic notes

These genera can be easily separated on the basis of: secretory ducts, laticifers, tannin cells, and leaf epidermal cell shapes. A case can be made for dividing them into separate tribes. There is also a large amount of variation within the genera. *Cercestis mirabilis* (*Rhektophyllum*) is distinctive on the basis of cuticle, epidermis, stomata, tertiary venation pattern and mesophyll tissue among other features. This falls short of a recommendation to reestablish *Rhektophyllum* until the range of anatomical variation of these two genera is more broadly understood.

This tribe belongs to the banded collenchyma lineage and has little in common with other Aroideae 'No alliance' genera of MBB which are nearly entirely of type Sv collenchyma anatomy. The banded collenchyma, type I vascular bundles, sclerenchyma, and secretory ducts place them at the base of the Philodendroideae. This supports the consensus tree of French *et al.* (1995).

Culcasia P.Beauv.

This genus of 20 species is distributed in tropical Africa. Plants are hemi-epiphytes, rhizomatous to somewhat scandent herbs, mostly with elongate internodes and with non-heteroblastic leaf development. Some species are occasionally terrestrial or epilithic. They occur mostly as understorey plants of primary or secondary forests, or in clearings and wet areas. The leaves examined are elliptic, ovate, cordate, or blunt-emarginate in *C. rotundifolia.* Chromosomes: $2n = 42$ or 84. Four species were available for study.

Leaf surface

Cuticle smooth or weakly to strongly striate on both surfaces, especially on abaxial midrib; rough on abaxial midrib in *C. obliquifolia.* **Epidermis**: adaxial cells polygonal-isohedral with straight anticlinal walls (1–1.5:1 *l*/*w*), abaxial cells polygonal (1–2:1 *l*/*w*). In *C. saxatilis*, adaxial epidermal cells sinuous at surface. **Stomata** brachyparacytic, evenly distributed on both surfaces, or mostly abaxial, or abaxial only. Stomata also brachy-para-tetracytic to -hexacytic in *C. mannii* (Fig. 59B). Inner subsidiary cells narrow or wide; often a second pair parallel with the first. **Venation**: midrib prominent. Secondaries pinnate, ± regular, arising broadly acute, curved and merging eucamptodromously at margins, or merging brochidodromously in submargins in *C. saxatilis*. Tertiaries and intersecondaries merging with each other in outer intercostal areas. Higher orders reticulate. In *C. rotundifolia*, basal secondaries widely acute, curved ascending, converging at lamina apex.

Leaf TS

Structure dorsiventral. **Specimen sizes**: *C. saxatilis*: 206 × 65 mm (*l*/*w*), *C. rotundifolia*: 26 × 39 mm (*l*/*w*), *C. mannii*: 180 × 100 mm (*l*/*w*) and *C. obliquifolia*: 122 × 78 mm (*l*/*w*). **Midrib** shallowly concave adaxially and broadly rounded-convex abaxially (Fig. 59A). **Cuticle** thick or thin. **Epidermis**: cells usually tall, flat on surface and rounded internally, or external and internal walls rounded. Cells adaxially papillate in *C. rotundifolia* (Fig. 1D). **Stomata** level with surface or slightly raised. **Mesophyll**: one palisade layer of short cells or T-shaped cells; palisade cells poorly or well differentiated from 3–8 layers of multi-armed spongy cells (type 2). **Air cavities** substomatal, forming up to 30–50% of spongy tissue; type 3 aerenchyma in midrib (Fig. 59A). **Collenchyma**: abaxial subepidermal band in midrib, 2–3 rows of angularly thickened cells (type B).

Vascular bundles in midrib scattered, 12–16, in two layers, arranged in bilateral symmetry (Fig. 59A); bundles well differentiated (type I). Xylem: 6–8 tracheary cells (up to four cells wide) forming straight boundary with broad semicircular phloem strand. **Sclerenchyma**: fibres completely ensheathing vascular bundles, 2–3 cell layers over xylem and phloem and one layer laterally, or sheath entirely 1-layered and thin-walled. Thickest-walled fibres capping phloem. **Secretory ducts**: one to several in midrib with 1–2 epithelial layers, except in *C. rotundifolia.* **Starch** granules common in midrib ground tissue of *C. mannii.* **Crystals**: raphides as single bundles in large oval cells, thick-walled, lignified and without terminal papillae, or thin-walled, rarely biforine-like (Fig. 59C). Other single bundles in short rounded cells. Occasional elongate cells with several overlapping crystal bundles. Raphides common or uncommon in midrib and mid-mesophyll, appearing randomly oriented near margin, otherwise aligned with secondary vascular bundles. Druse cells large and rounded, common throughout or restricted to palisade layer. Crystal sand common in upper mesophyll cells, in midrib ground tissue only, or common throughout, or absent.

Petiole TS

Outline crescent-shaped or semicircular proximally to subterete distally. **Size**: diameter 5–6 mm. **Cuticle** thick and striate. **Epidermis**: cells small and mostly squarish. **Collenchyma** band subepidermal on abaxial side, 3–5 layers with thin or thick angular thickenings (type B) (Fig. 59D). **Chlorenchyma**: chloroplasts in outer 8–10 ground tissue layers. **Ground tissue**: small cells compact at abaxial periphery grading to large cells with air cavities (type 2–3 aerenchyma) nearer adaxial side.

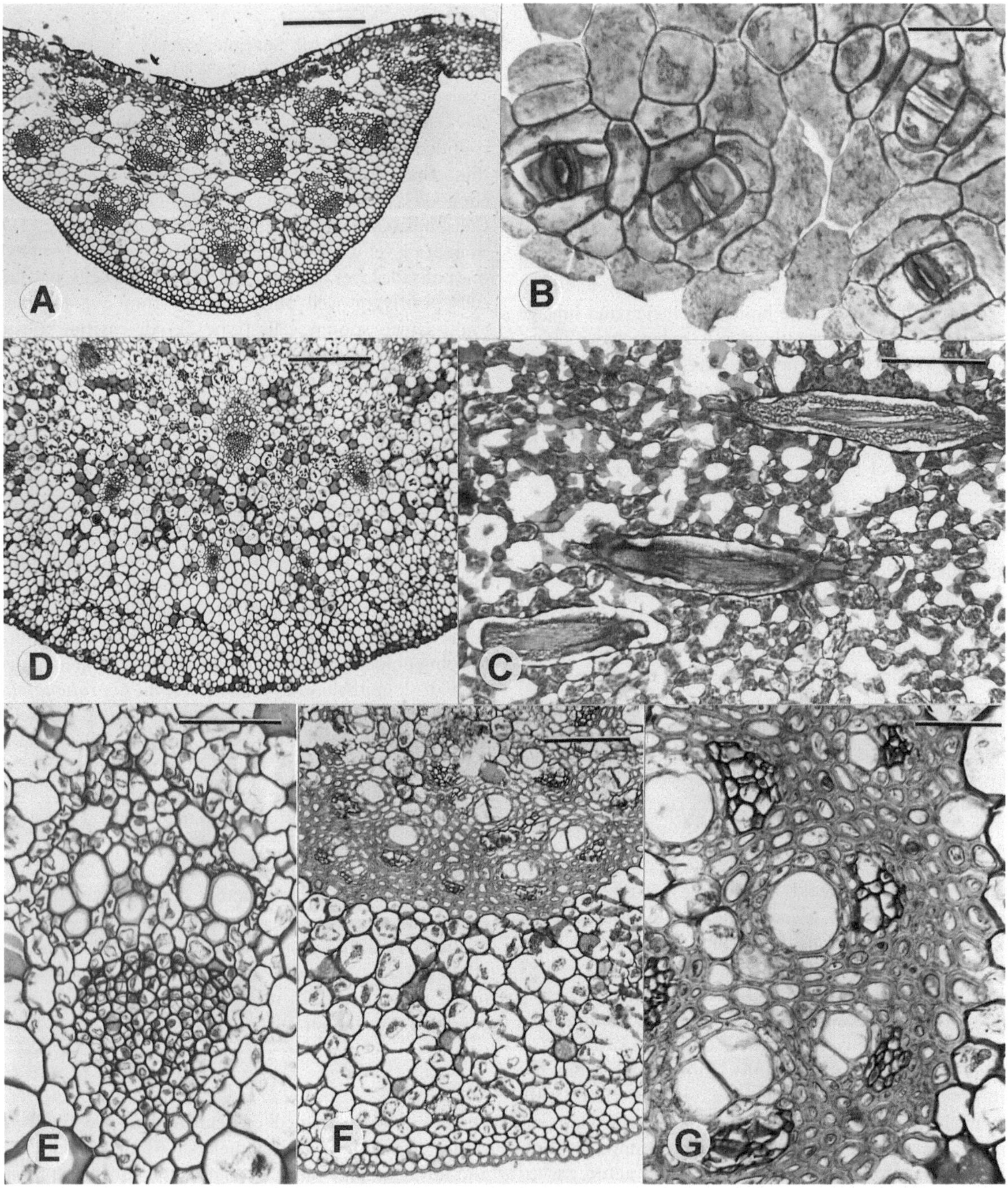

Fig. 59. *Culcasia* leaf, petiole and stem sections. (A) *C. saxatilis*; (B–E) *C. mannii*; (F,G) *C. rotundifolia*; (A) Midrib TS with type I vascular bundles and type 3 aerenchyma; (B) Lamina PS abaxial epidermis showing brachyparacytic and brachypara-tetracytic stomata; (C) Mesophyll PS spongy tissue and raphide cells; (D) Distal petiole TS, abaxial side; (E) Distal petiole TS, type I vascular bundle; (F) Stem TS showing cortex and central cylinder; (G) Detail of (F) showing confluent fibre tissue at central cylinder boundary with embedded phloem strands. Scale bars: A,D = 300 μm; B,E,G = 50 μm; C,F = 150 μm.

Vascular bundles numerous and scattered, well-defined (type I). Xylem: 1–2 narrow protoxylem lacunae and up to >12 metaxylem cells in several files of various sizes (Fig. 59E). Phloem: semicircular or elliptical strand at concave xylem boundary. **Sclerenchyma**: vascular bundles ensheathed proximally but not usually distally; sclerenchyma nearly absent in *C. saxatilis*. **Secretory ducts** small with two layers of small epithelial cells, common or uncommon at inner collenchyma boundary, in outer ground tissue. **Starch** granules common in central ground tissue. **Crystals**: raphide cells of two sizes. Thick-walled types without terminal papillae, uncommon in air spaces in central ground tissue. Thin-walled raphide cells common in outer ground tissue. Druses large and scattered, common in ground tissue.

Stem TS

Outline terete. **Size**: diameter 3–4 mm. **Cuticle** thick and striate. **Epidermis**: cells small, ± circular and outer walls lignified. **Hypodermis**: one layer of small cells. **Collenchyma**: subhypodermal band, 2–3 layers deep. **Cortex**: outer 12–14 cell layers of ground tissue. **Ground tissue**: cells small near periphery grading to large in centre. **Central cylinder**: mostly well demarcated by sclerenchyma boundary (Fig. 59F).

Vascular bundles scattered, most concentrated in outer part of central cylinder; many embedded in central cylinder sclerenchyma sheath (type I) (Fig. 59F). Xylem: 1–2 protoxylem lacunae and 6–12 metaxylem cells in several files. In *C. rotundifolia*, xylem with only 1–2 wide metaxylem cells (type II). Phloem strand small, circular or irregular (Fig. 59G). **Sclerenchyma**: thick-walled fibres forming boundary 3–12 layers deep, delineating central cylinder. Free vascular bundles within central cylinder ensheathed by fibres. **Secretory ducts**: small with two epithelial layers, uncommon within collenchyma layer. **Starch** granules common, scattered in central cylinder ground tissue or all ground tissue. **Crystals**: raphides in thin-walled cells, uncommon in outer ground tissue. Druses common in ground tissue.

Material examined

Culcasia mannii (Hook.f.) Engl., NYBG 11/16/78, Bogner 2340, cult. M; *C. obliquifolia* Engl., Bogner 654, Gabon, cult. M; *C. rotundifolia* Bogner, Bogner 749, Gabon, cult. M; *C. saxatilis* A.Chev., Knecht 42/381, Côte d'Ivoire.

Taxonomic notes

Venation patterns have distinctive characters in each species examined. In *C. obliquifolia*, several vascular bundles running as a single secondary rib, may originate by branching from different midrib bundles at different levels. In *C. mannii*, laminar vascular bundle origins resemble dicot branching patterns. Histological features vary little among species examined, although species-specific differences do occur.

Species reported on in the literature

Blanc (1978): developmental anatomy. • Engler (1905): several spp.: leaf. • French (1987b, c): several spp.: root. • French (1988): *C. mannii, C. saxatilis, C.* sp.: laticifers. • French and Tomlinson (1980, 1981c): *C. saxatilis, C.* sp.: stem vasculature. • Guttenberg (1968): root. • Knecht (1983): several spp.: leaf venation, secretory glands. • Lierau (1888): *C. mannii*: root. • Müller (1919): *C. scandens*: leaf tip. • Netolitzky (1929): crystals. • Ntépé-Nyamè (1984): several spp.: leaf venation. • Solereder (1919): several spp.: leaf. • Solereder and Meyer (1928): several spp.: all organs.

Additional literature

Barabé *et al.* 1987a; Barabé and Forget 1992; Domin 1911; French 1985a, 1986c, 1987a; Grayum 1990; Hegnauer 1986; Mayo 1986; Smolenski *et al.* 1974; Thanikaimoni 1969; Williams *et al.* 1981.

Cercestis Schott

This genus of 10 species is found in Guinea and Gabon, west Africa. *Rhektophyllum mirabile* has been incorporated into this description. The plants are rhizomatous terrestrial herbs or hemi-epiphytic climbers with appressed leaves, or scandent. Light requirements vary. Leaves are long-petiolate and ovate, digitately divided, hastate or sagittate. Chromosomes: 2n = 36, 42. Five species were available for study.

Leaf surface

Cuticle smooth on both surfaces. **Epidermis**: cells with straight to undulate or sinuous anticlinal walls on both surfaces. Cells sinuous to interlocking in *C. mirabilis* with 5–8 lobes on both surfaces. Cells polygonal with straight anticlinal walls on both surfaces (1–2:1 *l/w*) in *C. stigmaticus* (Fig. 60D). **Stomata** on both surfaces or more common abaxially; brachyparacytic or brachyparahexacytic (Fig. 60D). Subsidiary cells: 3–4 pairs adaxially, often only 1–2 pairs abaxially at slightly smaller stomata. Stomata brachyparacytic with very irregular subsidiary cells on both surfaces in *C. mirabilis*. Stomata abaxial only and brachyparacytic in *C. ivorensis*. **Venation**:

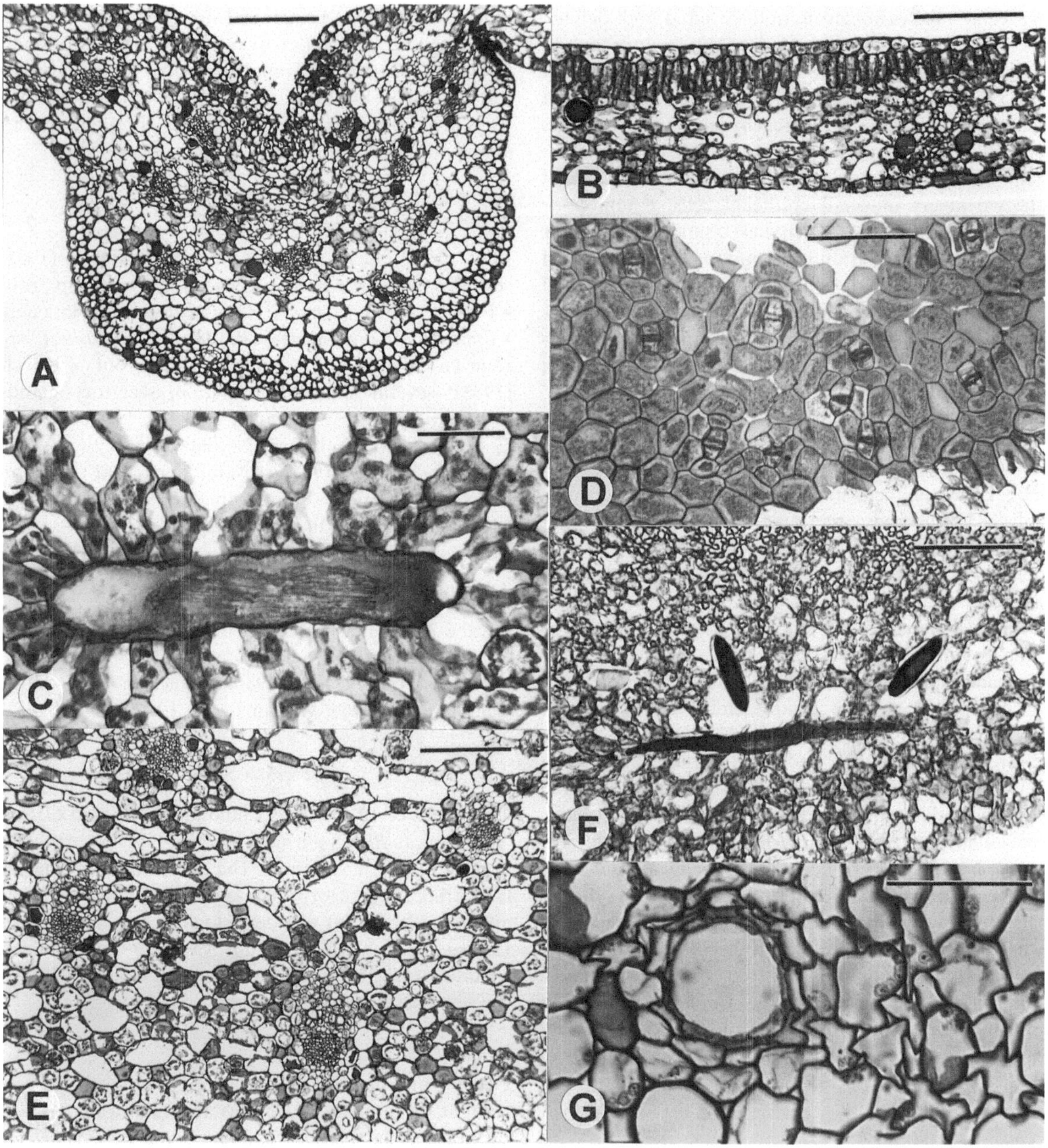

Fig. 60. *Cercestis* leaf, petiole and stem sections. (A) *C. afzelii*; (B,F,G) *C. congensis*; (C,D) *C. stigmaticus*; (E) *C. ivorensis*; (A) Midrib TS showing abaxial subepidermal collenchyma band; (B) Lamina TS; (C) Leaf mesophyll PS with raphide cell; (D) Abaxial leaf epidermis PS with mostly brachypara-hexacytic stomata; (E) Petiole TS centre with type I vascular bundles and type 3 aerenchyma; (F) Leaf mesophyll PS with biforine-like raphides; (G) Stem TS ground tissue with secretory duct. Scale bars: A,E = 200 μm; B,D,F = 150 μm; C,G = 50 μm.

midrib prominent with basal acrodromous basal lobe primaries in sagittate specimens, or having pedate branching. Secondaries arising infrequently and broadly acute from midrib, curving toward apex and merging eucamptodromously or brochidodromously. Tertiaries irregularly tending to follow secondaries, or reticulate as in *C. mirabilis*. Outer secondaries in *C. mirabilis* forming submarginal collectors for tertiaries. Higher orders reticulate.

Leaf TS

Structure dorsiventral. **Specimen sizes**: *C. afzelii:* 138 × 80 mm; *C. ivorensis*: 200 × 80 mm; *C. mirabilis*: 152 × 112 mm (*l/w*); lobe 90 × 58 mm; *C. stigmaticus*: 150 × 83 mm (*l/w*); *C. congensis*: 82 × 78 mm (*l/w*) for sagittate specimens. **Midrib** narrowly grooved adaxially, broadly and deeply convex-rounded abaxially (Fig. 60A). In *C. mirabilis*, midrib shallowly convex-rounded. **Cuticle** medium-thick on both laminar surfaces; thick abaxially on midrib. Cuticle thick at margins in *C. mirabilis*. **Epidermis** of squarish cells on both surfaces, flat externally, rounded internally. **Stomata** level with surface. **Mesophyll**: (Fig. 60B) one palisade layer and 5–6 spongy layers of armed cells. Spongy cells with well-organized or small irregular air cavities; type 1 in *C. mirabilis*, type 2 in *C. congensis* and *C. ivorensis*. Spongy cells large and rounded with inconspicuous lobes and corner air cavities in *C. stigmaticus* (type 1). **Air cavities**: 20% of spongy tissue in *C. stigmaticus*, in other species mostly substomatal and up to 40% of spongy layer in *C. afzelii*, up to 50% in *C. ivorensis*, or up to 90% in *C. mirabilis*. **Collenchyma**: band of 3–4 subepidermal layers on abaxial side of midrib (type B).

Vascular bundles: midrib with one large central bundle and 8–9 surrounding smaller bundles symmetrically disposed (type I) (Fig. 60A). Xylem: numerous metaxylem cells forming broad boundary with broad semicircular phloem strand; less robust in *C. mirabilis* and *C. stigmaticus*. **Sclerenchyma**: fibres ensheathing vascular bundles or as separate xylem and phloem caps on smaller bundles. **Laticifers** non-anastomosing and common, 1–2 lateral to vascular bundles. Cells with short processes extending within vein sheath ground tissue. **Secretory ducts** occasional in lamina with two layers of epithelium. **Crystals**: raphides common in mid-mesophyll as single bundles in short cells or as multiple bundles in tubular cells (Fig. 60C). Crystal cells randomly oriented or aligned with secondary veins. Raphide cells biforine-like in *C. stigmaticus, C. ivorensis*, and *C. congensis* (Fig. 60F). Druses, often sharp-pointed, in circular cells, common in mesophyll and midrib.

Petiole TS

Outline crescent-shaped proximally and terete distally. **Size**: *c.* 1 cm wide proximally. **Cuticle** medium-thick to thick and smooth. **Epidermis**: cells small and ± square. **Collenchyma**: subepidermal band (type B), *c.* 8 cells deep around abaxial periphery proximally and continuous distally. **Ground tissue**: mostly type 3 aerenchyma in central tissue (Fig. 60E), or small, cell-sized air cavities in *C. afzelii*. **Air space** up to 90%. **Vascular bundles** scattered and numerous or few, mostly type I to mostly type II. Metaxylem elements 3–10 opposing a semicircular or circular phloem strand (Fig. 6H). **Sclerenchyma**: fibre strands, *c.* 2 cells thick as xylem and phloem caps to peripheral bundles distally. **Secretory ducts** with single-layered epithelium, occasional in lateral and adaxial peripheral ground tissue. **Laticifers** non-anastomosing, lateral to phloem strands. **Tannin** cells occasional to common, scattered in ground tissue. **Starch** granules very common, 8–15 µm diameter, in ground tissue. **Crystals**: raphides common, biforine-like, mostly attached to partitions but not extending across partitions. Druses in small cells common subepidermally and throughout ground tissue.

Stem TS

Outline circular. **Size**: diameter *c.* 6 mm. **Cuticle** thick and smooth. **Epidermis**: cells small and ± square, interrupted irregularly by rounded protuberances formed of columnar epidermal and subepidermal cells. **Collenchyma**: 8–10 subepidermal layers around periphery. **Ground tissue** mostly types 2 and 3 aerenchyma. **Air space** in cortex about 50%. **Central cylinder** defined by abruptly greater density of vascular bundles and smaller, closely packed ground tissue cells.

Vascular bundles occasional in cortex, numerous in central cylinder, mostly types II or III. In central cylinder, protoxylem cells very wide, 1–2 cells per strand. Phloem strands semicircular with large-diameter sieve-tube elements. **Secretory ducts** occasional in cortex, with 1-layered epithelial lining (Fig. 60G). **Laticifers**: 1–2 cells lateral to phloem strands. **Tannin** cells common throughout ground tissue. **Crystals**: raphides biforine-like, scattered, oriented axially in cortex.

Root TS

Sclerotic hypodermis noted by French (1987c).

Material examined

Cercestis afzelii Schott, cult. M; *C. congensis* Engl., Bogner 353, cult. M; *C. ivorensis* A.Chev., Knecht 10/44, Côte d'Ivoire; *C. mirabilis* (N.E.Br.) Bogner, Bogner 620, Gabon, cult. M. (syn. *Rhektophyllum mirabile* N.E.Br.); *C. stigmaticus* N.E.Br., Bogner 632, Gabon, cult. M.

Species reported on in the literature

(*Rhektophyllum mirabile* cited as *C. mirabilis* throughout)

Engler (1911): *C. dinklagei, C. kamerunensis, C. mirabilis*: leaf, stem. • French and Tomlinson (1981a): *C. afzelii, C. congensis, C. stigmaticus*: stem. • French and Tomlinson (1986): *C. stigmaticus*: stem. • French (1987b): several spp.: root. • French (1987c): several spp.: root, hypodermis. • French (1988): several spp.: laticifers. • Guttenberg (1968): *C. mirabilis*: root. • Hotta (1971): *C. mirabilis*: leaf. • Knecht (1983): *C. afzelii, C. sagittatus, C. stigmaticus*: leaf venation. • Solereder and Meyer (1928): *C. mirabilis*: all organs; *C.* sp.: secretory tissue.

Additional literature

(Citations for *R. mirabile* alone = (R))

Blanc 1980 (R); French 1985b, 1986c, 1987a; Grayum 1990; Hegnauer 1986 (R); Mayo 1986; Thanikaimoni 1969; Williams *et al.* 1981.

Tribe Spathicarpeae Schott

A tribe consisting of 10 genera of the American tropics: *Bognera, Dieffenbachia, Spathantheum, Gorgonidium, Synandrospadix, Gearum, Spathicarpa, Asterostigma, Mangonia* and *Taccarum.*

Anatomical summary

Leaf surface

Cuticle smooth or striate; rugulate in *Dieffenbachia.* **Epidermis**: cells polygonal with straight anticlinal walls, or walls variable in *Dieffenbachia*, including sinuous; undulate to lobed in *Asterostigma* and *Gorgonidium.* **Trichomes** occasionally present on veins of *Dieffenbachia.* **Stomata** abaxial, occasionally on both surfaces or more frequent abaxially, mostly brachyparacytic. Subsidiary cells usually broad and irregular. **Venation** mostly with prominent midrib and brochidodromous or eucamptodromous margins.

Leaf TS

Structure dorsiventral and well or poorly differentiated. **Midrib** usually with very large rounded abaxial or subterete profile. **Cuticle** mostly thin, or thick at margins. **Epidermis**: cells usually tall. **Stomata** level with surface. **Mesophyll**: palisade cells poorly or well-differentiated; cells U-shaped in *Taccarum*, irregular in *Gearum.* Spongy tissue: types 1, 1a, or 2, often type 3 in midrib. **Collenchyma**: banded or interrupted bands (types B, Bi), subepidermal in abaxial midrib. **Vascular bundles** ranging from one to numerous in midrib; bundles mostly type II, occasionally types I or III. **Sclerenchyma**: fibres mostly absent, but forming bundle caps in *Dieffenbachia* or ensheathing bundles in *Asterostigma.* **Laticifers** reported as non-anastomosing, with colourless contents; not easily demonstrated. **Crystals**: raphides various. Single bundles in thin-walled cells in most genera, or multiple bundles grouped end to end in tubular cells, or side by side in wide cells. Biforines present in *Gearum* and *Dieffenbachia.* Single raphide bundles in cells aligned in articulated tubes in *Taccarum* and *Asterostigma.* Druses occasionally present.

Petiole TS

Shape crescent-shaped proximally, terete distally. **Cuticle** variously thickened and smooth or striate. **Epidermis**: cells mostly small and square. **Ground tissue** tending to be type 3 aerenchyma in large tissue expanses, type 2 in smaller organs. **Collenchyma**: type B, in peripheral bands, or discontinuous, type Bi, in *Dieffenbachia* and *Spathicarpa.* **Vascular bundles** mostly numerous and scattered, types II and III. **Crystals**: raphides few, as single bundles in small cells or multiple bundles in wide or elongate cells.

Taxonomic notes

While the genera show diagnostic variation, the tribe also has a number of coherent features. Laticifers have transparent, non-staining contents. Also, there is mostly type 1 spongy aerenchyma in the lamina while type 3 aerenchyma is common in the midrib and petiole. *Dieffenbachia* shows the largest variety of character states which also serves to isolate it. On closer inspection some of its diagnostic characters are shared, such as the presence of mucilaginous vesicles in air spaces, which is also found in *Gearum.* Biforines also occur in *Gearum* and type Bi collenchyma is also found in *Spathicarpa.*

A few rare conditions are present such as the articulated raphide tubes in *Asterostigma* and *Taccarum* and several cases of unusual palisade tissue, which also occurs sporadically in a few other tribes.

Bognera Mayo & Nicolson

Plants are terrestrial rhizomatous understorey herbs growing in leaf litter of sandy soil in dark humid forests. *Bognera recondita* (Madison) Mayo & Nicolson is the only species. Leaves are long-petiolate and ovate-elliptic. Chromosomes 2n = 34.

Material examined

No material available.

Additional literature

Grayum 1990; Hegnauer 1986.

Dieffenbachia Schott

This genus of about 50 species is distributed in the American tropics. Plants are light-seeking, caulescent, terrestrial understorey vines that may stand to a metre high before falling and growing horizontally toward the next light source in humid to wet forests. Some grow along edges of clearings along stream banks and can tolerate seasonal flooding. The leaves are simple, long-petiolate and ovate or elliptic. Chromosomes: 2n = 34, 36, 68. Three species were available for study.

Leaf surface

Cuticle: smooth or rugulate on both surfaces. **Epidermis**: cells with anticlinal walls undulate to deeply sinuous adaxially (1.5–2:1 *l*/*w*) and cells polygonal with walls straight to undulate abaxially (1–2:1 *l*/*w*). Cells polygonal with straight anticlinal walls on both surfaces in some species (Trivedi and Upadhyay 1984). **Trichomes** occasional, long, unicellular and striate on veins of *D. jenmanni*, *D. arvida* and *D. exotica* (Trivedi and Upadhyay 1984). **Stomata** abaxial or on both surfaces (Trivedi and Upadhyay 1984) and brachyparacytic to brachypara-tetracytic. Dunn *et al.* (1965) reported stomata length of 58 µm, very large for monocots. Subsidiary cells: 2–3 cell pairs, the outer cells often as a cycle of four; inner pair narrow or broad and usually rounded. **Venation**: midrib prominent. Secondaries pinnate, forming regular intercostal areas, widely-acute and curved-ascending. Tertiaries arising narrowly acute from secondaries, becoming parallel with them. Quaternaries often orthogonal to secondaries and tertiaries. Higher orders reticulate.

Leaf TS

Structure dorsiventral. **Size**: length up to 75 mm. **Midrib** broad, gradually and slightly convex on both surfaces (Fig. 61A). **Cuticle** medium-thick to thin on both surfaces, thick at margins. **Epidermis**: cells tall and square adaxially (Fig. 61B). Abaxial epidermal cells half as tall and of flattened cells. **Stomata** level with surface. **Mesophyll**: one layer of thin-walled, squarish palisade cells, poorly formed (Fig. 61B); *c.* $\frac{1}{10}$ mesophyll thickness. Spongy tissue of *c.* 10 layers of short-lobed cells with large aerenchyma spaces separated by uniseriate partitions (type 3). Solereder (1919) illustrated type 2 spongy aerenchyma. Lindorf (1980) illustrated mucilaginous vesicles suspended into spongy air space. **Collenchyma**: well-formed discrete strands, 6–10 cells deep, subepidermal on both sides of midrib (Fig. 61A,C) and closely spaced (type Bi). Breaks in collenchyma due to loosely packed spongy cells, often subtending stomata (Fig. 61C).

Vascular bundles scattered in midrib, often at corners of type 3 aerenchyma (type II). Xylem: usually one metaxylem element in midrib and several in laminar bundles. Phloem: circular or irregular strand. **Sclerenchyma**: fibres abaxial to phloem in laminar bundles, 2–6 cells deep and thin-walled. Phloem caps of collenchyma in most species. **Crystals**: raphides, one bundle per cell, in short or long cells in spongy mesophyll. Raphide cells occasionally suspended in partitions between air cavities and randomly oriented. Turpin (1836), Middendorf (1982, 1983) and Lindorf (1980) reported biforines and Solereder and Meyer (1928) described them as the largest seen in the family. In these specimens, well-developed biforines were found only in *D. lentii* and only in leaves. Druses in small cells, sharp-pointed and very common, most frequently in palisade layer and lower mesophyll.

Petiole TS

Outline: distally subterete, occasionally shallowly grooved at intervals around periphery; V-grooved adaxially. Proximally, crescent-shaped. **Size**: diameter 5–7 mm. **Cuticle** medium-thick and striate. **Epidermis**: cells small and ± isohedral. **Collenchyma**: discrete strands numerous and broad around periphery (Fig. 61D), defined by peripheral grooves when present (type Bi). **Chlorenchyma**: outer 3–4 ground tissue layers. **Ground tissue**: cells small at periphery grading to large in centre. Outer 15–20 layers of compact parenchyma, central type 3 aerenchyma with 1–2-celled partitions. **Air cavities**: small, width of several cell diameters.

Vascular bundles numerous, scattered (Fig. 61D) and small (types II, III). Xylem: one wide protoxylem cell, 2–8 narrow metaxylem cells in one ± irregular file, without sheath. Phloem: narrow strand, small and irregular. **Starch** granules few in ground tissue near phloem. **Crystals**: raphides uncommon in outer ground tissue, 1–3 bundles in undifferentiated thin-walled cells.

Stem TS

Outline terete. **Size**: diameter 1.5–2 cm. **Cuticle** thick and smooth. **Epidermis**: cells small and square. **Cork**: superficial thin-walled periderm, 8–10 cells deep, patchy on one side of stem. **Collenchyma**: peripheral band 12–15 cells deep, angularly-thickened, separated by a layer of parenchyma from epidermis (Figs 5E, 61E). **Ground tissue**: ± uniformly large cells internal to collenchyma. **Air cavities** small, at cell corners only. **Central cylinder**: obscure delimitation by vascular bundle type.

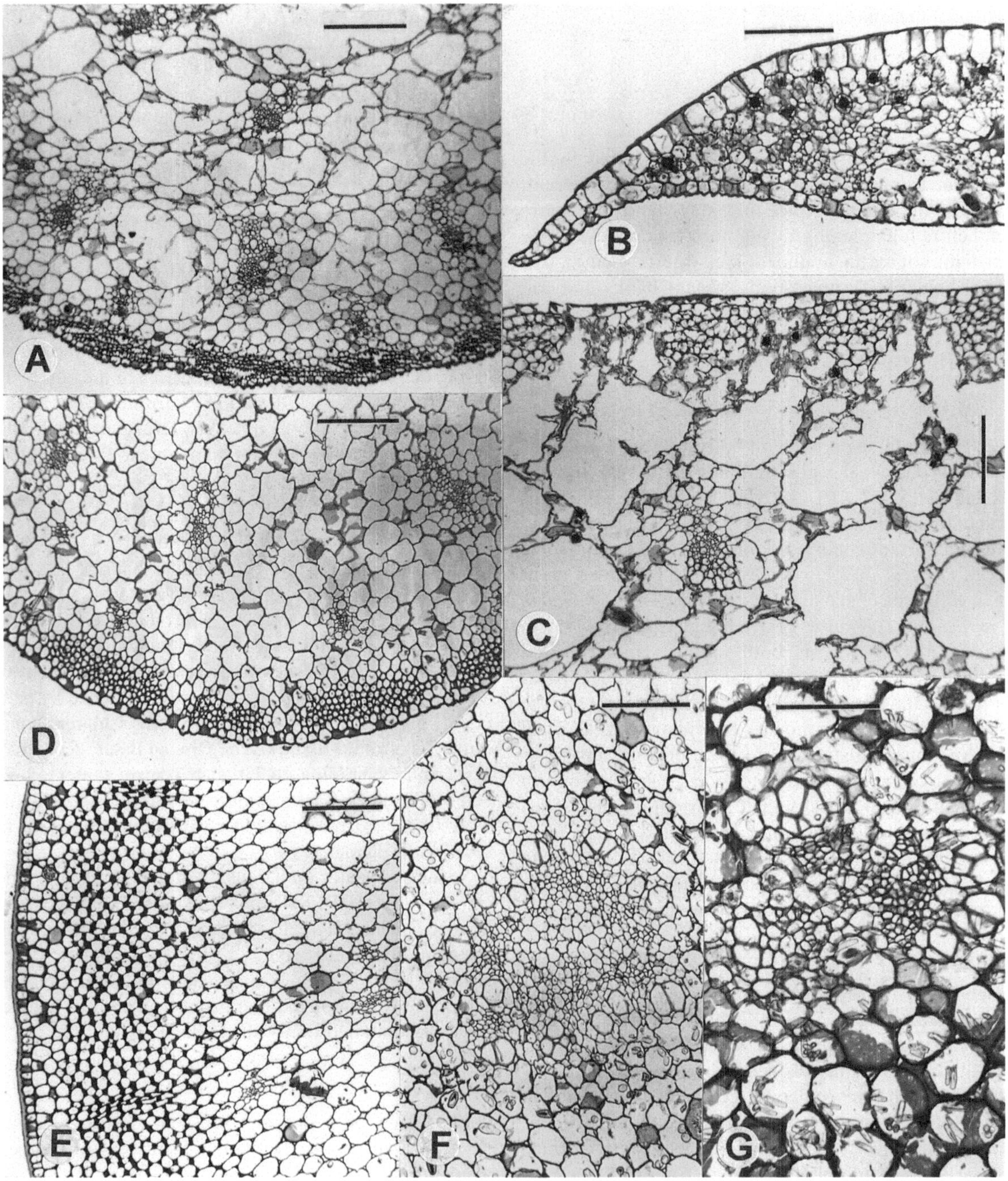

Fig. 61. *Dieffenbachia* leaf, petiole and stem sections. (A,D–F) *D. lentii*; (B,C,G) *D. longispatha var. creberapistila*. (A) Midrib TS, abaxial surface; (B) Lamina TS margin; (C) Midrib TS, adaxial surface; (D) Petiole TS, abaxial surface with type Bi collenchyma; (E) Stem TS, outer region with deep band of type B collenchyma in outer cortex; (F,G) Stems TS, central tissue with large compound vascular bundle. Scale bars: A,D,E = 300 μm; B,C,F = 200 μm; G = 100 μm.

Vascular bundles scattered, with some concentration in centre. Central bundles amphivasal or compound, 4–12 metaxylem strands surrounding 1–3 circular phloem strands (Fig. 61F,G). Outer cortical bundles small, collateral and simple, with irregular xylem and narrow phloem strands. Vessels reported present in stem xylem by Schneider and Carlquist (1998). **Starch** granules long (up to 40 μm), rod-like, common throughout ground tissue. **Crystals**: raphides common in central ground tissue, single wide bundles in thin-walled, unmodified ground tissue cells. Druses common in outer 1–4 layers around periphery.

Material examined

Dieffenbachia lentii Croat and Grayum, Croat 37479, Panama, MO; *D. longispatha* var. *creberapistila* Croat and Grayum, Croat 37361, Panama, MO; *D. oerstedtii* Schott, Croat 42645, Honduras, MO.

Taxonomic notes

This genus has a rich array of characters that show some relationship to the following tribe but it remains somewhat isolated from subsequent genera in this alliance. Collenchyma in this genus shows transitions from banded to closely spaced incipient strands. See next tribe for more comments.

Species reported on in the literature

Behnke (1969a,1972): *D. picta*: phloem. • Bergdolt (1955): *D. baussei, D. memoria, D. seguine*: leaf. • Chauveaud (1897): *D. costata*: root, phloem. • Czaja (1962): *D. fournieri* (hort.): leaf epidermis. • Dalitzsch (1886): *D. seguine*: leaf. • Dunn *et al.* (1965): *D. picta*: stomata. • El-Gazzar *et al.* (1977): *D. picta*: stomata. • Engler and Krause (1912): *D. parlatorei, D.* sp.: root, leaf, crystals. • French (1987c): several spp.: root. • French (1988): several spp.: laticifers. • French and Tomlinson (1981b): several spp.: stem vasculature. • French and Tomlinson (1986): stem. Gentner (1905): *D. picta:* leaf. • Genua and Hillson (1985): *D. maculata*: crystals. • Gulliver (1863–5): *D. maculata*: crystals. • Guttenberg (1968): root. • Hassack (1886): several spp.: leaf. • Kisser (1928): *D. picta, D. seguine, D. splendens*: leaf, petiole. • Leavitt (1904): *D. picta*: root. • Lierau (1888): several spp.: root. • Lindorf (1980): *D. seguine*: leaf. • Lohse (1924): *D. macrophylla, D. picta, D. sequine*: stem, tyloses. • Marcatili (1887): *D. seguine:* leaf. • Marlatt (1970): *D. amoena:* periderm. • Meyer (1962): *D. picta, D. seguine*: leaf. • Middendorf (1982): *D. picta*: leaf, crystals. • Middendorf (1983): crystals. • Mueller and Beckman (1979): root. • Müller (1919): *D. fournieri, D.* sp.: leaf tip. • Netolitzky (1929): *D. seguine*: crystals. • Noro *et al.* (1992): leaf. • Nyawuane and Gill (1990): *D. picta*: stomata. • Paliwal (1969): *D. sanguinea*: stomata. • Parthasarathy (1980): stem. • Petersen (1893): *D. seguine*: meristems. • Pirotta and Marcatili (1885): *D. seguine*: laticifers. • Reichert (1913): *D. illustris, D. seguine*: starch. • Rodrigue (1900): *D. imperialis, D. seguine*: leaf. • Roth (1949): leaf. Roth (1980, 1990): *D. seguine*: leaf. • Sakai and Nagao (1980): *D. maculata*: crystals. • Sakai *et al.* (1984): crystals. • Schneider and Carlquist (1998): *D. oerstedtii*: stem, root. • Schwendener (1882): root. • Shishkoff (1987): root. • Solereder (1919): *D. seguine*: leaf. • Solereder and Meyer (1928): *D. seguinae, D. picta*: all organs. • Stahl (1896): *D. picta*: leaf, petiole. • Stebbins and Khush (1961): stomata. • Sutter (1985): *D. maculata*: cuticle. • Swamy and Sivaramakrishna (1972): *D. picta*: stem. • Swamy and Sivaramakrishna (1976): root, ontogeny. • Tieghem (1866): *D. picta*: all organs. • Tieghem (1870–1): *D. picta*: root, secretory tissue. • Traquair (1987): *D. maculata*: crystals. • Trécul (1865, 1866): *D. picta, D. seguine*: vasculature. • Trivedi and Upadhyay (1984): several spp.: leaf. • Troll and Meyer (1955): *D. picta*: leaf. • Turpin (1836): *D. seguine*: leaf, crystals. • Uphof *et al.* (1962): hairs. • Webber (1960): *D. hoffmannii*: leaf, stomata. • Zettler and Rhodes (1975): *D. picta* : crystals. • Zinke (1924): *D. seguine*: leaf.

Additional literature

Arditti and Rodriguez 1982; Campbell 1900; Chouard 1936; Davis 1966; Domin 1911; Eller and Flach 1991; Erdtman 1952; Fox and French 1988; French 1985b, 1986c, 1987a; Grayum 1990; Hegnauer 1986; Henny 1983b; Lindinger 1908; Mayo 1986; Meusel 1951; Mookerjea 1955; Nahrstedt 1975; Ray 1988; Sharma 1967; Smolenski *et al.* 1972, 1973, 1974; Southorn 1964; Thanikaimoni 1969; Troll 1928; Vij *et al.* 1982; Walter and Khanna 1972; Williams *et al.* 1981; Young 1986, 1990.

Spathantheum Schott

This genus of two species is found in Andean Bolivia, Peru and northern Argentina. Plants are terrestrial tuberous herbs usually loosely rooted in the understorey, growing in humus deposits on rocks, near cliff faces, and along road banks in shady areas, or in mountain prairies up to 2400 m. The leaves are solitary, erect, long-petiolate with an ovate or pinnatifid lamina and slightly cordate base and acute apex. Chromosomes: 2n = 34. One species was available for study.

Leaf surface

Cuticle smooth on both surfaces. **Epidermis**: cells polygonal with straight anticlinal walls adaxially (1–2:1 *l/w*) (Fig. 62G) and straight to undulate abaxially (1–4:1 *l/w*). **Stomata** abaxial and brachyparacytic. Subsidiary cells mostly narrow, often in contact at one end, therefore unipolar (Fig. 62D). **Venation**: midrib prominent. Basal and suprabasal acrodromous secondaries leading to basal lobes. Secondaries apically curved in main lamina. Secondaries collected or merged eucamptodromously at margin. Tertiaries mostly parallel with secondaries. Higher orders reticulate (Fig. 62F).

Leaf TS

Structure dorsiventral. **Size**: lamina length 9 cm. **Midrib** shallowly concave adaxially, semicircular rounded-convex abaxially (Fig. 62C). **Cuticle** thin or not seen on both surfaces. **Epidermis**: cells large or slightly smaller abaxially; cells rounded externally and internally. **Stomata**: guard cells with birefringent walls and large ledges. **Mesophyll**: palisade cells in one loosely packed layer, up to 40% of lamina (Fig. 62B). Spongy tissue: 4–5 layers of long-armed cells (type 1–2) (Fig. 62E). **Air cavities** irregular, up to 80% of spongy tissue. **Collenchyma**: 1–2-layered subepidermal band on abaxial surface of midrib.

Vascular bundles: 2–3 in midrib (type II) (Fig. 62C). Xylem: 1–2 protoxylem elements and 2–3 narrow metaxylem elements. Phloem: irregularly shaped strand. **Starch** granules large and uncommon. **Crystals**: raphides common in single wide or elongate bundles, often in short rounded cells, randomly oriented in mesophyll. Some elongate raphide tubes with multiple bundles (Fig. 9A), not aligned with vascular bundles. Crystal sand common throughout mesophyll.

Petiole TS

Outline deeply grooved proximally with rounded abaxial surface. Distal section terete (Fig. 62A). **Size**: diameter 4–6 mm. **Cuticle** thin and smooth. **Epidermis**: cells medium-sized and rounded externally. **Collenchyma**: 1–2-layered subepidermal abaxial band in proximal section. **Ground tissue**: type 3 aerenchyma.

Vascular bundles scattered, types II and III, situated at corners of aerenchyma partitions (Fig. 62H), large proximally. Xylem: protoxylem lacunae wide, 1–2 narrower metaxylem cells facing small irregular phloem strand proximally. Distally, 4–8 metaxylem cells forming broad boundary with large or small phloem strand (type II) (Fig. 62H). **Starch** granules few and small distally, up to 7–8 μm diameter. **Crystals**: crystal sand sparse throughout.

Material examined

Spathantheum orbignyanum Schott, Rauh 26024, Peru, cult. M.

Species reported on in the literature

French (1987c): *S. orbignyanum:* root.

Additional literature

French 1986a, c; Davis 1966; Grayum 1990; Thanikaimoni 1969; Troll 1928.

Gorgonidium Schott

This genus of three species occurs in Peru, Bolivia and northern Argentina. Plants are terrestrial tuberous herbs from seasonally dry areas at 2500 to 3000 m. The leaves are erect, long-petiolate with a broad elliptic or pinnately lobed lamina. Chromosomes: 2n = 34. Two species were available for study.

Leaf surface

Cuticle smooth or finely striate. **Epidermis**: cell walls undulate, elongate adaxially (3:1 *l/w*) and sinuous-lobed to undulate abaxially (2:1 *l/w*). **Stomata** abaxial and brachyparacytic. Subsidiary cells mostly broad and rounded (Fig. 63E). **Venation**: midrib prominent. Secondaries irregularly spaced, narrowly acute, curved ascending and connecting eucamptodromously at submargin. In *G. mirabile*, secondaries connecting brochidodromously in distal lamina. Tertiaries irregularly parallel with secondaries or reticulate. Higher orders reticulate.

Leaf TS

Structure dorsiventral. **Specimen size**: 45 × 21 mm (*l/w*) to 80 × 35 mm (*l/w*). **Midrib** small, rounded-grooved adaxially and rounded-keeled abaxially (Fig. 63A). **Cuticle** very thin. **Epidermis**: cells somewhat larger adaxially, mostly flat or somewhat convex externally. **Mesophyll**: one palisade layer (3–4:1 *l/w*), up to 50% of lamina (Fig. 63B,C). Spongy tissue: 3–4 layers of irregular cells (type 1). **Air cavities** irregular, up to 80% air space.

Collenchyma: weakly developed subepidermal band abaxially in midrib. **Vascular bundles**: one large bundle in midrib, unsheathed (type II). Xylem: 2–6 metaxylem strands in 1–2 irregular files. Phloem: small irregular strand. **Tannin** cells occasional in mesophyll and unmodified. **Crystals**: raphides in rounded or elongated cells (Fig. 63D), in single or multiple bundles, common and

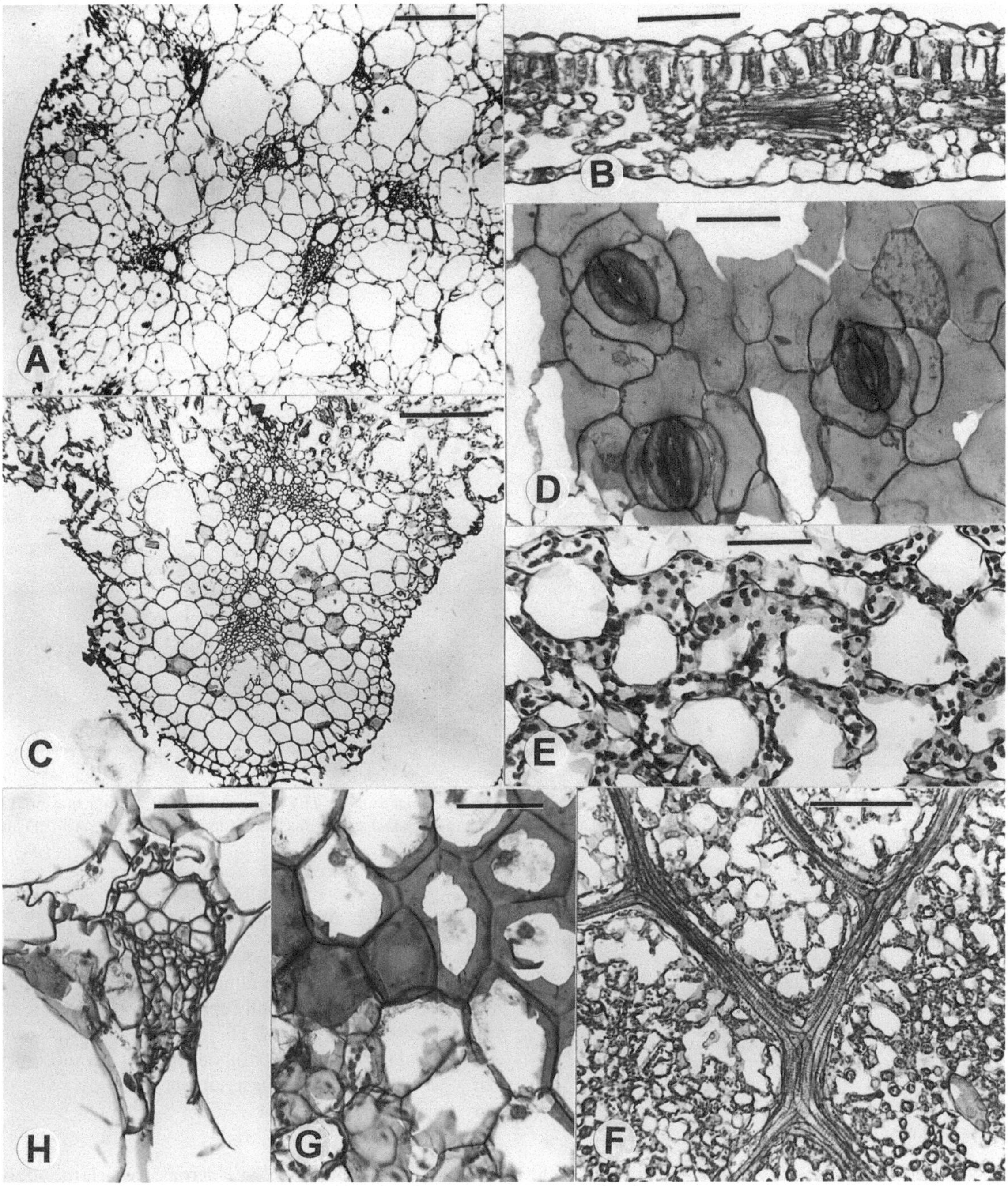

Fig. 62. *Spathantheum orbignyanum* leaf and petiole sections. (A) Distal petiole TS lateral margin; (B) Lamina TS; (C) Midrib TS, abaxial; (D) Abaxial epidermis PS showing two unipolar stomata; (E) Lamina PS showing type 2 aerenchyma; (F) Lamina PS showing venation and mesophyll; (G) Lamina adaxial epidermis PS; (H) Distal petiole TS showing type II vascular bundle. Scale bars: A = 300 μm; B,F,H = 150 μm; C = 200 μm; D,E,G = 50 μm.

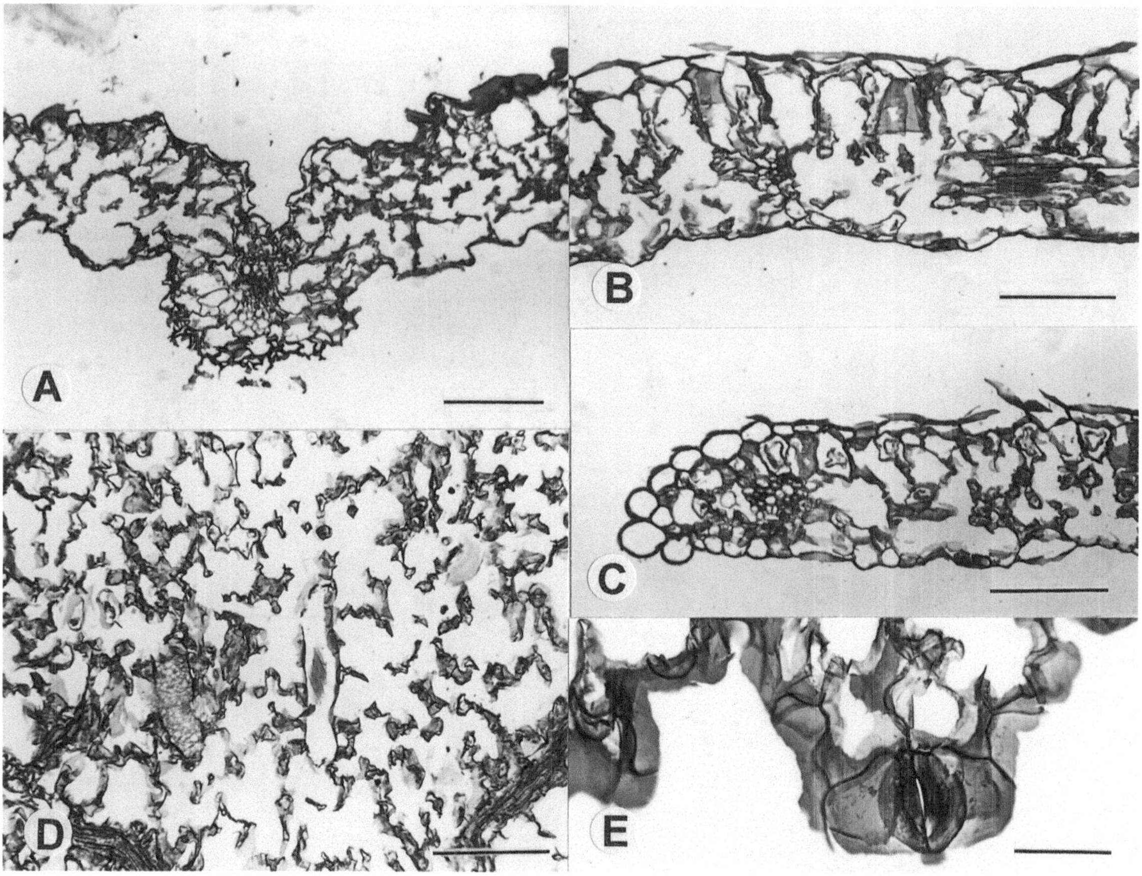

Fig. 63. ***Gorgonidium mirabile*** **leaf sections.** (A) Midrib TS with single vascular bundle; (B) Lamina TS; (C) Lamina margin TS; (D) Lamina PS showing elongated raphide cell; (E) Lamina PS showing abaxial epidermis and stoma. Scale bars: A = 200 μm; B–D = 150 μm; E = 50 μm.

randomly oriented in mesophyll. Crystal sand common throughout.

Material examined

Gorgonidium mirabile Schott, Bolivia, cult. M; *G. vargasii* Bogner & Nicolson, C.Vargas s.n., cult. M.

Species reported on in the literature

French (1987c): *G. vargasii*: root. • French (1988): *G. vargasii*: leaf, laticifers.

Additional literature

French 1986a, c, 1987a; Grayum 1990; Thanikaimoni 1969.

Synandrospadix Engl.

This genus of one species of cormose plants is found in Bolivia and northern Argentina to Paraguay. The plants are terrestrial or occasionally epiphytic, occurring in dry shrub forests up to 2000 m. They are tuberous herbs with long-petiolate, ovate or spatulate leaves. Chromosomes: 2n = 34. One specimen was available for study.

Leaf surface

Cuticle smooth or striate. **Epidermis**: cells large, polygonal with straight anticlinal walls (Fig. 64D,E) (1–1.5: *l*/*w*). **Stomata** on both surfaces; brachyparacytic with subsidiary cells broad and variably shaped (Fig. 64E). **Venation**: midrib prominent. Secondaries acute, running parallel and curving to eucamptodromous margin. Higher-order veins reticulate.

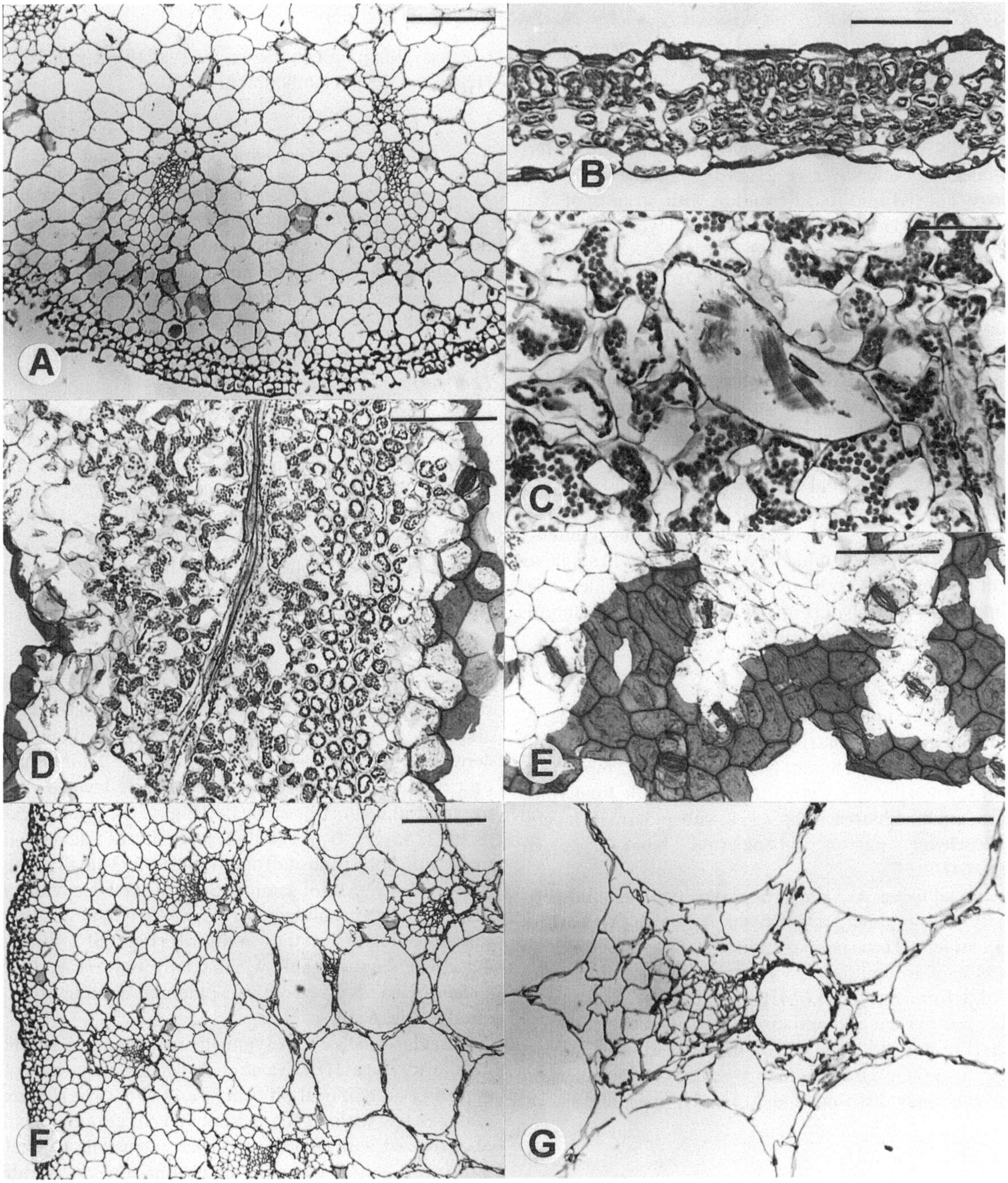

Fig. 64. *Synandrospadix vermitoxicus* leaf and petiole sections. (A) Midrib TS abaxial surface showing subepidermal type B collenchyma; (B) Lamina TS with stomata on both surfaces; (C) Lamina PS with raphide cell; (D) Lamina PS showing palisade and spongy mesophyll; (E) Abaxial epidermis PS and stomata; (F) Distal petiole TS with types 3 and 4 aerenchyma; (G) Distal petiole centre TS with type III vascular bundle. Scale bars: A = 200 μm; B,D,E,G = 150 μm; C = 50 μm; F = 300 μm.

Leaf TS

Structure dorsiventral. **Size**: lamina 20 × 9 cm (*l*/*w*). **Midrib** shallowly and broadly concave adaxially and deeply convex-rounded abaxially (Fig. 64A). **Cuticle** thin on both surfaces. **Epidermis**: cells large, rounded or flat internally or externally. **Mesophyll**: one weakly differentiated palisade layer (Fig. 64B); palisade extending across adaxial midrib, alternating with strands of collenchyma. In lamina, palisade a single layer of irregular cells. Spongy cells short-armed, in 3–4 layers of irregular size and shape (types 1, 1a) (Fig. 64D).

Air cavities: subepidermal, broad and conspicuous substomatal openings in palisade and spongy layers. Air space up to 40% of spongy layers. Midrib ground tissue type 3 aerenchyma. **Collenchyma**: 2-layered subepidermal band on abaxial side of midrib (Fig. 64A) and adaxial strands. **Vascular bundles**: midrib bundles poorly formed and scattered. Largest bundles with *c*. 6 narrow or wide metaxylem elements and rounded phloem strand flattened slightly at xylem boundary (type II). **Laticifers** non-anastomosing, 1–2 cells lateral to phloem strands; contents clear to very light-staining. **Crystals**: raphides in single or multiple bundles (side by side) in large wide cells, common in mesophyll (Fig. 64C). Crystal sand common in mesophyll and epidermis.

Petiole TS

Outline: proximally, rounded abaxially and deeply rounded-grooved adaxially. Distally, subterete and adaxially slightly concave. **Size**: diameter 1.2 mm proximally, 9 mm distally. **Cuticle** thin and smooth. **Epidermis**: cells small. **Chlorenchyma**: 1–2 subepidermal layers. **Collenchyma**: narrow subepidermal band (type B) (Fig. 64F).

Ground tissue: 4–5 layers of parenchyma around periphery, mostly type 3 aerenchyma in centre. **Air cavities** very large, *c*. 3 cells per partition (Fig. 64F). Air space up to 95% of ground tissue. **Vascular bundles** scattered and weakly formed (type II or III) (Fig. 64F,G). Xylem: 1–2 wide protoxylem cells and up to three narrow metaxylem cells (Fig. 64F). Phloem: narrow irregular strand. **Crystals**: raphides in single bundles in thin-walled cells, uncommon and randomly distributed and oriented in ground tissue.

Material examined

Synandrospadix vermitoxicus (Griseb.) Engl., Bogner 1480, Bolivia, cult. M.

Species reported on in the literature

French (1987c): root. • French (1988): laticifers.

Additional literature

Cocucci 1966; French 1986a, c, 1987a; Grayum 1990; Hegnauer 1986; Thanikaimoni 1969.

Gearum N.E.Br.

This genus consists of one species found in Goiás and Tocantins states in southern Brazil. Plants are terrestrial tuberous herbs of seasonally inundated areas. The leaves are palmately lobed or pedatisect with elliptic to linear lobes. Chromosome number unknown. Two specimens were available for study.

Leaf surface

Cuticle smooth, but striate on abaxial midrib. **Epidermis**: cells polygonal with straight anticlinal walls on both surfaces (1–2:1 *l*/*w*). **Stomata**: brachyparacytic on both surfaces. Subsidiary cells irregular and square-cornered (Fig. 65D). **Venation** pinnate. Basal secondaries of median sector steeply ascending, merging at straightened submarginal looped collector. Distal secondaries more broadly acute. Tertiaries and higher orders reticulate.

Leaf TS

Structure dorsiventral. **Size**: 70 × 9 mm (*l*/*w*). **Midrib** shallowly concave adaxially, deeply semicircular-convex abaxially (Fig. 65B). **Cuticle** thin to medium-thick. **Epidermis**: cells irregular in height, rounded or flat internally or externally. **Mesophyll**: palisade single-layered, cells irregular and short or elongate (Fig. 65C). Spongy cells in 10–12 layers, type 1. All mesophyll cells packed with numerous small (unusual) chloroplasts, 2–3 μm diameter. **Air cavities** *c*. 5% of lamina. Several type 3 cavities in adaxial midrib ground tissue (Fig. 65A). Numerous small vesicles lining air cavities in midrib (Fig. 65F). Substomatal air cavities shallow, abaxially only (Fig. 65E). **Collenchyma**: 2–3-layered subepidermal band in abaxial midrib (Fig. 65B).

Vascular bundles: four symmetrically arranged bundles in midrib (type II). Xylem: 1–2 wide metaxylem cells opposing an elliptical phloem strand (Fig. 65A). **Starch** granules few within each mesophyll cell; granules small, plastid-sized, with numerous lamellar extensions. **Crystals**: biforine raphides common in mid-mesophyll, parallel with laminar surface or arranged parallel to palisade cells in upper mesophyll (Fig. 65G). Druses common, mostly small, concentrated in upper mesophyll. **Unusual tissue**: mesophyll and midrib air cavities often lined with numerous vesicular cellular outgrowths. Vesicles empty or containing mucilaginous contents, walls thin and cellulosic,

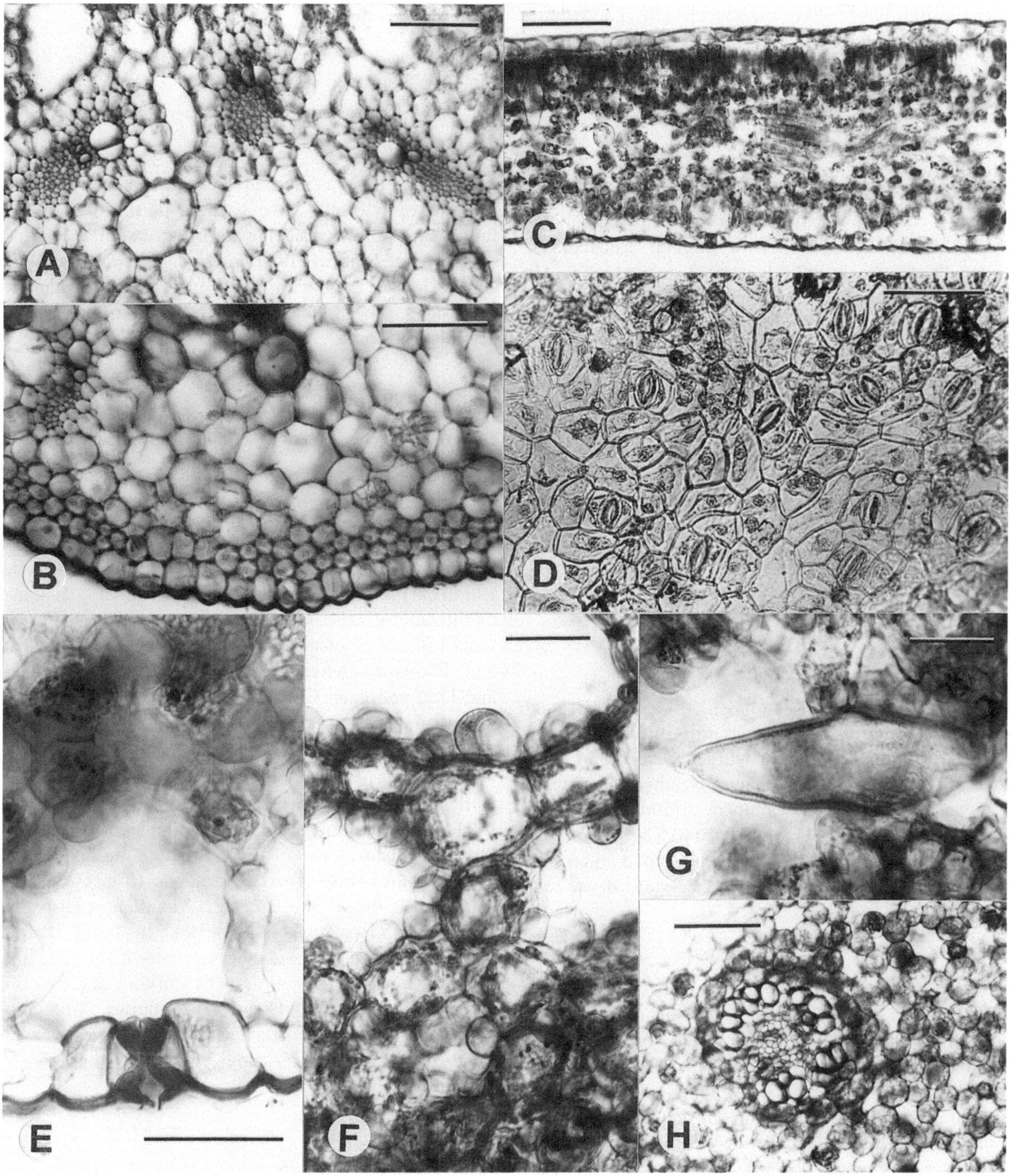

Fig. 65. *Gearum brasiliense* leaf and stem sections. Wet mounts: iodine in calcium chloride. (A) Adaxial midrib TS centre; (B) Abaxial midrib TS showing narrow subepidermal collenchyma band; (C) Lamina TS; (D) Abaxial epidermis PS showing brachyparacytic stomata; (E) Lamina TS showing abaxial epidermis and stoma; (F) Midrib aerenchyma TS showing mucilage vesicles bounding air cavities; (G) Mesophyll PS showing biforine raphide; (H) Tuber TS showing amphivasal bundle. Scale bars: A,C,H = 200 μm; B,D = 150 μm; E–G = 50 μm.

attaching soap-bubble-like to main cell walls, i.e. with no stalk or discrete attachment point (Fig. 65F). Nuclei large with numerous adherent organelles or vesicles.

Petiole base TS

Outline: semicircular to crescent-shaped, concave adaxially. **Size**: diameter 1.8 cm. **Cuticle** thin and smooth. **Epidermis**: cells square and small. **Collenchyma**: subepidermal band, 3–4-layered with angular thickenings. **Ground tissue**: nearly entirely type 3 aerenchyma. **Air cavities**: large, up to 95% of ground tissue. **Vascular bundles** scattered and small (type II). Xylem: 1–2 wide metaxylem cells surrounded by row of small cells. Phloem: laterally compressed elliptical strand.

Tannin cells rounded to slightly elongated in ground tissue, or cells elongated and narrow near vascular bundles, occasional in corners of aerenchyma partitions. **Starch** granules in single-celled arc of ground tissue capping phloem strand. **Crystals**: raphides uncommon in ground tissue, in wide bundles in rounded cells, or in multiple bundles in elongated cells. Crystal sand common in ground tissue.

Tuber TS

Outline subcircular. **Size** *c.* 2.2 cm diameter. **Cork** well-developed, *c.* 12 layers of radial tile-like cells. **Ground tissue**: cells ± uniform in size and rounded; loosely packed with numerous cavities larger in size than cells, up to *c.* 30% of tissue. **Vascular bundles**: innermost amphivasal with central circular phloem strand (Fig. 65H). Outer bundles collateral, type II. **Starch** granules subcircular, 21–29 μm diameter, closely packed in all ground tissue, up to eight granules per cell. **Crystals**: druse cells occasional in ground tissue.

Root TS

Terete. **Size**: diameter 2–3 mm. **Epidermis** a single layer of flattened cells. **Periderm** multi-layered, actively dividing beneath epidermis (Fig. 10E). **Cortex**: cells compact in outer 5–7 layers, loosely packed aerenchyma up to inner boundary. **Endodermis**: cylinder well-defined with broad casparian strips on anticlinal walls (Fig. 10C). **Pericycle**: single layer. **Vasculature**: protoxylem poles six, files of 1–3 small cells outside single wide metaxylem cells (Fig. 10C). Phloem strands six, on alternate radii, very small and irregular in outline. **Central region**: ground tissue. **Starch** granules in inner cortical ground tissue, surrounding stele. **Crystals**: raphides and druses in thin-walled cells, occasional, scattered mostly in outer ground tissue.

Material examined

Gearum brasiliense N.E.Br., Burchell 8598, Goiás, Brazil, (dry) K; Bogner 2254, Brazil, M.

Taxonomic notes

This species has several unique characters that isolate it. The mesophyll cells have very small plastids for the family and unique vesicular outgrowths similar to those described by Lindorf (1980) in *Dieffenbachia*. Mesophyll cell nuclei are large with numerous adherent organelles or vesicles. The presence of biforines is unique for the Spathicarpeae.

Additional literature

Grayum 1990.

Spathicarpa W.Hook.

This genus of seven species ranges from southern Brazil, Paraguay, and northern Argentina to the Seychelles Islands. The plants are terrestrial with tubers, rhizomes or rhizomatous tubers and they occur in understoreys of humid or marshy forests. One species occurs in dry forests. The several leaves are elliptic or cordate to sagittate and long-petiolate. Chromosomes: $2n = 34$. One species was available for study.

Leaf surface

Cuticle smooth to finely striate with a striate margin. **Epidermis**: cells polygonal with straight anticlinal walls on both surfaces (1–3:1 *l/w*). **Stomata** more frequent abaxially and simply brachyparacytic (Fig. 66F). Subsidiary cells broad and irregular. **Venation**: midrib prominent with irregular secondaries. In basal lamina, secondary laterals arising independently from petiole, at first wide-spreading then apically curving and merging eucamptodromously. Distal secondaries branching acutely from midvein. Tertiaries arising broadly acute or orthogonally, producing reticulate higher-order branching.

Leaf TS

Structure dorsiventral. **Size**: 70 × 50 mm (*l/w*), plus basal lobes (length 11 mm). **Midrib** small, deeply rounded-grooved adaxially, broadly convex-rounded abaxially (Fig. 66A). **Cuticle** thin on both surfaces. **Epidermis**: cells large on both surfaces, rounded or flat internally and externally. **Mesophyll**: one weakly differentiated palisade layer of squarish cells (Fig. 66B). Spongy tissue of

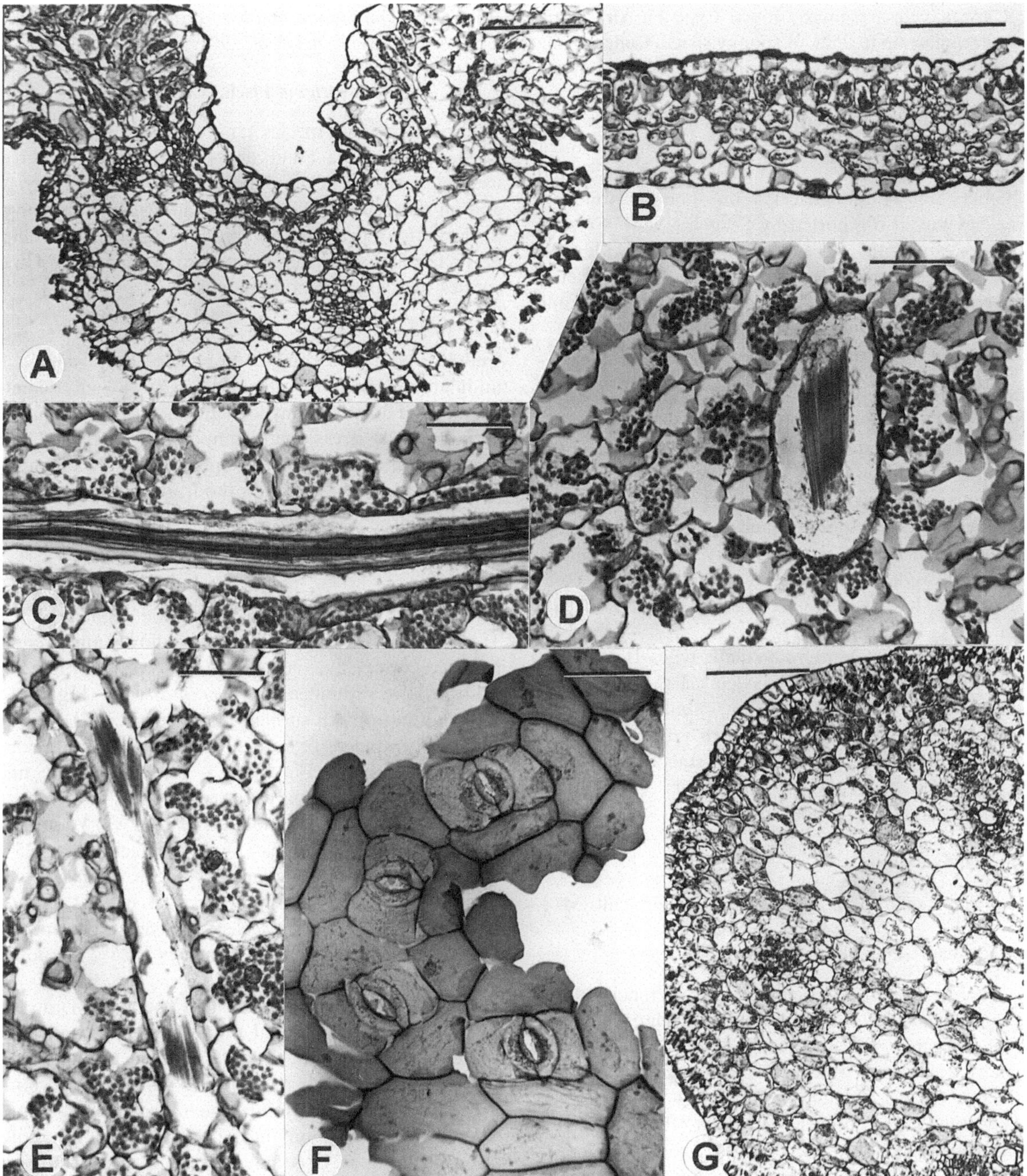

Fig. 66. *Spathicarpa sagittifolia* leaf and petiole sections. (A) Midrib TS; (B) Lamina TS; (C) Lamina PS showing vascular bundle and two adjacent laticifers; (D) Lamina PS with raphide cell; (E) Lamina PS with elongate raphide cell and multiple bundles; (F) Abaxial epidermis PS with stomata; (G) Petiole TS, abaxial surface at left. Scale bars: A,B,G = 150 μm; C–F = 50 μm.

c. 4 layers, cells irregularly lobed (type 1). **Air cavities** very irregular, up to 25% of spongy space. **Collenchyma**: abaxial, two-layered subepidermal band on abaxial surface of midrib.

Vascular bundles: midrib with one median and two lateral bundles (type II) (Fig. 66A). Xylem: 4–5 metaxylem elements. Phloem: a small semicircular strand. **Laticifers** not readily detected but non-anastomosing (Fig. 66C) as also reported by French (1988). **Crystals**: raphides abundant in single bundles in wide cells (Fig. 66D), or in multiple bundles in larger cells, or elongate cells (Fig. 66E). Crystal sand occasional, randomly distributed.

Petiole TS

Outline: proximally: deeply grooved adaxially, rounded abaxially. Distally: shallowly grooved, without laminar extensions, rounded abaxially (Fig. 66G). **Size**: diameter 3–5 mm. **Cuticle** thin and smooth or shallowly striate. **Epidermis**: cells small and flat externally. **Collenchyma**: 2–3 layers in discontinuous bands 1–2 layers inside epidermis. **Ground tissue**: outer 4–5 layers bearing chloroplasts. Cells gradually larger toward centre (Fig. 66G). **Air cavities**: few and cell-sized in central ground tissue (Fig. 66G).

Vascular bundles scattered, small; types II (Fig. 6B) and III. Xylem: 1–2 wide protoxylem cells and one or few narrow metaxylem cells. Phloem: circular to laterally elliptical strand, or irregular. **Starch** granules few and scattered, up to 5 μm diameter. **Crystals**: raphides few and scattered in axially elongated cells. Crystal sand proximally in adaxial epidermis.

Material examined

Spathicarpa sagittifolia Schott, Bogner 2335, cult. M.

Species reported on in the literature

Ertl (1932): *S. sagittifolia*: leaf venation. • French (1987c): *S. sagittifolia*: root. • French (1988): *S. sagittifolia*: laticifers. • French and Tomlinson (1983): *S. sagittifolia*: stem vasculature. • Fricke (1926): *S. sagittifolia*: leaf. • Lierau (1888): *S. sagittifolia*: root. • Löv (1925–6): *S. sagittifolia*: leaf ontogeny. • Müller (1919): *S. sagittifolia*: leaf tip. • Schwendener (1882): root. • Solereder (1919): *S. sagittifolia*: leaf, crystals. • Solereder and Meyer (1928): *S. sagittifolia*: anatomy. • Webber (1960): *S. sagittifolia*: epidermis, stomata, crystals.

Additional literature

Barabé and Chrétien 1985a; Campbell 1902, 1903; Erdtman 1952; French 1986a, c, 1987a; Grayum 1990; Ray 1988; Thanikaimoni 1969; Troll 1928; Ueno 1980; Uhlarz 1982; Williams *et al.* 1981.

Asterostigma Fisch. & C.A.Mey.

This genus of five species ranges from south Brazil to northern Argentina. Plants are terrestrial tuberous herbs, usually occurring in the understorey of primary forests, and on steep slopes of cool mountain valleys. The leaves are large, long-petiolate, and pinnately divided with elliptic or cordate segments. Chromosomes: 2n = 34. Three species were available for study.

Leaf surface

Cuticle smooth or rough and striate. **Epidermis**: cells undulate to lobed on both surfaces (3–6:1 *l*/*w*). **Stomata** abaxial and brachyparacytic. Subsidiary cell shape highly variable, cells occasionally unipolar (Fig. 67D). **Venation**: midrib prominent with pinnate, acutely ascending secondaries collecting eucamptodromously and irregularly at margin. Tertiary and higher order veins reticulate.

Leaf TS

Structure dorsiventral. **Size** of elliptic segment 120 × 31 mm (*l*/*w*). **Midrib** flat adaxially, shallowly and broadly rounded-convex abaxially in *A. riedelianum* (Fig. 2A). Midrib more prominent in *A. cryptostylum* (Fig. 67A). **Cuticle** thin on both surfaces. **Epidermis**: cells large and square to rounded on both surfaces, flat externally and rounded internally (Fig. 67B,C). **Trichomes** short, unicellular, sclerotic with constricted basal attachment, occasional on abaxial epidermis (Fig. 2A). **Mesophyll**: one well-differentiated layer of palisade cells, up to 25% of mesophyll, and *c.* 5 spongy layers in *A. riedelianum* (Fig. 67B). Layers weakly differentiated in *A. cryptostylum* (Fig. 67C). Spongy cells 5–6-layered, unarmed or short-armed, type 1 or 1a. **Air cavities** 5–25% of spongy layer, substomatal cavities shallow.

Vascular bundles: one median bundle in midrib (Fig. 67F), 1–2 additional bundles symmetrically disposed on each side (Fig. 67A) (largest type I, smaller type II). Xylem: metaxylem elements numerous, forming broad boundary with broad semicircular phloem strand. **Sclerenchyma**: fibres on abaxial side of vascular bundles or ensheathing all large bundles in midrib or lamina in *A. riedelianum* (Fig. 67F); fibres absent in smaller type II bundles of that species. **Laticifers**: elongate, non-anastomosing cells placed laterally along vascular bundles, contents colourless. **Crystals**: raphides, occasional in small cells, randomly distributed in mesophyll and abaxial side of midrib. In *A. riedelianum*, raphide cells also common and articulated forming elongated multi-celled

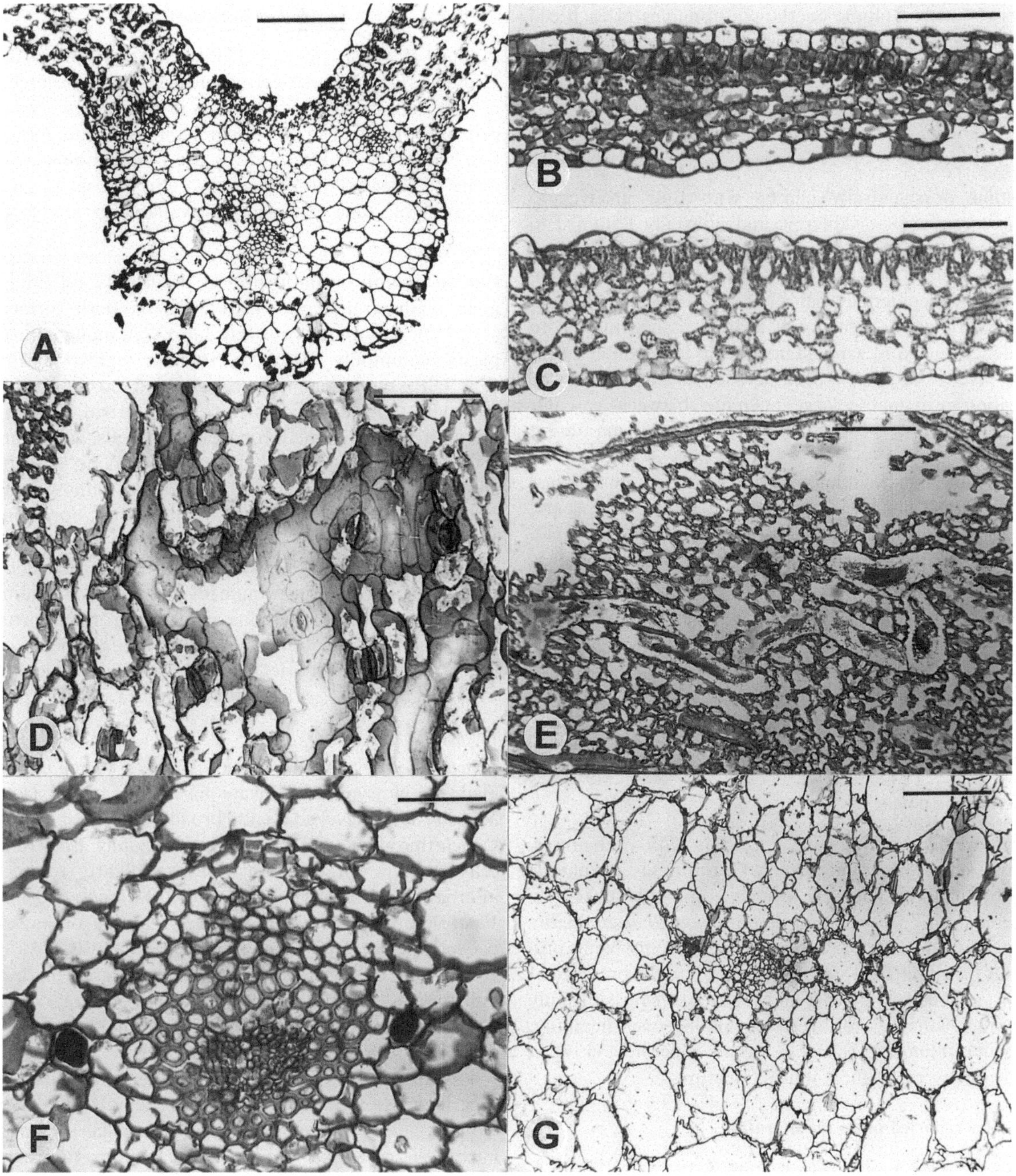

Fig. 67. *Asterostigma* leaf and petiole sections. (A,C) *A. cryptostylum*; (B,D–G) *A. riedelianum*; (A) Midrib TS; (B) Lamina TS; (C) Lamina TS; (D) Abaxial epidermis PS and stomata, with one unipolar stoma; (E) Mesophyll PS showing sinuous raphide 'vessel'; (F) Midrib TS, type I vascular bundle; (G) Petiole centre TS, type III vascular bundle. Scale bars: A,G = 200 μm; B–D = 150 μm; E = 300 μm; F = 50 μm.

'vessels' with multiple overlapping bundles in each cell (Figs 9D, 67E). Druses occasional in upper mesophyll. Crystal sand common in abaxial epidermis and as small prismatics in numerous mesophyll cells and midrib ground tissue.

Petiole TS

Outline dorsiventrally elliptic with four grooves in *A. riedelianum*, or subterete and adaxially flattened in *A. cryptostylum*. **Size**: diameter 9–15 mm. **Cuticle** thin and finely striate. **Epidermis**: cells small and rounded. **Collenchyma**: subepidermal band, 3–5 cells deep. **Chlorenchyma**: outer three layers of ground tissue, internal to collenchyma band. **Ground tissue**: cells large, outer 10–12 cell layers compact, type 3 aerenchyma in centre. **Air cavities**: proximally, large openings between vascular bundle islands, 1–3 partition cells. Aerenchyma absent distally.

Vascular bundles numerous and scattered, unsheathed (type III). Xylem: wide protoxylem lacuna or cell and 0–2 narrow metaxylem cells (Fig. 67G). Phloem: small, laterally narrowed strand. **Tannin** cells occasional in outer ground tissue. **Crystals**: raphides occasional in wide bundles in irregular, thin-walled cells.

Material examined

Asterostigma sp., Harley 18565, cult. K; *A. cryptostylum* Bogner, Bogner 1237, Brazil: Goias, cult. M; *A. riedelianum* (Schott) O. Ktze., Harley 18535, Brazil: Bahia, cult. K.

Taxonomic notes

Asterostigma species have a combination of features, which, if not unique individually for the family, are unique in combination. These are articulated raphide tubes, also shared with *Taccarum* and the unrelated *Biarum*, highly irregular and variable subsidiary cells, some being unipolar (in about 10–15% of stomata), and the presence of short abaxial trichomes. Finally, *Asterostigma* is the only member of the tribe to have fibres capping, or ensheathing, vascular bundles. *Taccarum* seems especially close in sharing several of these unusual features.

Species reported on in the literature

French (1987c): *A. lividum*: root. • French (1988): *A. lividum*, *A. riedelianum*: laticifers. • French and Tomlinson (1983): *A. lividum*: stem.

Additional literature

French 1986a, c, 1987a; Grayum 1990; Hay 1986; Hegnauer 1986; Thanikaimoni 1969; Williams *et al.* 1981.

Mangonia Schott

A genus of two species from tropical southern Brazil and Uruguay. Plants are terrestrial tuberous herbs occurring usually in seasonally dry habitats in the understorey of primary forests, or in more open habitats of cool mountain valleys to 2300 m. Leaves are long-petiolate with ovate to linear-elliptic blades. No chromosome counts are available. One specimen was available for study.

Leaf surface

Cuticle smooth on both surfaces. **Epidermis**: cells polygonal with straight anticlinal walls on both surfaces (1–2:1 *l*/*w*). Tanniniferous cells, also containing chloroplasts, commonly scattered. **Stomata** highly variable (Fig. 68B), more common abaxially and mostly brachyparacytic. Subsidiary cells narrow or broad, angular, often little differentiated from epidermal cells. Subsidiary cells often one large and one small, making stomata appear asymmetric. Up to 25% of stomata with two subsidiary cells on one side, one on the other, appearing nearly anomocytic. Stomata occasionally brachypara-tetracytic or brachypara-hexacytic with narrow subsidiary cells. **Venation**: midrib prominent with acute-ascending secondaries. Secondary veins eucamptodromous proximally and brochidodromous with straightened collectors distally. Tertiaries and higher orders reticulate with irregular areoles; free vein endings rare.

Leaf TS

Structure dorsiventral or weakly so. **Size**: young elliptic blade with asymmetric ovate base, 50 × 19 mm (*l*/*w*). **Midrib** robust, level adaxially, broadly convex abaxially. **Cuticle** thin abaxially, medium-thick adaxially. **Epidermis**: cells medium-tall on both surfaces; slightly rounded internally and externally. **Mesophyll** differentiated into 1 palisade layer (Fig. 68A) and 5–7 tightly or loosely packed spongy layers (type 1). **Air cavities**: 10–50% of lamina, up to 90% of type 3 aerenchyma in midrib. Substomatal cavities (Fig. 68C) small or larger. **Collenchyma**: 1–2-layered band, subepidermal on abaxial side of midrib.

Vascular bundles 5–7, symmetrically scattered, largest type I, others type II. Xylem: 7–16 metaxylem cells of various widths. Phloem: broadly semicircular strand or laterally narrowed strand in smaller bundles. **Laticifers**: not detected. **Tannin** cells mostly absent in ground tissue. When present in epidermis, chloroplasts also present. **Crystals**: raphides in large, often very elongated or wide cells. Cells mostly with multiple crystal bundles, occasionally discrete, often overlapping or tightly packed. Greatest raphide bundle complexity found in lamina marginal cells (Fig. 68D).

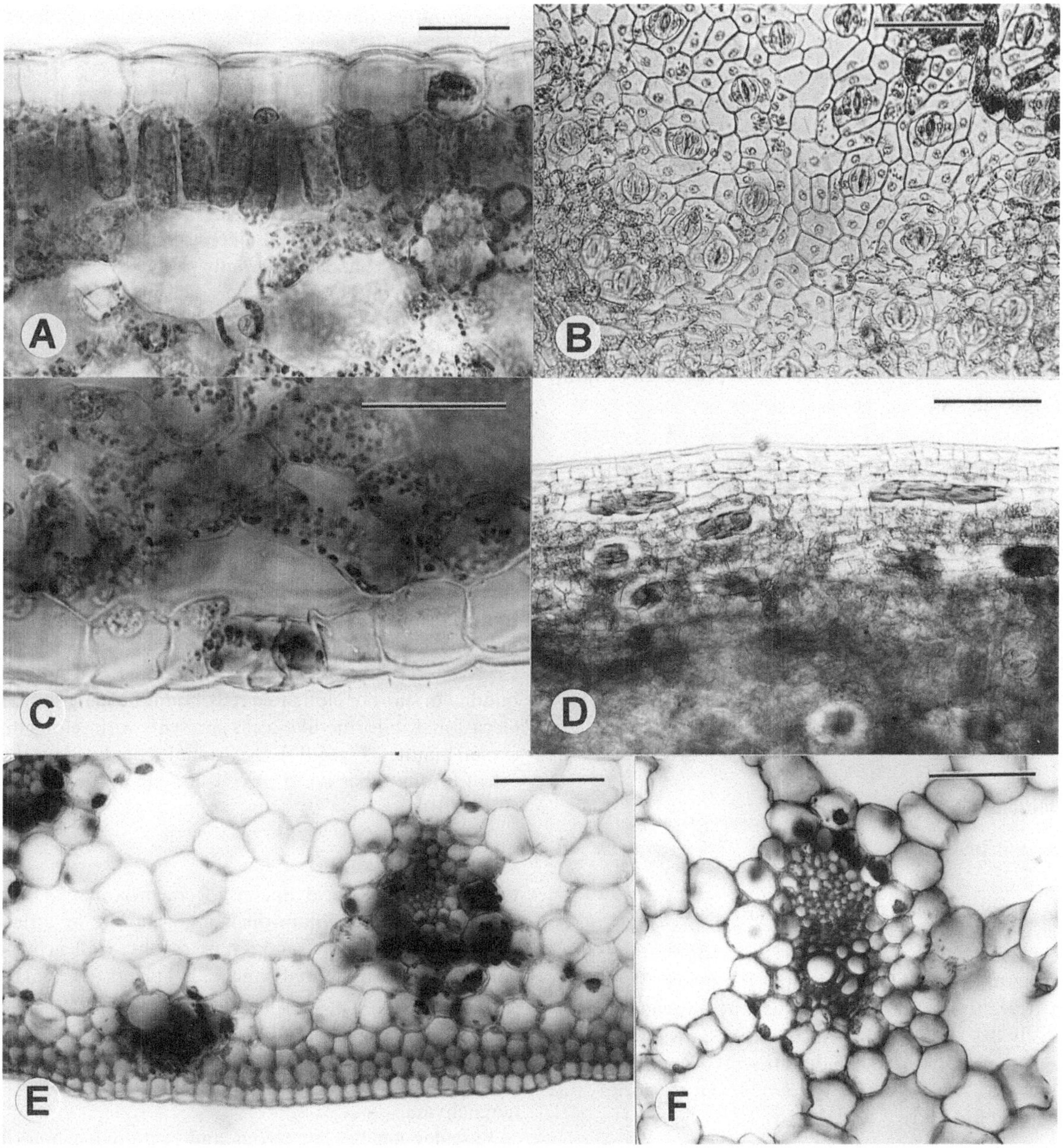

Fig. 68. *Mangonia tweedieana*, leaf and petiole hand sections. (A) Lamina TS, adaxial epidermis and palisade/spongy mesophyll boundary; (B) Lamina PS, abaxial epidermis and stomata; (C) Lamina TS, abaxial epidermis with stoma and shallow substomatal air cavity; (D) Lamina PS, margin and elongated raphide cells; (E) Proximal petiole TS, iodine-stained section. Note starch granules (black) and narrow band of subepidermal collenchyma; (F) Petiole TS, iodine-stained, large central type I vascular bundle suspended in type 3 aerenchyma. Scale bars: A,C = 50 μm; B,D–F = 150 μm.

Petiole TS

Outline crescent-shaped. **Size**: diameter 6 mm. **Cuticle** smooth, thick abaxially, thinner adaxially. **Stomata** occasional abaxially. **Collenchyma**: 2–3 angularly thickened subepidermal layers abaxially (Fig. 68E). **Ground tissue**: type 3 aerenchyma with 1–3 partition cells in central ground tissue. **Air cavities** up to 95% of aerenchyma.

Vascular bundles scattered with large type I (Fig. 68F) and smaller type II present. Xylem: 8–10 metaxylem cells with broad straight boundary with semicircular phloem strands. **Laticifers** not detected. **Starch**: granule-filled cells in discontinuous ring around most vascular bundles (Fig. 68E). Granules polygonal, small (diameter 4–5 μm). **Crystals** absent.

Material examined

Mangonia tweedieana Schott, Bogner 2376, Uruguay, cult M.

Taxonomic notes

The genus is distinct by virtue of high stomatal variation and the most robust vascular bundle formation for the tribe. It fits well within this anatomically variable tribe on collenchyma banding pattern, ground tissue configuration, raphide structure, and epidermal features.

Additional literature

Grayum 1990.

Taccarum Schott

This genus of five species ranges from southern Brazil to northern Argentina and possibly Paraguay. The plants are terrestrial and tuberous, usually growing in seasonally dry, open shrub forests. Leaves are solitary with long erect petioles. The blade is trisect, decompound and pinnatifid. Chromosomes: 2n = 34. One species was available for study.

Leaf surface

Cuticle smooth on both surfaces and slightly striate on abaxial midrib and near larger vascular bundles. **Epidermis**: cells polygonal with straight to undulate anticlinal walls (1–2:1 *l*/*w*). **Stomata** abaxial and brachyparacytic. Subsidiary cells narrow and elongate or broad and irregular (Fig. 69C). **Venation**: midrib prominent and straight or forked. Secondaries steeply acute with straightened marginal collectors and either brochidodromous or eucamptodromous connections. Tertiaries and higher orders reticulate.

Leaf TS

Structure weakly dorsiventral. **Size**: elliptic segment, broad at centre and tapering angularly to tip and base, 37 × 12 mm (*l*/*w*). **Midrib** large and subterete with lamina segments abruptly demarcated and arising close to adaxial side. **Cuticle** thin. **Epidermis**: cells large on both surfaces; cells rounded externally and internally. **Mesophyll**: poorly differentiated, one palisade layer and 3–4 spongy layers (Fig. 69A). Palisade cells rounded, often U-shaped, the short arms oriented toward epidermis. Spongy cells rounded and slightly lobed (type 1). **Air cavities** small and irregular, up to 30% of spongy tissue. Type 2 aerenchyma in central midrib. **Collenchyma**: band 1–2 cells thick, abaxial in midrib.

Vascular bundles numerous and scattered in midrib, type III; xylem: wide protoxylem lacuna, 1–2 narrow metaxylem cells opposing small, irregular phloem strand. Laminar bundles: 8–9 narrow metaxylem cells and circular phloem strand. Differentiated sheath cells absent. **Laticifers** reported as non-anastomosing by French (1988) but not convincingly demonstrated in this material. **Crystals**: raphides in single wide bundles or multiple bundles in short or elongated cells, common in mesophyll. Occasional biforine-like cells present with elongated crystal bundles (Fig. 69D). Occasional articulated raphide cells forming tubes up to six cells long in mid-mesophyll (Fig. 69B).

Petiole TS

Outline terete. **Size**: diameter 1.2 cm proximally, 8 mm distally. **Cuticle** medium-thick and smooth or striate. **Epidermis**: cells square and flat externally. **Collenchyma**: continuous subepidermal band of 2–4 layers of angularly thickened cells (Fig. 69E). **Ground tissue**: cells increasing in size toward centre. **Air cavities**: proximally, type 3 aerenchyma with prominent air cavities. Distally a few large, mostly cell-sized air cavities, type 2 aerenchyma.

Vascular bundles numerous and scattered in midrib, type II (Fig. 69E) or type III (Fig. 69F). Xylem: 1–2 wide protoxylem cells and 2–4 metaxylem cells. Phloem: an irregular strand. **Starch**: few small granules anywhere in ground tissue. **Crystals**: raphides few in rounded or elongated cells, randomly oriented in ground tissue.

Material examined

Taccarum weddellianum Brongn. ex Schott, Bogner 458, cult. M.

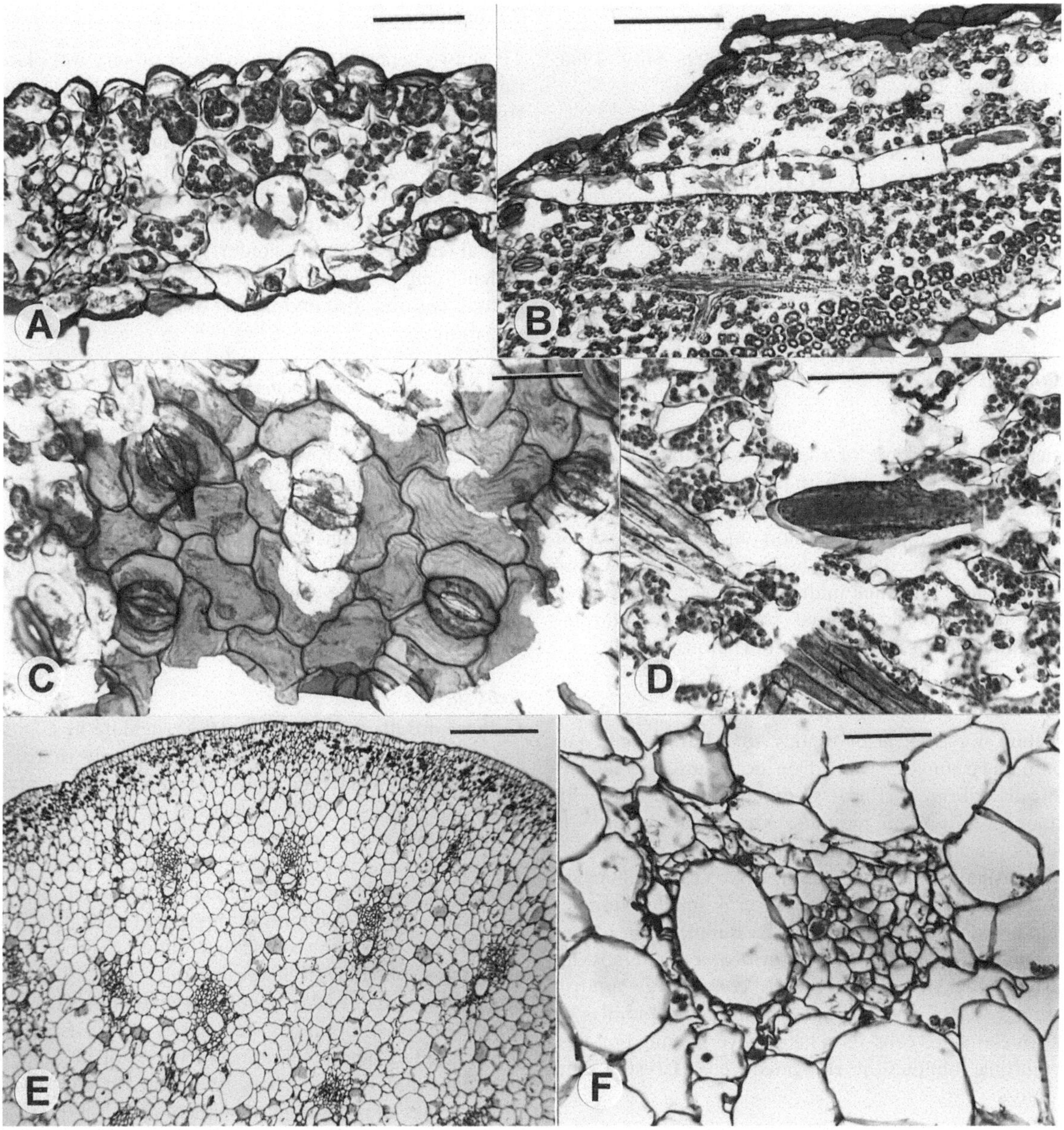

Fig. 69. *Taccarum weddellianum* leaf and petiole sections. (A) Lamina TS; (B) Leaf mesophyll PS showing articulated raphide tube with six cells in view; (C) Leaf abaxial epidermis PS showing brachyparacytic stomata and epidermal cells with undulate anticlinal walls; (D) Leaf mesophyll PS with biforine-like raphide cell; (E) Petiole TS showing subepidermal collenchyma band; (F) Petiole TS centre, with type III vascular bundle. Scale bars: A,C,D,F = 50 μm; B = 150 μm; E = 300 μm.

Taxonomic notes

Articulated raphide cells, rarely found in the family, link this genus to *Asterostigma*. See Taxonomic notes under that genus.

Species reported on in the literature

(All *T. weddellianum*)

French (1987c): root. • French (1988): laticifers. • French and Tomlinson (1983): stem vasculature.

Additional literature

French 1986a,c, 1987a; Grayum 1990; Hay 1986; Thanikaimoni 1969.

SUBFAMILY SCHISMATOGLOTTIDOIDEAE

Tribe Cryptocoryneae Blume

This tribe contains the Asian genera *Cryptocoryne* and *Lagenandra.*

Anatomical summary

Leaf surface

Cuticle smooth, striate or rugulate and thin to absent. **Epidermis**: cells polygonal with rounded or straight anticlinal walls. **Stomata** brachyparacytic or brachypara-hexacytic or -octocytic; common on both surfaces or more frequent abaxially. Subsidiary cells narrow and square-angled. **Venation**: midrib prominent with steeply ascending, regularly spaced secondaries. Margins eucamptodromous or basally acrodromous.

Leaf TS

Structure dorsiventral or weakly so. **Midrib** slightly or prominently convex-rounded on both surfaces. **Epidermis**: cells tall or short near stomata. **Stomata** level with surface or raised on mounds. **Mesophyll**: palisade 1-layered, often poorly differentiated. Spongy tissue type 1. **Collenchyma**: strands circular, type Sv. **Vascular bundles**: one large central bundle and several smaller bundles symmetrically arranged in midrib. Bundles types II or III, scattered, well- or poorly-formed. **Laticifers**: non-anastomosing or absent. **Starch** common in midrib ground tissue. **Crystals**: raphides in single bundles in wide or elongated cells, or in helical overlapping bundles, or multiple bundles in elongated cells. Crystal sand common.

Petiole TS

Crescent-shaped or semicircular. **Cuticle** thin and finely striate. **Epidermis**: cells small and square. **Collenchyma** strands circular, type Sv. **Ground tissue** compact or type 2 or 3 aerenchyma. **Vascular bundles** 12 or more, types II or III, may be arranged in concentric circles. **Laticifers** non-anastomosing. **Starch** granules occasional in ground tissue. **Crystals**: raphides in single or multiple bundles in short or elongated cells. Crystal sand present in ground tissue.

Taxonomic notes

These two genera both grow in water or very wet places but their anatomy has not been particularly modified from that of putative relatives. The genera are quite similar anatomically and are appropriately placed in the same tribe. The Cryptocoryneae share a number of similarities with the Schismatoglottideae, the probable nearest clade. Their epidermal, stomatal, and certain crystal features are quite similar. On the other hand, the collenchyma patterns are markedly different as are some crystal features, the condition of ducts, and aerenchyma variations.

Cryptocoryne Wydler

This genus of 50 species is found in India, Sri Lanka, south-east Asia and Malaysia. Plants are small rhizomatous herbs, primarily aquatic, living in swampy areas, gravelly stream beds, on rocks along streams, or in freshwater tidal zones of coastal rivers. Some live on seasonally dry sand flats and others are terrestrial. Plants have petiolate, long-elliptic to linear, ovate or cordate thin leaves. Chromosomes: $2n = 20, 22, 28, 30, 33, 34, 36, 42, 68, 85, 90, 102$. Nine species were available for study.

Leaf surface

Cuticle smooth or striate adaxially, rugulate in *C. pontederiifolia* (Fig. 70E); not seen or irregularly ridged in *C. annamica.* **Epidermis**: cells polygonal with straight anticlinal walls on both surfaces (1–2:1 *l/w*). **Stomata** common and brachyparacytic (Fig. 70F), to brachypara-tetra- and -hexacytic (Fig. 70E) on both surfaces. Subsidiary cells with straight anticlinal walls parallel to stomatal axis. Neighbouring epidermal cells variable, occasionally 4–6 elongate cells surrounding stomata and subsidiary cells. All stomatal apertures tending toward alignment with vascular bundles. **Venation**: midrib prominent. In most specimens secondaries steeply ascending to eucamptodromous margins. Tertiaries orthogonal or reticulate and higher orders reticulate. In *C. spiralis*, quaternaries may be aligned with midrib. In *C. affinis* and *C. pontederiifolia*, midrib basally prominent with secondaries basally or suprabasally acrodromous. Secondaries curved-ascending to leaf tip and tertiaries curved-orthogonal. Higher orders orthogonal to tertiaries or to midrib, or reticulate. Highest organization at all orders of venation found in *C. pontederiifolia.*

Leaf TS

Structure weakly dorsiventral. **Specimen sizes**: varying from 31×13 mm to 82×14 mm (*l/w*). **Midrib** broadly rounded-convex on both surfaces or broadly concave

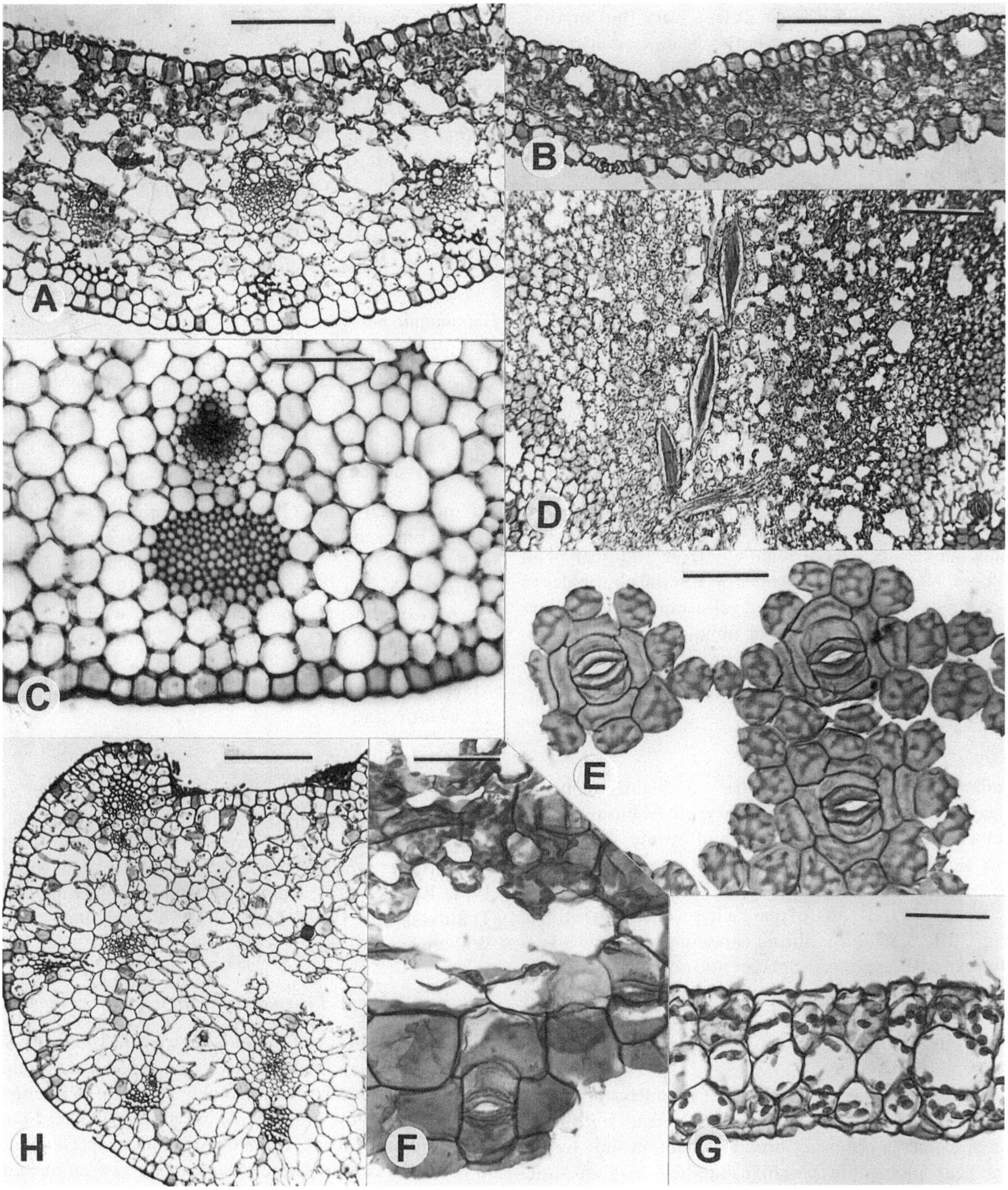

Fig. 70. *Cryptocoryne* leaf and petiole sections. (A,B,D,E) *C. pontederiifolia*; (C) *C. annamica;* (F) *C. spiralis*; (G) *C. crispatula* var. *balansae*; (H) *C. affinis*; (A) Midrib TS; (B) Lamina TS, note stomata on both surfaces; (C) Midrib abaxial margin with large type Sv collenchyma strand; (D) Mesophyll PS showing raphide cells; (E) Lamina epidermis PS showing brachypara-hexacytic stomata and rugulate cuticle; (F) Lamina PS showing brachyparacytic stomata; (G) Lamina TS showing undifferentiated mesophyll; (H) Distal petiole TS. Scale bars: A–C = 150 μm; D,H = 200 μm; E–G = 50 μm.

adaxially (Fig. 70A). **Cuticle** absent, very thin or thin. **Epidermis**: cells irregular on both surfaces, smaller near stomata. Cells flat or rounded externally and internally. **Stomata**: guard cells and subsidiary cells often raised on mounds above abaxial surface; mostly level with surface adaxially. **Mesophyll** poorly differentiated (Fig. 70B), one palisade layer and 4–9 spongy layers of irregular, rounded and unarmed cells (type 1). Mesophyll undifferentiated in *C. crispatula* var. *balansae* (Fig. 70G). **Air cavities** small and irregular, 10–40% of spongy layer. **Collenchyma**: several circular strands, subepidermal on abaxial side of midrib (Fig. 70C); aligned with vascular bundles but separated from them by several layers of ground tissue (type Sv).

Vascular bundles scattered in midrib, mostly poorly developed (type III), but well-formed in *C. annamica*. Xylem: several protoxylem elements near phloem strand. Phloem: small irregular or laterally compressed strand, except broad strand extending beyond edges of xylem strand in *C. annamica* and *C. pontederiifolia* (Fig. 70A). **Laticifers**: non-anastomosing, cells wide, lateral to vascular bundles, cells occasionally narrow and near or in phloem. Not found in *C. annamica*. **Crystals**: raphides in wide single bundles in rounded cells, randomly oriented in mesophyll, or in multiple or oblique bundles in elongated cells (Fig. 70D). Crystal sand uncommon and dispersed in mesophyll cells.

Petiole TS

Outline: rounded abaxially, flat to slightly concave adaxially (Fig. 70H). More deeply grooved basally. **Size**: diameter *c.* 3 mm. **Cuticle** thin and striate. **Epidermis**: cells square, rounded externally and internally. **Collenchyma**: strands circular, large or smaller, around periphery of organ, several rows of parenchyma beneath surface (Fig. 70H). Each large strand capping phloem of a vascular bundle (type Sv). Smaller strands not associated with bundles. **Chlorenchyma**: occasional plastids in outer ground tissue. **Ground tissue**: cell size uniform. **Air cavities** numerous, small and larger spaces separated by uniseriate partitions (type 3).

Vascular bundles: two circles of bundles, one near periphery and one internal (type II). Xylem: 1–4 protoxylem elements opposing broad phloem strand. **Laticifers** near phloem at vascular bundles; rare in some specimens and reported absent by French (1988). **Starch** rare, granules scattered in ground tissue. **Crystals**: raphides in single bundles, some in elongated cells suspended in partitions pointed toward air spaces on both sides; or in multiple bundles in elongated raphide tubes. Crystal sand common in ground tissue.

Material examined

Cryptocoryne affinis N.E.Br. ex Hook.f., Bogner 923, cult. M; *C. annamica* Serebryani, cult. M; *C. ciliata* (Roxb.) Fisch. ex Schott, Bogner 1327, cult. M; *C. crispatula* Engl. var. *balansae* (Gagnep.) Jacobsen, Bogner 468, cult. M; *C. lingua* Becc. ex Engl., Bogner 1436, Sarawak, cult. M; *C. pontederiifolia* Schott, Bogner 1093, Sumatra, cult. M; *C. retrospiralis* (Roxb.) Kunth, India, cult. M; *C. spiralis* (Retz.) Fisch. ex Wydler, cult. M; India, cult. M; *C. wendtii* de Wit, Bogner 470, cult. M.

Taxonomic notes

Leaf venation shows many species-specific characters with *C. pontederiifolia* being the most organized and distinct.

Species reported on in the literature

Anandakumar *et al.* (1982): *C. spiralis*: corm. • Blanc (1978): ontogeny. • Engler (1920b): *C. ciliata*: stem, leaf. • Ertl (1932): *C. nevillei*: leaf. • French (1987c): *C. retrospiralis, C. spiralis*: root. • French (1988): *C. ciliata, C. retrospiralis, C. spiralis*: laticifers. • French and Tomlinson (1983): *C. spiralis, C.* sp.: stem. • Roth (1949): leaf. • Shah and Gopal (1969): leaf. • Solereder and Meyer (1928): *C. wightii*: crystals. • Streitberg (1954): *C. bekettii*: leaf.

Additional literature

Anandakumar *et al.* 1981; Domin 1911; French 1986a, c, 1987a; Gernhard 1981; Grayum 1990; Hegnauer 1986; Kulkarni *et al.* 1990; Meusel 1951; Raynal-Roques 1981; Reumer 1984; Richter 1929; Subramanyam 1962; Thanikaimoni 1969; Troll 1933; Vogel 1962, 1990; Williams *et al.* 1981.

Lagenandra Dalzell

This genus of 14 species is found in western India and Sri Lanka. Plants are mainly aquatic and moderately small herbs with stoloniferous rhizomes. They live in swamps, wet places in low country, or on stream banks. Some species are terrestrial. The nine Sri Lankan species occur in rivers. Leaves are ovate to wide or narrow-elliptic and long-petiolate. Chromosomes (four species): $2n = 36$, *c.* 72. Five species were available for study.

Leaf surface

Cuticle smooth adaxially, striate abaxially and at margins or with rugulate knobs. **Epidermis**: cells rounded-polygonal (1:1 *l*/*w*) on both surfaces. **Stomata**: more common abaxially or common on both surfaces. Stomata usually uniformly brachypara-hexacytic (Fig. 71C); four narrow subsidiary cells forming a square surrounding the guard cells and first pair of narrow subsidiary cells. In *L. thwaitesii* and *L. meeboldii*, stomata appearing nearly brachypara-octocytic with additional subsidiary cells. **Venation**: midrib prominent. Secondaries steeply ascending, apically curved and regularly spaced; may become parallel to midrib. Secondaries merging eucamptodromously at margin or near apex. Tertiaries broadly acute, tending to be recurved from midrib, crossing secondaries on different vector. Leaves very organized: secondaries, intercostal area spacing, and tertiary spacing predictable. Higher orders orthogonal to tertiaries or reticulate.

Leaf TS

Structure dorsiventral or weakly so. **Size**: 26 × 12 mm to 72 × 29 mm (*l*/*w*). **Midrib** slightly convex on both surfaces (Fig. 71A). **Cuticle** thin on both surfaces. **Epidermis**: cells large on both surfaces, flat externally and rounded internally. Abaxially, cells rounded internally and especially externally. **Stomata** appearing to have two subsidiary cells per guard cell. Guard cells level with surface or raised slightly on mounds. **Mesophyll**: one palisade layer of square to elongate cells, tightly packed and often weakly differentiated (Fig. 71B). Spongy layers 3–6, large rounded cells with short lobes (type 1) and well packed. **Air cavities** up to 10–30% of spongy layer; largest spaces shallow substomatal cavities. **Collenchyma**: angularly thickened cells in discrete circular strands, 5–10 cells deep and 1–2 layers internal from epidermis, in abaxial midrib and opposite vascular bundles (type Sv).

Vascular bundles: one large median bundle and several abaxial symmetrically disposed smaller bundles in midrib (Fig. 71A). Bundles well-formed (type II). Xylem: narrow protoxylem lacuna and several narrow metaxylem elements (Fig. 71F). Phloem: wide strand separated by ground tissue from xylem (Fig. 71G). **Laticifers** non-anastomosing and wide or narrow, with dark granular contents, common and lateral to larger vascular bundles. Not present in *L. thwaitesii* and noted as not present in 4 species by French (1988). **Starch** granules very common in midrib ground tissue. **Crystals**: raphides in wide elongate cells, in single wide bundles (Fig. 71D) or spiral overlapping bundles, common in mid-mesophyll in most species and abundant in *L. bogneri*. Raphides mostly near leaf base or closest to midrib, or tending to be parallel with secondaries. Crystal sand common to occasional, fine to coarse in mesophyll, epidermis and midrib ground tissue.

Petiole TS

Outline crescent-shaped to semicircular (Fig. 71E). **Size**: diameter 1.5–4 mm. **Cuticle** thin and finely striate. **Epidermis**: cells small, squarish and rounded externally. **Collenchyma**: circular strands, 8–10 cells in diameter, separated by 1–2 ground tissue layers from epidermis beneath ridges around periphery (Fig. 71E,H). **Chlorenchyma**: plastids in outer 8–10 layers of ground tissue. **Ground tissue**: compact and large cells in *L. lancifolia*; type 2–3 aerenchyma in other species. **Air spaces**: from 10% to 80% of ground tissue.

Vascular bundles: *c.* 12 scattered bundles, unsheathed, usually type III. Xylem: one wide late protoxylem cell, 0–2 narrow metaxylem cells. Phloem: small circular strand. **Laticifers**: 2–3 lateral to all vascular bundles. **Tannin** cells occasional in ground tissue of most species. **Starch** granules occasional in outer ground tissue. **Crystals**: raphides occasional in outer ground tissue in thin-walled cells. Crystal sand occasional in ground tissue.

Material examined

Lagenandra bogneri de Wit, Bogner 1131, Ceylon, cult. M; *L. erosa* de Wit, Bogner 450, Ceylon, cult. M; *L. lancifolia* (Schott) Thwaites, Bogner 1126, Ceylon, cult. M; *L. meeboldii* (Engl.) C.E.C. Fischer, Bogner 515, India, cult. M; *L. thwaitesii* Engl., cult. 9/62 (K).

Species reported on in the literature

Blanc (1978): stem. • Engler (1920b) *L. ovata*: stem, leaf. • French (1987c): several spp.: root. • French (1988): several spp.: laticifers. • French and Tomlinson (1983): *L. lancifolia, L. meeboldii*: stem.

Additional literature

French 1986a,c, 1987a; Grayum 1990; Kulkarni *et al.* 1990; Panicker 1965; Thanikaimoni 1969.

Tribe Schismatoglottideae Nakai

This mostly Asian tribe comprises the genera *Phymatarum, Schismatoglottis, Aridarum, Piptospatha, Hottarum, Bucephalandra,* and *Heteroaridarum.*

Anatomical summary

Leaf surface

Trichomes in *Schismatoglottis* only. **Cuticle** smooth, rough, striate, centric-rugulate, or centric-knobbed.

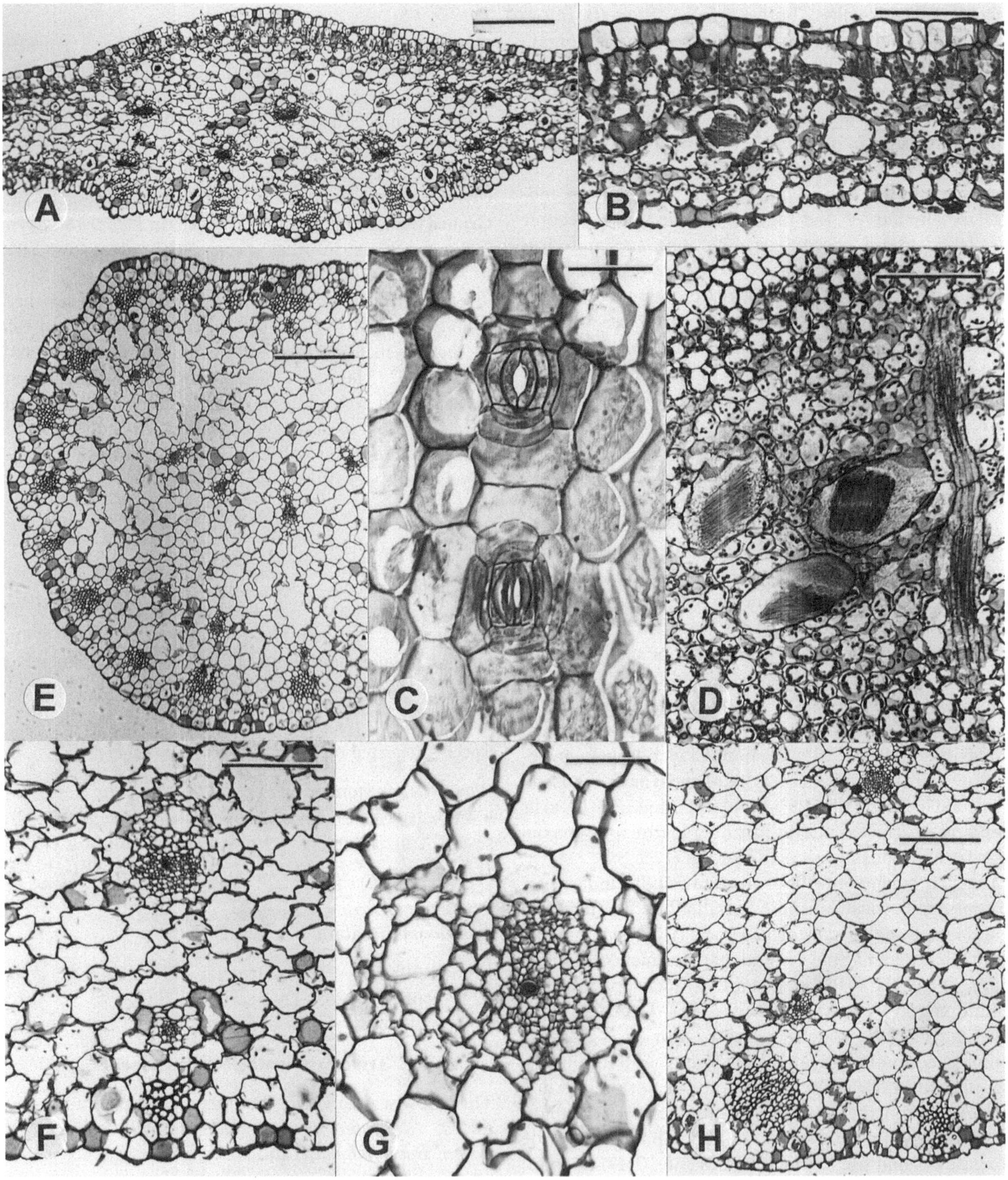

Fig. 71. *Lagenandra* leaf and petiole sections. (A,H) *L. bogneri*; (B,D,F,G) *L. erosa;* (C,E) *L. thwaitesii*; (A) Lamina TS midrib; (B) Lamina TS; (C) Abaxial leaf epidermis PS showing brachypara-hexacytic stomata; (D) Mesophyll PS showing wide raphide cells; (E) Distal petiole TS showing circular collenchyma strands near periphery; (F,G) Midrib abaxial margin TS; (G) Type II vascular bundle; (H) Proximal petiole TS. Note type Sv collenchyma strands. Scale bars: A,E,H = 300 μm; B,D,F = 150 μm; C,G = 50 μm.

Epidermis: cells polygonal, with sraight anticlinal walls on both surfaces (mostly 1–2:1 *l/w*). **Stomata** abaxial, equally numerous on both surfaces or fewer adaxially. Type brachyparacytic to brachypara-hexacytic, or brachypara-octocytic in *Phymatarum, Aridarum, Bucephalandra* and *Heteroaridarum*. **Venation** pinnate with usually narrowly acute ascending secondaries, usually ending eucamptodromously at margin.

Leaf TS

Structure strongly dorsiventral. **Midrib** shape usually strongly rounded-protruding abaxially, and of a unique type in *Bucephalandra*. **Epidermis**: cells mostly tall. **Hypodermis** adaxial at midrib in *Aridarum, Bucephalandra* and *Heteroaridarum*. **Mesophyll**: palisade cells 1–3 layers. Spongy type 1, 1a, 2 or 3 in lamina. **Collenchyma** banded or banded-interrupted (types B, Bi). **Vascular bundles** in midrib ranging from few to 30, randomly scattered or symmetrical. Bundles mostly type II, may be type III in *Schismatoglottis*. Fibre caps at phloem pole of vascular bundles in many genera. **Laticifers** non-anastomosing. **Secretory ducts** present in *Schismatoglottis*. **Tannin** cells common in ground tissue in most genera. **Starch**, often enlarged granules, present in most ground tissue. **Crystals**: raphides present as single bundles in thin-walled cells; as biforines in *Schismatoglottis, Aridarum* and *Piptospatha* and in multiple bundles or overlapping bundles in elongate cells in *Phymatarum*, in biforine-like cells in *Bucephalandra*. Druses and crystal sand usually present.

Petiole TS

Crescent-shaped at base to terete, subterete, or shallowly grooved distally, fluted in *Bucephalandra*. **Epidermis** of tall or short cells. **Ground tissue** compact or of type 3 aerenchyma; types 3 and 4 aerenchyma in *Schismatoglottis* and *Heteroaridarum*; type 2 in *Bucephalandra*. **Collenchyma** mostly in subhypodermal band, occasionally type Bi; in *Bucephalandra*, arranged in circular strands at three ridges (type Sv). **Vascular bundles** numerous, types II or III. **Secretory ducts** with 1–3 epithelial layers, with mucilage occasionally present in most genera. **Laticifers** non-anastomosing, 1–3 lateral to vascular bundles, or absent. **Crystals**: raphides as single bundles in wide cells, or biforines in type 3 aerenchyma in *Schismatoglottis*. Druses, crystal sand and starch granules usually present.

Stem TS (for *Schismatoglottis* and *Bucephalandra* only)

Cuticle thin or thick. **Epidermis**: cells small or large. **Collenchyma** absent or type Bi. **Ground tissue** ± compact, or large cells, with thickened primary walls in *Schismatoglottis*. Central cylinder not delimited except by starch cylinder or solid deposition in central cells. **Vascular bundles** scattered, collateral in outer tissue, mostly amphivasal in centre. Occasional endodermis development encircling some amphivasal bundles. **Cork** present at bract cicatrice zones in *Bucephalandra*. Irregularly shaped sclereids noted in *Schismatoglottis*. **Secretory cavities** may be present. **Laticifers** none or occasionally non-anastomosing. **Crystals**: raphides in single bundles per cell, occasionally as biforines, in *Schismatoglottis*; druses common.

Taxonomic notes

This group of genera is held together by a number of shared characters. Tendencies toward amphibrachyparacytic stomata, unusual cuticle development, transitions between banded and stranded collenchyma, presence of mucilage ducts are among characters that clearly isolate this tribe in the family. Within the tribe, *Bucephalandra* stands out by several exceptional characters as does *Schismatoglottis*, but each tends to share one or more states with at least one other genus.

Phymatarum M. Hotta

This genus of two species occurs in northern Borneo. Plants are small rheophytic herbs with short creeping rhizomes growing on steep banks of the forest floor near streams, in the understorey of primary riparian forests. The leaves are elliptic. Chromosomes: 2n = 26, 28. One species was available for study.

Leaf surface

Cuticle smooth adaxially, with some striate ridges abaxially and with folded-papillate ridges over midrib abaxially. **Epidermis**: cells polygonal (1–2(3):1 *l/w*). **Stomata** on both surfaces, brachypara-hexacytic (Fig. 72C), or -octocytic with up to four pairs of subsidiary cells. Subsidiary cells narrow and rounded. **Venation**: midrib prominent. Secondaries broadly acute, merging eucamptodromously in straightened marginal collector. Intersecondaries and tertiaries closely spaced and parallel to secondaries. Quaternaries irregular or cross-connecting tertiaries, orthogonal to midrib. Higher orders reticulate.

Leaf TS

Structure dorsiventral. **Size**: 101 × 38 mm (*l/w*). **Midrib** flat adaxially and rounded-semicircular abaxially (Fig. 72A). **Cuticle** thin adaxially and medium-thick abaxially. **Epidermis**: cells large and square or tall

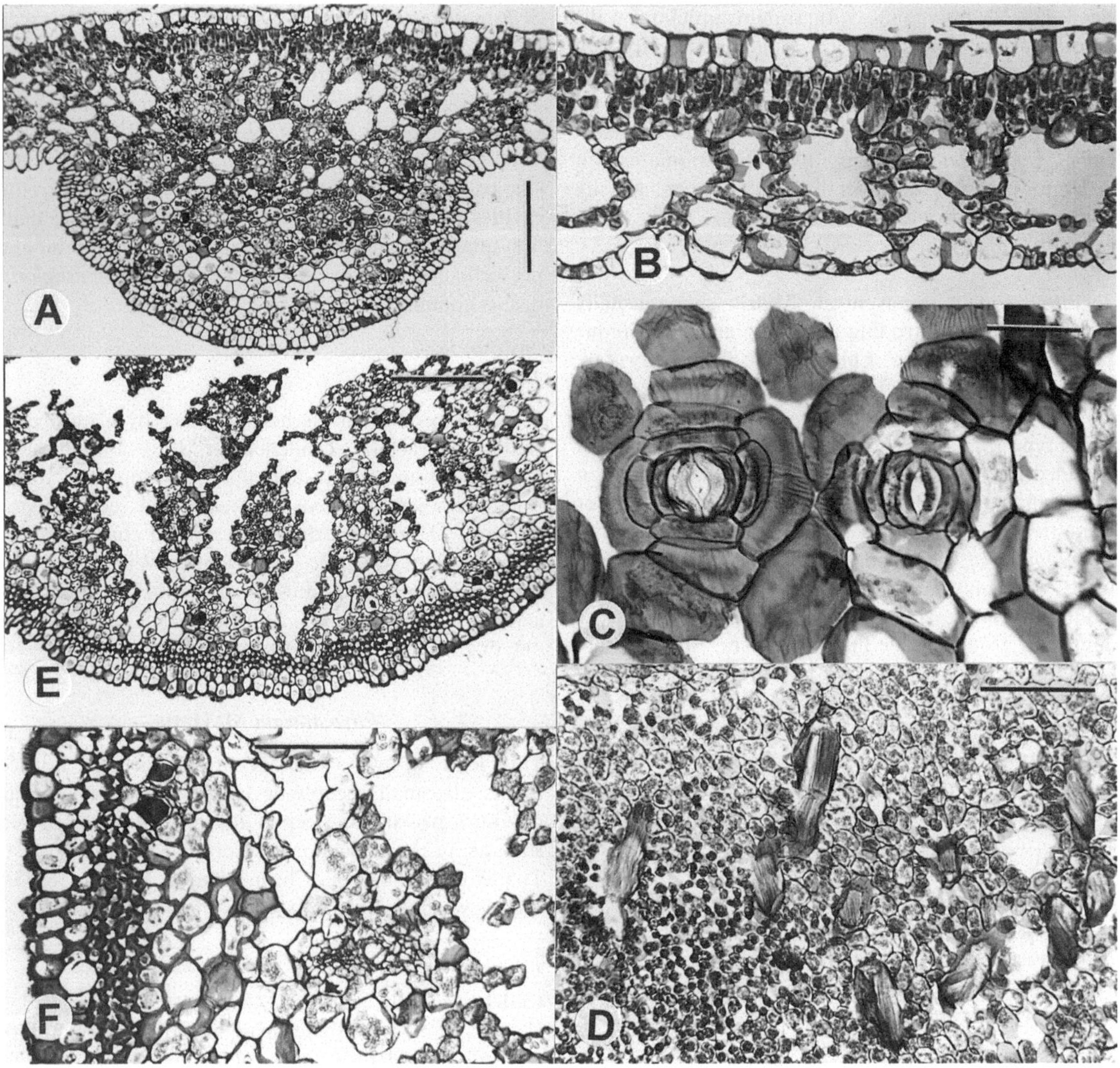

Fig. 72. *Phymatarum borneense* leaf and petiole sections. (A) Midrib TS; (B) Lamina TS showing type 3 spongy aerenchyma; (C) Lamina abaxial surface showing brachypara-hexacytic stomata; (D) Mesophyll PS showing abundant raphide cells; (E,F) Distal petiole TS showing peripheral collenchyma band on abaxial side. Scale bars: A,E = 300 μm; B,D,F = 150 μm; C = 50 μm.

adaxially, flat externally and rounded internally. Abaxial cells smaller, flattened or tall, very irregular in size; rounded internally and externally. **Stomata** level with epidermal surface and with markedly short subsidiary cells (Fig. 72B). Substomatal air cavities large and square-based adaxially, wide abaxially. **Mesophyll** well-differentiated. Palisade with two layers of short cells, up to 30% of mesophyll. Spongy layers forming an organized aerenchyma with large, tall substomatal air spaces with 1–2-seriate partitions (type 3) (Fig. 72B). **Air cavities**: *c*. 95% of spongy layer. **Collenchyma**: 2–3-layered subepidermal band around midrib periphery and discontinuous adaxially.

Vascular bundles: *c*. 9 in midrib, symmetrical in deltoid pattern and with one central bundle (Fig. 72A). Bundles small (type II). Xylem: 1 narrow protoxylem element and 2–5 narrow to medium-sized metaxylem elements. Phloem: strand small and laterally narrowed at constricted boundary with xylem. Some laminar vascular bundles larger than midrib bundles. **Sclerenchyma**: weakly lignified fibres as small phloem caps to vascular bundles. **Laticifers**: non-anastomosing and common, 1–2 lateral to vascular bundles. **Crystals**: raphides common to abundant in elongated cells (Fig. 72D), in multiple overlapping bundles, lateral and parallel to vascular bundles. Raphides also common in single bundles in short wide cells, mostly at palisade-spongy boundary. Druses common in mid-mesophyll or in palisade tissue. Crystal sand common in all mesophyll cells.

Petiole TS

Outline subterete, flattened and short-winged adaxially. **Size**: diameter 3–4 mm. **Cuticle**: thick and striate distally, or rugulate proximally. **Epidermis**: cells large and tall, rounded externally. **Collenchyma**: 2–4-layered peripheral band separated from epidermis by 1–2 ground tissue layers. Cells with distinct angular thickenings (Fig. 72E,F). **Chlorenchyma**: plastids in outer 4–5 layers of collenchyma and adjacent tissue. **Ground tissue**: cells large at collenchyma boundary, 2–5 layers of compact tissue. In centre, type 3 aerenchyma with partitions 1–3 cells long. **Air cavities**: up to 90% of aerenchyma.

Vascular bundles: three concentric rows; bundles numerous (type II). Xylem: one narrow or wide late protoxylem cell and 3–6 medium-width metaxylem cells (Fig. 72F). Phloem: small laterally narrowed strand. **Sclerenchyma**: fibres as small phloem caps to large vascular bundles; walls gelatinous distally, or thinly lignified proximally. **Laticifers**: large cells in pairs, lateral to phloem at all vascular bundles. **Starch** granules abundant in all ground tissue interior to collenchyma. **Crystals**: raphides common in single bundles, in short thin-walled cells in peripheral ground tissue, and in partition cells in centre. Druses and crystal sand occasional in central ground tissue.

Material examined

Phymatarum borneense M.Hotta, Bogner 1476, Sarawak.

Species reported on in the literature

French (1987c): root. • French (1988): laticifers. • French and Tomlinson (1981b): stem vasculature.

Additional literature

French 1985b, 1986c, 1987a; Grayum 1990; Hotta 1965, 1971.

Schismatoglottis Zoll. & Morr.

This genus of about 100 species is distributed mostly in Malaysia, a few species in south-east Asia and Burma, and 3 species in tropical South America. Most are terrestrial herbs but some are epilithic. Plants are short-stemmed or have stoloniferous rhizomes and occur as understorey plants of primary forests. Perhaps three species are true rheophytes. Leaves are long-petiolate and elliptic to cordate-sagittate. Chromosomes are medium-sized: 2n = 26, 52. Six species were available for study.

Leaf surface

Trichomes short, 1-celled, with striate cuticle on thick walls, common on abaxial midrib of Croat 53521 only. **Cuticle** smooth, rough or striate on both surfaces. Centric rugulate pattern often in centre of each epidermal cell. Abaxial cuticle in Croat 75532 with small, irregularly thickened cuticular papilla over centre of each cell. **Epidermis**: cells polygonal (1–2:1 *l*/*w*). **Stomata** more common abaxially, randomly oriented (Fig. 73D), brachyparacytic, brachypara-tetracytic, or brachypara-hexacytic. Subsidiary cells narrow or irregular and angular. **Venation**: midrib prominent. Secondaries pinnate, acute or broadly acute, closely spaced, straight, ending eucamptodromously at margin at small collector vein. Intersecondaries and tertiaries parallel with secondaries. Quaternaries orthogonal or irregular, when visible.

Leaf TS

Structure dorsiventral and well differentiated. **Sizes** up to 116 × 65 mm (*l*/*w*). Lobe length 15 mm. **Midrib** shallowly concave to slightly convex adaxially and strongly rounded-protruding abaxially (Fig. 73A). **Cuticle** thin on both surfaces (apart from papillae). **Epidermis**: cells tall and columnar to square on both surfaces, slightly rounded internally and externally. **Stomata** level with surface. **Mesophyll**: 2–3 irregular palisade layers; cells small, rounded to slightly elongated and packed with plastids. In *S. neoguineensis* and Croat 75532, palisade zone continuing across midvein as vertical ribs interrupting adaxial collenchyma band. Spongy cells in organized type 2 (Fig. 73E) to type 3 tissue forming tall air spaces (substomatal cavities) between abaxial epidermis and base of palisade layer, occupying *c.* 2/3 of mesophyll (Fig. 73B). **Air cavities**: up to *c.* 90% of spongy mesophyll. In midrib few to many irregularly spaced and sized air chambers (types 3 or 4). **Collenchyma**: continuous or discontinuous hypodermal band (Fig. 73A,C) in abaxial midrib, 3–8 layers deep, angularly thickened; gaps in collenchyma due to chlorenchyma ridges.

Vascular bundles: 15–30 randomly or symmetrically arranged bundles in midrib; very simple, unsheathed (types II, III). Xylem: 1–3 large or medium-sized protoxylem lacunae, 1–3 small metaxylem strands.

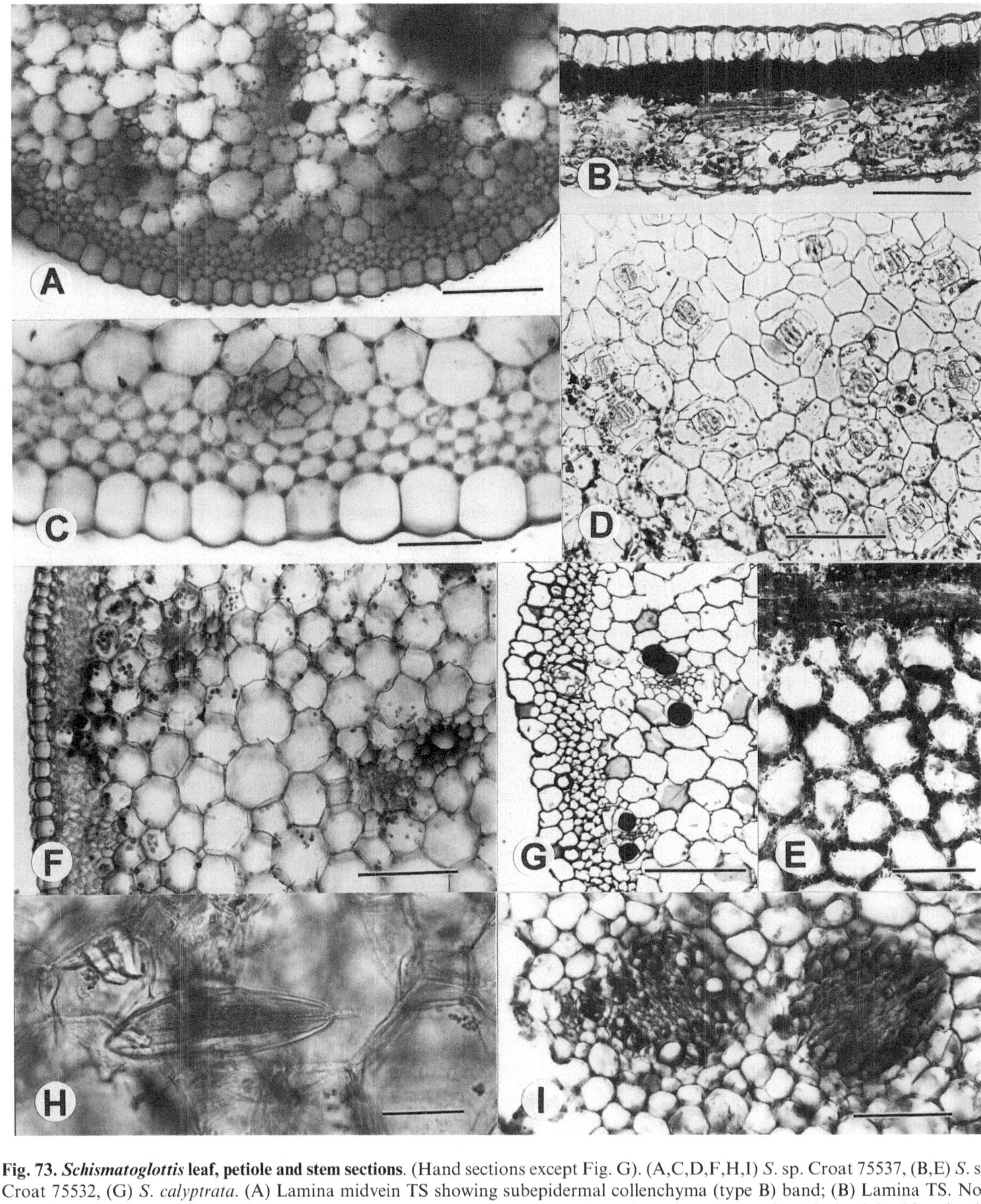

Fig. 73. *Schismatoglottis* leaf, petiole and stem sections. (Hand sections except Fig. G). (A,C,D,F,H,I) *S.* sp. Croat 75537, (B,E) *S.* sp. Croat 75532, (G) *S. calyptrata*. (A) Lamina midvein TS showing subepidermal collenchyma (type B) band; (B) Lamina TS. Note packed chloroplasts in palisade layer; (C) Midvein TS abaxial side. Note small duct in centre; (D) Lamina PS abaxial epidermis with stomata; (E) Lamina PS showing type 2 spongy aerenchyma; (F,G) Distal petiole TS showing collenchyma band and vascular bundles; (G) Note laticifer pairs (dark) lateral to vascular bundles; (H) Proximal petiole TS showing biforine raphide cell; (I) Stem TS with 2 amphivasal bundles. Scale bars: A,B,D,F,G,I = 150 μm; C,H = 50 μm; E = 200 μm.

Phloem: circular to laterally narrowed small strand. **Secretory ducts** small, with 1-layered epithelium, in peripheral ground tissue (Fig. 73C), common or rare. Ducts or idioblastic cells producing amorphous, colourless mucilaginous material. **Laticifers** non-anastomosing, in pairs, common, lateral to phloem strands along larger vascular bundles. **Tannin** cells common in ground tissue or vascular bundles. **Starch** granules present in some ground tissue cells and around some vascular bundles. **Crystals**: raphides common or occasional in double-ended biforines, one or both ends suspended in air cavities, randomly distributed. Also raphides in thin-walled irregular cells with very wide bundles. Raphides abundantly and widely spread during sectioning. Druses abundant throughout ground tissue and mesophyll in otherwise unspecialized cells, but often in small circular cells protruding into air cavities. In midribs, druses especially common in cells bounding air cavities. Crystal sand shaped as tiny raphides or rounded crystals, abundant or uncommon in almost all remaining cells, especially abundant in spongy tissue.

Petiole TS

Outline broadly sheathing at base, edges tapering to lamina-like margins. Distally: shape nearly terete, slightly flattened adaxially. **Size**: diameter 5–8 mm. **Cuticle** thin and striate or smooth. **Epidermis**: cells large and square or irregularly shaped. **Collenchyma**: subhypodermal or subepidermal abaxial band; or distally, beneath all epidermal layers (Fig. 73F,G). Cells angularly thickened, up to five layers deep, surrounding ground tissue of larger cells. Band may be interrupted laterally or adaxially by ridges of chlorenchyma. **Chlorenchyma** with sparse chloroplasts, in several layers internal to collenchyma layer, or as discrete zones interrupting collenchyma band. **Ground tissue**: cells large, mostly compact except for central zones of type 3 or 4 aerenchyma. **Air cavities** absent, or present in central ground tissue; present in adaxial half of basal section in Croat 75537. Cavities large or small with 1–2-seriate partitions.

Vascular bundles numerous and scattered (Fig. 73F,G) but variable in shape (type II). Xylem elements 1–6, usually protoxylem, opposing a small circular or laterally narrowed phloem strand. **Secretory ducts** absent or small, with 2–3 layers of epithelium, common in collenchyma layer or in outer ground tissue; exudate absent or, when present, mostly associated with large xylem elements. **Laticifers** absent, or 1–3 large and non-anastomosing tubes lateral to all vascular bundles (Fig. 73G). **Tannin** cells absent, or cells with dark contents common in epidermis or subepidermal layers. **Starch** granules large, sparse and scattered throughout ground tissue. **Crystals**: raphides, rare or common, in cells resembling ground tissue, randomly arranged, with single axially aligned crystal bundle, or cells large with abundant crystals in one wide bundle. Biforines occasional in Croat 75537 only (Fig. 73H), both ends suspended in adjacent type 3 air cavities. Druses common in proximal petiole sections, sharp-pointed, adjacent to air cavities or randomly distributed in ground tissue and near epidermis. In some species, druse cells often in sets of 2–3 adjacent cells. Crystal sand often present as small rods, common in ground tissue.

Stem TS and LS

Outline circular. **Size**: diameter 2 cm. **Cuticle** thin and smooth. **Epidermis**: cells small or large and irregular. **Collenchyma** not present, or incomplete subepidermal band, with 3–4 layers of angularly thickened cells. **Ground tissue** not differentiated into cortex and central cylinder; cells with thickened primary walls. **Air cavities**: loose undifferentiated spaces among some cells. **Vascular bundles**: outer bundles collateral, very small, with one narrow protoxylem or metaxylem element and small phloem strand. Central bundles compound or amphivasal (Fig. 73I) and possibly comprising 3–5 poles, but often with completely circular band of large metaxylem elements surrounding large homogeneous phloem strand. Sieve-cell end walls slightly oblique to transverse; cell length 150–200 μm. Outermost amphivasal bundles occasionally incompletely encircled by endodermis development. Casparian strips obscure and incompletely formed.

Sclerenchyma: occasional cells with lignified walls, forming apparent irregular sclereids. **Secretory ducts** not present, or cavities common and storing large quantities of mucilage within ground tissue without specialized epithelium. **Laticifers** not seen, or occasional and non-anastomosing. **Tannin** cells occasional in ground tissue or phloem. **Starch** granules packed into central cells of ground tissue or in cylinder 2–3 mm deep, dividing starch-free cortex from large starch-free central zone. Granules large and elongated, bulbous at one end; long axis oriented parallel to cross zone. **Crystals**: raphides uncommon to very common, in large cells with single very wide bundle, scattering crystals widely when slightly disturbed. Biforines occasional. Druses very common throughout, especially in small circular cells in outer ground tissue. **Adventitious roots**, when present, arising from no discernible pericycle or endodermis.

Material examined

Schismatoglottis calyptrata Zoll. & Morr., 75p220, 772049, MO; *S. concinna* Schott var. *immaculata* N.E.Br.

605/75, cult. M; *S. neoguineensis* N.E.Br., Bogor 17,6; *S.* sp., Croat 53521, Thailand; *S.* sp., Croat 75532, Malaysia; *S.* sp., Croat 75537, Malaysia.

Taxonomic notes

The six species examined here have a coherent core of characters that define the genus anatomically. However, many features vary significantly, making the genus an excellent subject for detailed study, especially given its disjunct distribution.

Species reported on in the literature

Chrysler (1904): *S. roebelinii*: stem ontogeny. • Dalitzsch (1886): *S. picta*: leaf. • Engler and Krause (1912): *S. ruprestris, S. lancifolia, S. lavalei*: leaf, root, secretory tissue. • French (1987c): several spp.: root. • French (1988): several spp.: laticifers. • French and Tomlinson (1981b): several spp.: stem vasculature. • Hassack (1886): *S. lavallei, S. picturata*: leaf. • Hotta (1971): leaf. • Janse (1897): root. • Lierau (1888): several spp.: root. • Müller (1919): *S. neoguineensis, S. rebelinum, S.* sp.: leaf tip. • Pant and Kidwai (1966): *S. wallichi*: leaf, epidermis. • Solereder and Meyer (1928): several spp.: all organs. • Tieghem (1872): secretory tissue. • Troll and Meyer (1955): *S. pulchra*: leaf ontogeny. • Webber (1960): *S. novoguineensis*: epidermis, stomata.

Additional literature

Blanc 1978; French 1985b, 1986b, 1987a; Grayum 1990; Hegnauer 1986; Kraus 1896; Mayo 1986; Mookerjea 1955; Thanikaimoni 1969; Williams *et al.* 1981.

Aridarum Ridl.

This genus of seven species occurs in northern Borneo. Plants are terrestrial or epilithic, often rheophytic herbs with short creeping rhizomes, occurring on wet rocks in or near shallow streams in shady forested areas. The leaf is long-petiolate with an elliptic lamina. Chromosomes: 2n = 24. One species was available for study.

Leaf surface

Cuticle smooth or rough adaxially and striate-wrinkled or reticulate abaxially. **Epidermis**: cells polygonal. **Stomata** abaxial, brachypara-hexacytic or amphi-brachypara-hexacytic (Fig. 74C), with up to 15 subsidiary cells, six in outer or third tier. **Venation**: midrib prominent with numerous closely-spaced, steeply-ascending secondaries and parallel tertiaries. Higher orders acute to orthogonal.

Leaf TS

Structure dorsiventral and well-differentiated. **Size**: 14 × 6 cm (*l*/*w*). **Midrib**: flat adaxially and rounded-convex abaxially (Fig. 74A). **Cuticle** thick on both surfaces. **Epidermis**: cells columnar adaxially (Fig. 74B), and 2–3 times as large as abaxial cells. **Hypodermis**: locally developed adaxially over larger veins (Fig. 74A,E). **Stomata** level with epidermal surface; tall substomatal air cavities (Figs 5C, 74B). **Mesophyll**: 1–2 palisade layers of short cells with densely concentrated chloroplasts. Spongy cells in *c.* 7 layers forming reticulate aerenchyma with large hexagonal air cavities (type 3) (Fig. 74B,D); corner cells larger; uniseriate partitions 2–3 cells long. **Collenchyma**: *c.* 2-layered subepidermal band (type B) in abaxial midrib.

Vascular bundles few and symmetrically scattered in midrib (Fig. 74A,E). Bundles type II with modified ground tissue sheath cells. Xylem: 2–3 protoxylem strands, all having spiral or annular thickenings. Phloem: circular or irregular strand. Sieve elements with horizontal or slightly oblique end walls. **Sclerenchyma**: fibre caps, 2–3 cells deep over phloem, only slightly lignified (Fig. 74E). **Laticifers** along vein sheaths, non-anastomosing, with dark amorphous contents (Fig. 74D). **Starch**: amyloplasts with large starch granules very common in spongy cells. **Crystals**: raphides as fine crystals in single bundles in short cells. One biforine seen (Fig. 74D). Crystal sand very fine, common in spongy layer.

Petiole TS

Outline subterete. **Size**: diameter 2 × 3 mm. **Cuticle** thick and smooth. **Epidermis**: cells small and flat externally. **Collenchyma**: continuous peripheral band 3–5 layers deep (Fig. 74F,G) separated from epidermis by 1–2 layers of ground tissue. **Chlorenchyma**: 3–4 layers inside collenchyma. **Ground tissue**: cells uniformly large. **Air cavities** numerous, 1–2 times cell size and separated by type 3 partitions. **Vascular bundles** numerous and scattered (Fig. 74F); bundles small and narrow (type III) (Fig. 74H). Xylem: wide protoxylem cell and 2–4 narrow metaxylem cells forming irregular boundary with small, irregular phloem strand. **Secretory tissue**: some air cavities filled with mucilaginous material. **Starch** granules common in ground tissue. **Crystals**: raphides in thin-walled cells common in ground tissue, especially near periphery.

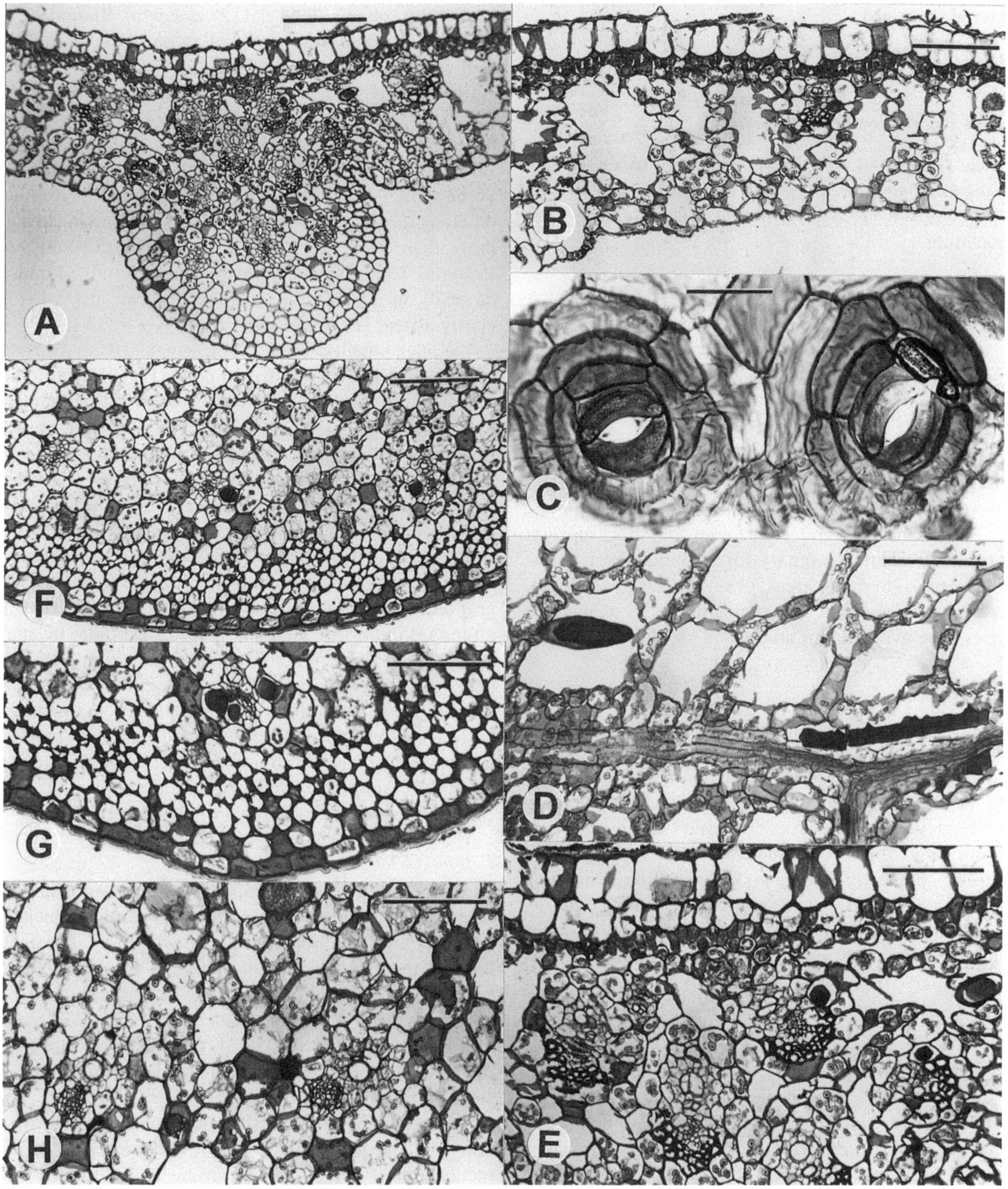

Fig. 74. ***Aridarum nicolsonii*** **leaf and petiole sections.** (A) Lamina TS midrib; (B) Lamina TS; (C) Lamina abaxial epidermis PS showing amphibrachypara-hexacytic stomata; (D) Mesophyll PS showing type 3 aerenchyma, biforine cell, and laticifer at vein; (E) Adaxial midrib TS showing small vascular bundles with fibre caps. (F,G) Proximal petiole TS showing abaxial collenchyma band and compact ground tissue; (H) Petiole central tissue TS showing type III vascular bundles. Scale bars: A = 300 μm; B,F = 200 μm; C = 50 μm; D,E,G,H = 150 μm.

Material examined

Aridarum nicolsonii Bogner, Bogner 1421, Sarawak, M.

Species reported on in the literature

French (1987c): *A. nicolsonii:* root. • French (1988): *A. annae, A. burtii, A. nicolsonii:* leaf, laticifers. • French and Tomlinson (1981b): *A. annae, A. nicolsonii*: stem vasculature.

Additional literature

French 1985b, 1986c, 1987a; Grayum 1990; Hotta 1965, 1971; Thanikaimoni 1969.

Piptospatha N.E.Br.

This genus of 10 species is distributed in Borneo and western Malaysia. Plants are small epilithic or rheophytic herbs with short creeping rhizomes or short upright stems. They occur as primary forest understorey plants along streams, in damp areas, or on well-drained slopes. The leaves are elliptic. Chromosomes: 2n = 26. Two species were available for study.

Leaf surface

Cuticle with striations perpendicular across cellular boundaries, or rugulate (*P. ridleyi*) or smooth (*P. elongata*). **Epidermis**: cells polygonal and of varying sizes (1–2:1 *l*/*w*). **Stomata** mostly abaxial and brachy-para-tetracytic to -hexacytic. Subsidiary cells narrow, 2–4 pairs (Fig. 75D). **Venation**: midrib prominent. Secondaries narrowly-acute and ascending, merging eucamptodromously at margins. Tertiaries closely spaced and parallel with secondaries. Quaternaries extending irregularly or orthogonally between secondaries and tertiaries.

Leaf TS

Structure dorsiventral and strongly differentiated. **Size**: 72 × 26 mm (*l*/*w*). **Midrib** flattened-convex adaxially, rounded-convex abaxially with lateral grooves delimiting the lamina (*P. ridleyi*) (Fig. 75A) or flattened-convex abaxially (*P. elongata*). **Cuticle** medium-thick to thick; on abaxial midrib, striate in tall irregular pattern (*P. ridleyi*) or very thin to not seen (*P. elongata*). **Epidermis**: adaxial cells very large, flat on surface and slightly rounded internally (*P. ridleyi*) or rounded both internally and externally (*P. elongata*). Abaxial cells large and irregularly shaped, rounded on surface and internally (*P. ridleyi*), or small and flat (*P. elongata*). **Stomata** level with surface. Subsidiary cells often narrower than adjacent epidermal cells (Fig. 75C). Guard cells with double cuticular ledges. **Mesophyll**: 1–2 loosely packed palisade layers, up to 25% of lamina (Fig. 75B). Spongy cell layers 6–10, cells mostly rounded, forming 1-seriate partitions surrounding small, tall substomatal air cavities (type 2). Some spongy tissue organization irregular, cells rounded or elongate and shallowly lobed (type 1). **Air cavities**: more than 50% of spongy layer, mostly tall and substomatal. **Collenchyma** in subepidermal bands in midrib (type B), up to three layers adaxially, two abaxially, or nearly absent (*P. elongata*).

Vascular bundles: *c*. 12 small, irregularly shaped and symmetrically placed midrib bundles (Fig. 75A). Bundles small (type II). Xylem: 2–7 narrow metaxylem cells, opposing an irregularly-shaped phloem strand. **Sclerenchyma**: thin-walled fibres, present as 1–2-layered phloem caps at larger vascular bundles, absent in *P. elongata*. **Laticifers** short-branched, non-anastomosing and robust, parallel to and on both sides of all phloem strands (Fig. 75F); cell length up to 270 μm. **Tannin** cells large and very common in palisade layer, common in spongy layer. **Starch** granules small and very common in mesophyll; large and abundant in midrib. **Crystals**: biforine raphides in single bundles in elongate, round-ended or papillate cells (Fig. 75E), or raphides in rounded or oval thin-walled cells (Fig. 75F). Raphides common, mostly in upper and mid-mesophyll. Druses small, occasional in palisade layer and near vascular bundles; larger druses occasional in midrib and abaxial mesophyll. Crystal sand very common in spongy mesophyll of *P. elongata*.

Petiole TS

Outline: subterete and shallowly V-grooved adaxially (Fig. 75H). **Size**: diameter 2–3 mm. **Cuticle** thick or irregular and smooth to striate. **Epidermis**: cells irregularly shaped and mostly flattened. **Collenchyma**: continuous band, 5–6 layers thick and separated from epidermis by 2–3 ground tissue layers (Fig. 75H); cells small and angularly thickened. **Ground tissue**: cells uniformly large, bounded by collenchyma. **Air cavities** cell-sized and numerous, up to 30% of ground tissue.

Vascular bundles: *c*. 17 scattered bundles, unsheathed and small (type II) (Fig. 75G). Xylem: 2–10 narrow metaxylem cells in 1–2 files; strand laterally narrowed. Phloem: narrow irregular strand. **Secretory ducts** occasional in central ground tissue, large and circular with 1-layered epithelium (Fig. 75H) and containing mucilaginous material. **Laticifers** non-anastomosing and of large diameter, in pairs, lateral to phloem strands (Fig. 75G). **Tannin** cells numerous throughout ground tissue. **Starch**

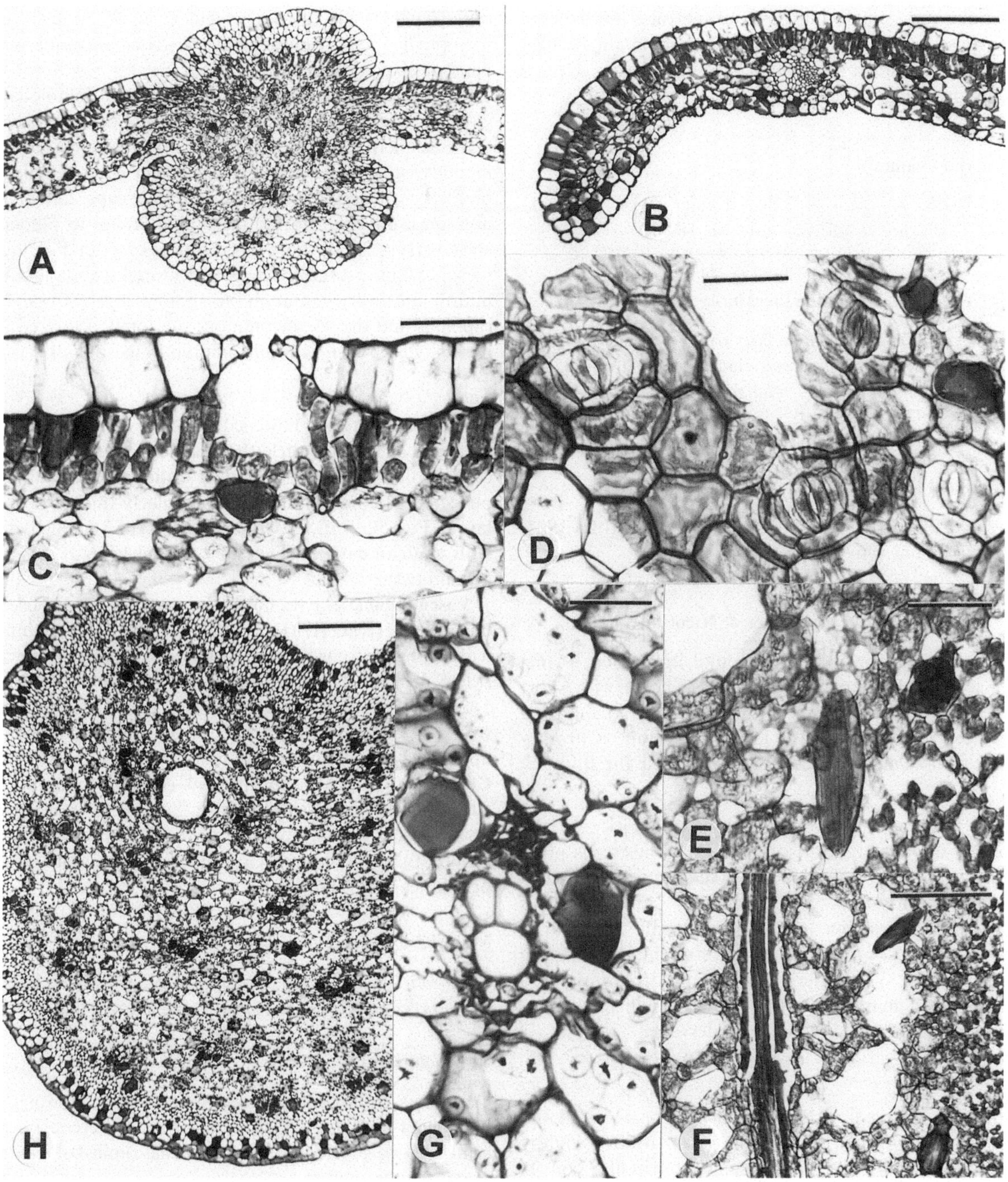

Fig. 75. *Piptospatha ridleyi* leaf and petiole sections. (A) Midrib TS; (B) Lamina and margin TS; (C) Lamina TS, adaxial stoma; (D) Abaxial epidermis PS showing brachypara-hexacytic stomata; (E) Mesophyll PS showing biforine-like raphide; (F) Mesophyll PS showing non-anastomosing laticifers at vascular bundle and two raphide cells; (G) Distal petiole TS, vascular bundle with two large laticifers with dark contents; (H) Proximal petiole TS. Note continuous collenchyma band at periphery and large secretory duct. Scale bars: A,G,H = 300 μm; B = 200 μm; F = 150 μm; C,D,E = 50 μm.

granules very common throughout central ground tissue. **Crystals**: raphides in wide bundles and in thin-walled cells, common in peripheral ground tissue. Druses common near collenchyma.

Material examined

Piptospatha elongata N.E.Br., Bogner 1478, Sarawak, cult. M; *P. ridleyi* N.E.Br., cult. M, 182/70.

Species reported on in the literature

French (1987c): *P. grabowskii*: root. • French (1988): *P. grabowskii, P. perakensis*: laticifers. • French and Tomlinson (1981b): *P. elongata*: stem vasculature. • Hotta (1971): leaf.

Additional literature

French 1985b, 1986b, 1987a; Grayum 1990; Hotta 1965; Thanikaimoni 1969.

Hottarum Bogner & Nicolson

This genus of four species is found in Borneo. Plants are terrestrial or epilithic, rheophytic herbs with short creeping rhizomes. They grow in and along streams and on steep mossy banks. The leaves are elliptic. Chromosomes: 2n = 26. Two species were available for study.

Leaf surface

Cuticle smooth on both surfaces. **Epidermis**: cells polygonal (1–4:1 *l*/*w*). **Stomata** abaxial and brachyparahexacytic with 2–3 pairs of narrow subsidiary cells (Fig. 76C). **Venation**: midrib prominent. Secondaries closely spaced, narrowly-acute and straight, then curving toward apex. Veins merging eucamptodromously into straightened marginal collector. Few small cross connections noted, the only higher orders seen.

Leaf TS

Structure dorsiventral. **Size**: *H. sarikeense*: 118 × 17 mm (*l*/*w*) and *H. lucens*: 79 × 12 mm (*l*/*w*). **Midrib** slightly convex adaxially and with a small semicircular convex rib abaxially (Fig. 76A). **Cuticle** thick on both surfaces, or slightly grooved abaxially. **Epidermis**: cells large and square adaxially, smaller and flat to rounded abaxially. **Stomata** level, with no obvious subsidiary cells. **Mesophyll**: palisade 1-layered comprising *c.* 30% of lamina (Fig. 76B). Spongy tissue: 6–7 layers of irregularly armed cells (type 1). **Air cavities** irregular, up to 30–50% of mesophyll area. **Collenchyma**: narrow subepidermal band at abaxial midrib (type B).

Vascular bundles: three in midrib, central bundle largest (Fig. 76A). Bundles (type II) with 3–4 tracheary cells and a semicircular phloem strand. **Sclerenchyma**: fibres in midvein arranged as broad phloem caps. Laminar bundles with large diameter abaxial fibre caps. **Laticifers** non-anastomosing, in mid-mesophyll lateral to phloem strands (Fig. 76D); short lobed processes extending into mesophyll interstitial space. **Starch** granules abundant in midrib and very common in spongy layer. **Crystals**: raphides and druses uncommon, randomly scattered in mesophyll. Crystal sand uncommon in mesophyll.

Petiole TS

Outline subterete and flattened adaxially (Fig. 76E). **Size**: diameter 2–3 mm. **Cuticle** thick and smooth. **Epidermis**: cells medium-sized and rounded. **Collenchyma**: continuous subepidermal band of 3–8 angularly thickened cell layers (Figs 5G, 76E,F). **Ground tissue**: type 3 aerenchyma with small air cavities (Fig. 76F). **Air space** up to 80% of ground tissue.

Vascular bundles: *c.* 12 bundles symmetrically disposed; unsheathed (type II) (Fig. 76G). Xylem: 0–1 medium-width late protoxylem elements and 2–8 narrow metaxylem elements. Phloem: small circular strand. **Laticifers**: 1–2 lateral to each vascular bundle; cell diameter wide with dark contents. **Starch** granules large and spheroidal, up to 12 µm diameter, abundant in ground tissue. **Crystals**: raphides in thin-walled cells, uncommon at edge of ground tissue. Druses large and randomly arranged in ground tissue, common, especially near ground tissue periphery. Crystal sand: occasional small prismatics in ground tissue.

Material examined

Hottarum lucens Bogner, Bogner 1439, Sarawak, cult. M; *H. sarikeense* Bogner & Hotta, Bogner 1553, cult. M.

Taxonomic notes

The genus *Hottarum* has been merged with *Piptospatha* by Bogner and Hay (2000). Only trivial differences separate the two genera and there is no anatomical objection to combining them.

Species reported on in the literature

French (1987c): *H. sarikeense*: root. • French (1988): *H. lucens*: laticifers. • French and Tomlinson (1981b): *H. lucens*: stem vasculature.

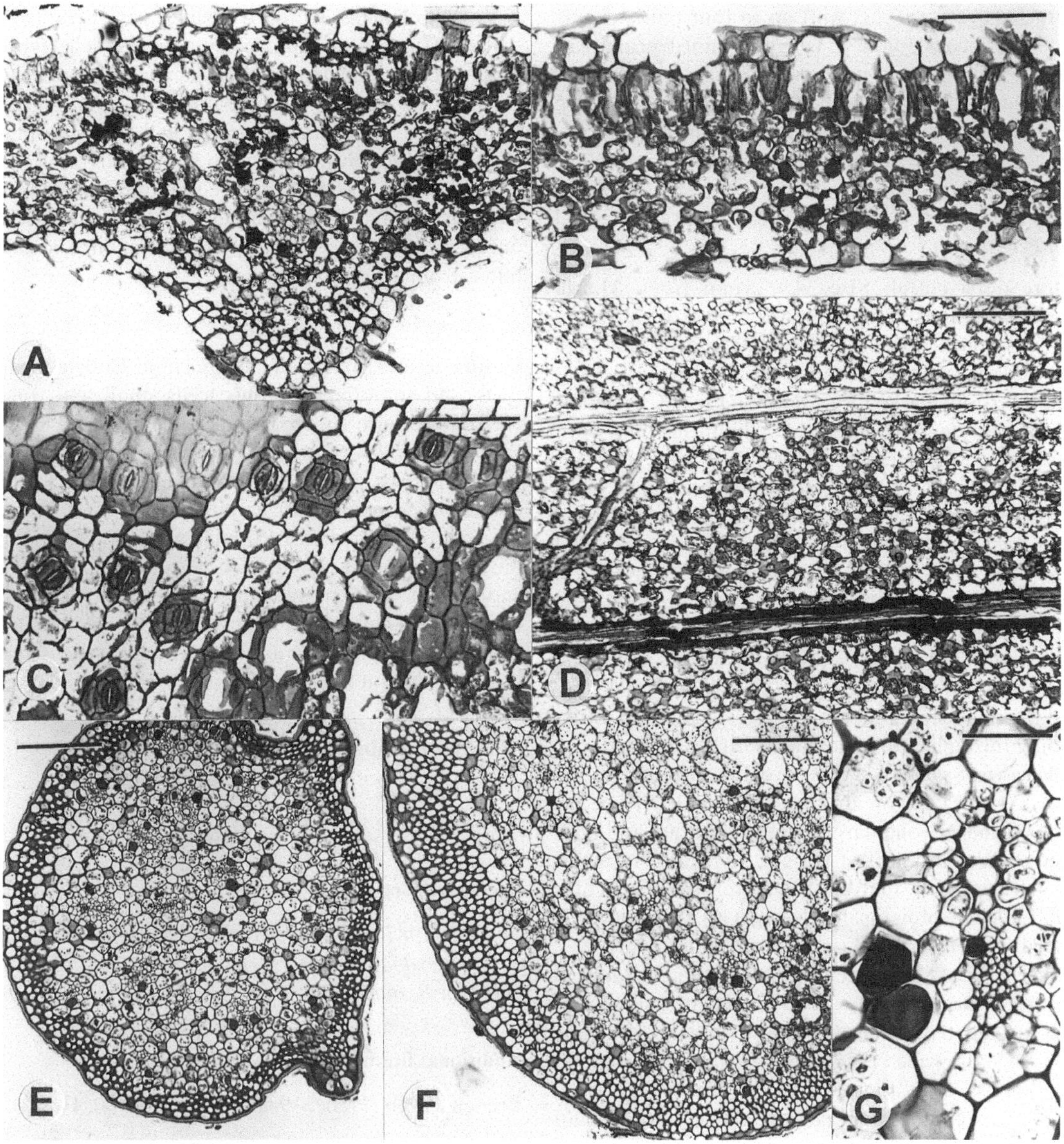

Fig. 76. *Hottarum lucens* leaf and petiole sections. (A) Midrib TS; (B) Lamina TS; (C) Abaxial epidermis PS with stomata; (D) Lamina PS with vascular bundles and non-anastomosing laticifers; (E) Distal petiole TS, adaxial surface on right; (F) Proximal petiole TS with deep collenchyma band beneath epidermis; (G) Proximal petiole TS with vascular bundle. Scale bars: A,D = 200 μm; B,C = 150 μm; E,F = 300 μm; G = 50 μm.

Additional literature

French 1985b, 1986c, 1987a; Grayum 1990.

Bucephalandra Schott

This genus, including *Microcasia* Becc., contains two species found in Borneo. Plants are small rheophytes growing in crevices in and along forest streams. The leaves are broadly or narrowly elliptic. Chromosomes: 2n = *c.* 26. One species was available for study.

Leaf surface

Cuticle markedly striate abaxially and smooth adaxially. **Epidermis**: cells polygonal (1–2:1 *l*/*w*). **Stomata** abaxial and brachypara-octocytic, with up to four pairs of narrow subsidiary cells (Fig. 4A). **Venation**: midrib prominent with narrow acute and steeply ascending secondaries and intersecondaries, becoming parallel, gradually merging with each other. Secondaries merging eucamptodromously in distal lamina. Tertiaries orthogonal to reticulate.

Leaf TS

Structure dorsiventral. **Size** 34 × 15 mm (*l*/*w*). **Midrib** shape unusual: adaxial side rounded-convex with lamina insertion sloping down toward abaxial side. Abaxial midrib rounded-convex, smaller than adaxial side and lamina insertion marked by furrows (Fig. 3C). **Cuticle** thick on margin and midrib, medium-thin on both lamina surfaces. **Epidermis**: adaxial cells large and columnar (Fig. 3C) or smaller; abaxial cells irregular, small or large. **Stomata** on slight mounds above epidermal surface. Subsidiary cells in up to three tiers of modified narrow cells (Fig. 77D). **Hypodermis**: 1–2 layers on adaxial side of midrib. **Mesophyll**: two columnar palisade layers, the second layer of shorter cells, up to *c*. 30% of lamina (Fig. 77C). Spongy layer of somewhat loosely packed, rounded cells (type 1a). **Air cavities** small and irregular or substomatal, or large, regular, and often extending through entire spongy layer (Fig. 77F).

Collenchyma: subepidermal, weakly formed broad strand, 2–3 cells deep on abaxial side of midrib (type B). **Vascular bundles**: one large central bundle and 2–3 smaller bundles in midrib (type II). Xylem: 8–10 metaxylem elements in midvein. Phloem: small circular strand. **Laticifers** non-anastomosing, with dark contents, at midrib and in pairs near smaller veins (Fig. 77F). **Tannin** cells with dark contents, common and random in mesophyll. **Starch** granules common in mesophyll and midrib ground tissue cells. **Crystals**: raphides in single bundles in short wide cells, in upper spongy mesophyll. Some raphide cells approaching biforine shape (Fig. 77E). Druses common in palisade layer and median midrib ground tissue; cells large, circular or narrow.

Petiole TS

Outline subterete and fluted, with eight ridges (Fig. 77A). **Size**: diameter 1–1.5 mm. **Cuticle** thick and striate. **Epidermis**: cells large, square, rounded externally. **Collenchyma**: circular or irregular strands, 2–3 cells deep, separated from epidermis by two ground tissue layers (Fig. 77B); strands tending to subtend small vascular bundles (type Sv). **Ground tissue**: cells large and compact in outer third of tissue, type 2 aerenchyma in centre. **Air cavities**: 70% of central aerenchyma.

Vascular bundles arranged in single peripheral row and scattered in centre; bundles weakly formed (type II). Xylem: one small protoxylem lacuna, 2–6 metaxylem elements of various sizes. Phloem: a circular strand. **Laticifers**: 2–3 non-anastomosing and lateral to phloem (Fig. 77B). **Tannin** cells few and randomly scattered in ground tissue. **Starch** granules common in ground tissue. **Crystals**: raphides uncommon, in thin-walled cells in outer ground tissue. Druses common in circular cells in outer ground tissue.

Stem TS

Outline terete. **Size**: diameter 3–4 mm. **Cuticle** medium-thick and smooth. **Epidermis**: cells small and rounded externally. **Cork**: subepidermal as cicatrice in vicinity of shrivelled bracts. **Ground tissue**: cells small in cortical region, largest in centre. **Vascular bundles** scattered, concentrated near centre; amphivasal (Fig. 77G). Xylem: ring surrounding central circular phloem strand. Outermost amphivasal bundles occasionally encircled by endodermis. Casparian strips obscure and incompletely formed. **Tannin** cells very common in ground tissue. **Starch** granules abundant in all ground tissue. **Crystals**: druses abundant in ground tissue, cells large and circular.

Material examined

Bucephalandra motleyana Schott, Bogner 1366, Sarawak, M.

Species reported on in the literature

French (1987c): *B. motleyana*: root. • French (1988): *B. motleyana*: laticifers. • French and Tomlinson (1981b): *B. motleyana*: stem.

Additional literature

French 1985b, 1986c, 1987a; Grayum 1990; Hotta 1971.

Heteroaridarum M. Hotta

This monotypic genus is found in Borneo. Plants are terrestrial or epilithic rheophytes, short-stemmed and rhizomatous, occurring as understorey herbs along streams. Leaves are long-petiolate and elliptic. No chromosome counts are available. One specimen was available for study.

Leaf surface

Cuticle finely striate adaxially and coarsely striate abaxially. **Epidermis**: cells polygonal (1–2:1 *l*/*w*). **Stomata**

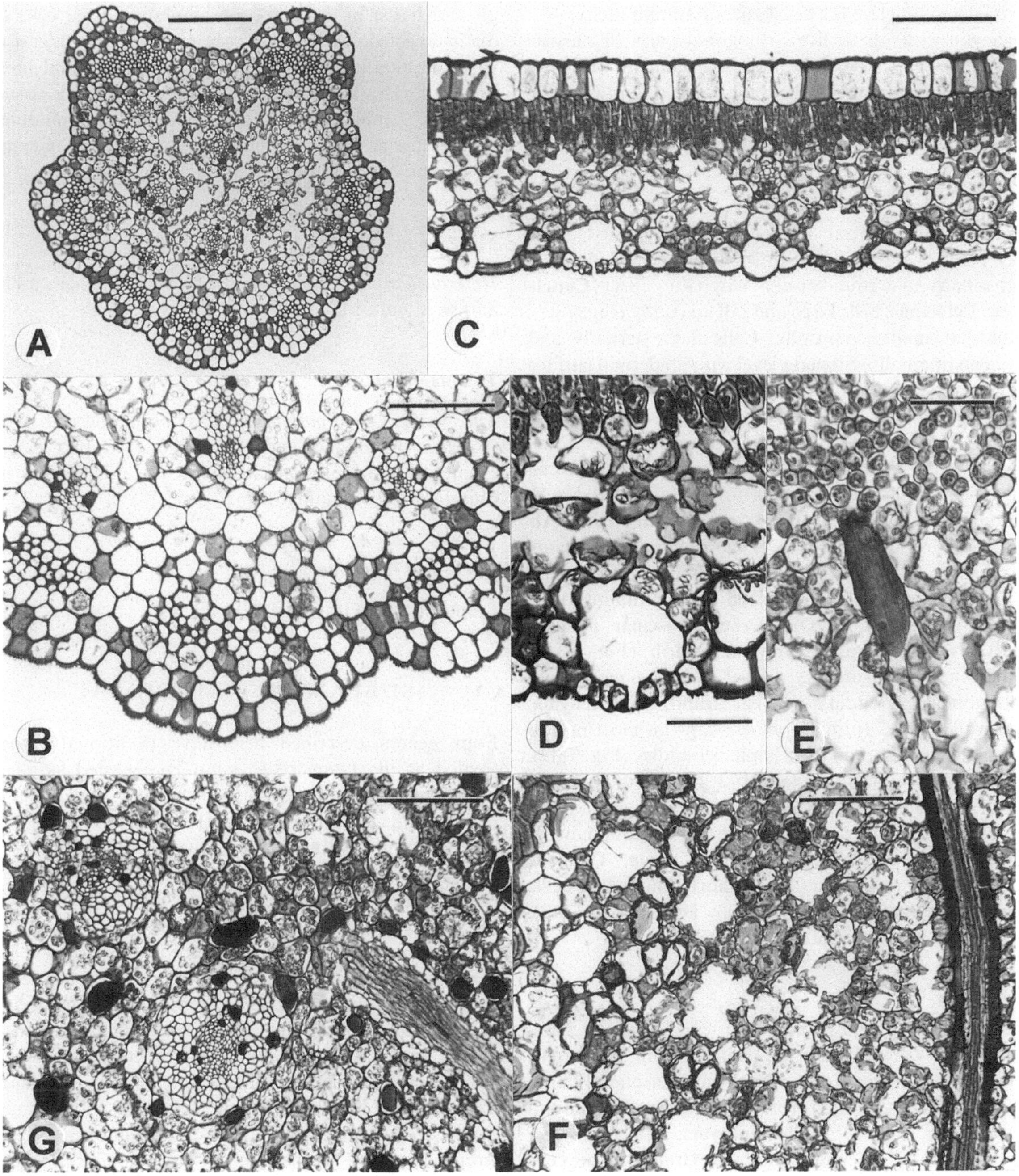

Fig. 77. *Bucephalandra motleyana* leaf, petiole and stem sections. (A) Petiole TS with ridged outline; (B) petiole TS, abaxial margin with ridges subtending irregular collenchyma strands; (C) Lamina TS showing tall adaxial epidermal cells and low substomatal air cavities; (D) Abaxial lamina TS showing guard cells surrounded by rows of small subsidiary cells; (E) Lamina PS showing raphide cell at boundary of palisade (above) and spongy tissue; (F) Lamina PS with ± circular substomatal air cavities and non-anastomosing laticifers at higher order vein; (G) Stem central cylinder TS with amphivasal bundles, starch-filled ground tissue, and tannin cells. Scale bars: A = 300 μm; B,C,F,G = 150 μm; D,E = 50 μm.

abaxial and brachypara-hexacytic, or amphibrachypara-hexacytic with up to five orthogonal pairs of narrow subsidiary cells. **Venation**: midrib prominent with secondaries arising acutely, ± evenly spaced, and merging at margin eucamptodromously. Tertiaries parallel with secondaries and connecting between secondaries.

Leaf TS

Structure dorsiventral and well differentiated. **Size** 16 × 7 cm (*l*/*w*). **Midrib** flat to slightly concave adaxially and semicircular-rounded abaxially (Fig. 78A). **Cuticle** thick. **Epidermis**: cells large and tall adaxially (Fig. 78C), somewhat smaller abaxially. Cells flat externally and rounded internally. **Stomata** level with epidermal surface and surrounded by 2–3 sets of distinguishable subsidiary cells. **Hypodermis**: one layer on adaxial side of midrib (Fig. 78A). **Mesophyll**: one layer of palisade (2:1–5:1 *l*/*w*). Spongy cells small and rounded, forming 8–10 layers, mostly as uniseriate partitions for well-organized substomatal aerenchyma (types 2 and 3) (Fig. 78B). **Air cavities** substomatal, up to 80%, dominating spongy layer.

Collenchyma weakly developed, discontinuous band beneath abaxial midrib epidermis. **Vascular bundles**: 9–10 arranged symmetrically in midrib (Fig. 78A). Bundles weakly formed (type II). Xylem: 1–6 metaxylem cells opposing a circular phloem strand. **Sclerenchyma**: thin-walled fibres forming phloem caps on most midrib vascular bundles and larger laminar bundles (Fig. 78D). **Laticifers** non-anastomosing, 1–2 lateral to vascular bundles. **Tannin** cells occasional in mesophyll and epidermis. **Starch**: large elongate granules abundant in spongy layers and midrib ground tissue. **Crystals**: raphides fine, single bundles in short thin-walled cells, common in palisade and spongy layers (Fig. 78B). Druses small, occasional in palisade cells.

Petiole TS

Outline subterete and shallowly flattened adaxially. **Size**: diameter 3–4 mm. **Cuticle** thick and smooth. **Epidermis**: cells small and irregular. **Collenchyma**: peripheral band, 2–3 cells deep, separated from epidermis by 1–2 ground tissue layers (Fig. 78E). **Chlorenchyma**: numerous plastids in up to six subepidermal layers. **Ground tissue**: cells large beneath collenchyma. **Air cavities**: 2–3 large cavities in centre; numerous small type 3 (Fig. 78G) or type 4 air cavities (up to twice cell diameter).

Vascular bundles numerous and scattered (type II) (Fig. 78F). Xylem: one protoxylem lacuna and 1–3 metaxylem cells in single file. Phloem: very small irregular strand. **Secretory tissue**: mucilage common in central ground tissue in distal sections; no specialized cells or ducts detected. **Laticifers** occasional, 1–2 cells lateral to vascular bundles. **Starch** granules numerous and large (up to >15 μm diameter) in central ground tissue. **Crystals**: raphides in single bundles in unmodified ground tissue cells, common in ground tissue. Crystal sand common in ground tissue.

Material examined

Heteroaridarum borneense M.Hotta, Bogner 1400, Sarawak, cult. M.

Taxonomic notes

The genus *Heteroaridarum* has been merged with *Aridarum* by Bogner and Hay (2000). The two genera essentially share anatomical characters and there is no objection to combining them.

Additional literature

French 1985b; Grayum 1990.

SUBFAMILY LEMNOIDEAE

Four genera described here have traditionally been treated as the Lemnaceae, a family regarded as allied with Araceae. However, recent evidence suggests that they belong within the Araceae (French *et al.* 1995). The species are mostly tiny floating aquatic plants common to relatively still fresh waters worldwide. In order of reduced size and complexity the genera are *Spirodela* (4–5 species); *Lemna* (13 species); *Wolffia* (11 species); and *Wolffiella* (11 species). The recently described genus *Landoltia* D.H. Les & D.J.Crawford contains one species previously assigned to *Lemna* or *Spirodela*. Thalloid in appearance, plants of all genera are small, circular to elliptic leaf-like organs commonly called fronds. Short stipes connect groups of ontogenetically related fronds. Vegetatively, daughter fronds originate within small cavities called pouches in the basal section of the parent frond. At least in the central USA, well-established colonies of lemnoids often consist of at least three genera and several species. Chromosomes have been counted mostly as 2n = 40 and 42, with other counts as 2n = 20, 30, 36, 44, 50, 60, 63, 64, 70, 80, 84, 126. Eight specimens representing four genera were available for study, with information added from Landolt (1998), which should be consulted for thorough review.

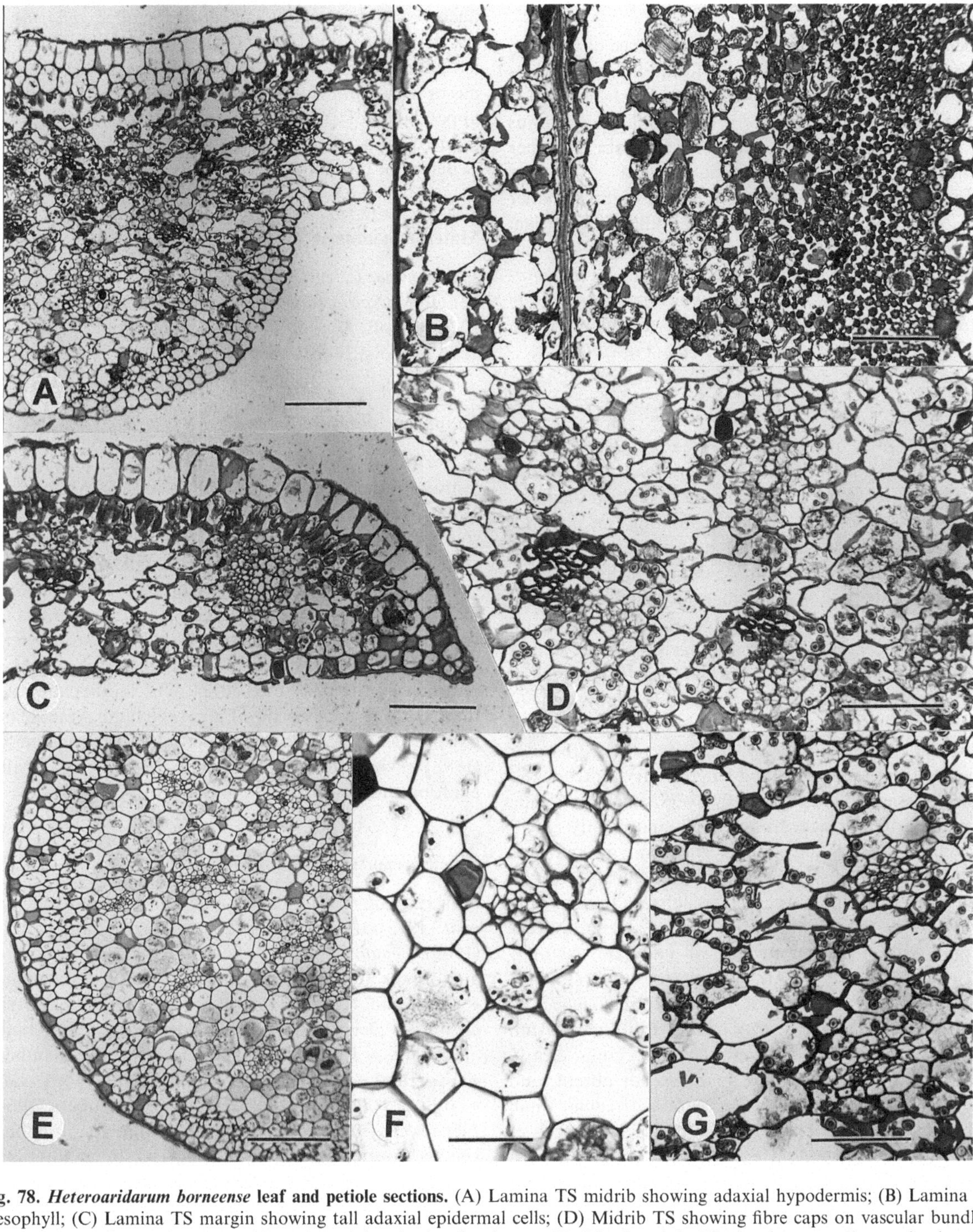

Fig. 78. *Heteroaridarum borneense* leaf and petiole sections. (A) Lamina TS midrib showing adaxial hypodermis; (B) Lamina PS mesophyll; (C) Lamina TS margin showing tall adaxial epidermal cells; (D) Midrib TS showing fibre caps on vascular bundles; (E) Distal petiole TS showing subhypodermal band of collenchyma; (F) Distal petiole TS showing type II vascular bundle; (G) Mid-petiole TS showing vasculature and aerenchyma. Scale bars: A,E = 300 μm; B,C = 200 μm; D,G = 150 μm; F = 50 μm.

Frond surface

Cuticle present and smooth on upper surface of floating, or absent in submerged fronds as found in some species of *Lemna* and *Wolffiella*. **Epidermis**: cells straight-sided polygonal in *Wolffia* and *Wolffiella*, undulate or sinuous in *Lemna* (Fig. 79F) and *Spirodela*. **Stomata** on upper surface, anomocytic or tending toward actinocytic having 4–7 (9) neighbouring cells, often slightly smaller than other epidermal cells and unmodified. **Venation**: acrodromous primary veins present in *Spirodela* and *Lemna*, the number species-specific: 5–16(–21) in *Spirodela* (Fig. 79A), 1–3(–4–5) in *Lemna* converging distally from point dividing distal and basal sections of frond. Veins absent in *Wolffia* and *Wolffiella* (Fig. 79G).

Frond TS

Structure: dorsiventrality defined clearly by aerenchyma structure but poorly when that is absent. **Midrib** and other primaries, when present, immersed within lamina. **Cuticle** thin on adaxial surface of floating but not submerged fronds. **Epidermis**: cells flat or tile-like. **Papules** or teeth, consisting of several epidermal cells, forming pointed protuberances, along margin of some species of *Lemna* and *Wolffiella*. Papilla-like cells occurring on adaxial surface, mostly along veins of *Spirodela* and *Lemna*. **Stomata** level with surface and functional in *Spirodela*, *Wolffiella* and *Wolffia*, and flowering cells of *L. trisulca*. Guard cells dead, lacking cytoplasmic structures, and always open in *Lemna*. Guard cells with outer cuticular ledges only; opening lengths varying from 18–34 µm. **Mesophyll**: one to >10 layers of parenchyma. Compact ground tissue around node and mostly type 3 aerenchyma through remainder of frond in *Spirodela* (Fig. 79B,C), *Lemna* (Fig. 79E) and in some species of other genera. In other species, aerenchyma present only around node. Variations are species-specific. Aerenchyma partitions often formed of undulate or interlocking cells (Fig. 79D). **Air cavities** very tall, up to 95% of frond; absent in turions of *Spirodela* and *Lemna*.

Vascular bundles small, with irregular outline. Xylem: tracheids few with spiral or annular thickenings; clearly evident in *Spirodela*, obscure in *Lemna* or absent, and absent in *Wolffia* and *Wolffiella*. Phloem: irregular small strand abaxial to xylem. **Tannin** cells (pigment cells of Landolt 1998) common throughout mesophyll (Fig. 79I). **Starch**: discoid granules, length up to 10 µm, throughout mesophyll tissues in *Lemna* fronds examined, or in proximal frond tissue in *Wolffiella* (Fig. 79H). **Crystals**: raphides present in elongated enlarged cells (Fig. 79F), single or overlapping bundles in *Lemna* and *Spirodela*, cells occasionally articulated with other raphide cells. Distributed throughout in *Spirodela* and *Lemna;* up to 4% dry weight of plant in *Lemna*, crystals usually elongated parallel to margin and veins. Druses present in small mesophyll parenchyma cells in *Spirodela*, mostly in upper layer of frond, or near nodes or veins. Druses isolated and occasional anywhere in mesophyll in *Lemna;* not present in turions.

Material examined

Lemna minor L., cult. M; Keating 2276, Illinois; *L. trisulca* L. cult. M; *Spirodela polyrrhiza* Schleid., cult. M; Keating 2276, Illinois; *Wolffia arrhiza* Wimm., cult. M; Keating 2276, Illinois; *Wolffiella* sp., cult M.

Taxonomic notes

In terms of anatomical complexity, the four genera examined here form a reduction series: *Spirodela* ⇒ *Lemna* ⇒ *Wolffiella* ⇒ *Wolffia* (cf. Les *et al.* 1997). In spite of their greatly reduced plant body, the genera are clearly similar to other Araceae. The cpDNA restriction site evidence developed by French *et al.* (1995) makes it impossible to regard these genera as a family, unless the Araceae themselves are to be regarded as paraphyletic. Thorne (2000) holds that their family status should be affirmed on the basis of their extreme modification and that resulting recognition of a paraphyletic Araceae is a reasonable cost. Subramanyam (1962) and Landolt (1998) review theories of family evolution and other anatomical literature.

Species reported on in the literature

Al-Rais *et al.* (1971): *Lemna minor:* crystals. • Anderson *et al.* (1973): *Wolffia*: fine structure. • Arnott and Pautard (1965): *Lemna*: raphides. • Bernard *et al.* (1990): *Wolffia:* anatomy. • Clark (1979): family: stomata, aerenchyma. • Cody and Horner (1983): *Lemna* sp., *Spirodela* sp.: crystals. • Den Hartog and Van der Plas (1970): family: all organs. • Efrat *et al.* (1977): *Lemna:* air chambers. • Fagerlind and Massalski (1974): *Lemna:* cell walls. • Gulliver (1863–5): 4 spp.: crystals. • Guttenberg (1968): *Lemna* (2), *Spirodela polyrrhiza, Wolffia* sp., *Wolffiella* sp.: root ontogeny. • Haccius and Lakshmanan (1966): *Lemna gibba*: shoot ontogeny. • Kirchner *et al.* (1934): *Lemna* (3), *Spirodela polyrrhiza, Wolffia arrhiza, Wolffiella* sp. • Korschgen (1981): *Lemna:* leaf. • Landolt (1998): family: all organs. • Ledbetter and Porter (1970): *Spirodela*: raphides. • Lemon and Posluszny (2000b): *Lemna minor, Spirodela polyrrhiza, Wolffia borealis*: frond and

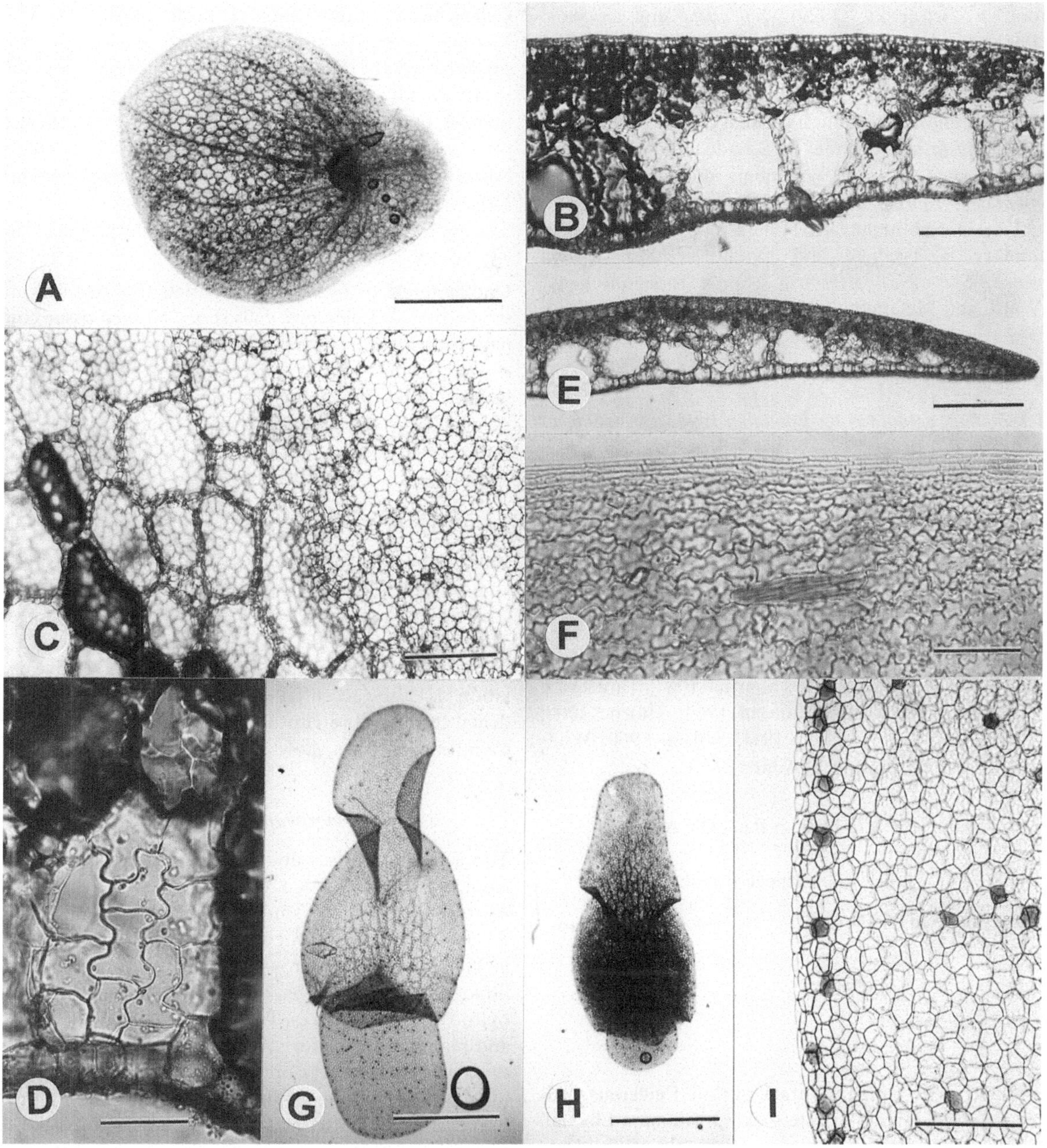

Fig. 79. Lemnoideae frond wet whole mounts and hand sections. (A–D) *Spirodela polyrrhiza*; (E,F) *Lemna minor*; (G–I) *Wolffiella* sp.; (A) Frond whole mount, ventral side; eight vascular bundles visible; (B) Frond TS with type 3 aerenchyma; (C) Frond PS showing aerenchyma partitions and polygonal subepidermal cells (right). Dark areas are air-filled cavities; (D) Frond TS showing undulate cells of aerenchyma partition; (E) Frond TS showing type 3 aerenchyma; (F) Frond margin, adaxial epidermal cells and raphide cell (centre); (G) Frond whole mount. Note type 3 aerenchyma in centre; (H) Frond whole mount showing IKI test for starch (dark area); (I) Frond whole mount showing lateral distal margin with scattered pigmented cells. Scale bars: A,G,H = 1500 μm; C,E = 200 μm; B,I = 150 μm; D,F = 50 μm.

shoot. • Netolitzky (1929): *Spirodela* sp.: crystals. • Pankey *et al.* (1965): *Spirodela:* starch. • Pant and Kidwai (1966): *Lemna minor, Spirodela polyrrhiza, Wolffia microscopica*: leaf. • Sakai and Hayashi (1973): *Lemna* sp.: chloroplast starch. • Sakai *et al.* (1984): *Lemna* sp., *Spirodela* sp.: crystals. • Schade and Guttenberg (1951): *Lemna* sp.: root. • Stebbins and Khush (1961): *Lemna* sp., *Spirodela* sp., *Wolffia* sp., *Wolffiella* sp.: stomata. • Subramanyam (1962): family: anatomical summary. • Tieghem and Douliot (1888): *Lemna trisulca, Spirodela polyrrhiza*: branch root ontogeny. • Walsh and Melaragno. (1976): *Lemna*: phloem.

Additional literature

(*A sampling, estimated by Landolt (1998) to include at least 4300 titles.*)

Arber 1920; Bate-Smith 1968; Bogner and Mayo 1998; Daghlian 1981; Daubs 1965; Davis 1966; Frölich and Barthlott 1988; Gunn and Ritchie 1988; Harborne 1974; Hegnauer 1986; Hofmeister 1858; Kuprianova and Tarasevich 1984; Landolt 1980; Les and Crawford 1999; Les *et al.* 1997; Maheshwari and Khanna 1956; Maheshwari 1958; McClure 1970; Meusel 1951; Mikesell 1990; Pott and Cervi 1999; Ravn 1894; Raynal-Roques 1981; Sargent and Wangermann 1959; Sharma 1967; Tarasevich 1990; Taylor 1990; Thanikaimoni 1969; Thorne 2000; Tillich 1985, 1992; Tsuchiya 1991; Wilson 1960; Wunderlich 1954; Zennie and McClure 1977.

SUBFAMILY AROIDEAE

Tribe Thomsonieae Blume

Comprising two Old World genera, *Amorphophallus* and *Pseudodracontium*.

Anatomical summary

Leaf surface

Cuticle striate or rough and thick or thin. **Epidermis**: cells polygonal to sinuous or interlocking. **Stomata** abaxial, brachyparacytic to brachypara-hexacytic. Subsidiary cells broad or narrow and rounded. **Venation** (of segments) pinnate with submarginal brochidodromous collector.

Leaf TS

Structure dorsiventral. **Midrib** concave adaxially, rounded convex or fluted abaxially. **Epidermis**: cells usually tall adaxially and short abaxially. **Mesophyll**: 1–3 palisade layers. Spongy tissue type 1 aerenchyma. **Collenchyma** strands circular to flattened (type Sv). **Vascular bundles** scattered in midrib; medium-sized or small (types II, III). **Laticifers** non-anastomosing with short lateral processes. **Crystals**: raphides in single bundles in short cells, randomly spaced, or vertical in palisade tissue, or in multiple bundles in elongated cells. Some crystals arranged as coarse 'styloid' bundles. Druses, prismatics, and crystal sand common.

Petiole TS

Outline terete or oval-shaped. **Cuticle** thin or thick and striate-ridged. **Epidermis**: cells short. **Ground tissue** compact near margin, type 3 aerenchyma in centre. **Collenchyma** strands subhypodermal (type Sv). **Vascular bundles** type III. **Laticifers** of large diameter near phloem. **Crystals**: raphides in single bundles in wide thin-walled cells.

Taxonomic notes

Amorphophallus and *Pseudodracontium* are very similar anatomically. They appear at the base of the clade constructed by French *et al.* (1995) leading to the Caladieae. They stand out from other members of this clade in not having anastomosing laticifers. Their relative isolation is also accentuated by their having no tendency toward fibre cap development as is common in Caladieae.

Amorphophallus Decne.

The genus comprises about 200 cormose species from tropical Africa (25), Asia, Madagascar, the Comores and Australia. Specimens studied as *Thomsonia* Wall. and *Plesmonium* Schott (Nicolson 1984; Croat 1988) are included here. The plants are terrestrial, sometimes epilithic, understorey herbs, or occasionally at the margins of forests or open places. The corms show long dormancy and plants flower before the single leaf (or occasionally two) develops. Leaves are trisected at the summit of a long petiole with main branches often dichotomously branched. Petiolule segments are pinnatifid to tripinnatifid and ultimate laminae are elliptic. Chromosomes are large (*c.* 15 species): 2n = 26, 28, 29, 36, 39. Seven species were available for study.

Leaf surface

Cuticle finely striate (Fig. 2D) or rough. **Epidermis**: cells large on both surfaces. Adaxial cells polygonal with straight to sinuous anticlinal walls (1–1.5:1 *l/w*) (Fig. 80E). Abaxial cell walls deeply undulate to interlocking (2:1 *l/w*). **Stomata** abaxial, brachyparacytic or brachypara-tetracytic (Fig. 80D). Subsidiary cells broadened laterally. **Venation**: segments mostly elliptic with prominent midrib and ± regular secondaries arising broadly acute and straight. Secondaries connecting abruptly at a submarginal brochidodromous collector. Tertiaries and intersecondaries tending to be parallel with secondaries. Higher orders reticulate. Smaller marginal and sometimes looping veins present.

Leaf TS

Structure dorsiventral. **Size** of segments 72 × 29 to 118 × 53 mm (*l/w*). **Midrib** deeply concave-grooved adaxially and rounded, fluted convex abaxially (Figs 1F, 80A). **Cuticle** thin or thick on both surfaces. **Epidermis**: adaxial cells medium-tall, rounded or flat internally and externally. Abaxial layer similar but shorter. **Stomata** level with surface. **Mesophyll**: 1–3 compact palisade layers occupying up to half mesophyll (Fig. 80B,C). Spongy layers 5–6, short-lobed cells with small irregular air spaces (type 1). **Air cavities** up to 60% of spongy layer. **Collenchyma** subepidermal with ± circular strands of angularly thickened cells beneath abaxial ridges of midrib (Fig. 80A) and beneath secondary vascular bundles; strands 8–10-seriate (type Sv).

Vascular bundles scattered in midrib (Figs 1F, 80A), symmetrically in *A. sumawongii* (*Thomsonia*), and weakly formed; type II. Xylem: wide or narrow protoxylem lacunae and several wide or narrow metaxylem cells adjacent to small rounded or irregularly shaped phloem strand. **Laticifers** non-anastomosing, at flanks of vascular bundles with short, irregular lateral processes extending between cells of ground tissue. **Tannin** cells uncommon in mesophyll in *A. sumawongii* (*Thomsonia*). **Photosynthate**: plastids store sugar rather than starch (Sakai and Hayashi 1973). **Crystals**: raphide bundles without visible sheaths, single or multiple, in short wide cells (Fig. 80F) or elongated cells, uncommon in mid-mesophyll. Raphide cells may occur vertically in palisade tissue. Most raphides fine, some coarse as bundles of styloids. Very small druses and prismatics common, especially in palisade tissue; large druses rare in mesophyll. Crystal sand common throughout mesophyll and ground tissue. Rounded prismatic bodies present in some tracheids of *A. sumawongii* (*Thomsonia*). Nuclei very large in mesophyll and epidermal cells of *A. su mawongii.*

Petiole TS

Outline terete. **Size**: diameter 8–20 mm. **Cuticle** thin and ± striate. **Epidermis** of squarish cells, rounded internally and externally. **Collenchyma**: strands ± circular (type Sv) separated from epidermis by *c.* 2 layers of ground tissue (Fig. 80G). **Ground tissue**: outer tissue of small compact cells grading to type 3 aerenchyma in centre; partition cells one (two) in *A. maculatus*, up to eight in *A. konjac*. In *A. variabilis*, large type 4 cavities intermixed with type 3.

Vascular bundles scattered in centre (type III); peripheral row of smaller bundles subtending collenchyma strands. Xylem of one wide protoxylem cell (Fig. 80H) and 1–2 narrow metaxylem cells. Phloem: a small, laterally narrowed strand. **Laticifers**: usually two large cells lateral to phloem. **Starch** granules scattered in ground tissue. **Crystals**: raphides in single bundles in thin-walled cells near periphery.

Material examined

Amorphophallus hildebrandtii (Engl.) Engl. & Gehrm., Bogner 168, Madagascar, cult. M; *A. konjac* K.Koch, Keating 2089, cult. Edwardsville, IL; *A. maculatus* N.E.Br., Bogner 600, Gabon, cult. M; *A. margaritifer* Kunth (*Plesmonium margaritiferum* (Roxb.) Schott), Haines 347, India, K; *A. sumawongii* (Bogner) Bogner & Mayo (*Thomsonia sumawongii* Bogner), Bogner 372, Thailand, cult. M; *A. sutepensis* Gagnep. Croat 71876, cult. MO; *A. variabilis* Blume, Indonesia, cult. M.

Taxonomic notes

Amorphophallus and the previously recognized genera *Plesmonium* and *Thomsonia* are very similar anatomically and their structure provides no grounds for objection to combining the latter two genera with *Amorphophallus*. Note: Dalitzsch (1886) illustrated an *A. rivieri (konjac)* brachyparacytic stoma with striate cells. This type of cell was not seen in specimens examined here.

Species reported on in the literature

Abranowicz (1912): *A. rivieri*: tuber. • Aisslinger (1907): fibres. • Barthlott and Lobin (1998): *A. titanum*: general survey. • Beck and Mannagetta (1912): crystals. • Borzi (1887–8): *A. bulbifer, A. rivieri:* root. • Bredemann and Meyer (1932): *A. titanum*: leaf fibres. • Dalitzsch (1886): *A. bulbifer, A. rivieri*: leaf. • Decrock (1909): *A. campanulatus, A. sativus*: starch. • Engler (1911): several spp.: root, stem, leaf. • Ertl (1932): *A. konjac*: leaf venation. • French (1987c): several spp.: root. • French (1988): several spp.: laticifers. • Gentner (1905): *A. rivieri*: leaf tip. • Genua and Hillson (1985): *Hydrosme rivieri*: crystals. • Giovannozzi (1911): *A. konjac*: tuber. • Guttenberg (1968): *A. rivieri*: root. • Hotta (1971): *A. kiusianus, A. titanum*: tuber, leaf. • Inaba (1984):

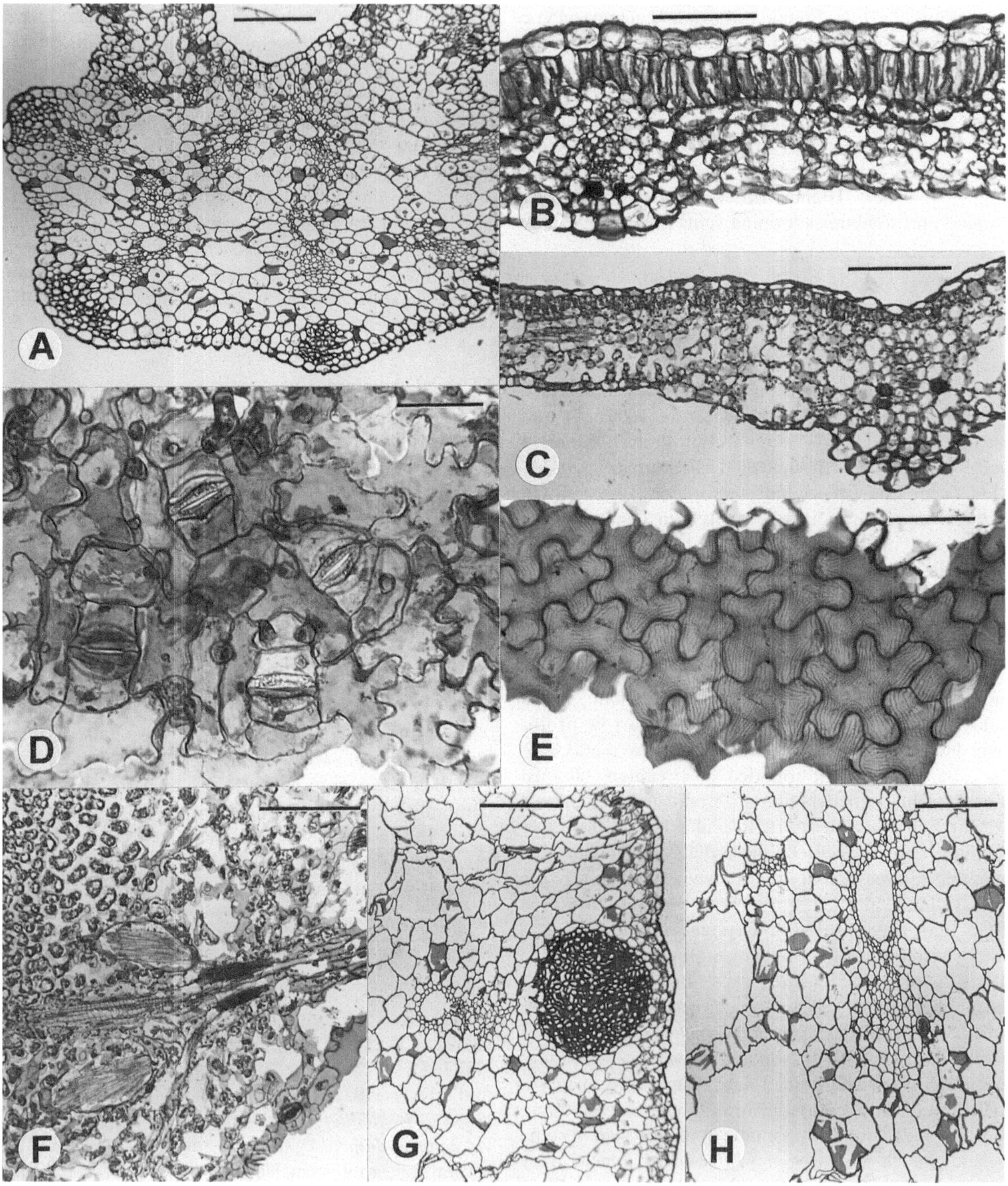

Fig. 80. *Amorphophallus* leaf and petiole sections. (A,D,G,H) *A. konjac*; (B) *A. hildebrandtii*; (C) *A. variabilis*; (E,F) *A. maculatus*: (A) Leaflet midrib TS; (B) Lamina TS; (C) Lamina TS; (D) Abaxial epidermis PS with stomata; (E) Adaxial epidermis PS, showing cuticular striations; (F) Lamina PS with two wide raphide cells; (G) Petiole TS margin with type Sv collenchyma strand subtending type III vascular bundle; (H) Petiole TS centre and type III vascular bundle. Scale bars: A,G,H = 300 µm; B,C,F = 150 µm; D,E = 50 µm.

A. konjac: leaf. • Knecht (1983): *A. abyssinicus, A. johnsonii, A. staudtii*: leaf venation. • Kroemer (1903): *Hydrosme rivieri*: root. • Lierau (1888): *A. campanulatus, Hydrosme rivieri*: root. • Majumdar (1929): *A. companulatus*: root. • Meyer (1962): *A. konjac*: leaf. • Möbius (1936): *A. bulbifer*: bulbil. • Motte (1932): *A konjac*: tuber. • Müller (1919): *A. rivieri*: leaf hydathode. • Namba *et al.* (1968): rhizome. • Netolitzky (1929): *A. rivieri*: crystals. • Noro *et al.* (1992): stem raphides. • Ohtsuki (1960): *A. konjac*: tuber starch. • Paliwal and Kavathekar (1972): *A. campanulatus*: stem. • Pant and Kidwai (1966): *A. campanulatus*: stomata. • Philipp (1923): *A. dubius*: abscission. • Planchon and Juillet (1909): *A. campanulatus, A. rivieri, A. sativus*: starch. • Reichert (1913): *A. rivieri*: starch. • Sakai and Hayashi (1973): *A. konjac*: leaf carbohydrates. • Sakai *et al.* (1984): raphides. • Schwendener (1882): root. • Shah and Gopal (1969): stomata. • Solereder and Meyer (1928): *A. variabilis, Thomsonia* sp.: all organs. • Stahl (1896): *A. variabilis*: leaf. • Van Fleet (1942): *A. giganteus*: stem, endodermis. • Wakabayashi (1957a): *A. kiusiana, A. konjac*: raphides. • Wakabayashi (1957b): *A. konjac*: corm, mucilage cells. • Widjaja and Lester (1987): *A. paeoniifolius*: raphides. • Zinke (1924): *A. rivieri*: leaf.

Additional literature

Chauhan and Brandham 1985; Engler 1911, *Thomsonia*; Erdtman 1952; French 1985b, 1986c, 1987a; Gandawijaja *et al.* 1983; Grayum 1990; Hay 1986; Hegnauer 1986; Huang 1972; Jos and Bai 1986; Koernicke 1937; Kulkarni *et al.* 1990; Lakowitz 1881; Meeuse and Raskin 1988; Meusel 1951; Pijl 1937; Plese *et al.* 1989; Sharma 1967; Smolenski *et al.* 1972, 1974; Thanikaimoni 1969; Williams *et al.* 1981.

Pseudodracontium N.E.Br.

This genus of seven species is distributed in Thailand and Indochina. Plants are tuberous herbs producing one or more leaves. They are found in tropical moist forests with dry seasons. Leaves are long-petiolate, trisect, and irregularly divided with few obovate or elliptic segments. Chromosomes: 2n = 26 for two species. One species was available for study.

Leaf surface

Cuticle striate. **Epidermis**: adaxial cells polygonal with straight anticlinal walls (1–2:1 *l/w*) and abaxial cells polygonal-undulate (1–3:1 *l/w*). **Stomata**: abaxial, usually brachypara-hexacytic (Fig. 81D), occasionally-brachypara-tetracytic. Subsidiary cells narrow and rounded. **Venation**: pinnate with secondaries regularly spaced. Secondaries merged at ± straightened submarginal brochidodromous collector. Tertiaries and higher orders ± parallel with secondaries but often reticulate.

Leaf TS

Structure dorsiventral. **Size**: segment 5 × 1.5 cm (*l/w*). **Midrib** shallowly and narrowly grooved adaxially, broadly and deeply convex-rounded abaxially (Fig. 81A). **Cuticle** medium-thick adaxially, thin abaxially. **Epidermis**: cells tall adaxially, flat or rounded on surface and rounded within. Abaxial epidermal cells mostly short, some tall, most flat on surface and within, some rounded. **Stomata** level with surface; guard cells with marked cuticular ledges. Substomatal air cavities shallow. **Mesophyll**: one well-differentiated palisade layer (Fig. 81C) and 5–6 spongy layers with small irregular air cavities (type 1). **Air cavities**: up to 50% of spongy tissue.

Collenchyma: five subepidermal strands, flattened, 2–3 layers thick, on abaxial side of midrib (type Sv). **Vascular bundles** weakly formed in midrib in two tiers; abaxial bundles aligned with collenchyma strands (Fig. 81A) (types II and III). Xylem: 2–3 wide metaxylem elements opposing a small circular or elliptic phloem strand (Fig. 81B). **Laticifers**: non-anastomosing, usually two cells lateral to vascular bundles (Figs 8B, 81B), in midrib and larger laminar bundles. **Crystals**: raphides in single or overlapping bundles, in irregular thin-walled, often elongate cells, common in mesophyll (Fig. 81E). Crystal sand, 2–3 per cell, common in epidermal layers, less so in mesophyll.

Petiole TS

Outline oval. **Size**: diameter 4 × 7 mm. **Cuticle** thick and striate-ridged. **Epidermis**: cells flat or rounded externally. **Collenchyma**: discrete circular strands, 10–15 cells wide, separated from epidermis by *c.* 2 layers of ground tissue (Fig. 81G), each subtending small vascular bundle (type Sv). **Chlorenchyma**: outer 6–8 ground tissue layers. **Ground tissue**: compact in outer 10–20 layers, irregular large cavities in centre (2–4 times cell diameter), mostly with type 3 partitions. **Air cavities**: up to 70% of central aerenchyma.

Vascular bundles: outer circle of bundles internal to collenchyma strands; inner bundles in 2–3 irregular circles. Bundles small (type III) (Fig. 81F). Xylem: 1–2 wide protoxylem elements. Phloem: small laterally narrowed strand. **Laticifers**: wide diameter cells lateral to phloem in pairs. **Crystals**: raphides in single bundles, in wide thin-walled cells, uncommon and scattered in ground tissue.

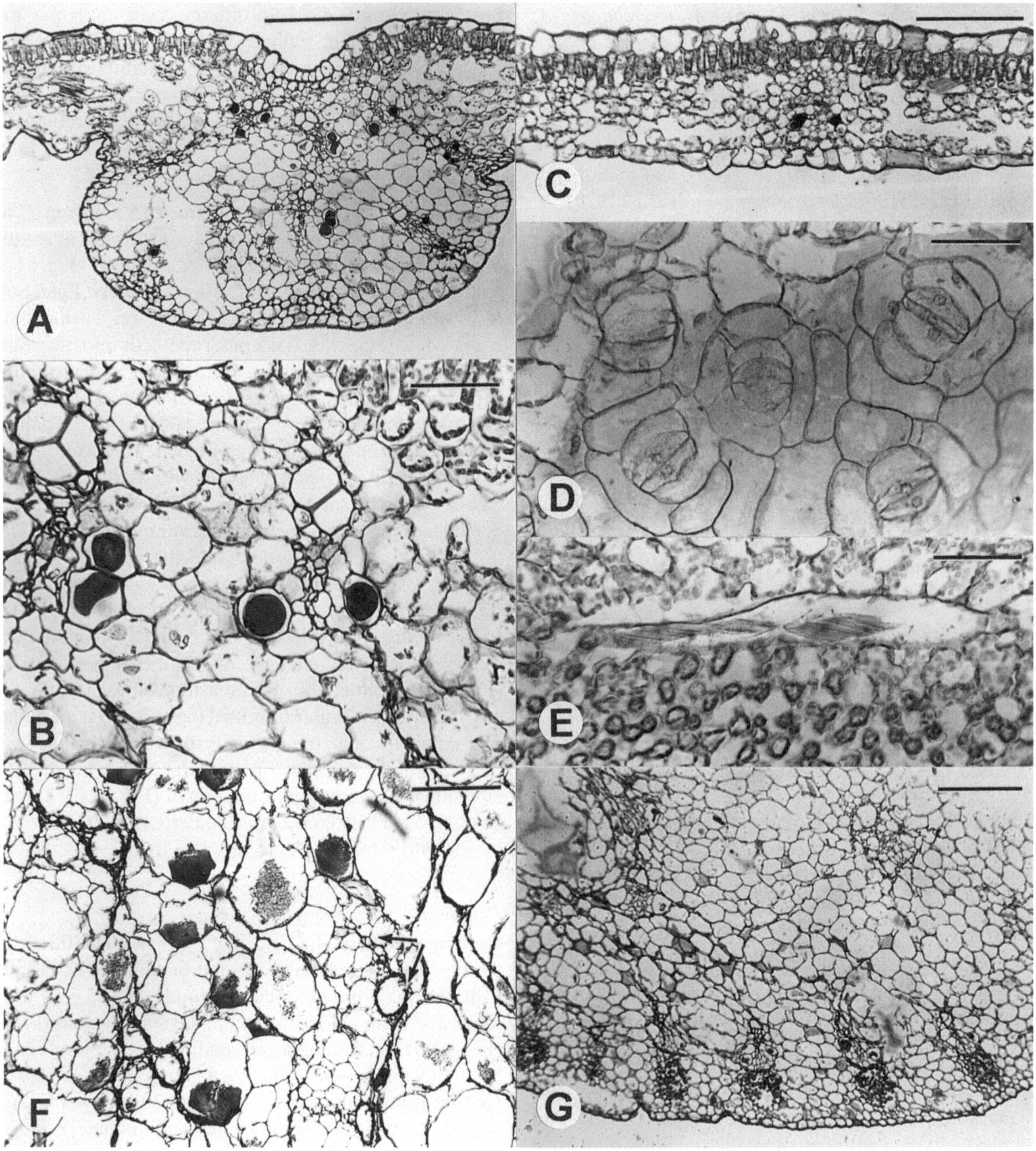

Fig. 81. *Pseudodracontium siamense* leaf and petiole sections. (A,B) Midrib TS: (B) Detail of vascular bundles and laticifers; (C) Lamina TS; (D) Abaxial epidermis PS and stomata; (E) Lamina PS showing elongated raphide cell; (F) Petiole TS centre showing weak type III vascular bundles (arrows); (G) Petiole TS lateral margin. Scale bars: A,F = 200 μm; B,D,E = 50 μm; C = 150 μm; G = 300 μm.

Material examined

Pseudodracontium siamense Gagnep., Kew 414-35-41401, ex Bogner (M).

Species reported on in the literature

French (1987c): *P. lacourii*: root. • French (1988): *P. siamense*: laticifers. • Solereder and Meyer (1928): secretory tissue.

Additional literature

French 1985b, 1986c, 1987a; Grayum 1990; Hay 1986; Thanikaimoni 1969.

Tribe Caladieae

Eleven genera constitute this tribe; *Hapaline* occurs in south-east Asia and the other ten are found in the New World tropics. They are *Syngonium, Xanthosoma, Chlorospatha, Ulearum, Filarum, Zomicarpella, Caladium, Scaphispatha, Jasarum*, and *Zomicarpa*.

Anatomical summary

Leaf surface

Cuticle smooth, striate, papillate or rugulate-ridged. **Epidermis**: cells mostly straight-sided polygonal, occasionally rounded polygonal, to undulate. **Stomata** mostly or entirely abaxial, brachyparacytic to brachyparahexacytic. Subsidiary cells narrow, broad and angular, or trapezoidal. **Venation** acrodromous or with prominent midrib. Secondaries brochidodromous, occasionally looped, or eucamptodromous, occasionally ending in straightened marginal collector vein.

Leaf TS

Structure dorsiventral. **Midrib** usually grooved adaxially, shallowly convex-rounded, semicircular, angular, or fluted abaxially. **Epidermis**: cells short, tall or papillate. Hypodermis present in *Syngonium schottianum* and *Ulearum*. **Mesophyll**: palisade cells in 1–3 layers. Spongy cells forming types 1, 1a, 2, or 3 aerenchyma. **Collenchyma**: usually type Sv in midrib, circular strands subtending vascular bundles and occasionally aligned with ridges. **Vascular bundles**: several, usually symmetrically disposed in midrib; bundle types II or III. **Sclerenchyma**: fibres usually absent, present as phloem caps in *Syngonium* and xylem and phloem caps in some *Xanthosoma* species. **Laticifers** anastomosing and often ramifying among ground tissue near vascular bundles. **Secretory tissue**: in *Ulearum*, no specialized histology but copious mucilage exuding from cut surfaces. **Crystals**: raphides in single bundles in broad thin-walled cells, or elongated tubular cells with 2–4 bundles. Also raphides common as biforines with one end attached or both ends free. Druses large in *Syngonium* and *Chlorospatha*. Prismatics small and numerous in *Jasarum*.

Petiole TS

Outline crescent-shaped or deeply V-grooved basally, circular distally. **Cuticle** thin and smooth or striate. **Epidermis**: cells small, often papillate in *Syngonium* and *Caladium*. **Ground tissue** variable: types 1 or 2, or types 3 or 4. **Collenchyma** type Sv mostly and type Bi distally in *Scaphispatha*. **Sclerenchyma**: fibres as phloem caps in *Scaphispatha* and *Chlorospatha*. **Vascular bundles** types II or III. **Laticifers** anastomosing, found in most genera. **Crystals**: raphides as single bundles in small thin-walled cells. Biforine-like cells attached at one end in many genera, or in multiple overlapping bundles in *Ulearum* and *Zomicarpella*.

Taxonomic notes

The tribe demonstrates significant morphological diversity ranging from terrestrial tuberous herbs to the immersed aquatic genus *Jasarum*. Leaf venation, stomatal type (or absence) epidermal cell variation, and mesophyll types all show diagnostic differences among genera. In spite of this, the genera form a coherent group based on the anastomosing laticifers, the presence of types 3 and 4 aerenchyma in the petiole, and the presence of biforines in mesophyll or petiole ground tissue. The genus *Ulearum* stands somewhat apart in having collenchyma as phloem caps, mucilage exudation, and raphides in multiple bundles.

Hapaline Schott

This genus of five species is distributed in the Himalayas, south-east Asia, and Malaysia. Plants are mostly terrestrial tuberous understorey herbs. One species is epilithic in limestone rock crevices. The leaves are long-petiolate, and elongate cordate or sagittate. Chromosomes: $2n = 26, 28$. One species was available for study.

Leaf surface

Cuticle smooth on lamina surfaces, striate-ridged on midrib. **Epidermis**: cells polygonal with straight anticlinal walls on both surfaces (2–3:1 *l*/*w*). **Stomata** more common abaxially and brachyparacytic with broadly-rounded subsidiary cells (Fig. 82D). **Venation**: midrib prominent and basal lobe primaries, acrodromous. Secondaries 1–2-pinnate from midrib, broadly acute and curved toward apex. Brochidodromous collector looped basally, more straightened distally. Secondaries and intersecondaries mostly parallel. Tertiaries and higher orders reticulate.

Leaf TS

Structure dorsiventral and well-differentiated. **Size**: 103 × 53 mm (*l*/*w* total length); midrib length 78 mm. **Midrib** with rounded adaxial groove and broadly convex abaxially (Fig. 82A). **Cuticle** thin on both surfaces, thicker at margin. **Epidermis**: cells larger adaxially, flat externally and rounded internally. **Stomata** level with surface. **Mesophyll**: one palisade layer (8–10:1 *l*/*w*) (Fig. 82C). Spongy tissue in 8–9 cell layers, cells horizontally extended, long-armed and irregular (type 2). **Air cavities** mostly regular and rounded, up to 80% of spongy tissue. Several large type 3 air cavities in midrib. **Collenchyma**: subepidermal strands abaxial in midrib (Fig. 82B), each capping a small vascular bundle (type Sv).

Vascular bundles symmetrically disposed in midrib, *c*. 4 large central bundles and *c*. 6 small abaxial ones; bundles small and unsheathed (type II). Xylem: one wide protoxylem cell and 4–5 narrow metaxylem elements. Phloem: a small laterally expanded strand. **Laticifers** anastomosing and with dark amorphous contents, several at vascular bundles; branching patterns complex at vascular bundle junctions. **Crystals**: raphides, uncommon, of two types. Biforine-shaped cells terminally papillate but with thin unlignified walls, uncommon in palisade layer. Larger, irregularly-shaped thin-walled cells uncommon in mid-mesophyll (Fig. 82E). Crystal sand common in mesophyll and epidermis.

Petiole TS

Outline subterete. **Size**: diameter 3 × 4 mm. **Cuticle** thin and striate. **Epidermis**: cells large and irregular or square. **Collenchyma**: subepidermal or subhypodermal strands around periphery, most subtending small vascular bundles (type Sv). **Ground tissue**: outer 8–10 layers compact, of large cells, separated by circle of large air cavities with type 3 or 4 partitions (Fig. 82F) from central, large-celled, compact tissue. **Chlorenchyma**: sparse plastids in outer 6–8 ground tissue layers.

Vascular bundles scattered throughout centre (Fig. 82G) plus single circle of bundles in outer ground tissue; bundles unsheathed (type II) (Fig. 82G). Xylem: one large protoxylem strand and 2–3 wide or medium-sized metaxylem cells. Phloem: small, circular or irregular strand. **Laticifers**: 1–3 large diameter cells, lateral to phloem at most vascular bundles, common throughout compact ground tissue. **Crystals**: raphides uncommon in peripheral ground tissue, in short thin-walled cells with single bundles. Crystal sand sparse throughout ground tissue.

Material examined

Hapaline brownii Hook. f., Burtt and Woods 1771, cult. M.

Species reported on in the literature

French (1987c): *H. brownii*: root. • French (1988): *H. brownii*: laticifers. • French and Tomlinson (1983): *H. brownii*: stem.

Additional literature

Fox and French 1988; French 1986a, c, 1987a; Grayum 1990; Loh *et al*. 2000; Thanikaimoni 1969.

Syngonium Schott

This genus of 36 species (including three formerly placed in *Porphyrospatha*, now section *Cordatum* of *Syngonium*) is distributed in Central and South America. Plants are usually hemi-epiphytic appressed climbers on understorey tree trunks of moist to wet forests, or they may be epiphytic or epilithic. They also occur in regrowth areas. Juvenile plants are terrestrial, growing skototropically toward the nearest large tree. Mature plants are rare in some areas, and revert easily to the juvenile form. Juvenile leaves are sagittate or hastate. Mature leaves are trisect or pedisect with ± elliptic segments. Chromosomes: 2n = 24, 26, 28. Eight species were available for study.

Leaf surface

Cuticle smooth or irregular on both surfaces, or rugulate-ridged abaxially. **Epidermis**: cells polygonal with straight anticlinal walls adaxially and straight or undulate walls abaxially (1–3:1 *l*/*w* on both surfaces). **Stomata**: mostly abaxial, brachypara-hexacytic (Fig. 83D) but with outer pairs of subsidiary cells unmodified except for position. **Venation**: median segment with prominent midrib. Secondaries arising acutely at regular intervals. Tertiaries arising acutely from secondaries and becoming parallel with secondaries, and often forming composite intersecondaries. Secondaries and tertiaries merging with two or more submarginal collector veins.

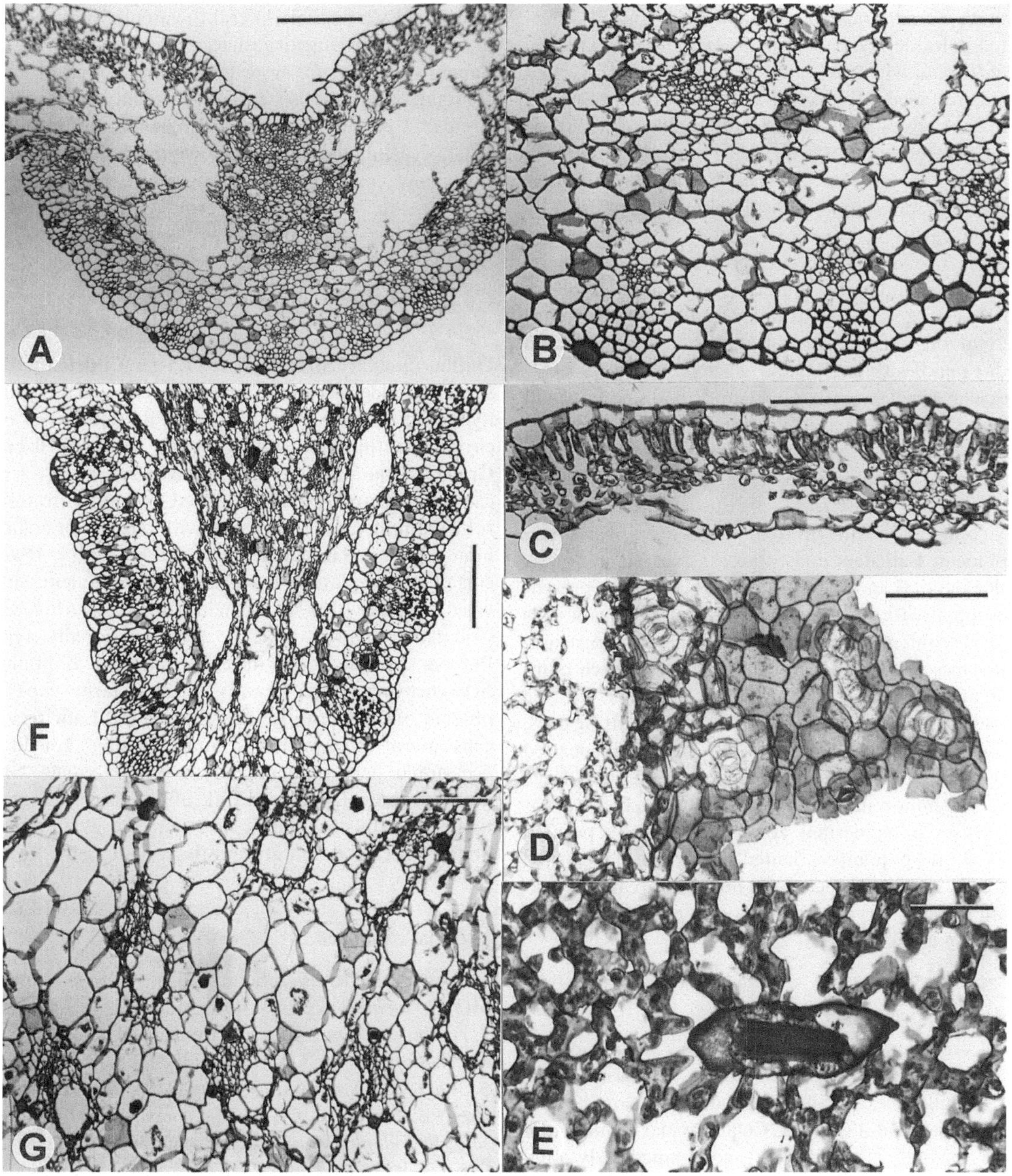

Fig. 82. *Hapaline brownii* leaf and petiole sections. (A,B) midrib TS with type Sv collenchyma; (C) Lamina TS; (D) Abaxial epidermis PS with brachyparacytic stomata; (E) Lamina PS with small raphide cell; (F) Petiole TS showing type 4 aerenchyma; (G) Petiole TS central tissue with compact parenchyma and type III vascular bundles. Scale bars: A,F = 300 μm; B–D,G = 150 μm; E = 50 μm.

Leaf TS

Structure dorsiventral. **Size**: segment 24 × 14 cm (*l*/*w*). **Midrib** flat adaxially, broadly and deeply semicircular-convex abaxially (Fig. 83A). **Cuticle** thick adaxially, medium-thick abaxially. **Epidermis**: cells somewhat larger adaxially, cells flat externally and rounded internally. **Stomata** level with surface. **Hypodermis** present adaxially in midrib and patchy in nearby lamina in *S. schottianum*. **Mesophyll**: 1–2 well-differentiated palisade layers, except poorly differentiated in *S. schottianum* where palisade present as 2–3 layers of compact, rounded cells packed with chloroplasts. Spongy cells in *c.* 11 layers, irregular and often long-armed (type 2) or with one partition cell (type 3). Most spongy cell plastids amyloplasts in *S. schottianum*. **Air cavities** mostly in organized high vertical columns, separated by uniseriate partitions (Fig. 83B), up to 95% of spongy layer. In *S. schottianum*, air cavities well developed in midrib, separated by uniseriate partitions (type 3). **Collenchyma**: ± circular strands 6–10 layers deep, subepidermal on abaxial side of midrib and subtending vascular bundles (type Sv).

Vascular bundles scattered in midrib; bundles small (type II). Xylem: 2–3 metaxylem elements opposing a broad oval strand of phloem. **Sclerenchyma**: 1–2 fibre layers abaxial to phloem, in some vascular bundles. In *S. schottianum*, fibres thin- or thick-walled, forming nearly complete sheaths up to 3–4 cells deep over xylem and phloem. **Laticifers** anastomosing, very common at vascular bundles and in phloem. **Secretory ducts** present in mesophyll with one layer of epithelium. **Tannin** cells with dark contents common, similar in form to neighbouring mesophyll and parenchyma cells. **Starch** common in abaxial mesophyll and midrib. **Crystals**: biforine raphides uncommon, in long bundles in tight-fitting cells, often with heavily birefringent walls, always suspended across spongy tissue partitions (Fig. 83C). Druses large and very common in midrib and lamina. In *S. schottianum*, druses of epidermal zones eccentrically pointed with 1–3 sharp points oriented toward surface; such druse cells superficially conical, not quite reaching epidermal surface. Crystal sand loose or clustered, very common in mesophyll.

Petiole TS

Outline crescent-shaped to subcircular. **Size**: diameter 5–14 mm. **Cuticle** thick and smooth. **Epidermis**: cells small, papillate externally. **Stomata** occasional, with shallow substomatal cavities. **Collenchyma**: semicircular strands abaxially (Figs 5I, 83F), not immediately adjacent to small vascular bundles (type Sv). **Chlorenchyma**: tissue weakly developed in outer ground tissue. **Ground tissue**: mostly type 3 aerenchyma, some 2-seriate partitions (type 4) (Fig. 83F). **Air cavities** up to 95% of ground tissue.

Vascular bundles: numerous and scattered, small near periphery; few large type I bundles in centre, mostly smaller, types II and III (Fig. 83E). Xylem: 1–10 metaxylem cells. Phloem: circular or semicircular strands. **Laticifers**: two or more lateral to vascular bundles, anastomosing, ramifying into adjacent ground tissue (Fig. 8C). **Tannin** cells few, in or near phloem. **Starch** absent or common in ground tissue; granules small and irregular or circular. **Crystals**: raphides in single bundles in biforine cells, occasionally across aerenchyma partitions or generally in ground tissue; some single wide bundles in large, circular, thin-walled cells in ground tissue. Druses and raphides suspended in air cavities in *S. macrophyllum* (Fig. 83E).

Stem TS

Outline circular. **Size**: diameter 1.5 cm. **Cuticle** thick and smooth. **Epidermis** and **hypodermis**: cells heavily lignified, hypodermal cells with U-shaped thickenings. **Collenchyma**: continuous subhypodermal band, 8–10 layers. **Ground tissue** cells mostly compact, some cell-sized air cavities. **Central cylinder** defined by concentration of robust amphivasal bundles, or with mixture of collateral bundles. **Vascular bundles** crowded in centre, fewer in cortex; only collateral bundles in cortex. Xylem: amphivasal bundles with 4–10 tracheid groups (Figs 6F, 83G,H). Collateral bundles type II and occasionally type I. Phloem strands circular in centre of amphivasal bundles. **Sclerenchyma**: fibre strands forming large caps over phloem of collateral bundles (Fig. 83H). **Laticifers**: 1–3 cells lateral to phloem and anastomosing. **Tannin** cells occasional near phloem and in ground tissue. **Starch** granules irregularly isohedral, abundant in all ground tissue. **Crystals**: raphides not seen, or as single wide bundles occasional in ground tissue.

Material examined

Syngonium angustatum Schott, Croat 42559, Honduras, MO; *S. auritum* (Vell.) Schott, Metcalfe 30, Jamaica, K; *S. erythrophyllum* Birdsey ex Bunting, Bogner 1883, cult. M; *S. hoffmannii* Schott, Croat 33508, Panama, MO; *S. macrophyllum* Engl., Croat 42528, Honduras, MO; Croat 33718, Panama, MO; Croat 43183, Costa Rica, MO; *S. podophyllum* Schott, Croat 33636, Panama, MO; Croat 42934, Nicaragua, MO; Croat 42567, Honduras; Keating 1004, cult. SEL; *S. salvadorense* Schott, Croat 43758, Guatemala, MO; *S. [Porphyrospatha] schottianum* Wendl. ex Schott, Croat 35984, Panama (MO).

Taxonomic notes

Section Cordatum, here represented by *S. schottianum*, shows several distinctive characters as noted in the leaf

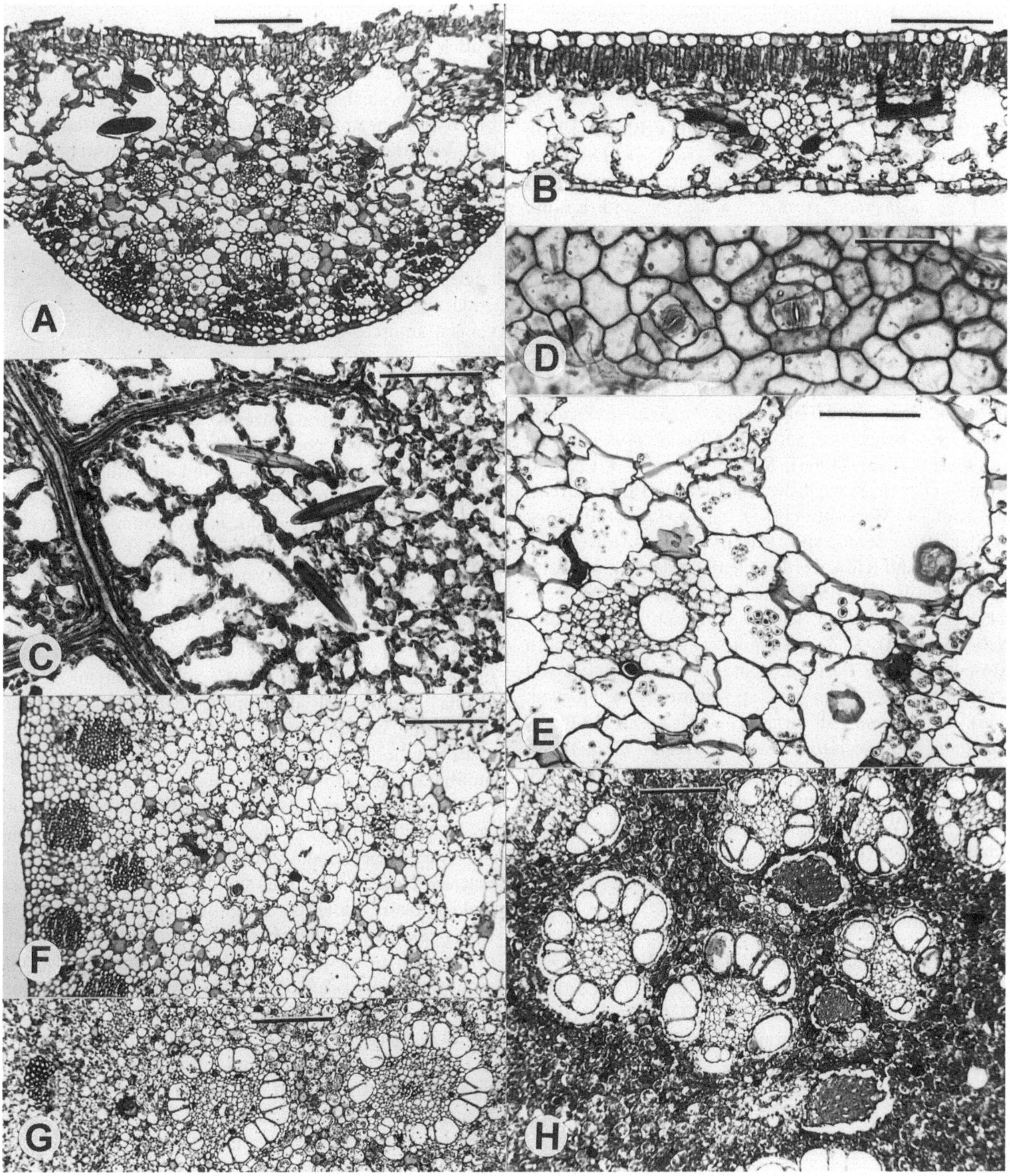

Fig. 83. *Syngonium* leaf, petiole and stem sections. (A,D) *S. auritum*; (B,G) *S. podophyllum*; (C) *S. angustatum*; (E,F,H); *S. macrophyllum*. Type 3 aerenchyma shown in (A,B,C,E). (A) Midrib TS, (B) Lamina TS showing anastomosing laticifers; (C) Mesophyll PS showing biforine cells; (D) Lamina abaxial epidermis showing brachypara-tetra- or -hexacytic stomata; (E) Proximal petiole TS showing type III vascular bundle; (F) Proximal petiole TS showing circular collenchyma strands; (G) Stem TS showing amphivasal vascular bundles; (H) Stem TS showing amphivasal vascular bundles, some collateral bundles with fibre caps and ground tissue with numerous starch granules. Scale bars: A = 200 μm; B,C,E = 150 μm; D = 50 μm; F–H = 300 μm.

descriptions. However, the genus remains coherent in general anatomy.

Species reported on in the literature

Berquam (1972): *S. podophyllum*: root. • Bloch (1926): *S. (Porphyrospatha) schottianum*: root. • Dalitzsch (1886): *S. auritum, S. riedelianum*: leaf. • Dunn *et al.* (1965): *S. neglectum*: stomata. • Engler and Krause (1920): *S. vellozianum*: leaf, secretory tissue. • Ertl (1932): *S. podophyllum*: leaf venation. • French (1987c): several spp.: root. • French (1988): several spp.: laticifers. • French and Tomlinson (1980): *S. podophyllum*: stem vasculature. • French and Tomlinson (1983): *S. erythrophyllum, S. schottianum, S. vellozianum*: stem vasculature. • Funke (1929): *S. podophyllum*: root. • Gentner (1905): leaf tip. • Genua and Hillson (1985): *S. podophyllum*: crystals. • Goebel and Sandt (1930): *S. podophyllum*: root. • Guttenberg (1968): *S. 'album'*: root. • Lee and Richards (1991): *S. rayii*: leaf. • Lierau (1888): several spp.: root. • Marcatili (1887): *S. auritum*: laticifers. • Müller (1919): several spp.: leaf tip. • Napp-Zinn (1953): root. • Pant and Kidwai (1966): leaf, epidermis. • Pirotta and Marcatili (1885): *S. auritum*: laticifers. • Roth (1949): *S. podophyllum*: leaf. • Rywosch (1909): root. • Scherer (1904): *S. podophyllum*: vasculature. • Smith and Lew (1970): *S. auritum*: stem, node. • Solereder and Meyer (1928): *S. auritum, S.* sp.: anatomy. • Tieghem (1866): *S. auritum*: all organs. • Tieghem (1870–1): *S. auritum*: root, secretory tissue. • Trécul (1865, 1866): *S. auritum, S. riedelianum*: vasculature. • Troll (1932a): *S. podophyllum*: leaf ontogeny. • Voronkina (1975): root ontogeny. • Webber (1960): several spp.: epidermis, stomata. • Weiss (1866): *S. decipiens*: root, laticifers. • Went (1895): *S. album, S. auritum*: root.

Additional literature

Blanc 1978; Croat 1981; DiBenedetto and Garcia 1992; Fox and French 1988; French 1986a, c, 1987a; Grayum 1986b, 1990; Murata 1990a; Ray 1986, 1987a, 1988, 1990, 1992a,b; Thanikaimoni 1969; Vágújfalvi 1971; Williams *et al.* 1981.

Xanthosoma Schott

This genus of 45 species is distributed in the American tropics. Plants are terrestrial rhizomatous or tuberous herbs of forest understoreys in moist to wet forests, or in marshy areas in full sun or partial shade. One species is epilithic. Stems of caulescent species recline over the forest floor. The leaves are elliptic, sagittate, hastate, or pedately compound. Chromosomes for eight species: 2n = *c.* 24, 26, 39. Seven species were available for study.

Leaf surface

Cuticle smooth or striate. **Epidermis**: cells polygonal with straight anticlinal walls on both surfaces. **Stomata** abaxial or on both surfaces and brachyparacytic (Fig. 84C) to brachypara-hexacytic. Subsidiary cells broad and rounded. **Venation**: prominent midrib and two basal primaries arising from summit of petiole. Widely spaced secondaries acute to narrowly acute, merging with brochidodromous, mostly straightened, submarginal collector. Tertiaries often closely spaced, arising at broadly acute angles from secondaries, becoming parallel with secondaries and merging into composite intersecondary veins in intercostal space. Higher orders reticulate.

Leaf TS

Structure dorsiventral. **Size**: 87 × 21 mm (average *l/w*). **Midrib** shallowly concave adaxially and broadly or deeply and abruptly convex-curved abaxially (Fig. 84A). **Cuticle** absent or thin on both surfaces. **Epidermis**: cells irregularly shaped and sized, mixed rounded or rectangular on both surfaces. **Stomata** level with surface. **Mesophyll**: 1–3 palisade layers, 30–40% of leaf (Fig. 84B). Spongy cells irregular, isodiametric or elongate in any direction (type 2 or 3) (Fig. 84E). **Air cavities** well developed in spongy layer (type 3), with 1-seriate partitions, 2–5 cells long, up to 80% of aerenchyma. **Collenchyma** strands well developed, discrete and circular, subtending vascular bundles on abaxial side of midrib (type Sv). In thin midribs, collenchyma in one thin strand on abaxial side of midrib.

Vascular bundles scattered in midrib. Xylem with wide protoxylem element or 1–2 wide metaxylem elements opposing large circular phloem strand (types I or II). **Sclerenchyma** absent, or as small fibre caps to both xylem and phloem. **Laticifers** anastomosing, 1–3 lateral to larger vascular bundles (Fig. 84D,E), or not in any special orientation to vascular bundles; cells may branch into mesophyll in various directions. **Tannin** cells with dark contents occasional in mesophyll. **Crystals**: raphides in rounded cells, up to 60 μm long (Sakai and Hanson 1974), or biforines, not common, high or low in lamina; biforines often arranged across aerenchyma partitions, pointing into two adjacent cavities. Druses absent or present, large or very small, sharp-pointed in circular cells, restricted to lower palisade layer or in any mesophyll or aerenchyma cell. Crystal sand absent, or common in aerenchyma partition cells.

Petiole TS and LS

Outline: crescent-shaped proximally; ± circular distally and slightly flattened adaxially. **Size**: diameter 4–5 mm

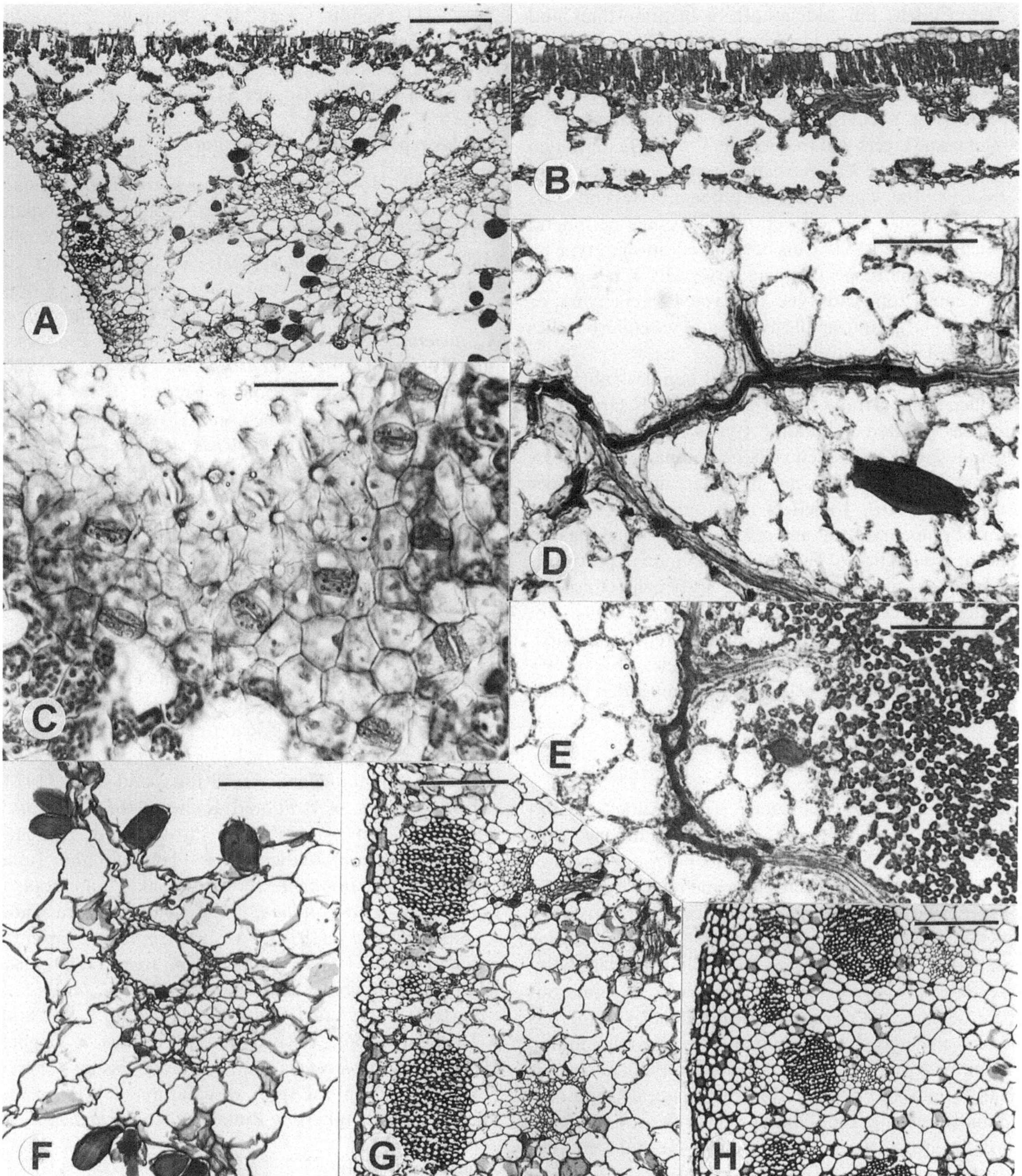

Fig. 84. *Xanthosoma* leaf, petiole and stem sections. (A–C,E–G) *X. sagittifolium*; (D) *X. pulchra*; (H) *X. robustum*; (A) Midrib TS, adaxial and lateral surfaces; (B) Lamina TS; (C) Abaxial epidermis PS showing epidermal papillae; (D) Lamina PS showing anastomosing laticifers and raphide cell; (E) Lamina PS and anastomosing laticifer; (F) Petiole TS centre with raphide cells suspended in cavities near type III vascular bundle; (G) Petiole TS margin showing type Sv collenchyma and type III vascular bundles; (H) Stem TS margin showing irregular arrangement of collenchyma strands. Scale bars: A,H = 300 μm; B,G = 200 μm; C = 50 μm; D–F = 150 μm.

to 2 cm. **Cuticle** thin and smooth or medium-thick and striate. **Epidermis**: cells large and square, flat externally and internally or shortly papillate. **Stomata** occasional, level with surface. **Collenchyma**: strands ± circular (type Sv) (Fig. 84G,H); separated from epidermis by 1–2 ground tissue layers and frequent in *X. mafaffa*, or in two peripheral rows in *X. robustum*. Strands semicircular in *X. robustum* but separated from phloem by several rows of ground tissue. Largest petiole with outer peripheral collenchyma band 3–4 layers deep, in addition to type Sv strands. **Ground tissue**: compact large cells in peripheral zone. Central zone with type 3 or type 4 aerenchyma, or mixed with large and small cavities and vascular bundles in corner islands of ground tissue.

Vascular bundles: small peripheral bundles subtended by collenchyma strands; central zone with larger and numerous scattered bundles (types II, mostly III). Xylem: 1–2 wide metaxylem cells, occasionally one narrow protoxylem lacuna. Phloem: large circular or broader strand (Fig. 84F). **Laticifers** anastomosing, 2–4 cells lateral to phloem strands and occasionally branching into adjacent ground tissue. **Tannin** cells common in partitions and other ground tissue in some specimens. **Crystals**: raphides, mostly in circular, short-elongate cells, not biforines. Cells attached by one end to aerenchyma partitions and very common in some species (Fig. 84F). Druses in small circular cells partially embedded in partitions, less frequent.

Stem TS

Outline circular. **Size**: diameter *c.* 9 mm. **Cuticle** discontinuous, medium-thick and smooth. **Epidermis**: cells mostly small. **Collenchyma**: 2–3 subepidermal layers of outer cortex. Also, circular or semicircular strands as phloem caps to peripheral bundles, or as 2–3 irregular rows of independent small strands in outer cortex (Fig. 84H). **Ground tissue**: cells small near periphery, larger in centre. **Air spaces**: large in centre as numerous type 4 aerenchyma cavities. **Vascular bundles** numerous and scattered, type III, plus numerous very small bundles. Xylem: one(two) wide protoxylem cells. Phloem strands mostly broad and circular. **Laticifers** lateral to phloem and anastomosing; ramifying into nearby ground tissue. **Starch** in ground tissue near phloem. Granules 3–4 µm diameter.

Material examined

Xanthosoma helleborifolium (Jacq.) Schott, Rauh 37175, cult. M; *X. mafaffa* (Schott) Engl., P. Tuley, s.n., 1963, Nigeria; *X. pubescens* Poepp., Bogner 2339, cult. M; *X. robustum* Schott, Croat 34243, Panama; *X. sagittifolium* (L.) Schott, Metcalfe s.n., Canal Zone; *X. striatipes* (Kunth & Buché) Madison, Bogner 1231, Brazil, cult. M; *X. weeksii* Madison, Bogner 1479, Ecuador, cult. M.

Species reported on in the literature

Cody and Horner (1983): *X. sagittifolium*: crystals. • Dalitzsch (1886): several spp.: epidermis, stomata. • Decrock (1909): *X. sagittifolium:* starch. • Deinega (1898): *X. belophyllum:* leaf. • Elisei (1941): *X. violaceum*: casparian strips. • Engler and Krause (1920): several spp.: root, leaf. • Ertl (1932): *X. atrovirens, X. lindenii, X. robustum*: leaf venation. • Franceschi and Horner (1980): *X. sagittifolium*: crystals. • French (1987c): several spp.: root. • French (1988): several spp.: laticifers. • French and Tomlinson (1983): *X. sagittifolium, X. tarapotense* : stem vasculature. • Fujimoto *et al.* (1990): *X. sagittifolium*: starch. • Gill and Karatela (1983): stomata. • Guttenberg (1968): root • Hanstein (1864): *X. sagittifolium*: laticifers. • Hassack (1886): *X. lindenii*: leaf. • Knecht (1983): *X. nigrum*: leaf venation. • Korner *et al.* (1983): *X. sagittifolium*: leaf. • Lierau (1888): several spp.: root. • Linsbauer (1930): epidermis. • Marcatili (1887): *X. roseum*: laticifers. • Melati and Scialabba (1984–5): *X. caracu*: leaf, corm, root. • Möbius (1936): *X. maximilianum*: bulbil. • Molisch (1899): *X. maximilianum*: tissues. • Müller (1919): leaf tip. • Netolitzky (1929): crystals. • Onken (1922): *X. maximilianum, X. violaceum*, *X.* sp.: laticifers. • Osisiogu and Uzo (1973): corm, raphides. • Pant and Kidwai (1966): *X. maximilianum*: leaf, stomata. • Pirotta and Marcatili (1885): *X. roseum*: laticifers. • Planchon and Juillet (1909): *X. sagittifolium*: starch. • Sakai and Hanson (1974): *X. sagittifolium, X. violaceum*: raphides. • Sakai *et al.* (1972): *X. sagittifolium*: raphides. • Sakai *et al.* (1984): raphides. • Scott (1889): leaf, laticifers. • Solereder and Meyer (1928): *X. lindenii, X. robustum, X. sagittifolium*: laticifers; anatomy. • Tieghem (1866): *X. violaceum*: all organs. • Tieghem (1870–1): *X. violaceum*: root, secretory tissue. • Trécul (1865, 1866): several spp.: vasculature. • Winton and Winton (1935): tubers. • Zinke (1924): *X. lindenii*: leaf.

Additional literature

Bahadur *et al.* 1977; Davis 1966; Erdtman 1952; Eyde *et al.* 1967; Fox and French 1988; French 1986a,c, 1987a; Grayum 1990; Hegnauer 1986; Kenten and Woods 1973; Loh *et al.* 2000; Mayo and Bogner 1988; Meusel 1951; Ray 1988; Richter 1929; Simmonds 1976; Thanikaimoni 1969; Troll 1928; Williams *et al.* 1981.

Chlorospatha Engl.

This formerly monotypic genus now contains about 15 species, including species formerly assigned to *Caladiopsis*. Species are found in northern tropical South America. They grow in wet ravines or boggy areas of forests, at middle to low elevations. The plants are terrestrial rhizomatous understorey herbs with leaves cordate to hastate to pedately compound. Chromosomes: 2n = 26. Three taxa were available for study.

Leaf surface

Cuticle smooth on both surfaces. **Epidermis**: cells polygonal with straight anticlinal walls on both surfaces (1–2:1 *l/w*). **Stomata** abaxial and brachyparacytic (Fig. 85D). Subsidiary cells broad or narrow, rounded or angular. **Venation**: midrib and basal lobe primaries arising from summit of petiole. Midrib prominent, secondaries acute or narrowly acute, curved and apically ascending, widely spaced or irregular. Tertiaries variable, sometimes parallel with secondaries or reticulate. Higher orders reticulate and merging marginally at ± straightened collector vein.

Leaf TS

Structure dorsiventral. **Size**: 48 × 26 mm (*l/w*). **Midrib** prominent, semicircular or ribbed abaxially (Fig. 85A). **Cuticle** thin. **Epidermis**: cells large adaxially and formed of rounded cells. Abaxial cells small to large, variably shaped and occasionally papillate. **Stomata** level with surface. **Mesophyll**: two columnar palisade layers (4–6:1 *l/w*) forming 25% of mesophyll area (Fig. 85C). Spongy zone of *c*. 5 layers of irregularly shaped, short-armed cells forming an aerenchyma of large, rounded, usually substomatal air cavities (type 2). Some aerenchyma cells with 1-seriate partitions, one partition cell between corner cells (type 3). **Air cavities** forming about 80–90% of spongy layers. **Collenchyma**: small strands subtending peripheral bundles (type Sv) (Fig. 85A,B).

Vascular bundles in midrib *c*. 12 around abaxial periphery and *c*. 8 scattered in centre. Bundles not large or well-differentiated (type III). Xylem: one protoxylem cell with one metaxylem cell occasionally present. Phloem: a circular strand. **Laticifers**: 1–2 anastomosing tubes lateral to each vascular bundle (Fig. 85E); branches may penetrate palisade tissue, ending at epidermis. **Tannin** cells with dark contents common in palisade. **Crystals**: raphides rare in single bundles in narrow, rounded-ended, thin-walled biforine-like cells oriented across aerenchyma partitions (Fig. 85G), or in rounded cells (Fig. 8E) and uncommon; raphides subepidermal in midrib. Druses large, very common in mid-mesophyll, often bordering large vascular bundles (Fig. 85F). Small druses very common in palisade cells. Crystal sand in abaxial guard cells of stomata.

Petiole TS

Outline semicircular at base, adaxially shallowly grooved distally. **Size**: 5–7 mm diameter. **Cuticle** thin and striate. **Epidermis**: cells large and rounded. **Collenchyma**: peripheral rounded strands, *c*. 12 layers deep, type Sv, 1–3 layers beneath epidermis. **Chlorenchyma**: chloroplasts sparse in outer 6–8 layers of ground tissue. **Ground tissue**: cells mostly compact, scattered type 3 or 4 cavities near centre. **Vascular bundles**: scattered, two sizes: largest type III, mixed with very small bundles comprised almost entirely of phloem. **Sclerenchyma**: fibres in 1–3 layers capping phloem of type III bundles in one specimen of *C. longipoda*. **Laticifers** anastomosing, 2–3 lateral to vascular bundles and ramifying throughout ground tissue. **Tannin** cells occasional in ground tissue. **Starch** granules occasional in outer ground tissue. **Crystals**: druses common in compact ground tissue and at margins of air cavities.

Material examined

Chlorospatha longipoda (K.Krause) Madison, 23-76-122, cult SEL, Plowman *et al.* 3979, Ecuador, cult. M; *C.* n. sp., Colombia, cult. SEL; *C.* sp., Madison 2527, cult. SEL

Taxonomic notes

Specimens originally obtained as *Caladiopsis* and *Chlorospatha* are nearly identical anatomically.

Species reported on in the literature

French (1987c): several spp.: root. • French (1988): several spp.: laticifers. • French and Tomlinson (1983): *C. castula*: stem, vasculature.

Additional literature

French 1986a,c, 1987a; Grayum 1990; Madison 1978.

Ulearum Engl.

This genus includes two species from the upper Amazon valley, Peru and Brazil. Plants are terrestrial understorey rhizomatous herbs growing on leaf litter of wet or humid

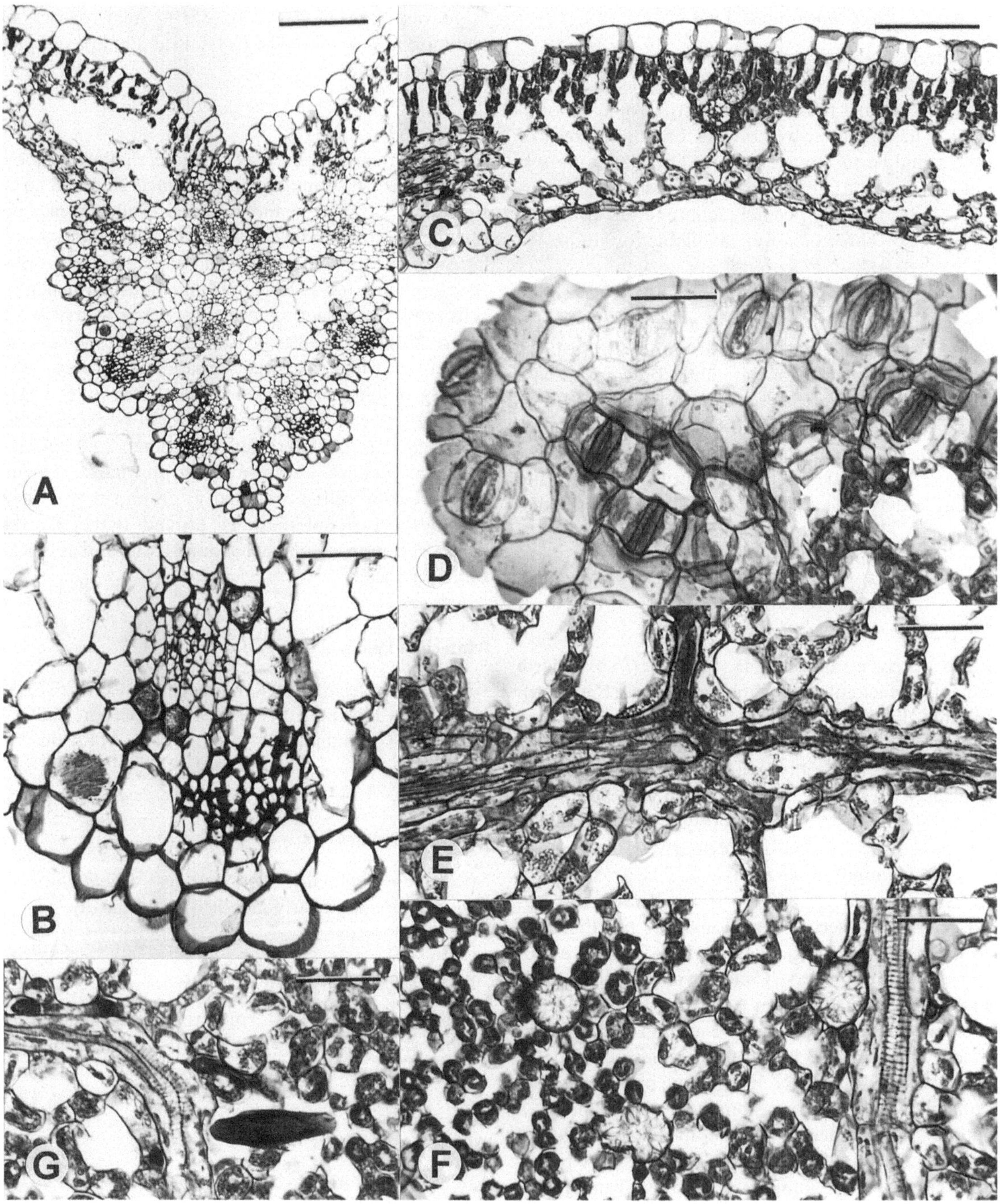

Fig. 85. *Chlorospatha* leaf sections. (A–C, E–G) *Chlorospatha* sp.; (D) *C. longipoda*. (A,B) Midrib TS; (B) Note type III vascular bundle subtended by collenchyma strand; (C) Lamina TS; (D) Leaf abaxial epidermis PS showing brachyparacytic stomata; (E,F,G) Mesophyll PS; (E) Anastomosing laticifers along vein; (F) Palisade/spongy tissue transition and druse cells; (G) Raphide cell near vein. Scale bars: A = 200 μm; B,D–G = 50 μm; C = 150 μm.

primary forests. The leaves are sagittate and long-petiolate, arising singly from the end of a rhizome. Chromosomes: 2n = 14. One species was available for study.

Leaf surface

Cuticle smooth or finely striate. **Epidermis**: cells polygonal with straight or slightly curved anticlinal walls on both surfaces (1–3:1 *l/w*). **Stomata**: abaxial and brachyparacytic (Fig. 86C). Subsidiary cells broad and variably shaped. **Venation**: midrib prominent, arising from summit of petiole with basal lobe primaries. Midrib secondaries brochidodromous; eucamptodromous nearer base. Higher orders reticulate.

Leaf TS

Structure dorsiventral. **Size**: lamina length 10 cm, petiole to 20 cm. **Midrib** rounded-grooved adaxially, broadly and shallowly convex abaxially (Fig. 86A). **Cuticle** thin. **Epidermis**: cells tall or square on both surfaces, flat to rounded externally and rounded internally. **Stomata** level with surface. **Hypodermis** occasional as patches where palisade layer becomes unpigmented adaxial layer, probably the cause of reported leaf variegation. **Mesophyll**: palisade single-layered, weakly differentiated from spongy layers; palisade cells flared, widest beneath adaxial epidermis (Fig. 86B). Numerous plastids attached in finger-like groups or mats appressed against palisade cell walls. Spongy cells in 4–5 layers, up to 75% of mesophyll; cells multi-armed, type 1a (Fig. 86D). **Air spaces**: *c.* 25% of spongy tissue. **Collenchyma**: *c.* 5 discrete subepidermal strands, forming caps over phloem of abaxial midrib vascular bundles (type Sv).

Vascular bundles in abaxial and adaxial rows in midrib (types II, III) (Fig. 86A). Largest bundles with one to several irregular strands of narrow protoxylem elements proximal to small irregular phloem strands. **Secretory tissue**: specialized cells not detected but dissected tissue exuding copious mucilage. **Laticifers**: anastomosing, following all vascular bundles except smallest, frequently ramifying among mesophyll cells and basal anticlinal walls of epidermal layers. **Starch** granules common in some plastids of all mesophyll and midrib ground tissue cells. **Crystals**: raphides in single bundles in small cells or in multiple overlapping bundles in elongate cells in upper mesophyll. Crystal sand occasional in adaxial epidermis.

Petiole TS

Outline: throughout length, deeply V-grooved adaxially (Fig. 86E), rounded abaxially. **Size**: 2–3 mm diameter. **Cuticle** thin, smooth or finely striate. **Epidermis**: cells medium-sized, rounded internally and externally. **Collenchyma**: subepidermal strands and as phloem bundle caps (Fig. 86E,F) to peripheral bundles (types Sv, Svc). **Chlorenchyma**: present in some outer parenchyma, especially clustered near peripheral vascular bundles. **Ground tissue** uniform; occasional corner air spaces; in LS, cells elongate to 6–8:1 *l/w*.

Vascular bundles: well-organized, forming row of peripheral bundles capped by phloem and collenchyma, and inner circle of bundles (Fig. 86E) plus two central bundles. Xylem of 5–10 protoxylem elements forming irregular clusters. Phloem forming irregular strands proximal to xylem, always oriented toward organ periphery (type II). **Secretory tissue**: no specialized cells; dissected organ exuding copious mucilage. **Laticifers** not detected (but see Leaf TS above); if present, then not near phloem. **Starch** granules densely present in uniseriate boundary between phloem and phloem cap, very common in all ground tissue. **Crystals**: raphides common in multiple overlapping bundles in long raphide tubes in ground tissue. Crystals sand occasional in ground tissue.

Stem TS

Outline terete. **Size**: 5–7 mm diameter. **Cork**: 4–6 layers of thin-walled cells. **Collenchyma** subepidermal, 1–2 layers of discrete strands and forming caps over phloem of peripheral bundles (types Sv, Svc). **Ground tissue** uniform with few air spaces.

Vascular bundles apparently randomly scattered, clustered on side nearest leaf base; collateral or in amphivasal clusters (Fig. 86G). Protoxylem and phloem forming irregular strands. **Secretory tissue**: no specialized cells detected but dissected tissue exuding mucilage. **Starch**: large spheroidal granules abundant in all ground tissue. **Crystals**: raphides in small single bundles in short cells near periphery, or as multiple bundles in elongate cells.

Root TS

Outline: terete, 1.5 mm diameter. **Periderm**: several thin-walled layers formed from periclinal divisions of epidermis. **Ground tissue** uniform, cells mostly in radial files. **Stele** delimited by endodermis with thin casparian strips in anticlinal walls. **Vascular tissue**: five xylem and phloem poles. Central ground tissue not sclerotic.

Material examined

Ulearum sagittatum Engl., Jangoux *et al.* 85–058 21–30. cult. M.

Taxonomic notes

This genus fits the anatomical pattern for the tribe, but is distinguished by an unusual volume of mucilaginous

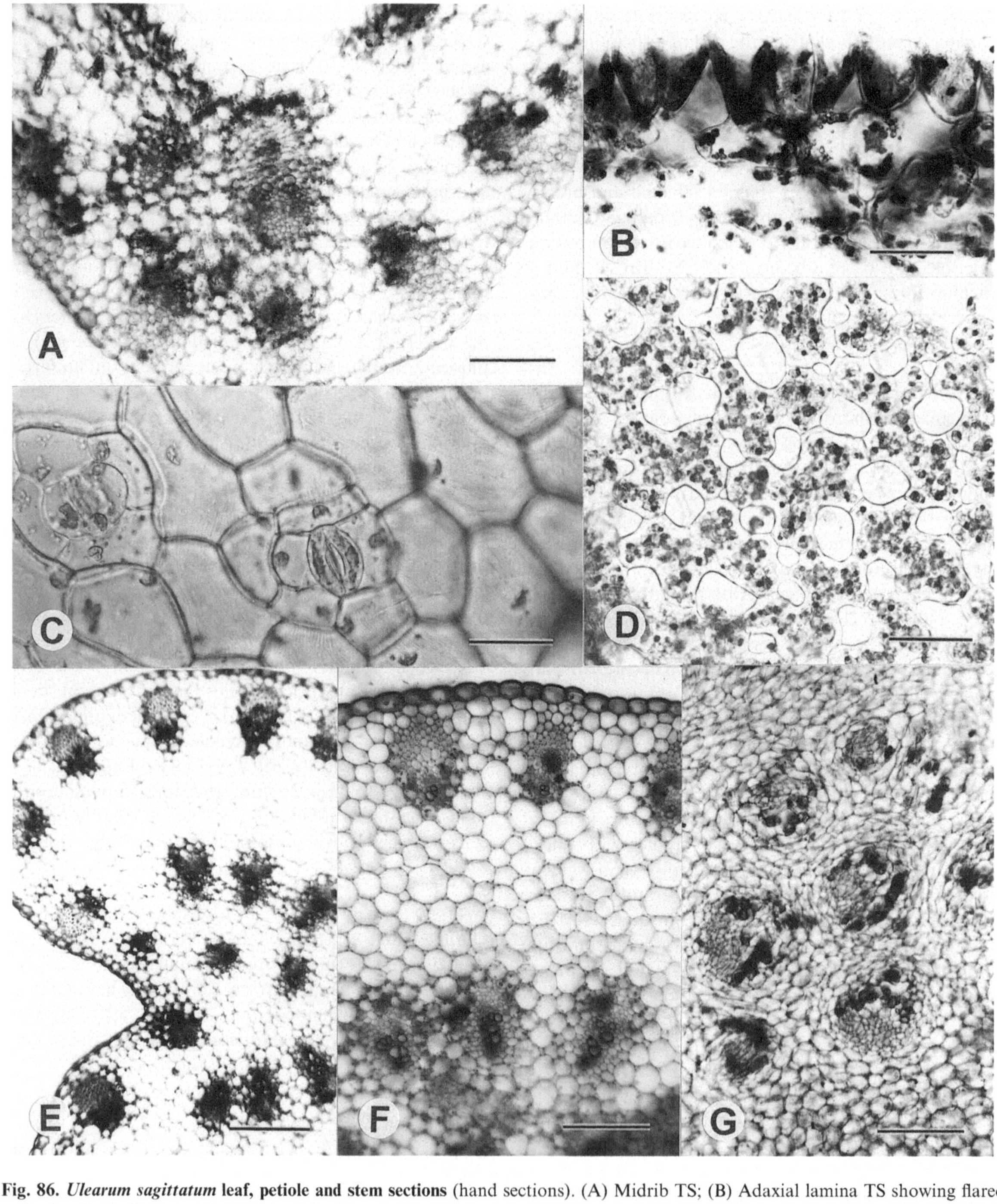

Fig. 86. ***Ulearum sagittatum* leaf, petiole and stem sections** (hand sections). (A) Midrib TS; (B) Adaxial lamina TS showing flared palisade cells (iodine mount); (C) Abaxial epidermis PS with stomata; (D) Mesophyll PS showing spongy tissue (type 1a); (E) Petiole TS, adaxial surface on left. Note distinct outer and inner rows of vascular bundles; (F) Petiole TS, lateral surface uppermost; (G) Stem TS central tissue showing amphivasal and collateral bundles. Scale bars: A,F,G = 200 μm; B–D = 50 μm; E = 300 μm.

exudation from the cut stem. It apparently originates from ground tissue since specialized ducts are absent.

Additional literature

Engler 1905; Grayum 1990.

Filarum Nicolson

This monotypic genus is found in Loreto, Peru. Plants are terrestrial tuberous understorey herbs of the primary forest. The leaves are entire or cordate and long-petiolate. Chromosomes: 2n = 28. Two young specimens were available for study.

Leaf surface

Cuticle smooth or finely striate. **Epidermis**: cells large and polygonal (Fig. 87D) with straight anticlinal walls on both surfaces (1–1.5:1 *l*/*w*). **Stomata** abaxial and brachyparacytic (Fig. 87E,F). Subsidiary cells broad and square-cornered or irregular. **Venation**: digitately veined from petiole summit; primaries five or more. Lateral primaries curving broadly, ending eucamptodromously near margin. Secondaries ± orthogonally branched or acute from midvein. Higher orders reticulate.

Leaf TS

Structure dorsiventral and weakly differentiated. **Size** *c*. 20 × 30 mm (*l*/*w*). **Midrib** slightly concave adaxially and rounded-convex-ribbed abaxially (Fig. 87A). **Cuticle** thin to absent, often patchy. **Epidermis**: cells large on both surfaces, flat externally and rounded internally. **Stomata** protruding or level with surface. **Mesophyll** poorly differentiated (Fig. 87B). Palisade: one layer of large cells, 30–50% of lamina. Spongy cells in 2–3 layers, type 1. Chloroplasts most dense in palisade cells. **Air cavities**: 30–40% of spongy layer. Substomatal cavities shallow. **Collenchyma** cells angularly thickened in strands abaxial to larger veins (type Sv).

Vascular bundles: 1–4 small midrib bundles (type II) (Fig. 87A). Xylem: 1–6 tracheary cells opposing small circular phloem strand (Fig. 87C). Secondary vascular bundles very small and widely spaced. **Laticifers** anastomosing, 1–2 lateral to larger vascular bundles (Fig. 87C). **Starch** granules common in mesophyll cells and midrib; most frequent surrounding veins. **Crystals**: raphides in elongated cells, one to several bundles per cell. Raphide cells also with dark granular contents. Crystal sand and small prismatics common in mesophyll and epidermal cells.

Petiole TS

Outline semicircular, shallowly grooved adaxially, abaxially fluted, ridges subtending collenchyma strands (Fig. 87G). **Size**: diameter 2 mm. **Cuticle** thin and finely striate. **Epidermis**: cells small and rounded. **Stomata** occasional and level with surface. **Collenchyma**: circular strands, up to eight layers deep, separated from epidermis by a layer of ground tissue; cells angularly thickened, each subtending small vascular bundle (type Sv). **Chlorenchyma**: chloroplasts sparse throughout ground tissue. **Ground tissue**: cells large throughout. **Air cavities** small at cell corners.

Vascular bundles: peripheral row of 10–12 bundles; central group of five symmetrically disposed bundles (Fig. 87G). Bundles small and unsheathed (type II). Xylem: 4–8 metaxylem elements in 1–2 irregular files. Phloem: small circular or irregular strand opposing xylem. **Laticifers**: anastomosing, 1–2 lateral to most phloem strands. Anastomosing plexuses extending into several layers of ground tissue near some vascular bundles (Fig. 87H). **Crystals**: raphides occasional in ground tissue cells near peripheral vascular bundles. Crystal sand clusters and small prismatics, common and random in ground tissue.

Material examined

Filarum manserichense Nicolson, Wurdack 2402, Peru, K; Sizemore, Peru, cult. M.

Additional literature

French 1986b; Grayum 1990; Thanikaimoni 1969.

Zomicarpella N.E.Br.

This genus contains two species found in Colombia. The plants are understorey herbs with creeping rhizomes that grow in unflooded areas of low primary forest. Leaves are long-petiolate and hastate or sagittate. Chromosomes: 2n = 26. Two specimens were available for study.

Leaf surface

Cuticle smooth on both surfaces. **Epidermis**: adaxial and abaxial cells polygonal with straight anticlinal walls (adaxial 1–2:1 *l*/*w*); abaxial cells often more elongated (1–3:1 *l*/*w*). **Stomata** abaxial, variably brachyparacytic (Fig. 88B) or brachypara-tetracytic. Subsidiary cells mostly broad and square-edged or occasionally irregular. **Venation**: multiple primaries emerging palmately from

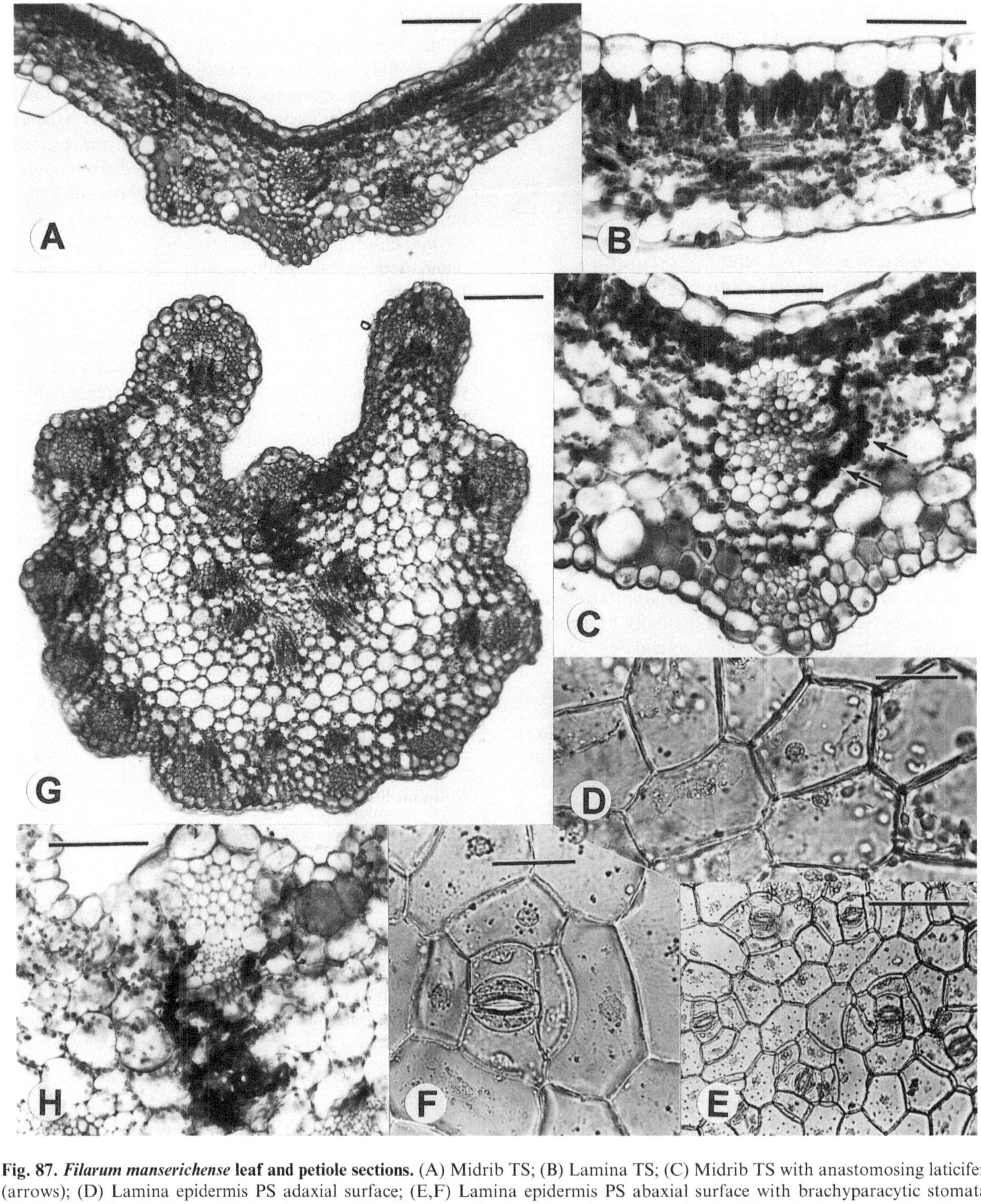

Fig. 87. *Filarum manserichense* leaf and petiole sections. (A) Midrib TS; (B) Lamina TS; (C) Midrib TS with anastomosing laticifers (arrows); (D) Lamina epidermis PS adaxial surface; (E,F) Lamina epidermis PS abaxial surface with brachyparacytic stomata; (G) Petiole TS; (H) Petiole TS showing adaxial central vascular bundle and anastomosing laticifers. Scale bars: A,G = 300 μm; B,C,E,H = 150 μm; D,F = 50 μm.

petiole apex, and joining a submarginal collector vein. Connections mostly brochidodromous, eucamptodromous at base. Smaller marginal veins also present. Higher order veins reticulate.

Leaf TS

Structure dorsiventral. **Size**: 75 × 45 mm (*l*/*w*). **Midrib** shallowly concave adaxially and strongly convex-rounded and ridged abaxially. **Cuticle** thin on both surfaces. **Epidermis**: cells medium-sized to large on both surfaces or smaller abaxially. Adaxial cells flat or rounded externally and internally; abaxial cells rounded on both surfaces. **Stomata** level with epidermal surface. Outer guard cell margins with weakly formed cuticular ledges. **Mesophyll** *c*. 6 layers, the uppermost a weakly defined palisade layer (Fig. 88A), cells widened beneath epidermis (1–2:1 *l*/*w*). Some mesophyll, mostly palisade, continuous beneath adaxial surface of midrib. Spongy cells short- or longer-armed, type 1, or type 2 especially in substomatal regions. **Air cavities** up to *c*. 10–30% of spongy tissue. **Collenchyma**: cells angularly thickened in *c*. 6 discrete subepidermal circular strands, abaxial in midrib; some directly subtending a vascular bundle (type Sv).

Vascular bundles: 7–9 scattered bundles in midrib (type II). Xylem: cluster of 2–3 wide metaxylem elements; several narrower metaxylem elements in largest bundles. Clusters surrounded by small ground tissue cells. Adjacent rounded phloem strand oriented toward nearest organ surface. **Laticifers** prominent, anastomosing, with contents dark or light and barely detectable; cells lateral to phloem, 1–2 on each side. **Crystals**: raphides of three types common in mesophyll: (1) biforine cells with crystals in narrow single bundles up to 100 μm long; (2) raphides in multiple overlapping bundles in irregular elongate cells; and (3) raphides in small broad bundles in short thin-walled cells. Druses common, especially immediately beneath adaxial epidermis. Crystal sand common, in most epidermal and mesophyll cells.

Petiole TS and LS

Outline: proximal end crescent-shaped abaxially; distal end terete, or flattened adaxially and rounded to regularly furrowed (Fig. 88C). **Size**: diameter 3 mm. **Cuticle** thin and smooth. **Epidermis**: cells small and rounded externally. **Collenchyma**: *c*. 20 large circular strands, 2–3 layers beneath epidermis, of *c*. 12 layers of angularly thickened cells. Strands aligned with ridges and subtending circle of peripheral bundles (type Sv) (Fig. 88C,D,F). **Ground tissue** more or less uniform; outer layers bearing some chloroplasts.

Vascular bundles: a central cluster of 9–11 scattered bundles, and peripheral row of bundles aligned with collenchyma strands; types II or III. Xylem: one wide protoxylem cell or cluster of several wide and narrower metaxylem cells, adjacent to a laterally compressed strand of phloem. Whole bundles appearing laterally narrowed. Vascular bundle cross connections frequent in petiole. **Laticifers** very common, lateral to phloem, cells anastomosing (Figs 8D, 88E,G). **Tannin cells** scattered in ground tissue. **Starch** granules infrequent in vascular bundle sheaths. **Crystals**: raphides in single bundles, cells more or less aligned longitudinally and common in ground tissue. Raphides also in multiple overlapping bundles in axially elongated tubular cells. Druses scattered in ground tissue. Crystal sand common in ground tissue.

Rhizome TS

Outline rounded. **Ground tissue** uniform, cells compact, mostly metabolite storage tissue. **Vascular bundles**, especially toward centre, apparently compound (amphivasal). **Starch** granules mostly rounded and irregular, densely packed in ground tissue except in outermost layers. **Crystals**: raphides in single bundles in elongate or broad cells, common in outer tissues.

Material examined

Zomicarpella amazonica Bogner, Croat 71763, cult. MO; Bogner 1985, Amazonas, Brazil, cult. M.

Additional literature

French 1986c; Grayum 1990.

Caladium Vent.

This genus of 17 species is distributed in tropical South America. *Aphyllarum tuberosum* is included here. Plants are terrestrial, occasionally epilithic, usually understorey tuberous herbs, especially common in semideciduous forests, open areas of forests, or creek banks. The plants have two leaves subtending the spathe. Leaves are long-petiolate, usually peltate, cordate, hastate or sagittate. Chromosomes are medium-sized: 2n = 22, 26, 28, 30, 48. Three species were available for study.

Leaf surface

Cuticle smooth on flat surfaces. **Epidermis**: cells polygonal with straight anticlinal walls on both surfaces (1–1.5:1 *l*/*w*). **Stomata** abaxial, or on both surfaces, usually

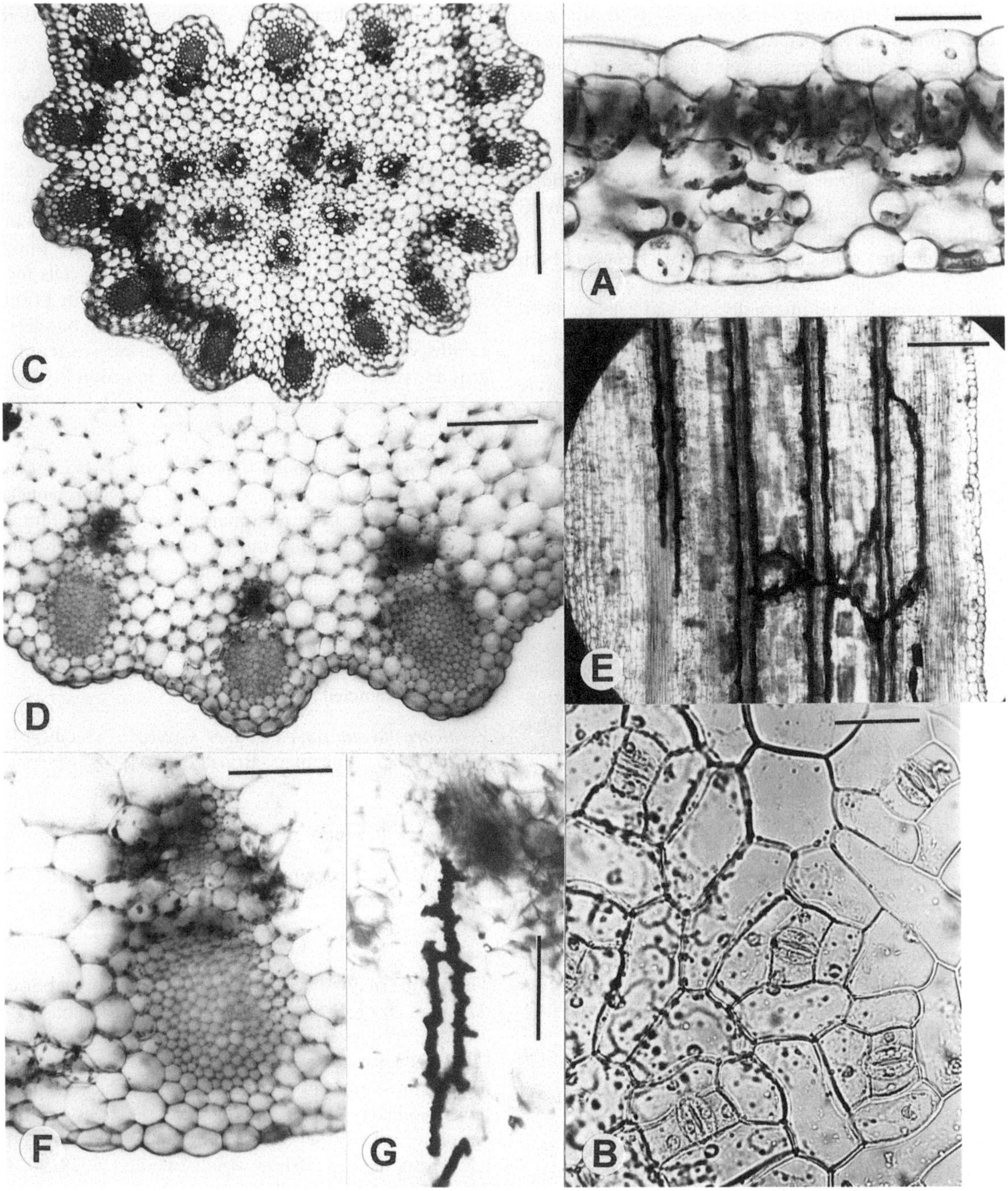

Fig. 88. *Zomicarpella amazonica* leaf and petiole sections. (A) Lamina TS; (B) Leaf abaxial epidermis PS showing brachyparacytic stomata; (C,D) Distal petiole TS showing fluted margin with vascular bundles and collenchyma strands (type Sv) within ridges; (E) Petiole LS showing anastomosing laticifers connecting between vascular bundles; (F) Petiole TS showing collenchyma strand and anastomosing laticifers in ground tissue and near bundle; (G) Petiole TS showing anastomosing laticifer in ground tissue. Scale bars: A,B = 50 μm; C,E = 300 μm; D = 200 μm; F,G = 150 μm.

brachyparacytic (Fig. 89F) with two or more pairs of subsidiary cells surrounding some stomata. Subsidiary cells narrow, broad or trapezoidal, often with square corners. **Venation**: midrib and two basal lobe primaries arising at petiole summit. Midrib prominent at base but dissipating as secondaries diverge laterally and regularly, base to apex. Resulting secondaries broadly acute or narrowly acute, ending at a straightened brochidodromous submarginal collector. Tertiaries tending to be parallel with secondaries or irregular, tending to merge in intercostal areas forming composite intersecondaries. Higher orders reticulate.

Leaf TS

Structure dorsiventral. **Size** 40 mm (midrib length) × 50 mm (*l*/*w*); lobes 40 × 30 mm (*l*/*w*). **Midrib** prominent, broadly and shallowly concave adaxially (Fig. 89D), broadly semicircular-convex abaxially (Fig. 89B). **Cuticle** thin or not detected. **Epidermis**: cells tall adaxially, less rounded internally. Abaxially, cells flat, elongate or occasionally large and rounded. **Stomata** level with surface. **Mesophyll** well-differentiated with palisade of 1–2 layers, up to 30% of mesophyll (Fig. 89A). Spongy tissue of 5–8 layers of irregularly armed type 1 cells. **Air cavities** large and irregular, mostly type 3 in midrib. **Collenchyma**: discrete subepidermal strands subtending small vascular bundles on abaxial side of midrib (type Sv).

Vascular bundles: bundles scattered asymmetrically in midrib (type II) (Fig. 89E), with parenchymatous sheath. Xylem: 1–2 large protoxylem lacunae, and 0–3 narrow metaxylem cells. Phloem a small circular strand. **Laticifers** anastomosing, having dark contents, lateral to phloem along most vascular bundles and ramifying into palisade or other mesophyll tissue (Fig. 89C). **Tannin** cells occasional in mesophyll. **Crystals**: raphides as single bundles, often obliquely arranged in *C. lindenii*, common in small cells in midrib and mid-mesophyll (Fig. 89G), or absent. Cells often suspended across partitions in midrib aerenchyma (Fig. 89I). Biforines suspended into air cavities in upper spongy mesophyll (Fig. 89H), occasional in *C. lindenii* and *C. tuberosum*. Occasional elongated raphide cells with small oblique crystal bundles in *C. tuberosum*. Druses small and occasional in mesophyll, especially in palisade layer. Crystal sand as coarse grains common in spongy cells.

Petiole TS

Outline semicircular or concave-grooved adaxially. **Size** 5–9 mm diameter. **Cuticle** thin and finely striate. **Epidermis**: cells small and level with surface. **Collenchyma** strands circular or elliptic around perimeter, 1–3 layers beneath epidermis (Fig. 89J). Strands 3–10 layers deep (type Sv), each strand subtending a small vascular bundle. **Chlorenchyma**: plastids few in outer 4–6 ground tissue layers. **Ground tissue**: outer 5–12 layers of compact cells; large, type 3 aerenchyma in centre.

Vascular bundles numerous and scattered. Ertl (1932) noted vasculature consisting of numerous bundles arranged in 2–3 concentric 'U's. The most adaxial and lateral bundles diverging near base, becoming secondaries, followed successively (abaxially) by more bundles. Central abaxial vascular bundle forming residual midrib at leaf tip. Bundles small, unsheathed, type III (Fig. 89J). Xylem: wide protoxylem cell, 0–1 narrow metaxylem cells. Phloem a small, irregular and laterally-narrowed strand. **Laticifers**: 2–3 lateral to vascular bundles, anastomosing and ramifying into adjacent ground tissue. **Starch** granules scattered throughout ground tissue. **Tannin cells** uncommon in all ground tissue in *C. macrotites*. **Crystals**: raphides common in ground tissue and partition cells in *C. macrotites*, absent in other specimens. Druses: large crystals in enlarged circular cells scattered throughout ground tissue, often common near vascular bundles.

Material examined

Caladium lindenii (André) Madison, Bogner 2338, cult M; *C. macrotites* Schott, Bogner 1279, Brazil/Bolivia boundary, M; *C. (Aphyllarum) tuberosum* (S.Moore) Bogner & Mayo, Harley 10477, Brazil, K.

Species reported on in the literature

Bergdolt (1955): *C. bicolor, C. hybridum*: leaf. • Bonnier (1904): *C. bicolor, C. odorum*: root. • Chrysler (1904): *C. bulbosum*: stem. • Dalitzsch (1886): *C. bicolor, C. duchartrei*: leaf. • Deinega (1898): *C. antiquorum*: leaf. • De La Rue (1866): *C. odorum*: leaf. • Engler (1920b): *C. marmoratum*: leaf. • Ertl (1932): *C. bicolor*: leaf. • Fenizia (1896): *C. violaceum*: leaf. • French (1987c): several spp.: root. • French (1988): *C. bicolor, C. macrotites*: laticifers. • French and Tomlinson (1983): *C. plowmannii, C. riopalenquensis*: stem. • Gentner (1905): *C. esculentum*: leaf. • Gulliver (1865): *C. viviparum*: crystals. • Guttenberg (1968): root. • Hanstein (1864): *C. sagittifolium*: laticifers. • Hassack (1886): *C. duchartrei*: leaf. • Kroemer (1903): *C. hybridum*: root. • Leavitt (1904): *C. esculentum*: root. • Lierau (1888): *C. bicolor*: root. • Linsbauer (1930): *C. bicolor*: epidermis. • Mangin (1882): root. • Marcatili (1887): *C. erubescens*: leaf, laticifer. • Molisch (1903): *C. nymphaeifolia*: leaf. • Müller (1919): *C. hastatum, C. odorum, C.* sp.: leaf tip. • Netolitzky

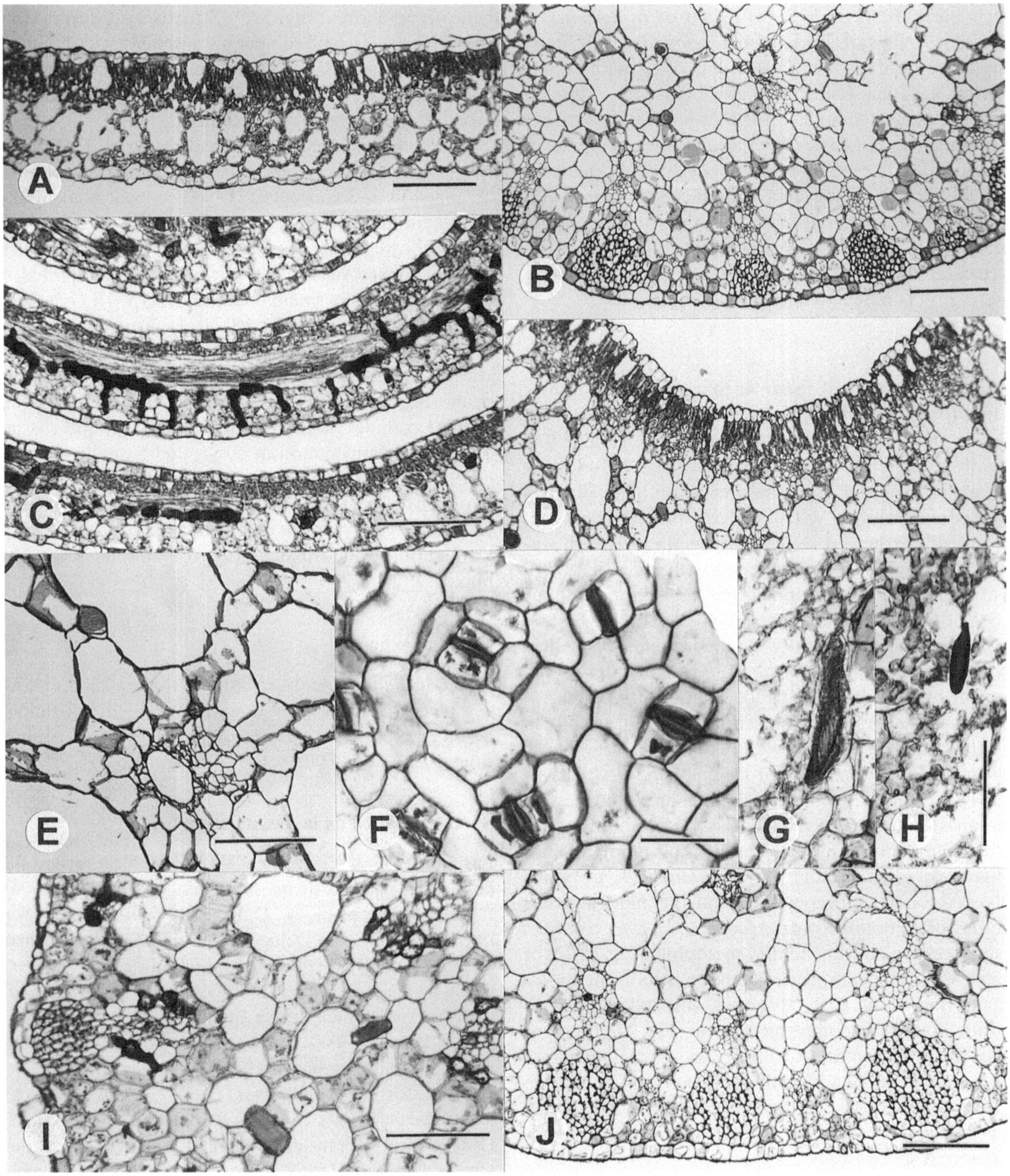

Fig. 89. *Caladium* leaf and petiole sections. (A–E,I,J) *C. macrotites*, (F–H) *C. lindenii*; (A) Lamina TS, note substomatal air cavities in palisade layer; (B) Midrib abaxial surface; (C) Young lamina TS showing anastomosing laticifers along vein; (D) Midrib TS showing adaxial surface and palisade layer; (E) Midrib TS with central type II vascular bundle; (F) Abaxial lamina epidermis PS with stomata; (G) Lamina PS showing elongated raphide cell; (H) Lamina PS showing biforine cell; (I) Midrib margin. Note raphide cells oriented across aerenchyma partitions; (J) Distal petiole TS showing collenchyma strands (type Sv) and type III vascular bundles. Scale bars: A,B,D = 300 μm; C,E,(G,H: one scale bar), I = 150 μm; F = 50 μm; J = 200 μm.

(1929): crystals. • Nyawuane and Gill (1990): *C. picturatum*: stomata. • Pant and Kidwai (1966): *C. bicolor*: leaf, stomata. • Pirotta and Marcatili (1885): *C. erubescens*: laticifers. • Roth (1949, 1952): leaf. • Schmidt (1882): *C. marmoratum*: laticifers. • Schwendener (1882): endodermis. • Shishkoff (1987) root. • Solereder and Meyer (1928): several spp.: secretory tissue, crystals. • Stahl (1896): *C. bicolor*: leaf. • Tieghem and Douliot (1888): *C. bicolor*: root. • Zinke (1924): *C. duchartrei*: leaf.

Additional literature

Domin 1911; Erdtman 1952; French 1986a,c, 1987a; Gertz 1916a; Gottsberger 1992; Grayum 1990; Loh *et al.* 2000; Mayo and Bogner 1988; Mookerjea 1955; Ray 1988; Thanikaimoni 1969; Tieghem 1907; Wheaton and Stewart 1970; Williams *et al.* 1981; Zettler and El-Nil 1979.

Scaphispatha Brongn. ex Schott

This monotypic genus is found in southern Peru and Brazil. Plants are terrestrial, tuberous herbs occurring in open grassy areas or temporary wet spots. The leaves are long-petiolate and hastate-peltate with rounded basal lobes. Chromosomes: 2n = 28. One specimen was available for study.

Leaf surface

Cuticle smooth on both surfaces. **Epidermis**: cells large, polygonal with straight anticlinal walls (1–2:1 *l*/*w*) on both surfaces (Fig. 2E). **Stomata**: abaxial and brachyparacytic (Fig. 90D) with broad, mostly angular subsidiary cells. **Venation**: acrodromous secondaries forming midrib and vascularization of basal lobes. Upper midrib prominent with pinnate secondaries. Tertiaries and higher orders reticulate.

Leaf TS

Structure dorsiventral. **Size**: 82 × 83 mm (*l*/*w*). Basal lobe length 45 mm. **Midrib** flat to shallowly concave adaxially, broadly and deeply rounded-convex abaxially (Fig. 90A). **Cuticle** thin on both surfaces. **Epidermis**: cells large adaxially and rounded externally and internally. Abaxial cells small and flat. **Stomata** level with surface. **Mesophyll**: palisade tissue complex, comprising well-differentiated palisade cells (6–8:1 *l*/*w*) nested among large cells without chloroplasts (Fig. 90B). Paradermally, large empty cells separated by uniseriate rows of palisade cells giving reticulate appearance (Fig. 90C). Spongy tissue of *c.* 4 layers of small lobed and irregular cells (type 1). **Air cavities** up to 40% of spongy layer.

Collenchyma strands up to three cells deep, subepidermal within fluted ridges on abaxial side of midrib (type Sv). **Vascular bundles**: midrib bundles scattered, some aligned with collenchyma strands; bundles weak (type III). Xylem: one wide functional protoxylem cell and 2–3 narrow metaxylem cells, bordering a small circular strand of phloem. **Laticifers** common, anastomosing, and lateral to vascular bundles (Fig. 90E). **Crystals**: uncommon; raphides in single bundles in biforine cells with lignified walls in mid-mesophyll (Fig. 90C), attached by one end to spongy cells. Druses rare.

Petiole TS

Outline crescent-shaped proximally, terete distally. **Size**: 3–5 mm diameter. **Cuticle** thin and smooth to finely striate. On one side of petiole, cuticle interpenetrating 3–4 layers of epidermis and adjacent ground tissue. **Epidermis**: cells small, flat externally. **Collenchyma**: flattened or irregular strands in proximal petiole, *c.* 2 layers deep, subtending some peripheral vascular bundles. Distally, occasionally interrupted hypodermal band, up to eight cells deep (Fig. 90F), opposite peripheral vascular bundles (type Bi). **Ground tissue**: 5–8 layers of large parenchyma cells at periphery of proximal petiole, grading to type 3 aerenchyma at centre. Aerenchyma not present distally.

Vascular bundles scattered throughout in proximal sections; distally arranged in three concentric circles (Fig. 90F); type III (Fig. 90G). Xylem: wide protoxylem lacuna and several metaxylem cells narrowly bordering laterally compressed strand of phloem. **Sclerenchyma**: fibre caps, 2–3 cells deep, to phloem proximally, absent distally. **Laticifers**: anastomosing, 2–3 strands lateral to phloem. **Starch**: distally, few large angular granules, to 18 μm diameter. **Crystals**: few proximally to rare distally. Raphides in single bundles in small cells or occasionally as biforines attached by one end and suspended in aerenchyma spaces. Druses in circular cells in outer ground tissue.

Material examined

Scaphispatha gracilis Brongn. ex Schott, Bogner 1211, Brazil, cult. M.

Species reported on in the literature

French (1988): laticifers.

Additional literature

French 1986a,c, 1987a; Grayum 1990; Thanikaimoni 1969.

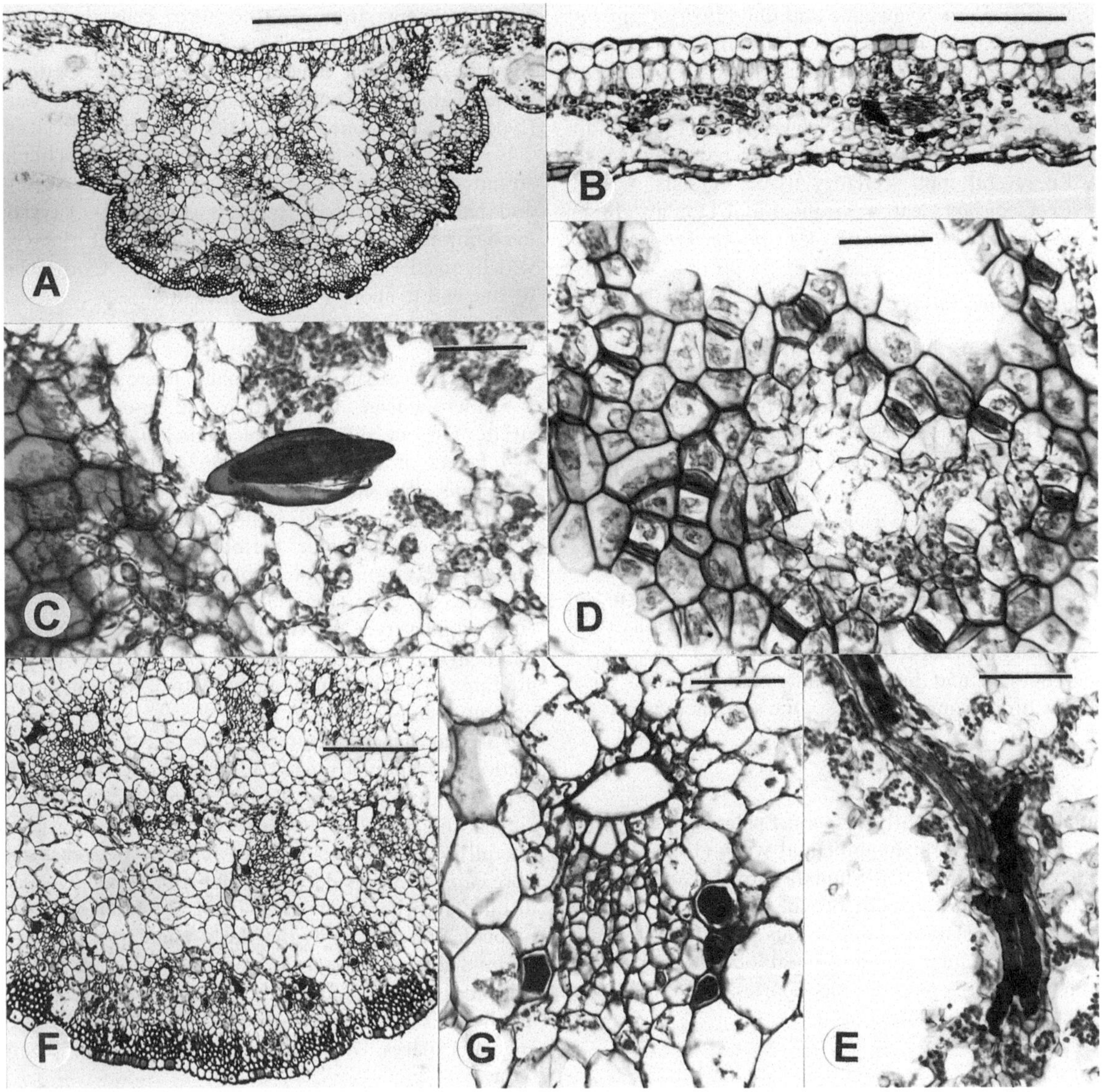

Fig. 90. *Scaphispatha gracilis* leaf and petiole sections. (A) Midrib TS; (B) Lamina TS with hypodermis; (C) Mesophyll PS showing biforine raphide; (D) Abaxial epidermis PS showing brachyparacytic stomata with broad angular subsidiary cells; (E) Mesophyll PS showing vein and anastomosing laticifer; (F,G) Distal petiole TS: (F) showing type Bi collenchyma; (G) showing type III vascular bundle. Scale bars: A = 300 μm; B = 150 μm; C–E,G = 50 μm; F = 200 μm.

Jasarum Bunting

This monotypic genus is endemic to the Guyana highland, Venezuela. Plants are corm-bearing and aquatic with vegetative parts entirely submerged in moderately slow moving black-water streams, in shade or full sunlight. The leaves are long-petiolate and have a narrow elliptic lamina. Chromosomes: $2n = 22$. One specimen was available for study.

Leaf surface

Cuticle smooth. **Epidermis**: cells rounded-polygonal (1–3:1 l/w) on both surfaces (Fig. 91D). **Stomata** not

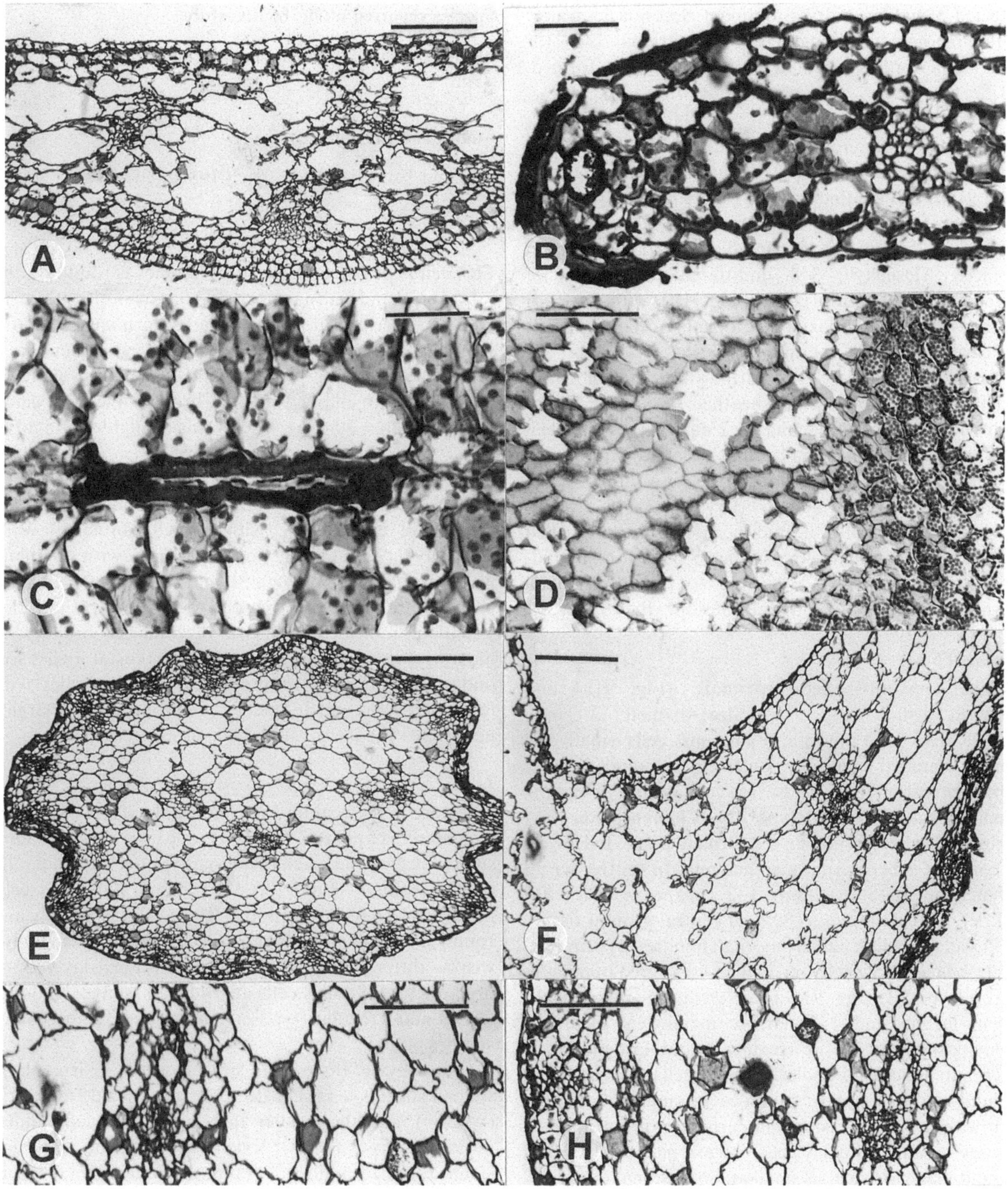

Fig. 91. *Jasarum steyermarkii* leaf and petiole sections. (A) Midrib TS; (B) Lamina margin TS. Mesophyll undifferentiated and stomata absent; (C) Mesophyll PS showing anastomosing laticifers; (D) Lamina epidermis PS; (E) Distal petiole TS showing type 3 aerenchyma; (F) Proximal petiole TS; (G) Distal petiole TS centre and vascular bundle; (H) Distal petiole TS margin and vascular bundle. Scale bars: A = 200 μm; B,C = 50 μm; D,G,H = 150 μm; E,F = 300 μm.

detected. **Venation**: midrib prominent. Secondaries widely acute, merging at straightened submarginal collector vein. Higher orders reticulate, sparse and poorly formed.

Leaf TS

Structure not differentiated. **Size**: 13 × 2 cm (*l*/*w*). **Midrib** flat adaxially, broadly and slightly convex abaxially (Fig. 91A). **Cuticle** thick at margin, patchy or very thin elsewhere. **Epidermis** of small, square to elongated cells, flat externally and flat or rounded internally on both surfaces. **Mesophyll**: 3–4 undifferentiated layers of rounded (isodiametric) cells (Fig. 91B) with rows of plastids packed on adaxial side of most cells (type 1a). **Air cavities** up to 2% of mesophyll.

Vascular bundles: 6–7 in midrib symmetrically arranged in two tiers; bundles unsheathed (type III). Xylem with wide protoxylem lacuna, 1–2 narrower metaxylem cells, opposing a small semicircular or laterally narrowed phloem strand. **Laticifers** mostly in pairs along veins, often anastomosing (Fig. 91C). **Crystals**: raphides common in single bundles, in mid-mesophyll in thin-walled cells. Druses in small circular cells, common in mid-mesophyll. Crystal sand grading to numerous small prismatics, common throughout mesophyll.

Petiole TS

Outline: crescent-shaped proximally (Fig. 91F) and subterete distally (Fig. 91E). **Size**: diameter 2–3 mm. **Cuticle** thick and smooth. **Epidermis**: cells small and square, rounded externally. **Collenchyma**: small circular subepidermal strands, up to five cells deep, subtending small vascular bundles (type Sv). **Chlorenchyma**: 3–4 peripheral subepidermal layers. **Ground tissue**: five layers of compact subepidermal parenchyma. In centre: type 3 aerenchyma with partitions 1–3 cells wide (Figs 5B, 91E,F). **Air cavities** up to 80% of central ground tissue.

Vascular bundles: numerous small bundles in peripheral circle, beneath collenchyma, and 3–4 scattered bundles in centre (type III) (Fig. 91G,H). Xylem: one wide protoxylem lacuna and 1–2 narrow metaxylem elements. Phloem: small circular or irregular strand. **Laticifers**: 1–2 lateral to vascular bundles. **Tannin** cells occasional in ground tissue. **Crystals**: raphides in single bundles in thin-walled cells, occasional in peripheral ground tissue. Druses in small circular cells, occasional in peripheral ground tissue. Prismatics small and occasional in ground tissue.

Material examined

Jasarum steyermarkii Bunting, Herker and Jesche s.n., Venezuela, cult. M.

Species reported on in the literature

French (1988): laticifers. • French and Tomlinson (1983): stem.

Additional literature

Fox and French 1988; French 1986a,c; Grayum 1990.

Zomicarpa Schott

This genus of three rhizomatous species is found on the subtropical and warm temperate Atlantic slopes of Brazil. Plants are terrestrial tuberous herbs dormant in the dry season. They occur in the seasonal forest understorey. The leaves are long-petiolate and hastate, trifoliate to pedatisect with elliptic leaflets. Chromosomes (2 species, 2 counts): 2n = 20. One species was available for study.

Leaf surface

Cuticle smooth or faintly striate. **Epidermis**: cells polygonal and isodiametric with straight anticlinal walls adaxially (1–1.5:1 *l*/*w*), and polygonal with undulate walls abaxially (2–3:1 *l*/*w*). **Stomata** abaxial and brachyparacytic. Subsidiary cells broad and irregular (Fig. 92E). **Venation**: midrib prominent. Secondaries broadly or narrowly acute forming irregular intercostal areas; secondaries connecting in brochidodromous arches with festooned marginal loops. Eucamptodromous margins near base. Tertiaries and higher orders reticulate.

Leaf TS

Structure very weakly dorsiventral. Leaf margin finely irregularly papillate. **Size**: 67 × 30 mm (*l*/*w*). **Midrib** shallowly grooved adaxially, broadly convex and laterally angled abaxially (Fig. 92A). **Cuticle** thin. **Epidermis**: cells medium-sized to tall on both surfaces, flat externally and rounded internally. **Stomata** level with surface. **Mesophyll** weakly differentiated (Fig. 92B). Palisade 1-layered, or not apparent. Spongy cells in *c*. 4 layers, somewhat long-armed and irregular (type 2). **Air cavities**: up to 90% of spongy tissue.

Collenchyma: three small subepidermal, ± irregularly shaped strands, *c*. 2 cells deep, on abaxial side of midrib (type Sv). **Vascular bundles**: three weakly formed bundles in midrib (types II, III). Xylem: 2–4 elements opposing small rounded phloem strand. **Laticifers** uncommon, anastomosing lobed cells lateral to most vascular bundles (Fig. 92C). **Crystals**: raphides common in single bundles in thin-walled mesophyll cells. Some elongated raphide cells with 3–4 short bundles. Biforines common in mid-mesophyll (Fig. 92D). Druses small, common in palisade cells. Crystal sand occasional in spongy layer.

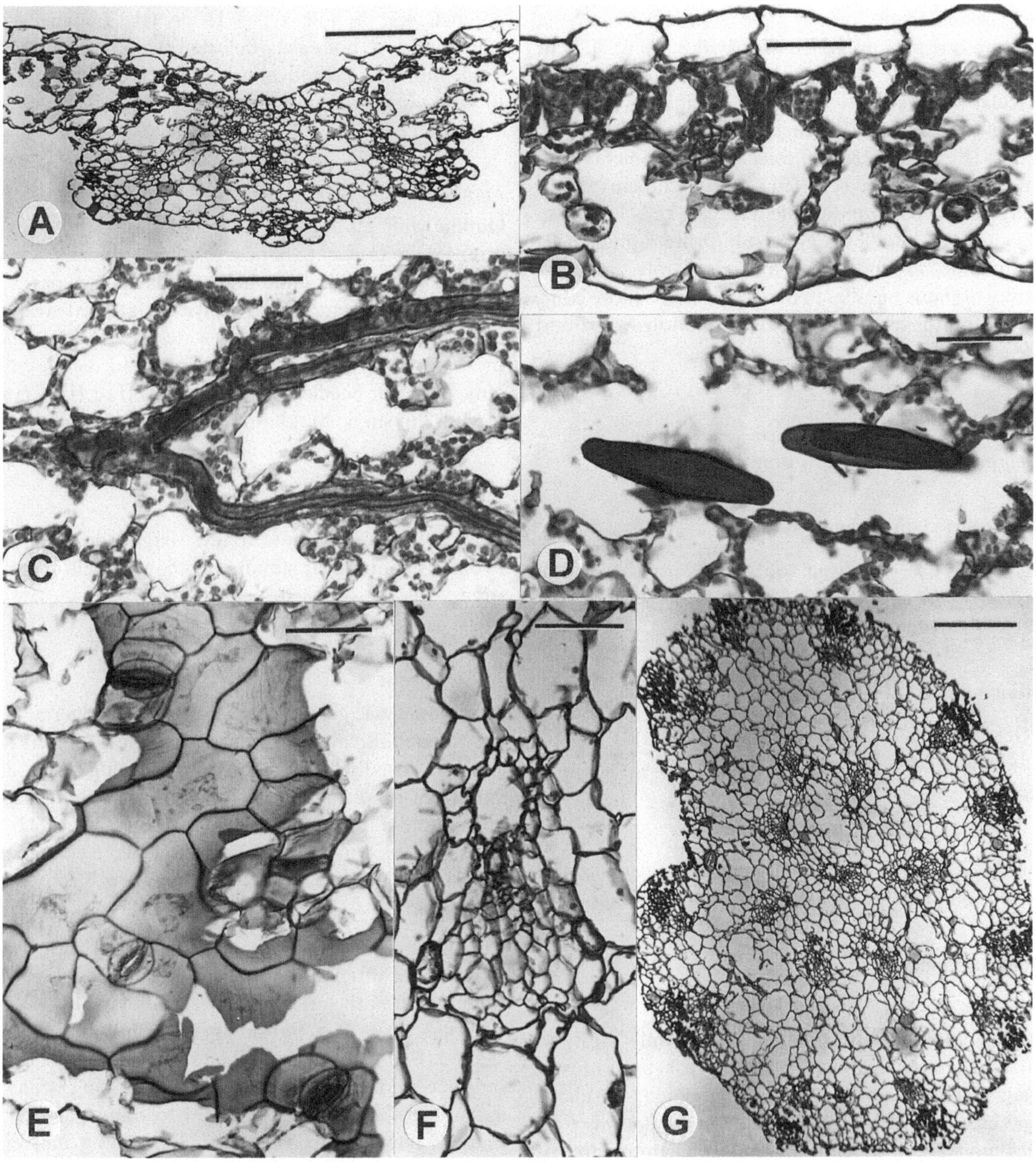

Fig. 92. *Zomicarpa riedeliana* leaf and petiole sections. (A) Midrib TS, abaxially ridged; (B) Lamina TS; (C) Mesophyll PS showing anastomosing laticifer; (D) Mesophyll PS showing 2 biforine raphides; (E) Abaxial epidermis PS showing brachyparacytic stomata with broad subsidiary cells; (F) Distal petiole TS showing type III vascular bundle; (G) Distal petiole TS showing marginal vascular bundles with type Sv collenchyma and central group of vascular bundles. Scale bars: A = 200 μm; B–F = 50 μm; G = 300 μm.

Petiole TS

Outline terete except for small rounded-concave adaxial groove (Fig. 92G). **Size**: diameter 2–3 mm. **Cuticle** thin and striate. **Epidermis**: cells medium-sized, rounded internally and externally. **Collenchyma**: rectangular or circular subepidermal strands directly adjacent to phloem of peripheral bundles (type Sv) (Fig. 92G). **Chlorenchyma**: *c.* 3 subepidermal layers of compact ground tissue. **Ground tissue**: mostly type 2 or 3 aerenchyma (with up to one partition cell). **Air cavities**: up to 90% of aerenchyma.

Vascular bundles: row of small peripheral bundles and nine symmetrically arranged larger central bundles (type III) (Fig. 92G). Xylem: one very wide metaxylem cell and 1–2 narrower cells (Fig. 92F). Phloem: small, laterally narrowed strand. **Laticifers**: 2–5 cells lateral to phloem at most bundles. **Crystals**: biforine raphide cells uncommon in ground tissue and randomly oriented. Raphides in single bundles in thin-walled cells more common near periphery. Crystal sand common in ground tissue, especially near bundles.

Material examined

Zomicarpa riedeliana Schott, Bogner 1213, cult. M.

Species reported on in the literature

French (1987c): *Z. riedeliana*: root, hypodermis. • French (1988): *Z. riedeliana*: laticifers. • Solereder and Meyer (1928) *Z. pythonium*: anatomy.

Additional literature

French 1987a; Grayum 1990; Mookerjea 1955; Thanikaimoni 1969.

Tribe Arisareae Dumort.

This tribe consists of the small genera *Ambrosina* and *Arisarum* found in the greater Mediterranean region.

Anatomical summary

Leaf surface

Cuticle smooth. **Epidermis**: anticlinal cell walls straight, undulate or sinuous. **Stomata** on both surfaces or mostly abaxial, brachyparacytic or lacking modified subsidiary cells. **Venation**: midrib prominent in *Arisarum* with acrodromous basal secondaries ending in brochidodromous looped margins. In *Ambrosina*, midrib prominent at base only and with acrodromous basal secondaries.

Leaf TS

Structure dorsiventral. **Midrib** flat adaxially, broadly convex abaxially. **Mesophyll**: palisade tissue in one layer of wide, rounded or lobed cells. Spongy layers 5–6, forming type 2 or 3 aerenchyma. **Collenchyma**: small flattened subepidermal strands, type Sv. **Vascular bundles** few in midrib, among type 3 aerenchyma or compact parenchyma; bundle types II or III. **Laticifers** non-anastomosing, not easily detected. **Crystals**: raphides in single bundles in elliptical or elongated cells in *Arisarum*. In *Ambrosina*, raphides in 1–several bundles present in elongated cells.

Petiole TS

Outline wide elliptical or subterete, adaxially shallowly grooved or crescent-shaped. **Cuticle** thin and smooth. **Ground tissue**: cells large and compact, or types 3 or 4 aerenchyma in centre. **Collenchyma**: flattened strands as though discontinuous band but widely spaced in *Arisarum*. In *Ambrosina*, type Sv strands found in distal sections only. **Vascular bundles** scattered, types II or III. **Crystals**: raphides in single bundles in small or wide cells.

Taxonomic notes

These two genera show a few diagnostic differences. From an anatomical standpoint they fit well in a single tribe.

Ambrosina Bassi

This monotypic genus occurs in Italy and Algeria. Plants are terrestrial, tuberous herbs found in Mediterranean climates, usually along coasts, north-facing slopes and a variety of other habitats. The plants have long petioles and simple elliptical leaves. Chromosomes: $2n = 22$. Two specimens yielding poor preparations were examined.

Leaf surface

Cuticle smooth. **Epidermis**: cells polygonal with straight to undulate anticlinal walls on both surfaces (1–2:1 *l*/*w*). **Stomata** on both surfaces, brachyparacytic, occasionally unipolar (Fig. 2F). Subsidiary cells broad, irregularly shaped (Fig. 93A). **Venation**: midrib prominent near base only. Secondaries, 4–5 on each side, arising near base, ascending steeply and curving, becoming parallel with margin, converging at leaf tip. Remaining midvein of tertiary size near leaf tip. Tertiaries curved, or orthogonal between secondaries, or reticulate. Higher orders reticulate. Areoles isohedral or elongate in any orientation.

Leaf TS

Structure weakly dorsiventral. **Midrib** flat adaxially, broadly and shallowly convex abaxially. **Size** 54 × 34 mm (*l*/*w*). **Cuticle** very thin. **Epidermis**: cells flat internally and externally. **Stomata** level with surface. **Mesophyll** with one layer of rounded palisade cells containing most of the

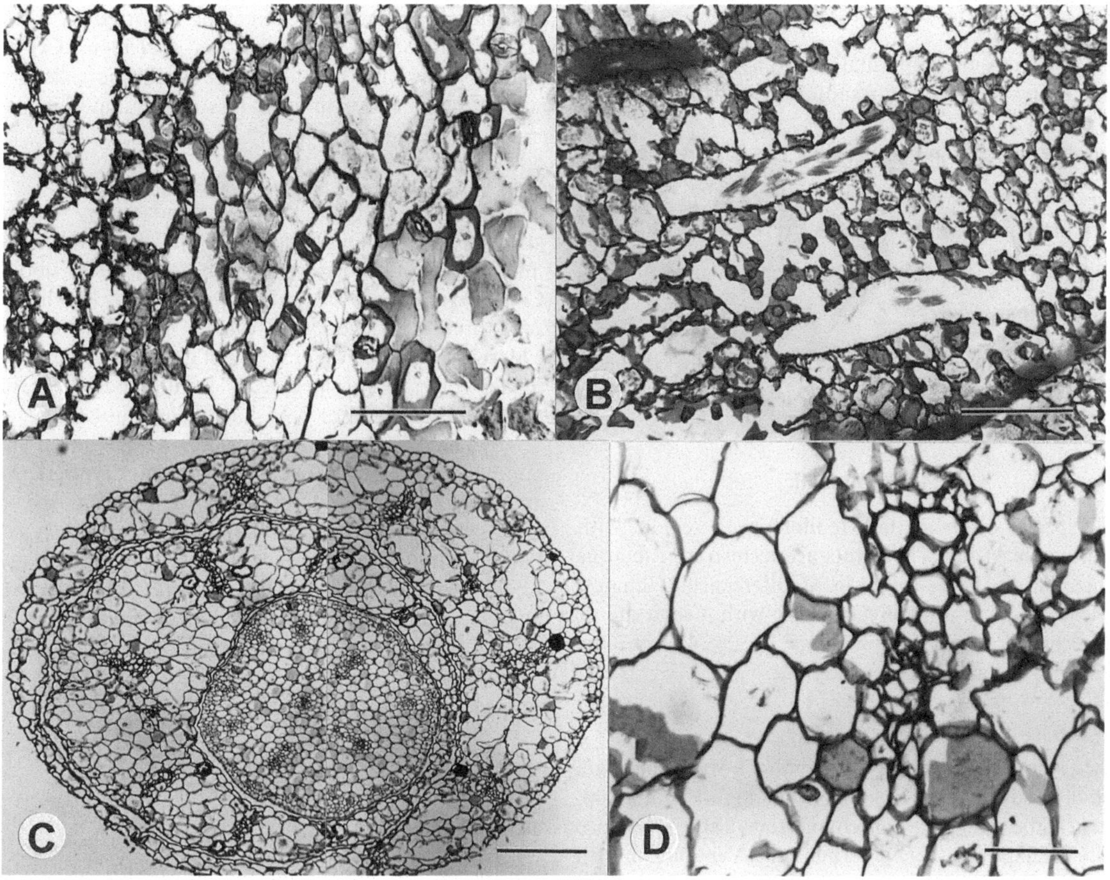

Fig. 93. *Ambrosina bassii* leaf and petiole sections. (A) Lamina PS showing adaxial epidermis with broad stomatal subsidiary cells; (B) Lamina PS showing elongate raphide cells bearing multiple raphide bundles; (C) Petiole bases TS distal to shoot tip. Note two crescent-shaped petiole bases and one distal subterete section in centre; (D) Petiole TS showing type II vascular bundle. Scale bars: A,B = 150 μm; C = 300 μm; D = 50 μm.

chloroplasts. Spongy cells gradually demarcated from palisade by lesser density of chloroplasts; size and shape similar (types 1, 2). **Air cavities**: up to 70% of spongy layer. **Collenchyma**: abaxial subepidermal strands subtending midrib vascular bundles (type Sv).

Vascular bundles: three large bundles in midrib, type II. Xylem: 6–10 metaxylem cells. Phloem: circular or irregular strand. **Laticifers** not found in this material but reported by French (1988) as non-anastomosing. **Crystals**: raphides in 1–several bundles, in cells much elongated beyond centrally concentrated bundles (Fig. 93B); cells in upper or lower region of lamina.

Petiole TS

Outline: proximally crescent-shaped and distally subterete (Fig. 93C). **Size** 1.5 × 2 mm diameter. **Cuticle** thin and smooth. **Epidermis**: cells irregularly shaped. **Collenchyma** in distal section: irregular strands subtending vascular bundles, 1–4 cells deep (type Sv). **Chlorenchyma**: scattered chloroplasts in outer layers. **Ground tissue**: large parenchyma cells throughout. **Air cavities** few and small, forming type 4 openings. **Vascular bundles**: 10 bundles scattered symmetrically; small, irregularly shaped (type II). Xylem: 2–5 narrow metaxylem cells in irregular strand (Fig. 93D). Phloem: small laterally compressed strand. **Crystals**: raphides in wide cells, few in outer ground tissue.

Material examined

Ambrosina bassii L.; Bogner 1521, Sardinia, cult. M; Wolfe s.n. (K) dry, Algiers.

Taxonomic notes

The genus has unexceptional anatomy not assignable more specifically beyond its subfamily affiliation.

Species reported on in the literature

Borzi (1887–8): root. • Ertl (1932): leaf. • French (1987c): root. • French (1988): laticifers. • French and Tomlinson (1983): stem vasculature.

Additional literature

Buscalioni and Lanza 1928; Domin 1911; French 1986a,b, 1987a; Grayum 1990; Hotta 1971; Irmisch 1874; Killian 1929; Thanikaimoni 1969; Vignoli 1939.

Arisarum Mill.

Three species occur in the Mediterranean region, with two in north Africa. The plants are terrestrial, tuberous or rhizomatous herbs found in Mediterranean climates. The leaves have a long, erect petiole with a sagittate or hastate blade. Chromosomes: 2n = 14, 28, 42, 56. One specimen was available for study.

Leaf surface

Cuticle smooth. **Epidermis** anticlinal cell walls undulate adaxially and sinuous abaxially. **Stomata** mostly abaxial, with neighbouring cells brachyparacytic and otherwise similar in shape to other epidermal cells. **Venation**: midrib prominent. Acrodromous basal secondaries with brochidodromous looped margins. Midrib secondaries broadly acute, apically curving, forming laminar brochidodromous connections. Marginal connections eucamptodromous. Higher orders reticulate.

Leaf TS

Structure dorsiventral. **Size**: 4 × 3.5 cm (*l*/*w*). **Midrib** flat adaxially and broadly and shallowly convex abaxially (Fig. 94A). **Cuticle** thin. **Epidermis**: cells small or large, rounded internally and flat externally. **Stomata** level with surface. **Mesophyll**: one palisade layer of short, wide, often lobed cells (Fig. 94D) and 5–6 spongy layers forming mostly type 2 (Fig. 94B) or some type 3 aerenchyma. **Air cavities**: *c*. 90% of spongy layer. **Collenchyma**: small, flat subepidermal strands, 2–3 cells deep, subtending vascular bundles (type Sv) (Fig. 94A,C).

Vascular bundles: one central and two lateral bundles in midrib among type 3 aerenchyma (Fig. 94A). Bundles small and laterally narrowed, type II or III. Xylem: wide protoxylem lacuna and one to several narrow metaxylem elements. Phloem: small irregular strand. **Laticifers**: non-anastomosing type reported by French (1988) but not detected in this specimen. **Crystals**: raphides as single bundles in small elliptic cells, some cells elongate and without dark-staining material.

Petiole TS

Outline wide elliptic (Fig. 94E), adaxially shallowly grooved. **Size** 3–4 mm wide, proximal to distal. **Cuticle** thin and smooth. **Epidermis**: cells small and cuboidal. **Collenchyma**: small subepidermal strands, 6–10 cells wide, 2–3 three cells deep, and widely spaced (Fig. 94F). **Chlorenchyma**: 3–5 outer layers of ground tissue. **Ground tissue**: compact, of large cells around periphery with transition to type 3 or type 4 aerenchyma in centre.

Vascular bundles scattered, outer row of seven small bundles subtended by collenchyma strands, types II or III, and a few central bundles (Fig. 94E,F). Xylem: one wide protoxylem lacuna and 3–6 narrow metaxylem cells. Phloem: small, laterally narrowed strand. **Starch**: occasional granules in single ground tissue cells adjacent to phloem strands. **Crystals**: raphides in single bundles in small, otherwise not differentiated cells.

Material examined

Arisarum vulgare Targ.Tozz., cult. M.

Taxonomic notes

This genus has simple anatomy. Beyond its subfamilial placement it has no characters distinctive enough to be used to further refine its position.

Species reported on in the literature

Borzi (1887–8): *A. vulgare*: root. • El-Kiey *et al.* (1967): *A. vulgare*: corm. • French (1987c): *A. vulgare*: root. • French (1988): *A. vulgare*: laticifers. • French and Tomlinson (1983): *A. vulgare*: stem. • Galil (1978): *A. vulgare*: stem, root. • Higashi and Kuo (1949): pharmacognosy. • Lierau (1888): *A. proboscidium*: root. • Morassi Bonzi and Fabbri (1975): *A. proboscidium, A. vulgare*: leaf. • Nicolas (1916): petiole. • Preda (1910): *A. vulgare*: leaf. • Solereder (1919): *A. vulgare*: leaf. • Solereder and Meyer (1928): *A. vulgare*: leaf, crystals, secretory tissue. • Webber (1960): *A. vulgare*: leaf, stomata.

Additional literature

Antony 1898; Batygina and Yakovlev 1990; Erdtman 1952; French 1986a,c, 1987a; French and Tomlinson 1983; Galil 1990; Grayum 1990; Halevy 1989; Hegnauer

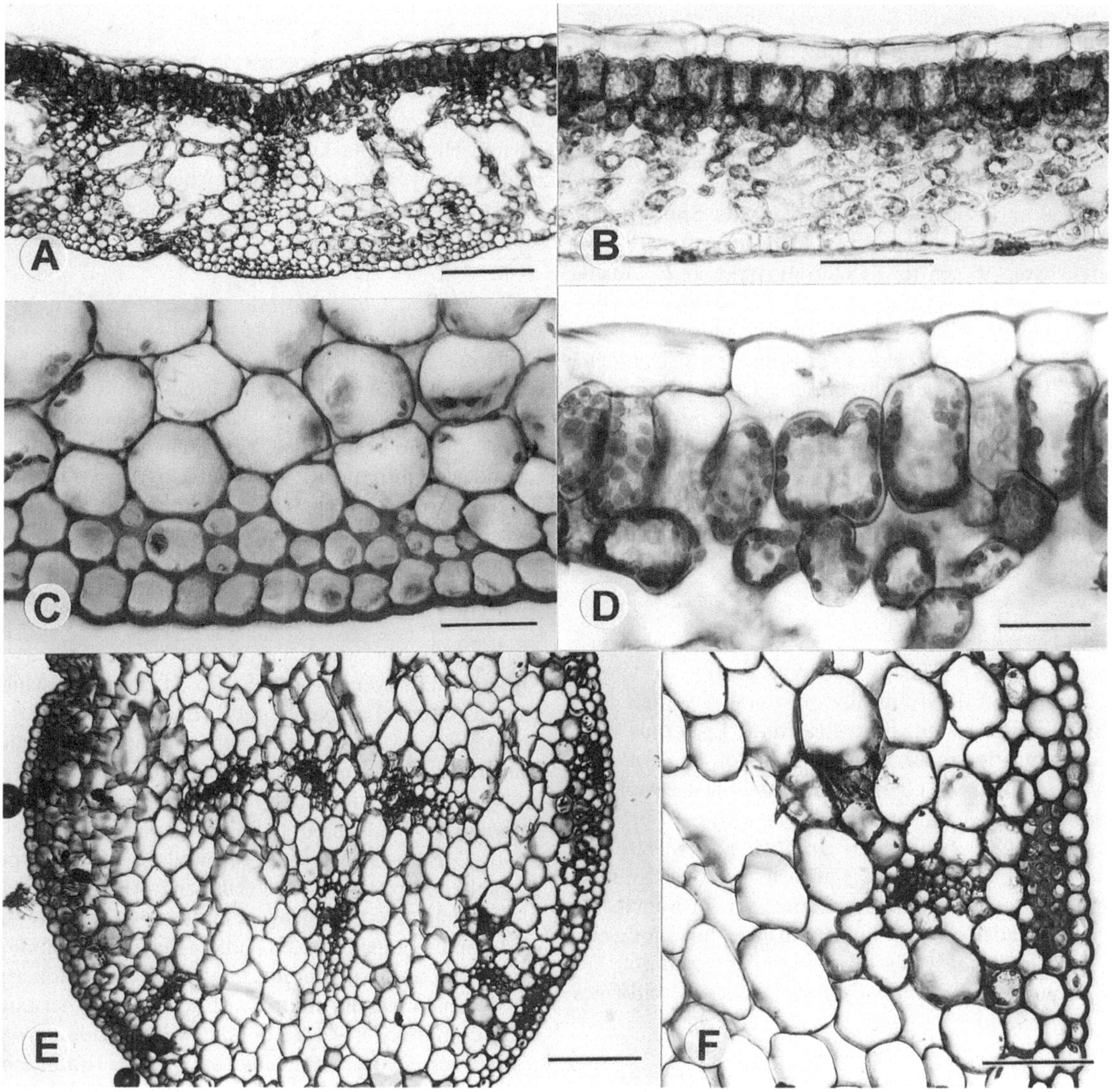

Fig. 94. *Arisarum vulgare* leaf and petiole hand sections. (A) Midrib TS; (B) Lamina TS; (C) Midrib TS, abaxial side showing small subepidermal collenchyma strand; (D) Lamina TS, adaxial side showing occasional lobed palisade cells; (E) Petiole TS showing small vascular bundles; (F) Petiole TS, lateral portion showing collenchyma strand and vascular bundle. Scale bars: A,E = 300 μm; B,F = 150 μm; C,D = 50 μm.

1986; Kirchner *et al.* 1934; Koach and Galil 1986; Pütz 1991; Thanikaimoni 1969; Vogel 1978.

Tribe Peltandreae Engl.

This tribe includes an unusual disjunction. *Peltandra* occurs in North America while the four genera *Typhonodorum, Colletogyne, Carlephyton* and *Arophyton* are restricted to Madagascar.

Anatomical summary

Leaf surface

Cuticle thin, smooth or reticulate. **Epidermis**: cells straight-sided polygonal, or cell walls curved. **Stomata** mostly abaxial or equally numerous on both surfaces and brachyparacytic, occasionally brachypara-tetracytic. Subsidiary cells mostly broad and rounded, angular or irregular. **Venation**: with strong midrib development or basally acrodromous. Secondaries mostly brochidodromous.

Leaf TS

Structure dorsiventral. **Midrib** small or larger, convex-curved abaxially. **Mesophyll**: palisade layers 1–2, cells cylindrical or T-shaped beneath epidermis. Spongy cells in 5–8 layers, up to 15 layers in *Typhonodorum*, with type 3 aerenchyma in centre, or entirely types 1–2. **Collenchyma** of type Sv strands in midrib. **Vascular bundles** types II and III. Fibres present in some phloem caps of *Peltandra*. **Laticifers** non-anastomosing, lateral to phloem strands. **Tannin** cells usually present. **Crystals**: raphides as single bundles in thin-walled cells, or biforine-like, without terminal papillae, or thin-walled cells with terminal papillae, or true biforines. Biforines attached by one end, suspended across aerenchyma partitions, or randomly distributed within mesophyll. Crystal sand present.

Petiole TS

Crescent-shaped proximally to terete distally, or deltoid and compressed dorsiventrally or laterally. **Cuticle** thin to thick, smooth, microstriate, or striate. **Epidermis**: cells variable, mostly small. **Stomata** common in *Typhonodorum*. **Ground tissue** compact and small near periphery, becoming type 3 aerenchyma in centre with 1–7 partition cells. Type 4 aerenchyma in *Arophyton humbertii* and *Carlephyton glaucophyllum*. **Collenchyma**: type Sv near periphery. **Vascular bundles** scattered, few or numerous, type III. **Laticifers** non-anastomosing, but not seen in *Typhonodorum*. **Crystals**: raphides common in single bundles in thin-walled cells, or as biforines (*Typhonodorum*).

Taxonomic notes

The genera *Peltandra* and *Typhonodorum* have long been considered related (see Tribe Peltandreae in MBB) and they present a remarkable disjunction, perhaps related to overlapping ranges in the past. In spite of the long isolation of *Peltandra* it appears to have diverged little in histology from the other four genera.

The other three genera form a coherent group (see Tribe Arophyteae in MBB) for which there is broad agreement in most anatomical details. They also fit well within the expanded Peltandreae. The presence of the uncommon type 4 aerenchyma in petioles of *Arophyton* and *Carlephyton* is an especially strong indicator of relationship.

Peltandra Raf.

This genus with two species occurs in Atlantic North America and the southern United States. Plants have tuberous rhizomes and extremely polymorphic leaf forms (Blackwell and Blackwell 1975) that may be narrow-ovate when young and sagittate when mature. Chromosomes are small: $2n = 56$, 112. Both species were available for study.

Leaf surface

Cuticle smooth or reticulate-ridged on both surfaces. **Epidermis**: cells polygonal with straight anticlinal walls on both surfaces (1–1.5:1 *l*/*w*). **Stomata** mostly abaxial and brachyparacytic (Figs 3G, 95C). Subsidiary cells broad, angular or irregular. **Venation**: midrib prominent, secondaries infrequent, connecting at ± straightened brochidodromous arch. Higher order marginal veins also present. Tertiaries closely spaced, parallel with secondaries. Quaternaries orthogonal. Higher orders reticulate. Areoles elongate and aligned with parallel tertiaries.

Leaf TS

Structure dorsiventral. **Size**: 18×11 cm (*l*/*w*). **Midrib**: convex-flattened adaxially and broadly rounded-convex abaxially (Fig. 95A). **Cuticle** very thin on both surfaces. **Epidermis**: cells squarish or flattened, usually flat externally and rounded internally. **Stomata** level with surface. **Mesophyll**: 1–2-layered palisade adaxially (Fig. 95B). Spongy tissue 5–7 cells deep, well organized type 3 aerenchyma; partitions 1–3 cells long. **Air cavities**: 90% of aerenchyma. **Collenchyma**: numerous discrete sub-epidermal strands on abaxial and adaxial sides of midrib (Fig. 95A), each subtending small vascular bundle (type Sv). **Vascular bundles**: midrib with small row of peripheral bundles and fewer, larger bundles within wide partition corners of central aerenchyma; bundles unsheathed. Xylem: 1–2 wide protoxylem tracheids and less frequently 2–3 narrow metaxylem elements (types II or III). Phloem: small irregular strand.

Sclerenchyma: fibres occasional as small caps over phloem; often not present. **Laticifers**: non-anastomosing, in pairs lateral to all vascular bundles. **Tannin**: cells with dark contents randomly distributed in mesophyll, common in aerenchyma partitions. **Starch** very common as large granules in most mesophyll cells. **Crystals**: biforine raphides in stalked lignified cells, circular or cylindrical, usually without terminal papillae, suspended in air spaces. Biforines suspended less frequently across partitions with papillate ends pointing into adjacent air cavities (Fig. 95D). Raphides also common in thinner-walled irregular cells, usually in single bundles. Raphide bundles commonly

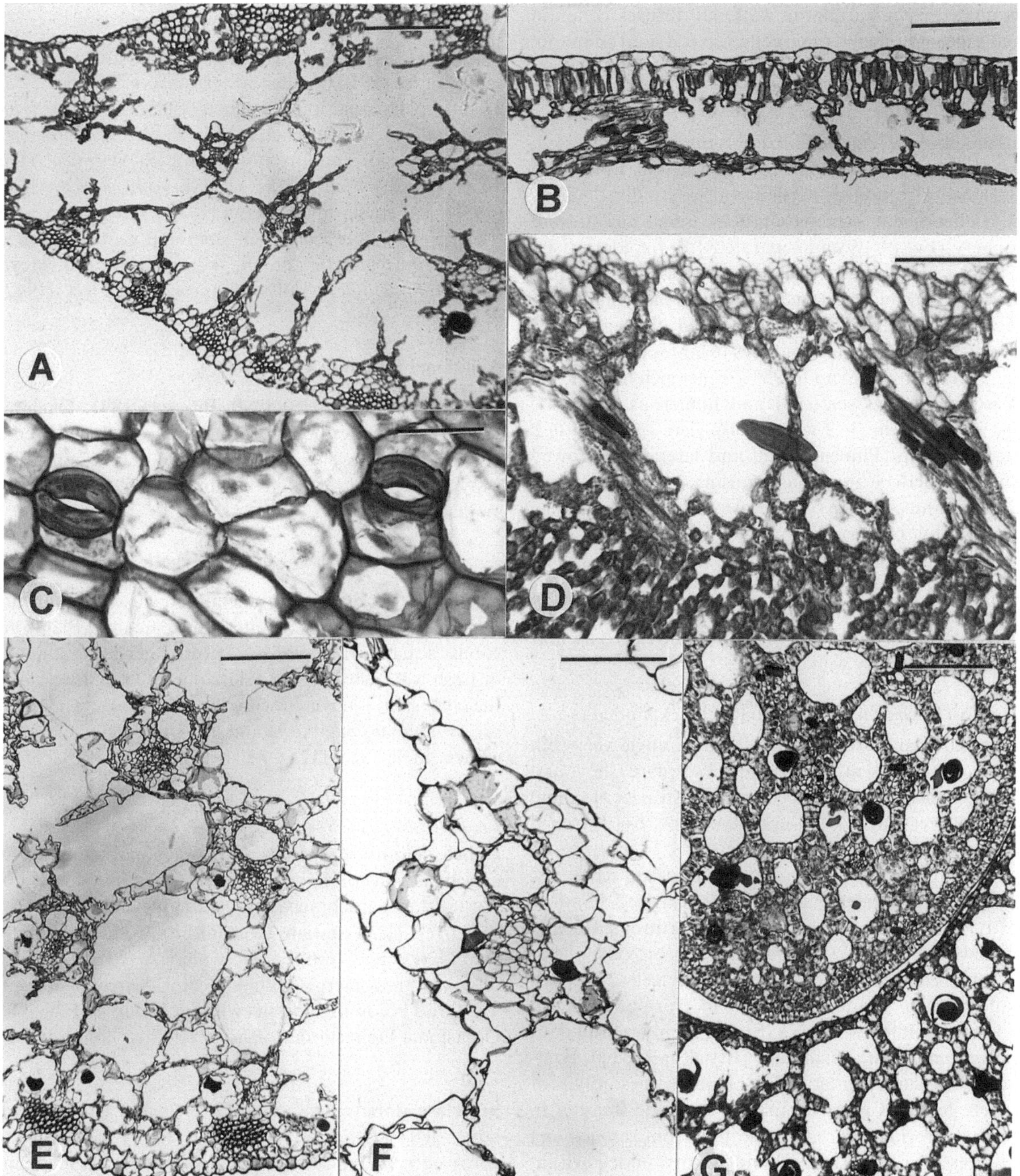

Fig. 95. *Peltandra* leaf, petiole and stem sections. (A–D,F,G) *P. virginica*; (E) *P. sagittifolia*; (A) Midrib TS; (B) Lamina TS; (C) Abaxial epidermis PS with stomata; (D) Lamina PS showing type 3 aerenchyma and biforine cell; (E) Distal petiole TS margin. Note type Sv collenchyma strand; (F) Proximal petiole TS, central type III vascular bundle; (G) Young stem (centre) surrounded by young leaf. Scale bars: A = 300 μm; B,E,G = 200 μm; C = 50 μm; D,F = 150 μm.

surrounded by mucilaginous material. Druses common in palisade and subepidermal cells. Crystal sand common, especially in vicinity of vascular bundles.

Petiole TS

Outline subterete, deltoid and dorsiventrally compressed. **Size**: diameter 4 mm. **Cuticle** thin and striate. **Epidermis**: cells isohedral, rounded internally and externally, or papillate. **Collenchyma**: strands circular, numerous around periphery (Fig. 95E) and separated from epidermis by 1–2 layers of ground tissue. Each strand subtending small vascular bundle (type Sv). **Chlorenchyma**: outer 3–5 layers of ground tissue. **Ground tissue**: type 3 aerenchyma with vascular bundles in large corner islands (Fig. 95G). **Air space**: up to 95% of aerenchyma.

Vascular bundles scattered and numerous (type III) (Fig. 95F). Xylem: 1–2 wide protoxylem cells and 0–2 metaxylem cells. Phloem: small and laterally narrowed strand. **Laticifers** non-anastomosing, 1–2 lateral to phloem strands. **Tannin** cells occasional in ground tissue. **Starch** granules occasional in ground tissue near phloem. **Crystals**: raphides in wide bundles, in large thin-walled biforine-like cells near periphery; cells suspended by one end in air space, or hypodermal and longitudinally aligned. Druses common in ground tissue near vascular bundles in air cavity walls.

Stem TS (*P. virginica*, young, partially expanded)

Outline circular. **Size**: diameter 2 mm. **Cuticle** very thin and smooth to absent. **Epidermis**: cells square, short and narrow. **Collenchyma**: small subcircular strands of young cells, not yet angularly thickened (type Sv). **Ground tissue**: outer 1–2 layers of compact cells. Centre mostly type 3 aerenchyma (Fig. 95G) with uniseriate partitions of 3–4 cells. **Vascular bundles**: 8–10 scattered bundles, mostly at corners of aerenchyma partitions; mostly procambium, type III. Xylem: narrow protoxylem lacunae and occasional mature metaxylem cells. Phloem: small irregular strand. **Laticifers**: in pairs adjacent to phloem, cells filled with latex. **Starch** granules abundant in all ground tissue, small and partially digested, compared with larger granules in mature mesophyll of adjacent leaves. **Crystals**: raphides as single bundles in full-sized biforine-like cells, projecting from aerenchyma partitions into air cavities; bundles surrounded within cells by thick, deep-staining mucilaginous layer.

Material examined

Peltandra sagittifolia (Michx.) Morong., Boufford 18467, Florida (MO); *P. virginica* (L.) Schott, Boufford and Wood 18862, North Carolina (MO).

Species reported on in the literature

Chrysler (1904): *P. virginica*: stem central cylinder. • Engler and Krause (1912): root, leaf, crystals. • Ertl (1932): *P. virginica*: leaf venation. • French (1987c): *P. sagittifolia, P. virginica*: root. • French (1988): *P. virginica*: laticifers. • French and Tomlinson (1981b): *P. virginica*: stem vasculature. • Guttenberg (1968): root. • Korschgen (1981): leaf epidermis. • Lierau (1888): *P. virginica*: root. • Ogden (1974): *P. virginica*: stem and root. • Reichert (1913): *P. undulata*: starch. • Solereder and Meyer (1928): *P. undulata*: all organs. • Van Fleet (1942): *P. virginica*: stem.

Additional literature

Batygina and Yakovlev 1990; Brandza 1891; Daghlian 1981; Davis 1966; Engler 1915; Eyde *et al.* 1967; French 1985b, 1986c, 1987a; Goldberg 1940; Grayum 1990; Mayo 1986; Ray 1988; Thanikaimoni 1969; Tillich 1985; Williams *et al.* 1981.

Typhonodorum Schott

This monotypic genus is found in Madagascar, Mauritius, Comores, Mascarenes, Pemba, and Zanzibar. Plants are robust aquatic, rhizomatous or tuberous herbs standing in fresh water swamps or tidal zones. On mature plants the erect stem is covered with dead leaf bases. The leaves are triangular, sagittate, and long-petiolate. Chromosomes small: $2n = 112$. One specimen was available for study.

Leaf surface

Cuticle smooth. **Epidermis**: cells polygonal with straight anticlinal walls on both surfaces (1–2:1 *l*/*w*). **Stomata** on both surfaces and brachyparacytic (Fig. 96D). Subsidiary cells irregular. **Venation**: large midrib producing parallel secondaries and tertiaries at widely-spaced ± regular intervals that merge with collector veins at margin. Tertiaries becoming parallel with secondaries (Fig. 96E). Occasional higher order orthogonal cross connections.

Leaf TS

Structure dorsiventral. **Size** 150 × 36 mm (*l*/*w*). **Midrib** large, flat adaxially (Fig. 96A), deeply semicircular-convex abaxially. **Cuticle** thin. **Epidermis**: cells small and square on both surfaces, flat externally and internally. **Stomata** level with surface, having double cuticular ledges. Stomatal apertures generally aligned parallel with vascular bundles. **Mesophyll**: palisade one, occasionally two layers (4:1–8:1 *l*/*w*). Spongy tissue: *c*. 15 layers composed of irregular cells near palisade and abaxial epidermis, central cells as uniseriate partitions for type 3 aerenchyma

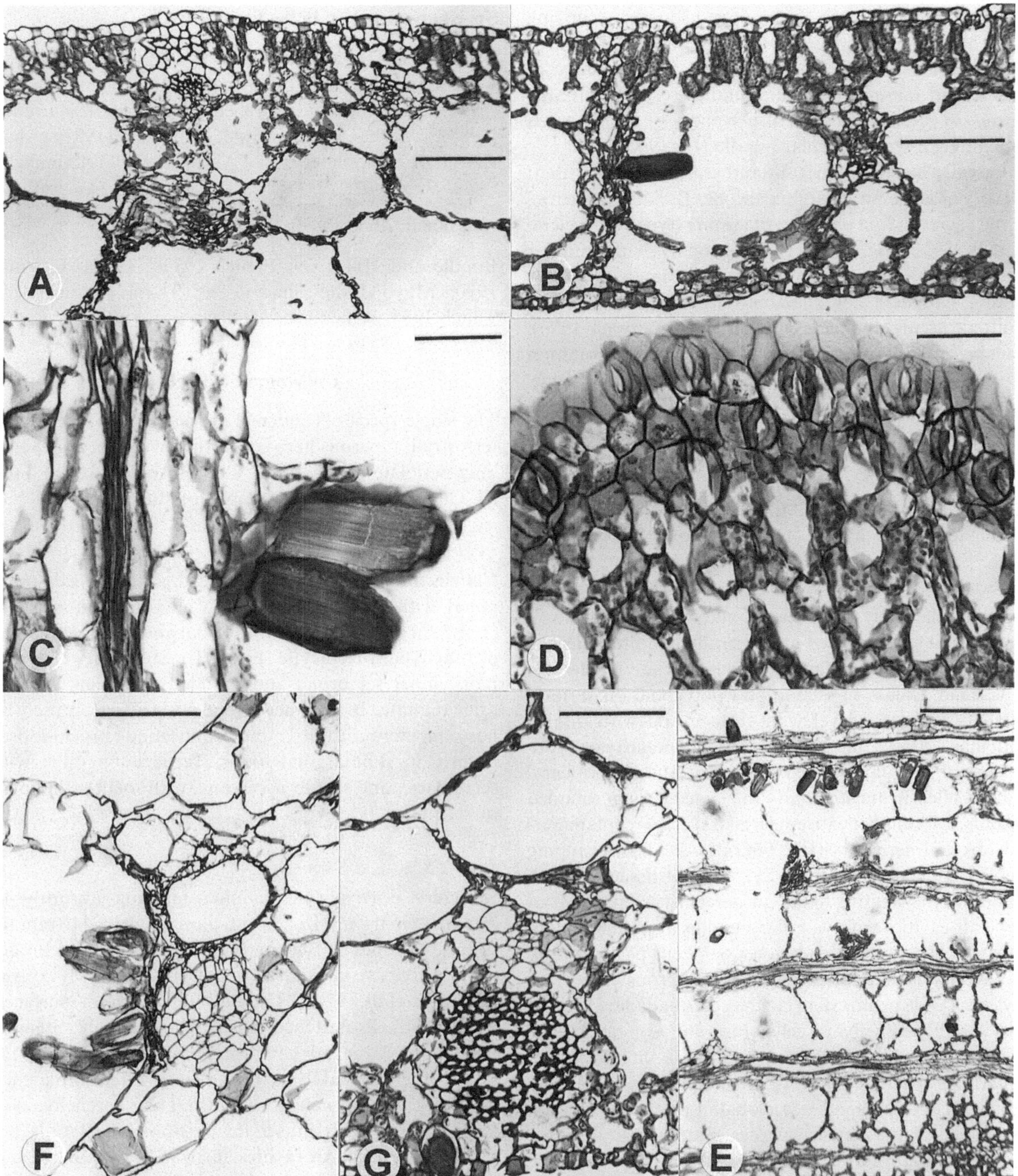

Fig. 96. *Typhonodorum lindleyanum* leaf and petiole sections. (A) Midrib TS showing type 3 aerenchyma; (B) Lamina TS; (C) Lamina PS showing biforine raphide cells; (D) Lamina PS abaxial surface with stomata; (E) Lamina PS showing parallel secondary vascular bundles; (F) Petiole TS showing large type III vascular bundles and raphide cells; (G) Petiole TS margin with type Sv collenchyma strand. Scale bars: A = 200 μm; B,F,G = 150 μm; C,D = 50 μm; E = 300 μm.

(Fig. 96B). **Air cavities** very large and square, occupying most of mesophyll space. **Collenchyma** often present as subepidermal strands on abaxial or adaxial sides of midrib and laminar vascular bundles (type Sv). Large strands at regular intervals at periphery of midrib, each subtending a small vascular bundle (Fig. 96A).

Vascular bundles numerous and scattered in midrib at corners of large air cavity partitions. Bundles occurring in each aerenchyma partition of lamina (type III). Xylem: usually 1–2 wide protoxylem elements with annular thickenings opposing an irregularly rounded phloem strand. **Laticifers** non-anastomosing and common laterally at vascular bundles (Goldberg 1940; Grayum 1990). **Tannin** cells idioblastic with dark contents, common throughout parenchyma tissue. **Starch** granules in groups in 1–2 cells near phloem of most bundles. **Crystals**: raphides abundant as biforine cells; single bundles in short narrow cells attached by one end to aerenchyma partitions, especially near vascular bundles, and suspended in air cavities (Fig. 96B,C,E). Crystal sand common throughout parenchyma of midrib, uncommon elsewhere.

Petiole TS

Outline: bilaterally symmetrical, adaxial side with single rib, abaxial side rounded and laterally compressed. **Size** 12 × 15 mm diameter. **Cuticle** thin and striate. **Epidermis**: cells small, square in section and polygonal on surface. **Stomata** common, level with surface and with double cuticular ledges on guard cells. **Chlorenchyma** subepidermal, *c.* 5 layers deep in discontinuous peripheral band. **Collenchyma**: angularly thickened cells in rounded strands, 2–3 cells beneath epidermis, at close regular intervals around periphery (Fig. 96G). Each strand capping small vascular bundle (type Sv). **Ground tissue**: beneath chlorenchyma, entire tissue an aerenchyma (type 3) of large air cavities with vascular bundles at points of convergence of uniseriate partitions; 6–7 cells per partition.

Vascular bundles scattered and numerous (type III). Xylem: a wide protoxylem lacuna surrounded by layer of very small parenchyma cells (Fig. 96F,G); metaxylem cells narrow and rare. Phloem strands ± circular, small in peripheral bundles and large in interior bundles. **Tannin** cells with dark contents occasional in partitions and not associated with vascular bundles. **Starch** granules concentrated in ground tissue. **Crystals**: biforines attached by one end to vascular bundle islands in aerenchyma and suspended in air cavities (Fig. 96F). Crystal sand common in most parenchyma cells.

Material examined

Typhonodorum lindleyanum Schott, cult. M.

Species reported on in the literature

Behnke (1972): leaf. • Claverie (1906, 1909): fibres. • Engler and Krause (1912): leaf, petiole, crystals. • Ertl (1932): leaf venation. • French (1987c): root. • French (1988): laticifers. • French and Tomlinson (1981b): stem vasculature. • Solereder and Meyer (1928): anatomy.

Additional literature

Boodle and Hill 1929; French 1985b, 1987a; Grayum 1990; Mayo 1986; Shaw 1990; Thanikaimoni 1969; Tillich 1985; Williams *et al.* 1981.

Colletogyne S.Buchet

The single species is endemic to Madagascar. Plants are terrestrial, tuberous herbs occurring in shade. Leaves are long-petiolate and cordate. Chromosomes: 2n = 44, 46, 54. One specimen was available for study.

Leaf surface

Cuticle smooth on both surfaces. **Epidermis**: cells polygonal with straight anticlinal walls on both surfaces (1–1.5 *l/w*). **Stomata** mostly abaxial and brachyparacytic or brachypara-tetracytic (Fig. 97E). Subsidiary cells 2–3 pairs; inner set broad and rounded. **Venation**: secondaries digitate, basally acrodromous, arising obtusely at base, acutely distally, and connecting brochidodromously in submarginal loops. Tertiaries parallel with secondaries and steeply ascending, with small orthogonal quaternaries.

Leaf TS

Structure dorsiventral. Lamina tapering abruptly to attenuate margin. **Size**: 85 mm along midrib, 110 mm tip to base of lobes × 102 mm (*l/w*). **Midrib** flat and broad, slightly concave adaxially, broadly and shallowly convex abaxially (Fig. 97A). **Cuticle** thin on both surfaces. **Epidermis**: cells flattened, and larger adaxially. **Stomata** subtending large air cavities. **Mesophyll**: one palisade layer of elongated cells (Fig. 97C), often broadened adaxially. Spongy cells forming irregular aerenchyma of large cavities with uniseriate partitions, 1–2 cells per partition (type 3). **Air cavities**: up to 95 % of spongy zone.

Collenchyma in midrib in up to 12 discrete subepidermal strands subtending vascular bundles abaxially (type Sv) (Fig. 97A). **Vascular bundles** numerous (type II), symmetrically arranged in midrib, in *c.* 3 rows. Xylem: 1–2 elements opposing irregular phloem strand. **Laticifers**: non-anastomosing, large-armed cells lateral to phloem strands (Fig. 97B). **Crystals**: raphides as single bundles in

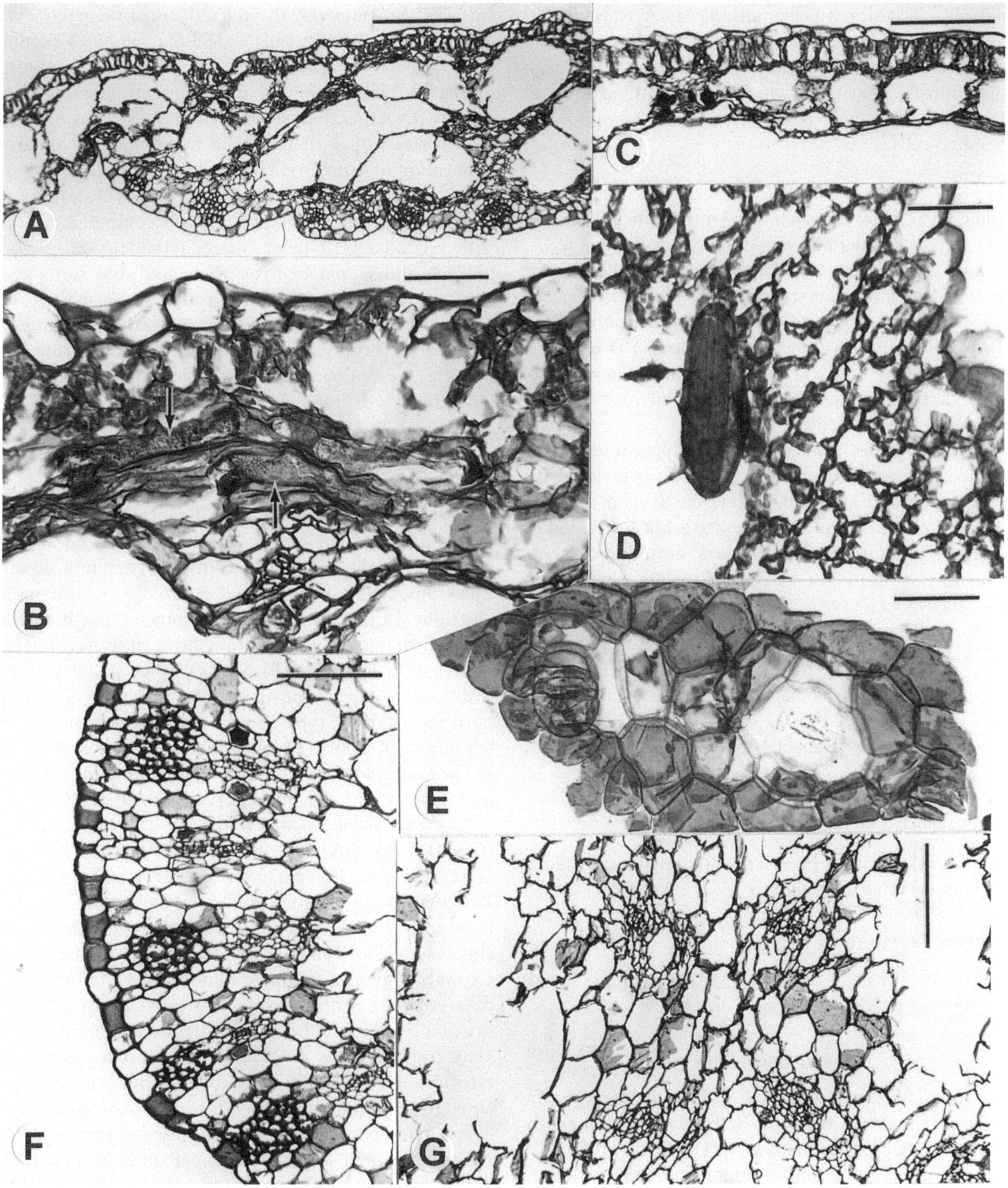

Fig. 97. *Colletogyne perrieri* leaf and petiole sections. (A) Midrib TS; (B) Part of midrib TS showing LS laticifers (arrows); (C) Lamina TS with type 3 spongy aerenchyma; (D) Lamina PS showing biforine raphide cell; (E) Lamina PS showing brachyparacytic and brachypara-tetracytic stomata; (F) Distal petiole TS showing type Sv collenchyma and small type III vascular bundles; (G) Distal petiole TS centre showing type III vascular bundles. Scale bars: A = 300 μm; B,D,E = 50 μm; C,F,G = 150 μm.

small cells, occasional near palisade layer. Raphides in biforine cells commonly suspended across spongy zone partitions; cell walls thin and papillae absent (Fig. 97D). Druses mostly near veins, small and few. Crystal sand surrounded by small dark membranes, common in mesophyll cells.

Petiole TS

Outline: oval. **Size**: diameter 4 × 3 mm. **Cuticle** thin and finely striate. **Epidermis** of square to polygonal cells, variously shaped externally and internally. **Collenchyma**: numerous small circular strands, two cell layers beneath surface, each strand up to 5–7 cells deep, each subtending phloem of small vascular bundle (type Sv) (Fig. 97F). **Ground tissue**: outer 5–8 layers of compact ground tissue, central tissue with numerous large air cavities (type 3) separated by 1–6-celled partitions.

Vascular bundles numerous, in peripheral row and scattered throughout (type III). Xylem: one wide protoxylem lacuna and occasionally one metaxylem element. Phloem: a small laterally compressed strand (Fig. 97G). **Laticifers**: 1–2 large cells with dark granular contents lateral to most vascular bundles. **Crystals**: raphides in thin-walled cells, occasional in peripheral ground tissue. Druses small and occasional in ground tissue. Crystal sand occasional in scattered ground tissue parenchyma.

Material examined

Colletogyne perrieri Buchet, Bogner 165, Madagascar, cult. M.

Species reported on in the literature

French (1987c): root. • French (1988): laticifers. • French and Tomlinson (1983): stem.

Additional literature

Bogner 1972; French 1986a,c, 1987a; Grayum 1990; Thanikaimoni 1969; Williams *et al.* 1981.

Carlephyton Jum.

This genus of three species is found in Madagascar. Plants are terrestrial tuberous herbs occurring in shady places, and adapted to a dry season. The leaves are usually solitary, occasionally two, cordate and long-petiolate. Chromosomes small: 2n = 54, 108. Three species were available for study.

Leaf surface

Cuticle smooth on both surfaces. **Epidermal** cells polygonal with straight anticlinal walls (2:1 *l*/*w*) on both surfaces. **Stomata** mostly abaxial and brachyparacytic (Fig. 98C). Subsidiary cells wide and rounded. **Venation**: *C. diegoense*: midrib prominent at base, secondaries obtuse at base becoming wide-acute to narrow-acute distally. Divergence of secondaries representing a loss of midrib size rather than bundle branching. Secondaries connecting at submarginal brochidodromous collector. Tertiaries arising acutely from secondaries, close, curved-ascending, merging with each other, becoming ± parallel with secondaries. Higher orders reticulate. *C. glaucophyllum* and *C. madagascariense*: basal and suprabasal acrodromous secondaries as prominent as midrib. All secondaries looping with brochidodromous submarginal collectors. Tertiaries arising narrowly acute from secondaries, becoming regularly parallel in intercostal areas. Quaternaries orthogonal with tertiaries.

Leaf TS

Structure dorsiventral and well-differentiated. **Size**: specimens 21 × 35 mm (*l*/*w*) to 55 × 70 mm (*l*/*w*). **Midrib** level with surface or slightly concave adaxially and slightly convex abaxially (Fig. 98A). **Cuticle** very thin or absent. **Epidermis**: cells flat or rounded externally and internally, and same size on both surfaces. **Stomata** level with surface. **Mesophyll**: one discrete palisade layer of loosely packed cells (4–8:1 *l*/*w*) (Fig. 98B). Spongy tissue an aerenchyma with large polygonal air cavities separated by uniseriate partitions (type 3; type 2 in *C. dieogense*); stomata subtending each cavity. **Collenchyma**: in *C. glaucophyllum*, strands subepidermal abaxially, subtending most midrib and laminar vascular bundles (type Sv) (Fig. 98D).

Vascular bundles in midrib 3–4, weakly formed (type II) (Fig. 98A). Xylem: 1–3 narrow protoxylem and metaxylem cells. Phloem: one irregular strand per bundle. **Laticifers** non-anastomosing, parallel with vascular bundles (Fig. 98F), with dark contents; occasionally 2–3 clustered cells. **Tannin** cells occasional in mesophyll. **Crystals**: raphides as single bundles in wide or narrow cells, common in mid-mesophyll, near veins, or randomly oriented, or near either epidermis. In *C. glaucophyllum* some raphide cells with terminal papillae, similar to biforines but with thin walls, distinct and randomly oriented in ground tissue (Fig. 98E); some suspended across aerenchyma partitions in *C. madagascariense* (Fig. 98F). Druses common, medium or small in vein sheaths or upper mesophyll. Crystal sand uncommon, scattered throughout ground tissue.

Petiole TS

Outline broadly oval, laterally compressed, or subterete. **Size**: diameter 3–4 mm. **Cuticle** thin and microstriate.

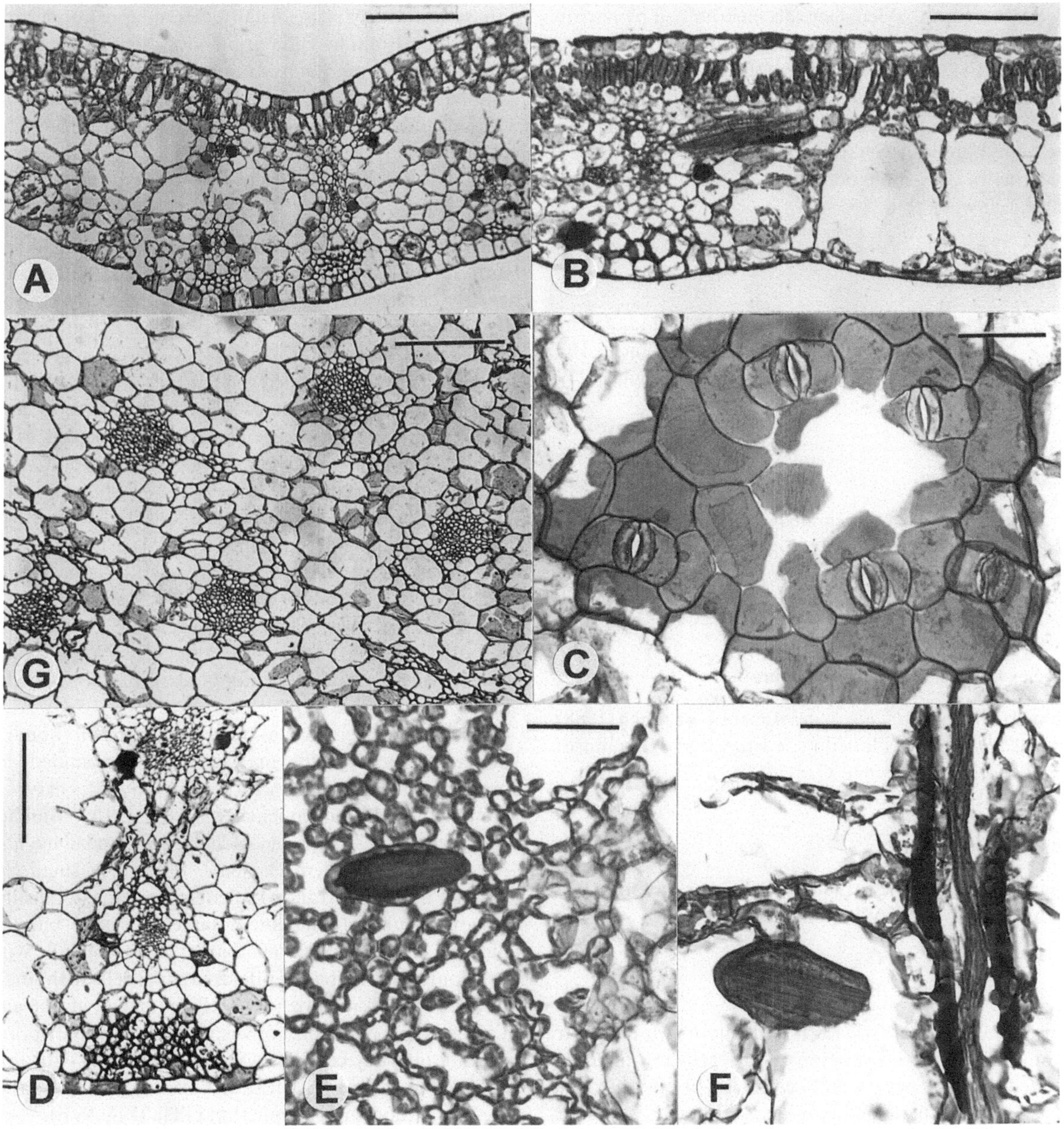

Fig. 98. *Carlephyton* leaf and petiole sections. (A) *C. diegoense*; (B–E,G) *C. glaucophyllum*; (F) *C. madagascariense*; (A) Midrib TS; (B) Lamina TS; (C) Abaxial epidermis PS showing brachyparacytic stomata; (D) Midrib abaxial margin TS; (E) Mesophyll PS showing biforine-like cell among palisade tissue; (F) Mesophyll PS showing laticifers and biforine-like cell suspended across type 3 aerenchyma partition; (G) Petiole TS showing type III vascular bundles. Scale bars: A = 200 μm; B,D,G = 150 μm; C,E,F = 50 μm.

Epidermis: cells small and flat or large and irregularly shaped. **Collenchyma**: circular strands, two cells beneath epidermis, strands subtending peripheral vascular bundles (type Sv). **Ground tissue**: outer tissue compact, cells large (Fig. 98G); central tissue type 3 aerenchyma. **Air cavities**: partitions 2–5 cells wide. Air cavities in *C. glaucophyllum* type 4 and small. Distal petiole sections with few large air spaces.

Vascular bundles forming outer circle and scattered smaller bundles in central ground tissue (type III).

Xylem: wide protoxylem lacunae, surrounded by narrow parenchyma sheath cells, and 1–2 narrow metaxylem elements (Fig. 98G). Phloem: small oval strand. **Laticifers**: 2–3 with dark contents, lateral to most vascular bundles. **Starch** granules common in ground tissue near vascular bundles. **Crystals**: raphides, common subepidermally or subhypodermally; bundles probably single, cell walls thin.

Material examined

Carlephyton diegoense Bogner, Bogner 234, Madagascar, cult. M; *C. glaucophyllum* Bogner, Bogner 167, Madagascar; *C. madagascariense* Jum., Bogner 279, Madagascar, cult. M.

Taxonomic notes

The three specimens examined exhibit species-specific variation that should be investigated further. It is possible that true biforines are developing or degenerating in this genus.

Species reported on in the literature

French (1987c): *C. madagascariense*: root. • French (1988): *C. madagascariense* : laticifers. • French and Tomlinson (1983): *C. glaucophyllum, C. madagascariense* : stem.

Additional literature

Bogner 1972; French 1986a,c, 1987a; Grayum 1990; Thanikaimoni 1969; Williams *et al.* 1981.

Arophyton Jum.

This genus of seven species is endemic to Madagascar. Plants are terrestrial tuberous herbs growing in forest understoreys, or occasionally epiphytic or epilithic. The leaves are long-petiolate, cordate, sagittate, hastate, trisect or pedately compound with elliptic leaflets. Chromosomes are small to medium-sized: 2n = 38, 40, 54, *c*.76. Five species were available for study.

Leaf surface

Cuticle smooth. **Epidermis**: cells polygonal with curved or straight anticlinal walls (1–2:1 *l/w*) on both surfaces. **Stomata** mostly abaxial, brachyparacytic (Fig. 99D). Subsidiary cells mostly broad with rounded sides. **Venation**: midrib prominent with pinnate secondaries, obtuse at base and widely acute distally. Secondaries connecting at brochidodromous submarginal, ± straightened collector. Smaller marginal straight collector originating from basal secondaries. Tertiaries tending to be parallel with secondaries or mostly reticulate. Higher orders reticulate.

Leaf TS

Structure dorsiventral. **Size**: lamina length including basal lobes 70–180 mm. **Midrib** small, flat adaxially, broadly curved-convex abaxially (Fig. 99A). Midrib in *A. humbertii* inconspicuous on both surfaces. **Cuticle** very thin or not detected. **Epidermis** of large square cells adaxially, flat or rounded on surface and internally. Abaxial epidermis similar or of smaller square or flattened cells. **Stomata** level with surface. **Mesophyll** with one, occasionally two layers of palisade, up to 20% of lamina. Palisade cells of *A. humbertii* ± T-shaped, flared facing epidermis (Fig. 99B). Spongy cells forming 5–8 layers of reticulate aerenchyma, types 1–2 near abaxial epidermis and palisade layer, type 3 at tissue centre (Fig. 99C). In *A. humbertii*, all spongy tissue of type 2 (Fig. 99B). **Air cavities** large and irregular, substomatal on both surfaces and up to 80% of spongy layers. **Collenchyma** in small subepidermal strands on abaxial side of midrib (type Sv).

Vascular bundles: abaxial outer row of midrib bundles subtended by collenchyma strands, surrounded by compact ground tissue; central scattered bundles in type 3 aerenchyma without collenchyma (Fig. 99A). Bundles small (type III). Xylem: one wide protoxylem lacuna and 1–3 narrow metaxylem elements. Phloem: a small circular or irregular strand. **Laticifers** non-anastomosing, on both flanks of phloem of larger vascular bundles (Fig. 99C). **Tannin** cells occasional in mesophyll, with dark contents. **Crystals**: raphides in small cells with single bundles, aligned with palisade cells, attached at one end in aerenchyma partitions, or suspended across partitions, pointing into adjacent air cavities. Some cells with thin cellulosic walls, others with thick-walled biforine structure but without thin terminal papillae (Fig. 99E).

Petiole TS

Outline C-shaped proximally, subterete (Fig. 99F) and distally flattened on adaxial side. **Size**: diameter 2–5 mm. **Cuticle** thin to medium-thick, smooth to striate. **Epidermis**: cells small and square. **Collenchyma**: strands circular and regularly spaced, within 2–4 ground tissue layers from epidermis (type Sv) (Fig. 99F). **Chlorenchyma**: some chloroplasts in up to 8–10 subepidermal cell layers. **Ground tissue**: cells small, compact in outer 5–10 layers,

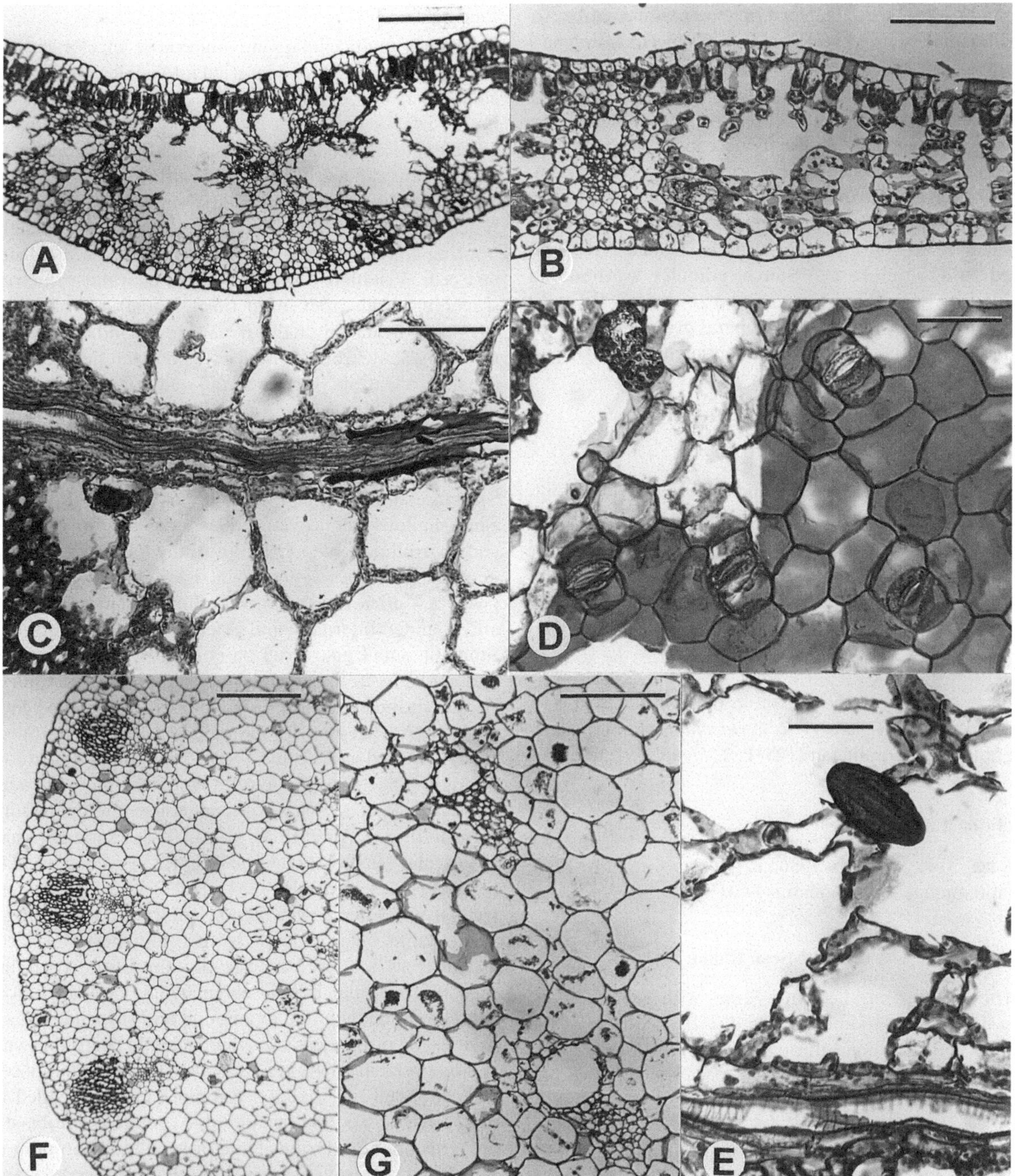

Fig. 99. *Arophyton* leaf and petiole sections. (A,D,E) *A. crassifolium*; (B,F,G) *A. humbertii*; (C) *A. buchetii*; (A) Midrib TS; (B) Lamina TS; (C) Mesophyll PS showing laticifers and raphide cell in type 3 aerenchyma; (D) Abaxial epidermis PS showing variable complexity of brachyparacytic stomata; (E) Mesophyll PS showing biforine raphide cell in aerenchyma; (F) Petiole TS showing type Sv collenchyma and compact ground tissue; (G) Petiole TS, central tissue showing type III vascular bundles. Scale bars: A,F = 300 μm; B,C,G = 150 μm; D,E = 50 μm.

expanding to type 3 aerenchyma in organ centre. **Air cavities**: partitions 3–5 cells long. *A. humbertii* with type 4 cavities, 2–3 times size of neighbouring ground tissue cells. Cavities up to 80% of central tissue.

Vascular bundles: peripheral row of small bundles mostly subtended by collenchyma caps (type Sv); scattered small bundles in petiole centre (type III). Xylem: one protoxylem lacuna and 0–3 narrow metaxylem cells (Fig. 99G). Phloem: a small irregular strand. **Laticifers**: 1–2 lateral to some phloem strands. **Tannin** cells scattered in ground tissue. **Starch** granules scattered in ground tissue cells near vascular bundles. **Crystals**: raphides most common in peripheral ground tissue; cell walls thin, cellulosic, not differentiated from ground tissue. Crystal sand common in ground tissue.

Material examined

Arophyton buchetii Bogner, Bogner 207, Madagascar, cult. M; *A. crassifolium* (Buchet) Bogner, Bogner 278, cult. M; *A. humbertii* Bogner, Bogner 160, Madagascar, cult. M; *A. tripartitum* Jum. var. *masoalaense* Bogner, Bogner 264, Madagascar, cult. M.

Species reported on in the literature

French (1987c): *A. humbertii*: root. • French (1988): *A. crassifolium, A. humbertii, A. rhizomatosum*: laticifers. • French and Tomlinson (1983): *A. humbertii*: stem.

Additional literature

Bogner 1972; French 1986a,c, 1987a; Grayum 1990; Thanikaimoni 1969; Williams *et al.* 1981.

Tribe Pistieae Blume

A tribe comprising the genus *Pistia*, with one species of free-floating aquatic plants found in fresh waters of the tropics and subtropics throughout the world.

Pistia L.

Plants occur at low elevation, in quiet fresh waters in a variety of life zones, and have well-developed vegetative propagation by stolon (Ntépé-Nyamè 1988). Plants have a short stem and a rosette of aerenchymatous, buoyant, obovate and emarginate leaves. The leaf bases are cuneate. Roots are numerous and hair-like. Chromosomes are small: 2n = 14, 28. Two specimens were available for study.

Leaf surface

Trichomes multicellular and uniseriate, of elongate or mostly rounded cells and inserted into epidermis as single enlarged rounded cells (Figs 2C, 100B); abundant on both surfaces forming a hydrophobic boundary layer. **Cuticle** smooth. **Epidermis**: cells regularly polygonal with straight anticlinal walls on both surfaces (1–1.5:1 *l/w*). **Stomata** more common adaxially, small and occurring in brachyparacytic relationship between corners of neighbouring epidermal cells, i.e. without specialized subsidiary cells. **Venation**: 6–7 evenly-spaced parallel primaries spreading from apex of petiole. Primaries connected eucamptodromously at distal margin. Secondaries arising narrowly-acute from primaries, 4–5 on each side, running parallel to leaf tip. Tertiaries and quaternaries reticulate or forming irregular cross connections.

Leaf TS

Structure dorsiventral. **Size**: 70 × 60 mm (*l/w*). **Midrib zone** shallowly concave-rounded adaxially and deeply convex-rounded V-shaped abaxially. Lamina arched between primaries resulting in deep abaxial floatation grooves. **Cuticle** not seen or very thin. **Epidermis**: cells tall and rounded internally and externally on both surfaces. **Stomata**: small guard cell pairs in crevices, at point of contact between the much larger epidermal cells. **Collenchyma** strands forming broad phloem caps (type Svc) (Fig. 100A). **Mesophyll**: one palisade layer of rounded, closely spaced cells, up to 10% of lamina thickness. Spongy tissue forming type 3 aerenchyma with very large air spaces with vertical uniseriate partitions (Fig. 100B–D). **Air cavities**: up to 99% of spongy tissue.

Vascular bundles: two large bundles in midrib, weakly formed (types II or III). Xylem: 1–2 wide protoxylem lacunae and often a wide metaxylem tracheid opposing a small semicircular phloem strand. **Laticifers** not seen. **Tannin** cells scattered among ground tissue, trichome cells, and vascular bundle parenchyma. **Crystals**: biforines common, anchored by one end and suspended in air cavities, or suspended across aerenchyma partitions with ends exposed in adjacent air cavities (Figs 9E, 100D); cell ends rounded or papillate. Raphides also embedded in mesophyll tissue in randomly oriented idioblastic cells, uncommon. Druses large or small and common in mesophyll, midrib, or epidermal layers. Crystal sand common in mesophyll or absent.

Petiole TS

Outline oval, dorsiventrally narrowed. **Size**: diameter 9 × 5 mm. **Cuticle** very thin or not seen. **Trichomes** uniseriate, multicellular, slightly lignified, common.

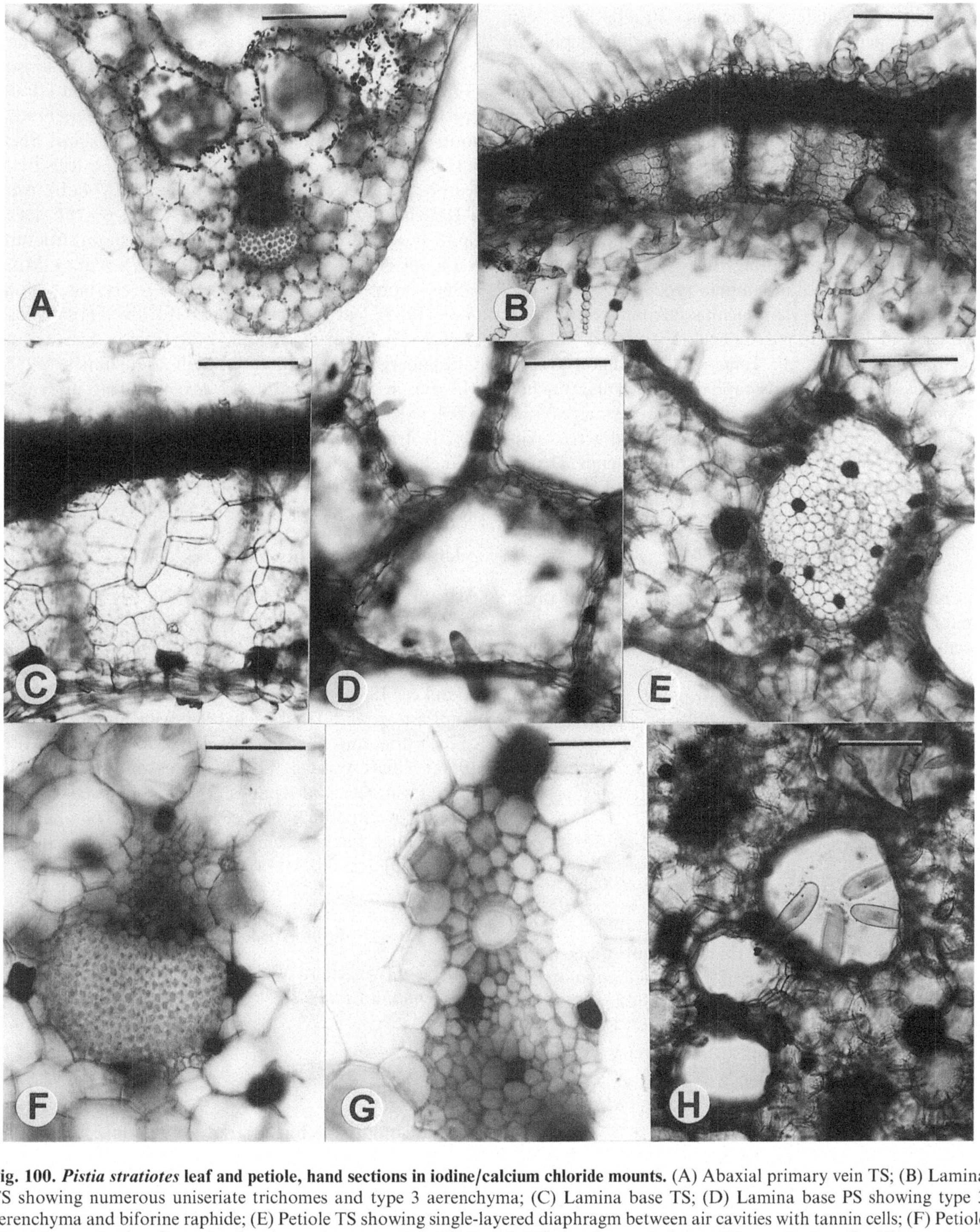

Fig. 100. *Pistia stratiotes* leaf and petiole, hand sections in iodine/calcium chloride mounts. (A) Abaxial primary vein TS; (B) Lamina TS showing numerous uniseriate trichomes and type 3 aerenchyma; (C) Lamina base TS; (D) Lamina base PS showing type 3 aerenchyma and biforine raphide; (E) Petiole TS showing single-layered diaphragm between air cavities with tannin cells; (F) Petiole TS with vascular bundle and collenchyma (near margin); (G) Petiole TS showing central type III vascular bundle; (H) Petiole TS showing raphide cells suspended in air cavity. Scale bars: A,D,H = 200 μm; B = 300 μm; C,E,F = 150 μm; G = 50 μm.

Epidermis: cells small. **Collenchyma** strands large, circular or semicircular (Fig. 100F), occurring as phloem caps (type Sv) around periphery, and separated from epidermis by 2–3 layers of ground tissue. **Ground tissue** compact except for type 3 or 4 aerenchyma cavities arranged in nine dorsiventral rows. One-layered transverse diaphragms common (Fig. 100E). **Air cavities**: up to 40% of ground tissue.

Vascular bundles arranged in dorsiventral rows, five per row, each row alternating with rows of aerenchyma cavities. Bundles representing ribs of acrodromous primaries ending in distal lamina; type II. Xylem: one medium-width protoxylem lacuna surrounded by small double or single row of narrow cells (Fig. 100G). Phloem: ± circular small strand. **Tannin** cells common among ground tissue cells. **Crystals**: raphides numerous, attached by one end and pointing into aerenchyma cavities (Fig. 100H). Some true biforines suspended across partitions. Most cells thin-walled wth rounded ends. Druses uncommon in ground tissue, some suspended halfway into cavities.

Root morphology

Discrete cap and continuous epidermis, one cell thick, separating cap from cortex. Roots large with large quiescent centres for their size (Clowes 1958, 1984). Clowes (1990) reported that major roots are adventitious; laterals arise from endogenous primordia within apical meristem of mother root; large numbers of primordia are found in sections of mother root tips.

Material examined

Pistia stratiotes L., cult. M; cult. MO.

Taxonomic notes

Pistia has many characters that fit comfortably within the Aroideae pattern in spite of its unusual apomorphies related to its floating aquatic habit. Subfamily status might be justified due to its unique features but placing it in its own family would appear unnecessary.

Species reported on in the literature

Arber (1922b): leaf. • Buscalioni (1895–6): crystals. • Charlton (1983): root. • Clowes (1958, 1984, 1985, 1990, 1992): root. • Datta and Biswas (1980): root, stolon, leaf. • Datta and Maiti (1968): mechanical tissue. • Dihoru (1986): root. • Engler (1920b): shoot, root. • Ertl (1932): leaf venation. • Fisher and French (1978): internodal meristems. • Flahault (1878): root. • Franceschi and Horner (1980): root, crystals. • French (1987c): root. • French (1988): laticifers. • French and Tomlinson (1983): stem vasculature. • Genua and Hillson (1985): crystals. • Gulliver (1865): crystals. • Guttenberg (1968): root. • Haccius and Lakshmanan (1966): shoot apex. • Hotta (1971): leaf. • Inamdar *et al.* (1983): leaf architecture. • Ito (1899): leaf. • Janczewski (1874a,b): root. • Knecht (1983): leaf venation. • Kubin (1878): development. • Lemon and Posluszny (2000a): shoot structure. • Lierau (1888): root. • Minden (1899): leaf. • Mitra (1966): morphology. • Netolitzky (1929): crystals. • Noda *et al.* (1985): leaf surface. • Pant and Kidwai (1966): leaf, stomata. • Sakai *et al.* (1984): crystals. • Schade and Guttenberg (1951): root. • Shah and James (1971): phloem. • Sifton (1945): *P. texensis*: leaf air space. • Solereder and Meyer (1928): all organs. • Staudermann (1924): trichomes. • Tieghem and Douliot (1888): root branch ontogeny. • Treub (1876): root meristem. • Uphof *et al.* (1962): hairs. • Voronkina (1975): root apex.

Additional literature

Arber 1920, 1922b; Batygina and Yakovlev 1990; Blanc 1978; Buscalioni and Lanza 1928; Buzgó 1994; Datta and Biswas 1970; Davis 1966; Domin 1911; French 1986a, c, 1987a; Gentner 1905; Grayum 1990; Hegelmaier 1874; Klercker 1888; Knecht 1983; Kulkarni *et al.* 1990; Kuprianova and Tarasevich 1984; McClure 1970; Mercado-Noriel and Mercado 1977–8; Meusel 1951; Mitra 1966; Nunes Caldeira 1967; Pettet and Pettet 1970; Ray 1988; Raynal-Roques 1981; Shadowsky 1931; Subramanyam 1962; Tarasevich 1990; Thanikaimoni 1969; Tillich 1985; Tsuchiya 1991; Wheaton and Stewart 1970; Williams *et al.* 1981; Zennie and McClure 1977.

Tribe Arisaemateae Nakai

A tribe of two genera of North America, Africa and Asia, *Arisaema* and *Pinellia*.

Anatomical summary

Leaf surface

Cuticle smooth to finely striate. **Epidermis**: cells polygonal-elongate adaxially, anticlinal walls undulate abaxially, or sinuous on both surfaces. **Stomata** abaxial and brachyparacytic to brachypara-hexacytic in *Arisaema*. Subsidiary cells broad or narrow and rounded or angular. **Venation** with strong midrib and secondaries ending in straightened brochidodromous loops.

Leaf TS

Structure dorsiventral. **Midrib** semicircular-convex and often ridged abaxially. **Epidermis**: cells irregular in size, tall or short. **Mesophyll** well-differentiated into one palisade layer and type 2 spongy aerenchyma, or irregular type 1. **Collenchyma** in circular abaxial strands in midrib (type Sv), subepidermal or separated from epidermis by 1–3 ground tissue layers. Collenchyma also present in abaxial lamina in *Arisaema*. **Vascular bundles** few or numerous and scattered in midrib; types II and III. **Laticifers** non-anastomosing, near phloem, or absent. Photosynthate more likely to be sugar than starch. **Crystals**: raphides in multiple bundles in elongate cells (*Pinellia*) or in multiple overlapping or single bundles (*Arisaema*); also in wide tubular cells in *Arisaema*.

Petiole TS

Outline subterete. **Cuticle** thin and smooth. **Epidermis**: cells small. **Ground tissue** compact and cells small near periphery grading to large in centre. Air cavities: type 4 in *Pinellia*, type 3 or 4 in *Arisaema*. **Collenchyma** type Sv. **Vascular bundles**: types II and III. **Crystals**: raphides as single bundles in wide cells.

Taxonomic notes

These genera show little variation and appear closely related. They share a tendency toward type 4 aerenchyma formation in the petiole, not a widespread feature in the family. Also the raphides in multiple bundles in elongate cells and the tendency toward sugar as a storage photosynthate are important shared character states.

***Arisaema* Mart.**

This genus of about 150 species is distributed in tropical Asia, Atlantic North America to Mexico, with four species in east Africa. Plants are terrestrial tuberous herbs, or rarely with rhizomatous tubers, occurring as understorey plants in open woodlands, occasionally in swampy areas, or in sunny open areas or steppes. Two species are infrequently epiphytic. Leaves are long-petiolate, and trifoliate to digitately or pedately compound. Chromosomes: 2n = 26, 28, 48, 52, 56, 58, 66–68, 72, 140. Four species were available for study.

Leaf surface

Cuticle smooth to finely striate. **Epidermis**: cells polygonal with straight anticlinal walls, or undulate or sinuous adaxially (1–3:1 *l*/*w*). Cells polygonal with straight anticlinal walls (Fig. 101E) or undulate or sinuous (Fig. 101F), and also isohedral to elongate abaxially (1–4:1 *l*/*w*). **Stomata**: abaxial, brachyparacytic to brachparahexacytic (Fig. 101E,F). Subsidiary cells broad and square or irregular, or narrow and rounded. **Venation**: prominent midrib within pinnate leaflet, acute and slightly curved secondaries ending at a brochidodromous submarginal collector. Tertiaries tending to be parallel with secondaries, or irregular and reticulate. Higher orders reticulate.

Leaf TS

Structure dorsiventral. **Size**: segments 15–18 × 8–10 cm (*l*/*w*). **Midrib** prominent and flat or shallowly concave adaxially, broadly and deeply semicircular-convex abaxially and often ridged (Fig. 101A,B). **Cuticle** not seen or thin. **Epidermis**: cells tall adaxially, 2–5 times taller than abaxial cells. Adaxial cells rounded on surface with rounded inner walls; abaxial cells short and flat externally and internally. **Stomata** level with surface. **Mesophyll**: one palisade layer, up to 20% of mesophyll, and 5–8 layers of long-armed spongy cells (type 2) or short, irregular type 1 cells (Fig. 101C). **Air cavities**: 20–90% of spongy layer. **Collenchyma** angularly thickened, midrib with 5–11 subepidermal abaxial strands and one central adaxial strand (Fig. 101A,B); abaxial strands subtending circle of vascular bundles (type Sv). Strands in lamina abaxial to large secondary bundles.

Vascular bundles numerous, symmetrically scattered in midrib, appearing reticulate paradermally in lamina (types II and III). Xylem: 1–2 wide protoxylem elements, 1–2 narrow metaxylem elements. Phloem: small circular strand. **Laticifers** non-anastomosing, noted only in *A. anomalum*, reported by French (1988) in *A. triphyllum* and *A. wrayii*. **Tannin** cells with dark contents absent or common; cells randomly scattered in mesophyll or near vascular bundles. **Photosynthate** storage reported as unstable in nine species, mostly more sugar than starch (Sakai and Hayashi 1973). **Crystals**: raphides in single or multiple overlapping bundles, uncommon or common, in wide and often tubular cells, often with radiating neighbouring spongy cells (Fig. 101D). Raphides often in palisade tissue and adaxial epidermis of *A. tortuosum* (Fig. 101C). Paradermally, raphide bundles often tending to be parallel with secondary vascular bundles. Crystal sand absent or occasional in midrib and lamina cells.

Petiole TS

Outline subterete. **Size**: diameters 3, 4, 12 mm. **Cuticle** thin and smooth. **Epidermis**: cells square, small, rounded

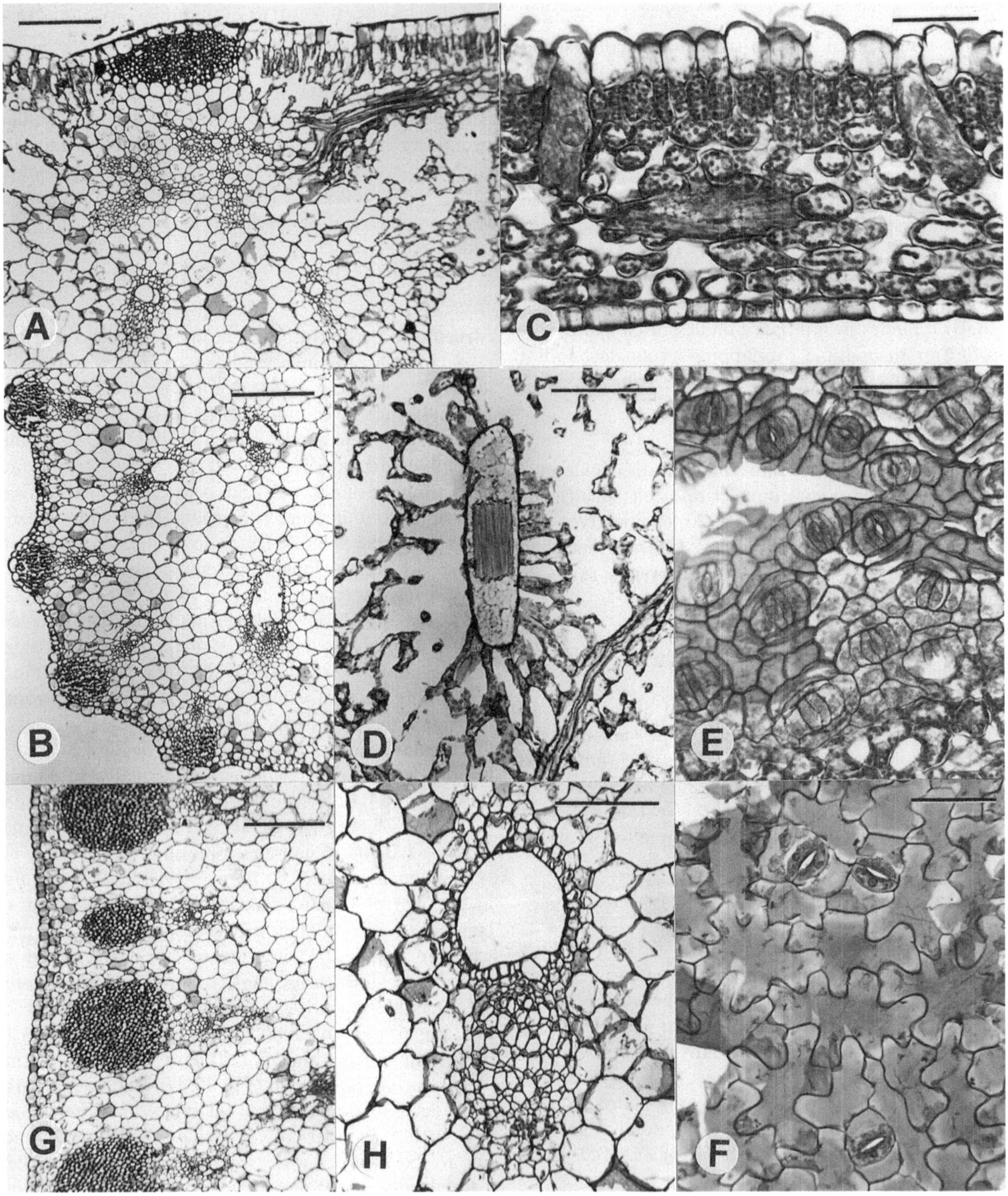

Fig. 101. *Arisaema* leaf and petiole sections. (A,D) *A. fargesii*; (B,C,E,G,H) *A. tortuosum*; (F) *A. anomalum*; (A) Midrib TS; (B) Midrib TS abaxial, lateral side; (C) Lamina TS with raphide cells horizontal and interpenetrating palisade and epidermal layer; (D) Lamina PS with raphide cell; (E) Abaxial epidermis PS with stomata; (F) Abaxial epidermis with stomata; (G) Petiole margin TS with type Sv collenchyma strands; (H) Petiole centre TS with type III vascular bundle. Scale bars: A,B,G = 300 μm; C,E,F = 50 μm; D,H = 150 μm.

internally and externally. **Collenchyma**: circular strands 10 to >20 cells deep around periphery, 2–4 cells beneath epidermis (Fig. 101G). Each strand subtending vascular bundle (type Sv). **Chlorenchyma**: plastids in outer 3–4 layers. **Ground tissue**: cells small near periphery grading to larger in centre. **Air cavities**: small, cell-sized and few, or large (3–4 times cell diameters), type 3 or 4 aerenchyma. Air space in centre 5–20%.

Vascular bundles in peripheral circle and with numerous scattered bundles in centre; bundles robust (types II and III). Xylem: 1–4 very wide protoxylem elements and 1–2 narrow metaxylem elements. Phloem: large circular strand (Fig. 101H). **Crystals**: raphides uncommon, in large thin-walled circular cells in peripheral ground tissue.

Material examined

Arisaema anomalum Hemsl., Bogner 1274, Borneo, cult. M; *A. fargesii* S. Buchet, cult. M; *A. tortuosum* Schott, cult. M, 75/75; *A. triphyllum* (L.) Schott, Boufford and Utech 18063, Illinois, USA, MO.

Taxonomic notes

A coherent genus but with much diagnostic variation among species. It should be examined in more detail.

Species reported on in the literature

Black (1918): *A. triphyllum*: corm, crystals. • Borzi (1887–8): root. • Boyd (1932): several spp.: seedling. • Brundrett and Kendrick (1988, 1990): *A. atrorubens* : root. • Chrysler (1904): several spp.: stem. • Dalitzsch (1886): *A ringens*: leaf. • Ertl (1932): *A. amurense*: leaf. • French (1987c): *A. candidissimum, A. triphyllum, A. wrayii*: root. • French (1988): *A. triphyllum, A. wrayii*: laticifers. • French and Tomlinson (1983): *A. triphyllum*: stem. • Gentner (1905): leaf. • Genua and Hillson (1985): *A. atrorubens*: crystals. • Higley (1880): *A. triphyllum*: crystals. • Holm (1911a): *A. triphyllum*: seedling. • Hotta (1964): several spp.: seedling. • Hotta (1971): *A. album, A. ringens, A. serratum*: tuber, root. • Ko *et al.* (1990): leaf, petiole, root. • Korschgen (1981): *A. triphyllum*: leaf. • Leavitt (1904): *A. triphyllum*: root. • MacDougal (1901): *A. dracontium, A. triphyllum*: seedling. • Mashima (1964): *A. serratum*: leaf. • Murata (1990a): Areae: leaf ontogeny. • Noro *et al.* (1992): leaf, stem. • Ohtsuki (1960): *A. atrorubens*: tuber. • Pant and Kidwai (1966): *A tortuosum*: stomata. • Periasamy and Muruganathan (1986): leaf ontogeny. • Pickett (1915): *A. triphyllum*: all organs. • Planchon and Juillet (1909): *A. pentaphyllum:* starch. • Qin (1988): rhizome. • Reichert (1913): *A. triphyllum*: starch. • Rennert (1902): *A. dracontium, A. triphyllum*: seedling. • Roth (1949, 1952): *A. ringens*: leaf and ontogeny. • Sakai and Hayashi (1973): several spp.: leaf. • Shah and Gopal (1969): leaf. • Solereder (1919): *A. enneaphyllum*: leaf. • Solereder and Meyer (1928): *A. enneaphyllum*: all organs. • Stebbins and Khush (1961): stomata. • Sun (1960): *A. consanguineum*: leaf. • Tieghem (1866): *A. atrorubens*: all organs. • Tieghem (1870–1): *A. atrorubens*: root, secretory tissue. • Troll (1932a): *A. amurense, A. consanguineum*: leaf ontogeny. • Van Fleet (1942): *A. dracontium, A. triphyllum*: stem. • Wakabayashi (1957a): *A. ringens, A. serratum*: crystals.

Additional literature

Antony 1898; Batygina and Yakovlev 1990; Blanc 1978; Campbell 1902; Daumann 1930, 1970; Davis 1966; Erdtman 1952; French 1986a, c, 1987a; Grayum 1990; Hay 1986; Hegnauer 1986; Huang 1972; Kulkarni *et al.* 1990; Kuprianova and Tarasevich 1984; Maheshwari and Khanna 1956; Mehra and Sachdeva 1976; Murata 1984, 1989, 1990b; Oh *et al.* 1990; Ohashi *et al.* 1983; Ray 1988; Sargant 1903; Sharma 1967; Smolenski *et al.* 1973, 1974; Tarasevich 1990; Thanikaimoni 1969; Tieghem 1907; Tillich 1985; Vogel 1962, 1978, 1990; Williams *et al.* 1981; Zavada 1983.

Pinellia Ten.

This genus of seven species is distributed in China and Japan. Plants are terrestrial, occasionally epilithic, tuberous herbs occurring mostly in shady wet areas along streams, or as understorey herbs. Leaves are long-petiolate and vary from a somewhat cordate primary leaf to basally lobed or digitately trilobed or pedately lobed lamina. Chromosomes: 2n = 26. One species was available for study.

Leaf surface

Cuticle smooth. **Epidermis**: cells polygonal-elongate adaxially with straight anticlinal walls (3–4:1 *l*/*w*) and undulate abaxially (4–5:1 *l*/*w*). **Stomata**: abaxial and brachyparacytic (Fig. 102F). Subsidiary cells broad and rounded or irregular. **Venation**: strong midvein (within lobes) having acute pinnate secondaries, looped near margin in straightened brochidodromous loops. Small higher order collector veins also present. Higher order venation reticulate.

Leaf TS

Structure dorsiventral. **Size**: 9 × 3 cm (*l*/*w*). **Midrib** large and flat adaxially, rounded semicircular-convex

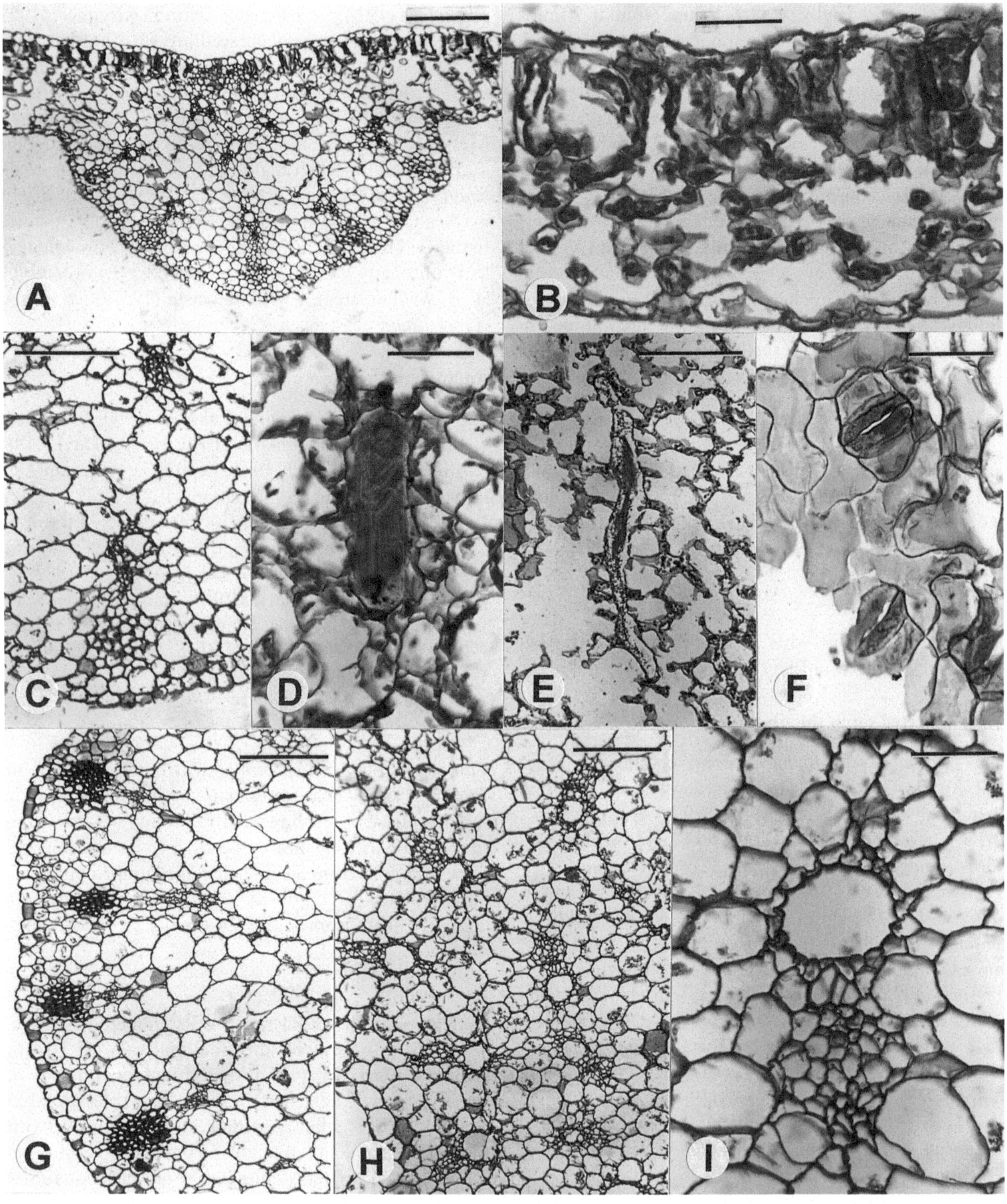

Fig. 102. *Pinellia tripartita* leaf and petiole sections. (A) Midrib TS; (B) Lamina TS; (C) Abaxial midrib TS; (D) Lamina PS with raphide cell; (E) Lamina PS showing raphide tube; (F) Abaxial epidermis PS with stomata; (G) Petiole TS margin with type Sv collenchyma strands; (H,I) Distal petiole centre TS (I) Detail of type III vascular bundle. Scale bars: A = 300 μm; B,D,F,I = 50 μm; C,E = 150 μm; G,H = 200 μm.

abaxially (Fig. 102A). **Cuticle** thin on both surfaces. **Epidermis**: cells mostly flattened, of irregular height on both surfaces. **Stomata** level with surface. **Mesophyll**: one palisade layer, up to about 40% of lamina (Fig. 102B); some tendency for subsurface lobing of palisade cells. Spongy tissue of up to *c.* 8 irregular cell layers with tendency toward uniseriate aligned partitions; cells long-armed (type 2). **Air cavities**: up to about 90% of spongy tissue. **Collenchyma** forming several discrete strands, circular or irregular (Fig. 102C), separated from epidermis by 1–3 layers of ground tissue and subtending vascular bundles abaxially in midrib (type Sv).

Vascular bundles: weakly formed, nine in two rows, scattered symmetrically (Fig. 102A); type III. Xylem: wide protoxylem lacuna and 2–3 narrow metaxylem elements. Phloem: small circular or irregular strand. **Laticifers**: infrequent and non-anastomosing, 0–2 along phloem strands. **Reserve photosynthate**: plastids storing sugar rather than starch (Sakai and Hayashi 1973). **Crystals**: raphides in multiple bundles in short or elongate cells, common in mesophyll (Fig. 102D,E). Crystal sand common in mesophyll.

Petiole TS

Outline subterete. **Size**: diameter 3–6 mm. **Cuticle** thin and smooth. **Epidermis**: cells small and square. **Collenchyma**: circular strands, 8–10 cells in diameter (Fig. 102G), around periphery (type Sv), separated from epidermis by 1–2 ground tissue layers. **Chlorenchyma**: outer 3–5 layers of ground tissue. **Ground tissue**: mostly compact, large parenchyma cells. **Air cavities**: 6–7 large type 4 cavities in circle between outer and inner rings of bundles.

Vascular bundles: outer circle of numerous small bundles subtended by collenchyma strands; two inner circles of fewer, larger bundles (Fig. 102H) (type III). Xylem: one wide protoxylem lacuna, 2–3 narrow metaxylem elements (Fig. 102I). Phloem: small irregular strand. **Crystals**: raphides in wide thin-walled cells, scattered and uncommon in ground tissue.

Material examined

Pinellia tripartita Schott, coll.1977, s.n., cult. M.

Taxonomic notes

Ertl (1932) noted and illustrated a 2-tiered vascular organization in the distal petiole. Multiple 'primary nerves' are shown entering the lamina from the petiole. A central 'secondary nerve' superimposes vascular branches over the primary nerves. If confirmed, this would be unique in the tribe.

Species reported on in the literature

Dalitzsch (1886): *P. tuberifera*: leaf. • Ertl (1932): *P. tuberifera*: leaf venation. • French (1987c): *P. tripartita*: root. • French (1988): *P. tripartita*: laticifers. • French and Tomlinson (1983): *P. tripartita*: stem. • Freundlich (1908): *P. tuberifera*: leaf vasculature. • Higuchi and Okada (1980): *P. cordata:* tuber. • Hotta (1971): leaf. • Kuster (1928–9): *P. tuberifera*: regeneration. • Li (1999): several spp.: leaf epidermis, stomata. • Lierau (1888): *P. tuberifera*: root. • Linsbauer (1934): *P. ternata*: bulbils • Löv (1925–6): *P. tuberifera*: leaf ontogeny. • Mannich and Yih (1926): *P. tuberifera*: tuber. • Möbius (1936): *P. ternata*: bulbil. • Netolitzky (1929): crystals. • Qin (1988): anatomy. • Sakai and Hayashi (1973): *P. ternata*: chloroplasts. • Solereder (1919): *P. tuberifera*: raphides. • Solereder and Meyer (1928): *P. tuberifera*: several organs. • Wakabayashi (1957a): *P. ternata, P. tripartita*: raphides. • Winkler (1907): *P. tuberifera*: petiole.

Additional literature

Batygina and Yakovlev 1990; Domin 1911; Eyde *et al.* 1967; French 1986a, b, c, 1987a; French and Tomlinson 1983; Grayum 1990; Hegnauer 1986; Huang 1972; Irmisch 1874; Murata 1990a; Nahrstedt 1975; Ray 1988; Schwendener 1874; Tarasevich 1990; Thanikaimoni 1969; Tieghem 1907; Tillich 1985; Uhlarz 1985, 1986; Vágújfalvi 1971; Williams *et al.* 1981.

Tribe Areae

A tribe of nine genera ranging from Europe across Asia to Australia. The genera are *Typhonium, Theriophonum, Sauromatum, Lazarum, Biarum, Arum, Eminium, Dracunculus* and *Helicodiceros*.

Anatomical summary

Leaf surface

Cuticle smooth or striate to rugulate (*Biarum*) and mostly thin. **Epidermis**: cells polygonal with straight or undulate anticlinal walls. **Stomata** brachyparacytic, infrequently brachypara-tetracytic, and abaxial or on both surfaces. Subsidiary cells broad or narrow, irregular, rounded or angular, or absent (i.e. stomata anomocytic: *Typhonium, Biarum, Dracunculus*). **Venation**: usually with prominent midrib and secondaries eucamptodromous (*Biarum, Arum*) or brochidodromous, usually at straightened collector vein.

Leaf TS

Structure dorsiventral or weakly so. **Midrib** usually prominent, broadly rounded abaxially or not differentiated

(*Helicodiceros*), or abaxially ridged (*Typhonium*, *Theriophonum* and *Sauromatum*). **Epidermis**: cells short or tall. **Stomata** level or on shallow mounds (*Biarum*, *Eminium*, *Dracunculus*). **Mesophyll**: palisade of 1–3 layers, or absent (*Biarum*). Spongy tissue types 1, 1a, 2 or 3. **Collenchyma**: circular or semicircular strands (type Sv) in abaxial midrib. **Vascularbundles** in midrib one, or few, or several; usually type II (type III in *Typhonium* and some in *Sauromatum*). Fibres present as phloem caps only in *Lazarum*. **Laticifers** reported as non-anastomosing for all genera but usually not seen in this material (seen only in *Theriophonum*, *Biarum* and *Eminium*). **Tannin** cells absent or infrequent. **Crystals**: raphides in small cells with single bundles or as 1–3 bundles or long overlapping bundles in large wide or elongated cells; raphide cells articulated into 'raphide tubes' in *Biarum*. Crystal sand and druse-like clusters occasionally present.

Petiole TS

Outline variable, often grooved adaxially, semicircular, or elliptical abaxially (fluted in *Theriophonum* and *Lazarum*). **Cuticle** thin to medium-thick and smooth to striate. **Epidermis**: cells small or large. **Stomata**, when present, sometimes arranged in rows between collenchyma strands (*Sauromatum*, *Biarum*, *Eminium*). **Collenchyma** type Sv. Ground tissue compact throughout or grading to type 3 aerenchyma in centre. **Vascular bundles** numerous and scattered, types II or III, rarely I (*Eminium*). **Crystals**: raphides usually as single bundles in thin-walled cells (1–several bundles in *Eminium*).

Taxonomic notes

This tribe demonstrates some diagnostic variation in its array of character states but the taxa are mostly coherent. One distinctive shared character of the group is the non-anastomosing laticifer that is difficult to detect. French (1988) has reported laticifers of this type in all of the nine genera but they were only observed at first hand in this study in *Theriophonum*, *Biarum*, *Eminium* and *Helicodiceros*. In this clade laticifers do not take up dyes and apparently lack the pigmented components found in most other laticifer-containing genera in the family. Special techniques for detecting latex as used by French were not used or found necessary for other genera in this survey.

Biarum appears to be the most distinct taxon in terms of midrib shape and aerenchyma, and its articulated raphide tubes, but it does share features of cuticle structure with *Helicodiceros*, stomatal placement with *Eminium*, spongy mesophyll type with *Arum*, *Dracunculus* and *Eminium*, leaf eucamptodromy with *Arum*, petiole aerenchyma type with *Arum*, *Dracunculus*, and *Helicodiceros*. Similar groupings can be made with other genera. *Lazarum*, *Sauromatum* and *Typhonium* share compact ground tissue in the petiole. The more one compares features in this group, the more it appears as a mosaic; there are no sets of characters that subdivide the taxa into discrete subgroups. The unmodified stomatal subsidiary cells found only in *Typhonium* (and at times in *Dracunculus* and *Biarum*) could turn out to be plesiomorphic for the group if the cpDNA tree of French *et al.* (1995) proves enduring.

Typhonium Schott

This genus of 25–40 species was originally distributed in India, Sri Lanka, the Himalayas, south-east Asia, China, Japan, Malaysia and Australia. One species is naturalized in the Comores, Ghana, Venezuela and Brazil. Plants are terrestrial tuberous perennial herbs occurring in a wide variety of habitats from near sea level to 4000 m. The leaves are long-petiolate and hastate or sagittate to pedately-lobed with elliptic segments. Chromosomes (13 species, 34 counts): 2n = 16, 18, 20, 26, 52, 54, 65, 118, 130, 152, 160, 168. Two species were available for study.

Leaf surface

Cuticle smooth on lamina, finely striate on midrib. **Epidermis**: cells polygonal with straight anticlinal walls adaxially (1–2: *l*/*w*) and straight to undulate abaxially. **Stomata** mostly abaxial with neighbouring cells arranged brachyparacytically (Fig. 103D), otherwise unmodified. **Venation**: midrib prominent. Basal lobe primaries dividing at petiole apex. Secondaries acute and straight, attaching to straightened brochidodromous collector. Tertiaries parallel with secondaries or reticulate. Higher orders reticulate.

Leaf TS

Structure dorsiventral. **Size**: 150 × 90 mm (*l*/*w*); basal lobes 65 × 35 mm (*l*/*w*). **Midrib** shallowly concave-rounded adaxially and broadly convex, ridged-angular abaxially (Fig. 103A). **Cuticle** thin on both surfaces. **Epidermis**: cells medium-sized on both surfaces and rounded or flat externally or internally. **Stomata** level with surface. **Mesophyll**: well-differentiated, single palisade layer, up to 40% of lamina (Fig. 103B), and 4–5 spongy layers of short-armed, irregularly lobed cells (types 1, occasionally 2). **Air cavities** circular or irregular, up to 40% of lamina. **Collenchyma**: 4–5 strands subtending vascular bundles (type Sv), up to five cells deep beneath abaxial ridges of midrib (Fig. 103A).

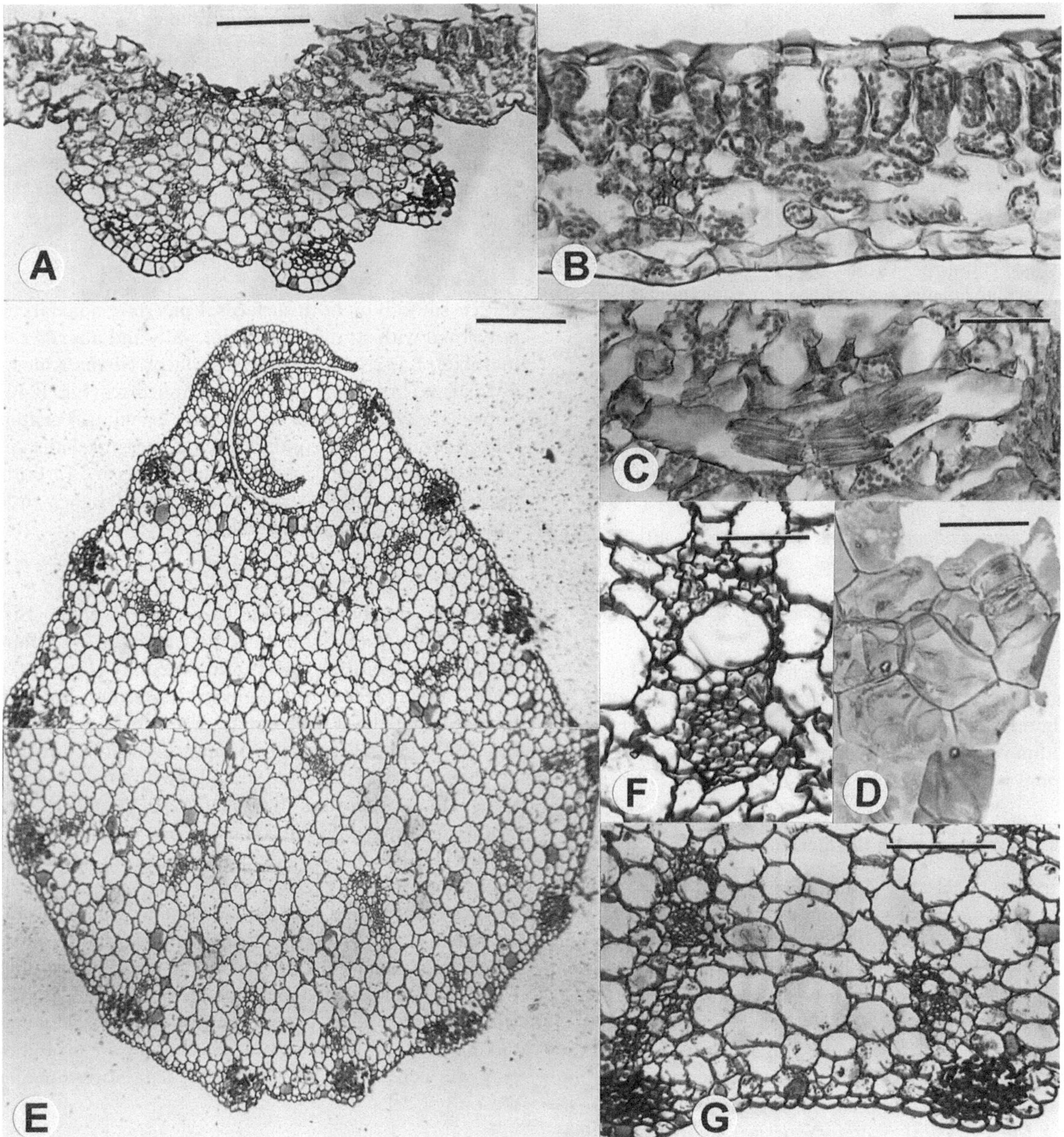

Fig. 103. ***Typhonium giraldii*** **leaf and petiole sections**. (A) Midrib TS; (B) Lamina TS; (C) Mesophyll PS showing raphide cell; (D) Adaxial epidermis PS with stoma; (E) Proximal petiole TS; (F) Distal petiole TS with type III vascular bundle; (G) Distal petiole TS lateral side showing type Sv collenchyma. Scale bars: A = 200 µm; B–D,F = 50 µm; E = 300 µm; G = 150 µm.

Vascular bundles: *c.* 4 bundles in midrib; bundles small, type III. Xylem one wide protoxylem lacuna and *c.* 3 narrow metaxylem cells opposing a small, laterally compressed strand of phloem. **Laticifers** reported as non-anastomosing by French (1988); not detected in this material. **Crystals**: raphides common in mid-mesophyll, ± aligned with secondary veins in multiple bundles in large, very elongated cells (Fig. 103C). Also, few raphides in single bundles in small isolated cells in mid-mesophyll.

Petiole TS

Outline: mostly terete-fluted and slightly laterally compressed with small adaxial groove (Fig. 103E). **Size**: 2–4 mm diameter. **Cuticle** smooth or finely striate. **Epidermis**: medium-sized cells, small and square. **Collenchyma**: *c*. 14 circular subepidermal strands, *c*. 8 cells deep and angularly thickened (Fig. 103E), distributed around periphery of petiole (type Sv). **Ground tissue** uniform with no zonation. **Air cavities** small, restricted to corners between cells.

Vascular bundles scattered, outer row aligned beneath collenchyma strands (Fig. 103E, G); types II and III. Xylem: 1–2 protoxylem lacunae and 1–3 narrow metaxylem cells adjacent to small laterally compressed phloem strand (Fig. 103F,G). **Crystals**: raphides in small cells, mostly concentrated near peripheral ground tissue.

Material examined

Typhonium blumei Nicolson & Sivadasan, 401/75, cult. M; *T. geraldii* (Baroni) Engl., Bogner 1108, cult. M.

Taxonomic notes

The genus *Typhonium* now includes *Lazarum* (Hay 1997) and *Sauromatum* (Hetterscheid and Boyce 2000). As originally envisioned, those genera are treated separately in this volume. See *Lazarum* and *Sauromatum* for additional notes.

Species reported on in the literature

Chrysler (1904): *T. divaricatum*: stem. • Ertl (1932): *T. divaricatum*: leaf venation. • French (1987c): *T. trilobatum*: root. • French (1988): *T. trilobatum*, *T.* sp.: laticifers. • French and Tomlinson (1983): *T. divaricatum*, *T. giraldii*: stem vasculature. • Giovannozzi (1911): *T. giganteum*: tuber. • Hotta (1971): leaf. • Lierau (1888): *T. divaricatum*: root. • Messeri (1925): *T. giganteum*: vasculature. • Murata (1990c): leaf ontogeny. • Pant and Kidwai (1966): *T. trilobatum*: leaf, stomata. • Solereder and Meyer (1928): secretory tissue. • Tieghem (1866): *T. trilobatum*: all organs. • Tieghem (1870–71): *T. trilobatum*: root, secretory tissue.

Additional literature

French 1986a, c, 1987a; Grayum 1990; Hegnauer 1986; Kulkarni *et al.* 1990; Mehra and Sachdeva 1976; Murata 1990a, b; Schnepf 1965a, b; Schnepf and Czygan 1966; Tarasevich 1990; Thanikaimoni 1969; Vogel 1962, 1990; Williams *et al.* 1981.

Theriophonum **Blume**

This genus of about seven species is found in peninsular India and Sri Lanka. The plants are terrestrial tuberous or cormose herbs occurring in moist shady areas, or grassy areas of dry country. Leaves are hastate or sagittate with an elliptic main lobe. Chromosomes (five counts, three species): 2n = 14, 16, 18. One species was available for study.

Leaf surface

Cuticle smooth on both surfaces. **Epidermis**: adaxial cells polygonal with straight anticlinal walls and abaxial cells undulate, 1.5–2:1 (*l*/*w*) for both surfaces. **Stomata** mostly abaxial and brachyparacytic on both surfaces (Fig. 104C). Subsidiary cells broad or narrow and irregular. **Venation**: midrib prominent. Secondaries acute, merging with acute angle at straightened submarginal collector. Tertiaries becoming ± parallel with secondaries. Higher order veins merging at marginal collectors.

Leaf TS

Structure dorsiventral and leaf very thin. **Size**: 110 × 33 mm (*l*/*w*). Basal lobes 50 × 16 mm (*l*/*w*). **Midrib** flat adaxially, convex and ridged abaxially (Fig. 104A). **Cuticle** thin. **Epidermis**: cells medium-sized on both surfaces. **Stomata** level with surface. **Mesophyll**: one layer of large rounded palisade cells. Spongy tissue a well-formed aerenchyma; cells large and irregular with small air spaces (type 1). **Collenchyma** strands abaxial at midrib, subtending vascular bundles (type Sv).

Vascular bundles in midrib few and scattered, weakly formed, type II. Xylem: a protoxylem lacuna and 2–3 metaxylem elements near a small circular phloem strand. **Laticifers** noted rarely, non-anastomosing and without contents. **Tannin** cells occasional in upper mesophyll. **Crystals**: raphides in single bundles oriented ± parallel with secondary veins, common in mid-mesophyll, or crystals as multiple bundles and overlapping in prominent elongate cells (Fig. 104B). Crystal sand occasional in mesophyll cells.

Petiole TS

Outline: elliptical, fluted and shallowly concave adaxially (Fig. 104D). **Size**: 2–4 mm diameter. **Cuticle** thin and smooth. **Epidermis**: cells large and rounded externally. **Collenchyma**: large circular strands, subepidermally beneath ridges and subtending vascular bundles (type Sv) (Fig. 104D). **Ground tissue** mostly type 3 aerenchyma. **Vascular bundles** scattered and numerous, type II. Xylem: 1–2 wide protoxylem cells and 2–3 narrow metaxylem cells, opposing small irregular phloem strand.

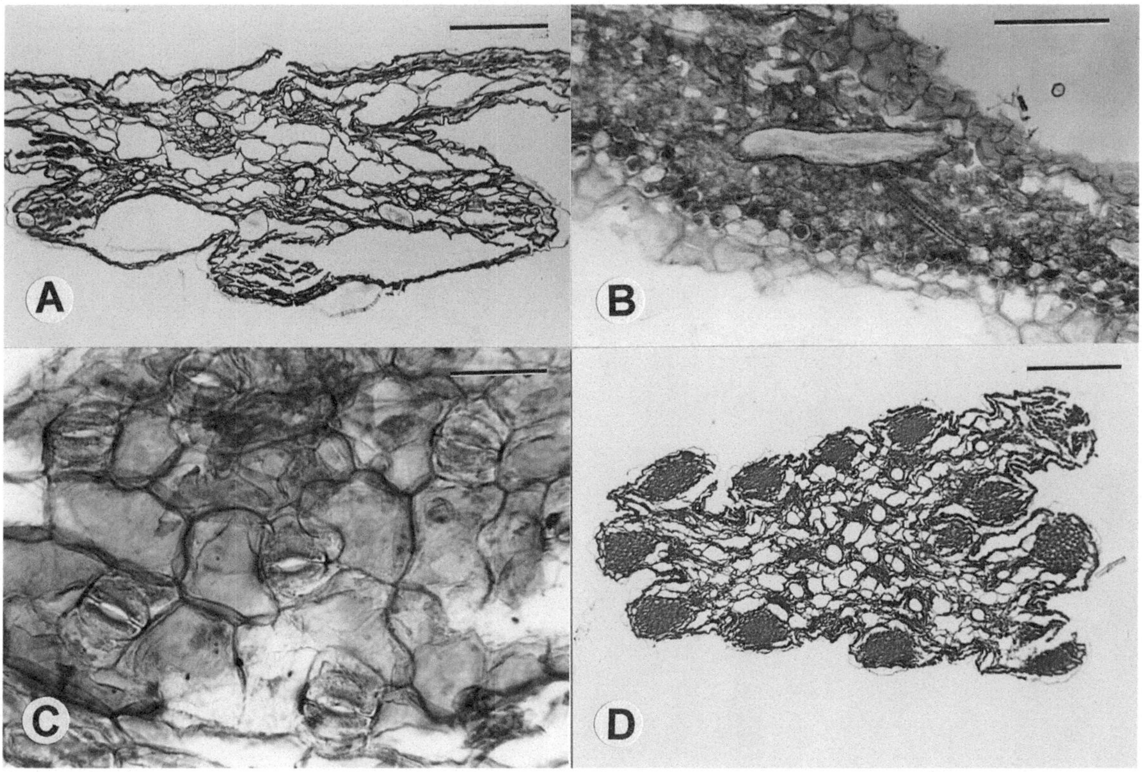

Fig. 104. *Theriophonum minutum* leaf and petiole, poorly rehydrated sections. (A) Midrib TS; (B) Lamina PS showing raphide cell; (C) Abaxial epidermis PS with stomata; (D) Petiole TS. Note large type Sv collenchyma strands. Scale bars: A = 200 μm; B = 150 μm; C = 50 μm; D = 300 μm.

Material examined

Theriophonum minutum Engl., Mooney 3963, India, K.

Species reported on in the literature

French (1988) *T. infaustum*: laticifers.

Additional literature

Davis 1966; French 1986a, c, 1987a; Grayum 1990; Kulkarni *et al.* 1990; Parameswaran 1959; Swamy and Krishnamurthy 1970; Thanikaimoni 1969.

Sauromatum (Blume) Schott

This genus of six species (Croat 1988: 2 spp.) occurs in India, the Himalayas to 1600 m, and western Malaysia. *Sauromatum venosum* is widespread in tropical Africa (naturalized; Croat 1988) as well as India. Plants are terrestrial tuberous herbs dormant in the dry season. They occur in forest understoreys in moist places, as well as in open sun. Leaves are hastate or hastately or pedately divided. Chromosomes: 2n = 26, 52, 104. Two species were available for study.

Leaf surface

Cuticle smooth. **Epidermis**: cells polygonal with straight anticlinal walls on both surfaces (1–1.5:1 *l*/*w*). **Stomata** abaxial, brachyparacytic (Fig. 105E) with one, occasionally two pairs of subsidiary cells. First pair of subsidiary cells broad or narrow, irregular, with angular outer corners. **Venation**: midrib prominent. Secondaries pinnate, acute, attaching to brochidodromous submarginal, ± straightened collector vein. Marginal brochidodromous veins collecting smaller veins. Intersecondaries common. Tertiaries and higher orders reticulate or tending to be elongate parallel with secondaries.

Leaf TS

Structure dorsiventral. **Size** (elliptic segment): 84 × 24 mm (*l*/*w*). **Midrib**: slightly concave adaxially, broadly

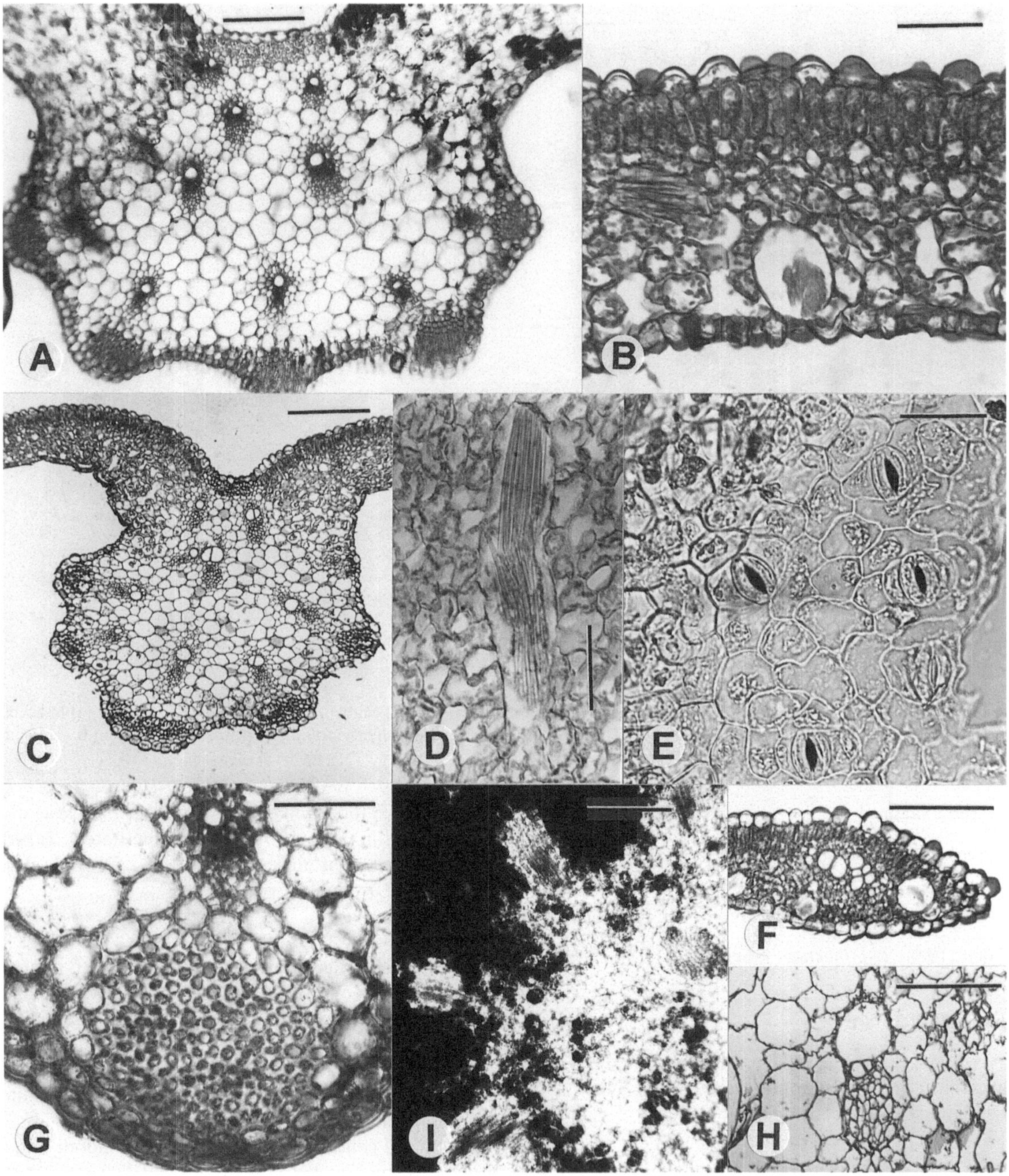

Fig. 105. *Sauromatum*: leaf, petiole and corm hand sections. (A,E,G,I) *S. guttatum*; (B–D,F,H) *S. venosum*. (A,C) Midrib TS showing fluted abaxial surface; (B) Lamina TS with papillate adaxial epidermis and raphide cell beneath abaxial surface; (D) Mesophyll PS showing elongated raphide cell; (E) Abaxial epidermis PS with undulate anticlinal walls and brachyparacytic stomata; (F) Lamina TS margin; (G) Petiole TS with large circular collenchyma strand subtending small vascular bundle; (H) Proximal petiole TS showing type III vascular bundle. (I) Corm TS showing cortical starch zone (black). Scale bars: A,C,I = 300 μm; F–H = 150 μm; B,D, E = 50 μm.

and deeply convex abaxially and fluted (Fig. 105A, C). **Cuticle** thin or not seen. **Epidermis**: cells medium-sized, rounded externally and internally, sometimes papillate (Fig. 105B,F). **Stomata** level with surface. **Mesophyll**: palisade 1-layered, well-differentiated, 4–6:1 (*l/w*). Spongy cells short-armed (type 1) in *S. venosum* (Fig. 105B), longer-armed (type 2) in *S. guttatum*. **Air cavities** up to 30% of spongy layers. **Collenchyma**: ± circular or broad subepidermal strands in abaxial midrib subtending fluted ridges (Fig. 105A,C); strands aligned with vascular bundles (type Sv). One broad strand of collenchyma beneath adaxial midrib epidermis.

Vascular bundles scattered, *c.* 9 bundles in midrib, small (types II, III). Xylem: one wide protoxylem cell or 2–3 narrower metaxylem cells opposing a small irregular phloem strand. **Laticifers** not detected but reported as non-anastomosing (French 1988). **Crystals**: raphides in multiple overlapping bundles in elongate cells, infrequent in mid-mesophyll (Fig. 105D). Crystal sand common in adaxial epidermal cells and abaxial subepidermal cells in *S. venosum*.

Petiole TS

Outline: deeply grooved basally, rounded distally. **Size**: 8–15 mm diameter. **Cuticle** thin, slightly longitudinally striate. **Epidermis**: cells short, rounded internally and externally. **Collenchyma**: cells angularly thickened, in circular strands (type Sv), separated from epidermis by one layer of ground tissue (Fig. 105G). **Chlorenchyma**: up to eight cell layers deep beneath surface, between collenchyma strands. Blotchy surface appearance (dark green to light green) due to variable chloroplast density ranging from densely packed in ground tissue cells, to cells with fewer chloroplasts, to none. **Ground tissue** cells polyhedral without significant air cavities, cells often larger in centre.

Vascular bundles scattered, those beneath collenchyma strands smaller than those deeper in ground tissue; bundles type III (Figs 6C, 105H). Xylem: wide protoxylem lacuna surrounded by circle of small ground tissue cells, 1–2 narrow metaxylem cells. Phloem: a small laterally compressed strand.

Corm TS (*S. guttatum* only)

Outline circular. **Size**: 15–50 mm diameter. **Ground tissue** mostly homogeneous, cells compact, with central and cortical vascular bundles. **Starch** zone forming a cylinder; outer cortex and central ground tissue without starch (Fig. 105I). Starch granules small, largest 7–10 μm diameter.

Material examined

Sauromatum guttatum Schott, cult. Edwardsville, IL; S. *venosum* (Ait.) Kunth, cult. M.

Taxonomic notes

There are no structural differences sufficient to object to the merger of this genus with *Typhonium* as proposed by Hetterscheid and Boyce (2000).

Species reported on in the literature

Abranowicz (1912): *S. guttatum*: tuber. • Beck and Mannagetta (1912): crystals. • Blanc (1978): ontogeny. • Dalitzsch (1886): *S. guttatum*: leaf. • Deinega (1898): *S. guttatum, S.* sp.: leaf ontogeny. • Engler (1920b): *S. guttatum*: stem, leaf. • Ertl (1932): *S. guttatum*: leaf venation. • French (1987c): *S. venosum*: root. • French (1988): *S. venosum*: laticifers. • Gentner (1905): *S. peltatum*: leaf tip. • Gravis *et al.* (1943): *S. guttatum*: seedling. • Lierau (1888): *S. venosum*: root. • Müller (1919): *S. peltatum*: leaf tip. • Netolitzky (1929): *S. guttatum*: crystals. • Petit (1889): *S. guttatum*: petiole. • Philipp (1923): *S. pedatum*: periderm. • Scott (1908): *S. guttatum*: root. • Shah and Gopal (1969): stomata. • Solereder (1919): leaf. • Solereder and Meyer (1928): *S. guttatum*: secretory tissue. • Troll (1932b): *S. guttatum*: leaf ontogeny.

Additional literature

Berger and Schnepf 1970; Dakwale and Bhatnagar 1985; Diamond 1989; French 1986a, c, 1987a; Froebe and Pütz 1988; Genau 1901; Grayum 1990; Hegnauer 1986; Kulkarni *et al.* 1990; Meeuse *et al.* 1984; Meeuse and Raskin 1988; Mehra and Sachdeva 1976; Murata 1990a, b; Pütz 1991, 1992; Skubatz and Kunkel 1999; Skubatz *et al.* 1991; Thanikaimoni 1969; Tillich 1985; Vágújfalvi 1971; Williams *et al.* 1981.

Lazarum A. Hay

This genus containing one species is found in Northern Territory, Australia, and Melville Island. Plants are deciduous, cormose geophytes. The leaves are solitary or paired and appear after flowering (Hay 1992b). They are long-petiolate with a broad elliptic lamina. The single small leaf specimen available for study was ovate-cordate. No chromosome counts are available.

Leaf surface

Cuticle smooth. **Epidermis**: cells large and polygonal with straight anticlinal walls on both surfaces (1–2:1 *l/w*).

Stomata brachyparacytic and on both surfaces. Subsidiary cells broad, irregular and angular. **Venation**: midrib giving off widely spaced, curved-ascending secondaries, obtuse near base, more acute distally. Connections brochidodromous in mid-lamina and eucamptodromous or looped at margin. Higher orders reticulate.

Leaf TS

Structure: weakly dorsiventral. **Size**: juvenile leaf: 1.5 × 1.5 cm (*l/w*). **Midrib** flat adaxially, slightly protruding abaxially. **Cuticle** thin. **Epidermis**: cells large and broad on both surfaces, rounded internally and externally. **Stomata** level with surface. Substomatal air cavities very shallow or absent. **Mesophyll** weakly differentiated; cells irregularly shaped, occasionally forming single palisade layer of square cells, but mostly horizontally extended spongy cells, very shallowly lobed, arranged without significant intercellular air spaces (type 1a).

Collenchyma: angularly thickened cells; strands subepidermal and abaxial to major veins (type Sv). **Vascular bundles**: one to several small bundles in midrib (type II). Xylem: 3–5 metaxylem strands. Phloem: small irregular strand. **Sclerenchyma**: possibly small strands over phloem of largest bundles. **Crystals**: raphides abundant, randomly oriented in mid-mesophyll within spongy zone; cells large to very large, broad or elongated with multiple wide raphide bundles, obliquely oriented to others within cell. Cells often appearing as duct-sized cavities that distend external leaf surface.

Petiole TS

Outline broad laterally, slightly grooved adaxially, with protruding ridges abaxially opposite vascular bundles. **Collenchyma** strands abaxial to each vascular bundle, in abaxial ridges (type Sv). **Ground tissue**: uniform large parenchyma cells. **Air cavities**: very small, at cell corners only.

Vascular bundles: 6–7 small bundles forming arc (type II). Xylem: 1–6 narrow to medium-sized metaxylem cells opposing small irregular phloem strand. **Crystals**: raphides in large cells, occasional and randomly oriented in ground tissue.

Material examined

Lazarum mirabile A.Hay, Bogner s.n., cult. M.

Taxonomic notes

Hay (1992b) suggested that *Lazarum* is apparently closely related to *Sauromatum* Schott, a genus of 6 species ranging from West Africa to the Himalayas. Also similar is *Biarum*, centred in Syria. See also notes for Tribe Peltandreae. Anatomy of this one small specimen with its enlarged epidermal cells and unusual mesophyll and raphide cells, may be due in part to its juvenile condition. There are no essential differences precluding the merger of *Lazarum* into *Typhonium* as proposed by Hay (1997).

Additional literature

Hay 1992b.

Biarum Schott

This genus of five species (or 21 spp.: Croat 1988) occurs in the Mediterranean region, and includes three species in north Africa. Plants are terrestrial, tuberous herbs of open areas in Mediterranean climates and become dormant in later spring. The leaves are long-petiolate with a narrow elliptic lamina. Chromosomes medium-sized: 2n = 16, 22, 24, 26, 32, 74, 96. One species was available for study.

Leaf surface

Cuticle striate to rugulate. **Epidermis**: cells polygonal with straight anticlinal walls on both surfaces (1–1.5:1 *l/w*). **Stomata** simple brachyparacytic, occurring on both surfaces (Fig. 106D) or anomocytic (Fig. 3F). Subsidiary cells narrow, broad or irregular. **Venation**: midrib prominent, secondaries steeply ascending and becoming parallel with midrib. Secondaries merging eucamptodromously, or brochidodromously inside distal margins. Tertiaries parallel or irregular near base, reticulate distally.

Leaf TS

Structure weakly dorsiventral. **Size**: lamina 76 × 10 mm (*l/w*). **Midrib** prominent, flat adaxially, rounded V-convex abaxially (Fig. 106E). **Cuticle** medium-thick on both surfaces. **Epidermis**: cells elongate, flat internally, rounded externally, appearing irregularly sized on both surfaces. **Stomata** and subsidiary cells raised above epidermal surface on mounds, subtending shallow substomatal air cavities. **Mesophyll**: cells smaller adaxially but without differentiation (Fig. 106A); cells irregularly polygonal and occasionally armed (type 1). **Air cavities**: up to 20% of mesophyll, mostly small and irregular, larger only immediately beneath stomata. Midrib with type 3 air cavities. **Collenchyma**: subepidermal strands on abaxial side of midrib subtending vascular bundles (type Sv) (Fig. 106B).

Vascular bundles few and scattered in midrib; bundles small (type II). Largest bundles with 1–2 wide protoxylem elements, 2–10 narrow metaxylem elements,

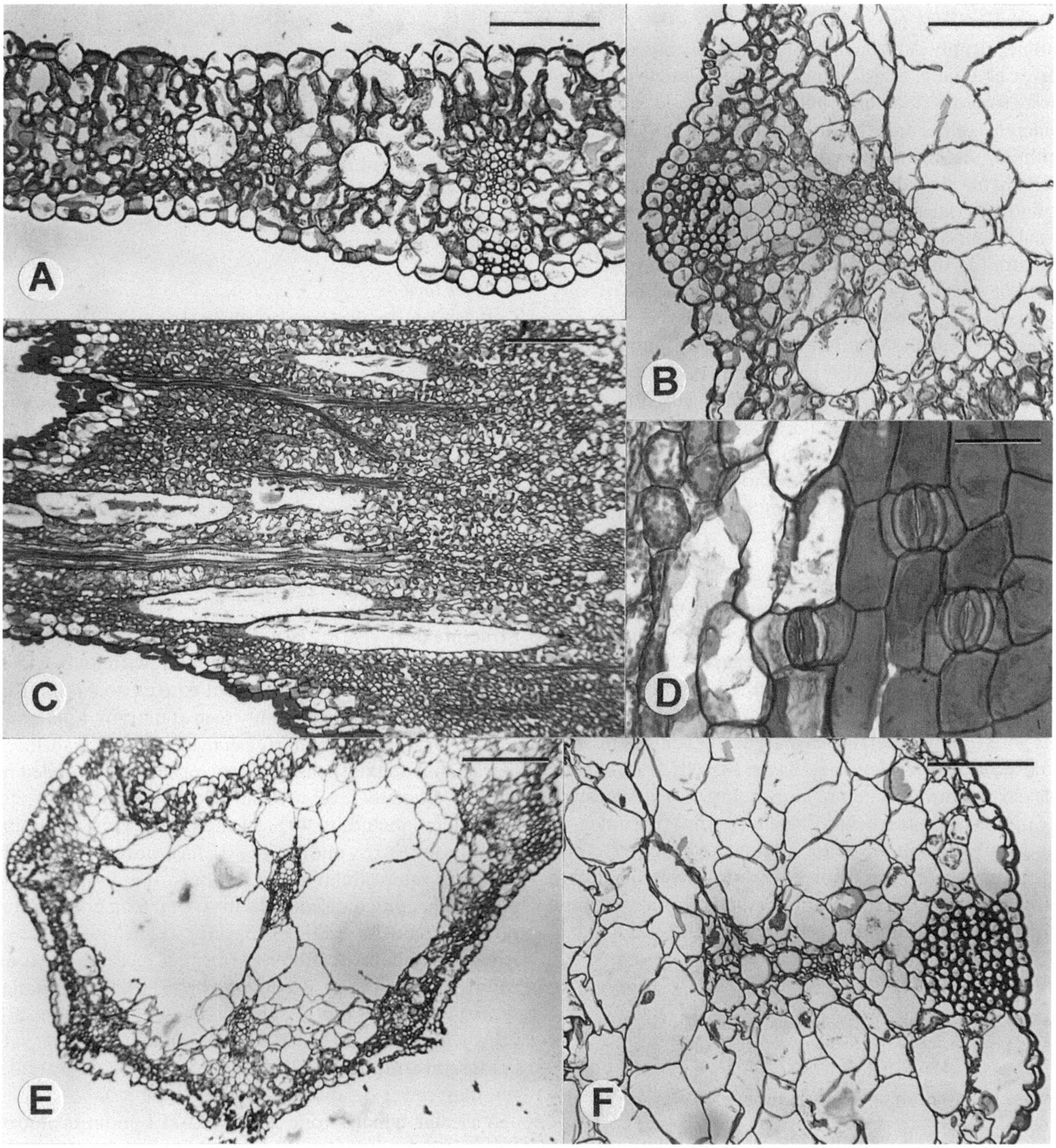

Fig. 106. *Biarum tenuifolium* leaf and petiole sections. (A) Lamina TS; (B) Midrib TS lateral margin; (C) Lamina PS showing elongate raphide cells; (D) Epidermis PS with brachyparacytic stomata; (E) Midrib TS with large aerenchyma cavities; (F) Proximal petiole margin TS with type Sv collenchyma strand. Scale bars: A,B,F = 150 μm; C,E = 300 μm; D = 50 μm.

and circular or irregular phloem strand. **Laticifers** non-anastomosing, lateral to phloem strands. **Crystals**: raphides in discontinuous multiple overlapping bundles in very wide and long duct-like cells, often articulated with end-to-end alignment (Fig. 106C). Cells abundant in mesophyll and parallel with major vascular bundles. Biforines reported by Solereder and Meyer (1928) but not seen in this material.

Petiole TS

Outline: deeply V-grooved adaxially, deeply rounded-convex abaxially. **Size**: diameter 3 mm. **Cuticle** medium-thick, smooth adaxially and finely striate abaxially. **Epidermis**: cells small, flat or rounded externally. **Stomata** common, on slightly raised epidermal mounds between collenchyma strands. **Collenchyma**: subepidermal strands around abaxial periphery (type Sv), each subtending a vascular bundle (Fig. 106F); cells angularly thickened with strands 10–12 cells deep. **Chlorenchyma**: outer 2–3 layers of ground tissue. **Ground tissue** grading to very large cells at centre. **Air cavities** few and large, type 3.

Vascular bundles scattered, numerous and weakly formed (type II) (Fig. 106F). Xylem: one wide protoxylem cell, several narrow metaxylem cells. Phloem: a laterally narrowed or irregular strand at narrow boundary with xylem. **Laticifers** common and lateral to phloem. **Starch** granules common in ground tissue near vascular bundles. **Crystals**: raphides common in cells in peripheral ground tissue.

Root TS

Outline circular. **Size**: 1.2 mm diameter. **Exodermis**: single row of peripheral cells, rounded externally with external surface having slightly birefringent cap-like thickenings. **Cortex**: cells smaller at periphery, internally mostly larger and compact with few corner air cavities. **Endodermis**: cells with thin and narrow casparian thickenings in anticlinal walls. **Pericycle**: single layer of cells. **Protoxylem** poles six. Central vascular region: metaxylem files meeting at large central tracheid. **Phloem** strands irregular or triangular; cells in LS with oblique to transverse end walls. **Tannin**: single cells or groups in cortex filled with granular contents, common. **Crystals**: small raphide bundles in non-idioblastic cells in outer ground tissue.

Material examined

Biarum tenuifolium (L.) Schott, cult. S. Mayo, London; cult. M.

Species reported on in the literature

Engler (1920c): *B. tenuifolium*: stem. • Flinck (1891): vegetative organs. • French (1987c): *B. tenuifolium*: root. • French and Tomlinson (1983): *B. tenuifolium*: stem. • Lierau (1888): *B. bovei*, *B. tenuifolium*: root. • Nicolas (1916): petiole. • Solereder and Meyer (1928): crystals, secretory tissue. • Tieghem (1866): *B. tenuifolium*: all organs.

Additional literature

French 1986a,c, 1987a; Grayum 1990; Talavera 1976; Thanikaimoni 1969; Williams *et al.* 1981.

Arum L.

This genus of 26 species (Croat 1988) ranges throughout the Mediterranean region of Europe with two species in northern Africa. Plants are terrestrial, tuberous herbs adapted to leafless dormancy and a broad range of climate and habitat requirements. The leaves are long-petiolate, broad elliptic, cordate, sagittate or hastate. Chromosomes: $2n = 28$, *c.* 32, 56, 70, 84. Two species were available for study.

Leaf surface

Cuticle smooth. **Epidermis**: cells polygonal with straight or undulate anticlinal walls on both surfaces (1.5-2:1 *l*/*w*). **Stomata** brachyparacytic (Fig. 107D) or brachyparatetracytic, more common abaxially or equally common on both surfaces. Subsidiary cells irregular, broad or narrow and angular. **Venation**: prominent midrib at lamina base, becoming smaller as pinnate lateral secondaries diverge into lamina. Acute secondaries ascending steeply, merging eucamptodromously at base and brochidodromously distally. Tertiaries and higher orders reticulate.

Leaf TS

Structure dorsiventral. **Size** 12–15 × 6–8 cm (*l*/*w*). **Midrib** prominent, flat adaxially, straight or with two widely spaced grooves; broadly rounded-convex abaxially (Fig. 107A,B). **Cuticle** thin and only seen at margin. **Epidermis**: cells short and irregularly shaped on both surfaces. Occasional adaxial cells papillate, or tall and rounded on surface. **Stomata** level with surface. **Mesophyll**: two palisade layers including 40% of mesophyll (Fig. 3D), and continuous across midrib (Fig. 107A). Carr *et al.* (1980b) reported intercellular pectic strands, 0.3–0.4 μm diameter, connecting palisade cells in *A. sanctum*; not detected in these paraffin sections. Spongy tissue: about seven irregularly shaped cell layers forming medium-sized air cavities (types 2, 3) occupying up to 50% of spongy space. In midrib, type 3 aerenchyma with partitions 3–5 cells wide (Fig. 107A, B). **Collenchyma**: discrete subepidermal strands in midrib subtending abaxial vascular bundles; fewer or absent adaxially (type Sv).

Vascular bundles: one large central bundle in midrib (Fig. 107C), and abaxial and adaxial rows of smaller bundles, symmetrically placed; type II. Xylem: one protoxylem cell, 8–12 metaxylem cells. Phloem: circular strand at narrow straight boundary with xylem. **Laticifers**: French (1988) reported non-anastomosing laticifers, but not detected in these specimens. **Crystals**: raphides common in small cells, oriented vertically in palisade layer, randomly in spongy layer or midrib ground tissue. Biforines reported by Solereder and Meyer (1928), not confirmed here.

Petiole TS

Outline subterete or semicircular (Fig. 107E). **Size**: 4–5 mm broad. **Cuticle** thin and smooth. **Epidermis**: cells small and square, rounded externally. **Stomata** occasional, level with surface. **Collenchyma**: circular subepidermal strands beneath shallow ridges around periphery (Fig. 107E) and mostly subtending small vascular bundles (type Sv). **Chlorenchyma**: chloroplasts in outer ground tissue. **Ground tissue**: outer 4–7 compact layers of large parenchyma cells. Central cells forming type 3 aerenchyma, partitions 1–3 cells wide. **Air cavities**: up to 95% of central aerenchyma.

Vascular bundles: peripheral circle of bundles capped by collenchyma strands; central semicircle of fewer, larger bundles (type II) (Fig. 107F). Xylem: one wide protoxylem cell, 8–15 narrow and medium-sized metaxylem cells. Phloem: a circular strand with short straight boundary at xylem. **Laticifers**: 1–2 large non-anastomosing cells lateral to phloem strands. **Crystals**: raphides in large, thin-walled cells in peripheral ground tissue and occasionally near vascular bundles in central ground tissue.

Material examined

Arum italicum Miller, cult. M; *A. maculatum* L., cult. M.

Taxonomic notes

The two species examined are very close anatomically. Raphide bundles and cells in this genus are the simplest and least diverse of any genus in the tribe, and are seldom this simple in the family.

Species reported on in the literature

Ambronn (1879–81): *A. maculatum*: collenchyma. • Arber (1922a): *A. italicum*: stem. • Arber (1922b): *A. italicum, A. maculatum*: leaf. • Arcangeli (1897): *A. italicum:* leaf. • Arzt (1937): *A. maculatum*: cuticle. • Azevedo and De (1970–1): *A. italicum, A. maculatum*: root mycorrhizae. • Beck and Mannagetta (1912): *A. maculatum:* crystals. • Behnke (1969a, 1972): *A. maculatum*: phloem. • Bohny (1906): *A. maculatum*: leaf. • Borzi (1887–8): *A. pictum:* root. • Boyd (1932): *A. italicum, A. hygrophyllum*: seedling. • Briens (1916): *A. maculatum*: all organs. • Brown and Holloway (1981): *A. maculatum*: cutin. • Bruce (1903): *A. maculatum*: leaf. • Brunzema (1928): *A. maculatum*: crystals. • Buscalioni (1895–6): *A. colocasia:* crystals. • Carr *et al.* (1980a): *A. sanctum*: leaf. • Carr *et al.* (1980b): leaf. • Catalano (1915): *A. italicum, A. maculatum:* contractile root. • Chauveaud (1911): *A. italicum*: stem. • Chrysler (1904): *A. italicum:* stem. • Czaja (1962): *A. italicum*: leaf epidermis. • Dalitzsch (1886): *A. italicum, A. maculatum, A. vulgare:* leaf. • Dangeard (1959–60): *A. italicum*: leaf. • Draheim (1929): *A. maculatum, A. ternatum:* root. • Dumann (1980): *A. maculatum*: leaf, crystals. • Ertl (1932): *A. hygrophilum, A. italicum, A. maculatum*: leaf. • Flahault (1878): *A. dracunculus:* root. • Flinck (1891): stem. • French (1987c): *A. italicum*: root. • French (1988): *A. dioscoridis, A. hygrophyllum:* laticifers. • French and Tomlinson (1983): *A. hygrophyllum*: stem. • Freundlich (1908): *A. maculatum*: leaf. • Gallaud (1905): *A. arisarum, A. maculatum*: root. • Giovannozzi (1911): *A. italicum:* tuber. • Gravis *et al.* (1943): *A. italicum*: seedling. • Guillot-Salomon *et al.* (1978): *A. maculatum*: leaf. Gulliver (1863–5): *A. maculatum*: crystals. • Guttenberg (1968): *A. italicum, A. maculatum*: root. • Hallé and Cusset (1969): *A. maculatum*: leaf. • Hather (1988, 1993): root, tuber. • Irmisch (1850): *A. maculatum*: tuber. • Kasapligil (1961): *A. orientale*: leaf. • Kirchner *et al.* (1934): *A. maculatum*: corm. • Kisser (1928): *A. maculatum*: leaf, petiole. • Knoll (1948): *A. italicum:* leaf ontogeny. • Kroemer (1903): root. • Lamant and Heller (1967): *A. italicum*: root. • Lazenby (1890): crystals. • Lierau (1888): *A. maculatum, A. orientale*: root. • Lignier (1915): *A. maculatum*: stem. • Linsbauer (1930): several spp.: epidermis. • Morassi Bonzi and Fabbri (1975): *A. italicum, A. maculatum, A. pictum*: leaf. • Morassi Bonzi and Fabbri (1978): *A. italicum*: phloem crystals. • Müller (1919): *A. italicum, A. maculatum A. peltatum*: leaf tip. • Netolitzky (1929): crystals. • Nicolas (1916): petiole. • Nougarède and Rondet (1981): *A. italicum*: rhizome. • Pethybridge (1903): *A. maculatum:* leaf. • Planchon and Juillet (1909): *A. esculentum, A. maculatum*: starch. • Potgieter and Van Wyk (1992): pectic protuberances. • Raunkiaer (1895–9): anatomy. • Reichert (1913): *A. cornutum, A. italicum, A. palaestinum*: starch. • Rimbach (1897): *A. maculatum*: root. • Roland (1967): *A. maculatum:* collenchyma • Scherer (1904): *A. maculatum*: root. • Scott and Sargent (1898): *A. maculatum*: ontogeny • Shishkoff (1987): *A.* sp.: root. • Solereder (1919): *A. maculatum:* leaf. • Solereder and Meyer (1928): *A. maculatum*: leaf, secretory tissue, crystals. • Stahl (1896): leaf. • Tieghem (1866): *A. italicum*: all organs. • Tobler (1957): *A. maculatum:* fibres. • Trécul (1865): *A. vulgare*: vasculature. • Turpin (1836): several spp.: leaf, crystals. • Van Fleet (1942): *A. crinitum, A. italicum, A. maculatum*: stem, endodermis. • Voronkina (1975): *A. maculatum*: root. • Walton (1964): *A. maculatum:* ontogeny. • Warncke (1911): *A. maculatum*: leaf. Webber (1960): *A. hygrophyllum,*

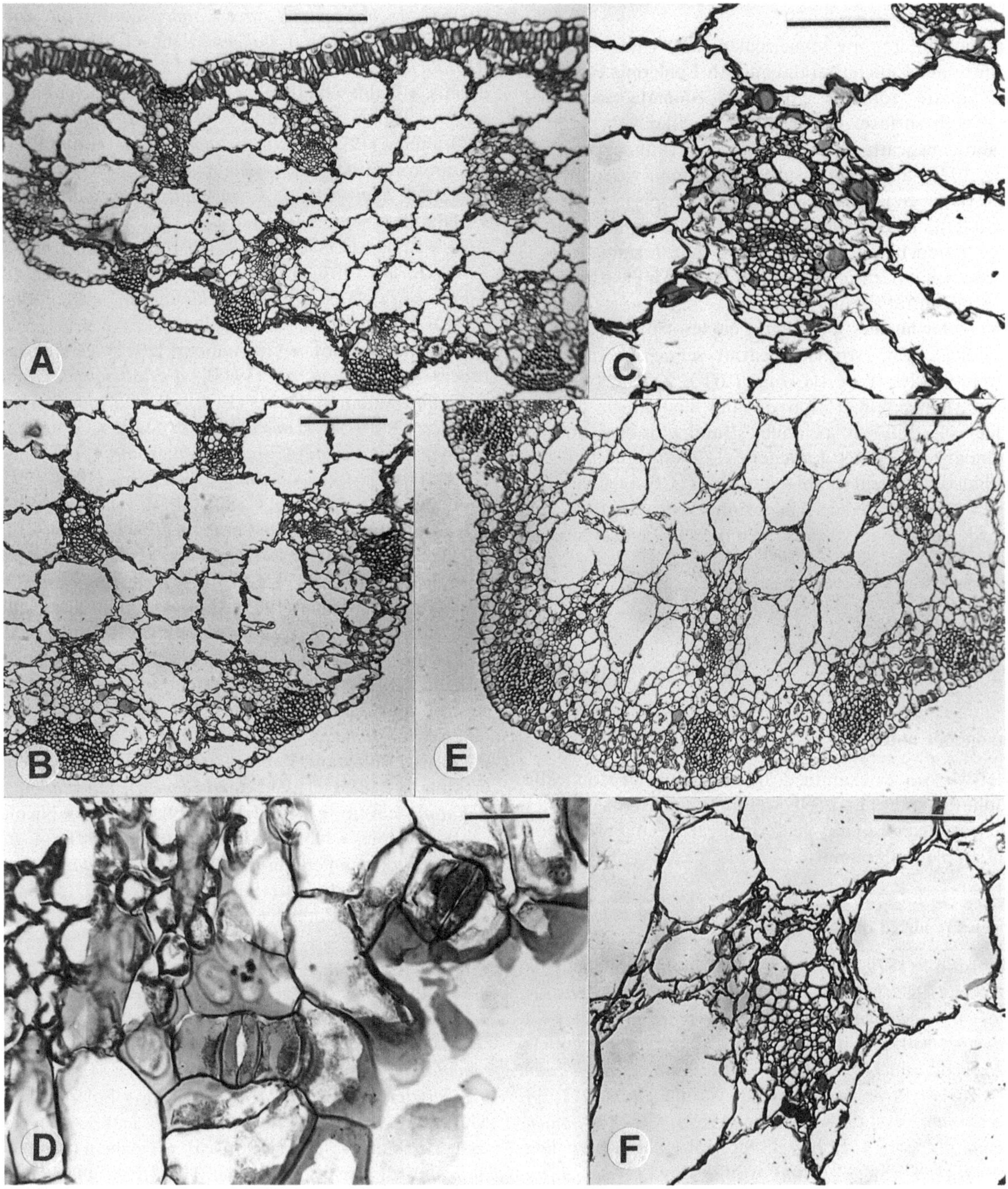

Fig. 107. *Arum* leaf and petiole sections. (A,C) *A. maculatum*; (B,D–F) *A. italicum*; (A–C,E,F) All showing ground tissue with strongly developed type 3 aerenchyma and stranded type Sv collenchyma: (A) Midrib TS with palisade continuous beneath adaxial surface. (B) Midrib TS abaxial side; (C) Lamina TS central type II vascular bundle; (D) Lamina PS showing 2 brachyparacytic stomata; (E) Petiole TS; (F) Petiole TS, central ground tissue with type II vascular bundle. Scale bars: A,E = 300 μm; B = 200 μm; C,F = 150 μm; D = 50 μm.

A. italicum: leaf, stomata. • Winton and Winton (1935): *A. maculatum, A. italicum*: tubers. • Winton and Winton (1945): *A. maculatum*: starch. • Ziegenspeck (1937–9): *A. maculatum*: stomata.

Additional literature

Antony 1898; Batygina and Yakovlev 1990; Berger and Schnepf 1970; Bochenska and Kozlowski 1974; Brandza 1891; Ceruti 1939; Daumann 1930, 1970; Davis 1966; Dihoru 1978; Domin 1911; Dormer 1960; Erdtman 1952; French 1986a, c, 1987a; Gerola 1963; Grayum 1990; Guéguen 1900, 1901; Guttenberg 1960; Halevy 1989; Hegnauer 1986; Henslow 1911; Hesse 1980; Hofmann 1930; Irmisch 1874; Jacobson-Paley 1920a, b; Knoll 1923; Lack and Diaz 1991; Mascré 1928; Meeuse and Raskin 1988; Meusel 1951; Murata 1990a; Nahrstedt 1975; Ohlendorf 1907; Pacini and Juniper 1983; Pritchard *et al.* 1993; Pütz 1991; Richter 1929; Salisbury 1916; Sargant 1903; Schnepf and Czygan 1966; Schulte 1969; Schwendener 1874; Signol and Lance 1975; Smolenski *et al.* 1974; Thanikaimoni 1969; Tieghem 1907; Vogel 1962, 1990; Weinert and Barckhaus 1974; Williams *et al.* 1981.

Eminium (Blume) Schott

This genus of six species ranges from the eastern Mediterranean to central Asia. The plants are terrestrial tuberous herbs of Mediterranean or monsoon seasonally dry climates. Leaves are elliptic to auriculate or hastate with twisted ascending basal lobes. Chromosomes: 2n = 28. Two species were available for study.

Leaf surface

Cuticle smooth or striate. **Epidermis**: cells polygonal with straight to slightly undulate anticlinal walls on both surfaces (1–4:1 *l*/*w*). **Stomata**: brachyparacytic on both surfaces (Fig. 108E,F). Subsidiary cells very broad, ends rounded or irregular. **Venation**: midrib prominent at base, becoming smaller distally. Secondaries narrowly acute and steeply ascending, the basal secondaries forming submarginal brochidodromous straightened collectors. Secondaries and tertiaries anastomosing brochidodromously, converging at leaf tip. Higher orders parallel to reticulate.

Leaf TS

Structure dorsiventral. **Size**: segment 73 × 31 mm (*l*/*w*). **Midrib** with slight adaxial groove, slight convex deformation abaxially. **Cuticle** thin. **Epidermis**: cells short and flat on both surfaces. **Stomata** raised on shallow mounds on both surfaces (Fig. 108C). **Mesophyll**: 1–2 palisade layers and 8–10 spongy layers comprising types 2 and 3 aerenchyma (Fig. 108C). Spongy cell partitions uniseriate between large irregular air spaces. **Air cavities**: up to 80% of spongy layer. Substomatal air cavities shallow. **Collenchyma**: strands large and circular, up to 15 cells deep, cells angularly thickened, abaxial to all larger vascular bundles (type Sv).

Vascular bundles: several large and small bundles in single or double line across broad midrib area; bundles small (type II). Xylem with 1–20 tracheary cells opposing a small circular phloem strand. **Laticifers**: large cells, not easily found, lateral to vascular bundles and uncommon (Fig. 108D). **Tannin** cells rounded or armed, with dark amorphous contents, occasional in mesophyll. **Crystals**: raphides in 1–3 bundles per cell, or long overlapping bundles, common in elongated mesophyll cells (Fig. 108B); raphide cells tending to be parallel with secondary veins. Crystal sand in small druse-like clusters common in mesophyll cells.

Petiole TS

Outline elliptic and shallowly fluted (Fig. 108A,G); adaxial groove may be present. **Size**: diameter 4 mm. **Cuticle** medium-thick and striate. **Epidermis**: cells large and isohedral. **Stomata** common between collenchyma strands. **Collenchyma**: large circular strands, up to >30 cells deep around periphery at outer ridges, subtending outer vascular bundles (type Sv). **Ground tissue**: outer 5–6 layers compact, inner tissue type 3 aerenchyma with 2–3 cells per partition. **Air cavities** large in centre, up to 95% of ground tissue.

Vascular bundles numerous and scattered (Fig. 108G), large and well developed (types I, II). Xylem: few up to >30 metaxylem cells opposing broad straight phloem boundary. Phloem: broad or narrow semicircular or elliptic strand. **Laticifers** present in *E. albertii*, uncommon and lateral to vascular bundles. **Crystals**: raphide cells with 1–several large bundles common in outer ground tissue.

Material examined

Eminium albertii (Regel) Engl., Wilson and Mewer 765, cult. K; *E. spiculatum* (Blume) Kuntze, Davis 6421, Lower Egypt, K; Bogner s.n., Israel.

Species reported on in the literature

El-Gazzar *et al.* (1977): *E. spicatum*: stomata. • El-Kiey *et al.* (1967–69): *E. spiculatum*: corm. • Lavrukhina and Akmedov (1983): *E. regelii*. Nechaeva (1943): *E. lehmannii*: corm, appendages. • Troll (1932b): *E. intortum*: leaf. • Vasilevskaya and Nechaeva (1964): *E. lehmannii*: corm.

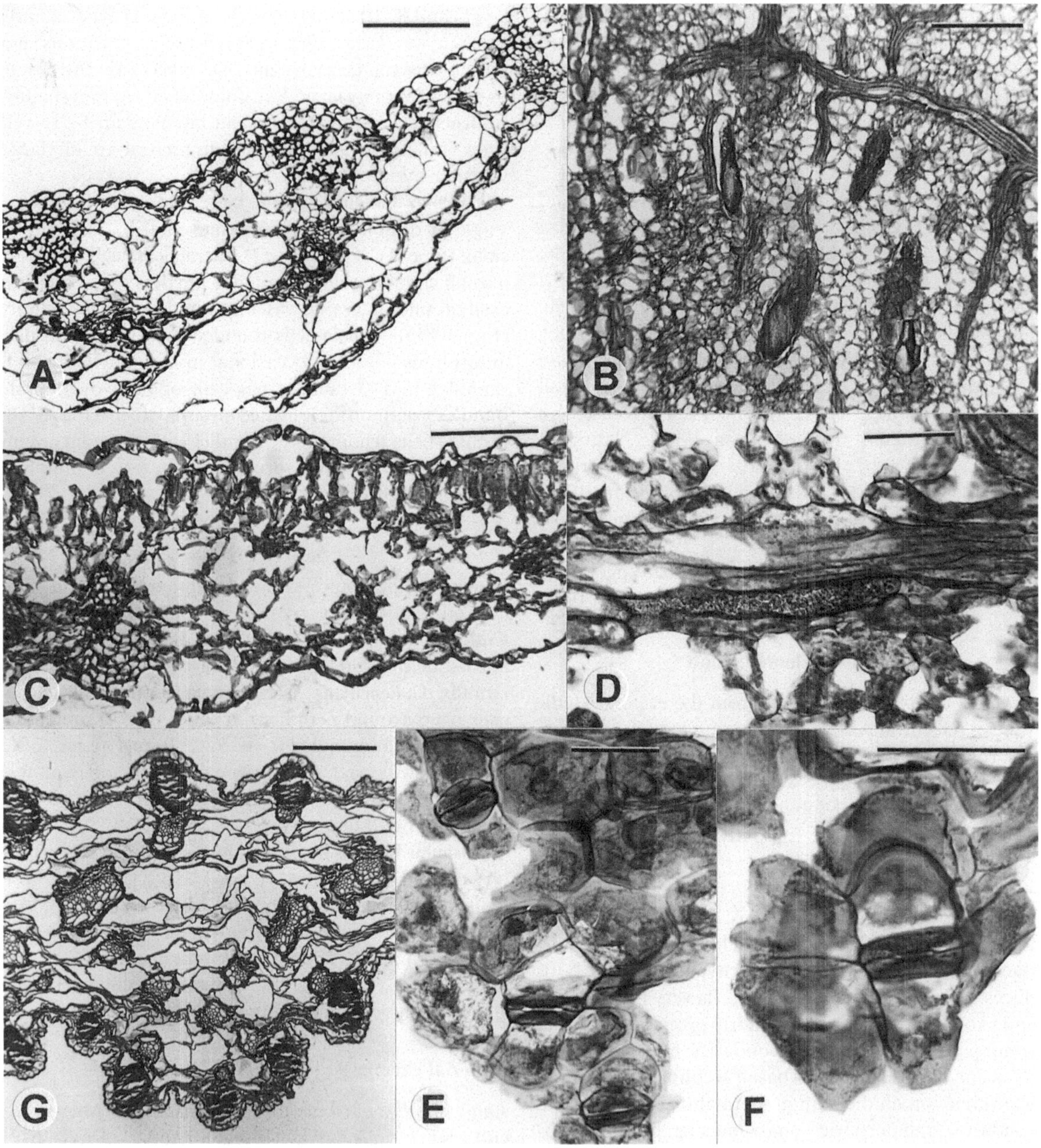

Fig. 108. *Eminium* leaf and petiole sections, rehydrated material. (A,C–E) *E. albertii*; (B,F,G) *E. spiculatum*; (A) Lateral petiole TS; (B) Lamina PS with veins and raphide cells; (C) Lamina TS; (D) Lamina PS showing vascular bundle and laticifer; (E,F) Lamina PS epidermis showing brachyparacytic stomata; (G) Petiole TS. Scale bars: A,C = 150 μm; B = 200 μm; D–F = 50 μm; G = 300 μm.

Additional literature

French 1986a,c, 1987a; Grayum 1990; Hegnauer 1986; Kandalova 1983; Popova *et al.* 1984; Smolenski *et al.* 1974; Tarasevich 1990; Thanikaimoni 1969.

Dracunculus Mill.

This genus of two species is distributed in the Mediterranean region including north Africa. Plants are terrestrial, tuberous herbs growing in shady Mediterranean habitats. The leaves are pedately decompound. Chromosomes in both species: 2n = 28. Both species were available for study.

Leaf surface

Cuticle smooth. **Epidermis**: cells polygonal to elongate with straight anticlinal walls or often undulate on both surfaces (1–4:1 *l/w*). **Stomata** common on both surfaces, brachyparacytic (Fig. 109D) or tending to be anomocytic. When present, subsidiary cells broad and irregular. **Venation**: basal and suprabasal acrodromous secondaries in *D. vulgaris;* tertiaries and higher orders mostly parallel or irregular between secondaries. Midrib prominent in *D. canariensis*; secondaries broadly acute and straight, connecting in mid-lamina at straightened brochidodromous arch; tertiaries parallel with secondaries or reticulate; higher orders reticulate.

Leaf TS

Leaf trifoliate. **Structure** poorly preserved, dorsiventral. **Size** (segment): 100 × 50 mm (*l/w*) to 34 × 14 mm (*l/w*), or median segment 15 × 6 cm (*l/w*). **Midrib** small, grooved adaxially and semicircular-convex abaxially (Fig. 109A). **Cuticle** thin. **Epidermis** probably bullate, of small short cells on both surfaces. **Stomata** raised on mounds. **Mesophyll**: 2–3 palisade layers, up to 30% of lamina thickness (Fig. 109B). Spongy cells in *c.* 10 layers of irregular small or long-armed cells forming aerenchyma of large spaces (type 2). Partitions of aerenchyma uniseriate, partition length one cell (type 1). **Air cavities**: up to 60% air space. **Collenchyma**: semicircular strand, 5–6 layers, subtending midvein (type Sv), separated from epidermis by a layer of ground tissue (Fig. 109A).

Vascular bundles: large bundle at centre of midrib; bundles type II. Xylem: up to four tracheids opposing a small semicircular phloem strand. **Laticifers**: absent in these specimens, reported non-anastomosing by French (1988). **Crystals**: raphides in 1–several bundles in large circular or slightly elongate thin-walled cells, very common in mesophyll (Fig. 109C) and often regularly spaced. Crystal sand very common in upper mesophyll.

Petiole TS

Outline: subterete with a few superficial grooves. **Size**: diameter 3–4 mm. **Cuticle** thin and smooth. **Epidermis**: cells elongate and short. **Collenchyma**: discrete circular strands, most subtending vascular bundles (type Sv) and separated from epidermis by a layer of ground tissue (Fig. 109E). **Ground tissue**: outer 5 layers of compact tissue. At centre large type 3 air cavities, *c.* 2 cells per partition (Fig. 109F).

Vascular bundles scattered, large and unsheathed. Xylem: 1–3 wide protoxylem cells, 1–2 narrow metaxylem cells (type II or III). Phloem: strand small and narrowly irregular (Fig. 109F). **Tannin** cells occasional in ground tissue. **Starch** granules large and few, scattered in ground tissue. **Crystals**: raphides common in thin-walled cells, throughout ground tissue.

Material examined

Dracunculus canariensis Kunth, Bogner 1519, Gran Canaria, cult. M; *D. vulgaris* Schott, cult. S. Mayo, London.

Species reported on in the literature

Borzi (1887–8): *D. vulgaris*: root. • French (1987c): *D. canariensis*: root. • French (1988): *D. canariensis*: laticifers. • Gravis *et al.* (1943): *D. canariensis*: seedling. • Guttenberg (1968): *D. vulgaris*: root. • Lierau (1888): *D.vulgaris*: root. • Morassi Bonzi and Fabbri (1975): *D. vulgaris*: leaf. • Planchon and Juillet (1909): *D. vulgaris*: starch. • Reichert (1913): *D. vulgaris*: starch. • Solereder and Meyer (1928): secretory tissue. • Tieghem (1866): *D. vulgaris*: all organs. • Tieghem (1870–1): *D. vulgaris:* root, secretory tissue. • Troll (1932a): *D. vulgaris*: leaf ontogeny.

Additional literature

Antony 1898; Arcangeli 1897; French 1986a, c, 1987a; Grayum 1990; Hegnauer 1986; Kirchner *et al.* 1934; Meeuse and Raskin 1988; Murata 1990a; Pütz 1991; Smolenski *et al.* 1974; Thanikaimoni 1969; Williams *et al.* 1981.

Helicodiceros Schott

A monotypic genus, populations are found in the Balearic Islands, Sardinia and Corsica. Plants are terrestrial tuberous herbs, dormant during hot dry season of Mediterranean climates. The leaves are deeply pedately lobed with twisted and erect basal lobes. Chromosomes: 2n = 56. One specimen was available for study.

Leaf surface

Cuticle smooth. **Epidermis**: cells polygonal with straight or undulate anticlinal walls (2–3:1 *l/w*) (Fig. 110C). **Stomata** brachyparacytic or occasionally unipolar on both surfaces. Subsidiary cells broad and rounded or irregular (Fig. 110C). **Venation**: midrib weak and formed of loosely associated vascular bundles in lower lamina; non-existent in distal lamina. Secondaries steeply

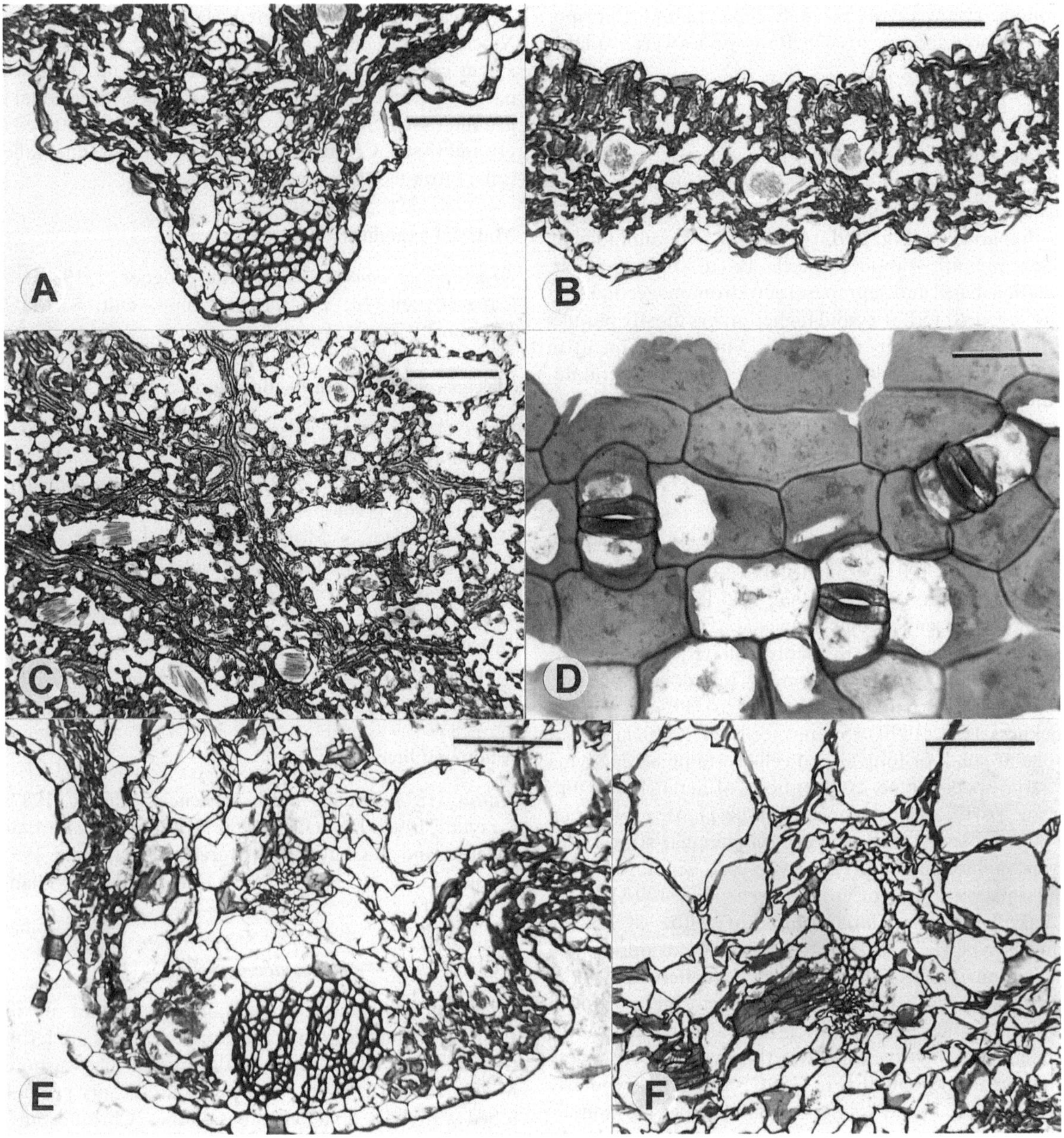

Fig. 109. *Dracunculus vulgaris* leaf and petiole sections, rehydrated material. (A) Midrib TS; (B) Lamina TS; (C) Mesophyll PS showing raphide cells; (D) Adaxial epidermis PS showing brachyparacytic stomata; (E,F) Proximal petiole TS: (E) type III vascular bundle and subtending type Sv collenchyma strand; (F) type 3 aerenchyma and type III vascular bundle. Scale bars: A,B,E, F = 150 μm; C = 200 μm; D = 50 μm.

ascending irregularly from base and higher on midrib. Secondaries merging brochidodromously in mid-lamina. Marginal collector also present. Tertiaries and higher orders reticulate.

Leaf TS

Structure dorsiventral, poorly differentiated and mechanically weak. **Size**: 55 × 19 mm (*l*/*w*). **Midrib** not differentiated from lamina. **Cuticle** thin. **Epidermis**: cells

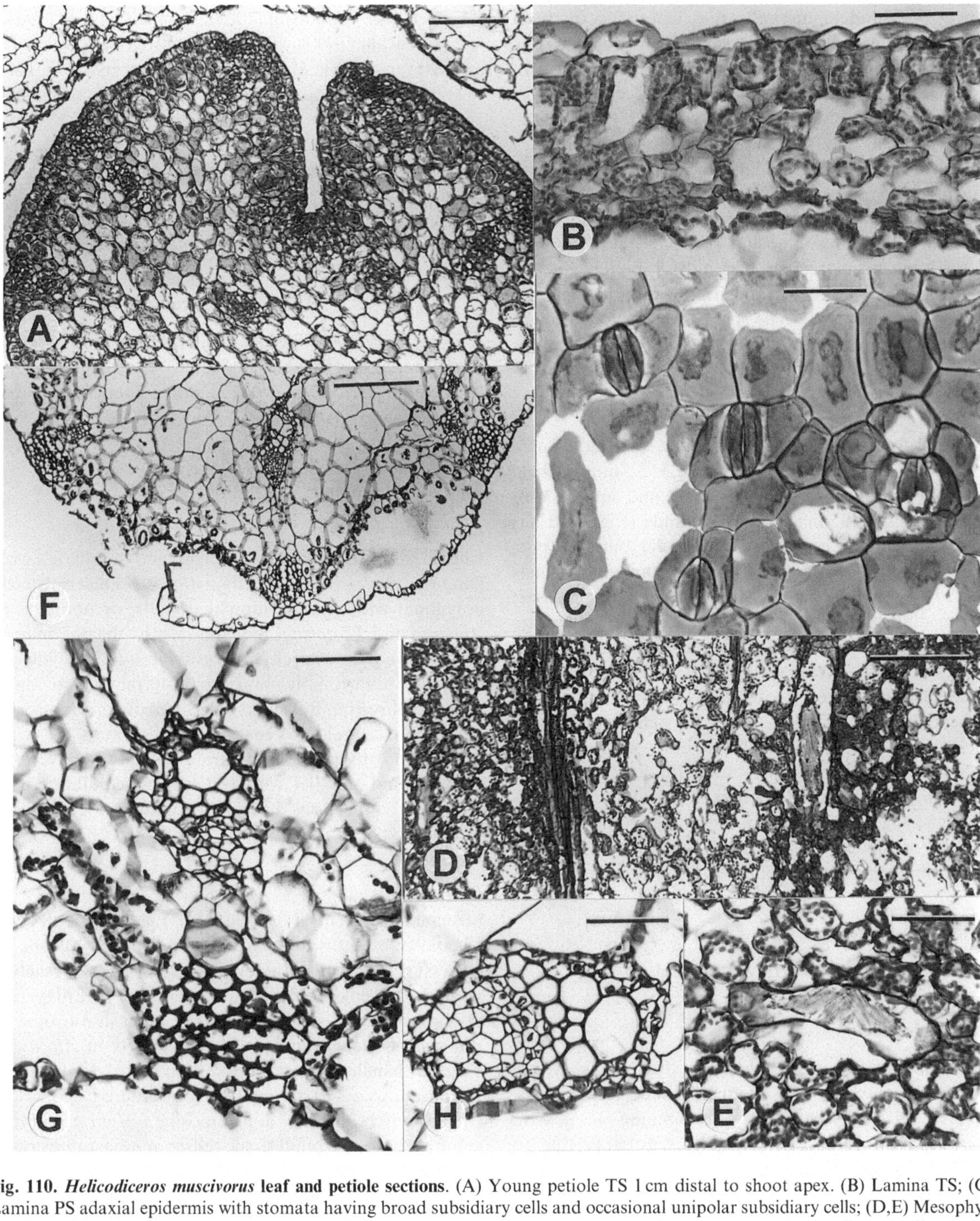

Fig. 110. *Helicodiceros muscivorus* leaf and petiole sections. (A) Young petiole TS 1 cm distal to shoot apex. (B) Lamina TS; (C) Lamina PS adaxial epidermis with stomata having broad subsidiary cells and occasional unipolar subsidiary cells; (D,E) Mesophyll PS with raphide cell; (F) Petiole TS showing collenchyma strands; (G) Petiole TS showing type II vascular bundle and collenchyma strand; (H) Petiole TS centre showing type II vascular bundle. Scale bars: A = 300 μm; B,C,E,G,H = 50 μm; D = 150 μm; F = 200 μm.

elongated, rounded internally and externally on both surfaces. **Stomata** level with surface. **Mesophyll**: one palisade layer of irregular cells (1–4:1 *l*/*w*) (Fig. 110B). Spongy layers 4–5; cells irregularly shaped with short arms (type 1, occasionally type 2).

Collenchyma: one small subepidermal strand in abaxial midrib (type Sv). **Vascular bundles**: one small weak midrib bundle (type II). Xylem: several narrow metaxylem cells bounding irregular phloem strand. **Laticifers**: reported as non-anastomosing (French 1988), but not detected in this specimen. **Crystals**: raphides fine or coarse, small or large, in single or overlapping bundles (Fig. 110D,E); raphide cells short to very elongate and common in mesophyll. Crystal sand uncommon in mesophyll.

Petiole TS

Outline: subterete, laterally narrowed, with adaxial V groove (Fig. 110A). **Size**: diameter 1.5–2.5 mm. **Cuticle** thin and striate. **Epidermis**: cells small and square. **Collenchyma**: subepidermal circular strands (Fig. 110F,G) usually subtending outer row of vascular bundles (type Sv); strands 5–7 cells deep. **Chlorenchyma**: chloroplasts in up to 3–4 layers of cells subepidermally. **Ground tissue**: cells medium-sized near epidermis and large in centre. **Air cavities**: central ground tissue mostly large type 3 air cavities with 1–2-celled partitions. Air space up to 95%.

Vascular bundles scattered and well developed; bundles unsheathed (type II) (Fig. 110H). Xylem: wide protoxylem cells with 8–15 medium-width or wide metaxylem cells in one irregular strand. Phloem: circular or irregular large strand. **Laticifers**: occasional and non-anastomosing, lateral to vascular bundles. **Crystals**: raphides in single bundles in thin-walled cells, common or uncommon in peripheral ground tissue.

Material examined

Helicodiceros muscivorus Engl., 1032/73, cult. M.

Taxonomic notes

Mabberley (1997) would include this genus within *Dracunculus*. By comparison of anatomical characters, there would be no objection to combining the genera. Leaf venation is poorly organized, probably due to simplification in ontogeny.

Species reported on in the literature

Arcangeli (1890): leaf. • Ertl (1932): leaf. • Figdor (1922): leaf. • French (1987c): root. • French (1988): laticifers. • French and Tomlinson (1983): stem. • Gentner (1905): leaf. • Murata (1990a, 1990c): leaf ontogeny. • Solereder and Meyer (1928): *H. crinitus*: secretory tissue. • Troll (1932b): leaf. • Troll and Meyer (1955): leaf ontogeny.

Additional literature

French 1986a, c, 1987a; Grayum 1990; Leinfellner 1954; Murata 1990a; Thanikaimoni 1969; Williams *et al.* 1981.

Tribe Colocasieae

A group of Old World genera with tuberous and cormose stems and mostly bearing anastomosing laticifers. It includes *Ariopsis, Alocasia, Remusatia*, *Colocasia*, *Steudnera* and *Protarum*.

Anatomical summary

Leaf surface

Cuticle smooth, striate, or rough and papillate; with flanges between epidermal cells in *Alocasia*. **Epidermis**: cells polygonal with straight anticlinal walls, or undulate in *Alocasia* and *Protarum*. **Stomata** mostly abaxial, mainly brachyparacytic; brachyparacytic to anomocytic in *Ariopsis*; occasionally brachypara-tetracytic to -hexacytic in *Alocasia*. Subsidiary cells narrow or broad, and curved or angular. **Venation** acrodromous or with prominent midrib, secondaries usually converging between and parallel with tertiaries. Secondaries with marginal connections brochidodromous or eucamptodromous.

Leaf TS

Structure dorsiventral or weakly so. **Epidermis** with large adaxial cells, short abaxial cells, or papillate on one or both surfaces. **Stomata** level with surface. **Mesophyll** strongly or weakly differentiated. Palisade 1(–2)-layered. Spongy tissue mostly types 2 or 3; types 1, 1a, or 2 in *Alocasia*. **Collenchyma** type Sv (Svc or Sv in *Alocasia*). **Vascular bundles** 6–8 or numerous in midrib (only one in *Remusatia vivipara*), symmetrical or scattered; types II or III. **Laticifers** mostly anastomosing, often ramifying among adjacent ground tissue cells; non-anastomosing in *Ariopsis*, *Alocasia 'port-o-dora'*, *Remusatia vivipara* and *Colocasia fallax*. **Tannin** cells occasional in mesophyll of *Ariopsis* and near diaphragms in two *Alocasia* species. **Crystals**: raphides in biforines or biforine-like cells usually present; also raphides in thin-walled cells as single bundles or two or multiple bundles in longer cells.

Petiole TS

Outline of various shapes with smooth or striate cuticle. **Epidermis** often of large cells, similar to adjacent ground tissue cells. **Ground tissue** of large, uniform, compact cells, or with type 3 air cavities; type 3–4 in *Colocasia* and type 4 in *Alocasia*. **Collenchyma** type Sv. **Vascular bundles** scattered, numerous in large petioles, types II or III. **Laticifers** anastomosing, except non-anastomosing in *Ariopsis*, *Alocasia 'port-o-dora'*, *A. plumbaea* and *Remusatia vivipara*. **Secretory ducts** in peripheral ground tissue in *Colocasia fallax* and in proximal petiole in *Remusatia*. **Crystals**: raphides in single bundles in thin-walled cells, and occasional biforines

Taxonomic notes

In spite of the existence of numerous diagnostic differences, this is a coherent group of genera. If it were not for the molecular analysis of the family by French *et al.* (1995), one would hypothesize that all genera with anastomosing laticifers form a monophyletic group. The same array of characters in the Caladieae and Colocasieae vary in the same ways. There are no structural trends, and no character or group of characters to separate them.

Ariopsis Nimmo

This monotypic genus of cormose plants is found in eastern and northern India and western Malaysia. Plants are terrestrial or epilithic, tuberous herbs, occurring in crevices or on forest soil. Corms are leafless during dry season. Leaves are mostly solitary, long-petiolate and ovate-peltate. Chromosomes small: 2n = 28, 80, 84 and 86. One specimen was available for examination.

Leaf surface

Cuticle smooth. **Epidermis**: cells polygonal with straight anticlinal walls (Fig. 111E) on both surfaces (1–1.5:1 *l*/*w*). **Stomata** abaxial, brachyparacytic to nearly anomocytic (Fig. 111C), with subsidiary cells narrow, irregular and often square-cornered. **Venation**: veins diverging acrodromously from summit of petiole, connecting eucamptodromously near proximal margins, or directly to collector veins distally. Higher orders reticulate or tending to be oriented parallel with secondaries.

Leaf TS

Structure dorsiventral and weakly differentiated. **Size**: lamina 10 × 9 cm (*l*/*w*). **Midrib** small and flat or slightly convex adaxially, broadly convex and slightly ridged abaxially (Fig. 111A). **Cuticle** very thin or absent. **Epidermis**: cells short-papillate on both surfaces (Fig. 111B). **Mesophyll** with one palisade layer and 4–5 spongy layers. Spongy cells long-lobed (type 2). **Air cavities**: up to 90% of spongy tissue, randomly spaced or columnar as substomatal cavities (Fig. 111B). **Collenchyma**: three subepidermal circular strands on abaxial side of midrib, subtending the three abaxial bundles (type Sv) and one subepidermal adaxial strand (Fig. 111A).

Vascular bundles: six symmetrically placed bundles in midrib (type III). Xylem: wide protoxylem lacuna and 1–2 metaxylem cells. Phloem: small circular strand. **Laticifers** non-anastomosing, uncommon and lateral to vascular bundles (Fig. 111D). **Tannin** cells occasional in mesophyll. **Crystals**: biforines common in mid-lamina (Fig. 111E); non-biforine raphides uncommon as single bundles in short cells, or as multiple bundles in elongate cells, randomly oriented in mesophyll. Druses small, uncommon in palisade layer and following vascular bundles.

Petiole TS

Outline subterete and fluted-ridged (Fig. 111F). **Size**: 4 × 2.5 mm diameter. **Cuticle** thin and smooth. **Epidermis**: cells flat externally and same size as ground tissue cells. **Collenchyma**: circular strands, capping phloem of peripheral vascular bundles (type Sv) (Fig. 111G). Strands 4–5 cells deep beneath epidermis. **Ground tissue** compact, cells large and uniform (Fig. 111G,H).

Vascular bundles scattered and numerous, arranged in central cluster and outer ring with collenchymatous phloem caps; bundles weak (type III) (Fig. 111H). Xylem: one wide protoxylem element. Phloem: small circular or irregular strand. **Laticifers**: 2–3 strands, non-anastomosing, lateral to phloem of most vascular bundles. **Crystals**: raphides in thin-walled cells, common in outer ground tissue.

Material examined

Ariopsis peltata J.Grah., Bogner 1481, cult. M.

Species reported on in the literature

Ertl (1932): leaf • French (1987c): root. • French (1988): laticifers. • French and Tomlinson (1983): stem. • Lierau (1888): root. • Müller (1919): leaf tip. • Roth (1949, 1952): leaf ontogeny. • Solereder and Meyer (1928): anatomy. • Troll (1932a): leaf.

Additional literature

Engler 1920b; Erdtman 1952; French 1986a,c, 1987a; Gowda 1980; Grayum 1990; Kulkarni *et al.* 1990; Thanikaimoni 1969; Tieghem 1907.

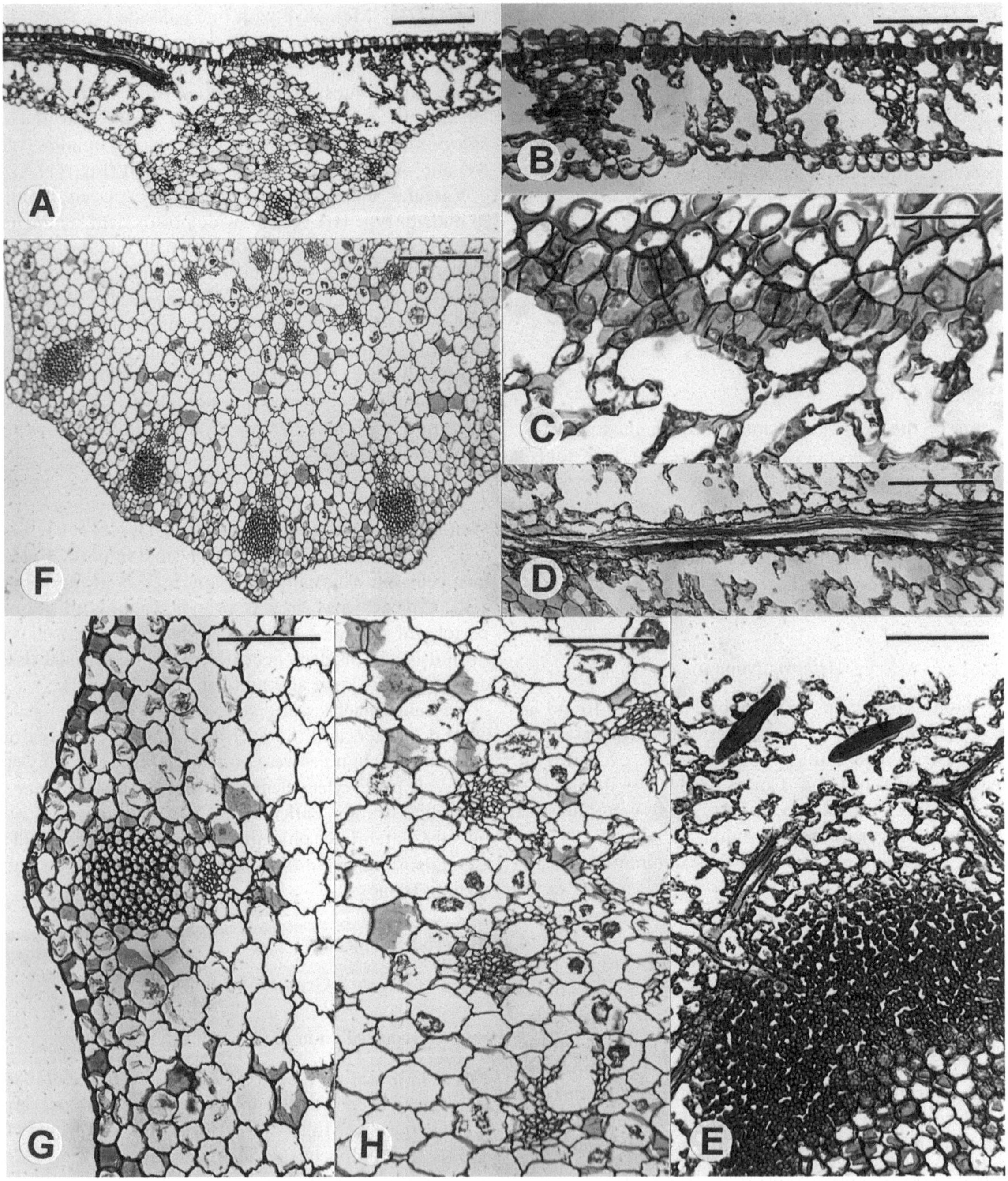

Fig. 111. ***Ariopsis peltata* leaf and petiole sections.** (A) Midrib TS; (B) Lamina TS; (C) Abaxial epidermis PS with stomata; (D) Lamina PS with vascular bundle and non-anastomosing laticifer; (E) Lamina PS showing adaxial epidermis, palisade and spongy tissue and biforine raphides; (F,G) Distal petiole TS with fluted margin, each ridge subtending circular type Sv collenchyma strand and vascular bundle; (H) Distal petiole TS, central tissue and type III vascular bundles. Scale bars: A,F = 300 μm; B,D,E,G,H = 150 μm; C = 50 μm.

Alocasia (Schott) G.Don

This genus of 70 species ranges from India, Sri Lanka, south-east Asia, Malaysia, southern China, Japan and the Caroline Islands, to Australia. This treatment also includes specimens obtained as *Xenophya*, now *A. lauterbachiana* (Engl.) A.Hay. Plants are caulescent or rhizomatous herbs of two types: large terrestrial herbs or smaller epilithic herbs. Terrestrial species tend to be found in open boggy areas. Lithophytic species are clustered, have repent stems, and grow in rock crevices. Corms may be produced from stolons as root system offshoots or by offshoots emerging from the side of the rhizome (Burnett 1984, cited in Croat 1988). Leaves are ovate, cordate, sagittate, or peltate. In *A. lauterbachiana*, leaves are long-petiolate, elliptic and have an orthogonally truncated base. Chromosomes are medium-sized, in most species $2n = 28$ but $2n = 40$, 42, 56, and 70 are also known. Seven specimens were examined.

Leaf surface

Cuticle smooth, rough or papillate. **Epidermis**: cells polygonal with straight or undulate anticlinal walls adaxially (1–3:1 *l/w*) and undulate abaxially (1–5:1 *l/w*). **Stomata** more frequent abaxially, variable in every species: brachyparacytic, occasionally brachypara-tetracytic to -hexacytic (Fig. 112B); uni- or bi-polar. Subsidiary cells broad with rounded edges, or narrow or irregular. **Venation**: midrib prominent with widely-spaced, broadly-acute lateral secondaries; secondaries narrowly-acute in *A. lauterbachiana*. Secondaries straight, ending orthogonally at marginal collector, or apically curving into brochidodromous arches. Tertiaries originating broadly-acute from both sides of secondaries, curving and becoming parallel with secondaries, anastomosing in centre of intercostal area. Higher orders orthogonal and irregular.

Leaf TS

Structure dorsiventral. **Size**: 10–40 × 6–47 cm (*l/w*). **Midrib** broadly convex adaxially, slightly convex to straight abaxially (Fig. 112A). **Cuticle** thin to medium-thick; forming flanges along anticlinal walls of epidermal cells. **Epidermis**: cells large on both surfaces, cells flat externally, slightly rounded internally. **Stomata** with shallow or deep substomatal air spaces. **Mesophyll**: palisade 1-layered, cells cylindrical (1:8 *l/w*). Spongy layers 5–6, up to nine layers in *A.* '*port-o-dora*', up to 12 layers in *A. lauterbachiana* (tissue types 1,1a, or 2). **Air cavities**: in lamina, 10% of spongy layers in *A. cuprea*, up to 30–70% in *A. cucullata* and *A.* '*port-o-dora*'; in midrib, type 4. Tall substomatal air cavities in latter two species. **Collenchyma**: discrete circular strands, subepidermal on both surfaces of midrib (Fig. 112A) and larger secondaries. Strands subtending vascular bundles as phloem caps (type Svc, Sv), or independent of bundles. **Vascular bundles** scattered in midrib (types II and III), not well developed in *A. lauterbachiana*. Xylem: mostly one wide protoxylem lacuna and often several additional metaxylem cells. Phloem: a circular, broadly elliptic or irregular strand.

Laticifers: 1–2 near phloem, in single non-anastomosing strands in *A.* '*port-o-dora*'. In other species, laticifers more frequent and extensively branching, ramifying and anastomosing among ground tissue near midvein and other larger vascular bundles. **Idioblasts**: in *A.* '*port-o-dora*', cells bearing orange rods common in palisade and mid-mesophyll; also yellow globular deposits in some epidermal cells. In same species, dark brown bodies, often with filamentous arms, one per cell, common among mesophyll parenchyma. **Tannin** cells found near diaphragms in *A. cuprea* and *A.* '*port-o-dora*'. **Crystals**: raphides in thin-walled, rounded-ended biforine-shaped cells, randomly oriented throughout mesophyll in all species, attached by one end at margin of midvein air cavities and pointed toward centre of cavity (Fig. 112A); cells abundant in *A. cuprea*, less frequent in other species. In *A.* '*port-o-dora*', raphides in thin-walled wide cells common as idioblasts among mesophyll. Druses large, common to uncommon in mesophyll of *A. cucullata* and *A.* '*port-o-dora*'; druses small, in adaxial mesophyll in *A. lauterbachiana*; druses absent in *A. cuprea*. Crystal sand occasional in ground tissue and mesophyll.

Petiole TS and LS

Outline rounded, flattened adaxially, or slightly laterally compressed or grooved. **Size**: 6–7 mm diameter, up to 20 mm in *A.* '*port-o-dora*'. **Cuticle** smooth, roughly ornamented or papillate, medium-thick. **Epidermis**: cells medium-sized, flat or rounded on surface and rounded internally. **Collenchyma** of discrete, ± circular strands, separated from epidermis by 1–several layers of ground tissue, alternating small and larger around periphery (Fig. 112C). Larger circular strands capping phloem of vascular bundles (type Sv), or not found juxtaposed with vascular bundles. **Chlorenchyma**: chloroplasts common within cells of outer ground tissue, except in *A. plumbaea*. **Ground tissue**: mostly uniform, large cells. **Air cavities** numerous, rounded, scattered, small or large, axially elongate, type 4 (Fig. 112F). Unicellular diaphragms present up to > 20 mm apart.

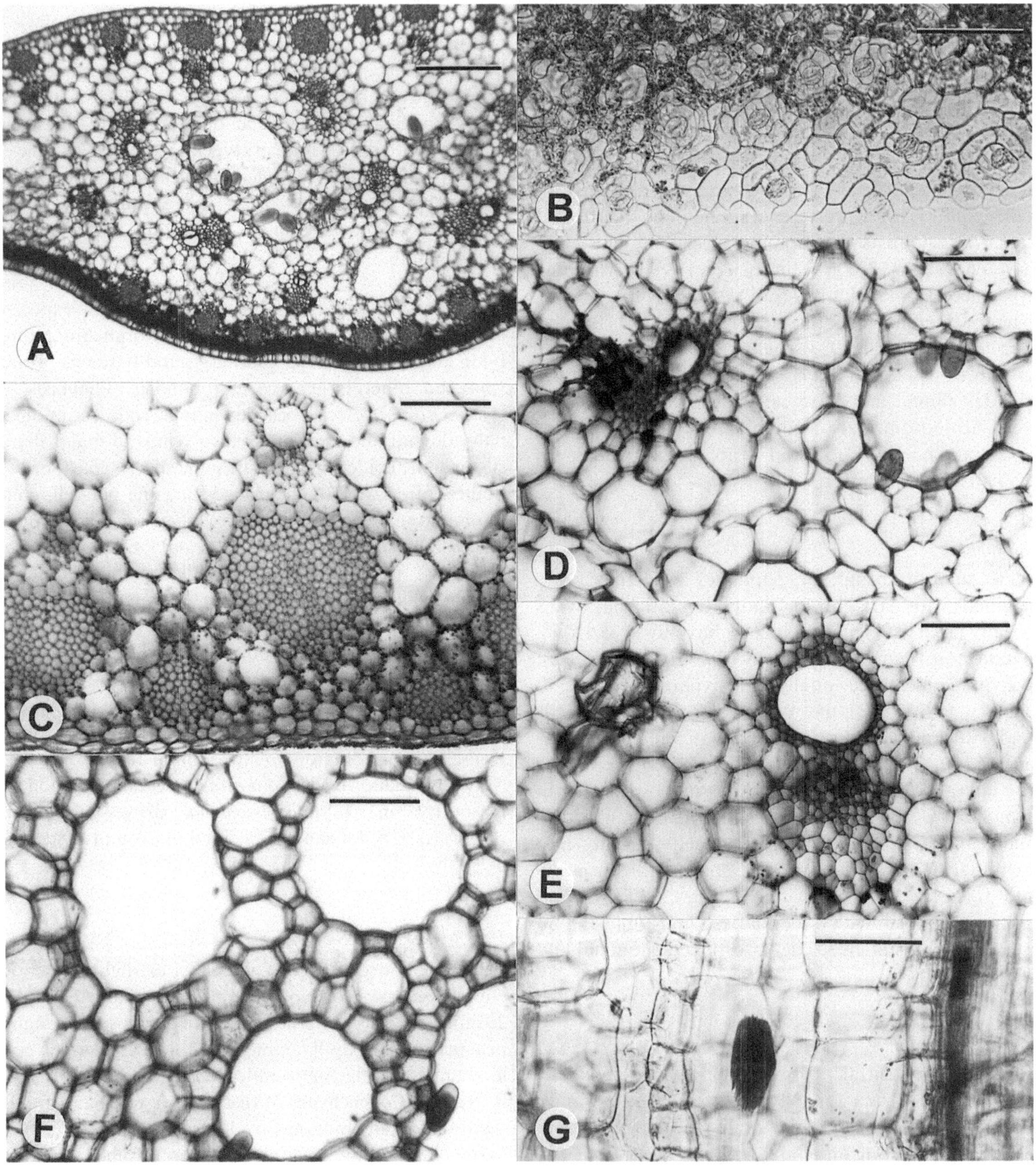

Fig. 112. *Alocasia* leaf and petiole sections. (A,D) *A. cuprea*; (B,F) *A. cucullata*; (C,E,G) *A. 'port-o-dora'*; (A) Midrib TS, abaxial surface uppermost. Note biforine-like raphides in aerenchyma cavities; (B) Abaxial epidermis PS with brachyparacytic and brachypara-tetracytic stomata; (C) Petiole TS showing large, nearly circular collenchyma strands of several sizes; (D) Petiole TS showing type 4 air cavity and type III vascular bundle with plexus of anastomosing laticifers. Biforine raphides suspended in large cavity; (E) Petiole TS with type III vascular bundle; (F) Petiole TS, central tissue with type 4 aerenchyma; (G) Petiole LS, ground tissue with small raphide cell. Scale bars: A = 300 μm; B–D,F = 200 μm; G = 150 μm; E = 50 μm.

Vascular bundles: ring of small bundles around periphery, numerous and scattered in centre; bundles weakly formed (type III). Xylem: usually a single protoxylem cell or lacuna, strongly outlined by a ring of small parenchyma cells, thereby resembling a duct (Fig. 112E). Metaxylem cells occasionally present, especially as cross-connecting minor veins. Phloem: small circular strand adjacent to xylem. **Laticifers**: in *A. cuprea* and *A. cucullata*, 3–6 adjacent to phloem, anastomosing and ramifying throughout ground tissue in apoplastic space, especially nearer periphery. In *A. 'port-o-dora'* and *A. plumbaea*, fewer non-anastomosing laticifers present. **Starch** granules in arc of cells abaxial to phloem, especially at peripheral bundles. **Crystals**: raphides in thin-walled biforine cells, attached at one end at periphery of air cavities with other end facing centre of cavity (Figs 9F, 112D); cells numerous, separated, or in clusters of 2–3. In *A. 'port-o-dora'* multiple bundles of raphides also common in large, elongate 'tubes' throughout ground tissue; also, single bundles in thin-walled cells scattered in ground tissue (Fig. 112G). Druses occasional in ground tissue in *A. plumbaea*. Crystal sand clusters often present in some ground tissue cells. **Nuclei**: unusually large nuclei visible in this genus with numerous large vesicles attached over surface.

Material examined

Alocasia beccarii Engl., Bogner 1342, Sarawak, cult. M; *A. cucullata* (Lour.) G.Don, cult. MO 894185; *A. cuprea* C.Koch, cult. MO 893385; *A. (Xenophya) lauterbachiana* (Engl.) A.Hay, Bogner 2337, cult. M; *A. plumbaea* C. Koch ex van Houtte, Croat 38233, Panama; *A. porphyroneura* H.Hallier ex Engl., cult. M; *A. 'port-o-dora'*, cult. MO 894203.

Taxonomic notes

The specimens examined here form a coherent group in spite of diagnostic differences noted above. *Alocasia lauterbachiana* (*Xenophya*) fits clearly within the anatomical variability of other *Alocasia* examined here.

Species reported on in the literature

Allen (1929): *A. macrorrhiza*: starch. • Alvim *et al.* (1969): *A. macrorrhiza*: stomata. • Ambronn (1879–81): *A. odora*: collenchyma. • Andersson (1888): *A. gigantea*: procambium. • Belin-Depoux (1978): several spp.: foliar glands. • Belin-Depoux (1989): leaf. • Carlquist and Schneider (1998): *A.* × *amazonica, A. macrorrhiza*: stem, root. • Chrysler (1904): *A. odorata*: stem. • Dalitzsch (1886): 5 spp.: leaf. • Engler (1920b): *A. esculenta, A. indica, A. macrorrhiza*: leaf, starch. • Ertl (1932): *A. indica, A. odora, A macrorrhiza*: leaf. • Franceschi and Horner (1980): *A. cucullata, A. macrorrhiza*: crystals. • French (1987c): several spp.: root. • French (1988): several spp.: laticifers. • French and Tomlinson (1983): several spp.: stem. • Gardiner (1889): leaf. • Gentner (1905): *A. van houtei*: leaf • Goodchild *et al.* (1971–2): *A. macrorrhiza*: leaf. • Guttenberg (1968): root. • Hotta (1965, 1971): *A. robusta*: leaf. • Ko *et al.* (1990): several spp.: tracheids: leaf sheath, petiole, roots. • Kundu (1942): *A. indica*: root. • Lierau (1888): several spp.: root. • Lohse (1924): *A. odora*: tyloses. • Marcatili (1887): *A. metallica*: laticifers. • Müller (1919): several spp.: leaf tip. • Namba *et al.* (1968): *A. odora*: rhizome. • Netolitzky (1929): crystals. • Onken (1922): *A. odora*: laticifers, calcium. • Pant and Kidwai (1966): *A. indica*: leaf, epidermis. • Pirotta and Marcatili (1885): *A. metallica*: laticifers. • Planchon and Juillet (1909): *A. indica*: starch. • Prychid and Rudall (2000): *A. micholitziana*: crystals. • Reichert (1913): *A. putzeysi*: starch. • Roth (1949): leaf ontogeny. • Roth (1952): *A. cucullata, A. macrorrhiza*: leaf ontogeny. • Sakai and Hanson (1974): *A. cucullata, A. macrorrhiza*: all organs, raphides. • Sakai *et al.* (1984): leaves, stems. • Schwendener (1874): *A. metallica*: petiole. • Scott (1889): leaf. • Sims and Pearcy (1992): *A. macrorrhiza*: leaf. • Sivaramakrishna and Vijayaraghavan (1979): root. • Solereder and Meyer (1928): several spp.: secretory tissue, crystals. • Swamy and Sivaramakrishna (1976): apical meristem, root. • Tieghem (1866): *A. metallica, A. odora*: all organs. • Tieghem (1870–1): *A. metallica, A. odora*: root, secretory tissue. • Tieghem and Douliot (1888): *A. boryi*: root. • Trécul (1865): *A. antiquorum, A. cucullata, A. zebrina*: vasculature. • Troll (1932a): several spp.: leaf. • Zinke (1924): *A. cuprea*: leaf.

Additional literature

Blanc 1978; Buchholz 1921; Chow *et al.* 1988; Clarke *et al.* 1978; Domin 1911; Erdtman 1952; Fox and French 1988; French 1986a,c, 1987a, 1988; French and Tomlinson 1983; Grayum 1990; Hegnauer 1986; Huang 1972; Irmisch 1874; Kraus 1896; Loh *et al.* 2000; Meusel 1951; Mookerjea 1955; Nahrstedt 1975; Ray 1988; Saha and Hussain 1983; Sharma 1967; Simmonds 1976; Tarasevich 1990; Thanikaimoni 1969; Troll 1928; Vogel 1962; Williams *et al.* 1981.

Remusatia Schott

This treatment includes species formerly assigned to *Gonatanthus*, now merged into *Remusatia* (Li and Hay 1992). Of the four species, one occurs from the Himalayas to Formosa. A second, *R. vivipara*, is widespread across tropical Asia, tropical Africa, Madagascar, and tropical Australia. These tuberous plants occur mostly in seasonally dry areas, often losing their leaves in the dry season. *R. vivipara* may grow as an epiphyte, among rocks in small streams, in rock crevices, in rocky forested areas, or pockets of soil on bare rock. Species may occur up to 2200 m in elevation. The leaves are long-petiolate, and peltate-cordate to lanceolate. The two species formerly of *Gonatanthus* occur from Thailand to southwest China, occurring in mossy, moist evergreen forests on deposits of humus on trees. Chromosomes: 2n = 28, 42, 56. Two species were available for study. Differences in *Remusatia pumila* (*Gonatanthus*) are noted below.

Leaf surface

Cuticle smooth on both surfaces, or striate on abaxial midrib in *R. pumila*. **Epidermis**: cells polygonal with straight anticlinal walls (1–2:1 *l*/*w*) on both surfaces, or rounded-polygonal abaxially (1–1.5:1 *l*/*w*) in *R. pumila*. **Stomata** mostly abaxial, brachyparacytic. Subsidiary cells broad or rounded or irregular and angular (Fig. 113C). **Venation**: basal acrodromy producing multiple primaries. Midrib prominent with widely spaced, broadly-acute secondaries. Tertiaries acute on both sides of secondaries and midrib; tertiaries closely spaced, parallel, curved-converging to centre of intercostal area and becoming parallel with secondaries. Higher orders reticulate or orthogonal.

Leaf TS

Structure dorsiventral, well-differentiated in *R. pumila*. **Size** 7–8 × 7–12 cm (*l*/*w*). **Midrib** slightly thicker than lamina, flat adaxially, slightly rounded-convex abaxially. In *R. pumila*, midrib more prominent and ribbed abaxially. **Cuticle** thin on both surfaces. **Epidermis**: cells large adaxially, rounded to irregular externally and internally. Abaxial cells shorter and flattened, or in *R. pumila* often tall and papillate. **Mesophyll** weakly differentiated in *R.vivipara*, one palisade layer (6–10:1 *l*/*w*) (Fig. 113A), except near margin (Fig. 113B). In *R. pumila* palisade layers often appearing as a hypodermis, empty of chloroplasts. Spongy layers 8–10 in well-formed type 3 aerenchyma with large cavities and irregular or vertical uniseriate partitions (Fig. 113D); type 1 spongy cells near abaxial epidermis. **Air cavities**: up to 95% of spongy layer. **Collenchyma**: large discrete strands in adaxial and abaxial sides of midrib (type Sv) in *R. pumila*.

Vascular bundles: one large bundle in midrib, or scattered and very small in *R. pumila;* bundles mostly large (type II), with irregular boundary with ground tissue. Xylem: 8–10 (1–4 in *R. pumila*) metaxylem elements opposing an irregular phloem strand, the whole surrounded by irregular sheath of parenchyma cells. **Laticifers** non-anastomosing (Fig. 113E) (anastomosing in *R. pumila)*, common, 1–2 near vascular bundles. Cells reported as anastomosing by French (1988). **Crystals**: raphides in wide single or double bundles common in pigmented sheaths in elongated cells (Fig. 113D,F); raphide cells aligned randomly and embedded among mesophyll cells. Occasional biforines oriented crosswise in aerenchyma partitions or attached by one end, with terminal papillae. Druses large and common, epidermal or subepidermal, or adjacent to vascular bundles. Crystal sand clustered in spongy cells or vascular bundle sheaths.

Petiole TS

Outline grooved proximally and terete distally. **Size**: 2–8 mm diameter. **Cuticle** thin and smooth, or thick and striate in *R. pumila*. **Epidermis**: cells rounded to papillate externally and flat internally. **Collenchyma**: discrete circular strands, small or large around periphery in *R. vivipara* (Fig. 113G) and separated from epidermis by 1–2 layers of ground tissue; most strands capping phloem of outermost vascular bundles (type Sv). Strands beneath ridges in *R. pumila* and 15–20 layers deep. **Ground tissue**: outer 5–6 compact layers bearing chloroplasts. Central ground tissue with large air cavities (type 3 aerenchyma).

Vascular bundles numerous and scattered, weak, mostly type III (Fig. 113H), or type II in *R. pumila*. Central bundles with wide protoxylem lacuna and adjacent circular phloem strand. Peripheral bundles: xylem with narrower protoxylem cell and 1–8 narrow metaxylem cells. **Secretory ducts**: 2–8 very wide ducts with 1-layered epithelium in ± abaxial half of proximal petiole, easily confused with aerenchyma cavities in some preparations. **Laticifers**: anastomosing in *R. pumila*, non-anastomosing in *R. vivipara*; 2–4 lateral to phloem strands in larger bundles. **Crystals**: raphides in axially elongated cells, mostly between collenchyma strands in *R. vivipara*, or common throughout in *R. pumila*. Biforines occasional in aerenchyma. Druses scattered and occasional.

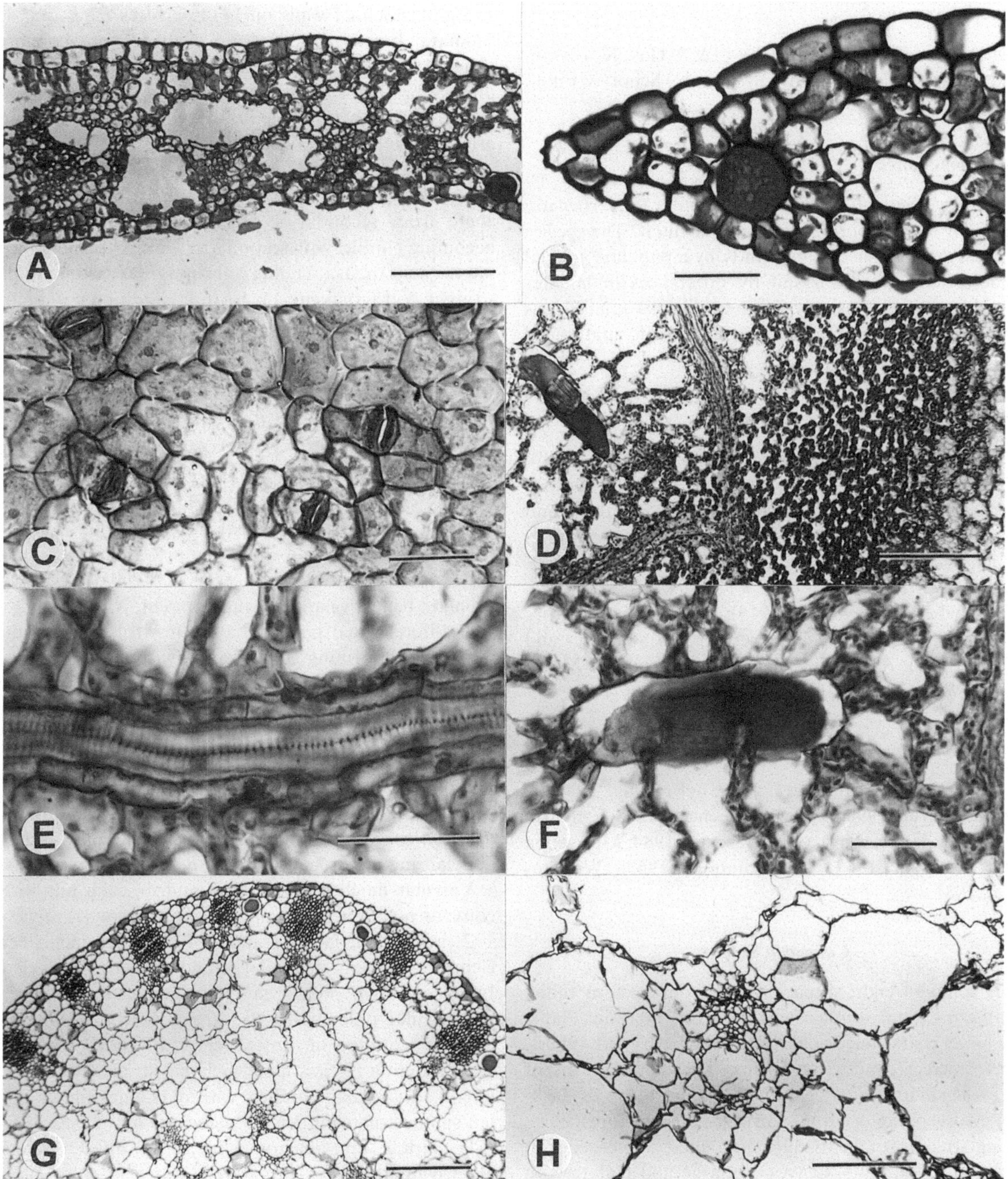

Fig. 113. *Remusatia vivipara* leaf and petiole sections. (A) Lamina TS; (B) Lamina TS margin; (C) Lamina PS abaxial epidermis and brachyparacytic stomata; (D) Lamina PS mesophyll and raphide cell; (E) Lamina PS higher order vascular bundle and lateral non-anastomosing laticifers; (F) Lamina PS showing raphide cell in spongy tissue; (G) Distal petiole TS showing type 3 aerenchyma and type Sv collenchyma; (H) Proximal petiole TS showing type III vascular bundle. Scale bars A,D,H = 150 μm; B,C,E,F = 50 μm; G = 300 μm.

Material examined

Remusatia pumila (D.Don) Heng Li & A.Hay, Kurtz s.n., s.d., Sikkim India, K; *R. vivipara* (Roxb.) Schott & Endl., K spirit collection.

Taxonomic notes

R. pumila (*Gonatanthus pumilus*) is similar to *R. vivipara*, especially as they both include petiole ducts. The species are best distinguished in this study by a papillate abaxial epidermis in *R. pumila* and by differences in laticifer morphology. As presently understood, this small genus shows more than the usual amount of anatomical variation.

Species reported on in the literature

Dalitzsch (1886): *R. vivipara*: leaf. • Engler (1920b): *G. sarmentosus:* branching. • French (1987c): *R. vivipara*: root. • French (1988): *R. vivipara*: laticifers. • French and Tomlinson (1983): *R. vivipara*: stem vasculature. • Gentner (1905): *R. vivipara*: leaf. • Lierau (1888): *G. sarmentosus, R. vivipara*: root. • Möbius (1935): *R. hookeriana*: bulbil. • Müller (1919): *R. vivipara*: leaf tip. • Philipp (1923): *R. vivipara*: periderm. • Solereder and Meyer (1928): *R. vivipara*: all organs. • Troll and Meyer (1955): *R. vivipara*: leaf.

Additional literature

Domin 1911; Erdtman 1952; French 1986a,c, 1987a; Grayum 1990; Jaeger 1968; Knecht 1983; Kulkarni *et al.* 1990; Kuprianova and Tarasevich 1984; Mookerjea 1955; Sharma 1967; Thanikaimoni 1969; Williams *et al.* 1981.

Colocasia Schott

This genus of eight species ranges from western India, southern China, southeastern Asia to Malaysia. Plants are large, coarse terrestrial or epilithic herbs with tubers, a short erect caudex, or corms. The edible corms of *C. esculenta* are widely grown as the crop taro or dasheen. Most species occur in shady places at the forest edge, along streams, or in shady humid sites. Some occur in steep rocky habitats. Leaves are large and peltate with cordate basal lobes. Chromosomes: 2n = 14, 22, 26, 28, 36, 42, 48. Three species were available for study.

Leaf surface

Cuticle smooth on both surfaces and striate on abaxial surface of midrib. **Epidermis**: cells polygonal with straight anticlinal walls on both surfaces (1–1.5:1 *l/w*). **Stomata**: brachyparacytic and slightly more common abaxially (Fig. 114D); in *C. gigantea* stomata abaxial only. Subsidiary cells broad and angular. **Venation**: basal acrodromy with prominent midvein among basal primaries and pectinal veins leading to basal lobes. Secondaries widely spaced and acute, producing regular intercostal areas. Tertiaries close, arising from midrib and broadly-acute from secondaries, curving toward margin and becoming parallel with secondaries. Tertiaries merging in intercostal areas, ending abruptly at submarginal straight collector vein; up to three additional submarginal collectors present. Higher orders orthogonal to reticulate.

Leaf TS

Structure dorsiventral. **Size**: 10–35 × 4–9 cm (*l/w*). **Midrib** slightly convex or flat adaxially, and slightly convex to broadly semicircular abaxially (Fig. 114A). **Cuticle** thin on both surfaces. **Epidermis**: cells square to elongate, rounded internally and externally in *C. gigantea*; markedly papillate on both surfaces in *C. fallax* (Fig. 114B). **Stomata** beneath papillae when present. **Mesophyll**: 1–2 well-differentiated palisade layers (6–10:1 *l/w*). Spongy layers 5–6, forming aerenchyma with large, vertical, substomatal air spaces with uniseriate partitions (type 3), 1–3 cells per partition between corners (Fig. 114D–F). **Air cavities** comprising 95% of spongy tissue; cavities large in midrib (types 3 and 4). **Collenchyma**: discrete subepidermal, ± circular strands of angularly thickened cells around midrib (Fig. 114A), or abaxial only in *C. gigantea*. Strands often forming bundle caps on outer abaxial ring of vascular bundles in *C. gigantea* (type Sv).

Vascular bundles scattered in midrib, each bundle at convergence of aerenchyma partitions (type II). Xylem: 1–2 protoxylem lacunae and 3–4 narrower metaxylem elements. Phloem: a circular strand abutting metaxylem. In *C. gigantea*, one large xylem lacuna and adjacent semicircular phloem strand (type III). Midrib vascular bundles scattered in central region in addition to peripheral bundles beneath collenchyma strands. **Laticifers**: lateral to vascular bundles, with dark amorphous contents; anastomosing in *C. gigantea* and *C. esculenta* (Fig. 114C); non-anastomosing in *C. fallax*. **Reserve photosynthate**: sugar, not starch, stored in chloroplasts (Sakai and Hayashi 1973). **Crystals**: raphides in single bundles common in large cells in mid-mesophyll (Fig. 114E). Biforine raphides in cylindrical cells up to 150 μm long, attached by one end to aerenchyma partitions, suspended in air cavities (Fig. 114F). Raphide cells common within aerenchyma partitions or corners.

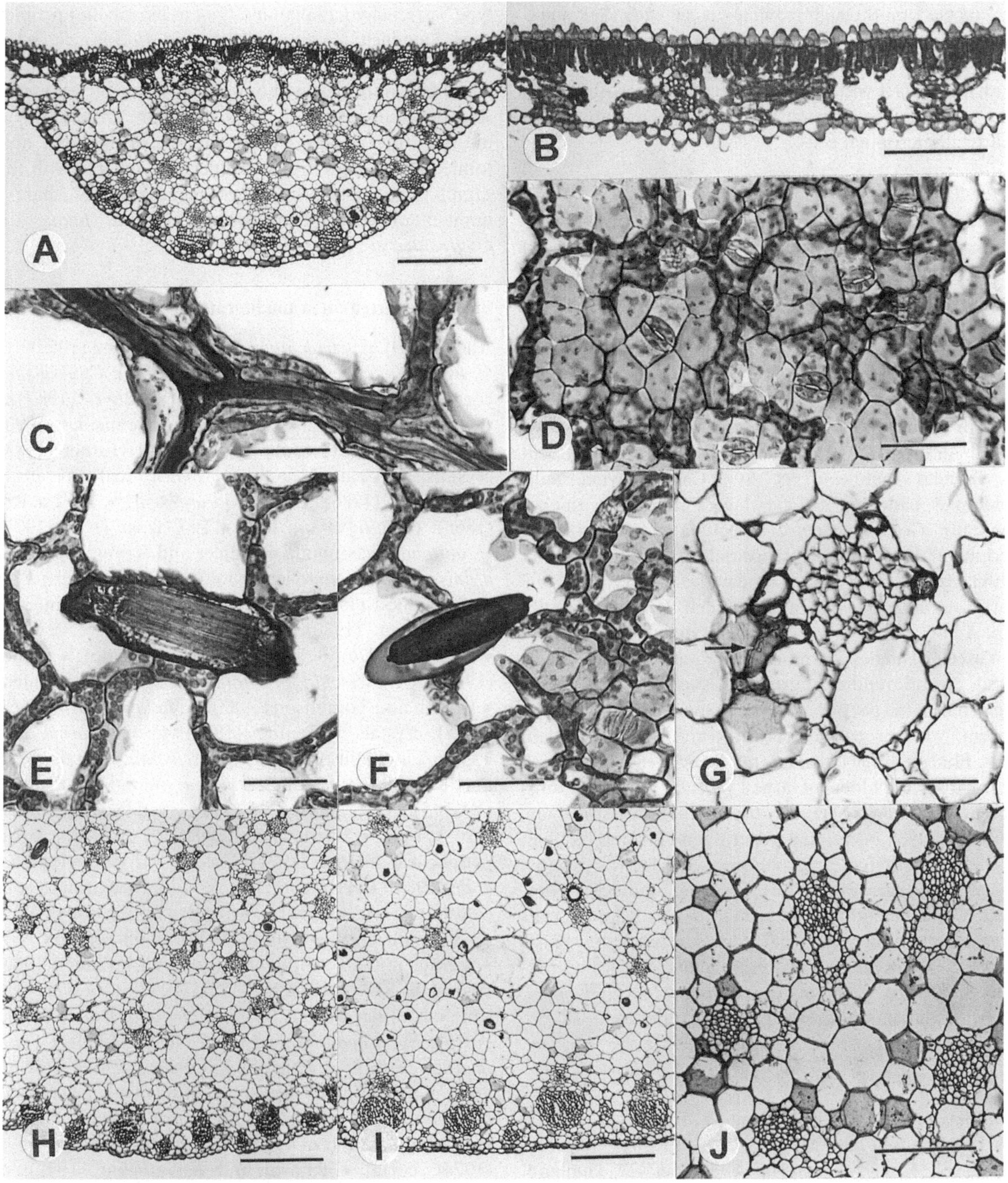

Fig. 114. *Colocasia* leaf and petiole sections. (A,B,D–F,I,J) *C. fallax;* (C) *C. gigantea*; (G,H) *C. esculenta* var. *illustris*. (A) Midrib TS; (B) Lamina TS; (C) Lamina PS with anastomosing laticifer; (D) Lamina PS, abaxial epidermis and brachyparacytic stomata; (E) Lamina PS, type 3 spongy aerenchyma and raphide cell; (F) Lamina PS showing biforine raphide; (G) Distal petiole TS showing central type III vascular bundle and anastomosing laticifer (arrow); (H) Distal petiole TS and (I) proximal petiole TS, both showing stranded (type Sv) collenchyma; (J) Distal petiole TS, central tissue with type III vascular bundles. Scale bars A,H,I = 300 μm; B,J = 150 μm; C–G = 50 μm.

Crystal ejection through papillae on cell ends (Sakai and Hanson 1974). Raphide cells often appearing with fused and eroded crystalline material. Raphide crystals H-shaped in TS with lateral grooves. Druses numerous and very small in mesophyll. Crystal sand common in nearly all mesophyll cells.

Petiole TS

Outline dorsiventrally flattened proximally, with adaxial laminar wings; distally rounded and irregularly grooved in *C. gigantea*. Petiole subterete in *C. fallax*. **Size**: diameters proximally and distally: 18 and 5 mm. **Cuticle** thin or thick and smooth or striate abaxially. **Epidermis**: cells square to elongate, somewhat rounded internally and externally. **Collenchyma**: cells angularly thickened, in strands 1–3 layers beneath epidermis. Strands abaxially only or around periphery; uniform or alternating large and small (Fig. 114H,I); most strands subtending phloem of vascular bundles (type Sv). **Chlorenchyma**: subepidermal band up to five layers adaxially, thicker abaxially. **Ground tissue**: compact parenchyma in 4–6 peripheral layers. Central ground tissue: aerenchyma with large or small type 3–4 air cavities. Uniseriate type 3 partitions in *C. gigantea*, up to five cells long. Air cavities type 4 in *C. fallax*, some containing mucilage.

Vascular bundles: peripheral ring and numerous scattered, central bundles in ground tissue islets of converging aerenchyma partitions (types II and III) (Fig. 114G,J). Xylem: wide protoxylem cavities and 0–4 metaxylem cells. Phloem: large circular strands. **Sclerenchyma**: small fibre caps over phloem of largest vascular bundles. Fibres absent in *C. fallax*. **Secretory tissue**: small ducts with one epithelial layer, occasional in peripheral ground tissue in *C. fallax*. **Laticifers** anastomosing, 3–4 lateral to most vascular bundles (Fig. 114G). **Crystals**: raphides in thin-walled cells, uncommon in ground tissue. Raphide cells occasionally attached by one end to aerenchyma partition in *C. gigantea*, but not in *C. fallax*. Druses numerous, small, scattered in ground tissue in *C. gigantea*. Crystal sand common, scattered in ground tissue.

Material examined

Colocasia esculenta (L.) Schott & Endl. var *illustris* (W.Bull.) Schott, cult. M; *C. fallax* Schott, Bogner 1139, India, cult. M; *C. gigantea* Hassk., Bogner 427, Thailand, cult. M.

Taxonomic notes

Comparisons between this genus and *Alocasia* reveal subtle differences. *Colocasia* species have more consistent type 3 aerenchyma in lamina (3–4 in midrib and petiole). Type 3 aerenchyma appears absent in *Alocasia* which has type 1, 1a or 2 in leaf, and type 4 in midrib and petiole. Abundant biforines, without papillae, line air cavities in *Alocasia* and other raphides occur in ground tissue; biforines in *Colocasia* have papillae. Otherwise, anatomical differences are not sufficiently reliable with this sample size to separate these genera. They share a number of common features. *Colocasia fallax* and *C. gigantea* are quite distinct.

Species reported on in the literature

Allen (1929): *C. esculenta*: starch. • Ambronn (1879–81): *C. esculenta*: collenchyma. • Black (1918): *C. esculenta*: corm. • Blanc (1978): ontogeny. • Borzi (1887–8): *C. antiquorum*: root. • Carlquist and Schneider (1998): *C. esculenta*: stem, root. • Cody and Horner (1983): crystals. • Czaja (1962): *C. indica*: leaf epidermis. • Dalitzsch (1886): *C. antiquorum*: leaf. • De La Rue (1866): *C. antiquorum*: leaf. • El-Gazzar *et al.* (1977): *C. antiquorum*: stomata. • Engler and Krause (1920): *C. affinis*, *C. antiquorum*: leaf, root. • Ertl (1932): *C. fallax*: leaf. • Flood (1919): *C. antiquorum*: leaf. • Franceschi and Horner (1980): *C. esculenta*: petiole, crystals. • French (1987c): *C. esculenta*, *C. fallax*: root. • French (1988): *C. esculenta*, *C. fallax*, *C. indica*: laticifers. • French and Tomlinson (1983): *C. fallax*: stem. • Fuchs (1898): crystals. • Gentner (1905): *C. antiquorum*: leaf. • Genua and Hillson (1985): *C. esculenta*: crystals. • Gill and Karatela (1983): *C. esculenta*: stomata. • Gulliver (1863–5): *C. antiquorum*, *C. odora*: crystals. • Guttenberg (1968): root. • Harris *et al.* (1992): *C. esculenta*: corm, petiole, leaf. • Hotta (1971): *C. esculenta*: leaf. • Inamdar *et al.* (1983): *C. esculenta*: leaf venation. • Karawya *et al.* (1980): *C. esculenta*: corm. • Kenten and Woods (1973): *C. esculenta*: corm. • Kundu (1942): *C. antiquorum*: root. • Lierau (1888): *C. antiquorum*: root. • Linsbauer (1930): epidermis. • Melati and Scialabba (1982): *C. antiquorum*: leaf. • Molisch (1903): *C. nymphaeifolia*: leaf. • Müller (1919): *C. antiquorum*: leaf tip. • Nasim (1990): *C. antiquorum*: corm, root. • Netolitzky (1929): crystals. • Nwufo and Fajola (1985): *C. esculenta*: corm. • Onken (1922): *C. antiquorum*, *C. multifida*: secretory tissue, calcium. • Osisiogu and Uzo (1973): starch. • Osisiogu *et al.* (1974): corm. • Paliwal and Kavathekar (1972): *C. esculenta*: corm. • Pant and Kidwai (1966): *C. antiquorum*: leaf. • Planchon and Juillet (1909): *C. antiquorum*, *C. esculenta*: starch. • Reichert (1913): *C. esculenta*: starch. • Rhee and Iwata (1982): *C. esculenta*: corm. • Roth (1952): *C. antiquorum*: leaf ontogeny. • Sakai and Hayashi (1973): *C. esculenta*: chloroplasts. • Sakai and

Hanson (1974): *C. esculenta, C.* cultivars: corm, petiole. • Schwendener (1874): *C. antiquorum*: petiole. • Smirnova *et al.* (1964): all organs. • Solereder (1919): *C. antiquorum*: leaf. • Solereder and Meyer (1928): *C. antiquorum*: all organs. • Stahl (1896): *C. antiquorum*: leaf. • Strauss (1983): *C. esculenta*: leaf, corm, root. • Sunell and Healey (1979): *C. esculenta*: corm • Sunell and Healey (1981, 1985): *C. esculenta*: leaf. • Swamy and Sivaramakrishna (1976): root. • Tieghem (1866): *C. antiquorum*: all organs. • Tieghem (1870–1): *C. antiquorum*: root, secretory tissue. • Tieghem and Douliot (1888): *C. odora*: root. • Tobler (1957): *C. antiquorum*: fibres. • Trécul (1866): *C. cucullata*: vasculature. • Troll and Meyer (1955): *C. antiquorum*: leaf. • Turpin (1836): as *Caladium esculenta*: leaf, biforines. • Uphof *et al.* (1962): hairs. • Van Fleet (1942): *C. esculenta*: stem, endodermis. • Wakabayashi (1957a): *C. antiquorum*: raphides. • Wakabayashi (1957b): corm. • Webber (1960): *C. antiquorum*: epidermis. • Winton and Winton (1935): *C. antiquorum, C. esculenta*: corm. • Yarbrough (1934): *C. antiquorum*: leaf. • Youngken (1919): *C. esculenta*: corm.

Additional literature

Batygina and Yakovlev 1990; Brandza 1891; Davis 1966; Erdtman 1952; Fox and French 1988; French 1986a,c, 1987a; Grayum 1990; Greenwell 1947; Hegnauer 1986; Huang 1972; Knecht 1983; Kraus 1896; Meusel 1951; Mookerjea 1955; Namba *et al.* 1968; Nyman *et al.* 1986; Saha and Hussain 1983; Sakai *et al.* 1984; Selvaraj and Subramanian 1987; Shaw 1975; Simmonds 1976; Smolenski *et al.* 1974; Thanikaimoni 1969; Tillich 1985; Wheaton and Stewart 1970; Williams *et al.* 1981.

Steudnera K.Koch

This genus of eight species ranges from northern India to south-east Asia, southern China and the Malay peninsula. Plants are terrestrial herbs with short ascending stems occurring in the understorey in wet areas along streams. The leaves are long-petiolate with an ovate-peltate lamina. Chromosomes: 2n = 28, 56. Two species were available for study.

Leaf surface

Cuticle smooth on both surfaces. **Epidermis**: cells large and polygonal with straight anticlinal walls (1–1.5:1 *l/w*) on both surfaces. **Stomata** abaxial and brachyparacytic (Fig. 115C). Subsidiary cells broad and irregular. **Venation**: midrib prominent. Secondary veins parallel and widely spaced. Tertiaries tending to form composite intersecondaries in centre of intercostal regions. Higher order veins reticulate.

Leaf TS

Structure dorsiventral. **Size**: lamina 15 × 9 cm (*l/w*). **Midrib** flat adaxially and broadly convex-rounded abaxially (Fig. 115A). **Cuticle** thin on both surfaces. **Epidermis**: cells large adaxially, flat externally and rounded internally. Cells markedly papillate abaxially (Fig. 115B). **Mesophyll**: one well-formed palisade layer of elongate cells (1:8 *l/w*). Spongy tissue: *c.* 6 layers of irregularly lobed or armed cells forming a large well-formed type 2 or 3 aerenchyma (Figs 5A, 115B,D–F). **Air cavities**: about 90% of spongy layer. **Collenchyma**: subepidermal strands on all sides of midrib, 8–10 cell layers deep; strands subtending phloem of some vascular bundles (type Sv).

Vascular bundles scattered in midrib, numerous and weakly formed (type II), in type 3 aerenchyma (Fig. 115A). Xylem: wide protoxylem lacuna and clusters of 6–8 narrow metaxylem tracheids. Phloem: small laterally compressed strand. **Laticifers** anastomosing and occasional at secondary vascular bundles, usually only one where found (but see two in Fig. 115F). Laticifers uncommon in mesophyll, subepidermal abaxially. **Crystals**: raphides rare, in single bundles or multiple bundles in wide (Fig. 115D) or elongated (Fig. 115E) cells in abaxial mesophyll. Druses large or small, common in mesophyll. Crystal sand very common in spongy mesophyll and midrib.

Petiole TS

Outline deeply U-shaped proximally and subterete distally. **Size**: 6–20 mm diameter. **Cuticle** thin and mostly smooth. **Epidermis**: cells short and flat externally. **Collenchyma**: circular strands (Fig. 115G) on abaxial side separated from epidermis by two layers of ground tissue and subtending vascular bundles as phloem caps (type Sv). **Ground tissue**: aerenchyma type 3 proximally, composed of large cells with few air cavities distally.

Vascular bundles scattered and well developed (type II) proximally (Fig. 115H) and type III distally (Fig. 115G). Xylem proximally with irregular strand of 8–12 metaxylem cells bounding circular phloem strand. Distally, one wide protoxylem lacuna with 2–4 narrow metaxylem cells bounding circular phloem strand. **Sclerenchyma**: thick-walled fibres in independent strands, or mostly capping phloem proximally (Fig. 115H); fibres absent distally. **Laticifers**: large anastomosing cells at vascular bundles and ramifying within ground tissue. **Crystals**: raphides and druses, few and scattered.

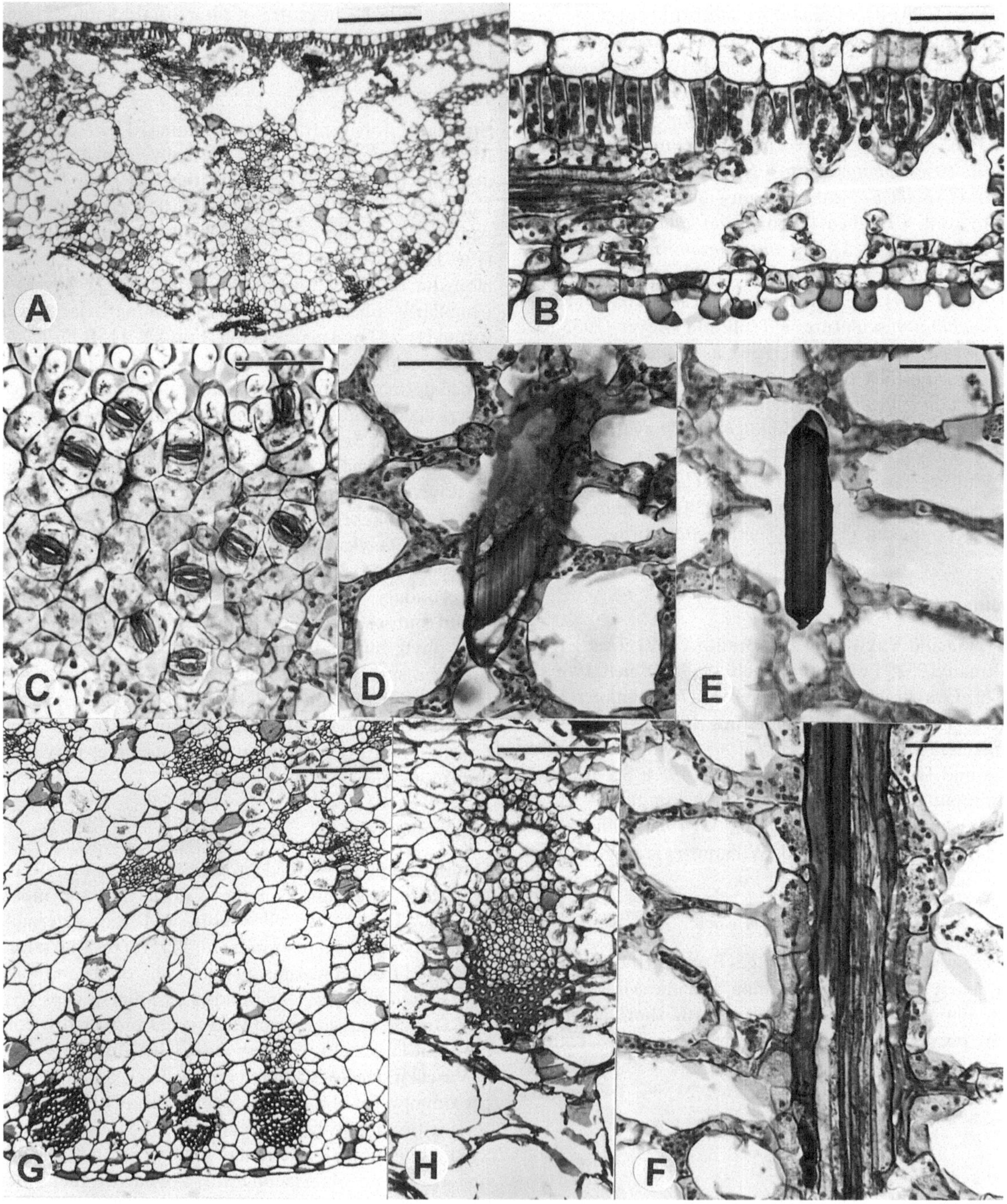

Fig. 115. *Steudnera colocasiifolia* leaf and petiole sections. (A) Midrib TS; (B) Lamina TS showing papillate abaxial epidermal cells; (C) Abaxial lamina PS showing stomata; (D,E) Mesophyll PS showing types 2 and 3 aerenchyma and raphide cells; (F) Mesophyll PS showing vascular bundle with laticifers on each side of it; (G) Distal petiole TS margin showing type Sv collenchyma strands and type III vascular bundles; (H) Proximal petiole TS showing type II vascular bundle. Scale bars: A = 300 μm; B–F = 50 μm; G = 200 μm; H = 150 μm.

Material examined

Steudnera colocasiifolia K.Koch, cult. M; *S. discolor* Bull, cult. M.

Species reported on in the literature

Deinega (1898): *S. colocasiifolia*: leaf. • Engler (1920b): *S. colocasiifolia*: root. • French (1987c): *S. colocasiifolia*: root. • French (1988): *S. virosa*: laticifers. • French and Tomlinson (1983): *S. colocasiifolia*: stem vasculature. • Lierau (1888): *S. colocasiifolia*, *S. discolor*: root. • Müller (1919): *S. griffithii*: leaf tip. • Solereder and Meyer (1928): *S. colocasiifolia*: anatomy. • Trécul (1865): *S. colocasiifolia*: vasculature.

Additional literature

French 1986a, c, 1987a; Grayum 1990; Meusel 1951; Thanikaimoni 1969; Tieghem 1907.

Protarum Engl.

The single species of this genus is found in the Seychelles Islands. Plants are terrestrial or epilithic, tuberous herbs adapted to the dry season of the forest understorey. The solitary leaves are digitately compound, pedisect or radiate, with elliptic segments. Chromosomes: $2n = 28$. One specimen was available for study.

Leaf surface

Cuticle smooth on both surfaces. **Epidermis**: cells polygonal with straight anticlinal walls or undulate on both surfaces (1.5–2.5:1 *l*/*w*). **Stomata** abaxial and brachyparacytic. Subsidiary cells broad and angular or rounded. **Venation**: leaflet midrib prominent. Secondaries broadly acute and straight, merging with straightened brochidodromous collector at margin. Tertiaries and higher orders reticulate.

Leaf TS

Structure dorsiventral. **Size**: (segment) 178×48 mm (*l*/*w*). **Midrib** broadly rounded-convex on both surfaces, somewhat ridged abaxially. **Cuticle** thin on both surfaces. **Epidermis**: cells very large adaxially, flat on surface and rounded internally. Cells short and flattened abaxially. **Stomata** with distinct large external ledges. **Mesophyll**: one palisade layer and *c*. 5 spongy layers. Spongy cells long-armed with large air spaces (type 2). **Air cavities**: up to 70% of spongy tissue. **Collenchyma** in large distinct ± circular strands, three subepidermal in adaxial midrib, five subepidermal abaxially; each strand subtending vascular bundle (type Sv).

Vascular bundles: eight small midrib bundles, weakly formed and symmetrically oriented (type II). Xylem: 2–3 narrow metaxylem cells and a small circular phloem strand. **Laticifers**: anastomosing, lateral to vascular bundles and ramifying into adjacent ground tissue. **Crystals**: biforine raphides in single bundles, in elongated thick-walled cells, common in any orientation in mid-mesophyll. Also, single bundles in thin-walled, irregularly shaped cells, common in mid-mesophyll. Raphide cells concentrated in spongy mesophyll. Druses in rounded cells, uncommon in palisade layer. Crystal sand common in palisade layer.

Material examined

Protarum sechellarum Engl., Gardiner 110, Seychelles, K.

Species reported on in the literature

French (1988): *P. sechellarum*: laticifers.

Additional literature

French 1986a,c, 1987a; Grayum 1990; Loh *et al.* 2000; Thanikaimoni 1969; Williams *et al.* 1981.

LISTS OF GENERA IN WHICH CERTAIN DIAGNOSTIC FEATURES OCCUR

The list includes genera where a character state was expressed in one or more species. Within genera with many species, character variation can be high. Also, a character state may not be uniformly present throughout an organ or in all species.

LEAF SURFACE: CUTICLE

Striate
Acorus, Amorphophallus, Anadendrum, Anchomanes, Anthurium, Anubias, Aridarum, Arisaema, Asterostigma, Biarum, Bucephalandra, Calla, Callopsis, Colocasia (midrib), *Cryptocoryne, Culcasia, Cyrtosperma, Dracontium, Eminium, Epipremnum, Filarum, Gearum, Gonatopus* (midrib), *Gorgonidium, Heteroaridarum, Holochlamys, Homalomena, Lagenandra, Lasia, Lasimorpha, Monstera, Montrichardia, Orontium, Philodendron* (subgenus *Philodendron), Phymatarum, Piptospatha, Pothos, Pseudodracontium, Pycnospatha, Remusatia, Rhaphidophora, Rhodospatha* (midrib), *Schismatoglottis, Scindapsus, Spathicarpa, Spathiphyllum, Stylochaeton* (midrib), *Symplocarpus, Synandrospadix, Taccarum, Typhonium, Ulearum, Urospatha, Xanthosoma, Zantedeschia, Zomicarpa.*

Rough or forming papillate mounds over midrib
Alocasia (on lamina), *Amorphophallus, Aridarum, Culcasia, Pseudohydrosme.*

Ridged, anastomosing or rugulate
Biarum, Cryptocoryne, Dieffenbachia, Dracontium, Furtadoa, Homalomena, Lagenandra, Monstera, Peltandra, Phymatarum Piptospatha, Schismatoglottis, Scindapsus, Spathiphyllum, Syngonium.

LEAF SURFACE: EPIDERMIS

Trichomes unicellular
Asterostigma, Dieffenbachia, Schismatoglottis.

Trichomes 2–3-celled
Lasia.

Trichomes multicellular
Homalomena, Pistia.

Cells elongated, parallel to leaf axis
Acorus, Gymnostachys.

Anticlinal cell walls sinuous or interlocking
Aglaonema, Alloschemone, Amorphophallus, Anaphyllum, Anchomanes, Anthurium, Arisaema, Arisarum, Callopsis, Cercestis, Culcasia, Cyrtosperma, Dieffenbachia, Dracontium, Gonatopus, Lasia, Lemna, Monstera, Nephthytis, Podolasia, Pseudohydrosme, Pycnospatha, Scindapsus, Spathiphyllum, Spirodela, Urospatha, Zamioculcas.

LEAF SURFACE: STOMATA

Anomocytic
Biarum, Dracunculus, Lemna, Orontium, Spirodela, Wolffia, Wolffiella.

Brachypara-hexacytic
Aglaodorum, Alloschemone, Alocasia, Amydrium, Anadendrum, Anaphyllum, Anthurium, Aridarum, Arisaema, Cercestis, Cryptocoryne, Culcasia, Dracontioides, Furtadoa, Heteroaridarum, Heteropsis, Holochlamys, Homalomena, Hottarum, Lagenandra, Mangonia, Monstera, Philodendron (subgenera *Philodendron, Meconostigma), Piptospatha, Phymatarum, Pothoidium, Pothos, Pseudodracontium, Rhaphidophora, Rhodospatha, Schismatoglottis, Scindapsus, Spathiphyllum, Stenospermation, Stylochaeton, Syngonium, Xanthosoma.*

Brachypara-octocytic or amphibrachy-paracytic
Aridarum, Bucephalandra, Furtadoa, Heteroaridarum, Lagenandra, Phymatarum.

Unipolar
Alocasia, Ambrosina, Asterostigma, Dracontium, Gonatopus, Helicodiceros, Holochlamys, Monstera, Podolasia, Pycnospatha, Scindapsus, Spathantheum.

LEAF TS: MIDRIB

Prominent adaxial ridge
Anthurium.

LEAF TS: CUTICLE

Irregularly thickened
Furtadoa, Rhaphidophora.

LEAF TS: EPIDERMIS

Papillate
Ariopsis, Arum, Chlorospatha, Colocasia, Culcasia, Lasia, Lemna, Lysichiton, Orontium, Remusatia, Sauromatum,

Schismatoglottis (cuticular), *Scindapsus*, *Spirodela*, *Steudnera*.

Stomata raised on mounds

Anthurium, *Biarum*, *Bucephalandra*, *Callopsis*, *Cryptocoryne*, *Dracunculus*, *Eminium*, *Lagenandra*.

LEAF TS: HYPODERMIS

In adaxial midrib

Aridarum, *Bucephalandra*, *Heteroaridarum*, *Homalomena*, *Philodendron* (subgenus *Philodendron*), *Remusatia*, *Syngonium*.

In adaxial lamina

Anthurium, *Scindapsus*, *Syngonium*, *Ulearum*.

LEAF TS: PALISADE LAYER

Weak or not differentiated

Acorus, *Ambrosina*, *Anaphyllum*, *Anthurium*, *Biarum*, *Callopsis*, *Cryptocoryne*, *Dieffenbachia*, *Filarum*, *Gymnostachys*, *Jasarum*, *Lazarum*, *Lemna*, *Pedicellarum*, *Spirodela*, *Wolffia*, *Wolffiella*, *Zomicarpa*, *Zomicarpella*.

Cells lobed, U- or H-shaped

Anaphyllum, *Anchomanes*, *Dracontium*, *Gonatopus*, *Holochlamys*, *Nephthytis*, *Pinellia*, *Pseudohydrosme*, *Pycnospatha*, *Scindapsus*, *Spathiphyllum*, *Taccarum*, *Zamioculcas*.

Cells T-shaped

Anthurium, *Arophyton*, *Culcasia*.

Large empty cells among uniseriate rows of palisade cells

Scaphispatha.

LEAF TS: SPONGY AERENCHYMA

Compact tissue or cell-sized air spaces, type 1, type 1a

Aglaodorum, *Aglaonema*, *Alocasia*, *Ambrosina*, *Amorphophallus*, *Anadendrum*, *Anaphyllopsis*, *Anchomanes*, *Anthurium*, *Anubias*, *Arisaema*, *Arophyton*, *Asterostigma*, *Biarum*, *Bucephalandra*, *Caladium*, *Callopsis*, *Cercestis*, *Cryptocoryne*, *Dracontioides*, *Dracontium*, *Dracunculus*, *Epipremnum*, *Filarum*, *Furtadoa*, *Gearum*, *Gonatopus*, *Gorgonidium*, *Gymnostachys*, *Helicodiceros*, *Heteropsis*, *Homalomena*, *Hottarum*, *Jasarum*, *Lagenandra*, *Lazarum*, *Mangonia*, *Monstera*, *Pedicellarum*, *Piptospatha*, *Pothoidium*, *Pseudodracontium*, *Pseudohydrosme*, *Remusatia*, *Rhaphidophora*, *Rhodospatha*, *Sauromatum*, *Scaphispatha*, *Scindapsus*, *Spathantheum*, *Spathicarpa*, *Spathiphyllum*, *Stenospermation*, *Synandrospadix*, *Taccarum*, *Theriophonum*, *Typhonium*, *Ulearum*, *Urospatha*, *Zamioculcas*, *Zomicarpella*.

Intercellular spaces 2–3 times cell diameters, type 2

Aglaonema, *Alloschemone*, *Alocasia*, *Ambrosina*, *Amydrium*, *Anchomanes*, *Anthurium*, *Anubias*, *Ariopsis*, *Arisaema*, *Arisarum*, *Arophyton*, *Arum*, *Carlephyton*, *Cercestis*, *Chlorospatha*, *Culcasia*, *Cyrtosperma*, *Dracontium*, *Dracunculus*, *Eminium*, *Epipremnum*, *Hapaline*, *Helicodiceros*, *Heteroaridarum*, *Heteropsis*, *Holochlamys*, *Homalomena*, *Lasia*, *Lasimorpha*, *Monstera*, *Montrichardia*, *Nephthytis*, *Philodendron* (subgenera *Philodendron*, *Pteromischum*), *Pinellia*, *Piptospatha*, *Pothos*, *Protarum*, *Pycnospatha*, *Rhaphidophora*, *Rhodospatha*, *Sauromatum*, *Schismatoglottis*, *Spathantheum*, *Steudnera*, *Syngonium*, *Taccarum*, *Typhonium*, *Xanthosoma*, *Zomicarpa*, *Zomicarpella*.

Intercellular spaces large with uniseriate partitions, type 3

Acorus, *Aglaodorum*, *Aglaonema*, *Anubias*, *Aridarum*, *Arisarum*, *Arophyton*, *Arum*, *Biarum*, *Caladium*, *Calla*, *Carlephyton*, *Chlorospatha*, *Colletogyne*, *Colocasia*, *Culcasia*, *Cyrtosperma*, *Dieffenbachia*, *Eminium*, *Gearum*, *Hapaline*, *Heteroaridarum*, *Holochlamys*, *Homalomena*, *Lemna*, *Lysichiton*, *Montrichardia*, *Orontium*, *Peltandra*, *Philodendron* (subgenera *Meconostigma*, *Philodendron*), *Phymatarum*, *Pistia*, *Podolasia*, *Remusatia*, *Schismatoglottis*, *Spirodela*, *Steudnera*, *Stylochaeton*, *Symplocarpus*, *Synandrospadix*, *Syngonium*, *Typhonodorum*, *Urospatha*, *Wolffia*, *Wolffiella*, *Xanthosoma*, *Zantedeschia*.

Intercellular spaces with multiseriate partitions, type 4 (mostly midribs)

Anaphyllum, *Colocasia*, *Cyrtosperma*, *Dracontium*, *Lasia*, *Lasimorpha*, *Podolasia*, *Pycnospatha*, *Schismatoglottis*.

LEAF TS: COLLENCHYMA

Banded, type B

Alloschemone, *Amydrium*, *Anadendrum*, *Anchomanes*, *Anthurium*, *Anubias*, *Aridarum*, *Bucephalandra*, *Cercestis*, *Culcasia*, *Cyrtosperma*, *Dracontium*, *Furtadoa*, *Gearum*, *Gorgonidium*, *Holochlamys*, *Homalomena*, *Hottarum*, *Lasia*, *Lasimorpha*, *Mangonia*, *Montrichardia*, *Nephthytis*, *Philodendron* (all subgenera), *Phymatarum*, *Piptospatha*, *Pycnospatha*, *Rhodospatha*, *Schismatoglottis*, *Scindapsus*, *Spathantheum*, *Spathicarpa*, *Spathiphyllum*, *Stylochaeton*, *Symplocarpus*, *Synandrospadix*, *Taccarum*, *Urospatha*, *Zamioculcas*.

Banded-interrupted, type Bi

Dieffenbachia, *Heteroaridarum*, *Nephthytis*, *Zantedeschia*.

Strands irregular, often flattened, type Sb

Aglaonema, *Aglaodorum*, *Nephthytis*, *Pseudohydrosme*.

Strands as phloem caps, type Svc

Alocasia, *Dieffenbachia*, *Lysichiton*.

Strands discrete, usually circular, type Sv

Alocasia, Ambrosina, Amorphophallus, Ariopsis, Arisaema, Arisarum, Arophyton, Arum, Biarum, Caladium, Carlephyton, Chlorospatha, Colletogyne, Colocasia, Cryptocoryne, Dracunculus, Eminium, Filarum, Hapaline, Helicodiceros, Lagenandra, Lazarum, Peltandra, Pinellia, Pistia, Protarum, Pseudodracontium, Remusatia, Sauromatum, Scaphispatha, Steudnera, Syngonium, Theriophonum, Typhonium, Typhonodorum, Ulearum, Xanthosoma, Zomicarpa, Zomicarpella.

LEAF TS: VASCULAR BUNDLES

Broad, phloem strand semicircular, type I

Acorus, Anadendrum, Anaphyllopsis, Anthurium, Anubias, Asterostigma, Calla, Callopsis, Cercestis, Culcasia, Cyrtosperma, Dracontioides, Dracontium, Epipremnum, Furtadoa, Gymnostachys, Heteropsis, Lasia, Lasimorpha, Lysichiton, Mangonia, Montrichardia, Orontium, Pedicellarum, Podolasia, Pothoidium, Pothos, Pycnospatha, Rhaphidophora, Rhodospatha, Scindapsus, Spathiphyllum, Stenospermation, Symplocarpus, Urospatha, Xanthosoma, Zamioculcas.

Narrow, phloem strand elliptical, type II

Aglaodorum, Aglaonema, Alloschemone, Alocasia, Ambrosina, Amorphophallus, Amydrium, Anaphyllum, Anchomanes, Aridarum, Arisaema, Arisarum, Arum, Asterostigma, Biarum, Bucephalandra, Caladium, Calla, Callopsis, Carlephyton, Colletogyne, Colocasia, Dieffenbachia, Dracontium, Dracunculus, Eminium, Filarum, Furtadoa, Gearum, Gonatopus, Gorgonidium, Hapaline, Helicodiceros, Heteroaridarum, Holochlamys, Homalomena, Hottarum, Lagenandra, Lazarum, Mangonia, Monstera, Montrichardia, Nephthytis, Peltandra, Philodendron (all subgenera), *Phymatarum, Piptospatha, Pistia, Protarum, Pseudodracontium, Pseudohydrosme, Remusatia, Rhaphidophora, Sauromatum, Schismatoglottis, Scindapsus, Spathantheum, Spathicarpa, Stenospermation, Steudnera, Stylochaeton, Synandrospadix, Syngonium, Theriophonum, Ulearum, Xanthosoma, Zomicarpa, Zomicarpella.*

Small, xylem a wide cell or lacuna, type III

Alocasia, Ariopsis, Arisaema, Arisarum, Arophyton, Chlorospatha, Colocasia, Cryptocoryne, Jasarum, Peltandra, Pinellia, Pistia, Pseudodracontium, Sauromatum, Scaphispatha, Schismatoglottis, Taccarum, Typhonium, Typhonodorum, Ulearum, Zantedeschia, Zomicarpa.

LEAF TS: SCLERENCHYMA FIBRES

Ensheathing vascular bundles

Acorus, Alloschemone, Anadendrum, Anthurium, Anubias, Asterostigma, Cercestis, Culcasia, Epipremnum, Gymnostachys, Heteropsis, Monstera, Pedicellarum, Philodendron (subgenera *Pteromischum, Meconostigma), Podolasia, Pothoidium, Pothos, Rhaphidophora, Rhodospatha, Scindapsus, Spathiphyllum, Stenospermation.*

Phloem caps

Acorus, Amydrium, Anthurium, Aridarum, Asterostigma, Callopsis, Cercestis, Cyrtosperma, Dieffenbachia, Dracontioides, Dracontium, Furtadoa, Heteroaridarum, Homalomena, Hottarum, Lasia, Lasimorpha, Lazarum, Montrichardia, Nephthytis, Orontium, Peltandra, Philodendron (all subgenera), *Phymatarum, Piptospatha, Podolasia, Pycnospatha, Rhaphidophora, Rhodospatha, Scindapsus, Syngonium, Urospatha, Xanthosoma.*

Xylem caps

Amydrium, Anthurium, Cercestis, Cyrtosperma, Dracontioides, Dracontium, Furtadoa, Homalomena, Lasia, Montrichardia, Nephthytis, Philodendron (all subgenera), *Podolasia, Pycnospatha, Scindapsus, Urospatha, Xanthosoma.*

Fibres showing concentric wall layering

Anthurium, Anubias, Heteropsis, Pedicellarum, Philodendron (subgenus *Philodendron*).

LEAF TS: SCLERENCHYMA TRICHOSCLEREIDS

Unaggregated

Epipremnum, Monstera, Rhaphidophora, Rhodospatha, Scindapsus, Spathiphyllum, Stenospermation.

Aggregated

Alloschemone, Holochlamys, Spathiphyllum.

LEAF TS: SECRETORY FEATURES

Oil cells in epidermis

Acorus.

Glands present at surface

Anthurium, Urospatha.

Ducts with epithelial cells

Cercestis, Culcasia, Furtadoa, Homalomena, Philodendron (all subgenera), *Rhaphidophora, Schismatoglottis, Syngonium.*

Copious exudates, no specialized ducts

Monstera, Rhodospatha, Scindapsus, Ulearum.

Laticifers non-anastomosing

Aglaodorum, Aglaonema, Alocasia, Ambrosina, Amorphophallus, Anchomanes, Anubias, Aridarum, Ariopsis, Arisaema, Arisarum, Arophyton, Arum, Asterostigma, Biarum, Bucephalandra, Calla, Callopsis, Carlephyton, Cercestis, Colletogyne, Colocasia, Cryptocoryne, Dracunculus, Eminium, Furtadoa, Helicodiceros, Heteroaridarum, Homalomena, Hottarum, Lagenandra,

Montrichardia, Nephthytis, Orontium, Peltandra, Philodendron (all subgenera), *Phymatarum, Pinellia, Piptospatha, Pseudodracontium, Remusatia, Rhodospatha, Sauromatum, Schismatoglottis, Scindapsus, Spathicarpa, Synandrospadix, Taccarum, Theriophonum, Typhonium, Typhonodorum, Zantedeschia.*

Laticifers anastomosing

Alocasia, Caladium, Chlorospatha, Colocasia, Filarum, Hapaline, Jasarum, Protarum, Remusatia, Scaphispatha, Steudnera, Syngonium, Ulearum, Xanthosoma, Zomicarpa, Zomicarpella.

LEAF TS: CALCIUM OXALATE CRYSTALS

Raphide cells elongated

Aglaonema, Alloschemone, Ambrosina, Amorphophallus, Amydrium, Anadendrum, Anaphyllopsis, Anaphyllum, Anchomanes, Anthurium, Ariopsis, Arisaema, Arisarum, Biarum, Callopsis, Cercestis, Culcasia, Cryptocoryne, Cyrtosperma, Dieffenbachia, Dracontioides, Dracontium, Dracunculus, Eminium, Epipremnum, Filarum, Gonatopus, Gorgonidium, Gymnostachys, Helicodiceros, Heteropsis, Holochlamys, Homalomena, Lagenandra, Lasia, Lasimorpha, Lazarum, Lemna, Mangonia, Monstera, Nephthytis, Orontium, Pedicellarum, Philodendron (subgenus *Philodendron*), *Phymatarum, Pinellia, Pothoidium, Pothos, Pseudodracontium, Pseudohydrosme, Pycnospatha, Remusatia, Rhaphidophora, Rhodospatha, Sauromatum, Scindapsus, Spathantheum, Spathicarpa, Spathiphyllum, Spirodela, Stenospermation, Steudnera, Symplocarpus, Taccarum, Theriophonum, Typhonium, Ulearum, Urospatha, Zantedeschia, Zomicarpa, Zomicarpella.*

Raphide cells articulated end to end

Asterostigma, Biarum, Lemna, Spirodela, Taccarum.

Raphides in oblique or spiral or multiple overlapping bundles

Alloschemone, Amydrium, Anadendrum, Anaphyllopsis (masses), *Anchomanes, Anthurium, Arisaema, Asterostigma, Biarum, Caladium, Cryptocoryne, Culcasia, Eminium, Epipremnum, Gonatopus, Helicodiceros, Heteropsis, Holochlamys, Lagenandra, Lasimorpha, Lazarum, Lemna, Mangonia, Monstera, Pedicellarum, Phymatarum, Pothoidium, Pothos, Pseudodracontium, Pseudohydrosme, Pycnospatha, Rhaphidophora, Rhodospatha, Sauromatum, Scindapsus, Spathiphyllum, Spirodela, Theriophonum, Ulearum, Zantedeschia, Zomicarpella.*

Raphide bundles 2 or more per cell (but not as above)

Ambrosina, Ariopsis, Biarum, Cercestis, Cyrtosperma, Dracontioides, Dracontium, Dracunculus, Eminium, Filarum, Gorgonidium, Gymnostachys, Homalomena, Lasia, Lazarum, Mangonia, Orontium, Pedicellarum, Pinellia, Podolasia, Remusatia, Sauromatum, Spathantheum, Spathicarpa, Stenospermation, Steudnera, Synandrospadix, Taccarum, Theriophonum, Typhonium, Urospatha, Zamioculcas, Zomicarpa.

Raphide cells vertical in palisade tissue

Amorphophallus, Arisaema, Arum, Montrichardia.

Raphides coarse

Amorphophallus, Gonatopus, Nephthytis, Zamioculcas.

Raphide cells biforine-like

Alocasia, Arophyton, Bucephalandra, Calla, Carlephyton, Cercestis, Chlorospatha, Culcasia, Cyrtosperma, Hapaline, Homalomena, Nephthytis, Philodendron (subgenus *Meconostigma*), *Taccarum, Zantedeschia.*

Raphide cells biforine

Aglaodorum, Aglaonema, Anchomanes, Anubias, Aridarum, Ariopsis, Arophyton, Caladium, Colletogyne, Colocasia, Dieffenbachia, Gearum, Peltandra, Philodendron (subgenus *Philodendron*), *Piptospatha, Pistia, Protarum, Pseudohydrosme, Remusatia, Scaphispatha, Schismatoglottis, Scindapsus, Stylochaeton, Symplocarpus, Syngonium, Typhonodorum, Xanthosoma, Zomicarpa, Zomicarpella.*

Druses coarse, bearing elongated, surface oriented, prismatic points

Philodendron (subgenus *Pteromischum*), *Pothoidium, Pothos, Syngonium.*

Prismatics in mesophyll and at vein sheaths

Acorus.

Prismatics subepidermal

Symplocarpus.

Prismatics along vascular bundles

Pedicellarum, Pothoidium, Pothos.

Styloids

Pycnospatha.

PETIOLE TS: OUTLINE

Abaxially ridged (fluted)

Ariopsis, Bucephalandra, Filarum, Theriophonum.

PETIOLE TS: CUTICLE

Striate

Alloschemone, Amorphophallus, Anadendrum, Arophyton, Asterostigma, Biarum, Bucephalandra, Caladium, Carlephyton, Chlorospatha, Colletogyne, Colocasia, Cryptocoryne, Culcasia, Dieffenbachia, Dracontium, Eminium, Filarum, Furtadoa, Gonatopus, Hapaline, Helicodiceros, Holochlamys, Homalomena, Lagenandra, Lasimorpha, Montrichardia, Peltandra, Phymatarum, Piptospatha, Pothos, Pseudodracontium, Remusatia, Sauromatum,

Scaphispatha, Schismatoglottis, Spathicarpa, Spathiphyllum, Stylochaeton, Taccarum, Typhonium, Typhonodorum, Ulearum, Urospatha, Xanthosoma, Zantedeschia, Zomicarpa.

Rugulate

Phymatarum (proximal), *Scindapsus.*

Ornamented or papillate

Aglaodorum, Alocasia, Caladium, Lysichiton, Monstera, Peltandra, Philodendron (subgenera *Philodendron, Pteromischum), Remusatia, Rhaphidophora, Rhodospatha, Syngonium, Xanthosoma.*

PETIOLE TS: CELLULAR SURFACE ELEVATIONS

Trichomes unicellular

Callopsis.

Trichomes multicellular

Pistia.

Spines or prickles

Cyrtosperma, Dracontium, Lasia, Lasimorpha, Podolasia, Pycnospatha.

PETIOLE TS: COLLENCHYMA

Banded, type B

Amydrium, Anaphyllum, Anthurium, Anubias, Aridarum, Asterostigma, Callopsis, Cercestis, Culcasia, Cyrtosperma, Dracontium, Epipremnum, Furtadoa, Gearum, Gonatopus, Heteroaridarum, Holochlamys, Homalomena, Hottarum, Lasia, Lasimorpha, Mangonia, Monstera, Philodendron (all subgenera), *Phymatarum, Piptospatha, Rhaphidophora, Schismatoglottis, Scindapsus, Spathantheum, Spathiphyllum, Stenospermation, Stylochaeton, Symplocarpus* (intermittent), *Synandrospadix, Taccarum, Urospatha, Zamioculcas.*

Banded interrupted, type Bi

Dieffenbachia, Scaphispatha, Spathicarpa, Zantedeschia.

Irregular flattened strands, non-aligned with vascular bundles, type Sb

Aglaodorum, Aglaonema, Anchomanes, Nephthytis.

Bundle caps, type Svc

Amydrium, Anadendrum, Anubias, Epipremnum, Gonatopus, Lysichiton, Monstera, Rhaphidophora, Scindapsus, Symplocarpus, Ulearum.

Strands aligned with vascular bundles, type Sv

Alocasia, Ambrosina, Amorphophallus, Ariopsis, Arisaema, Arisarum, Arophyton, Arum, Biarum, Bucephalandra, Caladium, Carlephyton, Chlorospatha, Colletogyne, Colocasia, Cryptocoryne, Dracunculus, Eminium, Filarum, Hapaline, Helicodiceros, Jasarum, Lagenandra, Lazarum, Peltandra, Pinellia, Pistia, Pseudodracontium, Remusatia, Sauromatum, Steudnera, Syngonium, Theriophonum, Typhonium, Typhonodorum, Ulearum, Xanthosoma, Zomicarpa, Zomiarpella.

Ensheathing vascular bundles

Anadendrum.

PETIOLE TS: WITH CENTRAL ZONE

Alloschemone, Amydrium, Anthurium, Monstera.

PETIOLE TS: GROUND TISSUE

Cells compact in centre

Ambrosina, Amydrium, Anadendrum, Anthurium, Anubias, Ariopsis, Callopsis, Epipremnum, Filarum, Furtadoa, Gonatopus, Heteropsis, Holochlamys, Homalomena, Lagenandra, Lazarum, Nephthytis, Pedicellarum, Pothoidium, Pothos, Philodendron (subgenus *Pteromischum), Rhaphidophora, Sauromatum, Scindapsus, Spathicarpa, Stylochaeton, Typhonium, Ulearum, Zamioculcas, Zomicarpella.*

Cells loosely packed (cell-sized cavities), type 1 and 2

Alloschemone, Bucephalandra, Cercestis, Culcasia, Lagenandra, Monstera, Piptospatha, Rhodospatha, Spathiphyllum, Stenospermation, Stylochaeton, Taccarum, Zomicarpa.

Uniseriate partitions separating cavities, type 3

Aglaodorum, Aglaonema, Amorphophallus, Anubias, Aridarum, Arisaema, Arisarum, Arophyton, Arum, Asterostigma, Biarum, Caladium, Calla, Carlephyton, Cercestis, Chlorospatha, Colletogyne, Colocasia, Culcasia, Cryptocoryne, Cyrtosperma, Dieffenbachia, Dracunculus, Eminium, Gearum, Hapaline, Helicodiceros, Heteroaridarum, Homalomena, Hottarum, Jasarum, Lagenandra, Lysichiton, Mangonia, Montrichardia, Orontium, Peltandra, Philodendron (subgenera *Philodendron, Pteromischum), Phymatarum, Pistia, Podolasia, Pothos, Pseudodracontium, Remusatia, Scaphispatha, Schismatoglottis, Scindapsus, Spathantheum, Spathiphyllum, Stenospermation, Steudnera, Stylochaeton, Symplocarpus, Synandrospadix, Syngonium, Taccarum, Theriophonum, Typhonodorum, Xanthosoma, Zantedeschia, Zomicarpa.*

Multiseriate partitions separating cavities, type 4

Alocasia, Ambrosina, Anaphyllum, Anchomanes, Arisaema, Arisarum, Arophyton, Caladium, Carlephyton, Chlorospatha, Colocasia, Cyrtosperma, Dracontioides, Dracontium, Hapaline, Heteroaridarum, Lasia, Lasimorpha, Monstera, Philodendron (subgenera *Meconostigma, Philodendron), Pinellia, Pistia, Podolasia, Pycnospatha, Schismatoglottis, Spathiphyllum, Syngonium, Urospatha, Xanthosoma.*

PETIOLE TS: VASCULAR BUNDLES

Large, type I
Alloschemone, Amydrium, Anadendrum, Anaphyllum, Anthurium, Calla, Callopsis, Cercestis, Culcasia, Cyrtosperma, Dracontium, Eminium, Epipremnum, Heteropsis, Lasia, Lasimorpha, Lysichiton, Mangonia, Orontium, Pedicellarum, Philodendron (subgenera *Philodendron, Meconostigma*), *Podolasia, Pothoidium, Pothos, Rhaphidophora, Rhodospatha, Scindapsus, Spathiphyllum, Stylochaeton, Symplocarpus, Urospatha.*
Medium-sized, type II
Aglaodorum, Aglaonema, Ambrosina, Amydrium, Anchomanes, Anubias, Arisaema, Arisarum, Arum, Biarum, Bucephalandra, Cercestis, Colocasia, Cryptocoryne, Dieffenbachia, Dracontioides, Dracontium, Dracunculus, Eminium, Filarum, Furtadoa, Gearum, Gonatopus, Hapaline, Helicodiceros, Heteroaridarum, Holochlamys, Homalomena, Hottarum, Lazarum, Mangonia, Monstera, Nephthytis, Philodendron (all subgenera), *Phymatarum, Piptospatha, Pistia, Pycnospatha, Remusatia, Schismatoglottis, Scindapsus, Spathantheum, Spathicarpa, Spathiphyllum, Stenospermation, Steudnera, Stylochaeton, Synandrospadix, Syngonium, Taccarum, Theriophonum, Typhonium, Ulearum, Xanthosoma, Zamioculcas, Zomicarpella.*
Small, with dominant single xylem element, type III
Aglaonema, Alocasia, Amorphophallus, Anubias, Aridarum, Ariopsis, Arisaema, Arisarum, Arophyton, Asterostigma, Caladium, Carlephyton, Chlorospatha, Colletogyne, Colocasia, Dieffenbachia, Dracunculus, Jasarum, Lagenandra, Montrichardia, Nephthytis, Peltandra, Philodendron (subgenera *Philodendron, Pteromischum*), *Pinellia, Pseudodracontium, Remusatia, Sauromatum, Scaphispatha, Spathantheum, Spathicarpa, Steudnera, Synandrospadix, Syngonium, Taccarum, Typhonium, Typhonodorum, Xanthosoma, Zantedeschia, Zomicarpa, Zomicarpella.*

PETIOLE TS: SCLERENCHYMA FIBRES

Band separating inner and outer ground tissue
Alloschemone, Monstera, Rhodospatha.
Ensheathing vascular bundles
Alloschemone, Anchomanes, Anthurium, Callopsis, Culcasia, Cyrtosperma, Heteropsis, Monstera, Pedicellarum, Philodendron (subgenera *Philodendron, Pteromischum*), *Pothoidium, Pothos, Rhodospatha, Spathiphyllum.*
Phloem caps
Anaphyllum, Anchomanes, Anthurium, Anubias, Cercestis, Colocasia, Cyrtosperma, Dracontioides, Dracontium, Gonatopus, Holochlamys, Homalomena, Lasia, Lasimorpha, Monstera, Montrichardia, Nephthytis, Philodendron (subgenera *Philodendron, Meconostigma*), *Phymatarum, Podolasia, Pycnospatha, Rhodospatha, Scaphispatha, Scindapsus, Spathiphyllum, Stenospermation, Steudnera, Urospatha.*
Xylem caps
Anchomanes, Cercestis, Cyrtosperma, Philodendron (subgenus *Philodendron*), *Podolasia, Pycnospatha, Spathiphyllum.*
Fibres showing concentric wall layering
Pedicellarum, Pothos.

PETIOLE TS: SCLERENCHYMA TRICHOSCLEREIDS

Unaggregated
Alloschemone, Epipremnum, Holochlamys, Monstera, Pothos, Rhaphidophora, Rhodospatha, Scindapsus.
Aggregated
Alloschemone, Holochlamys, Spathiphyllum.

PETIOLE TS: SECRETORY FEATURES

Ducts, mostly with epithelial cells
Cercestis, Colocasia, Culcasia, Furtadoa, Homalomena, Philodendron (all subgenera), *Piptospatha, Remusatia, Rhaphidophora* (no epithelium). *Schismatoglottis, Stenospermation.*
Mucilage in ground tissue cavities
Aridarum, Heteroaridarum, Rhodospatha (ducts, no epithelium), *Ulearum.*
Laticifers non-anastomosing
Aglaodorum, Aglaonema, Alocasia, Amorphophallus, Anchomanes, Anubias, Ariopsis, Arophyton, Arum, Biarum, Bucephalandra, Calla, Callopsis, Carlephyton, Cercestis, Colletogyne, Cryptocoryne, Eminium, Furtadoa, Helicodiceros, Heteroaridarum, Homalomena, Hottarum, Lagenandra, Nephthytis, Orontium, Peltandra, Philodendron (subgenera *Philodendron, Pteromischum*), *Phymatarum, Piptospatha, Pseudodracontium, Remusatia, Rhodospatha, Schismatoglottis, Scindapsus, Stenospermation, Zantedeschia.*
Laticifers anastomosing
Alocasia, Caladium, Chlorospatha, Colocasia, Filarum, Hapaline, Jasarum, Remusatia, Scaphispatha, Steudnera, Syngonium, Xanthosoma, Zomicarpa, Zomicarpella.

PETIOLE TS: STARCH

Compound crystalloids in central zone ground tissue
Amydrium.

PETIOLE TS: CALCIUM OXALATE CRYSTALS

Raphide cells elongated or tubular
Alocasia, Amydrium, Cryptocoryne, Cyrtosperma, Dracontium, Gearum, Monstera, Pedicellarum, Podolasia, Pycnospatha, Remusatia, Rhodospatha, Spathicarpa, Taccarum, Ulearum, Zomicarpella.
Raphide crystals in elongated and overlapping bundles
Amydrium, Cyrtosperma, Epipremnum, Ulearum, Urospatha, Zomicarpella.
Raphide bundles 2 or more per cell
Alocasia, Cryptocoryne, Dracontium, Gearum, Podolasia, Pycnospatha.
Raphide cells biforine
Aglaodorum, Aglaonema, Alocasia, Philodendron (subgenera *Philodendron, Meconostigma), Pistia, Remusatia, Scaphispatha, Schismatoglottis, Syngonium, Typhonodorum, Zantedeschia* (rare), *Zomicarpa.*
Raphide cells biforine-like
Cercestis, Colocasia, Dracontium, Peltandra, Philodendron (subgenus *Philodendron*), *Pistia.*
Druses bearing enlarged surface oriented points
Philodendron (subgenus *Pteromischum*), *Pothos.*
Prismatic crystals at bundle sheaths
Pedicellarum, Pothos.
Prismatic crystals aggregated in ground tissue
Alloschemone.

STEM ANATOMY (36 GENERA)

STEM SURFACE: CUTICLE

Smooth
Acorus, Aglaonema, Anadendrum, Anthurium, Bucephalandra, Cercestis, Dieffenbachia, Gonatopus, Heteropsis, Lasia, Monstera, Pedicellarum, Peltandra, Philodendron (subgenera *Philodendron, Pteromischum), Pothoidium, Pothos, Rhaphidophora, Schismatoglottis, Scindapsus, Spathiphyllum, Stenospermation, Symplocarpus, Xanthosoma, Zamioculcas.*
Striate or irregular
Alloschemone, Culcasia, Philodendron (subgenus *Philodendron*, rough).

STEM SURFACE: EPIDERMAL CELLS

Thick-walled (lignified)
Culcasia (outer walls), *Epipremnum, Montrichardia.*
Papillate
Acorus, Anadendrum, Philodendron (subgenus *Pteromischum*).

STEM TS: OUTLINE

Ridges around circumference
Epipremnum (3–5), *Pothoidium* (4).

STEM TS: HYPODERMIS

Thick-walled cells
Heteropsis, Stenospermation, Syngonium (U-shaped).
Separated from inner cortex by fibre band
*Philodendron (*subgenus *Philodendron).*

STEM TS: COLLENCHYMA

Caps over phloem (type Svc)
Symplocarpus, Ulearum.
Banded, at or near periphery (type B)
Anadendrum, Anthurium, Cercestis, Culcasia, Dieffenbachia, Philodendron (subgenera *Philodendron, Pteromischum), Symplocarpus, Syngonium, Xanthosoma.*
Band interrupted or incomplete
Aglaonema, Schismatoglottis.
Strands (type Sv)
Peltandra, Ulearum.

STEM TS: GROUND TISSUE

Compact
Aglaonema, Anadendrum, Anthurium, Bucephalandra, Culcasia, Dieffenbachia, Epipremnum, Gonatopus, Heteropsis, Lasia, Monstera, Montrichardia (outer ground tissue), *Pedicellarum, Peltandra* (outer ground tissue), *Podolasia, Pothoidium, Pothos, Pycnospatha, Rhaphidophora, Rhodospatha, Sauromatum, Scindapsus, Spathiphyllum, Syngonium, Ulearum, Zamioculcas, Zantedeschia, Zomicarpella.*
Types 1–2 aerenchyma (cell-sized cavities or cells loosely packed)
Alloschemone, Cercestis, Gearum, Lasia, Montrichardia (at centre), *Philodendron* (subgenera *Philodendron, Pteromischum), Podolasia, Schismatoglottis, Stenospermation.*
Type 3 aerenchyma in cortex or as noted
Acorus, Cercestis, Peltandra (central cylinder), *Spathiphyllum, Symplocarpus.*
Central ground tissue type 4 aerenchyma
Lasia, Xanthosoma.

STEM TS: FIBRES

Ensheathing vascular bundles

Acorus, Alloschemone, Anadendrum, Culcasia (in central cylinder), *Monstera* (in central cylinder), *Pedicellarum, Pothos, Rhodospatha* (cortex only), *Stenospermation* (cortex only).
Caps over phloem
Anadendrum, Anthurium, Epipremnum, Heteropsis, Monstera, Montrichardia, Philodendron (subgenus *Philodendron*), *Podolasia, Pothoidium, Rhaphidophora, Rhodospatha, Scindapsus, Syngonium, Zamioculcas.*
Strand in centre of compound bundles
Philodendron (subgenera *Philodendron, Pteromischum*), *Podolasia, Rhodospatha, Stenospermation.*
Fibres showing concentric wall layering (especially in central cylinder)
Anthurium, Heteropsis, Monstera, Pedicellarum, Philodendron, Pothos, Rhaphidophora, Scindapsus, Stenospermation.

STEM TS: SCLEREIDS IN CORTEX

Trichosclereids
Alloschemone, Epipremnum, Monstera.

STEM TS: CENTRAL CYLINDER

Absent or undefined
Aglaonema, Gearum, Lasia, Podolasia, Pycnospatha, Schismatoglottis, Symplocarpus, Ulearum, Xanthosoma, Zantedeschia, Zomicarpella.
Defined by vascular bundle concentration
Anadendrum, Anthurium, Bucephalandra, Cercestis, Montrichardia (uneven concentration of bundles), *Philodendron* (subgenera *Philodendron, Pteromischum*), *Rhaphidophora, Spathiphyllum* (plus starch cylinder), *Syngonium.*
Defined by lignified sclereid layer outside fibres
Monstera, Pothoidium, Rhaphidophora.
Defined by fibre layer
Alloschemone, Anadendrum (on one side), *Anthurium, Culcasia, Heteropsis, Monstera, Pedicellarum, Pothoidium, Pothos, Rhaphidophora* (on one side), *Rhodospatha, Scindapsus, Stenospermation.*
Defined by confluent fibre caps
Culcasia, Rhaphidophora.
Defined by starch layer
Gonatopus, Sauromatum (starch cylinder), *Schismatoglottis, Zamioculcas* (starch only in central cylinder), *Zantedeschia* (starch only in central cylinder).
Defined by central aerenchyma
Acorus, Peltandra.
Defined by endodermal casparian thickenings
Acorus, Epipremnum.

STEM TS: VASCULAR BUNDLES

Amphivasal in stem centre
Acorus, Bucephalandra, Dieffenbachia, Gearum, Lasia, Podolasia, Schismatoglottis, Spathiphyllum, Syngonium, Ulearum, Zantedeschia (discontinuously amphivasal).
Compound in stem centre
Aglaonema, Dieffenbachia, Montrichardia, Philodendron (subgenera *Philodendron, Pteromischum*), *Podolasia, Rhaphidophora, Rhodospatha, Schismatoglottis, Stenospermation, Zomicarpella.*
Bicollateral in stem centre
Alloschemone.
Type I in cortex or as noted
Acorus, Alloschemone (central cylinder), *Anadendrum* (central cylinder), *Anthurium, Culcasia* (central cylinder), *Epipremnum, Monstera, Pedicellarum* (throughout), *Pothos* (throughout), *Rhaphidophora* (throughout), *Rhodospatha, Spathiphyllum, Stenospermation, Symplocarpus* (throughout), *Syngonium, Zamioculcas* (central cylinder).
Type II in cortex or as noted
Aglaonema, Alloschemone (central cylinder), *Cercestis* (central cylinder), *Culcasia* (central cylinder), *Epipremnum, Gearum, Gonatopus* (throughout), *Heteropsis* (throughout), *Monstera* (central cylinder), *Philodendron* (subgenus *Pteromischum*), *Pothoidium* (throughout), *Pothos* (throughout), *Pycnospatha* (throughout), *Rhaphidophora* (throughout), *Scindapsus* (throughout), *Syngonium, Ulearum, Zamioculcas.*
Type III in cortex or as noted
Alloschemone (central cylinder), *Cercestis* (central cylinder), *Epipremnum, Montrichardia* (throughout), *Peltandra* (throughout), *Philodendron* (subgenus *Pteromischum*), *Pothoidium* (throughout), *Pothos* (throughout), *Rhaphidophora* (throughout), *Scindapsus* (central cylinder), *Xanthosoma.*

STEM TS: SECRETORY FEATURES

Cortical secretory ducts
Aglaonema, Cercestis, Culcasia, Gonatopus, Philodendron (subgenera *Philodendron, Pteromischum*), *Rhaphidophora, Schismatoglottis, Stenospermation.*
Oil cells
Acorus.
Laticifers non-anastomosing
Cercestis, Peltandra, Philodendron (subgenera *Philodendron, Pteromischum*), *Schismatoglottis.*
Tannin cells
Acorus, Aglaonema, Anthurium, Bucephalandra, Cercestis, Lasia, Montrichardia, Philodendron (subgenus

Philodendron), *Podolasia*, *Rhodospatha*, *Schismatoglottis*, *Stenospermation*, *Syngonium*.

STEM TS: CALCIUM OXALATE CRYSTALS

Prismatics in ground tissue
Acorus, *Pedicellarum*, *Pothoidium*, *Pothos*.
Prismatics associated with brachysclereids
Monstera, *Rhaphidophora*.
Prismatics associated with central cylinder fibre sheath
Pothos.
Raphides in single bundles in cells little modified from ground tissue
Aglaonema, *Alloschemone*, *Anadendrum*, *Anthurium*, *Culcasia*, *Dieffenbachia*, *Epipremnum*, *Gonatopus*, *Montrichardia*, *Pedicellarum*, *Philodendron* (subgenera *Philodendron* (elongate cells), *Pteromischum*), *Pothoidium*, *Pycnospatha*, *Rhaphidophora*, *Rhodospatha* (axially elongated), *Schismatoglottis*, *Scindapsus*, *Symplocarpus*, *Ulearum*, *Zomicarpella*.
Raphides in large circular cells surrounded by radiating parenchyma
Spathiphyllum.
Raphides in multiple bundles in elongated cells
Gonatopus (cells large, not elongated), *Heteropsis*, *Lasia*, *Rhodospatha*, *Ulearum*, *Zamioculcas* (rounded cells).
Raphides in oblique-overlapping bundles in elongated cells
Philodendron (subgenus *Philodendron*), *Scindapsus*, *Stenospermation* (single bundles).
Raphide cells biforines
Schismatoglottis.
Raphide cells biforine-like
Aglaonema, *Cercestis*, *Peltandra*.

BIBLIOGRAPHY

ABRANOWICZ, E. (1912). Über das Wachstum der Knollen von *Sauromatum guttatum* Schott und *Amorphophallus rivieri* Durieu. *Öst. bot. Z.* **62,** 449–458.

AISSLINGER, H. (1907). Beiträge zur Kenntnis wenig bekannter Pflanzenfasern. Diss., Zurich. 135 pp.

ALLEN, R. N. (1929). Photomicrographs of Philippine starches. *Philipp. J. Sci.* **38**, 241–258.

AL-RAIS, A. H., MYERS, A. and WATSON, L. (1971). The isolation and properties of oxalate crystals from plants. *Ann. Bot.* **35**, 1213–1218.

ALVIM, P. DE, MORRETES, T. and LANGE, B. DE (1969). Mecanismo estomatico de *Alocasia macrorhiza variegata* em relacao a anatomia foliar e ao papel desempenhado pelo parenquima clorofilado. *Bol. Fac. Filosof., Cienc., Letras Univ. São Paulo*, No. 331, Bot. No. **24**, 239–252. [English summary]

AMBRONN, H. (1879–81). Über die Entwickelungsgeschichte und die mechanischen Eigenschaften des Collenchyms. *Jb. wiss. Bot.* **12**, 473–541.

AMELUNXEN, F. and GRONAU, G. (1968). Elektronenmikroskopische Untersuchungen an den Ölzellen von *Acorus calamus* L. *Z. Pflanzenphysiol.* **60**, 156–168.

——and——(1969). Untersuchungen an den Gerbstoffzellen der Niederblätter von *Acorus calamus* L. *Cytobiologie* **1 (1)**, 58–59.

ANANDAKUMAR, A., RAJENDRAN, V., THIRUGNANASABHAN, P., BALASUBRAMANIUM, M. and MURALIDHARAN, R. (1981). 'Ativisha', its botanical identity. *J. Sci. Res. Pl. Med.* **2**, 43–46.

——,——,——,——and——(1982). The pharmacognosy of Nattu Atividayam - the corms of *Cryptocoryne spiralis* Fisch. *Ancient Sci. Life* **1**, 200–205.

ANDERSON, J. L., THOMSON, W. W. and SWADER, J. A. (1973). Fine structure of *Wolffia arrhiza*. *Can. J. Bot.* **51**, 1619–1622.

ANDERSSON, S. (1888). Om de primära kärlsträngarnes utveckling hos monokotyledonerna. *Bih. svenska Vet.-Akad. Handl.* **13**, No. 12, 23 pp.

ANGIOSPERM PHYLOGENY GROUP (1998). An ordinal classification for the families of flowering plants. *Ann. Missouri bot. Gard.* **85**, 531–553.

ANTONY, A. (1898). Sulla struttura e sulla funzione degli stomi nelle appendici del perianzio e nelle antere. *Bull. Soc. bot. ital.* 170–178.

ARAUS, J. L., ALEGRE, L., TAPIA, L., CALAFELL, R. and SERRET, M. D. (1986a). Relationships between photosynthetic capacity and leaf structure in several shade plants. *Amer. J. Bot.* **73**, 1760–1770.

——, SABIDÓ, J. and AGUILÁ, F. J. (1986b). Structural differences between green and white sectors of variegated *Scindapsus aureus* leaves. *J. Amer. Soc. hort. Sci.* **111**, 98–102.

ARBER, A. (1918a). The phyllode theory of the monocotyledonous leaf, with special reference to anatomical evidence. *Ann. Bot.*, Lond. **32**, 465–501.

——(1918b). Further notes on intrafascicular cambium in monocotyledons. *Ann. Bot.* **32**, 87–89.

——(1920). The vegetative morphology of *Pistia* and the Lemnaceae. *Proc. roy. Soc., Lond.*, B **91**, 96–103.

——(1921). The leaf structure of the Iridaceae, considered in relation to the phyllode theory. *Ann. Bot.* **35**, 301–226.

——(1922a). Studies on intrafascicular cambium in monocotyledons. V. *Ann. Bot.* **36**, 251–256.

——(1922b). On the nature of the 'blade' in certain monocotyledonous leaves. *Ann. Bot.* **36**, 329–351.

ARCANGELI, G. (1890). Sull' *Helicodiceros muscivorus* (L.fil.) Engler. *Nuovo G. bot. ital.* **22**, 467–472.

——(1897). Sull' *Arum italicum* e sopra la piante a foglie macchiate. *Bull. Soc. bot. ital.* 198–202.

ARDITTI, J. and RODRIGUEZ, E. (1982). *Dieffenbachia:* uses, abuses and toxic constituents: a review. *J. Ethnopharm.* **5**, 293–302.

ARESCHOUG, F. W. C. (1879). *Jemforande undersokningar ofver bladets anatomi.* Lund. 242 pp.

ARNOTT, H. J. and PAUTARD, F. G. E. (1965). Development of raphide idioblasts in *Lemna*. *Amer. J. Bot.* **52**, 618–619.

ARTSCHWAGER, E. (1921). Use of chloriodide of zinc in plant biology. *Bot. Gaz.* **71**, 400.

ARZT, T. (1937). Die Kutikula bei einigen Pteridophyten, Gymnospermen, und Monokotyledonen. *Ber. dt. bot. Ges.* **55**, 437–464.

AZEVEDO, N. F. DOS and DE, S. (1970–71). Micorrizas de plantas espontaneas e cultivadas. *Mem. Soc. brot.* **21**, 329–342.

BACHMANN, E. (1879–81). Ueber Korkwucherungen auf Blättern. *Jb. wiss. Bot.* **12**, 191–236.

BAHADUR, B., KUMAR, P. V., REDDY, N. P. and RAO, K. L. (1977). Bioisomerism in spathes of *Xanthosoma violaceum* Schott. *Curr. Sci.* **46**, 869–870.

BARABÉ, D. (1982). Vascularisation de la fleur de *Symplocarpus foetidus* (Araceae). *Can. J. Bot.* **60**, 1536–1544.

——(1987). La néotenie chez les Araceae (Angiospermae). *C. r. Acad. Sci., Paris*, sér. 3, *Sci. vie* **305**, 609–612.

——and CHRÉTIEN, L. (1985a). Anatomie florale de *Spathicarpa sagittifolia* (Araceae). *Beitr. Biol. Pfl.* **60**, 403–412.

——and——(1985b). Anatomie florale de *Monstera deliciosa* (Araceae). *Can. J. Bot.* **63**, 1423–1428.

——and——(1986). Vascularisation de la fleur de *Spathiphyllum wallisii*. *Can. J. Bot.* **64**, 1397–1401.

——,——and FORGET, S. (1986a). Vascularisation de la fleur pistillée d' *Anchomanes difformis* Engl. *Feddes Repert.* **97**, 831–835.

——,——and——(1987a). On the pseudomonomerous gynoecia of the Araceae. *Phytomorphology* **37**, 139–143.

——and FORGET, S. (1987). Analyse phylogénique des Calloideae (Araceae). *Nat. Canad.* **114**, 487–494.

——and——(1988a). Anatomie des fleurs fertiles et stériles de *Zamioculcas* (Araceae). *Bull. Mus. natn. Hist. nat., Paris*, sér. 4, **10**, B, *Adansonia*, 411–419.

BARABÉ, D. and FORGET, S. (1988b). Anatomie florale de l'*Aglaonema pictum* (Araceae). *Beitr. Biol. Pfl.* **63**, 453–462.

—— and —— (1992). Anatomie florale des *Culcasia* (Araceae). *Bull. Mus. natn. Hist. nat., Paris*, sect. B, *Adansonia* **14**, 263–278.

——, —— and CHRÉTIEN, L. (1986b). Sur les gynécées pseudo-monomères: cas de *Symplocarpus* (Araceae). *C. r. Acad. Sci., Paris*, sér. 3, **302**, 429–434.

——, —— and —— (1987b). Organogenèse de la fleur de *Symplocarpus foetidus* (Araceae). *Can. J. Bot.* **65**, 446–455.

—— and LABRECQUE, M. (1983). Vascularisation de la fleur de *Calla palustris* (Araceae). *Can. J. Bot.* **61**, 1718–1726.

—— and —— (1984). Vascularisation de la fleur de *Lysichitum camtschatcense* (Araceae). *Can. J. Bot.* **62**, 1971–1983.

—— and —— (1985). Vascularisation de la fleur d'*Orontium aquaticum* L. (Aracées). *Bull. Soc. bot. Fr.* **132**, *Lett. bot.* (2), 133–145.

——, —— and CHRÉTIEN, L. (1984). Vascularisation de la fleur d'*Anthurium lhotzkyanum* (Araceae). *C. r. Acad. Sci., Paris*, sér. 3, *Sci. Vie* **299**, 651–656.

—— and LACROIX, C. (2001a). The developmental floral morphology of *Montrichardia arborescens* (Araceae) revisited. *Bot. J. Linn. Soc.* **135**, 413–420.

—— and —— (2001b). Aspects of floral development in *Philodendron grandifolium* and *Philodendron megalophyllum* (Araceae). *Intn. J. Pl. Sci.* **162**, 47–57.

——, —— and JEUNE, B. (2000). Development of the inflorescence and flower of *Philodendron fragrantissimum* (Araceae): a qualitative and quantitative study. *Can. J. Bot.* **78**, 557–576.

BARAHONA CARVAJAL, M. E. (1977). Comparative morphological study of inflorescences of two species of Araceae: *Anthurium denudatum* Engler and *Philodendron radiatum* Schott. *Rev. Biol. trop.* **25**, 301–334. [Spanish, English summary]

BARANETZKY, J. (1897). Sur le développement des points végétatifs des tiges chez les Monocotylédones. *Ann. Sci. nat, Bot.*, sér. 8, **3**, 311–365.

BARLUND, H. (1939). Einige Beobachtungen über die Leitbündel von Rhizoma calami. *Suomen Apteskkariyhdist. Aikak.* **28**, 223–247.

BARTHLOTT, W. and LOBIN, W. (1998). *Amorphophallus titanum. Trop. subtrop. Pflanzenwelt* **99**, 225 pp.

—— and WOLLENWEBER, E. (1981). Zur Feinstruktur, Chemie und taxonomischen Signifikanz epicuticularer Wachse und ähnlicher Sekrete. *Trop. subtrop. Pflanzenwelt* **32**, 67 pp.

BATE-SMITH, E. C. (1968). The phenolic constituents of plants and their taxonomic significance. II. Monocotyledons. *Bot. J. Linn. Soc.* **60**, 325–356.

BATYGINA, T. B. and YAKOVLEV, M. S. (eds) (1990).*Comparative embryology of flowering plants. Monocotyledones. Butomaceae-Lemnaceae.* Nauka, Leningrad. 332 pp. [Russian only.]

BAY, D. (1995). Thermogenesis in the aroids. *Aroideana* **18**, 32–39.

BECK, V. and MANNAGETTA, G. (1912). Über die Ausbildung und das Vorkommen von oxalsäurem Kalk bei Araceen. *S. B. Lotos, Prag.* **60**, 192–193.

BECKER, R. and ZIEGENSPECK, H. (1931). Die Zytologie und Entwicklung der Raphidenzellen und die Entstehung ihres Inhaltes bei *Cissus gongylodes* und *Monstera deliciosa. Bot. Archiv* **33**, 81–96.

BEHNKE, H.-D. (1969a). Die Siebröhren-Plastiden der Monocotyledonen. Vergleichende Untersuchungen über Feinbau und Verbreitung eines charakteristischen Plastidentyps. *Planta* **84**, 174–184.

—— (1969b). Über den Feinbau und die Ausbreitung der Siebröhren-Plasmafilamente und über Bau und Differenzierung der Siebporen bei einigen Monocotylen und bei *Nuphar. Protoplasma* **68**, 377–402.

—— (1969c). Aspekte der Siebrohren-Differenzierung bei Monocotylen. *Protoplasma* **68**, 289–314.

—— (1972). Sieve-tube plastids in relation to angiosperm systematics—an attempt towards a classification by ultrastructural analysis. *Bot. Rev.* **38**, 155–197.

—— (1995). P-type sieve element plastids and the systematics of the Arales (*sensu* Cronquist 1988) - with S-type plastids in *Pistia. Pl. Syst. Evol.* **195**, 87–119.

BELIN-DEPOUX, M. (1978). Sur la présence de territoires foliaires glandularisés chez *Alocasia* Schott (Araceae). *Rev. gén. Bot.* **85**, 211–220.

—— (1989) [1990]. Des hydathodes aux nectaires foliaires chez les plantes tropicales. *Bull. Soc. bot. Fr.* **136**, *Act. bot.* (3–4), 151–168.

——, ROELAND, J. C. and SARTHOU, C. (1987). Aspects biologiques de "jardins de fourmis" en Guyane française. *Rev. Cytol. Biol. veg., Bot.* **10**, 185–210.

BENZING, D. H. (1990). *Vascular epiphytes. General biology and related biota.* Cambridge Univ. Press, Cambridge. 354 pp.

BERGDOLT, B. (1955). Anatomische und entwicklungsgeschichtliche Untersuchungen der Panaschierung und anderer Blattmuster bei Araceen. *Z. Bot.* **43**, 309–340.

BERGER, C. and SCHNEPF, E. (1970). Entwicklung und Altern der Spadix-Appendices von *Sauromatum guttatum* Schott und *Arum maculatum* L. I. Veränderungen der Feinstruktur. *Protoplasma* **69**, 237–251.

BERLYN, G. P. and MIKSCHE, J. P. (1976). *Botanical microtechnique and cytochemistry.* Iowa State Univ. Press, Ames. 326 pp.

BERNARD, F. A., BERNARD, J. M., and DENNY, P. (1990). Flower structure and life history of *Wolffia australiana* (Benth.) den Hartog & van der Plas. *Bull. Torrey bot. Club* **117**, 18–26.

BERQUAM, D. L. (1972). Size–structure correlation in developing roots of *Cissus* and *Syngonium. J. Minn. Acad. Sci.* **38 (1)**, 42–45.

BERRY, E. W. (1930). Revision of the Lower Eocene Wilcox flora of the southeastern states. *U. S. Geol. Surv. Prof. Pap.* **156**, 1–196.

BHANTHUMNAVIN, K. and MCGARRY, M. G. (1971). *Wolffia arrhiza* as a possible source of inexpensive protein. *Nature* **232**, 495.

BHAT, K. M., THULASIDAS, P. K. and MOHAMED, C. P. (1992). Strength properties of ten South Indian canes. *J. trop. for. Sci.* **5 (1)**, 26–34.

BHATTACHARYA, G. N. (1976). A cytological study on the tribe Anthurieae (Araceae). *Bull. bot. Soc. Bengal* **30**, 51–56.

Biermann, R. (1898). Über Bau und Entwicklung der Oelzellen und die Oelbildung in ihnen. Thesis, Bonn. 80 pp.

Black, O. F. (1918). Calcium oxalate in the Dasheen. *Amer. J. Bot.* **5**, 447–451.

Blackwell, W. H. and Blackwell, K. P. (1975). The taxonomy of *Peltandra* (Araceae). *J. Elisha Mitchell Sci. Soc.* **90**, 137–140.

Blanc, P. (1977a). Contribution à l'étude des Aracées. I. Remarques sur la croissance monopodiale. *Rev. gén. Bot.* **84**, 115–126.

——(1977b). Contribution à l'étude des Aracées. II. Remarques sur la croissance sympodiale chez l'*Anthurium scandens* Engl., le *Philodendron fenzlii* Engl. et le *Philodendron speciosum* Schott. *Rev. gén. Bot.* **84**, 319–331.

——(1978). Aspects de la ramification chez des Aracées tropicales. Thèse: Univ. Pierre & Marie Curie, Paris. 83 pp. + bibliog. + figs. (unpubl.).

——(1980). Observations sur les flagelles des Araceae. *Adansonia*, sér. 2, **20**, 325–338.

——(1986). Alternance de cataphylles et de feuilles assimilatrices chez les Araceae: implications écologiques. *Mem. Mus. natn. Hist. nat.*, sér. A, *Zool.* **132**, 263–272.

Bloch, R. (1926). Undifferenzierungen an Wurzelgeweben nach Verwundung. *Ber. dt. bot. Ges.* **44**, 308–316.

——(1944). Developmental potency, differentiation and pattern in meristems of *Monstera deliciosa*. *Amer. J. Bot.* **31**, 71–77.

——(1946). Differentiation and pattern in *Monstera deliciosa*. The idioblastic development of the trichosclereids in the air root. *Amer. J. Bot.* **33**, 544–551.

Bochenska, I. and Kozlowski, J. (1974). Comparative investigations of the structure of fruits of the family Araceae growing in Poland. *Herba polon.* **20**, 3–10. [Polish, English summary]

Bogner, J. (1972). Revision der Arophytae (Araceae). *Bot. Jb.* **92**, 1–63.

——(1976). Die systematische Stellung von *Acoropsis* Conwentz, einer fossilen Aracee aus dem Berstein. *Mitt. Bayer. Staatssamml. Paläont. hist. Geol.* **16**, 95–98.

——(1987). Morphological variation in aroids. *Aroideana* **10 (2)**, 4–16.

——and Hay, A. (2000). Schismatoglottideae (Araceae) in Malaysia II.–*Aridarum, Bucephalandra, Phymatarum* and *Piptospatha*. *Telopea* **9**, 179–222.

——and Mayo, S. J. (1998). Acoraceae, pp. 7–11 In: *Families and genera of vascular Plants: Monocotyledons* (ed. Kubitzki, K.). Springer, Berlin.

——and Nicolson, D. H. (1991). A revised classification of Araceae with dichotomous keys. *Willdenowia* **21**, 35–50.

Bohny, P. (1906). Über das Blatt von *Arum maculatum* L. und seine Verwechslung mit dem von *Paris quadrifolia* L. *Schweiz. Wochenschr. Chem. Pharm.* **44**, 89–96.

Bolli, M. (1953). Sui processi di cicatrizzazione delle ferite nei tuberi di patata. *Annali Bot.* **24**, 212–240. [English summary]

Bonde, S. D. (2000). *Rhodospathodendron tomlinsonii* gen. et sp. nov., an araceous viny axis from the Nawargaon intertrappean beds of India. *Palaeobotanist* **49**, 85–92.

Bonnier, G. (1904). Production accidentelle d'une assise génératrice intralibérienne dans des racines de Monocotylédones. *C. r. Acad. Sci., Paris* **138**, 1381–1384.

Boodle, L. A. and Hill, A. W. (1929). *Typhonodorum lindleyanum*: the development of the embryo and germination of the seed. *Ann. Bot.* **43**, 437–450.

Borzi, A. (1887–8). Formazione delle radici laterali nelle Monocotiledoni. *Malpighia* **1**, 391–413, 541–550 (1887). *Malpighia* **2**, 52–85, 394–402, 477–506 (1888).

Bown, D. (2000). *Aroids. Plants of the arum family*. ed. 2. Timber Press, Portland, Oregon. 392 pp.

Bown, T. M., Kraus, M. J., Wing, S. L., Fleagle, J. G., Tiffney, B. H., Simons, E. L. and Vondra, C. F. (1982). The Fayum primate forest revisited. *J. Human Evol.* **11**, 603–632.

Boyd, L. (1931, 1932). Evolution in the monocot. seedling. *Trans. bot. Soc. Edinb.* **30**, 286–303. Monocotylous seedlings. Ibid. **31**, 1–224.

Brandza, M. (1891). Développement des téguments de la graine. *Rev. gén. Bot.* **3**, 1–32, 71–84, 105–126, 150–165, 229–240.

Bredemann, G. and Meyer, H. (1932). Über die Blattstielfasern von *Amorphophallus titanum* Becc. *Faserforschung* **9**, 285–291.

Bretfeld, F. von (1879–81). Über Vernarbung und Blattfall. *Jb. wiss. Bot.* **12**, 133–160.

Briens, G. (1916). *Étude pharmacognosique de l' Arum maculatum.* La Cour d'Appel, Paris. 77 pp. [Also in : *Trav. Lab. Mat. med., Paris* **10** (1913–16) [1917] (3) 77 pp.]

Brown, G. A. and Holloway, P. J. (1981). Electron microscope histochemistry of plant cutins. *Micron* **12**, 193–194.

Brown, R. W. (1962). Paleocene Flora of the Rocky Mountains and Great Plains. *U. S. Geol. Survey Prof. Pap.* **375**, 1–119.

Bruce, W. B. (1903). The leaf markings of *Arum maculatum. Irish Naturalist* **12**, 161.

Bruhn, W. (1910). Beiträge zur experimentellen Morphologie, zur Biologie und Anatomie der Luftwurzeln. *Flora* **101**, 98–166.

Brundrett, M. C. and Kendrick, B. (1988). The mycorrhizal status, root anatomy, and phenology of plants in a sugar maple forest. *Can. J. Bot.* **66**, 1153–1173.

——and——(1990). The roots and mycorrhizas of herbaceous woodland plants. I. Quantitative aspects of morphology. II. Structural aspects of morphology. *New Phytol.* **114**, 457–468; 469–479.

Brunswik, H. (1921). Über Hesperidinsphärite im lebenden Hautgewebe von *Anthurium binotii* Linden. *Ber. dt. bot. Ges.* **39**, 208–212.

Brunzema, D. (1928). Die Entwickelung der Kalziumoxalatzellen mit besonderer Berücksichtigung der offizinellen Pflanzen. *Arch. Pharm.* **266**, 86–103.

Buchholz, M. (1921). Die Wasserleitungsbahnen in den interkalaren Wachstumszonen von monokotylen Sprossen. *Flora* **114**, 119–186.

Bugnon, F. (1981). Organogenèse foliaire commençante chez deux monocotylédones à feuilles végétatives supposées unifaciales: *Sansevieria cylindrica* Bojer et *Acorus calamus* L. *Rev. Cytol. Biol. vég., Bot.* **4**, 283–294.

BUINOVA, M. G. (1988). Anatomy and pigments of the leaf of plants of the Baikal area. *Novosibirsk Nauka, Sibir. Otdel.* 96 pp. [Russian only]

BUNTING, G. S. (1960). A revision of *Spathiphyllum* (Araceae). *Mem New York bot. Gard.* **10**, 1–53.

——(1968). Vegetative anatomy and taxonomy of the *Philodendron scandens* complex. *Gentes Herb.* **10**, 136–168.

BURGER, W. C. (1977). The Piperales and the monocots. Alternate hypotheses for the origin of monocotyledonous flowers. *Bot. Rev.* **43**, 345–393.

BUSCALIONI, L. (1895–6). Studi sui cristalli di ossalato di calcio. *Malphighia* **9**, 469–533 (1895) **10**, 3–67, 125–167 (1896).

——and LANZA, D. (1928). Sulla costituzione morfologica ed anatomica delle inflorescenze di *Ambrosinia bassii* L. e di *Pistia stratiotes* L. *Malpighia* **31**, 3–45.

BUZGÓ, M. (1994). Inflorescence development of *Pistia stratiotes* (Araceae). *Bot. Jb. Syst.* **115**, 557–570.

——(2001). Flower structure and development of Araceae compared with alismatids and Acoraceae. *Bot. J. Linn. Soc.* **136**, 393–425.

——and ENDRESS, P. K. (2000). Floral structure and development of Acoraceae and its systematic relationships with basal angiosperms. *Intn. J. Pl. Sci.* **161**, 23–41.

CAMAZINE, S. and NIKLAS, K. J. (1984). Aerobiology of *Symplocarpus foetidus*: interactions between the spathe and spadix. *Amer. J. Bot.* **71**, 843–850.

CAMPBELL, D. H. (1899). Notes on the structure of the embryo-sac in *Sparganium* and *Lysichiton*. *Bot. Gaz.* **27**, 153–166.

——(1900). Studies on the Araceae. *Ann. Bot.* **14**, 1–25.

——(1902). Recent investigations upon the embryo sac of angiosperms. *Amer. Nat.* **36**, 777–786.

——(1903). Studies on the Araceae. The embryo-sac and embryo of *Aglaonema* and *Spathicarpa*. *Ann. Bot.* **17**, 665–687.

——(1905). Studies on the Araceae, III. *Ann. Bot.* **19**, 329–349.

——(1912). The embryo-sac of *Aglaonema*. *Scott. bot. Rev.* **1** (2), 100–115.

CARLQUIST, S. and SCHNEIDER, E. L. (1997). Origin and nature of vessels in monocotyledons. 1. *Acorus*. *Int. J. Plant Sci.* **158**, 51–56.

——and——(1998). Origin and nature of vessels in monocotyledons. 5. Araceae subfamily Colocasioideae. *Bot. J. Linn. Soc.* **128**, 71–86.

CARR, D. J., CARR, S. G. M. and JAHNKE, R. (1980a). Intercellular strands associated with stomata: stomatal pectic strands. *Protoplasma* **102**, 177–182.

——OATES, K. and CARR, S. G. M. (1980b). Studies on intercellular pectic strands of leaf palisade parenchyma. *Ann. Bot.* **45**, 403–413.

CARVELL, W. N. (1989). Floral anatomy of the Pothoideae and Monsteroideae (Araceae). Doctoral thesis. Miami University, Oxford, Ohio. 566 pp. Univ. Microfilms #9015935.

CASADORO, G., RASCIO, N., PAGIUSCO, M. and RAVAGNAN, N. (1982). Flowers of *Orontium aquaticum* L.: membrane rearrangement in chloroplast-chromoplast interconversions. *J. ultrastruct. Res.* **81**, 202–208.

CATALANO, G. (1915). Intorno alla struttura e alla funzione di alcune radici contrattili. *Nuovo G. bot. ital.* (NS) **22**, 148–74.

CERUTI, A. (1939). Ricerche istofisiologiche sullo spadice di *Arum*. *Nuovo G. bot. ital.* (NS) **46**, 521–550.

CEVALLOS-FERRIZ, S. and STOCKEY, R. A. (1988). Permineralized fruits and seeds from the Princeton Chert (Middle Eocene) of British Columbia: Araceae. *Amer. J. Bot.* **75**, 1099–1113.

——,——and PIGG, K. B. (1991). The Princeton chert: evidence for in situ aquatic plants. *Rev. Palaeobot. Palynol.* **70**, 173–185.

CHARLTON, W. A. (1983). Patterns of distribution of lateral root primordia. *Ann. Bot.* **51**, 417–427.

CHASE, M. W. *et al.* (1993). Phylogenetics of seed plants: an analysis of nucleotide sequences from the plastid gene *rbc*L. *Ann. Missouri bot. Gard.* **80**, 528–580.

——, SOLTIS, D. E., SOLTIS, P. S., RUDALL, P. J., FAY, M. F., HAHN, W. H. *et al.* (2000). Higher level systematics of the monocotyledons: an assessment of current knowledge and a new classification. pp. 3–16 In: *Monocots. Systematics and evolution*. (eds Wilson, K. L. and Morrison, D. A.). CSIRO Publishing. Collingwood, Victoria, Australia.

CHAUHAN, K. P. S. and BRANDHAM, P. E. (1985). Chromosome and DNA variation in *Amorphophallus* (Araceae). *Kew Bull.* **40**, 745–758.

CHAUVEAUD, G. (1897). Recherches sur le mode de formation des tubes criblés dans la racine des monocotylédones. *Ann. Sci. nat., Bot.*, sér. 8, **4**, 307–381.

——(1911). L'appareil conducteur des plantes vasculaires et les phases principales de son évolution. *Ann. Sci. nat., Bot.*, sér. 9, **13**, 113–436.

CHEADLE, V. I. (1942). The occurrence and types of vessels in the various organs of the Monocotyledoneae. *Amer. J. Bot.* **29**, 441–450.

CHERNEJ, E. N. (1977). Structure anatomique de la spathe de l'inflorescence de diverses espèces d' Aroidées (Aracées). *Izv. Akad. Nauk moldav. SSR, biol. khim. Nauk* (2), 5–11. [Russian]

CHOUARD, P. (1936). La nature et le rôle des formations dites 'secondaires' dans l'édification de la tige des Monocotylédones. *Bull. Soc. bot. Fr.* **83**, 819–836.

CHOW, W. S., QIAN, L., GOODCHILD, D. J. and ANDERSON, J. M. (1988). Photosynthetic acclimation of *Alocasia macrorrhiza* (L.) G. Don to growth irradiance: structure, function and composition of chloroplasts. *Aust. J. Pl. Physiol.* **15**, 107–122.

CHRYSLER, M. A. (1904). The development of the central cylinder of Araceae and Liliaceae. *Bot. Gaz.* **38**, 161–184.

CLARK, H. L. (1979). LEMNACEAE. pp. 464–476 In: *Aquatic and wetland plants of the southeastern United States. Monocotyledons*. (eds Godfrey, R. K. and Wooten, J. M.). Univ. Georgia Press, Athens. 712 pp.

CLARKE, A. E., GLEESON, P. A., JERMYN, M. A. and KNOX, R.B. (1978). Characterization and localization of 3-lectins in lower and higher plants. *Aust. J. Pl. Physiol.* **5**, 707–722.

CLAUSEN, P. (1927). Ueber das Verhalten des Antheren-Tapetums bei einigen Monokotylen und Ranales. *Bot. Archiv* **18**, 1–27.

CLAVERIE, P. (1906). Étude morphologique et histologique du *Typhonodorum madagascariense*, textile de Madagascar. *Rev. gén. Bot.* **18**, 97–109.

——(1909). Contribution à l'étude anatomique et histologique des plantes textiles exotiques (Passiflorées, Musacées, Palmiers, Aroidées et Cyperacées). *Ann. Mus. colon. Marseille*, sér. 2, **7**, 11–207.

CLOWES, F. A. L. (1958). Development of quiescent centres in root meristems. *New Phytol.* **57**, 85–88.

——(1984). Size and activity of quiescent centres of roots. *New Phytol.* **96**, 13–21.

——(1985). Origin of epidermis and development of root promordia in *Pistia, Hydrocharis* and *Eichhornia*. *Ann. Bot.* **55**, 849–857.

——(1990). The discrete root epidermis of floating plants. *New Phytol.* **115**, 11–15.

——(1992). Regeneration of the discrete root epidermis of *Pistia stratiotes* L. after perturbation of the meristem. *New Phytol.* **120**, 209–213.

COCUCCI, A. E. (1966). Embriologia de *Synandrospadix vermitoxicus* (Araceae). *Kurtziana* **3**, 157–181.

CODY, A. M. and HORNER, H. T. (1983). Twin raphides in the Vitaceae and Araceae and a model for their growth. *Bot. Gaz.* **144**, 318–323.

COOK, C. D. K. (1999). The number and kind of embryo-bearing plants which have become aquatic: a survey. *Perspectives in Plant Ecology, Evolution and Systematics* **2**, 79–102.

COSTANTIN, J. (1885). Recherches sur l'influence qu'exerce le milieu sur la structure des racines. *Ann. Sci. nat., Bot.*, sér. 7, **1**, 135–182.

COSTERUS, J. C. (1875). Sur la nature des lenticelles et leur distribution dans le règne végétal. *Arch. neerl. Sci. exactes nat.* **10**, 385–400.

COURSEY, D. G. (1968). The edible aroids. *World Crops* **20**, 25–30.

CREPET, W. L. (1978). Investigations of angiosperms from the Eocene of North America: an aroid inflorescence. *Rev. Palaeobot. Palynol.* **25**, 241–252.

CROAT, T. B. (1979). The distribution of Arcaceae. pp. 291–308 In: *Tropical botany* (eds Larsen, K. and Holm-Nielsen, L. B.). Academic Press, London, New York, San Francisco.

——(1980). Flowering behavior of the neotropical genus *Anthurium* (Araceae). *Amer. J. Bot.* **67**, 888–904.

——(1981) [1982]. A revision of *Syngonium* (Araceae). *Ann. Missouri bot. Gard.* **68**, 565–651.

——(1983). A revision of the genus *Anthurium* (Araceae) of Mexico and Central America. *Ann. Missouri bot. Gard.* **70**, 211–420.

——(1985). Collecting and preparing specimens of Araceae. *Ann. Missouri bot. Gard.* **72**, 252–258.

——(1988). The ecology and life forms of Araceae. *Aroideana* **11**, 4–56.

——(1991). A revision of *Anthurium* section Pachyneurium (Araceae). *Ann. Missouri bot. Gard.* **78**, 539–855.

——(1997). A revision of *Philodendron* subgenus *Philodendron* (Araceae) for Mexico and Central America. *Ann. Missouri bot. Gard.* **84**, 311–704.

——(1998). History and current status of systematic research with Araceae. *Aroideana* **21**, 26–145.

CRONQUIST, A. (1968). *Evolution and classification of flowering plants*. Houghton-Mifflin, Boston.

——(1981). *An integrated system of classification of flowering plants*. Columbia Univ. Press, New York.

CZAJA, A. T. (1962). Die submikroskopische Struktur (Textur) der Aussenwände von Blattepidermen. *Planta* **57**, 669–698.

DAGHLIAN, C. P. (1981). A review of the fossil record of monocotyledons. *Bot. Rev.* **47**, 517–555.

DAHLGREN, G. (1989). An updated angiosperm classification. *Bot. J. Linn. Soc.* **100**, 197–203.

DAHLGREN, R. M. T. (1975). A system of classification of the angiosperms to be used to demonstrate the distribution of characters. *Bot. Notiser* **128**, 119–147.

——1980. A revised system of classification of the angiosperms. *Bot. J. Linn. Soc.* **80**, 91–124.

——and CLIFFORD, H. T. (1982). *The monocotyledons: a comparative study*. Academic Press, London.

——, ——and YEO, P. (eds) (1985). *The families of monocotyledons; structure, evolution, and taxonomy*. Springer, New York.

——and RASMUSSEN, F. N. (1983). Monocotyledon evolution, character and phylogenetic estimation. pp. 255–395 In: *Evolutionary biology* **16** (eds Hecht, M. K., Wallace, B. and Prance, G. T.). Plenum, New York.

——, ROSENDAL-JENSEN, S. and NIELSEN, B. J. (1981). A revised classification of the angiosperms with comments on correlation between chemical and other characters. pp. 149–204 In: *Phytochemistry and angiosperm phylogeny* (eds Young, D. A. and Siegler, D. S.). New York.

DAKWALE, S. and BHATNAGAR, S. (1985). Insect-trapping behaviour and diel periodicity in *Sauromatum guttatum* Schott (Araceae). *Curr. Sci.* **54**, 699–702.

DALITZSCH, M. (1886). Beiträge zur Kenntniss der Blattanatomie der Aroideen. *Bot. Centralbl.* **25**, 153–156, 184–187, 217–219, 249–253, 280–285, 312–318, 343–349.

DANGEARD, P. (1959–60) [1960]. Sur l'évolution de la structure dans les chloroplastes et dans les leucoplastes. *Botaniste* **43**, 1–58.

DANIEL, L. (1899). Greffe de quelques Monocotylédones sur elles-memes. *C. r. Acad. Sci., Paris* **129**, 654–656.

DATTA, P. C. and MAITI, R. K. (1968). Stereome disposition of paddy field weeds of West Bengal. *Bull. bot. Soc. Bengal* **22**, 221–236.

DATTA, S. C. and BISWAS, K. K. (1970). Germination pattern and seedling morphology of *Pistia stratiotes* L. *Phyton*, Argentina **27**, 157–161.

——and ——(1980). The biology of *Pistia stratiotes* L. var. *cuneata* Engl. *Bull. bot. Soc. Bengal* **34**, 15–26.

DAUBS, E. H. (1965). A monograph of Lemnaceae. *Illinois Biol. Monogr.* **34**, 1–118. Univ. Illinois Press, Urbana.

DAUMANN, E. (1930). Nektarabscheidung in der Blütenregion einiger Araceen. *Planta* **12**, 38–48.

——(1970). Das Blütennektarium der Monocotyledonen unter besonderer Berücksichtigung seiner systematischen und phylogenetischen Bedeutung. *Feddes Repert.* **80**, 463–590.

DAVIE, R. C. (1917). On the leaf-trace in some pinnate leaves. *Trans. roy. Soc. Edinb.* **52**, 1–36.

DAVIS, G. L. (1966). *Systematic embryology of the angiosperms.* John Wiley, New York, London, Sydney. 528 pp.

DAVIS. J. I. (1995). A phylogenetic structure for the monocotyledons, as inferred from chloroplast DNA restriction site variation, and a comparison of measures of clade support. *Syst. Bot.* **20**, 503–527.

DECAPITE, L. and MILLETTI, G. (1967). Anatomia delle radici aeree e terrestri in *Anthurium andreanum* Lindl. *Ann. Fac. Agr. Univ. Perugia* **22**, 307–315.

—— and —— (1968). Modificazioni indotte dall' acido N-dimetilamino succinammico sul rizoderma delle radici aeree di *Monstera deliciosa* Libm. *Ann. Fac. Agr. Univ. Perugia* **23**, 65–72.

DECROCK, E. (1909). Sur quelques fécules des colonies et en particulier de l'Indo-Chine. *Ann. Mus. colon. Marseille*, sér. 2, **7**, 325–359.

DEINEGA, V. (1898). Beiträge zur Kenntniss der Entwickelungsgeschichte des Blattes und der Anlage der Gefässbündel. *Flora* **85**, 439–498.

DE LA RUE, E. (1866). Zur Anatomie und Physiologie des Blattes der Aroideen. *Bot. Ztg.* **24**, 317–323.

—— (1869). Ueber Krystalldrusen bei einigen Pflanzen. *Bot. Ztg.* **27**, 537–539.

DEN HARTOG, C., and VAN DER PLAS, F. (1970). A synopsis of Lemnaceae. *Blumea* **18**, 355–368.

DENISSOVA, G. A. (1975). Attempt of classification of terpenoid-containing conceptacles in plants. *Bot. Zh. SSSR* **60**, 1698–1706. [Russian, English summary]

DESHPANDE, B. D. (1956). Velamen in terrestrial Araceae. *Sci. Cult.* **21**, 686–687.

DEY, D. and DAS, M. N. (1982). Pharmacognostic studies of *Acorus calamus* and its adulterants. *Acta bot. indica* **10**, 28–35.

—— and —— (1988). Pharmacognosy of antidysenteric drugs of Indian medicine. *Acta bot. indica* **16**, 216–226.

DIAMOND, J. M. (1989). Hot sex in voodoo lilies. *Nature* **339** (25 May), 258–259.

DIBENEDETTO, A. H. and COGLIATTI, D. H. (1990). Effects of light intensity and quality on the obligate shade plant *Aglaonema commutatum*. I. Leaf size and leaf shape. *J. hort. Sci.* **65**, 689–698.

—— and GARCIA, A. F. (1992). Adaptation of ornamental aroids to their indoor light environments. I. Spectral and anatomical characteristics. *J. hort. Sci.* **67**, 179–188.

DICKISON, W. C. (2000). *Integrative plant anatomy.* Harcourt Academic Press, San Diego. 533 pp.

DIHORU, A. (1978). Cercetări chemotaxonomice la unele specii de *Arum* şi *Iris* din flora României. *Stud. cerc. Biol.*, ser. *Biol. veg.* **30 (1)**, 15–18. [English summary]

DIHORU, G. (1986). New data on the anatomy of the root of *Pistia stratiotes* and *Eichhornia crassipes. Rev. roum. Biol., ser. Biol. veg.* **31**, 85–90.

DILCHER, D. L. and DAGHLIAN, C. P. (1977). Investigations of angiosperms from the Eocene of southeastern North America: *Philodendron* leaf remains. *Amer. J. Bot.* **64**, 526–534.

DOMIN, K. (1911). Morphologische und phylogenetische Studien über die Stipularbildungen. *Ann. Jard. bot. Buitenzorg*, sér. 2, **9**, 117–326.

DONOGHUE, M. J. and SANDERSON, M. J. (1992). The suitability of molecular and morphological evidence in reconstructing plant phylogeny. pp. 340–368 In: *Molecular systematics of plants* (eds Soltis, P. S., Soltis, D. E. and Doyle, J. J.). Chapman and Hall, New York.

DORMER, K. (1960). The truth about pollination in *Arum. New Phytol.* **59**, 298–301.

DOYLE, J. A. (1973). Fossil evidence on early evolution of the monocotyledons. *Quart. Rev. Biol.* **48**, 399–413.

DRAHEIM, W. (1929). Beiträge zur Kenntnis des Wurzelwerks von Iridaceen, Amaryllidaceen und Liliaceen. *Bot. Archiv* **23**, 385–440. [English summary.]

DRING, J. V., KITE, G. C., NASH, R. J. and REYNOLDS, T. (1995). Chemicals in aroids: a survey, including new results for polyhydroxy alkaloids and alkylresorcinols. *Bot. J. Linn. Soc.* **117**, 1–12.

DUDLEY, M. G. (1937). Morphological and cytological studies of *Calla palustris*. *Bot. Gaz.* **98**, 556–571.

DUKE, J. A. and VASQUEZ, R. (1994). *Amazonian ethnobotanical dictionary.* CRC Press, Ann Arber, Michigan.

DUMANN, H. (1980). Oxalatkristalle in Pflanzenzellen. IV. Kristallsand und Raphiden. *Mikrokosmos* **69**, 165–167.

DUNN, D. B., SHARMA, G. K. and CAMPBELL, C. C. (1965). Stomatal patterns of dicotyledons and monocotyledons. *Amer. Midl. Nat.* **74**, 185–195.

DUTT, B. S. M. (1956). Occurrence of longitudinal lenticels on the aerial roots of *Pothos* L. *Sci. Cult.* **22**, 163–164.

DUVAL-JOUVE, J. (1873). Diaphragmes vasculifères des Monocotylédones aquatiques. Mém. Acad. Sci. Montpellier **8**, 157–176.

DUVALL, M. R. (2000). Seeking the dicot sister group of the monocots. pp. 25–32 In: *Monocots: Systematics and Evolution.* (eds Wilson, K. L. and Morrison, D. A.). CSIRO Publishing. Collingwood, Victoria, Australia.

—— (2001). An anatomical study of anther development in *Acorus* L.: phylogenetic implications. *Pl. Syst. Evol.* **228**, 143–152.

——, LEARN, G. H., EGUIARTE, L. E. and CLEGG, M. T. (1993). Phylogenetic analysis of *rbc*L sequences identifies *Acorus calamus* as the primal extant monocotyledon. *Proc. Nat. Acad. Sci. U.S.A.* **90**, 4641–4644.

EAMES, A. J. and MACDANIELS, L. H. (1947). *An introduction to plant anatomy.* ed. 2. McGraw-Hill, New York. 427 pp.

EFRAT, Y., PIETERSE, A. H., and BOUMAN, F. (1977). A comparative study of developing air chambers in flat and gibbous fronds of *Lemna gibba* L. *Acta bot. neerl.* **26**, 343–347.

EL-GAZZAR, A., BADAWI, A. A. and HAMZA, M. K. (1977). The distribution of associated stomata in angiosperms, I. Monocotyledoneae. *Publ. Cairo Univ. Herb.*, Nos. 7–8, 47–55.

EL-HADIDI, M. N. (1969). Observations on the wound-healing process in some flowering plants. *Mikroskopie* **25**, 54–69.

ELISEI, F. G. (1941). Ricerche microfluoroscopiche sui punti di Caspary. *Atti Ist. bot. Pavia*, ser. 4, **13**, 3–66.

EL-KIEY, M. A., EL-WAHAB, S. M. A., ZAKI, A. Y. and WASSEL, G. M. (1967) [1969]. A pharmacognostical study of the corms of *Eminium spiculatum* (Blume) and *Arisarum vulgare* (Schott) growing in Egypt. I. Macro and micro-morphology. *J. pharm. Sci. U.A.R.* **8**, 147–157.

ELLER, B. M. and FLACH, B. M. T. (1991). Optische Eigenschaften chlorophyllfreier und grüner Blätter. *Flora* **185**, 39–46. [English summary]

ELSIK, W. C. (1968). Palynology of a Paleocene Rockdale Lignite, Milam County, Texas. 1. Morphology and taxonomy. *Pollen et Spores* **10**, 263–314.

ENGLER, A. (1877). Vergleichende Untersuchungen ber die morphologischen Verhältnisse der Araceae. *Nova Acta Leop. Carol.* **39**, 133–232.

——(1884). Beiträge zur Kenntniss der Araceae V. 12. Über den Entwicklungsgang in der Familie der Araceen und über die Blütenmorphologie derselben. *Bot. Jb.* **5**, 141–188; 287–336.

——(1889). Araceae. *Die natürlichen Pflanzenfamilien* 2, **3**, 102–153.

——(1905), (1908). Araceae - Pothoideae *Das Pflanzenreich* IV; **23B**; 330 pp. (1905). Addit. ad Araceae-Pothoideae. *Das Pflanzenreich* IV, **23B**, 3 pp. (1908).

——(1911). Araceae-Lasioideae. *Das Pflanzenreich* IV, **23C**, 130 pp.

——(1915). Araceae - Philodendroideae - Ambiadeae ad Peltandreae. *Das Pflanzenreich* IV, **24D**[c], 78 pp.

——(1920a). Araceae, pars generalis. *Das Pflanzenreich* IV, **23A**, 71 pp.

——(1920b). Addit. ad Araceae - Philodendreae, 2 pp.; Araceae - Colocasioideae. *Das Pflanzenreich* IV, **23E**, 139 pp.

——(1920c). Araceae - Aroideae & Araceae - Pistoideae. *Das Pflanzenreich* IV, **23F**, 274 pp.

——(1990). Comparative studies on the morphology of the Araceae. II. On leaf placement and shoot organization of Araceae. (1887 paper translated and annotated by Ray, T. S. and Renner, S. S.). *Englera* **12**, 1–139.

——and Krause, K. (1908). Araceae-Monsteroideae. *Das Pflanzenreich* IV, **23B**, 4–13.

——and——(1912). Araceae-Philodendroideae-Philodendreae. *Das Pflanzenreich* IV, **23Da**, 134 pp.

——and ——(1920). Araceae-Colocasioideae. *Das Pflanzenreich* IV, **23E**, 139 pp.

ERDTMAN, G. (1952). *Pollen morphology and plant taxonomy. Angiosperms.* Chronica Botanica, Waltham, MA 539 pp.

ERTL, P. O. (1932). Vergleichende Untersuchungen über die Entwicklung der Blattnervatur der Araceen. *Flora* **126**, 115–248.

ESAU, K. (1960). *Anatomy of seed plants.* ed. 1. Wiley, New York.

——(1965). *Plant anatomy.* ed. 2. Wiley, New York.

——(1977). *Anatomy of seed plants.* ed. 2. Wiley, New York.

ESCHRICH, W. (1983). Phloem unloading in aerial roots of *Monstera deliciosa. Planta* **157**, 540–547.

EYDE, R. H., NICOLSON, D. H. and SHERWIN, P. (1967). A survey of floral anatomy in Araceae. *Amer. J. Bot.* **54**, 478–497.

FAEGRI, K. and PIJL, L. VANDER (1971). *The principles of pollination ecology.* Pergamon, Oxford.

FAGERLIND, F. and MASSALSKI, A. (1974). The development of cell walls and intercellulares in the root of *Lemna minor* L. *Svensk bot. Tidskr.* **68**, 64–93.

FAHN, A. (1990). *Plant anatomy.* ed. 4 Pergamon Press, Oxford. 588 pp.

FENIZIA, C. (1896). La papille cérifère de *Caladium violaceum. Rivista ital. Sci. nat. Siena* **16**, 105–107.

FIGDOR, W. (1922). Über die Entwicklung der Wendeltreppenblätter von *Helicodiceros muscivorus* Engl. *S. B. Akad. Wiss. Wien, math.-naturw. Kl.*I, **131**, 233–241.

FISHER, J. B. and FRENCH, J. C. (1976). The occurrence of intercalary and uninterrupted meristems in the internodes of tropical monocotyledons. *Amer. J. Bot.* **63**, 510–525.

——and——(1978). Internodal meristems of monocotyledons: further studies and a general taxonomic summary. *Ann. Bot.* **42**, 41–50.

FLAHAULT, C. (1878). Recherches sur l'accroissement terminal de la racine chez les Phanérogames. *Ann. Sci. nat., Bot.*, sér. 6, **6**, 1–168.

FLEISCHER, E. (1885). *Die Schutzeinrichtungen der Pflanzenblätter gegen Vertrocknung.* Dobeln.

FLETCHER, F. and ARNOTT, H. J. (1963). The venation of *Lemna minor* in sterile culture. *Amer. J. Bot.* **50**, 621.

FLINCK, J. A. (1891). *Om den anatomiska byggnaden hos de vegetativa organen för upplagsnäring.* Akad. Afh. Helsingfors. 140 pp.

FLOOD, M. G. (1919). Exudation of water by *Colocasia antiquorum. Notes Bot. School, Dublin* **3 (1)**, 59–65.

FOX, M. G. and FRENCH, J. C. (1988). Systematic occurrence of sterols in latex of Araceae: subfamily Colocasioideae. *Amer. J. Bot.* **75**, 132–137.

FRANCESCHI, V. R. and HORNER, H. T. (1980). Calcium oxalate crystals in plants. *Bot. Rev.* **46**, 361–427.

FRANCINI, E. (1946) [1947]. La funzione dell' amido nella cuffia radicale. *Nuovo G. bot. ital.* (NS) **53**, 679–93.

FRANCK, D. H. (1975). Early histogenesis of the adult leaves of *Darlingtonia californica* (Sarraceniaceae) and its bearing on the nature of epiascidiate foliar appendages. *Amer. J. Bot.* **62**, 116–132.

FRENCH, J. C. (1985a). Patterns of endothecial wall thickenings in Araceae: subfamilies Pothoideae and Monsteroideae. *Amer. J. Bot.* **72**, 472–486.

——(1985b). Patterns of endothecial wall thickenings in Araceae: subfamilies Calloideae, Lasioideae, and Philodendroideae. *Bot. Gaz.* **146**, 521–533.

——(1986a). Patterns of endothecial wall thickenings in Araceae: subfamilies Colocasioideae, Aroideae, and Pistioideae. *Bot. Gaz.* **147**, 166–179.

——(1986b). Patterns of stamen vasculature in the Araceae. *Amer. J. Bot.* **73**, 434–449.

——(1986c). Ovular vasculature in Araceae. *Bot. Gaz.* **147**, 478–495.

——(1987a). Structure of ovular and placental trichomes of Araceae. *Bot. Gaz.* **148**, 198–208.

FRENCH, J. C. (1987b). Systematic survey of resin canals in roots of Araceae. *Bot. Gaz.* **148**, 360–371.

——(1987c). Systematic occurrence of a sclerotic hypodermis in roots of Araceae. *Amer. J. Bot.* **74**, 891–903.

——(1988). Systematic occurrence of anastomosing laticifers in Araceae. *Bot. Gaz.* **149**, 71–81.

——(1997). Vegetative anatomy, pp. 9–24 In: *The genera of Araceae* (eds Mayo, S. J., Bogner, J. and Boyce, P. C.). Royal Botanic Gardens, Kew.

——, CHUNG, M. and HUR, Y. (1995). Chloroplast DNA phylogeny of the Ariflorae. pp. 255–275 In: *Monocotyledons: systematics and evolution* (eds Rudall, P. J., Cribb, P. J., Cutler, D. F. and Humphries C. J.). vol. 1. Royal Botanic Gardens, Kew.

——and TOMLINSON, P.B. (1980). Preliminary observations on the vascular system in stems of certain Araceae. pp.105–116 In: *Petaloid monocotyledons* (eds Brickell, C. D., Cutler, D. F. and Gregory, M.). Academic Press, London.

——and——(1981a). Vascular patterns in stems of Araceae: subfamilies Calloideae and Lasioideae. *Bot. Gaz.* **142**, 366–381.

——and ——(1981b). Vascular patterns in stems of Araceae: subfamily Philodendroideae. *Bot. Gaz.* **142**, 550–563.

——and ——(1981c). Vascular patterns in stems of Araceae: subfamily Pothoideae. *Amer. J. Bot.* **68**, 713–729.

——and ——(1981d). Vascular patterns in stems of Araceae: subfamily Monsteroideae. *Amer. J. Bot.* **68**, 1115–1129.

——and ——(1983). Vascular patterns in stems of Araceae: subfamilies Colocasioideae, Aroideae and Pistioideae. *Amer. J. Bot.* **70**, 756–771.

——and ——(1984). Patterns of stem vasculature in *Philodendron. Amer. J. Bot.* **71**, 1432–1443.

——and ——(1986). Compound vascular bundles in monocotyledonous stems: construction and significance. *Kew Bull.* **41**, 561–574.

FREUNDLICH, H. F. (1908). Entwicklung und Regeneration von Gefässbündeln in Blattgebilden. *Jb. wiss. Bot.* **46**, 137–206.

FRICKE, G. (1926). Über die Beziehungen der Hochblätter zu den Laubblättern und Blüten. *Planta* **2**, 249–294.

FROEBE, H. A. and PÜTZ, N. (1988). Orientierende Versuche zur Verlagerung pflanzlicher Organe im Erdboden durch definierte Kräfte. *Beitr. Biol. Pfl.* **63**, 81–100. [English summary]

FRÖLICH, D. and BARTHLOTT, W. (1988). Mikromorphologie der epicuticularen Wachse und das System der Monokotylen. *Trop. subtrop. Pflwelt.* **63**, 1–135. [English summary]

FRYNS-CLAESSENS, E. and VAN COTTHEM, W. (1973). A new classification of the ontogenetic types of stomata. *Bot. Rev.* **39**, 71–138.

FUCHS, P. C. A. (1898). Untersuchungen über den Bau der Raphidenzelle. *Öst. bot. Z.* **48**, 324–332.

FUJIMOTO, S., YAMANAKA, O., SUGANUMA, T. and NAGAHAMA, T. (1990). Starches from Akaza-kazura (*Boussingaultia baselloides*), Amerika-sato-imo (*Xanthosoma sagittifolium*) and Syokuyo-kanna (*Canna edulis*). *Bull. Fac. Agric. Kagoshima Univ.*, **(40)**, 49–54. [Japanese, English summary]

FUNKE, G. L. (1929). Einige Bemerkungen über das Wachstum und die Wurzelbildung bei *Syngonium podophyllum. Ann. Jard. bot. Buitenzorg* **40**, 75–86.

——(1931). On the biology and anatomy of some tropical leaf joints II. *Ann. Jard. bot. Buitenzorg* **41**, 33–64.

GAISER, L. O. (1923). Intracellular relations of aggregate crystals in the spadix of *Anthurium. Bull. Torrey bot. Club* **50**, 389–398.

GALIL, J. (1978). Morpho-ecological studies on *Arisarum vulgare* Targ.-Tozz. *Israel J. Bot.* **27**, 77–89.

——(1990). Pollen germination in *Arisarum vulgare* Targ.-Tozz. *Flora* **184**, 51–61.

GALLAUD, I. (1905). Études sur les mycorhizes endotrophes. *Rev. gén. Bot.* **17**, 1–48, 66–85, 123–136, 223–229, 313–325, 423–433, 479–500.

GANDAWIJAJA, D., IDRIS, S., NASUTION, R., NYMAN, L. P. and ARDITTI, J. (1983). *Amorphophallus titanum* Becc.: a historical review and some recent observations. *Ann. Bot.* **51**, 269–278.

GARDINER, W. (1889). On the occurrence of secreting glandular organs on the leaves of some aroids. *Proc. Camb. phil. Soc.* **6**, 83–84.

GENAU, K. (1901). Physiologisches über die Entwicklung von *Sauromatum guttatum* Schott. *Öst. bot. Z.* **51**, 321–325.

GENTNER, G. (1905). Über die Vorläuferspitzen der Monokotylen. *Flora* **95**, 327–383.

GENUA, J. M. and HILLSON, C. J. (1985). The occurrence, type and location of calcium oxalate crystals in the leaves of fourteen species of Araceae. *Ann. Bot.* **56**, 351–361.

GERNHARD, K. (1981). Einige Beobachtungen an *Cryptocoryne affinis. Inform. Zentr. Arbeitsgr. Wasserpfl.* **11**, 2–3.

GEROLA, F. M. (1963). Ricerche sulle infrastrutture cellulari dello spadice di *Arum. G. bot. ital.* **70**, 177–183.

GERTZ, O. (1916a). Untersuchungen über septierte Thyllen nebst anderen Beiträgen zu einer Monographie der Thyllenfrage. *Acta Univ. Lund.* (NS) II, **2 (12)**, 45 pp.

——(1916b). Om septerade thyllbildningar. *Bot. Notiser* 43–44.

GILL, L. S. and KARATELA, Y. Y. (1983). Structure and ontogeny of stomatal complex in leafy vegetables of Nigeria. *Feddes Repert.* **94**, 529–541.

GIOVANNOZZI, U. (1911). Intorno al sughero delle Monocotiledoni. *Nuovo G. bot. ital.* **18**, 5–79.

GLÜCK, H. (1901). Die Stipulargebilde der Monokotyledonen. *Verh. naturhist.-med. Ver., Heidelberg* (NS) **7**, 1–96.

GOEBEL, K. (1878). Ueber Wurzelsprosse von *Anthurium longifolium. Bot. Ztg.* **36**, 645–648.

——(1922). Erdwurzeln mit Velamen. *Flora* **115**, 1–26.

——and SANDT, W. (1930). Untersuchungen an Luftwurzeln. *Bot. Abhandl.* **17**, 1–124.

GOLDBERG, B. (1940). Life history of *Peltandra virginica. Bot. Gaz.* **102**, 641–662.

GOLDEN, K. E. (1894). Growth in length and thickness of the petiole of *Richardia. Proc. Indiana Acad. Sci.* 235–237.

GONÇALVES, E. G., PAIVA, E. A. S. and COELHO, M. A. N. (2002). A preliminary survey of petiolar collenchyma in the Araceae. *Mon. Syst. Bot. Missouri Bot. Gard.* (submitted)

GOODCHILD, D. J., BJORKMAN, O. and PYLIOTIS, N. A. (1971–72) [1972]. Chloroplast ultrastructure, leaf anatomy, and content of chlorophyll and soluble protein in rainforest species. *Carnegie Instn. Yearbk* **71**, 102–107.

Gottsberger, G. (1992). Diversität der Bestäubung und Reproduktionsbiologie von ursprünglichen Angiospermen. *Stapfia* **28**, 11–27.

——and Amaral, A. Jr. (1984). Pollination strategies in Brazilian *Philodendron* species. *Ber. dt. bot. Ges.* **97**, 391–410.

Gould, S. J. (1977). *Ontogeny and phylogeny*. Belknap Press, Cambridge, MA. 501 pp.

Gowda, M. T. G. X. (1980). Endosperm in *Ariopsis peltata* Nimmo., Araceae. *Curr. Sci.* **49**, 601–603.

Graham, A. (1976). Studies in neotropical paleobotany. II. The Miocene communities of Veracruz, Mexico. *Ann. Missouri bot. Gard.* **63**, 787–842.

Grau, A. (1983). Las epidermis foliares de las Araceas cultivadas en la ciudad de Tucuman y sus alrededores. *Lilloa* **36**, 15–41.

Gravis, A., Monoyer, A., and Fritsché, E. (1943). Observations anatomiques sur les embryons et les plantules. *Lejeunia* (*Rev. de Bot.*) **3**, 1–180 + plates.

Grayum, M. H. (1984). Palynology and phylogeny of the Araceae. Doctoral thesis, Univ. Massachusetts, Amherst. 852 pp. Univ. Microfilms #8418886.

——(1985). Evolutionary and ecological significance of starch storage in pollen of the Araceae. *Amer. J. Bot.* **72**, 1565–1577.

——(1986a). Phylogenetic implications of pollen nuclear number in the Araceae. *Pl. Syst. Evol.* **151**, 145–161.

——(1986b). Correlations between pollen morphology and pollination biology in the Araceae, with some implications for angiosperm evolution. pp. 313–327 In: *Pollen and spores: form and function* (eds Blackmore, S. and Ferguson, I. K.). Academic Press, New York.

——(1987). A summary of evidence and arguments supporting the removal of *Acorus* from the Araceae. *Taxon* **36**, 723–729.

——(1990). Evolution and phylogeny of the Araceae. *Ann. Missouri bot. Gard.* **77**, 628–697.

——(1991). Systematic embryology of the Araceae. *Bot. Rev.* **57**, 167–203.

——(1992). Comparative external pollen ultrastructure of the Araceae and putatively related taxa. *Monog. Syst. Bot. Missouri Bot. Gard.* **43**, 1–167.

——(1996). Revision of *Philodendron* subgenus *Pteromischum* (Araceae) for Pacific and Caribbean Tropical America. *Monog. Syst. Bot. Missouri Bot. Gard.* **47**, 1–233.

Grear, J. W. (1973). Observations on the stomatal apparatus of *Orontium aquaticum* (Araceae). *Bot. Gaz.* **134**, 151–153.

Greenwell, A. B. H. (1947). Taro – with special reference to its culture and uses in Hawaii. *Econ. Bot.* **1**, 276–289.

Greger, J. (1920). Über Wurzelhaubenneste an den Luftwurzeln von *Monstera deliciosa* Liebm. *Lotos* 67–68, 93–97.

Gregor, H. J. and Bogner, J. (1984). Fossile Araceen Mitteleuropas und ihre rezenten Vergleichsformen. *Documenta Naturae* **19**, 1–12.

——and——(1989). Neue Untersuchungen an tertiären Araceen II. *Documenta Naturae* **49**, 12–22.

Gronegress, P. (1974). The structure of chromoplasts and their conversion to chloroplasts. *J. Microsc. Fr.* **19**, 183–192.

Grubert, M. (1981). *Mucilage or gum in seeds and fruits of angiosperms. A review*. Minerva Publikation, München. 397 pp.

Guéguen, F. (1900). Recherches sur le tissu collecteur et conducteur des phanérogames. *J. de Bot.* **14**, 140–148; 165–172.

——(1901). Anatomie comparée du tissu conducteur du style et du stigmate des Phanérogames (I. Monocotylédones, Apétales et Gamopétales). *J. de Bot.* **15**, 265–300.

Guillaud, A. J. (1878). Recherches sur l'anatomie comparée et le développement des tissus de la tige des Monocotylédones. Thèse, Paris. (Also publ. in *Ann. Sci. nat.*, sér. 6, **5**, 5–176.)

Guillot-Salomon, T., Tuquest, C., Lubac, M. de, Hallais, M. F. and Signol, M. (1978). Comparative analysis of ultrastructure and lipid composition of plastids from sun and shade plants. *Cytobiologie* **17**, 442–452.

Guinaudeau, H., Leboeuf, M. and Cave, A. (1975). Aporphine alkaloids. *Lloydia* **38**, 275–338.

Gulliver, G. (1863–5). On raphides in British plants. *Ann. Mag. nat. Hist.*, ser. 3, **11**, 13–15 (1863). On raphides and sphaeraphides of Phanerogamia; with a notice of the crystal prisms of Iridaceae. Ibid. **12**, 226–229. Observations on raphides. Ibid. **12**, 365–367. Observations on raphides and sphaeraphides. Ibid. **12**, 446–447.

Gunn, C. R. and Ritchie, C. A. (1988). Identification of disseminules listed in the federal noxious weed act. *USDA, Agric. Res. Serv., Tech. Bull.* No. 1719, 313 pp.

Guttenberg, H. von (1940). *Der primäre Bau der Angiospermenwurzel.* [Handbuch der Pflanzenanatomie vol. VIII]. Borntraeger, Berlin. 369 pp.

——(1960). Embryologische Studien an Monokotyledonen. III. Die Embryogenese von *Triglochin maritimum* L., *Arum maculatum* L. und *Typha latifolia* L. *Flora* **149**, 243–281.

——(1968). *Der primäre Bau der Angiospermenwurzel.* [Handbuch der Pflanzenanatomie VIII (5)]. 2nd edn. Borntraeger, Berlin. 472 pp.

Haccius, B. and Lakshmanan, K. K. (1966). Vergleichende Untersuchung der Entwicklung von Kotyledo und Sprossscheitel bei *Pistia stratiotes* und *Lemna gibba*, ein Beitrag zum Problem der sogenannten terminalen Blattorgane. *Beitr. Biol. Pfl.* **42**, 425–443.

Halevy, A. H. (ed.) (1989). *CRC handbook of flowering*. VI. CRC Press, Boca Raton, Florida. 753 pp.

Hallé, F., Oldeman, R. A. A. and Tomlinson, P. B. (1978). *Tropical trees and forests. An architectural analysis.* Springer-Verlag, New York. 441 pp.

Hallé, N. and Cusset, G. (1969). Anomalie remarquable chez une feuille d'*Arum maculatum* L. (Araceae). *Adansonia*, sér. 2, **9**, 451–455.

Hanstein, J. (1864). *Die Milchsaftgefässe und die verwandten Organe der Rinde.* Wiegandt and Hempel, Berlin. 92 pp.

Hara, N. (1957). Study of the variegated leaves, with special reference to those caused by airspaces. *Jap. J. Bot.* **16**, 86–101.

Harborne, J. B. (1974). Flavonoids as systematic markers in the angiosperms. pp. 103–15 In: *Chemistry in botanical classification* (eds Bendz, G. and Santesson, J.). Academic Press. New York.

HARRIS, P. J., FERGUSON, L. R., ROBERTON, A. M., MCKENZIE, R. J. and WHITE, J. B. (1992). Cell-wall histochemistry and anatomy of taro (*Colocasia esculenta*). *Aust. J. Bot.* **40**, 207–222.

——and HARTLEY, R. D. (1980). Phenolic constituents of the cell walls of monocotyledons. *Biochem. Syst. Ecol.* **8**, 153–160.

HASSACK, C. (1886). Untersuchungen über den anatomischen Bau bunter Laubblätter, nebst einigen Bemerkungen, betreffend die physiologische Bedeutung der Buntfarbung derselben. *Bot. Zbl.* **28**, 84–85, 116–121, 150–154, 181–186, 211–215, 243–246, 276–279, 308–312, 337–341, 373–375, 385–388.

HATHER, J. G. (1988). The morphological and anatomical interpretation and identification of charred vegetative parenchymatous plant remains. PhD thesis, Inst. Archaeology, Univ. London. 2 vols. 640 pp.

——(1993). *An archaeobotanical guide to root and tuber identification.* Volume 1. *Europe and South West Asia.* Oxbow Monograph 28. 154 pp.

HAY, A. (1986). *Cyrtosperma* Griffith and the origin of the aroids. Thesis, Univ. Oxford. 442 pp. (unpubl.).

——(1992a). Tribal and subtribal delimitation and circumscription of the genera of Araceae tribe Lasieae. *Ann. Missouri bot. Gard.* **79**, 184–205.

——(1992b). A new Australian genus of Araceae, with notes on generic limits and biogeography of the Areae. *Bot. J. Linn. Soc.* **109**, 427–434.

——(1997). Two new species and a new combination in Australian *Typhonium* (Araceae Tribe Areae). *Edinburgh J. Bot.* **54**, 329–336.

——and MABBERLEY, D. J. (1991). 'Transference of function' and the origin of aroids: their significance in early angiosperm evolution. *Bot. Jb.* **113**, 339–428.

HAYDON, G. F. and SHAW, D. E. (1991). Zinc accumulation in *Monstera deliciosa* near Brisbane. *Aust. Pl. Pathol.* **20 (4)**, 125–129.

HEGELMAIER, F. (1874). Zur Entwicklungsgeschichte monokotyledoner Keime nebst Bemerkungen über die Bildung der Samendeckel. *Bot. Ztg.* **32**, 631–639, 648–656, 657–671, 673–686, 689–700, 705–719.

HEGNAUER, R. (1963). *Chemotaxonomie der Pflanzen. 2. Monocotyledoneae.* Birkhäuser Verlag, Basel and Stuttgart. 540 pp.

——(1986). *Chemotaxonomie der Pflanzen.* vol. 7. *Nachträge zu Band 1 und Band 2.* BirkhäuserVerlag, Basel 804 pp.

——(1987). Phytochemistry and chemotaxonomy of the Araceae. *Aroideana* **10**, 17–19.

HENNY, R. J. (1983a). Inheritance of foliar variegation in three *Aglaonema* species. *J. Hered.* **74**, 475–476.

——(1983b). Inheritance of the white foliar midrib in *Dieffenbachia* and its linkage with the gene for foliar variegation. *J. Hered.* **74**, 483–484.

——(1986). Single locus, multiallelic inheritance of foliar variegation in *Aglaonema. J. Hered.* **77**, 214–215.

——(1992). Inheritance of the foliar variegation pattern from *Aglaonema nitidum* (Jack) Kunth 'Ernesto's Favorite'. *HortScience* **27**, 274.

HENSLOW, G. (1912). The origin of monocotyledons from aquatic dicotyledons. *J. roy. hort. Soc.* **37**, 289–294.

HERR, J. M., JR. (1992). New uses for calcium chloride solution as a mounting medium. *Biotech. Histochem.* **67**, 9–13.

HESSE, M. (1980). Entwicklungsgeschichte und Ultrastruktur von Pollenkitt und Exine bei naheverwandten entomophilen und anemophilen Angiospermensippen der Alismataceae, Liliaceae, Juncaceae, Cyperaceae, Poaceae und Araceae. *Pl. Syst. Evol.* **134**, 229–267.

——, BOGNER, J., HALBRITTER, H. and WEBER, M. (2001). Palynology of the perigoniate Aroideae: *Zamioculcas*, *Gonatopus* and *Stylochaeton* (Araceae). *Grana* **40**, 26–34.

HETTERSCHEID, W. L. A. and BOYCE, P. C. (2000). A reclassification of *Sauromatum* Schott and a new species of *Typhonium* Schott (Araceae). *Aroideana* **23**, 48–55.

HICKEY, L. J. (1977). Stratigraphy and paleobotany of the Golden Valley Formation (Early Tertiary) of western North Dakota. *Geol. Soc. Amer. Mem.* **150**, 181 pp.

——(1979). A revised classification of the architecture of dicotyledonous leaves. pp. 25–39 In: *Anatomy of the dicotyledons* (eds Metcalfe, C. R. and Chalk, L.), ed. 2, vol. 1. Clarendon Press, Oxford.

HIGAKI, T., RASMUSSEN, H. P. and CARPENTER, W. J. (1980a). Calcium deficiency of *Anthurium andreanum* Lind. spathes. *J. Amer. Soc. hort. Sci.* **105**, 438–440.

——, ——and——(1980b). Color breakdown in *Anthurium (Anthurium andreanum* Lind.) spathes caused by calcium deficiency. *J. Amer. Soc. hort. Sci.* **105**, 441–444.

——, ——and——(1984). A study of some morphological and anatomical aspects of *Anthurium andreanum* Lind. *Res. ser. Hawaii Inst. Trop. Agric. Hum. Resour. Univ. Hawaii* No. 30, 12 pp.

HIGASHI, J. and KUO, J. (1949). A pharmacognostical study on Manchurian *Arisarum. Jap. J. Pharm.* **3**, 6–10.

HIGLEY, W. K. (1880). On the microscopic crystals contained in plants. *Amer. Nat.* **14**, 720–725, 776–782.

HIGUCHI, M. and OKADA, M. (1980). On the tuber of *Pinellia cordata* and the Chinese medicinal "Sui-hange". *J. Jap. Bot.* **55**, 314–317. [Japanese only]

HINCHEE, M. A. W. (1981). Morphogenesis of aerial and subterranean roots of *Monstera deliciosa*. *Bot. Gaz.* **142**, 347–359.

——(1983). The quantitative distribution of trichosclereids and raphide crystal cells in *Monstera deliciosa. Bot. Gaz.* **144**, 513–518.

HOFMANN, E. (1930). Vorkommen, Verteilung und Funktion der Spaltöffnungen an den Blütenorganen. *Beih. bot. Zbl.* **47 (1)**, 139–168.

HOFMEISTER, W. (1858). Neuere Beobachtungen über Embryobildung der Phanerogamen. *Jb. wiss. Bot.* **1**, 82–188.

HOLFERT, J. (1889). Über die primäre Anlage der Wurzeln und ihr Wachstum. *Arch. Pharm. Berl.* **27**, 481–509.

HOLM, T. (1911a). Medicinal plants of North America 49. *Arisaema triphyllum* (L.) Tocr. *Merck's Rep.* **20**, 66–69.

——(1911b). Medicinal plants of North America 56. *Acorus calamus* L. *Merck's Rep.* **20**, 277–281.

HOLTTUM, R. E. (1955). Growth habits of monocotyledons. *Phytomorphology* **5**, 399–413.

HOTTA, M. (1964). On the juvenile plants of the genus *Arisaema*. *Acta Phytotax. Geobot.* **21**, 9–16.

HOTTA, M. (1965). Notes on Schismatoglottidinae of Borneo, I. *Mem. Coll. Sci., Univ. Kyoto* B **32**, 19–30.

——(1971). Study of the family Araceae. General remarks. *Jap. J. Bot.* **20**, 269–310.

HOTTON, C. L., LEFFINGWELL, H. A. and SKVARLA, J. J. (1994). Pollen ultrastructure and the fossil genus *Pandaniidites*. pp. 171–191 In: *Ultrastructure of fossil spores and pollen* (eds Kurmann, M. H. and Doyle J. A.). Royal Botanic Gardens, Kew.

HOWARD, R.A. (1969). The ecology of an elfin forest in Puerto Rico 8. Studies of stem growth and form and of leaf structure. *J. Arnold Arbor.* **50**, 225–267.

HUANG, T.-C. (1972). *Pollen flora of Taiwan*. Nat. Taiwan Univ. Bot. Dept. Press. 279 pp.

HUYNH, K. L. (1974) [1975]. Quelques phénomènes de polarité du pollen à plis multiples du genre *Spathiphyllum* (Araceae). *Beitr. Biol. Pfl.* **50**, 445–456.

IBRAHIM, R. K. (1965). The phenolic acid pattern of some pollen grains in relation to phylogeny. *J. Palynol.* **1**, 27–31.

INABA, K. (1984). Effect of shading on leaf anatomy in Konjak plant (*Amorphophallus konjac* K. Koch). *Jap. J. Crop Sci.* **53**, 243–248. [Japanese, English summary]

INAMDAR, J. A., SHENOY, K. N. and RAO, N. V. (1983). Leaf architecture of some monocotyledons with reticulate venation. *Ann. Bot.* **52**, 725–735.

IRMISCH, T. (1850). *Zur Morphologie der monokotylischen Knollen- und Zwiebelgewächse*. Reimer, Berlin. 286 pp.

——(1874). Beiträge zur vergleichenden Morphologie der Pflanzen. 5. Über einige Aroideen. *Abhandl. nat. Ges. Halle* **13 (2)**, 159–206.

ITO, T. (1899). Floating-apparatus of the leaves of *Pistia stratiotes*. *Ann. Bot.* **13**, 466.

ITTENBACH, S. and BOECKER, M. (1997). Untersuchungen zur Anatomie der Infloreszenzachse afrikanischer *Amorphophallus*-Arten (Monocotyledoneae, Araceae). *Bot. Jahrb. Syst.* **119**, 386–406.

JACKSON, B. P. and SNOWDON, D. W. (1968). *Powdered vegetable drugs*. J & A. Churchill, London. 203 pp.

——and ——(1990). *Atlas of microscopy of medicinal plants, culinary herbs and spices*. Belhaven Press, London. 257 pp.

JACOBSON-PALEY, R. (1920a). Sur le haustorium et la formation de l'albumen dans l'*Arum maculatum* L. *Bull. Soc. bot. Genève*, sér. 2, **12**, 55–64.

——(1920b). Le periplasmodium dans les anthères de l'*Arum maculatum* L. *Bull. Soc. bot. Genève*, sér. 2, **12**, 306–318.

JAEGER, P. (1968). Note sur la multiplication végétative du *Remusatia vupara* Schott (Aracées) et de l'*Acidanthera aequinoctialis* Baker (Iridacées). *Bull. Inst. fond. Afr. noire* **30A**, 1289–1301.

JANCZEWSKI, E. DE (1874a). Recherches sur l'accroissement terminal des racines dans les phanérogames. *Ann. Sci. nat., Bot.*, sér. 5, **20**, 162–201.

——(1874b). Recherches sur le développment des radicelles dans les phanérogames. *Ann. Sci. nat., Bot.*, sér. 5, **20**, 208–233.

JANSE, J. M. (1897). Les endophytes radicaux de quelques plantes javanaises. *Ann. Jard. bot. Buitenzorg* **14**, 53–201.

JAYALAKSHMI, S. K. and RAMACHANDRAN, K. (1993). Pollen morphology of *Anthurium* from South India. *J. Palynol.* **29**, 1–10.

JOHRI, B. M., AMBEGAOKAR, K. B. and SRIVASTAVA, P. S. (1992). *Comparative embryology of angiosperms*. 2 vols. Springer-Verlag, New York and Berlin. 1221 pp. (Araceae, Lemnaceae, vol 2, 953–968.)

JOS, J. S. and BAI, K. V. (1986). Seed set and polyembryony in *Amorphophallus campanulatus* Blume. *J. Indian bot. Soc.* **65**, 178–184.

JUEL, H. O. (1915). Untersuchungen über die Auflösung der Tapetenzellen in den Pollensacken der Angiospermen. *Jb. wiss. Bot.* **56**, 337–364 + Plates IV-V.

KÄLLERSJÖ, M., FARRIS, J. S., CHASE, M. W., BREMER, B., FAY, M. F., HUMPHRIES, C. J. *et al.* (1998). Simultaneous parsimony jackknife analysis of 2538 *rbc*L DNA sequences reveal support for major clades of green plants, land plants, seed plants, and flowering plants. *Pl. Syst. Evol.* **213**, 259–287.

KANDALOVA, G. T. (1983). Long-term life cycle of *Eminium lehmannii* (Araceae) in the eastern Karakums. *Bot. Zh. SSSR* **68**, 581–593. [Russian, English summary]

KANERVA, L., ESTLANDER, T. and AALTO-KORTE, K. (2000). Occupational protein contact dermatitis and rhinoconjunctivitis caused by spathe (*Spathiphyllum*) flowers. *Contact Dermatitis* **42**, 369–370.

KAPLAN, D. R. (1968). Interpopulational variations in foliar histogenesis in *Acorus calamus* (Araceae; Pothoideae). *Amer. J. Bot.* **55**, 715–716. [Abstract]

——(1970). Comparative foliar histogenesis in *Acorus calamus* and its bearing on the phyllode theory of monocotyledonous leaves. *Amer. J. Bot.* **57**, 331–361.

——(1973a). Comparative developmental analysis of the heteroblastic leaf series of axillary shoots of *Acorus calamus* L. (Araceae). *Cellule* **69**, 251–290.

——(1973b). The monocotyledons: their evolution and comparative biology: VII. The problem of leaf morphology and evolution in the monocotyledons. *Quart. Rev. Biol.* **48**, 437–57.

——(1975). Comparative developmental evaluation of the morphology of unifacial leaves in the monocotyledons. *Bot. Jb.* **95**, 1–105.

——(1984). Alternative modes of organogenesis in higher plants. pp. 261–300 In: *Contemporary problems in plant anatomy* (eds White, R. A. and Dickison, W. C.). Academic Press, Orlando.

——and BRUCK, D. K. (1980). Leaf (botany). In: *McGraw-Hill Yearbook Science and Technology*, McGraw-Hill. 4 pp.

KAPOOR, L. D., SINGH, A., KAPOOR, S. L. and SRIVASTAVA, S. N. (1975). Survey of Indian plants for saponins, alkaloids and flavonoids. 4. *Lloydia* **38**, 221–224.

KARAGYOZOVA, M. D. (1968–9) [1971]. Localisations des cristaux d'oxalate de calcium chez *Anthurium scherzerianum* Schott. *Sofii Univ., biol. Fak., God.* **63**, 43–53.

KARAWYA, M. S., WASSEL, G. M., BAGHDADI, H. H. and AMMAR, N. M. (1980). Mucilagenous contents of certain Egyptian plants. *Planta med.* **38**, 73–78.

KASAPLIGIL, B. (1961). Foliar xeromorphy of certain geophytic monocotyledons. *Madroño* **16**, 43–70.

KATTEIN, A. (1897). Der morphologische Werth des Centralcylinders der Wurzel. *Bot. Zbl.* **72**, 55–61, 91–97, 129–139.

KEATING, R. C. (1996). Anther investigations: a review of methods. pp. 255–271 In: *The anther, form, function and phylogeny* (eds D'Arcy, W. G. and Keating, R. C.). Cambridge Univ. Press, New York.

——(1997). Vegetative anatomy. pp. 340–341 In: A revision of *Philodendron* subgenus *Philodendron* (Araceae) for Mexico and Central America (by Croat, T. C.). *Ann. Missouri bot. Gard.* **4 (3)**, 311–704.

——(2000). Collenchyma in Araceae: trends and relation to classification. *Bot. J. Linn. Soc.* **134**, 203–214.

——(2002a). Systematic occurrence of raphide crystals in Araceae. *Missouri Bot. Gard. Monographs* (in press).

——(2002b). A revised outline of the genera of Araceae. *Missouri Bot. Gard. Monographs* (in press).

KENTEN, R. H. and WOODS, R. D. (1973). Viruses of *Colocasia esculenta* and *Xanthosoma saggitifolium*. *PANS* **19**, 38–41.

KILLIAN, C. (1929). Développement et biologie de l'*Ambrosinia bassii* L. 1. *Bull. Soc. Hist. nat. Afr. N.* **20**, 257–278.

KIMURA, K., HATA, K. and NISHIOKA, I. (1963). Pharmacognostical studies on Chinese drug "Ch'ang P'u", Chinese *Acorus*-rhizomes. *Syoyakugaku Zasshi* **17 (1–2)**, 25–35. [Japanese, English summary]

KIMURA, Y. and NAGAMACHI, T. (1935). Ueber die japanischen Arzneidrogen, ihren Anbau, ihre Einsammlung und Zubereitung etc. (VII). Ueber die japanische Kalmus und Wurzel von *Acorus gramineus*. *J. Jap. Bot.* **11**, 58–67. [Japanese]

KING, H. W. (1894). Physiology of *Monstera deliciosa*. *J. Quekett microsc. Cl,.* ser. 2, **5**, 125–136.

KIRCHNER, O., LOEW, E. and SCHRÖTER, C. (1904–38). In: *Lebensgeschichte der Blütenpflanzen Mitteleuropas*. Stuttgart. [Issued in parts.] [1934; I, Abt. 3, 1–56, *Arum, Dracunculus, Arisarum*. 57–80, Lemnaceae]

KISSER, J. (1928). Untersuchungen über das Vorkommen und die Verbreitung von Pektinwarzen. *Jb. wiss. Bot.* **68**, 206–232.

KLEBAHN, H. (1884). Die Rindenporen. Ein Beitrag zur Kenntniss des Baues und der Function der Lenticellen und der analogen Rindenbildungen. *Jen. Z. Naturwiss.* **17**, 537–592.

KLEIN, G. (1923). Zur Atiologie der Thyllen. *Z. Bot.* **15**, 417–439.

KLERCKER, J. E. F. AF (1888). Studien über die Gerbstoffvakuolen. *Bih. svenska Vet.-Akad. Handl.* **13 (8)**, 1–63.

KNECHT, M. (1983). Contribution à l'étude biosystématique des représentants d'Aracées de la Côte d'Ivoire. *Phanérogam. Mon.* 17. J. Cramer, Vaduz. 290 pp.

KNOLL, F. (1923). Über die Luckenepidermis der *Arum*-spatha. *Öst. bot. Z.* **72**, 246–254.

——(1948). Bau, Entwicklung und morphologische Bedeutung unifazialer Verläuferspitzen an Monokotylenblättern. *Öst. bot. Z.* **95**, 163–193.

KNOTH, R. (1981). Ultrastructure of lycopene containing chromoplasts in fruits of *Aglaonema commutatum* Schott (Araceae). *Protoplasma* **106**, 249–259.

KO, S. C., OH, B. U., LEE, H. S. and KIM, Y. S. (1990). A phylogenetic study of *Arisaema* by anatomical and palynological characters. I. Anatomical characters. *Korean J. Pl. Taxon.* **20**, 9–35. [Korean, English summary]

KOACH, J. and GALIL, J. (1986) [1987]. The breeding system of *Arisarum vulgare* Targ.-Tozz. *Israel J. Bot.* **35**, 79–90.

KOERNICKE, M. (1937). Über den Bonner *Amorphophallus titanum* Becc. *Ber. dt. bot.Ges.* **55**, 427–436.

KONSTANTY, E. C. (1926). Ueber der Entstehung der Kristallzellenreihen mit besonderer Berücksichtigung der Drogenpflanzen. *Bot. Archiv* **15**, 131–186.

KORNER, C., ALLISON, A. and HILSCHER, H. (1983). Altitudinal variation of leaf diffusive conductance and leaf anatomy in heliophytes of montane New Guinea and their interrelation with microclimate. *Flora* **174**, 91–135.

KORSCHGEN, L. J. (1981). Study No. 32. Histological characteristics of plant leaf epidermis and related structures as an aid in food habits. Job No. 1. Histological characteristics of plant leaf epidermis and related structures as an aid in food habits studies. (Monocotyledons other than grasses and sedges.) Performance Report. Missouri, Federal Aid Project No. W-13-R-35. 162 pp.

KOVACS, A. and RAKOVAN, J. N. (1975). Development of raphide idioblasts in the aerial root of *Monstera deliciosa* Liebm. *Acta agron. hung.* **24**, 39–52.

KRAUS, G. (1896). Physiologisches aus den Tropen. Über Blüthenwärme bei Cycadeen, Palmen und Araceen. *Ann. Jard. Bot. Buitenzorg* **13**, 217–275.

KRAUSE, K. (1908). Araceae-Calloideae. *Das Pflanzenreich* IV, **23B**, 140–155.

——(1913). Araceae-Philodendrinae. *Das Pflanzenreich* IV, **23Db**, 1–143.

KROEMER, K. (1903). Wurzelhaut, Hypodermis und Endodermis der Angiospermenwurzel. *Bibl. Bot.* **12**, Heft 59, 151 pp.

KUBIN, E. (1878). Die Entwickelung von *Pistia stratiotes*. *Bot. Abhandl. Bonn* **3**, 1–30.

KUBITZKI, K. (ed.) (1998). *The families and genera of vascular plants. IV. Flowering plants – monocotyledons.* Springer-Verlag, Berlin.

KULKARNI, A. R., DOSI, D. and MANOJ, V. M. (1990). Fruit and seed structure in Araceae. *Proc. Indian Acad. Sci., Pl. Sci.* **100**, 61–69.

KUNDU, B. C. (1942). On the multiperforate plates occurring in the xylem vessels of some monocotyledonous roots. *Proc. Indian Acad. Sci.*, B **16**, 180–189.

KUPRIANOVA, L. A. and TARASEVICH, V. F. (1984). The ultrastructure of the surface of pollen grain wall in some genera of the family Lemnaceae and the related genera of the family Araceae. *Bot. Zh. SSSR* **69**, 1656–1662 + 2 plates. [Russian]

KUSTER, E. (1928–9). Regenerationserscheinungen bei *Pinellia tuberifera*. *Ber. oberhess. Ges. Nat. Heilk.* **12**, 3–10.

KVAČEK, Z. (1995). *Limnobiophyllum* Krassilov (Araceae)—a fossil link to the Lemnaceae. *Aquat. Botany* **50**, 49–61.

LACK, A. J. and DIAZ, A. (1991). The pollination of *Arum maculatum* L.—a historical review and new observations. *Watsonia* **18**, 333–342.

LAKOWITZ, C. W. (1881). *Amorphophallus rivieri* Dur. und*A. campanulatus* Bl. Diss., Breslau. 41 pp.

LAMANT, A. and HELLER, R. (1967) [1968]. Sur la contraction des racines d'*Arum italicum*. *Bull. Soc. fr. Physiol. vég.* **13**, 179–193.

LANDOLT, E. (1980). Bibliographie der Familie der Lemnaceae. *Veröff. Geobot. Inst. ETH, Stiftung Rübel Zurich* **70**, 142–204.

——(1998). Anatomy of the Lemnaceae (duckweeds). pp. 1–127 In: *Extreme adaptations in angiospermous hydrophytes* (eds Landolt, E., Jäger-Zürn, I. and Schell, R. A. A.). Borntraeger, Berlin.

LANGLET, O. (1927). Zur Kenntnis der polysomatischen Zellkerne in Wurzelmeristem. *Svensk bot. Tidskr.* **21**, 397–422.

LAUX, W. (1887) [1888]. Ein Beitrag zur Kenntnis der Leitbündel im Rhizom monokotyler Pflanzen. Diss. Berlin. 49 pp. (1887). Also *Verh. bot. Ver. Brandenburg* **29**, 65–111.

LAVRUKHINA, L. F. and AKMEDOV, U. A. (1983). Anatomical and morphological study of *Eminium regelii*. *Uzbek. biol. Zh.* **(6)**, 30–33. [Russian]

LAZENBY, W. R. (1890). Notes upon the crystals in certain species of the *Arum*family. *Proc. Amer. Ass. Adv. Sci.* **39**, 332.

LEAF ARCHITECTURE WORKING GROUP. (1999). *Manual of leaf architecture*. Morphological description and categorization of dicotyledonous and net-veined monocotyledonous angiosperms. Smithsonian Institution. 65 pp.

LEAVITT, R. G. (1904). Trichomes of the root in vascular cryptogams and angiosperms. *Proc. Boston Soc. nat. Hist.* **31**, 273–313.

LEBLOIS, A. (1887). Recherches sur l'origine et le développement des canaux sécréteurs et des poches sécrétrices. *Ann. Sci. nat., Bot.*, sér. 7, **6**, 247–330.

LEDBETTER, M. C. and PORTER, K. R. (1970). *Introduction to the fine structure of plants*. Springer-Verlag, New York.

LEE, D. W. and RICHARDS, J. H. (1991). Heteroblastic development in vines. pp. 205–243 In: *The biology of vines* (eds Putz, F. E. and Mooney, H. A.). Cambridge Univ. Press, Cambridge.

LEHMANN, N. L. and SATTLER, R. (1992). Irregular floral development in *Calla palustris* (Araceae) and the concept of homeosis. *Amer. J. Bot.* **79**, 1145–1157.

LEICK, E. (1915). Die Erwärmungstypen der Araceen und ihr blütenbiologische Deutung. *Ber. dt. bot. Ges.* **33**, 518–536.

LEINFELLNER, W. (1954). Die petaloiden Staubblätter und ihre Beziehungen zu den Kronblättern. *Öst. bot. Z.* **101**, 373–406.

LEITGEB, H. (1864). Die Luftwurzeln der Orchideen. *Denkschr. Akad. Wiss. Wien,math.-naturw. Kl.* **24**, 179–222.

LEMON, G. D. and POSLUSZNY, U. (2000a). Shoot development and evolution in *Pistia stratiotes* (Araceae). *Intn. J. Pl. Sci.* **161**, 721–732.

——and ——(2000b). Comparative shoot development and evolution in the Lemnaceae. *Intn. J. Pl. Sci.* **161**, 733–748.

LEOPOLD, E. B. (1969). Miocene pollen and spore flora of Eniwetok Atoll, Marshall Islands. *Geol. Surv. Prof. Pap.* **260–II**, 1133–1185.

LES, D. H. and CRAWFORD. D. J. (1999). *Landoltia* (Lemnaceae), a new genus of duckweeds. *Novon* **9**, 530–533.

——, LANDOLT, E., and CRAWFORD, D. J. (1997). Systematics of the Lemnaceae (duckweeds): inferences from micromolecular and morphological data. *Pl. Syst. Evol.* **204**, 161–177.

LESICA, P. and ANTIBUS, R. K. (1990). The occurrence of mycorrhizae in vascular epiphytes of two Costa Rican rainforests. *Biotropica* **22**, 250–258.

LI, H. (1986). The ecological phytogeography and origin of the Araceae. *Acta bot. yunnan.* **8**, 363–382. [In Chinese, English summary].

——and HAY, A. (1992). Notes on the classification of genera *Remusatia* and *Gonatanthus* in Araceae. *Acta bot. yunnan.*, Suppl. 5, 27–33. [In Chinese].

LI, L. (1999). Micromorphological characters and enzyme analysis of some Chinese Pinellias. *Acta bot. yunnan.* **21**, 442–448 +2 plates [Chinese, English summary]

LI, W.-H., WOLFE, K. H. and SHARP, P. (1993). Origin of angiosperms inferred from DNA sequence data. *Recent Adv. Bot. (Inst. Bot.; Acad Sinica Monog. Ser.* **13**), 153–162.

LIERAU, M. (1888). Über die Wurzeln der Araceen. *Bot. Jb.* **9**, 1–38.

LIGNIER, O. (1915?) (n. d.). Nouveau cas de cambium chez les monocotylédones. *C. r. Ass. fr. Avanc. Sci.*, Session 43, Le Havre 1915, 487–490.

LINDINGER, L. (1905–06). Zur Anatomie und Biologie der monokotylen Wurzel.*Beih. bot. Zbl.* **19 (1)**, 321–358.

——(1908). Die Bewurzelungsverhältnisse grosser Monocotylenformen und ihre Bedeutung für den Gärtner. *Gartenflora* **57**, 281–291, 308–318, 367–377.

LINDORF, H. (1980). Leaf structure of 15 shade monocotyledons of the cloud forest of Rancho Grande: 1. Bifacials: Araceae, Marantaceae, Musaceae. *Mem. Soc. Cienc. nat. La Salle* **40 (113)**, 19–72. [Spanish, English summary]

LINSBAUER, K. (1924). Über Teilungsanomalien und metaplastische Chlorophyllbildung in der Epidermis von *Monstera*. *Z. Pflanzenkrankh.* **34**, 220–223.

——(1930). *Die Epidermis.* [Handbuch der Pflanzenanatomie. Abt. 1, Teil 2, vol. 4]. Gebrüder Borntraeger, Berlin. 283 pp.

——(1934). Die Bulbillenbildung bei *Pinellia ternata* und die Frage der Hemmungstoffe. *Ber. dt. bot. Ges.* **52**, 574–580.

LIPAYEVA, L. I. (1989). On the anatomy of petals in angiosperms. *Bot. Zh. SSSR* **74**, 826–830. [Russian only]

LOH, J. P., KIEW, R., HAY, A., KEE, A., GAN, L. H. and GAN, Y. Y. (2000). Intergeneric and interspecific relationships in Araceae tribe Caladieae and development of molecular markers using amplified fragment length polymorphism (AFLP). *Ann. Bot.* **85**, 371–378.

LOHSE, R. (1924). Entwurf einer Kritik der Thyllenfrage mit Ergebnissen eigener Versuche. *Bot. Archiv* **5**, 345–380.

LOPEZ-PORTILLO, J., EWERS, F. W., ANGELES, G. and FISHER, J. B. (2000). Hydraulic architecture of *Monstera acuminata*: evolutionary consequences of the hemiepiphytic growth form. *New Phytol.* **145**, 289–299.

LÖV, L. (1925–6). Zur Kenntnis der Entfältungszellen monokotylen Blätter. *Flora* **120**, 282–343.

LOVETT DOUST, L., LOVETT DOUST, J. and TURI, K. (1986). Fecundity and size relationships in Jack-in-the-pulpit, *Arisaema triphyllum* (Araceae). *Amer. J. Bot.* **73**, 489–494.

LUGARDON, B., LOBREAU CALLEN, D. and LE THOMAS, A. (1987–8) [1988]. Structures polliniques chez les Araceae. I, Tribu des Spathiphylleae. *J. Palynol.* **23–24**, 51–57.

LUHAN, M. (1957). Das Verhalten der Rhizomgewebe einiger Wasser- und Sumpfpflanzen bei Vitalfärbung. *Ber. dt. bot. Ges.* **70**, 361–370.

MABBERLEY, D. J. (1997). *The plant book.* Cambridge Univ. Press, Cambridge. 858 pp.

MACDOUGAL, D. T. (1901). Seedlings of *Arisaema. Torreya* **1**, 2–5.

MADISON, M. (1977). A revision of *Monstera* (Araceae). *Contrib. Gray Herb.*, **207**, 3–100.

——(1978). A synopsis of *Caladiopsis* (Araceae). *Contrib. Gray Herb.* **208**, 95–98.

——(1979). Additional observations on ant-gardens in Amazonas. *Selbyana* **5**, 107–115.

——and TIFFNEY, B. H. (1976). The seeds of the Monstereae: their morphology and fossil record. *J. Arnold Arbor.* **57**, 185–204.

MAERCKER, U. (1965). Über das Vorkommen von Stomata in der Epidermis bunter Perianthblätter. *Z. Pflanzenphysiol.* **53**, 422–428.

MAHESHWARI, S. C. (1958). *Spirodela polyrrhiza*: the link between the aroids and the duckweeds. *Nature* **181**, 1745–1746.

——and KHANNA, P. P. (1956). The embryology of *Arisaema wallichianum* Hook. f. and the systematic position of the Araceae. *Phytomorphology* **6**, 379–388.

MAI, D. H. and GREGOR, H. J. (1982). Neue und interessante Arten aus dem Miozän von Salzhausen im Vogelsberg. *Feddes Repert.* **93**, 405–435 + plates XVII–XXIII.

MAILLEFER, A. (1916–17). Anatomie de la racine d'*Acorus calamus. Bull. Soc. Vaud. Sci. nat.* **51**, *Proc. Verb.* 136.

——(1920). Sur la présence d'une assise dans la racine d' *Acorus calamus. Bull. Soc. Vaud. Sci. nat.* **53**, 77–79.

MAJUMDAR, G. P. (1929). Secondary growth in thickness in the roots of *Amorphophallus campanulatus* Bl. *Proc. 16th Indian Sci. Congr. Assn.*, Madras **16**, 229–230.

——and PAL, P. (1961). Developmental studies: VI. The morphology of the so-called stipule of *Piper* (Piperaceae) from a comparative ontogenetical study of this structure in *Piper betle, Piper longum, Piper nigrum, Piperornatum* and two other species of *Piper* and the morphology of the leaf-sheath of *Scindapsus officinalis* (Araceae). *Proc. Nat. Inst. Sci. India* B **27**, 26–39.

MANGIN, L. (1880). Relations anatomiques entre la tige, la feuille et l'axe floral de l'*Acorus calamus. Bull. Soc. Sci. Nancy*, 31 pp.

——(1882). Origine et insertion des racines adventives et modifications corrélatives de la tige chez les Monocotylédones. *Ann. Sci. nat., Bot.*, sér. 6, **14**, 216–363.

MANNICH, C. and YIH, S. (1926). Über die Knollen von *Pinellia tuberifera.* pp. 158–61 In: *Festschrift für Alexander Tschirch.* Tauchnitz, Leipzig.

MANNING, J. C. (1996). Diversity of endothecial patterns in the angiosperms. pp. 136–158 In: *The anther. Form, function and phylogeny* (eds D'Arcy, W. G. and Keating, R. C.). Cambridge Univ. Press, New York.

MANYA-CHERNEJ, E. N. (1978). Anatomical and morphological studies of inflorescence in some species of Araceae. *Bot. Zh. SSSR* **63**, 510–522. [Russian, English summary]

MARCATILI, L. (1887). Il vasi laticiferi ed il sistema assimilatore. *Ann. R. Ist. Bot. Roma* **3**, 17–42.

MARCHANT, C. J. (1970). Chromosome variation in Araceae: I. Pothoeae to Stylochitoneae. *Kew Bull.* **24**, 315–322.

——(1971a). Chromosome variation in Araceae. II: Richardieae to Colocasieae. *Kew Bull.* **25**, 47–56.

——(1971b). Chromosome variation in Araceae III. Philodendreae to Pythonieae. *Kew Bull.* **25**, 323–329.

——(1972). Chromosome variation in Araceae IV: Areae. *Kew Bull.* **26**, 395–404.

——(1973). Chromosome variation in Araceae: V. Acoreae to Lasieae. *Kew Bull.* **28**, 199–210.

MARLATT, R. B. (1970). Suberin and periderm formation in *Dieffenbachia. J. Amer. Soc. hort. Sci.* **95**, 32–33.

MASCRÉ, M. (1928). Sur le tapis staminal et le grain de pollen de l *Arum maculatum* L. *C. r. Acad. Sci., Paris* **186**, 1642–1644.

MASHIMA, M. (1964). Studies of the stomatal complex in *Arisaema serratum* Scott forma *thunbergii* Makino, with special reference to its pattern of organization and developmental sequence. *Mem. Fac. liberal Arts Educ. Shiga Univ., nat. Sci.* **14**, 49–64.

MAYO, S. J. (1986). Systematics of *Philodendron* Schott (Araceae) with special reference to inflorescence characters. PhD thesis, Reading Univ. 2 vols. 972 pp.

——(1989). Observations of gynoecial structure in *Philodendron* (Araceae). *Bot. J. Linn. Soc.* **100**, 139–172.

——(1991). A revision of *Philodendron* subgenus *Meconostigma* (Araceae). *Kew Bull.* **46**, 601–681.

——and BOGNER, J. (1988). A new species of *Caladium* (Araceae) with notes on generic delimitation in the Colocasioideae-Caladieae. *Willdenowia* **18**, 231–242.

——, ——and BOYCE, P. C. (1997). *The genera of Araceae.* Royal Botanic Gardens, Kew. 370 pp.

——, ——and——(1998). Araceae. pp. 26–74 In: *Families and genera of vascular plants: Monocotyledons* (ed. Kubitzki, K.). Springer-Verlag, Berlin.

MBAYA, N. (1976). Recherches sur les feuilles unifaciales chez les monocotylédones. Thèse, Dijon. 82 pp.

MCCLURE, J. W. (1970). Secondary constituents of aquatic angiosperms. pp. 233–268 In: *Phytochemical phylogeny* (ed. Harborne, J. B.). Academic Press, New York.

MEEUSE, B. J. D. (1975). Thermogenic respiration in aroids. *Ann. Rev. Pl. Physiol.* **26**, 117–126.

——and RASKIN, I. (1988). Sexual reproduction in the arum lily family, with emphasis on thermogenicity. *Sexual Pl. Reprod.* **1**, 3–15.

——, SCHNEIDER, E. L., HESS, C. M., KIRKWOOD, K. and PATT, J. M. (1984). Activation and possible role of the "food-bodies" of *Sauromatum* (Araceae). *Acta bot. neerl.* **33**, 483–496.

MEHRA, P. N. and SACHDEVA, S. K. (1976). Cytological observations on some W. Himalayan monocots. V. Araceae. *Cytologia* **41**, 55–61.

MEISCHNER, V. and SCHRÖDER, P. (1990). Sauerstoffzufuhr zu den Wurzeln von Pflanzen aus Feuchtgebieten. *Phyton* (Horn) **30**, 37–45.

MELATI, M R. and SCIALABBA, A. (1982). Indagini sulla *Colocasia antiquorum* (Schott) var. *esculenta* (L.) Schott. *Atti Accad. Sci. Lett. Arti Palermo*, ser. 4, **40**, 161–167.

——and ——(1984–85) [1986]. Contributo allo studio di *Xanthosoma caracu* C. Koch et Bouché: struttura e potenzialita. *Atti Accad. Sci. Lett. Arti Palermo* , ser. 5, **5**, 195–215.

MELVILLE, R. (1971). Some general principles of leaf evolution. *S. Afr. J. Sci.* **67**, 310–316.

——and WRIGLEY, F. A. (1969). Fenestration in the leaves of *Monstera* and its bearing on the morphogensis and colour patterns of leaves. *Bot. J. Linn. Soc.* **62**, 1–16.

MERCADO-NORIEL, L. R. and MERCADO, B. T. (1977–8). Floral anatomy and seed morphology of water lettuce (*Pistia stratiotes*). *Philipp. Agric.* **61**, 281–290.

MESSERI, E. (1925). Ricerche sullo sviluppo del sistema vascolare in alcune monocotiledoni. *Nuovo G. bot. ital.* (NS) **32**, 317–362.

METSÄVAINIO, K. (1931). Untersuchungen über das Wurzelsystem der Moorpflanzen. *Ann. Bot. Soc. zool.-bot. fenn. Vanamo* **1 (1)**, 1–422.

MEUSEL, H. (1951). Die Bedeutung der Wuchsform für die Entwicklung des natürlichen Systems der Pflanzen. *Feddes Repert. Spec. Nov. Regni Veget.* **54**, 137–172.

MEYER, F. J. (1925). Untersuchungen über den Strangverlauf in den radialen Leitbündeln der Wurzeln. *Jb. wiss. Bot.* **65**, 88–97.

——(1959). Beiträge zur Kenntnis der Assimilationsgewebe II. Isodiametrische Assimilationszellen. *Ber. dt. bot. Ges.* **72**, 25–36.

——(1962). *Das trophische Parenchym. A. Assimilationsgewebe.* (Handb. Pflanzenanat. IV, 7A). ed. 2. Borntraeger, Berlin. 88 pp.

MIDDENDORF, E. A. (1968). Plants with blowguns. *Turtox News* **46**, 162–164.

——(1982). The remarkable shooting idioblasts. *Carolina Tips* **45**, 25–26.

——(1983). The remarkable shooting idioblasts. *Aroideana* **6**, 9–11.

MIKESELL, J. (1990). Anatomy of terminal haustoria in the ovule of plantain (*Plantago major* L.) with taxonomic comparison to other angiosperm taxa. *Bot. Gaz.* **151**, 452–464.

MINDEN, M. VON (1899). Beiträge zur anatomischen und physiologischen Kenntnis wasser-secernierender Organe. *Bibl. Bot.* **9**, Heft 46, 1–76.

MITRA, E. (1966) [1968]. Contributions to our knowledge of Indian freshwater plants. 5. On the morphology, reproduction and autecology of *Pistia stratiotes* Linn. *J. Asiatic Soc.* **8**, 115–135.

MITTMANN, R. (1888). Beiträge zur Kenntnis der Anatomie der Pflanzen-stacheln. *Verh. bot. Ver. Brandenburg* **30**, 32–72.

MÖBIUS, M. (1885). Die mechnischen Scheiden der Secretbehälter. *Jb. wiss. Bot.* **16**, 262–301.

——(1887). Ueber das Vorkommen concentrischer Gefässbündel mit centralem Phloëm und peripherischem Xylem. *Ber. dt. bot. Ges.* **5**, 2–24.

——(1899). Ueber Bewegungsorgane an Blattstielen. pp. 37–62 In: *Botanische Untersuchungen S. Schwendener zum 10 Februar (1899) dargebracht.* Borntraeger, Berlin.

——(1935). Über Brutknospen bei Araceen. *Bot. Jb.* **67**, 123–31.

MOFFLER, M. D. (1983). A commentary on the anatomy of a neotropical species of *Homalomena. Ariodeana* **6**, 19–24.

MOLE, S. (1993). The systematic distribution of tannins in the leaves of angiosperms: a tool for ecological studies. *Biochem. Syst. Ecol.* **21**, 833–846.

MOLISCH, H. (1899). Über Zellkerne besonderer Art. *Bot. Z.* **57**, 177–191.

——(1903). Das Hervorspringen von Wassertropfen aus der Blattspitze von *Colocasia nymphaeifolia* Kth. (*Caladium nymphaeifolia* Hook.). *Ber. dt. bot. Ges.* **21**, 38–90.

MOLL, J. W. and JANSSONIUS, H. H. (1923). *Botanical penportraits.* Martinus Nijhoff, The Hague. 472 pp.

MOLLENHAUER, H. H. and LARSON, D. A. (1966). Developmental changes in raphide-forming cells of *Vanilla planifolia* and *Monstera deliciosa. J. ultrastruct. Res.* **16**, 55–70.

MOOKERJEA, A. (1955). Cytology of different species of aroids with a view to trace the basis of their evolution. *Caryologia* **7**, 221–291.

MORASSI BONZI, L. and FABBRI, F. (1975). Chloroplast protrusions in *Arisarum proboscideum* (L.) Savi. *Caryologia* **28**, 407–426.

——and ——(1978). Paracrystalline inclusions in the chloroplasts of sieve parenchyma cells of *Arum italicum* Mill. Preliminary report. *Caryologia* **31**, 129–136.

MOREL, P. (1910). Salsepareille. Notes sur quelques falsifications et substitutions. *Trav . Lab. Mat. med., Paris* **6 (4)**, 8 pp.

MOTLEY, T. J. (1994). The ethnobotany of Sweet Flag, *Acorus calamus* (Araceae). *Econ. Bot.* **48**, 397–412.

MOTTE, J. (1932). Le Konnyaku du Japon. *Ann. Mus. colon. Marseille* , sér. 4, **10 (3)**, 1–22.

MUELLER, W. C. and BECKMAN, C. H. (1979). Isotropic layers in the secondary cell walls of fibers in the roots of banana and other monocotyledons. *Can. J. Bot.* **57**, 2776–2781.

MÜHLDORF, A. (1925). Über den Ablösungsmodus der Gallen von ihren Wirtspflanzen nebst einer kritischen Übersicht über die Trennungserscheinungen im Pflanzenreiche. *Beih. bot. Centralbl.* **42 (1)**, 1–110.

——(1927). Beiträge zur Anatomie und Physiologie der Lentizellen besonder bei den Gymnospermen und den Araceen. *Bull. Fac. Stiinte Cernauti* **1**, 326–329.

MULLER, J. (1981). Fossil pollen records of extant angiosperms. *Bot. Rev.* **47**, 1–142.

MÜLLER, L. (1919). Über Hydathoden bei Araceen. Wien. 28 pp. Also *S. B. Akad. Wiss. Wien, math.-naturw. Kl.* I, **128**, 665–692.

MURASHIGE, T. (1974). Plant propagation through tissue cultures. *Ann. Rev. Pl. Physiol.* **25**, 135–166.

MURATA, J. (1984). An attempt at an infrageneric classification of the genus *Arisaema* (Araceae). *J. Fac. Sci. Univ. Tokyo, Sect. III, Bot.* **13**, 431–482.

——(1989). Diversity in the stem morphology of *Arisaema*, Araceae. *Pl. Species Biol.* **2**, 57–66.

——(1990a). Developmental patterns of pedate leaves in tribe Areae (Araceae-Aroideae) and their systematic implication. *Bot. Mag., Tokyo* **103**, 339–343.

MURATA, J. (1990b). Present status of *Arisaema* systematics. *Bot. Mag., Tokyo* **103**, 371–382.

——(1990c). Diversty of shoot morphology in *Typhonium* (Araceae). *Amer. J. Bot.* **77**, 1475–1481.

NAHRSTEDT, A. (1975). Triglochinin in Araceen. *Phytochemistry* **14**, 2627–2628.

NAMBA, T., YONEDA, K., KUBO, M. and TAKAHASHI, S. (1968). Pharmacological studies on the Chinese crude drugs Lang-tu and Ta-chi. I. On the marketed Lang-tu. *Syoyakugaku Zasshi* **22 (2)**, 115–124. [Japanese, English summary]

NAPP-ZINN, K. (1953). Studien zur Anatomie einiger Luftwurzeln. *Öst. bot. Z.* **100**, 322–330.

NASIM, G. (1990). Vesicular-arbuscular mycorrhizal endophytes in roots, scale leaves and epidermal cells of the rhizomatous tissues of *Colocasia antiquorum Schott. Sci. Khyber* **3 (2)**, 183–194.

NECHAEVA, N. T. (1943). Scientific notes. Observations on *Eminium lehmannii* (Bge) O. Ktze. in the KaraKum. *Sovet. Bot.* **(3)** 28–30. [In Russian]

NEGER, F. W. (1922). Beiträge zur Kenntnis des Baues und der Wirkungsweise der Lentizellen. II. *Ber. dt. bot. Ges.* **40**, 306–313.

NETOLITZKY, F. (1929). *Die Kieselkörper. Die Kalksalze als Zellinhaltskörper*. pp. 1–80 In: Handbuch der Pflanzenanatomie vol. III (1a). Gebrüder Borntraeger, Berlin.

NICOLAS, G. (1916). Sur l'existence d'un cambium chez quelques Aracées. *Bull. Soc. Hist. nat. Afr. noire* **7**, 106–108.

NICOLSON, D. H. (1960). The occurrence of trichosclereids in the Monsteroideae (Araceae). *Amer. J. Bot.* **47**, 598–602.

——(1969). A revision of the genus *Aglaonema* (Araceae). *Smithsonian Contrib. Bot.*, No. 1, 69 pp.

——(1984). *Amorphophallus konjac* vs. *A. rivieri* (Araceae). *Aroideana* **7**, 7–8.

NIKLAS, K. J. (1991). Effects of tissue volume and location on the mechanical consequences of dehydration of petioles. *Amer. J. Bot.* **78**, 361–369.

NODA, K., PRAKONGVONGS, C. and CHAIWIRATNUKUL, L. (1985). *Topography of the seeds and leaves of tropical weeds; with scanning electron microscope*. Project manual No. 2. National Weed Science Research Institute Project, Bangkok, Thailand. 159 pp.

NORO, Y., HISATA, Y., OKUDA, K. and KAWAMURA, T. (1992). Applications of scanning electron microscopy for pharmacognostical studies (1). New preparations for scanning electron microscope and fine details of needle crystals in the monocotyledon. *Shoyakugaku Zasshi* **46**, 115–124. [Japanese, English summary]

NOUGARÈDE, A. and RONDET, P. (1981). Fonctionnement sympodial continu et multiplication végétative chez l'*Arum italicum*. *Can. J. Bot.* **59**, 238–250.

NTÉPÉ-NYAMÈ (1984). Cinq nouvelles espèces pour le genre *Culcasia* P. Beauv. (Araceae). *Bull. Mus. natn. Hist. nat., Paris* , sér. 4, **6**, sect. B, *Adansonia* **(3)**, 313–323.

NUNES CALDEIRA, G. C. (1967). Contribuição para o conhecimento da distribuição dos compostos fenólicos nas Monocotiledóneas. *Rev. Est. ger. Univ. Moçambique, Lourenço Marques*, ser. II, **4**, 271–334.

NWUFO, M. I. and FAJOLA, A. O. (1985). Histological and biochemical changes in corms of *Colocasia esculenta* infected by *Botryodiplodia theobromae* and *Sclerotium rolfsii*. *J. Food Biochem.* **9 (1)**, 15–26.

NYAWUANE, H. G. K. and GILL, L. S. (1990). Structure and ontogeny of stomata in some tropical ornamental plants (Monocotyledons). *G. bot. ital.* **124**, 249–258.

NYMAN, L. P., WEBB, E. L., GU, Z. and ARDITTI, J. (1986). Structure and in vitro growth of zygotic embryos of taro (*Colocasia esculenta* var. *antiquorum*). *Ann. Bot.* **57**, 623–630.

O'BRIEN, T. P. and MCCULLY, M. E. (1981). *The study of plant structure. Principles and selected methods.* Termarcarphy, Melbourne.

OGDEN, E. C. (1974). Anatomical patterns of some aquatic vascular plants of New York. *N.Y. State Mus. Sci. Serv., Bull.* **424**, 133 pp.

OH, B. U., KO, S. C., HONG, W. P. and KIM, Y. S. (1990). A phylogenetic consideration of *Arisaema* by anatomical and palynological characters. II. Palynological characters. *Korean J. Pl. Taxon* . **20 (1)**, 37–52. [Korean, English summary]

OHASHI, H., MURATA, J. and TAKAHASHI, M. (1983). Pollen morphology of the Japanese *Arisaema* (Araceae). *Sci. Rep. Tohôku Univ.*, ser. 4, *Biol.* **38**, 219–251.

OHLENDORF, O. (1907). *Beiträge zur Anatomie und Biologie der Früchte und Samen einheimischer Wasser- und Sumpfpflanzen.* Diss., Erlangen. Also: A. Liesecke, Osnabrück. 109 pp.

OHTSUKI, T. (1960). On main reserve polysaccharide in tubers of *Amorphophallus rivieri* and *Arisaema atrorubens*. *J. Jap. Bot.* **35**, 310–314. [Japanese, English summary]

OLIVIER, L. (1880). Note sur le système tégumentaire des racines chez les phanérogames. *Bull. Soc. bot. Fr.* **27**, 233–241.

——(1881). Recherches sur l'appareil tégumentaire des racines. *Ann. Sci. nat., Bot.*, sér. 6, **11**, 5–133.

ONKEN, A. (1922). Ueber die Bedeutung des Milch- und Schleimsaftes für die Beseitigung des überschüssigen Calziums, ein Beitrag zur Exkretphysiologie der höheren Pflanzen. *Bot. Archiv* **2**, 281–333.

OSISIOGU, I. U. W. and UZO, J. O. (1973) [1974]. Industrial potentialities of some Nigerian yam and cocoyam starches. *Trop. Sci.*, London **15**, 353–359.

——, ——and UGOCHUKWU, E. N. (1974). The irritant effects of cocoyams. *Planta med.* **26**, 166–169.

PACINI, E. and FRANCHI, G. G. (1982). Germination of pollen inside anthers in some non-cleistogamous species. *Caryologia* **35**, 205–215.

——and JUNIPER, B. E. (1983). The ultrastructure of the formation and development of the amoeboid tapetum in *Arum italicum* Miller. *Protoplasma* **117**, 116–129.

PAIS, M. S. S. (1981). Regreening of *Zantedeschia aethiopica* spathe during fructification. Cytochemical characterization of peroxisomes and activities of catalase and glycolate-oxidase. *Ann. Sci. nat., Bot.*, sér. 13, **2–3**, 109–114.

PALANDRI, M. (1967). Modificazioni ultrastrutturali presentate dai plastidi della spata nel corso dell'inverdimento in *Spathiphyllum wallisii* Regel. *Caryologia* **20**, 273–285.

PALIWAL, G. S. (1969). Stomatal ontogeny and phylogeny. 1. Monocotyledons. *Acta bot. néerl.* **18**, 654–668.

—— and KAVATHEKAR, A. K. (1972). Anatomy of vegetative food storage organs. II. Stems. *Acta agron. Acad. Sci. Hung.* **21**, 313–318.

PANICKER, T. K. B. (1965). Female gametophyte and endosperm in *Lagenandra ovata* Thw. *Curr. Sci.* **34**, 614–615.

PANKEY, R. D., DRAUDT, H. N., and DESROSIER, N. W. (1965). Characterization of starch of *Spirodela polyrrhiza*. *J. Food Sci.* **30**, 627–631.

PANT, D. D. (1965). On the ontogeny of stomata and other homologous structures. *Plant. Sci. Ser. (Allahabad)* **1**, 1–24.

—— and KIDWAI, P. F. (1966). Structure of leaves and stomatal ontogeny in some Pandanales and Spathiflorae. *Senck. biol.* **47**, 309–333.

PARAMESWARAN, N. (1959). A contribution to the embryology of *Theriophonum minutum* Engl. *Proc. Indian Acad. Sci.* B, **50**, 15–25.

PARMELEE, J. A. and SAVILE, D. B. O. (1954). Life history and relationships of the rusts of *Sparganium* and *Acorus*. *Mycologia* **46**, 823–836.

PARTHASARATHY, M. V. (1980). Mature phloem of perennial monocotyledons. *Ber. dt. bot. Ges.* **93**, 57–70.

PATTERSON, C. (1982). Morphological characters and homology. pp. 21–74 In: *Problems of phylogenetic reconstruction* (eds Joysey, K. A. and Friday, A. E.). Systematics Association special volume **25**, Academic Press, London.

PELT, J. M. (1966). Apports récents de la biochemie a la taxinomie. *Nat. Monspel.*, sér. Bot. **17**, 157–188.

PERIASAMY, K. and MURUGANATHAN, E. A. (1986). Ontogeny of palmately compound leaves in angiosperms: 3. *Arisaema* spp.*Proc. Indian Acad. Sci., Pl. Sci.* **96**, 475–486.

PERUMALLA, C. J., PETERSON, C. A. and ENSTONE, D. E. (1990). A survey of angiosperm species to detect hypodermal Casparian bands. I. Roots with a uniseriate hypodermis and epidermis. *Bot. J. Linn. Soc.* **103**, 93–112.

PETERSEN, G. (1989). Cytology and systematics of Araceae. *Nordic J. Bot.* **9**, 119–166.

—— (1993a). New chromosome numbers in Araceae. *Willdenowia* **23**, 239–244.

—— (1993b). Chromosome numbers of the genera of Araceae. *Aroideana* **16**, 37–46.

PETERSEN, O. G. (1893). Bemaerkninger om den monocotyledone Staengels Tykkelsevaert og anatomiske Regioner. *Bot. Tidsskr.* **18**, 112–126. (French summary).

PETERSON, C. A. and PERUMALLA, C. J. (1990). A survey of angiosperm species to detect hypodermal Casparian bands. II. Roots with a multiseriate hypodermis or epidermis. *Bot. J. Linn. Soc.* **103**, 113–125.

PETHYBRIDGE, G. H. (1903). The leaf-spots of *Arum maculatum* . *Irish Nat.* **12**, 145–151.

PETIT, L. (1889). Nouvelles recherches sur le pétiole des phanérogames. *Actes Soc. linn. Bordeaux* **43**, 11–60.

PETTET, A. and PETTET, S. J. (1970). Biological control of*Pistia stratiotes* L. in Western State, Nigeria. *Nature* **226**, 282.

PFEIFFER, H. (1925). Über die Wasserstoffionenkonzentration (H') als Determinationsfaktor physiologischer Gewebegeschehen in der sekundären Rinde der Pflanzen. *New Phytol.* **24**, 65–98.

PFITZER, E. (1872). Beiträge zur Kenntnis der Hautgewebe der Pflanzen. *Jb. wiss. Bot.* **8**, 16–72.

PFITZER, P. (1962). Apomixis und Embryosackentwicklung in der Gattung *Aglaonema* Engl. *Portugal. Acta Biol.* A **6**, 279–293.

PHILIPP, M. (1923). Über die verkorkten Abschlussgewebe der Monokotylen. *Bibl. Bot.* **23**, Heft 92, 1–30.

PICKETT, F. L. (1915). A contribution to our knowledge of *Arisaema triphyllum*. *Mem. Torrey bot. Cl.* **16**, 1–55.

PIGNEUR, H. (1947). L'ontogenèse du tube criblé chez *Cucurbita* et *Anthurium*. *Cellule* **51**, 259–278.

PIJL, L. VANDER. (1937). Biological and physiological observations on the inflorescence of *Amorphophallus*. *Rec. Trav. bot. neerl.* **34**, 157–167.

—— (1969). *Principles of dispersal in higher plants* . Springer-Verlag, Berlin.

PINKERTON, M. E. (1936). Secondary root hairs. *Bot. Gaz.* **98**, 147–158.

PIROTTA, R. and MARCATILI, L. (1885) [1886]. Ancora sui rapporti tra i vasi laticiferi ed il sistema assimilatore. *Ann. R. Ist. bot. Roma* **2**, 156–159.

PLANCHON, L. and JUILLET, A. (1909). Étude de quelques fécules coloniales. *Ann. Mus. colon. Marseille*, sér. 2, **7**, 411–556.

PLESE, N., HORVAT, G. and WRISCHER, M. (1989). Etiology of virus disease of the aroid *Amorphophallus rivieri*. *Acta bot. croat.* **48**, 15–18.

PLOWMAN, T. (1969). Folk uses of new world aroids. *Econ. Bot.* **23**, 97–122.

POHL, F. (1931a). Anatomische und ökologische Untersuchungen am Blutenstände von *Philodendron selloum* Schott, mit besonderer Berücksichtigung der Harzkanale und der Beschaffenheit der Pollenkittstoffe. *Planta* **15**, 506–529.

—— (1931b). Das Bewegungsgewebe in der Spatha von *Philodendron selloum* Schott. *Planta* **15**, 530–539.

POLITIS, I. (1914). Sulla presenza di amiloide nelle cellule cristallofore del *Philodendron melanochrysum* Linden., e del *Philodendron oxycardium* Schott. *Atti Ist. bot. Pavia*, ser. 2, **14**, 397–402.

POPOVA, O. F., SLEMNEV, N. N., POPOVA, I. A. and MASLOVA, T. G. (1984). The content of plastid pigments in the plants of Gobi and Karakum deserts. *Bot. Zh. SSSR* **69**, 334–344. [Russian, English summary]

PORSCH, O. (1905). *Der Spaltöffnungsapparat im Lichte der Phylogenie*. Gustav Fischer, Jena. 196 pp.

—— (1911). Anatomie den Nahr- und Haftwurzeln von *Philodendron selloum* C. Koch. *Denkschr. Akad. Wiss. Wien* **79**, 390–451.

POTGIETER, M. J. and VAN WYK, A. E. (1992). Intercellular pectic protuberances in plants: their structure and taxonomic significance. *Bot. Bull. Acad. Sin.* **33**, 295–316.

POTIGUARA, R. C. V. and NASCIMENTO, M. E. (1994). Contribuição a anatomia dos orgãos vegetativos de *Heteropsis jenmanii* Oliv. (Araceae). *Bol. Mus. para. Emilio Goeldi, Bot.* **10**, 237–247.

POTT, V. J. and CERVI, A. C. (1999). A família Lemnaceae Gray no Pantanal (Mato Grosso e Mato Grosso do Sul), Brasil. *Revta bras. Bot.* **22**, 153–174.

POULSEN, V. (1875). Om Korkdannelse paa Blade. *Vidensk. Medd. naturh. Foren. Kjobenhavn* 44–58. [French summary 6–9.]

PRAY, T. R. (1957). Marginal growth of leaves in monocotyledons: *Hosta, Maranta* and *Philodendron*. *Phytomorphology* **7**, 381–387.

PREDA, A. (1910). Una nuova forma di *Arisarum vulgare* Targ. Tozz. *Bull. Soc. bot. ital.* 147–149.

PRIESTLEY, J. H. and WOFFENDEN, L. M. (1922). Physiological studies in plant anatomy. V. Causal factors in cork formation. *New Phytol.* **21**, 252–268.

PRITCHARD, H. W., WOOD, J. A. and MANGER, K. R. (1993). Influence of temperature on seed germination and the nutritional requirements for embryo growth in *Arum maculatum* L. *New Phytol.* **123**, 801–809.

PRYCHID, C. J. and RUDALL, P. J. (2000). Distribution of calcium oxalate crystals in monocotyledons. pp. 159–162 In: *Monocots: systematics and evolution* (eds Wilson, K. L. and Morrison, D. A.). CSIRO Publishers, Collingwood, Australia.

PURSEGLOVE, J. W. (1972). *Tropical crops. Monocotyledons*. vol 1. Longman, London.

PÜTZ, N. (1991). Die Zugbewegungstypen bei den Monokotylen. *Bot. Jb.* **112**, 347–364.

——(1992). Measurement of the pulling force of a single contractile root. *Can. J. Bot.* **70**, 1433–1439.

QIN, M. (1988). Comparative studies on histology of Rhizoma Arisaematis and Pinelliae. *Bull. chin. mat. med.* **13 (4)**, 6–8, 61. [In Chinese]

QIU, Y.-L., CHASE, M. W., LES, D. H. and PARKS, C. R. (1993). Molecular phylogenetics of the Magnoliidae: Cladistic analyses of nucleotide sequences from the plastid gene *rbc*L. *Ann. Missouri bot. Gard.* **80**, 587–606.

RADA, F. and JAIMEZ, R. (1992). Comparative ecophysiology and anatomy of terrestrial and epiphytic *Anthurium bredemeyeri* Schott in a tropical Andean cloud forest. *J. exp. Bot.* **43**, 723–727.

RAJAGOPAL, R., ULVSKOV, P., MARCUSSEN, J., ANDERSEN, J. M. and ALLERUP, S. (1987). Hormonal and phenolic changes accompanying and following UV-C induced stress in *Spathiphyllum* leaves. *J. Pl. Physiol.* **130**, 291–306.

RAKOVAN, J. N., KOVACS, A. and SZUJKO-LACZA, J. (1973). Development of idioblasts and raphides in the aerial root of *Monstera deliciosa* Liebm. *Acta biol. Acad. Sci. hung.* **24**, 103–118.

RAMBOUR, S. (1965). Quelques observations sur les cellules à raphides de *Spathiphyllum wallisii*. *Bull. Soc. bot. Nord Fr.* **18 (1)**, 27–31.

RAO, T. A. (1954). Nature and occurrence of septa in foliar sclereids of *Scindapsus* sp. *Curr. Sci.* **23**, 365–366.

——(1964). On the multiseptate trichosclereids in the genus *Scindapsus*. *Proc. Indian Acad. Sci.* B **59**, 66–69.

——(1977). The typology of foliar sclereids in taxonomic problems of the Monsteroideae of the Araceae. *Proc. Indian Acad. Sci.*, B **86**, 73–80.

——(1991). *Compendium of foliar sclereids in angiosperms*. Wiley Eastern, New Delhi.

——and BHATTACHARYA, J. (1978). A review of foliar sclereids in angiosperms. *Bull. bot. Surv. India* **20**, 91–99.

RASKIN, I., TURNER, M. and MELANDER, W. R. (1989). Regulation of heat production in the inflorescences of an *Arum* lily by endogenous salicylic acid. *Proc. Nat. Acad. Sci.* **86**, 2214–2218.

RASMUSSEN, H. (1983). Stomatal development in families of Liliales. *Bot. Jb.* **104**, 261–287.

RAUNKIAER, C. (1895–9). *De danske blomsterplanters Naturhistorie*. vol. I. Kjobenhavn. pp. 282–300.

RAVEN, P. H. and AXELROD, D. I. (1974). Angiosperm biogeography and past continental movements. *Ann. Missouri bot. Gard.* **61**, 539–673.

RAVN, F. K. (1894). Om flydeevnen hos frøene af vore Vandog sumpplanter. *Bot. Tidsskr.* **19**, 143–188. [French summary 178–188]

RAY, T. S. (1986). Growth correlations within the segment in the Araceae. *Amer. J. Bot.* **73**, 993–1001.

——(1987a). Cyclic heterophylly in *Syngonium* (Araceae). *Amer. J. Bot.* **74**, 16–26.

——(1987b). Leaf types in Araceae. *Amer. J. Bot.* **74**, 1359–1372.

——(1987c). Diversity of shoot organization in the Araceae. *Amer. J. Bot.* **74**, 1373–1387.

——(1988). Survey of shoot organization in the Araceae. *Amer. J. Bot.* **75**, 563–584.

——(1990). Metamorphosis in the Araceae. *Amer. J. Bot.* **77**, 1599–1609.

——(1992a). Landmark eigenshape analysis: homologous contours: leaf shape in *Syngonium* (Araceae). *Amer. J. Bot.* **79**, 69–76.

——(1992b). Foraging behaviour in tropical herbaceous climbers (Araceae). *J. Ecol.* **80**, 189–203.

RAYNAL-ROQUES, A. (1981). Contribution à l'étude biomorphologique des angiospermes aquatiques tropicales—essai d'analyse de l'évolution. Thèse, Univ. Sci. Tech. Languedoc. 2 vols. 382 + 450 pp.

REFFSTRUP, T. and BOLL, P. M. (1985). Allergenic 5-alkyl- and 5-alkenylresorcinols from *Philodendron* species. *Phytochemistry* **24**, 2563–2565.

REICHERT, E. T. (1913). *The differentiation and specificity of starches in relation to genera, species, etc.* Parts I, II. Publ. 173. Carnegie Institution, Washington, DC.

——(1919). *A biochemic basis for the study of problems of taxonomy, heredity, evolution etc., with especial reference to the starches and tissues of parent-stocks and hybrid-stocks and the starches and hemoglobins of varieties, species, and genera.* Carnegie Institution, Washington, DC 2 vols. 834 pp.

REINHARDT, M. O. (1885). Das leitende Gewebe einiger anomal gebauten Monocotylenwurzeln. *Jb. wiss. Bot.* **16**, 336–366.

RENNERT, R. J. (1902). Seeds and seedlings of *Arisaema triphyllum* and *Arisaema dracontium*. *Bull. Torrey bot. Cl.* **29**, 37–54.

REUMER, J. W. F. (1984). Cytotaxonomy and evolution in *Cryptocoryne* (Araceae). *Genetica* **65**, 149–58.

RHEE, J. K. and IWATA, M. (1982). Histological observations on the chilling injury of taro tubers during cold storage. *J. Jap. Soc. hort. Sci.* **51**, 362–368.

RICHTER, A. (1901). Physiologisch-anatomische Untersuchungen über Luftwurzeln mit besonderer Berücksichtigung der Wurzelhaube. *Bibl. Bot.* **10**, Heft 54, 50 pp.

RICHTER, S. (1929). Über den Öffnungsmechanismus der Antheren bei einigen Vertretern der Angiospermen. *Planta* **8**, 154–184.

RIEDL, H. (1980). The importance of ecology for generic and specific differentiation in the Araceae-Aroideae. *Aroideana* **3**, 49–54.

RIMBACH, A. (1897). Ueber die Lebensweise des *Arum maculatum*. *Ber. dt. bot. Ges.* **15**, 178–182.

RITTERBUSCH, A. (1971). Morphologische Untersuchungen zur Wuchsform von *Philodendron*. *Bot. Jb.* **90**, 527–549.

RODRIGUE, A. (1900). Les feuilles panachées et les feuilles colorées. (Rapports entre leurs couleurs et leur structure.) *Mém. Herb. Boissier*, No. 17, 11–75.

ROLAND, J.-C. (1967). Recherches en microscopie photonique et en microscopie électronique sur l'origine et la différenciation des cellules du collenchyme. *Ann. Sci. nat., Bot.*, sér. 12, **8**, 141–214.

ROSANOFF, S. (1867). Ueber Krystalldrusen in den Pflanzenzellen. *Bot. Ztg.* **25**, 41–42. [Translated from Russian]

ROSENDAHL, C. O. (1911). Observations on the morphology of the underground stems of *Symplocarpus* and *Lysichitum*, together with some notes on geographical distribution and relationship. *Minnnesota Bot. Studies* **4**, 137–152.

ROST, L. C. M. (1978). Biosystematic investigations with *Acorus* L. (Araceae). 1. Cytotaxonomy. *Proc. kon. Ned. Acad. Wet. Amsterdam*, C **81**, 428–441.

——(1979a). Biosystematic investigations with *Acorus* L. (Araceae). II: Essential oil contents.*Proc. kon. Ned. Akad. Wet. Amsterdam*, C **82**, 113–126.

——(1979b). Biosystematic investigations with *Acorus*. 4. A synthetic approach to the classification of the genus. *Planta med.* **37**, 289—307.

——and BOS, R. (1979). Biosystematic investigations with *Acorus* L. 3. Constituents of essential oils. *Planta med.* **36**, 350–36l.

ROTH, I. (1949). Zur Entwicklungsgeschichte des Blattes, mit besonderer Berücksichtigung von Stipular- und Ligularbildungen. *Planta* **37**, 299–336.

——(1952). Beiträge zur Entwicklungsgeschichte der Schildblätter. *Planta* **40**, 350–376.

——(1980). Blattstruktur von Pflanzen aus feuchten Tropenwäldern. *Bot. Jb.* **101**, 489–525.

——(1990). *Leaf structure of a Venezuelan cloud forest in relation to the microclimate*. (Handbuch der Pflanzenanatomie XIV, 1). Gebrüder Borntraeger, Berlin and Stuttgart. 44 pp.

——and FAST, G. (1954). Über die Entstehung dorsaler Blattschläuche bei *Anthurium scherzerianum* Schott. *Planta* **44**, 543–561.

——and MERIDA DE BIFANO, T. (1979). Morphological and anatomical studies of leaves of the plants of a Venezuelan cloud forest. II. Stomatal density and stomatal patterns. *Acta biol. venez.* **10**, 69–107.

RYWOSCH, S. (1909). Untersuchungen über die Entwicklungsgeschichte der Seitenwurzeln der Monocotylen. *Z. Bot.* **1**, 253–283.

SAHA, B. P. and HUSSAIN, M. (1983). A study of the irritating principle of aroids. *Indian J. agric. Sci.* **53**, 833–836.

SAKAI, T. and HAYASHI, K. (1973). Studies on the distribution of starchy and sugary leaves in monocotyledonous plants. *Bot. Mag., Tokyo* **86**, 13–25.

SAKAI, W. S. and HANSON, M. (1974). Mature raphid and raphid idioblast structure in plants of the edible aroid genera *Colocasia, Alocasia*, and *Xanthosoma*. *Ann. Bot.* **38**, 739–748.

——,——and JONES, R. C. (1972). Raphides with barbs and grooves in *Xanthosoma sagittifolium* (Araceae). *Science* **178**, 314–315.

——and NAGAO, M. A. (1980). Raphide structure in *Dieffenbachia maculata*. *J. Amer. Soc. hort. Sci.* **105**, 124–6.

——, SHIROMA, S. S. and NAGAO, M. A. (1984). A study of raphide microstructure in relation to irritation. *Scanning Electron Microscopy* 1984/2, 979–986.

SALISBURY, E. J. (1916). The emergence of the aerial organs in woodland plants. *J. Ecol.* **4**, 121–128.

SAMUELS, J.-A. (1913). Études cytologiques sur les relations existant entre le noyau et le développement des cristaux dans les cellules parenchymateuses du périanthe d'*Anthurium*. *C. r. Acad. Sci., Paris* **156**, 1275.

SANFORD, R. L., Jr. (1987). Apogeotropic roots in an Amazon rain forest. *Science* **235**, 1062–1064.

SARGANT, E. (1903). A theory of the origin of monocotyledons, founded on the structure of their seedlings. *Ann. Bot.* **17**, 1–92.

SARGENT, J. A. and WANGERMANN, E. (1959). The effect of some growth regulators on the vascular system of *Lemna minor*. *New Phytol.* **58**, 345–363.

SATTLER, R. (1973). *Organogenesis of flowers. A photographic text-atlas*. Univ. Toronto Press. 207 pp.

SAVCHENKO, M. I. and MANIA, E. N. (1970). Comparative anatomical studies on the spathe and the rachis in certain Araceae. *Bot. Zh. SSSR* **55**, 406–421. [Russian, English summary]

SAVILE, D. B. O. (1971). Coordinated studies of parasitic fungi and flowering plants. *Nat. canad.* **98**, 535–552.

SAXTON, W. T. (1913). The leaf-spots of *Richardia albo-maculata* Hook. *Trans. roy. Soc. S. Afr.* **3**, 135–138.

SCARAMELLA PETRI, P. and SPADONI, R. (1968). Sulla coltura in vitro di tre monocotiledoni di interesse farmacologico. *Studi Urbinati* **41C (11)**, 5 pp.

SCHADE, C. and GUTTENBERG, H. von (1951). Über die Entwicklung des Wurzelvegetationspunktes der Monokotyledonen. *Planta* **40**, 170–198.

SCHANTZ, M. VON (1958). Über das ätherische Öl bei Kalmus, *Acorus calamus* L.; pharmakognostische Untersuchung. *Acta Bot. Fennica* **59**, 1–138.

SCHERER, P. E. (1904). Studien über Gefässbündeltypen und Gefässformen. *Beih. bot. Zbl.* **16**, 67–110.

SCHIMPER, A. F. W. (1884). Über Bau und Lebensweise der Epiphyten Westindiens. *Bot. Zbl.* **17**, 192–195, 223–227, 253–258, 284–294, 319–326, 350–359, 381–389.

SCHIMPER, A. F. W. (1888). Die epiphytische Vegetation Amerikas. *Bot. Mitt. aus den Tropen*, Jena, Heft 2, 1–162.

SCHLOTMANN, A. (1933). Untersuchungen über die Struktur pflanzlicher Haare und Fasern. *Planta* **19**, 313–334.

SCHMID, P. C., HOLMAN R. T. and SOUKUP, V. G. (1997). 13-phenyltridecanoic acid in seed lipids of some aroids. *Phytochemistry* **45**, 1173–1175.

SCHMIDT, E. (1882). Ueber den Plasmakörper der gegliederten Milchröhren. *Bot. Ztg.* **40**, 435–448 and 451–456.

SCHNEIDER, E. L. and CARLQUIST, S. (1998). Origin and nature of vessels in monocotyledons. 4. Araceae subfamily Philodendroideae. *J. Torrey bot. Soc.* **125**, 253–260.

SCHNEPF, E. (1965a). Morphologie der Duftölausscheidung bei *Typhonium divaricatum* (Araceae). *Planta* **66**, 374–376. [English summary]

——(1965b). Die Morphologie der Sekretion in pflanzlichen Drüsen. *Ber. dt. bot. Ges.* **78**, 478–483.

——and CZYGAN, F., C. (1966). Feinbau und Carotinoide von Chromoplasten im Spadix-Appendix von *Typhonium* und *Arum*. *Z. Pflanzenphysiol.* **54**, 345–355. [English summary]

SCHOUTE, J. C. (1902). Über Zellteilungsvorgänge im Cambium. *Verh. kon. Akad. Wet. Amsterdam*, Sect. II, **9** (4), 1–60.

SCHULTE, H. (1969) [1970]. Die Infloreszenzkolben von *Arum maculatum*. Ein Beitrag zur Histologie. *Ber. dt. bot. Ges.* **82**, 643–650.

SCHULTES, R. E. (1963). Plantae Colombianae XVI. Plants as oral contraceptives in northwest Amazon. *Lloydia* **26**, 67–74.

——(1970a). The botanical and chemical distribution of hallucinogens. *Ann. Rev. Pl. Physiol.* **21**, 571–598.

——(1970b). De plantis toxicariis e mundo novo tropicale commentationes. 7. Several ethnotoxicological notes from the Colombian Amazon. *Bot. Mus. Leafl. Harvard Univ.* **22**, 345–347.

——and FARNSWORTH, N. R. (1980). Ethnomedical, botanical and phytochemical aspects of natural hallucinogens. *Bot. Mus. Leafl. Harvard Univ.* **28**, 123–214.

——and HOFMANN, A. (1973). *The botany and chemistry of hallucinogens*. Charles C. Thomas, Springfield, Illinois. 267 pp.

SCHWARZ, F. (1878). Über die Entstehung der Löcher und Einbuchtungen an dem Blatte von *Philodendron pertusum* Schott. *S. B. Akad. Wiss. Wien, math.-nat. Kl.* (I) **77**, 367–374.

SCHWENDENER, S. (1874). *Das mechanische Princip im anatomischen Bau der Monocotylen mit vergleichenden Ausblicken auf die übrigen Pflanzenklassen*. W. Engelmann, Leipzig. 179 pp.

——(1882). Die Schutzscheiden und ihre Verstärkungen. *Abh. kon. Akad. Wiss. Berlin*, 75 pp.

SCOTT, D. H. (1889). The distribution of laticiferous tissue in the leaf. *Ann. Bot.* **3**, 445–448.

——(1908). The contractile roots of the aroid *Sauromatum guttatum*. *Rept. British Assoc. Adv. Sci. Dublin.* (1908), 910–911.

SCOTT, R. and SARGENT, E. (1898). On the development of *Arum maculatum* from the seed. *Ann. Bot.* **12**, 399–414.

SCRIBAILO, R. W. and TOMLINSON P. B. (1992). Shoot and floral development in *Calla palustris* (Araceae-Calloideae). *Intn. J. Pl. Sci.* **153**, 1–13.

SELVARAJ, T. and SUBRAMANIAN, G. (1987). Vesicular-arbuscular mycorrhizal fungi in roots and scale-like leaves of *Acorus calamus* Linn. and *Colocasia esculenta* Linn. *Curr. Sci.* **56**, 1112–1114.

SEROBJAN, P. A. (1974). On anatomy and function of aerial roots of *Philodendron pertusum*. *Biol. Zh. armen.* **27 (6)**, 77–81. [Armenian]

SEUBERT, E. (1993). *Die Samen der Araceen*. Koeltz Sci. Books, Koenigstein. 433 pp.

——(1997a). Sclereids of Araceae. *Flora* **192**, 31–37.

——(1997b). A comparative study of the seeds of Lasieae (Araceae). *Bot. Jahrb. Syst.* **119**, 407–426.

SEYMOUR, R. S., BARTHOLOMEW, G. A. and BARNHART, M. C. (1983). Respiration and heat production by the inflorescence of *Philodendron selloum* Koch. *Planta* **157**, 336–343.

SHADOWSKY, A. E. (1931). Einige Angaben über die Embryogenie von *Pistia stratiotes* L. *Ber. dt. bot. Ges.* **49**, 350–356.

SHAH, G. L. and GOPAL, B. V. (1969) (rcd. 1974). Observations on the stomata and their ontogeny in some monocotyledons. pp. 55–56 In: *Seminar on morphology, anatomy and embryology of land plants* (eds Johri, B. M., Kapil, R. N. and Rashid, A.). Delhi.

SHAH, J. J. and JAMES, M. R. (1971). Some aspects of phloem structure in *Pistia stratiotes*. *Proc. Indian natn. Sci. Acad.*, B, **37**, 131–139.

SHARMA, M. (1967) [1968]. Pollen morphology of Indian monocotyledons. *J. Palynol.*, special vol. 1–96.

SHARMA, S., JAIN, N. and SHAH, A. K. (1985). Occurrence of *Azotobacter chroococcum* in *Pothos scandens*. *Curr. Sci.* **54**, 142–143.

SHAW, D. E. (1975). Illustrated notes on flowering, flowers, seed and germination in taro (*Colocasia esculenta*). *Dept. Agric. Stock Fish , P. Moresby, Res. Bull.* **13**, 39–59.

——(1990) [1992]. Germination of *Typhonodorum lindleyanum* at Brisbane, Australia. *Aroideana* **13**, 14–20.

——(1992). The occurrence and frequency of stomata of leaves of *Monstera deliciosa* (Araceae). *Aroideana* **15**, 8–16.

SHIRLEY, J. and LAMBERT, C. A. (1918). The stems of climbing plants. *Proc. Linn. Soc. N. S. W.* **43**, 600–609.

SHISHKOFF, N. (1987). Distribution of the dimorphic hypodermis of roots in angiosperm families. *Ann. Bot.* **60**, 1–15.

SIEDLER, P. (1892). Über den radialen Saftstrom in den Wurzeln. *Beitr. Biol. Pfl.* **5**, 407–442.

SIFTON, H. B. (1945). Air-space tissue in plants. *Bot. Rev.* **11**, 108–143.

SIGNOL, M. and LANCE, C. (1975). Relation entre l'évolution ultrastructurale des mitochondries et l'activité respiratoire du spadice d'*Arum maculatum* L. au cours de son développement. *Physiol. vég.* **13**, 549–562.

SIMMONDS, N. W. (ed.) (1976). *Evolution of crop plants*. Longman, London and New York. 339 pp.

SIMS, D. A. and PEARCY, R. W. (1992). Response of leaf anatomy and photosynthetic capacity in *Alocasia macrorrhiza*

(Araceae) to a transfer from low to high light. *Amer. J. Bot.* **79**, 449–455.

SINGH, H. (1968). Foliar sclereids in some species of *Raphidophora* and *Scindapsus*. *Gardens' Bull.*, Singapore **22**, 461–467.

SINNOTT, E. W. and BLOCH, R. (1946). Comparative differentiation in the air roots of *Monstera deliciosa*. *Amer. J. Bot.* **33**, 587–590.

SIVARAMAKRISHNA, D. and VIJAYARAGHAVAN, P. (1979) [1980]. Development and histological differentiation in roots after bisection of the apex. pp. 258–265 In: *Histochemistry, developmental and structural anatomy of angiosperms: a symposium* (ed. Periasamy, K.). P & B Publications, Tiruchirapalli, India.

SKUBATZ, H. and KUNKEL, D. D. (1999). Further studies of the glandular tissue of *Sauromatum guttatum* (Araceae) appendix. *Amer. J. Bot.* **86**, 841–854.

——, NELSON, T. A., MEEUSE, B. J. D. and BENDICH, A. J. (1991). Heat production in the voodoo lily (*Sauromatum guttatum*) as monitored by infrared thermography. *Pl. Physiol.* **95**, 1084–1088.

SMIRNOVA, E. S., Kantor, T. S. and FURST, G. G. (1964). Biology of *Colocasia antiquorum* (L.) Schott. *Bull. glav. bot. Sada* **52**, 52–67. [Russian only]

SMITH, C. W. and LEW, L., F. (1970) [1971]. Cellular arrangement in the node of various angiosperms. *Bot. Gaz.* **131**, 269–272.

SMOLENSKI, S. J., SILINIS, H. and FARNSWORTH, N. R. (1972). Alkaloid screening. I. *Lloydia* **35**, 1–34.

——, —— and —— (1973) [1974]. Alkaloid screening III. *Lloydia* **36**, 359–389.

——, —— and —— (1974). Alkaloid screening. IV, V. *Lloydia* **37**, 30–61, 506–536.

——, —— and —— (1975) [1976]. Alkaloid screening. VIII. *Lloydia* **38**, 497–528.

SOLEREDER, H. (1919). Beiträge zur Anatomie der Araceen. *Beih. bot. Zbl.* **36** (1), 60–77.

—— and MEYER, F. J. (1928). *Systematische Anatomie der Monokotyledonen*. Heft III, pp. 100–169. Borntraeger, Berlin.

SOLTIS, D. E., SOLTIS, P. S., NICKRENT, D. L., JOHNSON, L. A., HAHN, W. J., HOOT, S. B. *et al.* (1997). Angiosperm phylogeny inferred from 18S ribosomal DNA sequences. *Ann. Missouri bot. Gard.* **84**, 1–49.

SOUTHORN, W. A. (1964). A complex sub-cellular component of widespread occurrence in plant latices. *J. exp. Bot.* **15**, 616–621.

SPJUT, R. W. (1994). A systematic treatment of fruit types. *Mem. New York bot. Gard.* **70**, 1–182.

SPRATT, A. V. (1920). Some anomalies in monocotyledonous roots. *Ann. Bot.* **34**, 99–105.

STAHL, E. (1896). Über bunte Laubblätter. Ein Beitrag zur Pflanzenbiologie II. *Ann. Jard. bot. Buitenzorg* **13 (2)**, 137–215.

STAUDERMANN, W. (1924). Die Haare der Monokotylen. *Bot. Archiv* **8**, 105–184.

STEBBINS, G. L. and KHUSH, G. S. (1961). Variation in the organization of the stomatal complex in the leaf epidermis of monocotyledons and its bearing on their phylogeny. *Amer. J. Bot.* **48**, 51–59.

STEENIS, C. G. G. J. VAN (1981). *Rheophytes of the world.* Sythoff & Noordhoff, Alphen a/d Rijn. 407 pp. [Araceae, pp. 183–196.]

—— (1987). Rheophytes of the world: Supplement. *Allertonia* **4**, 267–330.

STEVENSON, D. W., DAVIS, J. I., FREUDENSTEIN, J. V., HARDY, C. R., SIMMONS, M. P., and SPECHT, C. D. (2000). A phylogenetic analysis of the monocotyledons based on morphological and molecular character sets, with comments on the placement of *Acorus* and Hydatellaceae. pp. 17–24 in: *Monocots. Systematics and Evolution*. (eds Wilson, K. L. and Morrison, D. A.). CSIRO Publishing. Collingwood, Victoria, Australia.

STEWART, R. N. and DERMEN, H. (1979). Ontogeny in monocotyledons as revealed by studies of the developmental anatomy of periclinal chloroplast chimeras. *Amer. J. Bot.* **66**, 47–58.

STOCKEY, R. A., HOFFMAN G. L. and ROTHWELL, G. W. (1997). Fossil monocot *Limnobiophyllum scutatum:* resolving the phylogeny of Lemnaceae. *Amer. J. Bot.* **84**, 355–368.

STRAUSS, M. S. (1983). Anatomy and morphology of taro, *Colocasia esculenta* (L.) Schott. pp. 20–33 In: *Taro. A revision of Colocasia esculenta and its potentials* (ed. Wang, J.-K.). Univ. Hawaii Press, Honolulu.

STREITBERG, H. (1954). Über die Heterophyllie bei Wasserpflanzen mit besonderer Berücksichtigung ihrer Bedeutung für die Systematik. *Flora* **141**, 567–597.

STRONG, D. R. and RAY, T. S. (1975). Host tree location behaviour of a tropical vine (*Monstera gigantea*) by skototropism. *Science* **190**, 804–806.

STUDER, C. and BRANDLE, R. (1984). Sauerstoffkonsum und Versorgung der Rhizome von *Acorus calamus* L., *Glyceria maxima* (Hartmann) Holmberg, *Menyanthes trifoliata* L., *Phalaris arundinacea* L., *Phragmites communis* Trin. und *Typha latifolia* L. *Bot. helv.* **94**, 23–31.

SUBRAMANYAM, K. (1962). *Aquatic angiosperms.* A systematic account of common Indian aquatic angiosperms. Bot. Mon. No. 3. C. S. I. R., New Delhi. 190 pp.

SUEUR, H. (1866–7). Des cellules consolidantes du *Scindapsus pertusus* Schott. *Adansonia* **7**, 292–296.

SUN, C. N. (1960). The effect of excising certain structures on foliar determination in *Arisaema consanguineum* Schott. *Phytomorphology* **10**, 37–42.

SUNELL, L. A. and HEALEY, P. L. (1979). Distribution of calcium oxalate crystal idioblasts in corms of taro (*Colocasia esculenta*). *Amer. J. Bot.* **66**, 1029–1032.

—— and —— (1981). Scanning electron microscopy and energy dipersive X-ray analysis of raphide crystal idioblasts in taro. *Scanning Electron Microscopy* 1981/3, 235–244.

—— and —— (1985). Distribution of calcium oxalate crystal idioblasts in leaves of taro (*Colocasia esculenta*). *Amer. J. Bot.* **72**, 1854–1860.

SUTTER, E. G. (1985). Morphological, physical and chemical characteristics of epicuticular wax on ornamental plants regenerated in vitro. *Ann. Bot.* **55**, 321–329.

SWAMY, B. G. L. and KRISHNAMURTHY, K. V. (1970) [1971]. On the so-called endothelium in the monocotyledons. *Phytomorphology* **20**, 262–269.

——and SIVARAMAKRISHNA, D. (1972) [1974]. Wound healing responses in monocotyledons. I. Responses in vivo. *Phytomorphology* **22**, 305–314.

——and——(1976). Wound healing responses in monocotyledons: III. Development of split apical meristems. *Phytomorphology* **26**, 119–123.

SZENDEL, A. J. and WERYSZKO, E. (1973). Investigations on structure of flowers and inflorescences of the *Anthurium andreanum* hort., non Linden. *Acta agrobot. Warsaw* **26**, 251–256.

SZUJKO-LACZA, J., RAKOVAN, J. N. and KOVACS, A. (1974). Rafid-idioblastztok es Ca-oxalat tukristalyok kialakulasa a *Monstera deliciosa* Liebm. leggyokereben. [Development of idioblast and raphides in the aerial root of *Monstera deliciosa* Liebm.] *Bot. Kozlem.* **61**, 1–11.

TAKHTAJAN, A. (ed.) (1985). *Comparative anatomy of seeds.* Vol. I. Izdat. Nauka, Leningrad. 318 pp. [Russian.]

——(1997). *Diversity and classification of flowering plants.* Columbia University Press, New York.

TALAVERA, S. (1976). Revision de las especies españolas del genero *Biarum* Schott. *Lagascalia* **6**, 275–296.

TARASEVICH, V. F. (1990). Palynological evidence of the position of the Lemnaceae family in the system of flowering plants. *Bot. Zh. SSSR* **75**, 959–965 + 2 plates. [Russian only]

TAYLOR, D. W. (1990). Paleobiogeographic relationships of angiosperms from the Cretaceous and Early Tertiary of the North American area. *Bot. Rev.* **56**, 279–417.

THALER, I. (1955–6). Eiweisskristalle in der Epidermis von *Scindapsus aureus*. *Phyton* (Austria) **6**, 89–91.

THANIKAIMONI, G. (1969). Esquisse palynologique des Aracées. *Trav. Sect. sci. tech. Inst. fr. Pondichery* **5 (5)**, 1–31 + 20 plates.

THIELKE, C. (1948). Beiträge zur Entwicklungsgeschichte unifazialer Blätter. *Planta* **36**, 154–177.

——(1960). Die Panaschure von *Acorus*. *Planta* **54**, 210–220.

THORNE, R. F. (2000). The classification and geography of the monocotyledon subclasses Alismatidae, Liliidae and Commelinidae, pp. 75–124 In: *Plant systematics for the 21st century.* (eds Nordenstam, B., El-Ghazaly, G. and Kassas, M.). Portland Press, London. 366 pp.

TIEGHEM, P. VAN (1866). Recherches sur la structure des Aroidées. *Ann. Sci. nat., Bot.*, sér. 5, **6**, 72–210.

——(1870–1). Sur la symétrie de structure des plantes vasculaires. *Ann. Sci. nat., Bot.*, sér. 5, **13**, 5–314.

——(1872). Mémoire sur les canaux sécréteurs des plantes. *Ann. Sci. nat., Bot.*, sér. 5, **16**, 96–201.

——(1891). Sur les tubes criblés extralibériens et les vaisseaux extraligneux. *J. de Bot.* **5**, 117–128.

——(1905). Sur les diverses sortes de meristeles corticales de la tige. *Ann. Sci. nat., Bot.*, sér. 9, **1**, 33–44.

——(1907). Remarques sur l'organisation florale et la structure de l'ovule des Aracées. *Ann. Sci. nat., Bot.*, sér. 9, **5**, 312–320.

——and DOULIOT, H. (1888). Recherches comparatives sur l'origine des membres endogènes dans les plantes vasculaires. *Ann. Sci. nat., Bot.*, sér. 7, **8**, 1–656.

TILLICH, H.-J. (1985). Keimlingsbau und verwandtschaftliche Beziehungen der Araceae. *Gleditschia* **13**, 63–73.

——(1992). Bauprinzipien und Evolutionslinien bei monocotylen Keimpflanzen. *Bot. Jb.* **114**, 91–132.

——(1995). Seedlings and systematics in monocotyledons. pp. 303–352 In: *Monocotyledons: systematics and evolution* (eds Rudall, P. J., Cribb, P. J., Cutler, D. F. and Humphries, C. J.). Royal Botanic Gardens, Kew.

TOBLER, F. (1922). Aningafaser. *Faserforschung* **2**, 153–156.

——(1957). *Die mechanischen Elemente und das mechanische System.* [Handbuch der Pflanzenanatomie IV (Teil 6)]. Borntraeger, Berlin. 60 pp.

TOMLINSON, P. B. (1970). Monocotyledons–towards an understanding of their morphology and anatomy. *Adv. bot. Res.* **3**, 207–292.

——(1974). Development of the stomatal complex as a taxonomic character in the monocotyledons. *Taxon* **23**, 109–128.

——(1982). *Anatomy of the Monocotyledons. VII. Helobiae (Alismatidae).* Clarendon Press, Oxford. 522 pp.

——(1984). Development of the stem conducting tissues in monocotyledons. pp. 1–51 In: *Contemporary problems in plant anatomy* (eds White, R. A. and Dickison, W. C.). Academic Press, Orlando.

TRAQUAIR, J. A. (1987). Backscattered electron imaging as a tool for histochemically localizing calcium oxalate with the scanning electron microscope. *Can. J. Bot.* **65**, 888–892.

TRÉCUL, A. (1854). Note sur la formation des perforations que présentent les feuilles de quelques Aroidées. *Ann. Sci. nat., Bot.*, sér. 4, **1**, 37–40.

——(1865). Des vaisseaux propres dans les Aroïdées. *C. r. Acad. Sci., Paris* **61**, 1163–1167.

——(1866). Vaisseaux propres dans les Aroidées. *C. r. Acad. Sci., Paris* **62**, 29–33, 248.

——(1866–70). Recherches sur les vaisseaux laticifères. *Adansonia* **7**, 100–212, 318–342 (1866–7); **8**, 91–131 (1867–8); **9**, 81–106, 135–145 (1868–70).

TREUB, M. (1876). *Le méristème primitif de la racine dans les Moncotylédones.* Musée Bot. Leyde **2**, E. J. Brill, Leiden. 78 pp.

TRIVEDI, B. S. and UPADHYAY, N. (1984). Epidermal structures of Araceae: 1. *Dieffenbachia* Schott. *Geophytology* **14**, 143–153.

TROLL, W. (1928). Über *Spathicarpa sagittifolia* Schott. Mit einem Anhang über die Stellung der Spathicarpeen im System der Araceen. *Flora* **123**, 286–316.

——(1932a). Morphologie der schildformigen Blätter. *Planta* **17**, 153–314.

——(1932b). Über Diplophyllie und verwandte Erscheinungen in der Blattbildung. *Planta* **15**, 355–406.

——(1933). Beiträge zur Morphologie des Gynaeceums. III. Über das Gynaeceum von *Nigella* und einiger anderer Helleboreen. *Planta* **21**, 266–291.

——(1939). *Vergleichende Morphologie der höheren Pflanzen.* Band I. *Vegetationsorgane.* Teil 2. Borntraeger, Berlin.

———and MEYER, H. J. (1955). Entwicklungsgeschichtliche Untersuchungen über das Zustandekommen unifazialer Blattstrukturen. *Planta* **46**, 286–360.

TSCHIRCH, A. (1894). Ueber Secrete und Secretbildung. *Bot. Zbl.* **60**, 289–293.

——(1909–10), (1912–17), (1923–25). *Handbuch der Pharmakognosie*. Leipzig. Vol. 1, 1072 pp. (1909–10). vol. 2, 1625 pp. (1912–17). Vol. 3, 1176 pp. (1923–25).

——(1913). Die Gerbstoffzellen des Kalmusrhizons. *Schweiz. Wochenschr. Chem. Pharm.* **51**, 269.

TSUCHIYA, T. (1991). Leaf life span of floating-leaved plants. *Vegetatio* **97**, 149–160.

TURLIER, M. F. (1968). Observations sur l'organogenèse foliaire précoce chez une Aracée: *Scindapsus argyraea* Engl. *C. r. Acad. Sci., Paris,* D, **266**, 2067–2070.

TURPIN, P. J. F. (1836). Observations sur les Biforines, organes nouveaux situés entre les vésicles du tissu cellulaires des feuilles dans un certain nombre d'espèces végétales appartenant à la famille des Aroïdées. *Ann. Sci. nat., Bot.* II, **6**, 5–27.

UENO, J. (1980). On the spiraperturate pollen grains. *Jap. J. Palynol.* **25**, 33–47. [Japanese, English summary].

UHLARZ, H. (1982) [1983]. Typologische und ontogenetische Untersuchungen an *Spathicarpa sagittifolia* Schott (Araceae); Wuchsform und Infloreszenz. *Beitr. Biol. Pfl.* **57**, 389–429. [English summary].

——(1985) [1986]. Ist *Pinellia tripartita* (Blume) Schott (Araceae, Aroideae) habituell anemophilgeitonogam? *Beitr. Biol. Pfl.* **60**, 277–291.

——(1986). Zum Problem des "blattlosen Sprosses": Morphologie und Anatomie der Infloreszenz von *Pinellia tripartita* (Blume) Schott (Araceae, Aroideae). *Beitr. Biol. Pfl.* **61**, 241–282. [English summary]

ULLISCH, A. (1930). Die Gewebespannungen beim Entfälten monokotyler Blätter. *Bot. Archiv* **30**, 195–234. [English summary]

UPHOF, J.T., HUMMEL, K. and STAESCHE, K. (1962). *Plant hairs.* [Handbuch der Pflanzenanatomie vol. IV (5)]. Gebrüder Borntraeger, Berlin. 292 pp.

VÁGÚJFALVI, D. (1971). Die Alkaloidakkumulation im Latex. *Acta bot. Acad. sci. hung.* **17**, 217–241.

VAN FLEET, D. S. (1942). The development and distribution of the endodermis and an associated oxidase system in monocotyledonous plants. *Amer. J. Bot.* **29**, 1–15.

VASILEVSKAYA, V. K. and NECHAEVA, N. T. (1964). Materiali k biologii *Eminium lehmannii* (Bge.) Ktze. *Izvest. Akad. Nauk Turk. SSR* **(2)**, 35–40.

VIGNOLI, L. (1939). Gametofiti e cromosomi di *Ambrosinia bassii* L. *Lav. r. Ist. bot. Palermo* **10**, 54–80.

VIJ, S. P., SHARMA, M. and CHAUDHARY, J. D. (1982). Cytogenetical investigations into some garden ornamentals III. Chromosomes in some monocot taxa. *Cytologia* **47**, 649–663.

VOGEL, S. (1962). Duftdrüsen im Dienste der Bestäubung. Über Bau und Funktion der Osmophoren. *Akad. Wiss. Lit., Abh. mat.-nat. Kl., Mainz*, **10**, 601–763.

——(1978). Pilzmuckenblumen als Pilzmimeten. Fortsetzung und Schluss. *Flora* **167**, 367–398.

——(1990). *The role of scent glands in pollination.* Amerind, Delhi. 230 pp. [Translation of earlier German version].

VOLKENS, G. (1882, 1883). Über Wasserausscheidung in liquider Form an den Blättern höherer Pflanzen. Diss. Berlin (1882). Also: *Jb. bot. Gart. Berlin* **2**, 166–209 (1883).

VORONKINA, N. V. (1975). Histogenesis in root apices of angiospermous plants and possible ways of its evolution. *Bot. Zh. SSSR* **60**, 170–87. [Russian, English summary]

WAGNER, P. (1977). Vessel types of the monocotyledons: a survey. *Bot. Notiser* **130**, 383–402.

WAKABAYASHI, S. (1957a). On the form and formative stage of raphidian cells in some araceous and orchidaceous plants. *J. Jap. Bot.* **32**, 268–274. [Japanese, English summary]

——(1957b). Studies on the structure and the formation of mucilage cells in corms of *Amorphophallus konjac*. *J. Jap. Bot.* **32**, 337–346. [Japanese, English summary]

WAKKER, J. H. (1888). Studien über die Inhaltskörper der Pflanenzelle. *Jb. wiss. Bot.* **19**, 423–496.

WALSH, M. A. and MELARAGNO, J. E. (1976). Ultrastructural features of developing sieve elements in *Lemna minor* L. The sieve plate and lateral sieve areas. *Amer. J. Bot.* **63**, 1174–1183.

WALTER, W. G. and KHANNA, P. N. (1972). Chemistry of the aroids. I. *Dieffenbachia seguine, amoena* and *picta*. *Econ. Bot.* **26**, 364–372.

WALTON, A. (1964). A morphogenetic study of *Arum maculatum* L. *Ann. Bot.* (NS) **28**, 271–282.

WANNAKRAIROJ, S. and KAMEMOTO, H. (1990). Histological distribution of anthocyanins in *Anthurium* spathes. *HortScience* **25**, 809.

WARNCKE, F. (1911). Neue Beiträge zur Kenntnis der Spaltöffnungen. *Jb. wiss. Bot.* **50**, 21–66.

WATSON, J. B., KRESS, W. J. and ROESEL, C. S. (1987) [1988]. A bibliography of biological literature on vascular epiphytes. *Selbyana* **10**, 1–23.

WATTENDORFF, J. (1974). Ultrahistochemical reactions of the suberized cell walls in *Acorus, Acacia*, and *Larix*. *Z. Pflanzenphysiol.* **73**, 214–225.

WEBBER, E. E. (1960). Observations on the epidermal structure and stomatal apparatus of some members of the Araceae. *Rhodora* **62**, 251–258.

WEIDNER, M., FRANZ, A. and NAPP-ZINN, K. (1985). Plastid ultrastructure and photosynthesis in greening petaloid hypsophylls. *Planta* **163**, 164–174.

WEINERT, H. and BARCKHAUS, R. (1974). Der Kolben des Aronstabes und seine Erwärmung zur Blütezeit. Elektronenmikroskopische Untersuchungen. *Mikrokosmos* **63**, 169–176.

WEISS, A. (1866). Milchsaftgefässe in den Luftwurzeln von *Syngonium decipiens* Karstens. *Bot. Untersuch. phys. Lab. landwirt. Lehranstalt Berlin* **1 (2)**, 170–173.

WEISSE, A. (1897). Ueber Lenticellen und verwandte Durchlüftungseinrichtungen bei Monocotylen. *Ber. dt. bot. Ges.* **15**, 303–320.

WENT, F. A. F. C. (1895). Ueber Haft- und Naehrwurzeln bei Kletterpflanzen und Epiphyten. *Ann. Jard. bot. Buitenzorg* **12**, 1–72.

WHEATON, T. A. and STEWART, I. (1970). The distribution of tyramine, N-methyltyramine, hordenine, octopamine, and synephrine in higher plants. *Lloydia* **33**, 244–254.

WICHTL, M. (ed.) (1989). *Teedrogen. Ein Handbuch fur die Praxis auf wissenschaftlicher Grundlage.* 2nd edn. Wissenschaftliche Verlagsgesellschaft, Stuttgart. 568 pp.

WIDJAJA, E. A. and LESTER, R. N. (1987). Morphological, anatomical and chemical analyses of *Amorphophallus paeoniifolius* and related taxa. *Reinwardtia* **10**, 271–280.

WIESNER, J. (1875). Über das Vorkommen von Haaren in den Intercellulargängen des Mesophylls von *Philodendron pertusum*. *Öst. bot. Z.* **25**, 4–7.

WILDE, V. (1989). Untersuchungen zur Systematik der Blattreste aus dem Mitteleozän der Grube Messel bei Darmstadt (Hessen, Bundesrepublik Deutschland). *Cour. Forsch.-Inst. Senckenberg* **115**, 1–213.

WILLIAMS, C. A., HARBORNE, J. B. and CLIFFORD, H. T. (1971). Flavonoid patterns in the monocotyledons. Flavonols and flavones in some families associated with the Poaceae. *Phytochemistry* **10**, 1059–1063.

——, —— and MAYO, S. J. (1981). Anthocyanin pigments and leaf flavonoids in the family Araceae. *Phytochemistry* **20**, 217–234.

WILLIAMS, K. A. (1919). A botanical study of skunk cabbage, *Symplocarpus foetidus*. *Torreya* **19**, 21–29.

WILLIAMS, M. (1994). Leaf anatomical investigation into *Philodendron* Schott. (Araceae Juss.). Thesis, Univ. Reading. 42 pp.

WILLIS, J. (1949). The classification and distribution of the Araceae. pp. 248–295 In: The birth and spread of plants. *Boissieria* **8**, 1–561.

WILSON, C. A. and PETERSON, C. A. (1983). Chemical composition of the epidermal, hypodermal, endodermal and intervening cortical cell walls of various plant roots. *Ann. Bot.* **51**, 759–769.

WILSON, K. A. (1960). The genera of the Arales in the southeastern United States. *J. Arnold Arbor.* **41**, 47–72.

WINKLER, H. (1907). Über die Umwandlung des Blattstieles zum Stengel. *Jb. wiss. Bot.* **45**, 1–82.

WINTON, A. L. and WINTON, K. B. (1935). *The structure and composition of foods.* Vol. II. Wiley, New York. 904 pp.

—— and —— (1945). *The analysis of foods.* Wiley, New York.

WITSCH, H. VON (1939). Zum Feinbau der Zellwand in Wurzeln. *Planta* **29**, 409–418.

WITTLIN, J. (1896). Über die Bildung der Kalkoxalat-Taschen. *Bot. Centralbl.* **67**, 65–73, 97–102, 129–133.

WOLFE, J. A. (1975). Some aspects of plant geography of the northern hemisphere during the Late Cretaceous and Tertiary. *Ann. Missouri Bot. Gard.* **62**, 264–279.

WOLFE, K. H., GOUY, M., YANG Y.-W., SHARP, P. M. and LI, W.-H. (1989). Date of monocot dicot divergence estimated from chloroplast DNA sequence data. *Proc. Nat. Acad. Sci.* **86**, 6201–6205.

WOLLENWEBER, E. F. (1897). *Vergleichende Anatomie der Schwimmblätter.* Diss., Albert-Ludwigs Univ. zu Freiburg im Breisgau. Bonn. 34 pp.

WULFF, H. D. (1954). Zur Zytologie, geographischen Verbreitung und Morpholgie des Kalmus. *Arch. Pharm. Ber. dt. pharm. Ges.* **287**, 529–541.

WUNDERLICH, R. (1954). Über das Antherentapetum mit besonderer Berücksichtigung seiner Kernzahl. *Öst. bot. Z.* **101**, 1–63.

YARBROUGH, J. A. (1934). Stomatal count of a single leaf of *Colocasia*. *Proc. Iowa Acad. Sci.* **41**, 71–73.

YOUNG, H. J. (1986). Beetle pollination of *Dieffenbachia longispatha* (Araceae). *Amer. J. Bot.* **73**, 931–944.

—— (1990) [1991]. Pollination and reproductive biology of an understory neotropical aroid. pp. 151–164 In: *Reproductive ecology of tropical forest plants* (eds Bawa, K. S. and Hadley, M.). [Man and the biosphere series, Vol. 7]. Parthenon, New Jersey.

YOUNGKEN, H. W. (1919). Notes on the Dasheen and Chayote. *Amer. J. Bot.* **6**, 380–386.

ZACHARIAS, E. (1879). Ueber Secret-Behälter mit verkorkten Membranen. *Bot. Ztg.* **39**, 617–628 and 633–645.

ZAVADA, M. S. (1983). Comparative morphology of monocot pollen and evolutionary trends of apertures and wall structures. *Bot. Rev.* **49**, 331–379.

ZENNIE, T. M. and MCCLURE, J. W. (1977). The flavonoid chemistry of *Pistia stratiotes* L. and the origin of the Lemnaceae. *Aquat. Bot.* **3**, 49–54.

ZETTLER, F. W. and EL-NIL, M. M. A. (1979). Mode of inheritance of foliage color in *Caladium*. *J. Hered.* **70**, 433–435.

—— and RHODES, B. B. (1975). Calcium oxalate variability in *Dieffenbachia* seedlings. *HortScience* **10**, 481–482.

ZIEGENSPECK, H. (1937–39). Die Micellierungen der Turgeszenzmechanismen. I. Die Spaltöffnungen (mit phylogenetischen Ausblicken). *Bot. Archiv* **39**, 268–309.

—— (1949). Zur Phylogenie der Hydathoden. *Phyton* (Horn) **1**, 302–318.

ZIMMERMANN, J. G. (1932). Über die extrafloralen Nektarien der Angiospermen. *Beih. bot. Zbl.* **49 (1)**, 99–196.

ZINKE, W. (1924). Über das Assimilationsgewebe der Monocotylen und seine Verwendung in der Frage ihres systematischen Anschlusses. *Bot. Archiv* **5**, 74–91.

INDEX

Note. *Where genera have been specially studied, the page numbers are given in bold type. Italic page numbers refer to illustrations. Most cell and tissue types are not indexed for the generic descriptions.*